Günter Kahl

The Dictionary of
Gene Technology

Genomics, Transcriptomics, Proteomics

Third Edition

Translated into Chinese

Related Titles from WILEY-VCH:

Stefan Lorkowski, Paul Cullen
Analysing Gene Expression
A Handbook of Methods. Possibilities and Pittfalls, Possibilities and Pitfalls
ISBN 3-527-30488-6
2002

Dev Kambhampati
Protein Microarray Technology
ISBN 3-527-30597-1
2004

Jennifer E. van Eyk, Michael J. Dunn
Proteomic and Genomic Analysis of Cardiovascular Disease
ISBN 3-527-30596-3
2003

Jean-Charles Sanchez, Garry L. Corthals, Denis F. Hochstrasser
Biomedical Applications of Proteomics
ISBN 3-527-30807-5
2004

Rolf D. Schmid, Ruth Hammelehle
Pocket Guide to Biotechnology and Genetic Engineering
ISBN 3-527-30895-4
2003

Günter Kahl

The Dictionary of Gene Technology

Genomics, Transcriptomics, Proteomics
Third Edition

WILEY-VCH Verlag GmbH & Co. KGaA

Professor Dr. Günter Kahl
Plant Molecular Biology, Biocenter
Johann Wolfgang Goethe University
Marie-Curie-Straße 9
D–60439 Frankfurt am Main
Germany
Phone: +49(069)–7982–9267
Fax: +49(069)–7982–9268
e-mail: kahl@em.uni-frankfurt.de

Visiting Professor
Center of Pharmacy
University of Vienna
Althanstraße 14
A–1090 Wien
Austria

This book was carefully produced. Nevertheless, author and publisher do not warrant the information contained therein to be free of errors. Readers are advised to keep in mind that statements, data, illustrations, procedural details or other items may inadvertently be inaccurate.

First Edition 1994
Second Edition 2001
Corrected Reprint 2001
Third Edition 2004

Cover Illustration:
The Title Page shows an atomic force microscopic (AFM) image of human chromosome 7, a low-density expression microarray after hybridisation with cDNAs labeled with different fluorophores, a multi-color fluorescence *in situ* hybridisation (FISH) karyotype of a human being, and an atomic force microscopic image of a cosmid molecule with bound *Eco*RI restriction endonucleases. The underlying text is taken from the description of "microarray" in this Dictionary. The photos were kindly provided by Dr. Stefan Thalhammer (Ludwig-Maximilians-University, Munich, Germany), Dr. Ilse Chudoba (MetaSystems GmbH, Altlußheim, Germany) and Dr. Peter R. Hoyt (Molecular Imaging Group, The Oak Ridge National Laboratory, Oak Ridge, Tennessee, USA). For details see Acknowledgements.

Library of Congress Card No. applied for

A catalogue record for this book is available from the British Library

Bibliographic information published by Die Deutsche Bibliothek
Die Deutsche Bibliothek lists this publication in the Deutsche Nationalbibliografie; detailed bibliographic data is available in the Internet at ⟨http://dnb.ddb.de⟩.
ISBN 3–527–30765–6

Printed on acid-free paper.

Composition: pagina GmbH, Tübingen
Printing: betz-druck GmbH, Darmstadt
Bookbinding: Großbuchbinderei J. Schäffer GmbH & Co. KG, Grünstadt

Printed in the Federal Republic of Germany

Preface

To see what is in front of one's nose
requires a constant struggle!

(Novelist and social critic George Orwell, 1946)

The Dictionary of Gene Technology: Genomics, Transcriptomics, Proteomics in its completely updated and expanded third edition now contains a total of 9000 different technical terms, 2500 more than the second edition, many of them originating from most recent developments and brand-new technologies. It therefore remains the most comprehensive collection of descriptions of molecular processes and techniques worldwide, and is in fact the only one. The contents could no longer be accomodated in a single volume and now appears in two volumes. However, this doubling in size only faintly and inadequately reflects the tremendous growth in facts, insights and techniques in the fundamentally important field of gene technologies over the past three years. The increasing contributions by physicists and technical engineers and the invention of novel materials and design of technologies of ever-increasing sophistication mark a new era in gene technology. Traditional PCR and its multiple variants still dominates the field with a total of more than 150 entries, but is challenged by so called chip technologies ("microarrays": about 180 entries) with dramatic miniaturization using atomic force microscopy and far-reaching automation, and also by novel developments such as single molecule tracking and sequencing as well as label-free detection and DNA computing. The numerous genome sequencing projects brought a plethora of new insights into the structure and function of genomes and genes, and as a consequence, several hundred new technical terms. Synthetic DNA and RNA variants are presently changing concepts and expanding applications. In short, it is still valid, more than ever before: "All is flux, nothing stays still" (Greek philosopher Herakleitos of Ephesos, 544–483 B.C.)

The Dictionary of Gene Technology: Genomics, Transcriptomics, Proteomics has warmly been welcomed and sold as a bestseller in its previous two editions. The author hopes, that the third edition will also be of benefit for all those, who are interested in genome and gene technologies, especially the Chinese readers (the third edition also appears in a Chinese translation) and the Spanish speaking scientists (a translation into Spanish is planned). The author appreciates the hospitality of his colleagues at the Biocenter of Frankfurt University, the successful research and research fund-raising of Dr. Peter Winter, Mr. Carlos Molina Medina and Mrs. Ruth Jungmann, the patient artist work of Mr. Uwe Kahl and the cooperation of the Wiley-VCH team. The latest technologies have been added while being visiting professor at the Pharma Center (University of Vienna, Austria), for which I am grateful to Professors Dr. Christian Noe (Institute of Pharmaceutical Chemistry) and Dr. Brigitte Kopp (Institute of Pharmacognosy).

Frankfurt am Main, December 2003   Günter Kahl

Preface to the Second Edition

παντα ρει

**All is flux –
nothing stays still**

**(Herakleitos of Ephesos,
Greek Philosopher, 544 – 483 B.C.)**

The Dictionary of Gene Technology in its completely revised and up-dated new version is the most comprehensive collection of terms of this extremely rapidly evolving science. A total of more than 2500 new entries have been added to the already existing 4000 terms of the first edition. This dramatic increase in the number of terms reflects the unprecedented and literally explosive development of a multitude of different techniques, their wide applications, the achieved results – and their world-wide marketing. The past five years have witnessed the inexorable change of traditional gene technology, involving molecular cloning and bacterial transformation, to polymerase chain reaction (PCR)-based technologies, that revolutionized and still revolutionize nearly all branches of modern molecular biology. Today, we can amplify single DNA molecules, detect DNA sequence variation in traces of trash, isolate full-length genes in a fraction of time required for the «old« techniques, and even do not need to spend intensive and frustrating bench work to do so: virtual cloning emerged. PCR also catalyzed miniaturization and automation throughout gene technology. Nano-technologies and super high-throughput procedures allow the simultaneous expression analysis of literally hundreds of thousands of genes on chips as little as a Cent coin. A whole nanoindustry led the transition from genes to screens, and from genetic backwaters to genomic boom, adding to the diversification of gene technology into many disciplines that were hitherto completely unknown: the *omics* era began.

As a consequence, the chaos in terminology grew and is growing. On one hand, companies coin their own techniques, though they are probably different from already existing ones by a mere change in PCR conditions. On the other, different laboratories use their own pet name for a technique, a DNA sequence or a protein, and refuse to accept other names. For example, a tyrosine kinase receptor, part of a signal cascade in the brain, was first described as Cek 5, then Nuk, later on Sek 3, Drt, Tyro 6, Qek 5, Erk, and Eph B2. And this holds for many other proteins as well, not to speak of DNA sequence elements and molecular techniques.

The Dictionary of Gene Technology aims at ordering the chaos a bit. It focusses on nucleic acids, their structure, modification, processing, transfer, and function analysis, but includes many terms from other disciplines if they are considered necessary or helpful for an understanding. So, proteins are described, if they function in concert with genes, for example. Moreover, it tries to portray newest technologies and newest discoveries. However, the author is fully aware of the fact, that a single person's capacity and competence is limited, so that misinterpretations and errors are inevitable. These considerations also held for the first edition, which was nevertheless warmly welcomed and reviewed extremely favorable, and sold as a bestseller.

The Dictionary of Gene Technology in its new version would not have been possible without the patience and dedicated research of all coworkers in the Plant Molecular Biology Department at the Biocenter of Frankfurt University, the never-tiring help of Mrs. S. Kost and Mrs. L. Mark-Abdelhamid, many funding agencies supporting our research, and the hospitality of foreign institutions, notably the Iwate Biotechnology Research Center at Kitakami, Iwate, Japan (Director: Dr. Hitoshi Enei), as well as the intensive care of Naoko and Ryohei Terauchi (Kitakami, Iwate, Japan). The cooperation with the team in the publishing house Wiley-VCH (Achim Kraus, editorial; Hans Jochen Schmitt, production, and Elina Niskanen, marketing) is highly appreciated. Surprising as it is, Sigrid still knows me.

Frankfurt am Main, January 2001   Günter Kahl

Preface to the First Edition

The **Dictionary of Gene Technology** is the most modern and most comprehensive collection of all terms of this modern science. With a volume of more than 4000 entries it reflects the importance of gene technology for present-day biology. It also documents myriads of acronyms, a serious obstacle for clearness, and a swamp of jargons, a deterrence for students and other newcomers. While acronyms may well be a help in daily laboratory work, the numerous synonyms are indeed annoying, though they are characteristic of the discipline of gene technology. Sometimes they differ from each other by only a word more or less, or may be minor expression variants. Wherever possible, this dictionary stresses the most commonly used term, and treats inferior terms secondarily.

The **Dictionary of Gene Technology** targets at students in the fields of molecular biology and biotechnology, and seasoned researchers in other fields who are keen on making themselves familiar with the vocabulary of gene technology, specifically bioengineers, biochemists, biologists, chemical engineers, chemists, geneticists, medics, microbiologists and pharmacists, working at universities, in laboratories of industry or public institutions. It also offers a guide for reporters, scientific journalists and politicians in the stormy sea of public dispute over the pros and cons of gene technology. This book also serves as reference work for the active researcher in biotechnology, genetic engineering, and molecular genetics. Especially for this group the most recently introduced techniques and terms were included, which are embedded in a net of terms from allied sciences such as bacteriology, biochemistry, biophysics, biotechnology, cell biology, chemistry, cytogenetics, genetics, immunology and virology.

Gene Technology is an extremely rapidly expanding science. It is therefore inevitable that new terms will soon be created, new techniques will be introduced and this dictionary will have omissions. Also, I have striven to avoid errors, ambiguities and misinterpretations, and to be as complete as ever possible. Nevertheless, I am sure that certain inadequacies will be discovered, and I apologize for them at this stage. Sometimes the definition of an entry might look lengthy. However, I frequently felt that a brief definition would be inadequate to convey the essence of the entry.

A **Dictionary of Gene Technology** brings an author to the utmost limits of his capacity. It is therefore a pleasure to acknowledge the numerous supports of many colleagues, the patience of my coworkers in the Plant Molecular Biology Group at Frankfurt University, the exhaustive help of Mrs. S. Kost, and the cooperation of the VCH editor, Dr. H. J. Kraus, and, last but not least, the expert editing of Dr. P. Falkenburg.

I honestly appreciate the hospitality of various institutions in different countries, in which I have been working on this book, especially the Research Institute for Bioresources, Okayama University (Japan), the Department of Biology and Molecular Biology, University of California Los Angeles (USA), the International Center for Agricultural Research in the Dry Areas, Aleppo (Syria) and the Centro Agronomico Tropical de Investigacion y Ensenanza, Turrialba (Costa Rica). Last but not least I would like to thank Sigrid that she still knows me.

Frankfurt am Main, September 1994   Günter Kahl

Contents

Instructions for Users

- All the entries are arranged in strict alphabetical order, letter by letter. For example, "mismatched primer" precedes "mismatch gene synthesis", and this is followed by "mismatch repair". Or, "photo-digoxygenin" precedes "photo-footprinting", which in turn precedes "photo-reactivation". In case an entry starts with or contains a Roman, Greek or Arabic numeral, it has first to be translated into Latin script. A few examples illustrate the translation:

cI	: c-one
exonuclease VII	: exonuclease seven
exonuclease III	: exonuclease three
5'	: five prime
G 418	: G fourhundred and eighteen
λ	: lambda
P1	: p-one
ΦX 174	: phi X one-seven-four
Qβ	: q-beta
RP 4	: RP four

For help, the user may consult the Roman numerals and the Greek alphabet below.

- The main entry title, printed in bold type, is followed by synonyms in parentheses. Italicized letters in titles (and text) of entries indicate use of these letters for abbreviations.

- Cross referencing is either indicated by an arrow, or the words "see", "see also", and "compare".

Greek Alphabet and Roman Numerals

Greek alphabet:

Capital	Lower case	Name		Capital	Lower case	Name
A	α	alpha		N	ν	nu
B	β	beta		Ξ	ξ	xi
Γ	γ	gamma		O	o	omicron
Δ	δ, δ	delta		Π	π	pi
E	ε	epsilon		P	ρ	rho
Z	ζ	zeta		Σ	σ, ς	sigma
H	η	eta		T	τ	tau
Θ	θ, ϑ	theta		Y	υ	ypsilon
I	ι	iota		Φ	φ	phi
K	κ	kappa		X	χ	chi
Λ	λ	lambda		Ψ	ψ	psi
M	μ	mu		Ω	ω	omega

Roman numerals:

I	II	III	IV	V	VI	VII	VIII	IX	X
1	2	3	4	5	6	7	8	9	10

XX	XXX	XL	L	LX	LXX	LXXX	XC	IC	C
20	30	40	50	60	70	80	90	99	100

CC	CCC	CD	D	DC	DCC	DCCC	CM	XM	M
200	300	400	500	600	700	800	900	990	1000

Abbreviations and Symbols

a	– atto (10^{-18})
A	– adenine or adenosine, absorbance
Å	– Ångstrom unit (1Å = 0.1nm)
~	– approximately
≅	– approximately equal
A/D	– analog-to-digital
Ap	– ampicillin
ATP	– adenosine triphosphate
BAC	– bacterial artificial chromosome
Bis	– N, N'-methylenebisacrylamide
BLAST	– basic local alignment search tool
bp	– base pair(s)
Bq	– Becquerel
BSA	– bovine serum albumin
c	– centi (10^{-2})
C	– cytosine or cytidine
^{14}C	– radioactive carbon
°C	– centigrade (degrees Celsius)
Ca	– calcium
CCD	– charge-coupled device
cDNA	– complementary DNA
CE	– capillary electrophoresis
Ci	– Curie
cm	– centimeter(s)
Cm	– chloramphenicol
CO_2	– carbon dioxide
cpm	– counts per minute
CTAB	– cetyltrimethylammonium bromide
Cy	– cyanine
D, Da	– Dalton
DAF	– DNA amplification fingerprinting
dATP	– deoxyadenosine triphosphate
dCTP	– deoxycytosine triphosphate
ddNTP	– 2',3'-dideoxynucleotide triphosphate
DGGE	– denaturing gradient gel electrophoresis
dGTP	– deoxyguanosine triphosphate
DMF	– N, N'-dimethylformamide
DMSO	– dimethyl sulfoxide
DMT	– dimethyloxytrityl
DNA	– deoxyribonucleic acid
DNase	– deoxyribonuclease
dNTP	– deoxynucleotide triphosphate
ds	– double-stranded
dT	– deoxythymidine
DTT	– dithiothreitol, Cleland's reagent
dTTP	– deoxythymidine triphosphate

dUTP	– deoxyuridine triphosphate
EC	– enzyme classification number
ECL	– enhanced chemiluminescence
E. coli	– *Escherichia coli*
EDTA	– ethylenediaminetetraacetic acid
e.g.	– for example
ELISA	– enzyme-linked immunosorbent assay
ESI	– electrospray ionization
EST	– expressed sequence tag
EtBr	– ethidium bromide
EtOH	– ethanol
f	– femto (10^{-15})
FIGE	– field inversion gel electrophoresis
FITC	– fluorescein isothiocyanate
fmol	– femto mol
5'	– carbon atom 5 of deoxyribose
g	– gram(s) or gravity
G	– guanine or guanidine, giga (10^9)
Gb	– gigabase
GC	– gas chromatography
GFP	– green fluorescent protein
Gm	– gentamycin
GMO	– genetically modified organism
>	– greater than
h	– hour(s)
HAC	– human artificial chromosome
^{3}H	– tritium, radioactive hydrogen
HCl	– hydrochloric acid
HEPES	– N-(2–hydroxyethyl)piperazine-N'-(2–ethanesulfonic acid)
HIV	– human immunodeficiency virus
HPCE	– high-performance capillary electrophoresis
HPLC	– high pressure liquid chromatography
HRP	– horseradish peroxidase
HTE	– high Tris-EDTA buffer
H_2O	– water
H_2O_2	– hydrogen peroxide
HVR	– hypervariable region
i.e.	– that is
IEF	– isoelectric focusing
Ig	– immunoglobulin
IVS	– intervening sequence, intron
k	– kilo (10^3)
kb	– kilobase(s)
KB	– kilobyte
kD (kDa)	– kilo Dalton
kg	– kilogram(s)
Km	– kanamycin
l	– liter(s)

<	– less than
LC	– liquid chromatography
LiCl	– lithium chloride
LIF	– laser-induced fluorescence
LTE	– low Tris-EDTA buffer
mAb	– monoclonal antibody
MALDI-MS	– matrix-assisted laser desorption/ionization-mass spectrometry
m	– meter(s) or milli (10^{-3})
μ	– micro (10^{-6})
μg	– microgram(s)
μl	– microliter(s)
M	– molar or mega (10^{6})
Mb (Mbp)	– megabase pairs
MB	– megabyte
MCS	– multiple cloning site
mg	– milligram(s)
Mg	– magnesium
$MgCl_2$	– magnesium chloride
$MgSO_4$	– magnesium sulfate
min	– minute(s)
ml	– milliliter(s)
mm	– millimeter(s)
mM	– millimolar
mmol	– millimole
mol	– mole
M_r	– relative molecular mass (no dimension)
mRNA	– messenger RNA
MS	– mass spectrometry
MS/MS	– tandem mass spectrometry
mtDNA	– mitochondrial DNA
MW	– molecular weight
n	– number or nano (10^{-9})
NaCl	– sodium chloride
Na_2EDTA	– disodium-EDTA
NC	– nitrocellulose
ng	– nanogram(s)
NH_4Cl	– ammonium chloride
NH_4OAc	– ammonium acetate
nm	– nanometer(s)
NMR	– nuclear magnetic resonance
nt	– nucleotide
OD	– optical density
OH	– hydroxyl
oligo	– oligonucleotide(s)
ORF	– open reading frame
p	– pico (10^{-12})
P	– phosphorus
P_i	– inorganic phosporus

^{32}P	– radioactive phosphorus
PAGE	– polyacrylamide gel electrophoresis
PBS	– phosphate buffered saline
PCR	– polymerase chain reaction
PEG	– polyethylene glycol
PFGE	– pulsed field gel electrophoresis
pfu	– plaque forming unit
pg	– picogram(s)
pH	– logarithm of reciprocal of hydrogen (H) ion concentration
pI	– isoelectric point
PMS	– phenazine methosulfate
PMSF	– phenylmethylsulfonyl fluoride
PNA	– peptide nucleic acid
pp	– page(s)
ppm	– parts per million
PSD	– post-source decay
PTFE	– polytetrafluoroethylene
PVDF	– polyvinylidene difluoride
PVP	– polyvinyl pyrolidone
RAPD	– random amplified polymorphic DNA
RFL	– restriction fragment length
RFLP(s)	– restriction fragment length polymorphism(s)
RIA	– radioimmunoassay
RNA	– ribonucleic acid
RNase	– ribonuclease
RP	– reversed phase
rpm	– revolutions per minute
rRNA	– ribosomal RNA
RT	– room temperature (also reverse transcriptase)
RT-PCR	– reverse transcriptase PCR
^{35}S	– radioactive sulfur
SAGE	– serial analysis of gene expression
SD	– standard deviation
SDS	– sodium dodecyl sulfate, lauryl sulfate
SE (SEM)	– standard error (standard error of the mean)
sec	– second(s)
Σ	– sum of
Sm	– streptomycin
SNP	– single nucleotide polymorphism
ss	– single-stranded
SSC	– sodium chloride sodium citrate (saline sodium citrate)
SSCP	– single-strand conformation polymorphism
ssDNA	– single-stranded DNA
SSO	– sequence-specific oligonucleotide
SSP	– sequence-specific probe
SSPE	– sodium chloride-sodium phosphate-EDTA
STR	– short tandem repeat
T	– thymine or thymidine, tera (10^{12})

$\tau_{1/2}$	– half-life
TAE	– Tris-acetate-EDTA
TBE	– Tris-borate-EDTA
TBS	– Tris-buffered saline
Tc	– tetracycline
TE	– Tris-EDTA-buffer
TEMED	– N, N, N', N'-tetramethylethylene diamine
3'	– carbon atom 3 of deoxyribose
TLC	– thin-layer chromatography
T_m	– melting temperature
TOF	– time of flight
Tp	– trimethoprim
Tris	– tris (hydroxymethyl) aminomethane
tRNA	– transfer RNA
U	– unit(s)
U	– uracil or uridine
UV	– ultraviolet
V	– voltage, volt(s)
VNTR	– variable number of tandem repeats
vol	– volume
v/v	– volume/volume
w/v	– weight/volume
$\bar{X}$	– mean
χ^2	– chi squared
YAC	– yeast artificial chromosome
yr	– year(s)

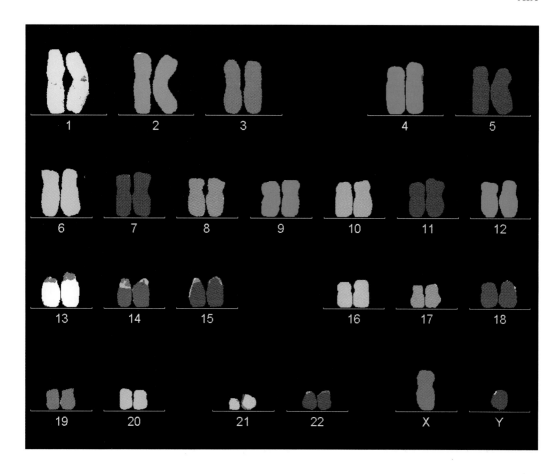

Multicolor fluorescence *in situ* **hybridization (multicolor FISH)**

Multicolor karyotype of a male human, visualized by → *in situ* hybridisation of a mixture of 24 whole → chromosome libraries labeled with combinations of five → fluorochromes to metaphase spreads. All 24 chromosomes are simultaneously detected by a single hybridisation. Note the → translocation onto chromosomes 13 and 14.
(Kind permission of Dr. Ilse Chudoba, MetaSystems, Altlußheim, Germany)

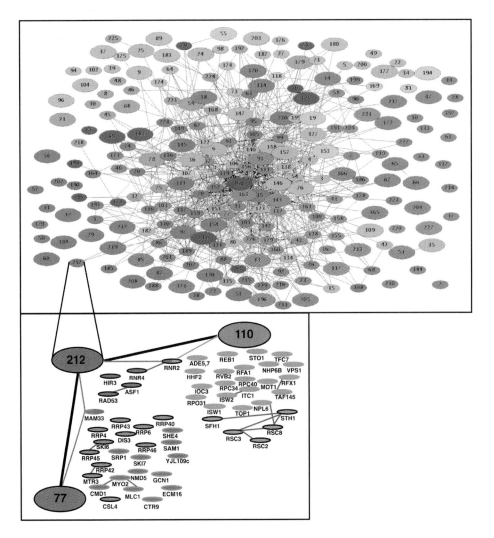

Protein-protein interaction map

A higher order network map of part of the → proteome of yeast (*Saccharomyces cerevisiae*), as detected by → tandem affinity purification (TAP) tagging and → matrix-assisted laser desorption-ionization time-of-flight mass spectrometry (MALDI-TOF-MS).

Upper panel: The functions of the individual protein complexes in cell metabolism are encoded in colors (*dark blue*: → transcription, DNA maintenance, → chromatin structure; *light blue*: membrane synthesis and membrane trafficking; *red*: cell cycle; *dark green*: signalling; *pink*: protein and RNA transport; *orange*: RNA metabolism; *light green*: protein synthesis and protein turnover; *brown*: cell polarity and structure; *violet*: intermediate and energy metabolism).

Lower panel: Details of the connection between one yeast protein complex (212) and two other complexes (77 and 110). Red lines: physical interactions.

(Kind permission of Dr. Monika Blank, Cellzome, Heidelberg, Germany; see Acknowledgements).

See → protein atlas, → protein complex, → protein interaction map, → protein interactome, → protein network, → protein-protein interactions, → proteome map.

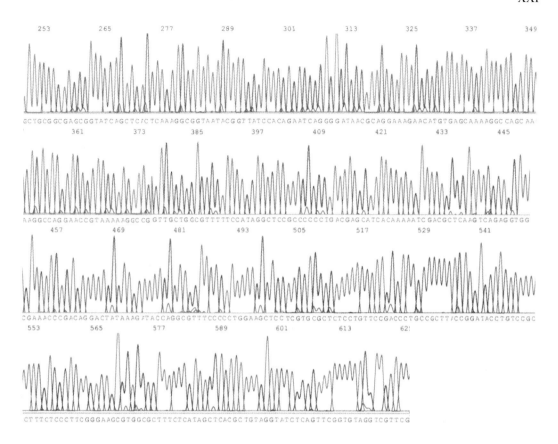

GCTGCGGCGAGCGGTATCAGCTCTCTCAAAGGCGGTAATACGGT TATCCACAGAATCAGGGG ATAACGCAGGAAAGAACATGTGAGCAAAAGGC CAGC AA

AAGGCCAGGAACCGTAAAAAGGCCG GTTGCTGGCGTTTTTCCATAGGCTCCGCCCCCTG ACGAGCATCACAAAAATC GACGCTCAAG TCAGAGGTGG

CGAAACCCGACAGGACTATAAAG TACCAGGCGTTTCCCCCTGGAAGCTCC TCGTGCGCTCTCCTGTTCCGACCCTGCCGCTTTCCGGATACCTGTCCGC

CTTTCTCCCTTCGGGAAGCGTGGCGCTTTCTCATAGCTCACGCTGTAGGTATCTCAGTTCGGTGTAGGTCGTTCG

Sanger sequencing

The sequence of bases in part of a → plasmid DNA, as it appears on a → fluorogram, where green symbolizes adenine (A), red thymine (T), blue cytosine (C) and black guanine (G), respectively. The different → dideoxynucleotides in this case are labeled with different → fluorochromes.

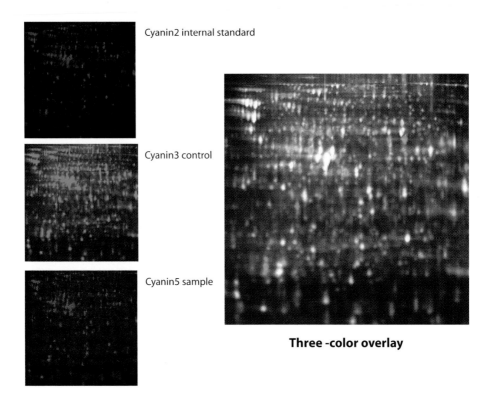

Cyanin2 internal standard

Cyanin3 control

Cyanin5 sample

Three -color overlay

Two-dimensional difference gel elctrophoresis (2–D DIGE)

A composite graphical image of a two-dimensional gel electrophoresis experiment, consisting of three originally separate images of one and the same protein separation gel probed with three different fluorochromes (e.g. cyanin 5, cyanin 3 and cyanin 2, respectively), that are merged *in silico*. The proteins from e.g. two different tissues are differentially labeled with cyanin 3 and cyanin 5, respectively, the pre-labeled protein samples mixed and co-separated by two-dimensional SDS polyacrylamide gel electrophoresis. The gel is imaged for the two fluorophores (and a control fluorophore, cyanin 2), and the resulting, perfectly overlapping image analysed and proteins detected, that are only present in one sample.
(Kindly provided by Dr. Joachim Häse, Amersham Biosciences, Germany)

See → green-red overlay, → two-color overlay.

M

M:
a) Abbreviation for either → adenine or → cytosine (a*M*ino in large groove), used in sequence data banks.
b) See → mismatch, → perfect match.

^mA (mA): Abbreviation for adenine carrying a methyl group (e. g. at N^6).

MAAP: See → arbitrarily amplified DNA.

mAB: See → *m*onoclonal *a*nti*b*ody.

MAB: See → marker-assisted breeding.

MAC:
a) See → *m*ammalian *a*rtificial *c*hromosome
b) See → *m*ap-*a*ssisted *c*loning.

Macroarray (nylon macroarray): Any nylon or nitrocellulose membrane (also plastic support), onto which several hundreds to a few thousand target molecules (e.g. cDNAs, oligonucleotides) are regularly spotted ("gridded"). In contrast, → microarrays contain thousands, hundred thousands, or millions of spots. Macroarrays are typically hybridised to radioactively labelled → probes, and the hybridization and washing procedures are identical to the development of a → Southern blot. Also, detection of the hybridisation event occurs via → autoradiography. Macroarrays are optimal for the expression analysis of a limited set of genes, as e.g. genes encoding enzymes of a particular metabolic pathway.

Macroautoradiography: See → autoradiography.

Macromolecule: Any molecule whose molecular weight exceeds a few thousand daltons (e.g. polysaccharides, proteins, nucleic acids).

Macronucleus: The larger of the two nuclei of certain protozoan species (ciliatae), that actively transcribes its genes during asexual growth, replicates during asexual reproduction, but is destroyed and re-formed during sexual reproduction. Therefore, macronuclei do not transmit genetic information to sexual offspring. Destruction of the macronucleus is followed by the development of a new macronucleus, which starts with multiple rounds of → DNA replication and leads to the formation of → polytene chromosomes. The extent of → polyploidization varies with the species, but it reaches ploidy levels of up to 64. The polytenic chromosomes undergo fragmentation, vesicle-like structures form and enclose the different chromosome fragments: short macronuclear chromosomes appear. The vesicles persist, but large amounts of DNA are eliminated (see → DNA deletion). In some cases, up to 90% of the micronuclear genome vanishes. Finally, the vesicles decay, and multiple rounds of DNA replication produce the ultimate ploidy level of the mature macronucleus. Compare → gene-sized DNA, → micronucleus, → nuclear dimorphism.

Macro-restriction map: A graphical description of the linear arrangement of an ordered set of large DNA fragments generated by → rare-cutting → restriction endonucleases of → genomic DNA. This type of → physical map spans long DNA stretches and can be used to localize interesting sequences (e. g. genes) by hybridization of fluorescent or radiolabeled → probes to the different cloned restriction fragments. See → ordered clone map.

Macrosatellite: A somewhat imprecise term for a → satellite DNA that exceeds a certain size (which is not precisely fixed). See → megasatellite, → microsatellite, → minisatellite.

Macrosynteny: The conserved order of large genomic blocks (in the megabase range) in the genomes of related (but also unrelated) species, as detected by e. g. → chromosomal *in situ* suppression hybridization, → fluorescent *in situ* hybridization. See → homosequential linkage map, → microsynteny, → synteny.

MAD-DNA: See → moderately affected Alzheimer disease DNA.

MADGE: See → *m*icroplate *a*rray *d*iagonal *g*el *e*lectrophoresis.

MADS box: A 56 bp conserved motif of → transcription factors that function in the regulation of various genes (e. g. *MCM1* of yeast, *ag*amous homoeotic gene *AG* of *Arabidopsis thaliana*, *DEF A* [*def*icient flower] gene in *Antirrhinum majus*).

MADS box gene: Anyone of a series of genes encoding → transcription factors containing a highly conserved domain of 56 amino acids (→ MADS box domain) functioning as a DNA-binding site.

***Magnet-assisted substraction technique* (MAST):** A method for the detection of sequences expressed in only one of two cell types. In short, total → RNA is separately isolated from both cell types, and each RNA separately chromatographed over oligo(dT)$_n$ fixed to → magnetic beads. The → polyadenylated RNA, including most → messenger RNAs, is bound to the oligo(dT) and can easily be separated from the → poly (A)$^-$-RNA by magnetic force and washing. Then the poly(A)$^+$-RNA from cell type A is converted to → cDNA, using → reverse transcriptase, and the resulting mRNA-cDNA complexes denatured such that the cDNA remains attached to the paramagnetic beads ("driver cDNA"). The same procedure produces cDNA from mRNA of cell type B ("tracer cDNA"). Now all cDNAs, that are present in equal amounts in both cell types, are removed using a 25fold excess of driver cDNA. Only those cDNAs will remain that are expressed in cell type B specifically. Compare → subtractive hybridization.

Magnetic bead (*paramagnetic particle*, PMP): Paramagnetic materials (e.g. iron oxide), coated with → polyacrylamide and → agarose and packaged into submicron-sized particles that have no magnetic field but form a magnetic dipole when exposed to a magnetic field. Magnetic beads serve as a solid-phase support for the separation of DNA or RNA molecules from complex mixtures of biomolecules. Specific binding is usually achieved via specifically designed DNA fragments (e.g. → oligonucleotides) coupled to the magnetic beads. See → magnetic crosslinking, → magnetic polyvinyl alcohol, → paramagnetic particle technology.

Magnetic crosslinking: A somewhat incorrect term for a simple technique to separate contaminating → cloning vector sequences from labeled (e.g. by → nick-translation) DNA inserts, that uses → magnetic beads to which single-stranded capture DNA sequences are covalently attached. These capture sequences will hybridize to the undesired vector DNA, and can be removed together with the magnetic support by centrifugation or magnetic separation. Compare → paramagnetic particle technology.

Magnetic *polyvinyl alcohol* (M-PVA) microparticle: A somewhat misleading term for → magnetic beads, composed of the hydrogel matrix polyvinyl alcohol with encapsulated magnetite, charged with a functional group (e. g. –COOH, –NH$_2$, –NHR, –CHO, → streptavidin, → oligo d(T), or → protein A) that allows interaction with and binding of the corresponding site on e. g. a protein. Since M-PVA shows only minimal unspecific protein adsorption compared to other carrier media, it is used for immunoassays, affinity separations, → messenger RNA isolation, and detection of → DNA- or RNA-binding proteins.

Magnetic *resonance imaging* (MRI): A technique for the creation of an *in vivo* image of gene expression, that uses an MRI contrast agent with the ability to indicate → reporter gene expression. For example, (1–[2–(β-galactopyranosyloxy)propyl]–4,7,10–tris(carboxymethyl)1,4,7,10– tetra-azacyclododecane)gadolinium(III); EgadMe), in which access of water to the first coordination sphere of a chelated paramagnetic gadolinium ion (Gd^{3+}) is blocked by a galactopyranose residue, is such a contrast agent. The galactopyranose cap is released by cleavage catalyzed by → beta-galactosidase, which exposes the Gd^{3+} to water, allowing modulation of water proton relaxation times and increase in magnetic resonance signal intensity. Regions of a cell or an organelle with higher MR image intensity therefore correlate with higher expression of the reporter gene.

Magnetofection: An *in vitro* and *in vivo* technique for the rapid and efficient transfer of any nucleic acid (e.g. a → plasmid vector and its → insert) into target cells by associating superparamagnetic nanoparticles with the DNA (using salt-induced colloid aggregation). In short, the magnetic particle is first coated with a polyelectrolyte (e.g. *polyethyleneimine*, PEI), then mixed with naked DNA in a salt-containing buffer. The DNA binds to or co-aggregates with the particles. Target cells are then incubated with the particle-DNA cocktail and exposed to a magnetic gradient field, that attracts the particles toward the cells. Magnetofection increases the number of transfected cells in comparison to other techniques of → artificial gene transfer.

Magnification: The increase in 18S and 28S → ribosomal RNA genes, that occurs in germ-line cells of rDNA-deficient *Drosophila* flies. The → amplification process probably occurs via unequal sister chromatid exchange (see → unequal crossing-over).

Main band: A broad band of genomic DNA that appears after → isopycnic centrifugation in cesium chloride density gradients in the presence of ethidium bromide. It contains most of the cellular DNA, including → cryptic satellites.

Maintenance gene: Any one of a set of genes, that are turned on early in fetal development and remain active throughout the lifetime of the organism. For example, ATP synthase of mitochondria, elongation factor EF–1 α, histone deacetylase, RNA polymerase II, and ubiquitin-conjugating enzymes are encoded by such maintenance genes.

Major gene: Any → gene, whose contribution to the expression of a particular polygenic trait is superior to the contribution(s) of other → minor gene(s).

Major groove: The indentation on the surface of a DNA → double helix molecule, formed by the sugar phosphate backbones and the edges of the base pairs (linked by → Watson-Crick base pairing forces), that contain the N6, N7, O6 (in → purines) or O4 and N4 atoms (in → pyrimidines). See → double helix, → minor groove.

Major or
wide groove

X: helix axis
The oxygen of the guanine deoxyribose ring lies above, the oxygen of the cytosine deoxyribose ring below the level of the base pair.

MALDI: See → matrix-assisted laser desorption-ionization mass spectrometer.

MALDI-MS: See → matrix-supported laser desorption-ionization mass spectrometry.

MALDI post source decay mass spectrometer (MALDI-PSD-MS): A specially designed mass spectrometer that allows to determine the masses of peptide fragments, generated by ionization of isolated proteins. The mass spectrometer contains a reflector ("reflectron"), that diverges the ions from their normally linear flight, such that their speed is first slowed down and then their direction and speed of flight are changed. After reflection, they reach the detector according to their mass-charge ratio. Since fragmentation occurs at the reflector (i. e. after the acceleration), this type of analysis is called post source decay (PSD) mass spectrometry.
As a result of MALDI-PSD analysis, a spectrum of peptide fragment ions becomes available, that can be compared to the theoretically expected fragment ions of the known proteins, or peptide sequence accessions in appropriate data banks, so that proteins and their post-translational modifications can be identified, using software packages as e. g. SEQUEST. See → tandem mass spectrometer.

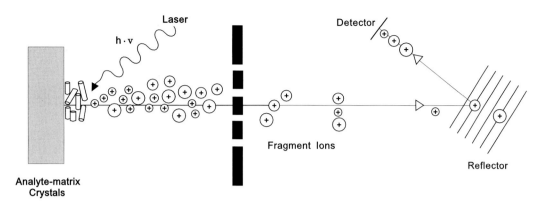

MALDI

MALDI-PSD-MS: See → MALDI *post* source *decay* mass *s*pectrometer.

MALDI-TOF: See → matrix-assisted laser desorption ionisation-time of flight.

Maltose binding protein (MBP): A protein whose gene is used in gene fusion experiments. See → protein fusion and purification technique.

Mammalian artificial chromosome (MAC): A high-capacity → cloning vector for mammalian cells that contains a mammalian → origin of replication, → telomeres, → centromeres, and other sequences necessary for its function in mammals. Since MACs are not integrated into the host cell genome, but nevertheless stably maintained at one copy per cell, they may be used in → gene therapy. See → bacterial artificial chromosome, → human artificial chromosome, → plant artificial chromosome, → P1 cloning vector, → *Schizosaccharomyces pombe* artificial chromosome, → transformation-competent artificial chromosome vector, → yeast artificial chromosome.

Mammalian vector: Any → cloning vector that functions in mammalian cells.

Mammalian-wide interspersed repeat (MIR): Any one of about 120,000 to 300,000 copies of → transfer RNA-derived → short interspersed nuclear elements (SINEs) of the primate genome, that can transpose either in → sense or → antisense orientation, also into genic sequences. The original MIR seems to be a 260 bp SINE, fragments of which are found as 70–100 bp elements. A central core region of about 25 bp is conserved in the MIRs of different mammals.

Mannopine: An amino acid derivative that is synthesized in plant cells transformed by the soil bacterium *Agrobacterium tumefaciens*. Mannopine belongs to the so-called → opines. See also → crown gall.

CH_2OH
|
$(CHOH)_4$
|
CH_2
|
NH
|
$HOOC - CH - (CH_2)_2 - CO - NH_2$

Mant nucleotide: See → N-Methylanthraniloyl nucleotide.

Map: A graphical description of genetically or physically defined positions on a circular (e.g. → plasmid) or linear DNA molecule (e.g. → chromosome) and their relative locations and distances. A map may show the distribution of specific → restriction sites (→ restriction map), genes (→ gene map), markers (→ marker map, chromosome markers (→ chromosome map), or the distance between two loci (e.g. a marker and a gene) in base pairs (→ physical map) or → centiMorgans (→ genetic map). The term is now also used for the illustration of peptide-peptide-, peptide-protein-, and protein-protein interaction networks in a cell or an organelle, and for the intracellular distribution of low-molecular weight cellular compounds (metabolites). See → BAC map, → base pair map, → biallellic genetic map, → bit map, → cDNA map, → cell map, → chromosome expression map, → chromosome features map, → chromosome map, → circular linkage map, → circular restriction map, → Cleveland map, → clone-based map, → contact map, → content map, → contig map, → cytogenetic map, → deletion map, → denaturation map, → diallelic map, → difference map, → diversity map, → DNA map, → doublet frequency map, → epitope map, → EST map, → expression map, → expression imbalance map, → fine-structure map, → frequency distance map, → functional map, → functional map atlas, → gene expression map, → gene expression terrain map, → gene map, → genetic map, → genome control map, → genome fingerprint map, → genome map, → haplotype map, → high density genetic map, → high-density map, → high resolution genetic map, → high resolution physical map, → homology map, → homosequential linkage map, → *in silico* map, → integrated map, → integrated physical-genetic map, → interactome map, → landmark map, → linkage map, → long-range restriction map, → macro-restriction map, → map, → marker map, → metabolic map, → microsatellite map, → nucleotide diversity map, → ordered clone map, → peptide map, → pharmacophore map, → physical map, → protein expression map, → protein interaction map, → protein linkage map, → protein-protein interaction map, → proteome map, → quantitative chromosome map, → radiation hybrid map, → recombinational map, → recombination frequency map, → response regulation map, → restriction map, → RN map, → SAGE map, → segregation map, → self-organizing map, → sequence map, → sequence-tagged sites map, → SNP map, → telomere map, → transcript map, → transcriptome map, → two-dimensional gel map, → ultra-high density map, → YAC map.

mAP: See → *m*essenger *a*ffinity *p*aper.

Map-assisted cloning: See → positional cloning.

Map-based cloning: See → positional cloning.

Map-based sequencing: See → clone-by-clone sequencing.

Map distance: The distance between two genes on a linear DNA molecule, expressed as → map units or centiMorgans (see → Morgan unit).

MAP kinase: See → mitogen-activated protein kinase.

Mapmaker: An interactive computer program for the construction of genetic → linkage maps that allows the estimation of the most likely order of specific genetic loci (e.g. → RFLP), and recombination frequencies between them. Calculations presuppose extensive data on meiotic segregation.

*M*apped *r*estriction *p*olymorphism (MRP): See → mapped restriction site polymorphism.

*M*apped *r*estriction *s*ite *p*olymorphism (MRSP; *m*apped *r*estriction *p*olymorphism, MRP): A variant of the → restriction *f*ragment *l*ength *p*olymorphism (RFLP) technique for the genetic fingerprinting of individual genomes, that is based on the amplification of target sequences (e.g. genes) with → primers complementary to conserved parts of these sequences. In short, genomic DNA is first isolated and primers directed against conserved gene sequences used to amplify these genes in a conventional → polymerase chain reaction in the presence of a ^{32}P-labeled deoxynucleoside triphosphate (e.g. dCTP). Frequently employed primers span conserved regions in e.g. 16S rRNA (*rrs*) and 23S rRNA (*rrl*) genes of eubacteria. After amplification, the products are restricted with → restriction endonucleases that cleave frequently in the target sequence. The restriction fragments are then separated by native → polyacrylamide gel electrophoresis and visualized by → autoradiography. Differences in the electrophoretic mobility of bands represent differences in the distribution of restriction sites along the target gene(s).

Mapping:
a) The plotting of gene positions or other defined sites along a strand of DNA. See also → acetylation mapping, → antigenic mapping, → association mapping, → bottom-up mapping, → cell mapping, → centromere mapping, → chromosome mapping, → clinical mapping, → comparative gene mapping, → comparative mapping, → compositional mapping, → contact mapping, → contig mapping, → cosmid insert restriction mapping, → cross-mapping, → deletion mapping, → denaturation mapping, → domain mapping, → epitope mapping, → exon-intron mapping, → expressed sequence tag mapping, → expressed sequence tag polymorphism mapping, → expression mapping, → fine mapping, → function mapping, → gene mapping, → genetic mapping, → genome mapping, → haplotype mapping, → HAPPY mapping, → heteroduplex mapping, → high density mapping, → high resolution physical mapping, → H-mapping, → homozygosity mapping, → *in silico* mapping, → integrative mapping, → interphase mapping, → intron-exon mapping, → localisome mapping, → long-range restriction mapping, → map, → megabase mapping, → nucleotide analogue interference mapping, → nucleotide mapping, → optical mapping, → pathway mapping, → peptide mapping, → protein expression mapping, → protein-protein interaction mapping, → proteome mapping, → QTL mapping, → radiation hybrid mapping, → receptor mapping, → restriction mapping, → retentate mapping, → saturation mapping, → Smith-Birnstiel mapping, → S1 mapping, → STS content mapping, → telomere mapping, → top-down mapping, → visual mapping. Compare → epitope mapping.
b) MAPPing, see → *m*essage *a*mplification *p*henotyping.

MAPREC: See → *m*utant *a*nalysis by *P*CR and *r*estriction *e*nzyme *c*leavage.

MAPS: See → *m*inisatellite-primed *a*mplification of *p*olymorphic *s*equences.

Map unit: One centiMorgan (cM). See → Morgan unit.

MAR: See → scaffold-associated region.

Mariner: Anyone of a family of animal → transposons (originally detected in insects and related arthropods, but also present in the genomes of other animals, including man).

Marker:
a) A → genetic marker.
b) Any protein, RNA or DNA molecules of known size or molecular weight that serve to cali-brate the electrophoretic or chromatographic separation of proteins, RNAs and DNAs, re-spectively. See → binning marker, → ladder, → molecular weight standard.

Marker-assisted breeding **(MAB):** The use of → molecular markers for the development of new animals and plant varieties, e.g. by → marker-assisted selection. See → marker-assisted introg-ression.

Marker-assisted introgression: A technique to facilitate → introgression of desirable genes into target organisms, that is based on the detection of → molecular markers closely linked to the gene encoding the trait of interest, and the monitoring of their fate in the progeny of sexual crosses. Marker-assisted introgression therefore avoids lengthy evaluation processes (e.g. the continuous monitoring of the phenotype of plants in the field over several years). See → marker-assisted breeding, → marker-assisted selection.

Marker-assisted selection: See → marker-based selection.

Marker-assisted selection (*marker-based selection*, MAS; *marker-mediated selection*, MMS): A technique to select individual organisms (bacteria, fungi, plants, animals) carrying a desirable gene with the aid of genetic → markers linked to the gene. For example, marker-based selection allows to screen for pathogen-resistant plants via closely linked markers without exposing them to the pathogen.

Marker-based selection: See → marker-assisted selection.

Marker bracket: The location of two or more → genetic or → molecular markers in the vicinity of a → gene, so that it is tagged both → upstream and → downstream ("bracketed").

Marker exchange: See → homogenization.

Marker map: Any → genetic or → physical map that is either based on phenotypic (→ mor-phological) or → molecular markers. See also → chromosome expression map, → chromosome map, → cytogenetic map, → denaturation map, → diversity map, → expression map, → fine-structure map, → frequency distance map, → gene map, → genetic map, → integrated map, → landmark map, → linkage map, → macro-restriction map, → map, → nucleotide diversity map, → ordered clone map, → quantitative chromosome map, → recombinational map, → re-combination frequency map, → response regulation map, → restriction map, → RN map, → se-quence map, → SNP map, → ultra-high density map.

Marker-mediated selection **(MMS):** See → marker-assisted selection.

Marker rescue:
a) The survival of gene(s) from an irradiated, inactive bacteriophage, by recombination with an unirradiated active bacteriophage. If a bacterial host is infected with two genetically marked phages (mixed infection) of which only one type is irradiated (and hence inactivated), rare recombination processes occur between both phage types. Thus recombinants can be found that contain genes from the irradiated parent. These are referred to as "rescued".

b) The re-isolation of a → genetic marker from a transgenic host, into which it has been transferred e.g. by → direct gene transfer techniques. Marker rescue allows the detection of marker alterations (e.g. truncations, → deletions, → inversions, generally → rearrangements) which have occurred during its transfer to the host and/or its integration into the host's genome.

MAS:
a) See → *mar*ker-based *s*election.
b) See → *m*askless *a*rray *s*ynthesizer.

MASA: See → *m*utant *a*llele-specific *a*mplification.

Masked *m*essenger *RNA* (masked mRNA): An inactive, stable and longlived → messenger RNA that has to be unmasked before its translation. Such masked messages occur in such diverse systems as unicellular algae (e. g. *Acetabularia*), angiosperm seeds, and echinoderm oocytes. Masking is brought about by RNA-binding proteins ("mRNA masking proteins"), that probably need phosphorylation for their activity. Activation of masked messenger RNAs is catalyzed by → polyadenylation.

Masked mRNA: See → masked messenger RNA.

*M*askless *a*rray *s*ynthesizer (MAS): A computerized instrument for the light-directed synthesis of high-resolution oligonucleotide microarrays, using a digital micromirror array generated on a computer to form virtual masks, instead of the conventional chrome/glass photolithographic masks. In short, microscope slides are first silanized. Then the photolabile protecting group (R, S)-1-(3,4-(*m*ethylene-dioxy)-6–*n*itro*p*henyl) ethyl *c*hloroformate (MeNPOC) is attached to the nucleotides and *h*exaethyleneglycol (HEG) as a spacer molecule. The photoprotected HEG is converted to a phosphoramidite, which in turn is covalently bound to the silanized slide. This procedure produces a microscope slide covered with a monolayer of spacer molecules containing hydroxyl groups protected by photolabile MeNPOC groups. These protective groups are conventionally removed by UV-light revealing all free hydroxyl groups. This deprotection does not occur at random, if MAS is used. Instead, a high-resolution pattern of UV light produced with the MAS and projected onto the microscope slide reproduces an identical pattern of free hydroxyl groups on its surface. Coupling of nucleotides then occurs at the free hydroxyl groups. Such MAS-produced microarrays accomodate about 100,000 oligonucleotides on spaces of 16 μm^2 with the potential to discriminate single-base → mismatches in thousands of genes simultaneously. Do not confuse with → marker-based selection.

*M*assively *p*arallel *s*ignature *s*equencing (MPSS): A high-throughput technique for the sequencing of millions of → cDNAs conjugated to oligonucleotide tags on the surface of 5 µm diameter microbeads, that avoids separate cDNA isolation, template processing and robotic procedures. In short, 32mer capture oligonucleotides are attached to the surface of separate microbeads (diameter: 5µm) by combinatorial synthesis, such that each microbead has a unique tag for its complementary cDNA. Then → messenger RNA is reverse transcribed into → cDNA using oligo(dT) primers, restricted at both ends with e.g. *Dpn* I, complements of the capture oligonucleotides are attached to the poly(A) tail of each cDNA molecule and the construct cloned into an appropriate vector containing PCR handles, which serve as primer-binding sites for → polymerase chain reaction based amplification of the tagged cDNA. The cDNA is now amplified with a → fluorochrome-labeled primer, denatured, and the single-stranded address tag-containing fragments annealed ("cloned") to the surface of microbeads containing address tag sequences as hybridization anchors, and then ligated ("in vitro cloning"). Each microbead displays about 100,000 identical copies of a particular cDNA ("microbead library"). The fluorescent microbeads (all containing a cDNA) are then separated from non-fluorescent ones (not containing a cDNA) by a *f*luorescence-*a*ctivated *c*ell *s*orter (FACS). Each single microbead in the library harbors multiple copies of a cDNA derived from different mRNA molecules. If a particular mRNA is highly abundant in the original sample, its sequence is represented on a large number of microbeads, and vice versa. In the original version of MPSS, 16–20 bases at the free ends of the cloned templates on each microbead are sequenced ("signature sequences"). First, millions of template-containing microbeads are assembled in a densely packed planar array at the bottom of a flow cell such that they remain fixed as sequencing reagents are pumped through the cell, and their fluorescence can be monitored by imaging. Then the fluorophore at the end of the cDNA is removed, and the sequence at the end of the cDNA determined in repetitive cycles of ligation of a short → adaptor carrying a restriction recognition site for a class IIS → restriction endonuclease (binding within the adaptor and cutting the cDNA remotely, producing a four nucleotide overhang; e.g. *Bbv*I). Next, a collection of 1,024 specially encoded adaptors are ligated to the overhangs, and the coded tails interrogated by the successive hybridization of 16 different fluorescent decoder oligonucleotides. This process is repeated several times to determine the signature of the cDNA on the surface of each bead in the flow cell. The abundance of each mRNA in the original sample is estimated by counting the number of clones with identical signatures. Compare → serial analysis of gene expression.

Massively parallel single molecule sequencing: A technique for the parallel sequencing of hundred thousands of DNA, oligonucleotide, cDNA, messenger RNA or genomic DNA molecules spotted on a → DNA chip. In short, the probe molecules are first immobilized on a chip surface optimised for single molecule detection in a distance of about 400 nm from each other. After → priming, a DNA polymerase starts sequencing reactions at all spotted DNAs simultaneously, using a selected "temporarily terminating" and fluorescently labeled nucleotide (structure not disclosed), which leads to a reversible chain termination. This socalled "pausing" of the polymerase allows to detect all incorporated bases on the complete chip surface with a fluorescence microscope. Then the modification (not disclosed) of the incorporated base is removed with a proprietary technology, which leads to the liberation of the 3'-ends of the DNA molecules and allows continuation of the process with another labeled nucleotide. Once a sequence has been determined over 15–35 nucleotides, it can be compared to entries in the databanks.

MAST: See → *m*agnet-*a*ssisted *s*ubstraction *t*echnique.

Master circle: The idealized → restriction map of the → mitochondrial DNA of a plant cell. Since a single cell, and even a single mitochondrium contain mtDNAs of different size, composition and gene order, it is impossible to isolate **the** mitochondrial genome per se. Instead, the total mtDNA of a plant species is restricted and the restriction fragments arranged in the socalled master circle.

Master gene: Any → gene that controls one or more other genes.

Master gene: See → source gene.

Master mix: A laboratory slang term for a pre-mixed solution consisting of a suitable buffer, optimised Mg^{2+} concentrations, all four dNTPs and a heat-stable DNA polymerase (e.g. → *Taq* DNA polymerase). This master mix is usually made for several (mostly 10, or 100) → polymerase chain reactions, is therefore first aliquoted, and each aliquot pipetted once into a reaction tube. Then → primers and → template DNA are added to start the reaction. Variants of the master mix exist, but in each case such mixes avoid multiple pipetting steps and multiple pipetting errors.

MAT: See → mating type.

Matching gene: Any host gene, that possesses a pathogen gene counterpart (and *vice versa*) in a gene-for-gene interaction. For example, socalled virulence gene(s) of a pathogenic fungus encode peptides or proteins, that produce a usually low molecular weight substances. These elicitors are recognized by a receptor protein anchored in the host cell membrane or located in the cytoplasm, and as a consequence of the interaction, a signalling cascade is incited leading to the activation of host genes and host defense reactions. The particular gene in the fungal genome "matches" the corresponding receptor gene in the host's genome.

Mate pair: See → paired-end sequence.

Maternal inheritance: See → cytoplasmic inheritance.

Maternal messenger RNA (maternal messenger, maternal mRNA): Any mRNA that is transcribed from the maternal genome during the oogenesis of animals. Maternal mRNA may be deposited in the oocyte and is needed for early embryogenesis. See → maternal effect genes.

Maternal mRNA: See → maternal messenger RNA.

Mates: A pair of DNA sequence reads with overlapping end sequences, that are randomly sampled from a genomic library and assembled with a special computer program. Mates are critical for → whole genome shotgun sequencing.

Mating: See → conjugation.

Mating type: Any one of two different cell types of → yeast (*Saccharomyces cerevisiae*) that allows → conjugation with the respective other type. In short, meiosis in yeast leads to the production of four haploid cells from an original diploid mother cell. This tetrad remains in a sac (ascus), formed by the cell wall of the mother cell, and is composed of two α and two *a* cells. Conjugation can only occur between an α and an *a* cell, never between cells of the same mating type, and leads to a diploid cell. A single gene locus (MAT, *ma*ting *ty*pe) regulates the formation of the cell type. Its allele α is necessary for the generation of the α cell, its allele *a* for the *a* cell type.

*M*atrix-*a*ssisted *l*aser *d*esorption-*i*onization mass spectrometer (MALDI; gene balance): An instrument that allows to determine the mass of a gene (generally, a DNA sequence). Basically, the gene balance is a mass spectrometer (*m*atrix-*a*ssisted *l*aser *d*esorption-*i*onization mass spectrometer, MALDI). The DNA sample is first embedded into a matrix, which is evaporated by a short laser pulse. This releases the DNA molecules into the gas phase, where they are ionized by collision with matrix molecules. These ionized molecules are accelerated into a field-free channel. DNA molecules with differing base sequences (i. e. different masses) reach a detector at different times, which allow to calculate their precise masses. The gene balance can be used to e. g. determine the different masses of → alleles.

Matrix-*assisted laser desorption ionisation- time of flight* **(MALDI-TOF):** See → matrix-supported laser desorption-ionization mass spectrometry.

Matrix attachment region: See → scaffold-associated region.

Matrix comparative genomic hybridisation (matrix CGH, array CGH, CGH chip): A variant of the conventional → comparative genomic hybridization technique, that works with genomic fragments in the range from 30–200 kb rather than metaphase chromosome spreads. These fragments (in the form of → bacterial artificial chromosome [BAC] clones) are spotted onto chip supports (e.g. surface-modified glass or plastics), and the resulting chips hybridised to differentially labeled → cDNAs from two sources (e.g. normal versus tumorous tissues). Subsequently a laser scanner detects the fluorescence signals on the chip, and the resulting data are accessible for analysis. Matrix CGH increases the relatively poor resolution of conventional CGH (with chromosomes) by several orders of magnitude. The technique can also be combined with → cDNA microarray experiments, so that genes expressed under certain conditions (e.g. in tumors) can first be selected and then spotted onto the matrix CGH array. For example, the combination of both techniques led to the molecular differentiation of two subtypes of liposarkomas (differentiated liposarkomas, DLs, versus polymorphic liposarkomas, PLs). Matrix CGH arrays covering the whole human genome, consisting of about 30,000 BAC clones, are used to detect small mutated regions of the genome.

Matrix-*supported laser desorption-ionization* **(MALDI) *mass* spectrometry (MALDI-MS; MALDI-TOF-MS):** A technique for the production and mass analysis of intact gas phase ions from a wide variety of biomolecules (e. g. peptides, proteins, oligonucleotides, carbohydrates, or glycolipids, to name few). The various analytes are prepared for MALDI-MS analysis by dissolving them in a solution containing a matrix compound, that absorbs at the wave-length of the employed laser light (UV-laser, $\lambda = 337$ nm, for example). The matrix compounds are either cinnamic or benzoic acid derivatives (e. g. 2,5-dihydroxybenzoic acid), that additionally function to individualize the analyte molecules. The solvent is then evaporated and the resulting analyte crystals irradiated with a short pulse of laser light to destroy the crystal structure, to desorb and ionize the analyte molecules, thereby creating a burst of ions. These ions in the particle cloud are then accelerated in the electric field of the mass spectrometer (voltage: 20-30 kV) and directed towards a detector. The time of flight (TOF) of the ions from the original location to the detector is measured and transformed into ion masses. Advanced variants of MALDI-MS work with a socalled delayed extraction (DE): the acceleration tension does not act on the ionized particles at the time of ionization, but a few hundreds of nanoseconds later. This delay allows the ions to move into the acceleration channel, driven by the surplus energy of ionization. Therefore the ions are no more fully accelerated by the separation tension, which altogether leads to an improved resolution. Another improvement is the use of ion reflectors in → MALDI post source decay mass spectrometry.

The raw data are collected, processed and analyzed. Usually the range of mass resulution is not unlimited, because the kinetic energy of the laser-produced ions is too widely distributed. MALDI-MS is increasingly being used in → proteome research, allowing the analysis of e. g. 100-200 kDa proteins and the determination of the molecular weights of the resulting peptide fragments in the fmol range (with an accuracy of few ppm).

For peptide analysis, the solubilized protein is pipetted onto a carrier, whose surface is densely packed with either one (e. g. trypsin) or several immobilized proteases (e. g. trypsin, α-chymotrypsin and V8 protease). The fixation of these proteases prevents their autolysis, but allows the digestion of the protein analytes into a series of peptides. After limited proteolysis, the reaction is terminated by the addition of a socalled acidic matrix solution, dried at room temperature, and laser irradiation started. Basically the same technique can be applied to DNA analysis (with immobilized phosphodiesterases) or oligosaccharides (immobilized exoglycosidases).

Mass spectrometry therefore replaces the whole repertoire of traditional fragmentation of the analyte molecule (by e. g. → restriction) and the gel electrophoretic separation of the fragments. See also → electrospray ionization mass spectrometry, → electrospray ionization time-of-flight, → parent-ion-scan technique, → tandem mass spectrometer.

Maturation:
a) of proteins, see → post-translational modification.
b) of RNA, see → post-transcriptional modification and → RNA editing.

Mature protein (ligated protein, spliced protein): The product of → protein splicing. A mature protein consists of → exteins, combined by peptide bond formation after the → cleavage of extein-intein junctions in the → precursor protein and the joining of the free exteins.

Mature RNA: Any RNA, that underwent one or several → post-transcriptional modification(s). For example, → pre-messenger RNA is first synthesized, then trimmed by → splicing (removal of → introns and joining of → exons), → polyadenylated at its 3'-end and → capped at its 5'-end. Only after these modification is the then mature → messenger RNA transported to the cytoplasm and translated on cytoplasmic ribosomes. See → RNA precursor.

Maxam-Gilbert sequencing: See → chemical sequencing.

Maxi-cells: *E. coli* or *B. subtilis* cells (*recA*, *uvrA*) irradiated with UV light, which leads to an extensive degradation of the chromosomal DNA and to cessation of chromosomal DNA synthesis. Plasmids contained in these cells are not damaged by UV light and therefore continue to replicate and to express their genes. Thus maxi-cells can be used to study cloned genes in a system without appreciable chromosomal background (in vivo transcription-translation system).

Maxizyme: A short allosterically regulatable synthetic → ribozyme, consisting of one molecule that binds to the substrate region, and another that cleaves the substrate RNA at the sequence NUX (where N=any nucleotide; X=A,C or U). The name derives from *m*inimized, *a*ctive, *x*-shaped [functions as dimer], *i*ntelligent [allosterically regulatable] ribo*zyme*.

Mb: See → *megabase.*

ᵐC (mC): Abbreviation for cytosine carrying a methyl group (e. g. at C5).

Mc: See → mis-cleavage.

MCAC: See → immobilized metal affinity chromatography.

Mcm: See → *m*ini*c*hromosome *m*aintenance.

Mcr system: See → *m*odified *c*ytosine *r*estriction system.

MCS: See → *m*ultiple *c*loning *s*ite.

MDA: See → multiple displacement amplification.

MDE: See → *m*utation *d*etection *e*lectrophoresis gel.

M-DNA (metal DNA): A complex of double-stranded DNA and divalent ions (e.g. Zn2+, Co2+, Ni2+) formed above pH 8.0, in which the imino proton of each base in the duplex is substituted by a metal ion. Therefore, the DNA is coated with metal ions, and consequently possesses special conductive properties not owned by normal DNA (for example, an electron transfer can proceed along the molecule). M-DNA is also called a molecular wire ("nanowire"). See → A-DNA, → B-DNA, → C-DNA, → D-DNA, → E-DNA, → ε-DNA, → G-DNA, → G4–DNA, → H-DNA, → P-DNA, → V-DNA, → Z-DNA. See → DNA wire.

MEA: See → microelectronic array.

MeCP: See → *m*ethyl-*C*p*G*-binding *p*rotein.

Mediator (adaptor): A multi-protein complex of about 20 largely conserved proteins necessary for the transcriptional activation in a fully reconstituted → RNA polymerase II transcription system. The mediator proteins of yeast fall into three broad categories: Sin 4/Rgr 1 proteins (function in repression as well as activation), Srb (*s*uppressor of *R*NA polymerase *B*), and Med proteins. The complex itself interacts with the non-phosphorylated → *c*arboxy-*t*erminal *d*omain (CTD) of the large subunit of → RNA polymerase II (B) to form a 1.5 Md holoenzyme. During this interaction the mediator unfolds and envelops the globular polymerase molecule. Mediator complexes act as interface between → activators (definition a) and the enzyme (i.e. between → enhancer sequences and → promoters).

Medical genomics: The detection, isolation and characterization of genes and the encoded proteins with medical relevance. See → behavioral genomics, → comparative genomics, → environmental genomics, → epigenomics, → functional genomics, → genomics, → horizontal genomics, → integrative genomics, → nutritional genomics, → pharmacogenomics, → phylogenomics, → proteomics, → recognomics, → structural genomics, → transcriptomics, → transposomics. Compare → clinical mapping.

Medium copy plasmid: Any → plasmid, that is present in 40 to 60 copies per bacterial host cell. For example, → pBR322 is such a medium copy plasmid.

Medium density chip: A laboratory slang term for a → DNA chip, onto which from 100 –10,000 → probes are spotted. Compare → high density chip, → low density chip.

Medium overlap: The number of bases matched between two clones (e.g. → bacterial artificial chromosome clones), that are not matched using the strictest criteria, but are matched using less strict criteria. See → strong overlap, → total overlap, → weak overlap.

Megabase (Mb): One million nucleotides or nucleotide pairs; 1000 kb. See → base pair. See also → megabase mapping, → megabase marker.

Megabase cloning: A technique for the → cloning of extremely large fragments of DNA (in the range from one to several megabases) into suitable vectors.

Megabase mapping: The establishment of a linear → gene map using markers that are separated from each other by one million bases (a megabase).

Megabase marker: A series of DNA fragments that range in size from about 50 to more than 1000 kb, used in → pulsed-field gel electrophoresis as size markers for the estimation of the molecular weight of large DNA molecules. For this purpose, → lambda phage → concatemers can be used that cover a molecular weight range of 48.5 kb to 1.2 Mb.

Megadalton (Md): Equivalent to 10^6 → daltons.

Megagene: Any unusually large gene whose length exceeds 10-20 kb (e. g. the X-linked Duchenne muscular dystrophy [DMD] gene with about 1000 kb, or the dystrophin gene with a total length of 2300 kb and 100 introns).

Megalinker (megalinker I-*Sce*I): The oligodeoxynucleotide → linker 5'-GATCCGC*TAGGGA-TAACAGGGTAAT*ATA-3' that contains a unique → meganuclease I-*Sce*I site. This linker permits the insertion of the unique I-*Sce*I recognition sequence into any *Bam* HI site of a → cloning vector or, generally, target DNA.

Megalinker I-*Sce* I: See → megalinker.

Meganuclease (meganuclease I-*Sce*I; omega nuclease, omega transposase): An → endonuclease encoded by a mobile group I → intron of yeast mitochondria that catalyzes the cleavage of the 18 base pair recognition sequence 5'-TAGGGATAA/CAGGGTAAT-3' to generate 3' → cohesive ends. Since an 18 bp recognition sequence will statistically occur only once in $6.9 \cdot 10^{10}$ bp of genomic DNA, the enzyme represents an extreme → rare cutter that can be used for the cloning and mapping of artificially inserted sequences in pro- and eukaryotic genomes, the specific fragmentation of whole chromosomes, and the mapping of large DNA fragments in genome analysis.

Meganuclease I -*Sce*I: See → meganuclease.

Megaplasmid: An imprecise term for any → plasmid whose size exceeds 200 kb.

Megaprimer method: See → megaprimer PCR mutagenesis.

Megaprimer mutagenesis: See → megaprimer PCR mutagenesis.

Megaprimer PCR mutagenesis (megaprimer mutagenesis, megaprimer method): A technique for the introduction of site-specific mutations (i. e. single base exchanges) into a target DNA. In short, the template is first amplified in a conventional → polymerase *chain reaction* (PCR) by using a flanking → primer (A or B) and an internal mutagenesis primer (M1 or M2). M1 as well as M2 should be designed such that the mutational mismatch is about 10 – 15 bases away from the 3' terminus to allow for normal amplification. Also, a mixture of two thermostable DNA polymerases is employed, → *Taq* DNA polymerase without proofreading, but an → extendase activity, and → *Pfu* DNA polymerase with a 3'→5' exonuclease function. This combination of enzymes reduces the extendase activity of *Taq* polymerase, which would otherwise lead to undesirable additional mutations in the final product. The amplification leads to a socalled megaprimer that contains the desired mismatch mutation. After electrophoretic purification and extraction of this megaprimer from the gel matrix, a second PCR with the flanking primers A and B and the product of the previous PCR (A-M1, or B-M2) are used to introduce the mutation into the target gene.

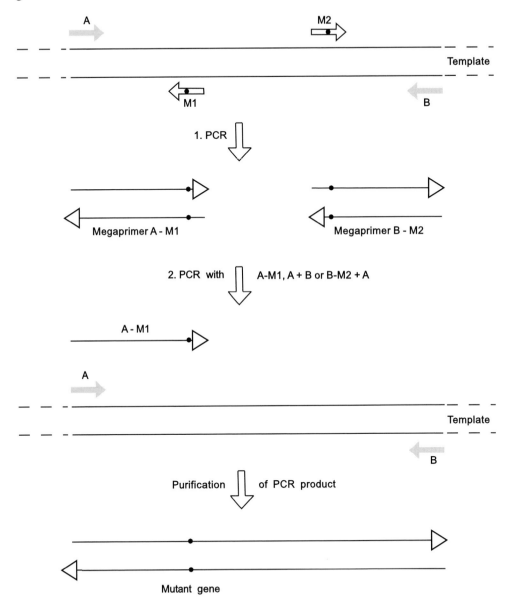

Megaprimer PCR mutagenesis

Megasatellite (megasatellite repeat): Any → satellite DNA with a repeat size of more than 1kb, that is tandemly arranged with other repeats of the same size to form large domains at specific sites in mammalian chromosomes. For example, a 4.7kb human megasatellite repeat, which is arranged with other repeats in head-to-tail tandem clusters of 50–70 copies per haploid genome, even contains a → promoter and an → open reading frame encoding a deubiquitinating enzyme. See → macrosatellite, → microsatellite, → minisatellite, → satellite.

Megasequencing (megabase sequencing): The determination of the primary sequence of DNA fragments (see → DNA sequencing) of at least 1 Mb (1 million bases). Compare → kilosequencing.

Mega-Yac: See → mega-yeast artificial chromosome.

Mega-yeast artificial chromosome: Any → Yac clone that contains an insert of more than one million base pairs.

Meiotic drive: The preferential transmission of a particular → allele (or chromosome) of a heterozygous pair to the progeny, occuring in natural populations of fungi, plants, insects, and mammals. Meiotic drive does not conform with classical Mendelian genetics. See → segregation distorter.

MEK kinase (MAP kinase kinase kinase, MKKK, MAP3K, or MEKK): Any one of a family of kinases, that phosphorylate and thereby activate MAP kinase kinases or MKKs). The phosphorylated MAP kinase kinases in turn phosphorylate and activate MAP kinases. For example, MEK kinase–1, MEK kinase–3, MEK kinase–4, MEK kinase–5 (also called ASK 1) Raf–1, Raf.B, and Mos are such MEK kinases.

MELK: See → multi-epitope ligand cartography.

Melting (DNA melting, RNA melting): The dissociation of the complementary strands of double stranded DNA or RNA, as well as of DNA-RNA heteroduplex molecules to form single strands. In the laboratory melting is usually achieved by heating, while in vivo various nucleic acid binding proteins catalyze strand-separation in e.g. DNA → replication, or RNA → translation. See also → melting curve, → melting temperature, and compare → C$_0$t analysis. Also → denaturation, → denatured DNA, → G + C content.

Melting curve (DNA melting curve): The graphical display of the dissociation of strands in a DNA duplex molecule to form single strands as a function of temperature. Compare → C$_0$t curve, see → melting temperature.

Melting point: See → melting temperature.

Melting protein: See → DNA topoisomerase I, also → helix-destabilizing protein.

Melting temperature (T$_m$, t$_m$, t$_{1/2m}$; melting point): The temperature at which fifty percent of existing DNA duplex molecules are dissociated into single strands. For measurement of T$_m$, a DNA solution is heated and its absorbance at 260 nm is continuously monitored. Transition from double- to single-stranded DNA occurs over a narrow temperature range and shows a characteristic increase in absorbance at 260 nm, so that a sigmoidal (S-shaped) curve results. T$_m$ is defined as the temperature at the midpoint of the absorbance increase, that is, the temperature at which fifty percent of the molecule(s) are dissociated.

Melting temperature (T_m) calculation:

a) Simplified calculation: $T_m = [2°C \text{ x } (\#A + \#T)] + [4°C \text{ x })\#G + \#C)]$
For example, the melting temperature of the 10mer oligonucleotide ACG TAC GTA C is: $[2°C \times (3 + 2)] + [4°C \times (2+3)] = 30°C$

b) Alternative calculation:
$T_m = 81.5°C - 16.6 + [41 \text{ x } (\#G + \#C)]/\text{oligonucleotide length} - (500/\text{oligo length})$
For example, the melting temperature of the 10mer ACG TAC GTA C is: $81.5°C - 16.6 + [41 \times (5)]/10 - (500/(10) = 35.4°C$

meltMADGE: See → programmable melting display microplate-array diagonal gel electrophoresis.

Membrane-based two-hybrid system: See → split ubiquitin two hybrid system.

Membrane microarray:
a) Any → microarray, that contains target sequences (e.g. → cDNAs, → oligonucleotides, peptides, proteins) on a membrane support (e.g. nylon or→ nitrocellulose). Other microarrays are made of glass or quartz (also plastic) supports.
b) A misleading laboratory slang term for a → microarray, onto which a series of membrane-bound proteins are spotted.

Membrane slide: A microscope slide carrying a thin layer of a microporous polymer with high capacity to bind either DNA or oligonucleotides, RNA or proteins (e.g. antibodies). Such membrane slides accomodate up to several thousand spotted probes, that are e.g. crosslinked by UV, and serve as → microarrays (which can even be produced manually with the aid of an appropriate → arrayer [" MicroCaster™"]).

***Membrane-translocating sequence* (MTS):** A short (e.g. 12 amino acids long) hydrophobic peptide sequence at the C-terminus of proteins, that mediates their translocation across the cellular membrane. Such MTSs are used to deliver cargo proteins into cells for functional tests. For example, an MTS from the h region of the signal sequence of the Kaposi fibroblast growth factor, if fused to the C-terminus of reporter proteins, efficiently imported these proteins into fibroblasts and also other cells.

Memory suppressor gene (long-term memory suppressor gene): Any one of a series of genes encoding proteins, that function to inhibit memory formation and long-term memory storage by e. g. decreasing synaptic strength and forcing neurons to learn only salient features. For example, long-term memory is at least partly a consequence of synaptic plasticity (i.e. the ability of neurons to alter the strength of their synaptic connections with prolonged activity and experience), which is controlled by a series of protein kinase signalling cascades and positive regulators of transcription as e.g. *cyclic adenosine monophosphate* (cAMP) *response element binding protein 1* (CREB1) and C/EBP. Activation of these positive regulators is essential for the consolidation of short-term memory into long-term memory. However, the removal of negative, inhibitory elements is equally important. In *Aplysia*, for example, the cAMP-dependent *protein kinase A* (PKA) pathway, mediated by CREB, stimulates the growth of new synaptic connections between sensory and motor neurons of the gill-withdrawal reflex after repeated exposure to serotonin (or behavioral training). Now, CREB2 is a repressor of these morphological and also functional changes, because an anti-CREB2 injection replaces serotonin functionally. The gene encoding this CREB protein (in *Aplysia*, ApCREB2) belongs to the family of memory suppressor genes. Compare → tumor suppressor gene.

MEMS: See → micro-electromechanical system sequencing machine.

MEPS: See → *minimum efficient processing segment*.

Mercaptoethanol: See → β-mercaptoethanol.

Mercaptopurine (6-mercaptopurine): A purine analogue that blocks the conversion of → inosine to → adenine (as well as the biosynthesis of 4-aminoimidazol-5-carboxamide ribotide).

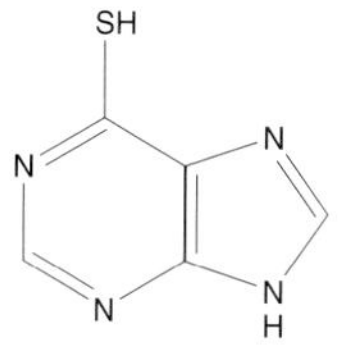

6-Mercaptopurine

Merging genes: A misleading term for the combination of two gene names into one, after experimental evidence (e.g. the isolation of a full-length cDNA) proved, that the two genes are representing only one single locus. The new name corresponds to the locus with the majority of sequences, the abandoned name is kept associated to the locus of the merged gene. See → splitting genes.

Message amplification phenotyping (**MAPPing**): A rapid and sensitive technique to analyze multiple mRNAs present in a single cell or a small population of cells simultaneously. In short, mRNA is isolated by a guanidinium thiocyanate/cesium chloride microscale procedure, reverse transcribed into → cDNA, primed with amplification → primers (amplimers) specific for the target messages (which may for example be derived from sequence data), and amplified in the → polymerase chain reaction (PCR). The figure shows MAPPing results obtained with cytokine primers.
Figure see page 642.

Messenger affinity paper (**mAP**): A diazo-thiophenyl paper to which poly(U) chains of more than 100 nucleotides in length are covalently bound. This paper is used to isolate polyadenylated mRNA that binds to poly(U) via hydrogen bonds (→ base-pairing).

Messenger ribonucleoprotein (**mRNP**): The fully processed → messenger RNA molecule, complexed with a series of proteins, representing the transport form of mRNA. mRNPs may also be associated with → translational control RNA.

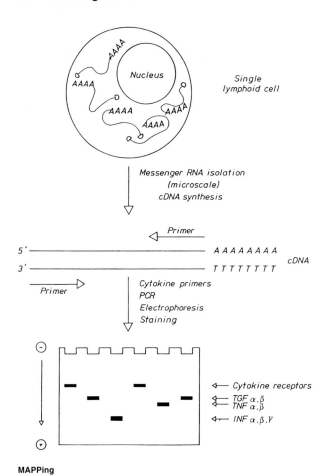

MAPPing

Messenger RNA **(mRNA):** A single-stranded RNA molecule synthesized by → RNA polymerase (RNA polymerase II or B in eukaryotic organisms) from a protein-encoding gene template (→ structural gene) or several adjacent genes (→ polycistronic mRNA). An mRNA specifies the sequence of amino acids in a protein during the process of → translation.

Messenger RNA circularisation **(mRNA circularisation, transcript circularization):** The interaction between the 3'-end (see → trailer) and the 5'-end (see → leader sequence) of a eukaryotic → messenger RNA (mRNA) molecule, mediated by protein-protein interactions, that lead to the formation of a loop structure ("closed loop", "circular structure"). For example, the *poly(A)–binding protein* (PABP), once bound to the → poly(A)-tail of a particular mRNA, contacts → translation initiation factor eIF4G, which in turn interacts with the → cap-binding protein eIF4E, thereby effectively and physically circularising the mRNA in a head-to-tail loop. Transcript circularisation increases translational efficiency, which can be compromised by the intervention of a protein (or proteins) bound at the trailer.

Messenger RNA display **(mRNA display, *in vitro* virus, mRNA-protein fusion):** A technique for the *in vitro* discovery and → directed molecular evolution of new peptides and proteins from combinatorial libraries, in which the → messenger RNA molecules are covalently attaches to the peptide or protein they encode. In short, a synthetic oligonucleotide containing → puromycin at its 3'-end is first enzymatically ligated or photochemically attached to the *in vitro* transcribed messenger RNA. This mRNA is then *in vitro* translated by e.g. a → rabbit reticulocyte lysate. The ribosome reads the message in the 5'→ 3'direction, and puromycin as a chemically stable, small mimic of aminoacyl → transfer RNA binds to the ribosomal A-site and attaches the mRNA to the C-terminus of the nascent peptide. The resulting covalently linked mRNA-peptide complex is isolated, reverse-transcribed and used for *in vitro* selection experiments. After binding to a target molecule (e.g. a drug), the fused molecule complex is eluted and the mRNA recovered by RT-PCR. Therefore, phenotype and genotype are elegantly linked. See → phage display, → ribosome display.

Messenger RNA expression array: See → cDNA expression array.

Messenger RNA-interfering complementary RNA: See → countertranscript.

Messenger RNA isoform: Any one of a series of → messenger RNAs, that all originate from one single gene, but differ in the combination of their exons. Isoforms are generated by → alternative splicing.

Messenger RNA **profiling (mRNA profiling):** The simultaneous detection of thousands of → messenger RNAs (indicative for the transcription of thousands of genes) upon developmental, physiological, environmentally influenced or pathological processes. Profiling can be achieved by → cDNA expression arrays, → massively parallel signature sequencing, or → serial analysis of gene expression, to name only few techniques.

Messenger RNA **scanning (mRNA scanning):** The movement of a → ribosome along a → messenger RNA, bound to the ribosome by its methylated → cap, until the → initiation codon 5'-AUG-3' is reached, where → translation starts.

Messenger RNA translation state: The number of → messenger RNAs in a given cell at a given time, that are actually translated into their cognate proteins. Since not all mRNA transcripts are also translated, and since proteins, not mRNAs determine the → phenotype of a cell, estimation of the messenger RNA translation state informs about the protein potential of the cell, and can be measured by → translation state array analysis.

MeST: See→ methylated sequence tag.

Metabolic engineering: The use of → genetic engineering technology to transfer, stably integrate, and express foreign genes in a host organism to shift a metabolic pathway towards overproduction of its products, or to rechannel metabolites of a pathway into another one.

Metabolic fingerprinting: See → metabolic profiling.

Metabolic map: The graphical depiction of (preferably) all metabolites of a cell, showing their quantitative relationships among each other at a specific point of time.

Metabolic phenomics: Another vague term of the → omics era, describing the analysis, interpretation and prediction of genotype-phenotype relationships from genomic data. See → phenome, → phenomic fingerprint.

Metabolic profiling (chemical profiling, metabolic fingerprinting, metabolite profiling):
a) The isolation of (preferably) all metabolites of a cell, their separation (by e.g. liquid chromatography, capillary electrophoresis, gas chromatography, or → matrix-assisted laser desorption/ ionization) and identification (by e.g. matching of the mass of each compound to reference masses or using internal standards) to establish a metabolic map (an inventory of all metabolites of a cell at a given time), or the cataloguing of up- or down-regulated compounds as a result of intrinsic or environmental stimuli. Metabolic profiling allows to monitor entire pathways simultaneously.
b) In a more specific sense, metabolic fingerprinting encircles the identification of a sample on the basis of the profile (i.e. the pattern and concentration) of a selected series of metabolites, that are indicative for specific metabolic pathways.

Metabolome: The complete set of low molecular weight compounds (metabolites) in a given cell and its organelles at a given time. The thousands of metabolites (*E. coli*: about 1,200) are extracted, separated by e.g. two-dimensional thin layer chromatography, and identified by various detection techniques. Conveniently the target cells are fed with ^{14}C-labeled precursors (e.g. ^{14}C-glucose), and the newly synthesized compounds extracted, separated, and detected by → phosporimaging. The result is termed a "metabolite profile". Compare → genome, → proteome, → transcriptome.

Metabolomics: The whole repertoire of techniques to study the → metabolome, the complete set of metabolites of a cell. The competing term "metabonomics" is virtually identical to metabolomics.

Metabolon: A series of tightly connected protein complexes (many of the proteins being enzymes), that catalyse the highly coordinated and cooperative processing of a substrate to a product (in some cases, an endproduct).

Metabonomics: The technologies to monitor changes of the → metabolome in response to stress.

Metagenome: The entirety of the nucleic acid material in a soil, water, or rock sample, resembling the genomes of an extremely complex mixture of a natural, mostly bacterial community. See → environmental genomics, → trash sequencing.

Metagenomic DNA: The total DNA isolated from a → metagenome.

Metagenomics: The analysis of the genomes of whole living communities in water, soil or rocks. See → environmental genetics, → metagenome.

Metal-chelate affinity chromatography: See → immobilized metal affinity chromatography.

Metal DNA: See → M-DNA.

Metallothionein: Any one of a series of highly conserved, low molecular weight, cysteine-rich proteins, that bind heavy metals such as cadmium, zinc, copper, mercury, and others. See → metallothionein gene.

Metallothionein gene (MT gene): Any member of a small gene family that codes for the synthesis of → metallothioneins, cysteine-rich proteins with the potential to bind heavy metals (e.g. zinc).

The promoter regions of these genes contain a highly conserved → consensus sequence of 15 bp (→ metal regulatory element), which causes activation of the adjacent genes in the presence of heavy metals. In the mouse, MT genes are selectively amplified in the presence of heavy metals. See also → heavy metal resistance.

Metal regulatory element (metal responsive element, MRE): A short (15 bp) sequence element in the → promoter region of → metallothionein genes that specifies → heavy metal resistance in animal and human cells. It is highly conserved (consensus sequence 5'-CTNTGCPuC-PyCGGCCC-3') and occurs in multiple copies in a metallothionein gene promoter. The insertion of synthetic MREs into heterologous promoters (e.g. the HSV thymidylate kinase promoter) renders the adjacent gene inducible by heavy metals. See also → heavy metal resistance gene promoter.

Metal responsive element: See → metal regulatory element.

Methidium: An intercalating dye (see → intercalating agent), used for → DNA capture procedures.

Methotrexate (Mtx, amethopterin, 4-amino-10-methylfolic acid): An analogue of dihydrofolate that inhibits → dihydrofolate reductase and consequently purine synthesis. See also → methotrexate resistance.

Methotrexate^r: See → methotrexate resistance.

Methotrexate resistance (*methotrexate^r*, Mtx^r): The ability of an organism to grow in the presence of the dihydrofolate analogue → methotrexate. The drug inhibits → dihydrofolate reductase (DHFR) and consequently purine biosynthesis. Resistance against methotrexate is usually based on the → amplification of the DHFR gene (→ gene dosage effect) but may also be a consequence of DHFR gene mutation. Methotrexate resistance is used as → selectable marker in cloning experiments with animal cells, but has also been used in plant cells.

Methylase: See → methyltransferase.

Methylase-limited partial digestion: The incomplete restriction of a DNA sequence by a particular → restriction endonuclease, caused by a simultaneously acting DNA modification methyltransferase, that methylates cytosine residues within the → recognition site of the endonuclease. This technique is used to partially digest DNA in → agarose plugs for → pulsed field gel electrophoresis.

Methylated adenine recognition and restriction (Mrr) system (modified adenine recognition and restriction system): A series of → restriction endonucleases of *E. coli* that recognize DNA sequences containing methylated adenine residues (such as G^{N6m}AC and C^{N6m}AG). Compare → methylated cytosine recognition and restriction system.

Methylated cap: See → cap.

Methylated cytosine recognition and restriction (Mcr) system (modified cytosine restriction system): A series of restriction endonucleases of *E. coli* that recognize DNA sequences containing methylated cytosine residues, and cleave them. Among these systems, Mcr A restricts the sequence C^{5m}CGG, Mcr B the sequence Pu^{me}C (where three different cytosine modifications are recognized: 5-methylcytosine, N-4-methylcytosine, and 5-hydroxymethylcytosine).

Methylated sequence tag (MeST): Any DNA sequence whose → cytosine (and adenosine) → methylation is diagnostic for a particular cellular situation (e.g. cancer). For example, specific MeSTs are expected to allow early detection of colon cancer.

Methylation: The transfer of a methyl group from a methyl donor (e.g. → S-adenosyl-L-methionine) to a methyl acceptor molecule (e.g. a protein, RNA or DNA) by a → methyl transferase. → DNA methylation is described in more detail, see there, and also → restriction-modification system. For an example of RNA methylation, see → methylated cap. RNA methylation is a → post-transcriptional modification.

Methylation assay (DNA methylation assay): A technique for the detection of methylated nucleotides within → recognition sequences of → restriction endonucleases in genomic DNA, using methyl-sensitive endonucleases, or pairs of endonucleases recognizing the same sequence but differing in methylation sensitivity (→ heteroprostomers). For example, the endonucleases *Mbo* I and *Hpa* II recognize and cut the same cleavage site (5'-CCGG-3'). *Msp* I also recognizes this site, if the internal cytosine is methylated (i.e. 5'-CC^{m}GG-3'), whereas *Hpa* II does not. By comparison of the cleavage pattern obtained from the same genomic DNA with either *Msp* I or *Hpa* II differences in methylation of CCGG-sequences can be detected. Since methylation of specific bases in → promoters may influence their activity (see → DNA methylation), such methylation assays allow to correlate promoter methylation with transcription of the adjacent gene. See also the table "Methyl-sensitivity of restriction endonucleases" of the Appendix.

Methylation-free island: See → CpG-rich island.

Methylation induced premeiotically (MIP): The extensive methylation of cytosyl residues in naturally or artificially duplicated DNA segments (regardless of their endogenous or exogenous origin) in the filamentous fungus *Ascobolus immersus*. MIP inactivates the genes located on the duplicated segments. Once established, both the C-methylation and gene silencing are stably maintained through vegetative and sexual reproduction, even after segregation of the duplicated segments. The minimum duplicate size for MIP-induced methylation is about 300-400 bp (i. e. smaller duplications are not methylated). MIP represents a special type of epimutation, and probably functions to shelter the genome against invasion by mobile elements and recombination of ectopic repeats that could be lethal.

Methylation interference (methylation interference assay, methylation interference analysis, methylation interference footprinting): A method to test the specificity of binding interaction(s) between a specific DNA sequence and a sequence-specific binding protein. In short, the DNA sequence is partially methylated in vitro at purine residues by dimethyl sulfate, mixed with a nuclear extract or a purified nuclear binding protein, and tested for its binding properties in a → mobility-shift DNA binding assay. Methylation of purine residues within the target DNA interferes with the binding of the specific protein that readily binds to the identical non-methylated sequence.

Methylation-mediated gene silencing: The down-regulation of the → transcription of genes, whose → promoters carry cytosines with 5-methyl groups at strategic positions. These methylated cytosines may either sterically prevent the binding of → transcription factors to their respective recognition sites, or bind → methyl-CpG-binding proteins recruiting → histone deacetylases. In both cases, the adjacent gene is silenced.

Methylation pattern (histone methylation pattern): The specific distribution of methylated side chain residues in → histones within the → chromatin of eukaryotic cells, that is continuously changing during the life cycle of a eukaryote. Methylation predominantly occurs on lysine and arginine residues in histone H3 and H4. For example, lysine residue 4 and 9 (K9) in histone H3 and lysine 20 in histone H4 are methylated by histone methyltransferase SU(VAR)$_{3-9}$ (in mammals) or Clr4 (in yeast). This methylated lysine is the only binding site for → heterochromatin protein HP$_1$, that is associated with silent heterochromatic regions of a genome. See → histone code.

Methylation protection: The masking of specific → restriction endonuclease → recognition sites (e.g. *Eco* RI sites) within a clonable → genomic DNA fragment or → cDNA by specific methylation of C or A residues using site-specific → methyltransferases (e.g. → *Eco* RI methylase). The protected DNA can then be modified e.g. by → linker tailing and restricted without internal → cuts. Compare → restriction-modification system.

Methylation-sensitive amplification polymorphism (MSAP): A variant of the conventional → amplified fragment length polymorphism (AFLP) technique, that uses the → isoschizomers *Hpa* II and *Msp* I instead of *Mse* I as the frequently cutting → restriction endonuclease (the rare cutter *Eco* RI remains unchanged) in order to detect the extent and pattern of cytosine methylation in a target genome. In short, genomic DNA is isolated and divided into two reactions, one of which is digested with *Eco* RI and *Hpa* II (recognizes and cleaves the sequence 5'-CCGG-3', is inactive if one or both cytosines are fully methylated [both strands methylated] and cleaves the hemimethylated sequence [only one strand methylated]). The other part of DNA is restricted with *Msp* I (cleaves 5'-C^{5m}CGG-3, but not 5'-^{5m}CCGG-3') instead of *Hpa* II. Then → adaptors (*Eco* RI- and *Hpa* II- *Msp* I adaptor) are ligated to the restriction fragments using → T4 DNA ligase, and the products pre-amplified with adaptor-specific primers (no selective bases). The pre-amplified products are diluted and amplified with the *Eco* RI and *Hpa* II-*Msp* I primers in a → touchdown → polymerase chain reaction as described for → AFLP. The *Hpa* II-*Msp* I primer was end-labeled with [γ-^{32}P]-ATP. The denatured fragments are separated on a denaturing → polyacrylamide gel, and the labeled fragments visualized by → autoradiography. See → methylation-specific polymerase chain reaction.

Methylation-sensitive single nucleotide primer extension (Ms-SNuPE): A technique for the detection and quantitation of cytosine methylation at specific CpG sites in → genomic DNA. In short, genomic DNA is first treated with sodium bisulfite, which converts unmethylated cytosine to uracil, but leaves methylcytosine, since it is resistant to deamination. In the subsequent → polymerase chain reaction amplification of the target sequence, using → primers specific for bisulfite-converted template DNA, the uracil is relicated as thymine and the methylcytosine as cytosine. After → agarose gel electrophoresis of the amplification product and its isolation, it is used as template in a primer extension reaction using appropriate internal Ms-SNuPE primers (that terminate immediately 5' of the single nucleotide under investigation), and ^{32}P-labeled dNTPs (either ^{32}PdCTP or ^{32}PdTTP). The radiolabeled products are than separated by denaturing 15 % → polyacrylamide gel electrophoresis and visualized by → autoradiography or → phosphorimaging. The ratio of methylated versus unmethylated cytosine (C versus T) at the genomic CpG sites can then be determined. Ms-SNuPE avoids → restriction endonucleases and can be multiplexed.

Methylation-specific polymerase chain reaction (MSP): A technique for the detection of the extent and pattern of cytosine methylation in a target genome. In short, unmethylated cytosines are first chemically converted to uracil, using the → bisulfite technique, which leaves the methylated cytosines unchanged. Then two types of → primers are employed for amplification of unmethylated versus methylated DNA, respectively. These primers are designed such that → mismatches are created that prevent → mispriming between primer and undesirable target sequences. The unmethylated sequence can only be amplified with the *U* primer set (*u*nmethylated sequence detection), whereas the methylated sequence can be amplified only with the *M* primer set (*me*thylated sequence detection). MSP allows the precise mapping of cytosine methylation in all CpG residues of a target DNA (e.g. → CpG-rich islands). Compare → methylation-sensitive amplification polymorphisms.

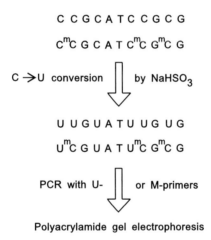

C C G C A T C C G C G

C^mC G C A T C^mC G^mC G

C →U conversion   by $NaHSO_3$

U U G U A T U U G U G

U^mC G U A T U^mC G^mC G

PCR with U-   or M-primers

Polyacrylamide gel electrophoresis

	Unmethylated/modified template:	Methylated/modified template:	Unmodified template:
U-Primer:	T T G T A T T T G T	T T G T A T T T G T	T T G T A T T T G T
	U U G U A T U U G U G	*U C G U A T U C G C G*	*C C G C A T C C G C G*
M-Primer:	T **C** G T A T T **C G C**	T C G T A T T C G C	T C G T A T T C G C

Methylation-specific polymerase chain reaction

Methyl-CpG-binding domain column (MBD column): An affinity matrix with an attached DNA-binding domain of the → methyl-CpG-binding protein, that is used for the isolation of DNA regions with methylated cytosines. DNA fragments with many mCpG sites own a higher affinity to the column than fragments containing less CpG sites. Therefore, the former are eluted at high salt (sodium chloride) concentrations, the latter at lower salt concentrations. The relative salt concentration reflects the overall methylation status of different DNA fragments with otherwise identical nucleotide sequence. MBD column chromatography suffers from insensitivity to small variations in methylation, and the fact that the methylation status of DNA fragments with identical or similar elution profiles can be heterogeneous.

M*e*thyl-C*p*G-binding *p*rotein (MeCP):

M*e*thyl-C*p*G-*bind*ing protein (MBD, C*p*G-binding protein, MeCP): Any one of a family of nuclear proteins, that bind to methyl-CpG rich regions of a genome (see → CpG-rich islands) and mediate transcriptional silencing associated with → DNA methylation (see → methylation-mediated gene silencing). For example, rat nuclear protein MeCP2 binds symmetrically to methylated CpG (mCpG) sites in the genome and has a DNA-binding domain of 85 amino acids. MeCPs probably prevent the demethylation of methylated CpG residues and may thus be involved in long-term silencing of transcription of whole → chromatin domains, especially if located in → promoters.See → methyl-CpG-binding domain column.

Methyl filtration (methyl filtration genome sequencing, methyl filtering): A somewhat misleading term for a technique to separate methylated from less or non-methylated regions of a genome. Most of the → repetitive genomic DNA (i. e. → retrotransposons, → satellite DNAs, → transposons) is heavily methylated, whereas → genic DNA is under- or unmethylated. For the separation of both classes of DNA, shotgun libraries of the target genome are established in genetically engineered bacterial strains that restrict methylated DNA. Thereby the library is enriched for non-methylated (genic) DNA, which can directly be sequenced to discover the genes. In short, genomic DNA is isolated and size-selected fragments generated by either → restriction or mechanical shearing. The relatively small fragments are then cloned in the methylation-restrictive *E. coli* host strains JM 101, JM 107, or JM 109. The enzyme encoded by gene *McrBc* of these strains restricts methylated DNA, requiring two 5'-Pu-^{m}C-3' dinucleotides separated by 40-80 bp for restriction. Therefore, methylated DNA is under-represented ("filtered") from these libraries, and as a consequence, → genic DNA (which is less or not at all methylated) is enriched several-fold.

Methyl filtration genome sequencing: See → methyl filtration.

Methyl guanosine: 1-methyl guanosine, and N²-dimethyl guanosine, both → rare bases.

Methyl inosine: 1-methyl inosine, a → rare base.

Methylmercuric hydroxide: The toxic chemical CH_3HgOH that reacts with imino bonds of uridine and guanosine in RNA and thus prevents the formation of secondary structures. The chemical is used to denature RNA completely before its electrophoresis in → agarose gels.

Methylome: The pattern of methylation of cytosyl (also adenyl) residues of a genome or parts of it (e.g. → promoters) at a given time. See → methylomics.

Methylomics: The whole repertoire of techniques for the quatitative determination of the methylation pattern of cytosyl or adenyl residues in target DNA. See → methylome.

Methylosome: A 20S protein complex consisting of methyltransferases and associated proteins, that symmetrically dimethylate arginine residues in arginine- and glycine-rich domains of socalled Sm proteins. For example, the methyltransferase JBP1 produces methylated SmD1 and

SmD3, which drastically increase their affinity for the *survival motor neuron* (SMN) complex, that in turn plays a decisive role in the assembly of → *small nuclear* (sn) RNA-protein core particles. See → pre-spliceosome, → spliceosome.

Methylphosphonate: A hydrophobic non-ionic nucleic acid analogue, that contains nuclease-resistant methylphosphonate linkages instead of the naturally occurring negatively charged → phosphodiester bonds. Methylphosphonates form duplex hybrids with complementary DNAs by standard → Watson-Crick base pairing and are effective → antisense molecules, that prevent virus and cellular → messenger RNA translation (antisense inhibition of herpes simplex virus replication, triplex-directed inhibition of chloramphenicol acetyltransferase mRNA expression in cell cultures, and inhibition of human collagenase IV mRNA expression). Methylphosphonate-RNA duplexes are not attacked by → RNase H.

Methylphosphonate

Methyltransferase **(MTase, DNA methyltransferase, DNA methylase, methylase, modification methylase):** An enzyme that catalyzes the transfer of a methyl group from → S-adenosyl-L-methionine to a substrate (e. g. a protein or nucleic acid). The C-terminal domain of mammalian methyltransferase displays two activities: maintenance methylation of the symmetrical cytosine in a hemimethylated 5'-CpG-3' doublet, and de novo methylation of unmethylated CpG dinucleotides. This domain also contains conserved amino acid motifs, that are characteristic of all bacterial and eukaryotic methyltransferases, some of which methylate CpNpG trinucleotide sequences. Multi-specific phage methyltransferases methylate more than one specific DNA target, including CpNpG and CpG sites. For an example of RNA methylation, see → cap. Methylation of DNA is described in more detail, see → DNA methylation and → restriction-modification system, also → Dam methylase, → Dcm methylase, → *Eco* RI methylase, → heteroprostomer, → isoprostomer, → modification methylase.

MFI: See → CpG-rich island.

M-FISH: See → *m*ultiplex *f*luorescent *in situ h*ybridization.

MFLP: See → microsatellite-anchored fragment length polymorphism.

MGB: See → *m*inor *g*roove *b*inding probe.

mhpDAF: See → *m*ini-*h*airpin *p*rimed *D*NA *a*mplification *f*ingerprinting.

MIAME: See → minimal information about a microarray experiment.

***M*icellar *e*lectrokinetic *c*apillary *c*hromatography (MECC):** A variant of the → capillary electro-phoresis technique, that allows to separate uncharged molecules. In short, surface-active sub-stances are first added to the separation buffer in concentrations above the *critical micelle conc*-entration (CMC), leading to the formation of socalled micelles. In an electro-osmotic flow, negatively charged micelles migrate to the cathode, but slower than the running buffer. Molecules are distributed within and out of the micelles according to their lipophilic character, and different molecules are separated from each other by their different affinity towards the micelles.

micRNA: See → countertranscript.

Microamplification: The amplification of DNA from a single band of a polytenic chromosome. The target band is first microdissected from a polytenic chromosome preparation, the underlying DNA digested with the restriction endonuclease *Sau*3A, then oligonucleotide → adaptors ligated to the resulting fragments, and → primers complementary to these adaptors used to amplify the fragments in a conventional → polymerase chain reaction. See → microcloning.

Microarray: Any microscale solid support (e.g. nylon membrane, nitrocellulose, glass, quartz, silicon, or other synthetic material), onto which either DNA fragments, → cDNAs, → oligo-nucleotides, → genes, → open reading frames, peptides or proteins (e.g. antibodies) are spotted in an ordered pattern ("array") at extremely high density. Such microarrays (laboratory jargon: "chips") are increasingly used for high-throughput → expression profiling. See → antibody chip, → antibody array, → antigen array, → antigen microarray, → antisense genome array, → ap-tamer chip, → aptazyme array, → BAC DNA microarray, → BAC microarray, → bead array, → bead-based array, → bioarray, → bioelectronic array, → biological array, → Brownian rat-chet, → cancer cell profiling array, → cantilever array, → capillary chip, → cDNA array, → cDNA expression microarray, → cDNA microarray, → cell-based microarray, → cell bio-chip, → cell chip, → cell microarray, → cellular biochip, → cellular chip, → cellular microarray, → chemical microarray, → chip, → combinatorial protein array, → cryoarray, → dendrimer-based microarray, → designer microarray, → diffusion sorting array, → DNA array, → DNA chip, → DNA microarray, → double-stranded DNA microarray, → electrochemical microar-ray, → electronic biochip, → electronic microarray, → electrophoresis chip, → entropic trap array, → EST array, → exon array, → expression array, → fiber bead array, → fiber-optic DNA array, → filter array, → flow-through biochip, → flow-through chip, → format I microarray, → format II microarray, → 4D array, → 4D chip, → functional protein array, → gel pad array, → gene array, → gene chip, → gene expression microarray, → genome array, → genome chip, → genomic array, → genomic microarray, → genomic tiling array, → genomic tiling path mi-croarray, → glycochip, → gold microarray, → haplotype chip, → high-density chip, → high den-sity colony array, → high-density oligonucleotide array, → high-density protein array, → hi-stological chip, → HLA chip, → human endogenous retrovirus chip, → human leucocyte antigen

chip, → human single nucleotide polymorphism probe array, → hybridization array, → hydrogel-based microarray, → immobilized microarray of gel elements, → *in situ* array, → interaction chip, → ion channel array, → lab-on-a-chip, → live cell microarray, → living chip, → living microarray, → LNA microarray, → low density array, → low density chip, → lymphochip, → macroarray, → medium density chip, → membrane microarray, → microarray Western, → microcantilever array, → microchip, → microelectronic array, → microelectrophoresis chip, → microfluidics chip, → microtube microarray, → modular array, → modular microarray, → *Mu* array, → multiallergen chip, → multi-functional biochip, → multiplex hybridisation array, → nanoarray, → non-living array, → nucleic acid microarray, → nucleic acid-programmable protein array, → nylon macroarray, → oligonucleotide array, → oligonucleotide chip, → oligonucleotide microarray, → one-chip-for-all, → ordered array, → pathochip, → pathway slide, → peptide array, → peptide chip, → phenotype array, → photoaptamer array, → planar array, → PNA array, → population-specific array, → printed microarray, → programmable chip, → protein biochip, → protein chip, → protein domain array, → protein *in situ* array, → protein microarray, → protein-protein interaction chip, → proteome array, → proteome chip, → proteome microarray, → recombinant protein array, → retroarray, → retrochip, → retrovirus chip, → reverse format array, → RNA biochip, → RNA chip, → SELDI chip, → separation chip, → sequencing array, → single base extension tag array on glass slides, → single molecule array, → single nucleotide polymorphism chip, → sipper chip, → small molecule microarray, → splice oligonucleotide array, → spotted array, → spotted microarray → subarray, → substrate chip, → suspension array, → tandem array, → theme array, → tissue array, → tissue microarray, → tissue-specific microdissection coupled with protein chip array technology, → transcript array, → transgene chip, → 2D/3D biochip, → ultra-high density microarray, → universal array, → universal microarray, → universal protein array, → whole genome oligonucleotide array, → whole proteome microarray. Compare → microarray architecture, → microarray noise.
See color plate 4.

Microarray architecture: A laboratory slang term for the overall layout of all components of a microarray system, such as the design of the microarray itself, the hybridization chamber, the detector with all the filters, light sources (e.g.lasers), optics and other hardwares.

Microarray noise: An undesirable contribution of → microarray parameters such as background or substrate fluorescence or cross-reactivity of the probe to the readings of the fluorescence detection instrument. See → background subtraction, → dark current, → electronic noise, → optical noise, → sample noise, → substrate noise.

Microarray Western: Any → microarray, onto which target proteins (for example, in the form of cellular extracts) are immobilized, that can be used to screen with → antobodies raised against specific proteins. If such a specific antibody recognizes and binds its cognate protein on the chip, the complex can be detected by a secondary antibody labeled with e.g. a → fluorochrome and active against the first antibody. The chip can then be scanned by a laser. Compare → Western blot, → Western blotting.

Microautoradiography: A variant of → autoradiography, which uses a liquid photoemulsion into which a sample (e. g. a tissue section whose RNA was labeled with ^{3}H-uridine) is embedded. The generated silver grains can be visualized on a sensitive film. See → macroautoradiography.

Microbial cell-surface display (cell-surface display): A technique for the display of peptides or proteins on the surface of bacterial or lower eukaryote cells, that is based on the expression of a fused gene encoding N- or C-terminal sequences of so called carrier proteins (usually cell surface proteins or their fragments) and sequences encoding the target protein. In short, the sequence encoding the target peptide or protein (→ "passenger protein") are first fused to either the N- or C-terminus of the carrier, or inserted into the center of the carrier (→ sandwich fusion), cloned into an → expression vector and expressed in an appropriate bacterial host cell, which should be compatible with the displayed protein and deficient from cell wall-associated or extracellular proteases (as e.g. certain *E.coli*, *Bacillus* and *Staphylococcus* strains). Distinct strains of *Saccharomyces cerevisiae* are also used, because they are considered as safe (e.g. for food or pharmaceutical applications), and possess protein folding and secretory systems similar to other eukaryotes (e.g. mammals). The expressed → fusion protein is then transported to the membrane (preferentially to the outer membrane) and exposed on the surface of the cell. The carrier protein should possess an efficient → signal peptide (or transport signal) and a strong anchoring motif to prevent detachment of the fusion protein from the cell surface, should remain stable after fusion and be resistant towards proteases of the periplasmic space. For example, bacterial fimbriae, S-layer proteins, ice nucleation proteins and some *E.coli* outer membrane proteins (e.g. TraT) are such efficient carrier proteins (especially for immunostimulation and the development of recombinant vaccines). The passenger protein sequence also influences the efficiency of display and can even prevent it. For example, a passenger containing four phenylalanine residues is only inefficiently displayed on *Staphylococcus xylosus* cells, but their replacement by serine residues allows efficient display. Substantial improvement of the display system can be introduced by spacers of appropriate (experimentally proven) lengths, that permit correct folding of both carrier and passenger proteins, prevent functional interference between both, or between passenger and cell surface. Microbial cell-surface display is widely applied for e.g. vaccine development (by exposure of heterologous epitopes on human commensal or attenuated pathogenic bacterial cells to elicit antigen-specific antibody responses), bioremediation (e.g. the development of efficient bioadsorbents for the removal of toxic chemicals or heavy metals from the environment), whole-cell biocatalysis (by e.g. immobilization of enzymes on the surface), biosensor design (by e.g. anchoring enzymes, receptors, or other signal-sensitive compounds for diagnostic or environmental purposes) and mutation screening (e.g the detection of single amino acid changes in target peptides after → random mutagensis). See → peptide display, → phage display, → ribosome display. Compare → differential display.

Microbiome: The entirety of all microorganisms in a certain environment. For example, all the symbiotic and commensal, but also parasitic and temporal microorganisms, that populate a human body (including skin and hair), are considered a microbiome. On the extreme, the mitochondria (in animals and plants) and the plastids (in plants) are also part of the microbiome.

Microcantilever: Any microfabricated silicon support, onto which a gold monolayer is deposited, which in turn serves as docking substrate for the covalent immobilization of synthetic 5'thio-modified oligonucleotides. Any hybridisation of an unlabeled → probe to the immobilized targets leads to a difference in surface tension between the functionalized gold and the non-functionalized silicium surface, which bends the microcantilever. The bending force can be transduced into measurable electric signals. See → cantilever array, → nanomechanical transduction.

Microcantilever array: Any → microarray, that contains hundreds or thousands of → microcantilevers. Microcantilever arrays are used for the label-free detection of DNA-DNA-, DNA-RNA-, DNA-protein-, RNA-protein-, protein-protein-, and peptide protein interactions. See → nanomechanical transduction.

Microcell-*m*ediated gene *t*ransfer (MMGT): A method for the transfer of single chromosomes from one mammalian somatic cell to another, using so-called microcells. In brief, donor cells are treated with colcemid to block mitoses. This leads to a reorganization of the nuclear membrane which engulfs single or small groups of chromosomes (micronuclei). Addition of cytochalasin B and centrifugation of these multinucleated cells produce microcells (micronuclei surrounded by plasma membrane) which can be fused to recipient normal-sized cells with the aid of → polyethylene glycol.

Microchip ("chip"):
a) A packaged computer circuitry ("integrated circuit") of minute dimensions, that is manufactured from silicon and produced for program logic ("microprocessor chip") or for computer memory (memory or RAM chip).
b) Any miniaturized solid support (e.g. of nylon, nitrocellulose, glass, quartz, silicon or other synthetic material), onto which socalled target molecules are spotted at a low, medium or high density (see → microarray), or into which nanochannels are microfabricated (see → microfluidic chip). Such microchips are used for → microarrays.

Microchromosome: Any extremely small chromosome (e. g. in birds).

Microcin: Any low molecular weight → colicin.

Microcloning: The → cloning of specific subchromosomal regions produced by microdissection (i.e. the removal of parts of a metaphase chromosome by physical means). Microdissection and microcloning procedures are used to generate → markers from specific chromosome regions that can serve as starting points to clone more extended regions of the chromosome. See → chromosome hopping, → chromosome walking.

Micrococcal nuclease (staphylococcal nuclease, micrococcus nuclease, nuclease S7, EC 3.1.31.1): An endonucleolytic enzyme from *Staphylococcus aureus* catalyzing the Ca^{2+}-dependent nucleolytic cleavage of → linker DNA between adjacent → nucleosomes in → chromatin. The enzyme is preferentially used to isolate nucleosome monomers (monosomes) and linker-free particles (→ core particles). The complete digestion of DNA by this nuclease leads to 3' mononucleotides.

Micrococcus nuclease: See → micrococcal nuclease.

Microdissection: A technique to fragment a chromosome by physical microsurgery (e.g. by a laser beam). Subchromosomal fragments can then be used to establish → subgenomic gene libraries.

Microdissection PCR: See → microdissection *p*olymerase *c*hain *r*eaction.

Microdissection *p*olymerase *c*hain *r*eaction (microdissection PCR): A method to amplify DNA fragments obtained by the dissection of specific regions of chromosomes. In brief, squashed chromosomes are microdissected, the fragment is extracted and the DNA of the fragment digested to completion (e.g. with Mbo I). Then → adaptor DNA sequences are ligated to the termini of the Mbo I-fragments. These adaptors serve as → primers for the → polymerase chain reaction that allows the amplification of the microdissected DNA in some 35-40 cycles to amounts of 100 ng or more.

Microdrop *in situ* hybridization (MISH): A technique for the detection of mutations in DNA of isolated nuclei, chromosomes, or RNA in single cells, which uses micro-encapsulation of the test material in gel droplets, where the → fluorescence *in situ* hybridization takes place. Subsequently → flow cytometry allows to reveal gross aberrations in chromosomes.

Micro-*e*lectromechanical *s*ystem sequencing machine (MEMS): An instrument for the → sequencing of DNA, consisting of a silicon chip, that contains engraved in its surface all components necessary for a sequencing reaction (e.g. a thermal cycler for → cycle sequencing, purification of the product, electrophoresis and a detector).

Micro*e*lectronic *a*rray (MEA, microelectronic chip, microelectrode array, bioelectronic array): Any → hybridization array, onto which the target sequences (e.g. nucleic acids or proteins) are directed to specific locations by programmable sets of microelectrodes. Each microelectrode (diameter: " 100 μm) is covered by a thin layer of → agarose and generates a controllable electric current, that forces the target sequences to specific preprogrammed location on the chip surface.

Microelectrophoresis chip: A variant of the → microfluidics chip, that contains multiple channels for the sequencing of nucleic acids and the separation of amplified DNA fragments in short periods of time.

Microexon: Any extremely short → exon that encodes only one or up to a dozen amino acids.

Microfluidics: A series of technologies for the management and control of flow of minute volumes of liquids (and gasses) in miniaturized systems (e.g. a → microchip). These technologies encircle the production of glass or silicon chips ("microprocessors") with microchannel systems, appropriate valves, and devices to apply and move nanoliters of various samples and reagents in parallel, and the detection of interactions between target and probe molecules or the generation of products. See → microfluidic chip, → microfluidics technology.

Microfluidics chip (microfluidics-based chip): A microfabricated silicon chip, whose interior contains a network of micrometer channels, in which fluids can be moved by controlled pressure-driven flow. Microfluidic chips can be used to monitor cellular parameters (e.g. cell density, cell shape, apoptotic cells, transformed cells), allow highly parallel and *h*igh-*t*hroughp*u*t (HTP) analyses, and help to reduce the costs for chemicals and test materials. See for example → continuous-flow polymerase chain reaction, → flow-through biochip.

Microfluidics technology: The whole repertoire of techniques combining liquid chromatography and microfabricated chip application. Such glass, quartz, silicon, or plastic chips contain interconnected gel-filled microchannels for the molecular sieving of nucleic acid molecules. In short, each sample is transported from its well to a separation capillary system, into which it is injected. Separation of the nucleic acid molecules in the capillaries occurs according to their size. The differently sized nucleic acids are in-capillary loaded with an intercalating fluorochrome (e. g. → ethidium bromide) and the complexes detected by fluorescence. Appropriate software plots fluorescence intensity versus time and displays the data as an electropherogram. See → microfluidics.

Microgel electrophoresis: See → single cell electrophoresis.

Microgene: Any gene that encodes a → microRNA.

Microgenomics: The whole repertoire of techniques to study the → genome, → transcriptome and → proteome of single cells on a microscale. Single cells are first isolated (by e.g. → laser microdissection) and the DNA, RNA or total protein extracted with techniques specifically adapted to extremely small quantities. For example, → messenger RNA is isolated from such cells, reverse transcribed into → cDNA, and the cDNA amplified by → *in vitro* transcription. The amplified cDNA is then labeled and used in → microarray experiments to detect cell-specific transcription or transcriptional responses of single cells to external stimuli (e.g. drugs). The term microgenomics is also employed for expression profiling of tissue sections.

Microhomology: Any → homology between two nucleic acid strands (e.g. DNA-DNA, DNA-RNA, RNA-RNA), that is based on only one or few complementary nucleotides.

Microinjection: A technique for the mechanical transfer of genes into living cells, that uses a → micromanipulator consisting of a glass capillary with a blunt end ("holding capillary") to position the recipient cell and to hold it under slight suction, and a finely drawn glass capillary into which the injection fluid (e.g. a DNA solution) is sucked under slight vacuum ("injection capillary"). The injection capillary is then moved with mechanical or hydraulic devices ("chop-sticks") towards the fixed cell, until it penetrates the cell membrane. It can also be directed to enter the nucleus (A). The injection capillary is now emptied and releases the solution (frequently mixed with an indicator dye such as red oil or the fluorescent Lucifer-Yellow). The recipient cell may also be glued to a glass surface with poly-L-lysine (B) or partly embedded in a thin layer of agarose (C). This technique of → direct gene transfer is comparably accurate and highly efficient, but needs experienced experimentors.
Figure see page 657.

Microlesion: See → point mutation.

Micromachining: The patterning of glass or silicon surfaces such that three-dimensional micro-structures are generated, as for example in → DNA microchip production.

Micromanipulator: An instrument for the injection of subcellular particles or molecules (e.g. DNA) into cells, see → microinjection. It also allows the isolation of single cells or → protoplasts.

Micron (μ, μm): A unit of length, equivalent to 10^{-6} meter. Frequently used in contour-length determination of DNA (or RNA) molecules (1 μ = 3 kb).

Micronucleus:
a) The smaller, generative nucleus of certain protozoan species (ciliatae), that contains the complete genome in the form of typical eukaryotic chromosomes with associated → histones, divides by mitosis, and is transcriptionally silent during asexual growth of the ciliate. However, it becomes active during sexual reproduction and is responsible for the genetic continuity of the protozoon ("germ-line nucleus"). Compare → gene-sized DNA, → macronucleus, → nuclear dimorphism.
b) Any one of several structures formed from acentric chromosome fragments and the nuclear membrane in cells treated with colcemid. See → microcell-mediated gene transfer.

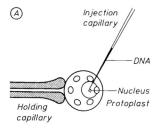

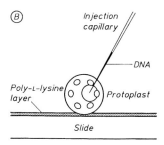

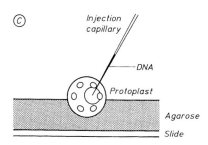

Microinjection

Microplate *a*rray *d*iagonal *g*el *e*lectrophoresis (MADGE): A high-throughput micro-format technique to separate minute amounts of proteins and nucleic acids. In short, a microplate format slot former with 9 mm pitch between wells is placed in an appropriate tray with the teeth upward. Then a → polyacrylamide (PA) gel is poured into the tray, and a sticky silane-coated glass plate placed onto the arrangement. After about 5 minutes, the glass plate is lifted with the open-faced microplate-compatible, 2 mm thick, 96-well PA gel attached to it. The array of wells is slightly turned on a diagonal angle (18.4° to the axis of the rows) such that the track lengths for electrophoresis are extended (i. e. the samples of a particular well are electrophoresed in between two wells of the next array beneath). The gel can be loaded with multi-channel pipettes, and running times be reduced to less than an hour.

Microplate-based PCR: See → microplate-based *p*olymerase *c*hain *r*eaction.

Microplate-based *p*olymerase *c*hain reaction (microplate-based PCR): A variant of the → polymerase chain reaction technique that allows the amplification of DNA sequences directly in lysed bacterial colonies or phage plaques without laborious DNA preparation. In short, bacterial colonies or phage plaques are transferred to the wells of a microplate using a toothpick. A microplate consists of a thin flexible polycarbonate mold that provides good thermal transfer properties (i.e. does not warp at high temperatures nor leach any potentially inhibitory organic compounds). Then a PCR reaction mixture containing deoxynucleotides, primers (→ amplimers) and → *Thermus aquaticus* DNA polymerase is added, and overlayered with light mineral oil. The bacterial cells or the phages are lysed and their DNA is denatured by heating. Then amplification cycles are started. This technique facilitates the rapid and simultaneous characterization of large numbers of clones and the production of single-stranded DNA templates for → Sanger sequencing, using appropriate primers (e.g. M13-based amplimers). If biotinylated nucleotides (see → biotinylated dATP, → biotinylated dUTP) are used for PCR amplification of the target DNA, it is possible to recover the amplified products easily. Following PCR, → streptavidin-coated magnetic beads are added to the microplate wells, again covered with mineral oil. The biotinylated strand is bound to the magnetic beads, the non-biotinylated strand eliminated (e.g. by alkali treatment) and the remaining single strand can be prepared for sequencing with e.g. fluorescently labeled → dideoxynucleotides directly in the microplate well.

Microprotein: Any one of a series of naturally occurring peptides of a few dozen amino acids in length, that possess extraordinary stability, a distinct and highly ordered tertiary structure and a good affinity for target proteins. For example, the socalled cystine knot microproteins are highly selective inhibitors of targets, e.g the microprotein EE-TI-II from *Ecballium elaterium* (a Mediterranean plant of the Cucurbitaceae) inhibits trysin efficiently, and the neuroactive conotoxins from deep sea snails of the genus *Conus* block ion channels in the membranes of neurons. Other microproteins exhibit hemolytic, antiviral, antimicrobial or uterotonic effects. Artificial microproteins can be derived from a native precursor (e.g. a cystine knot microprotein) by permutation. Libraries of permutated variants can then be screened for interaction partners by expressing them in *E.coli*, presenting them on the surface of the host cells, where interaction with binding ligands can be monitored by e.g.cytometric methods. The interacting complex can then be identified. Such selected microproteins are lead structures for the development of novel pharmaca.

Micro-representational difference analysis (micro-RDA): A variant of the conventional → representational difference analysis, that eliminates the high proportion of → ribosomal RNA by employing the → phenol emulsion reassociation kinetics technique (PERT) during the subtractive hybridisations.

Micro-ribonucleoprotein (miRNP): A 15S → ribonucleoprotein particle, that consists of several proteins (e.g. the survival of motor neurons [SMN] proteins Gemin 3 [a DEAD-box RNA helicase], 4, 5 and 6, and the Argonaute protein and eukaryotic translation initiation factor eIF2C2 as major constituents) in a complex with at least 40 → microRNAs, ranging in size between 16 and 24 nucleotides. The complexity of microRNAs reflects the ability to recognize a wide range of diverse target RNAs for degradation via the → RNA interference pathway. MiRNPs are probably involved in the maturation and activity of microRNAs and → small temporal RNAs.

MicroRNA (miRNA, also tiny RNA): Any one of a class of hundreds of ubiquitous, usually single-stranded, evolutionary conserved, non-coding, 16–24 nucleotides long non-coding, regulatory, eukaryotic → RNAs, that are processed by → Dicer from longer transcripts (70–171 nucleotides) carrying a stem-loop structure and are associated with proteins to form socalled → micro-ribonucleoprotein (RNP) complexes. One of the proteins of this RNA-protein complex is the eukaryotic translation initiation factor eIF2C2, others are Gemin3 and 4 (components of the survival of motor neurons [SMN] complex). Some of the miRNAs (e.g. *Lin*–4 and *Let*–7) are also called → small temporal RNAs, because their mutational inactivation affects developmental timing in *Caenorhabditis elegans*. MicroRNAs inhibit the translation of target mRNAs containing 3'-untranslated region (3'-UTR) sequences with partial complementarity, and are probably involved in the development of spinal muscular atrophy, a hereditary neurodegenerative disease of (predominantly) children. The SMN complex is involved in the assembly of and restructuring of diverse → ribonucleoprotein machines, as e.g. the → spliceosomal small nuclear RNPs (snRNPs), the → small nucleolar RNPs (snoRNPs), the → heterogenous nuclear RNPs (snRNPs), and the → transcriptosomes. MicroRNAs should not be confused with → short interfering RNAs, though the two RNA species are both generated by Dicer from longer precursors. However, siRNAs are not encoded by discrete genes, microRNAs are. See → cell cycle RNA, → non-coding RNA, → short hairpin RNA, → short interfering RNA, → small RNA, → small endogenous RNA, → small non-messenger RNA, → small regulatory RNA, → small temporal RNA, → spatial development RNA, → stress response RNA, → tiny RNA.

MicroRNA (*mir*) gene: Any one of a family of evolutionary conserved eukaryotic genes, that encode → microRNAs, are partly arranged in tandem gene clusters, and are e. g. coexpressed in the germ line and early embryo of *Caenorhabditis elegans* and *Drosophila melanogaster.*

MicroSAGE: A variant of the original → serial analysis of gene expression (SAGE) technique for the global analysis of gene expression patterns that requires only minute quantities of starting material (e.g. bioptic material or microdissections). MicroSAGE is run in a single → streptavidin-coated PCR tube (to which the RNA or cDNA remains immobilized) from RNA isolation to the release of tags, thus avoiding step-by-step losses. Also, re-amplification of excised → ditags is reduced to only 8–15 cycles. In between different steps, enzymes from the previous reactions are removed by heat inactivation and disposal, so that after washing the reaction buffer and all ingredients for the next step can easily be added. MicroSAGE also uses total RNA rather than → polyadenylated RNA, because the poly(A)$^+$-fraction is directly bound to the strepavidin-coated wall of the tube via a biotinylated oligo(dT)primer, that also serves as primer in subsequent cDNA synthesis. See → SAGE-Lite, → SAR-SAGE.

Microsatellite (*short tandem repeat*, STR; *repetitive simple sequence*, RSS; *simple repetitive sequence*, SRS; *simple sequence repeat*, SSR): Any one of a series of very short (2-10 bp), → middle repetitive, tandemly arranged, highly variable (hypervariable) DNA sequences dispersed throughout fungal, plant, animal and human genomes. For example, the microsatellite sequence $(TG)_n$ is present in 5-10 · 10^4 copies per human genome, spaced at intervals of 50-100 kb. Such microsatellites arise by → slipped-strand mispairing in combination with point mutations and → unequal crossing over of sister chromatids or homologous chromosomes during meiosis. See also → hypervariable region, → simple repetitive sequence, compare → minisatellite and → variable number of tandem repeats.

Microsatellite-anchored *fragment length polymorphism* (MFLP): A technique for the fingerprinting of genomes, that is based on a combination of the → amplified fragment length polymorphism and → microsatellite-anchored primer technique. In short, → genomic DNA is digested with a → restriction endonuclease (e.g. *Mse* I) and an *Mse* I-adaptor ligated onto the restriction fragments. Then an *Mse* I-adaptor → primer and a microsatellite-anchor primer are used to amplify the intervening sequences. Usually over 100 fragments are amplified with MFLP, using conventional → polymerase chain reaction techniques, many of which are polymorphic between individuals. The sequence polymorphisms are partly caused by mutations in the *Mse* I site (→ restriction *fragment length polymorphisms*, RFLPs), in the microsatellite itself (→ *variable number of tandem repeats*, VNTRs), or in the internal sequence.

Microsatellite expansion (triplet repeat expansion): The increase in numbers of a specific → microsatellite at a particular genomic locus. Such expansions probably occur at various locations in genomes, but in most cases the resulting → mutations (here: → insertion mutations) remain neutral, i. e. without phenotypic effect. However, in a series of human disorders such microsatellite expansion causes the onset of a disease. Some of the more important triplet expansion diseases are detailed:

*H*untington's *d*isease (HD) is an autosomal, dominantly inherited neurodegenerative disorder with uncontrolled movements (chorea), general motoric impairment, psychiatric abnormalities (personality changes) and dementia, which usually starts in the third or fourth decade of life and affects one in 10,000 individuals of European origin. The symptoms progressively worsen over the next 15-20 years and lead to death, associated by neuronal death and astrogliosis (especially in the caudate and putamen, but later on throughout the cerebral cortex). The underlying human gene IT 15 is 170 kb in length, consisting of 67 exons, located on chromosome 4 p 16.3, and mutated in HD chromosomes by a CAG microsatellite expansion at its 5'-end. The normal range of CAG repeats is from 6-34 triplets, in HD patients from 37 to over 100. An inverse correlation exists between age of disease onset and repeat length. IT 15 encodes a 348 kDa protein (huntingtin), which in HD suffered a polyglutamine expansion. The gene is widely expressed in human tissues, with highest expression levels in the brain. The huntingtin protein is localized in the cytoplasm.

The HD gene is highly conserved throughout vertebrates (murine-human sequence identity on the peptide level: 91 %). The generally smaller (23 kb) homologous gene from the pufferfish *Fugu rubripes* contains all 67 exons, is highly conserved, and serves as model to decipher the disease mechanism(s). It is most probable that the disease is caused by a gain-of-function (e. g. the stimulation or inhibition of some unrelated target gene by the mutated protein, most likely a → transcription factor).

Moderate expansions of glutamine-encoding CAG repeats are also underlying other neurological disorders. The so-called *d*enta*r*ubral-*p*allido*l*uysian *a*trophy (DRPLA), a rare autosomal dominant disease with progressive dementia, epilepsy, gait disturbance and involuntary movements (chorea and myoclonus) is linked to gene CTG-B37 on the short arm of chromosome 12, that contains CAG repeats whose number in normal individuals ranges from 7-23, but expands to 49-75 repeats in DRPLA patients. Again, the number of CAG repeats is inversely correlated with the age of disease onset and is clearly associated with the severity of clinical symptoms.

Another disease, the so-called **spino***c*erebellar *a*taxia type 1 (SCA 1), an autosomal dominant disorder with ataxia, progressive motor deterioration and severe loss of cerebellar Purkinje neurons, is likewise caused by an expansion of CAG repeats within a gene. Also here, the age of disease onset and severity is highly correlated with the size of the CAG repeat island. Both the normal and expanded alleles are transcribed in lymphoblasts of SCA 1 patients. The SCA 1 gene product, ataxin-1, is localized in nuclei of neurons from various cortical regions, caudate, putamen, globus pallidus, pons, and dentate nucleus of the cerebellum, but predominantly in the cytoplasm of Purkinje cells. The expanded glutamine stretch of the mutant protein probably leads to a gain-of -function.

Other CAG microsatellite expansion diseases are *s*pino*b*ullar *m*uscular *a*trophy (SBMA), Kennedy's disease; affected is a gene encoding the *a*ndrogen *r*eceptor, AR) and *M*achado-Joseph *d*isease (MJD). However, other microsatellite motifs may also expand and lead to disorders, e. g. fragile X syndrome (caused by an expanding CGG repeat in a large open reading frame, which turns off the transcription of the adjacent gene) and *m*yotonic *d*istrophy (MD, caused by an expanding CTG repeat in the 3'-untranslated region of an mRNA encoding a protein kinase). See → dynamic mutation, → microsatellite instability. Compare → loss of heterozygosity.

*Microsatellite in*stability (MIN): The expansion or contraction of the number of → microsatellite repeats at a given locus of a → genome. For example, repeated CAG codons within an → open reading frame of the *H*untington's *d*isease gene (HD gene) on human chromosome 4p16.3 are stable, if the repeat number stays below a threshold of about 40 triplets. Above this threshold, the repeat number becomes instable. As a consequence of this microsatellite instability there is a significant probability that the length of the CAG island will increase when transmitted from one generation to the next. In such cases, the carriers with such a → dynamic mutation will develop a serious neurodegenerative disorder, Huntington's chorea. See → microsatellite expansion.

Microsatellite map: A → genetic map, that is solely based on single-locus, codominant → microsatellite markers.

Microsatellite *o*btained from *B*AC (MOB): Any → microsatellite sequence that has been cloned into a → bacterial artificial chromosome (BAC) and recovered by either hybridization with microsatellite-complementary → probes, or amplification in a → polymerase chain reaction using → primers complementary to BAC sequences or sequences flanking the microsatellite. Do not confuse with → *mob*.

Microsatellite *o*btained *u*sing *s*trand *e*xtension (MOUSE): A fast and effective method to enrich → genomic libraries for → clones containing → microsatellite sequences. In short, blunt-ended genomic → restriction fragments are first size selected (350-550 bp) on agarose gels, ligated to double-stranded, linearized → M13 vector DNA, and then transformed into electro-competent bacterial host cells. The bacteria are plated, → plaques develop, and M13 phage particles are eluted from the plate with → LB medium. Single-stranded M13 DNA is then isolated and used for the enrichment procedure, which starts by adding single-stranded → biotinylated microsatellites (e. g. [CA]$_{20}$). These anneal to the complementary microsatellites of the M13 clones and are extended by the → Klenow fragment, which displaces the newly formed strand from the template strand (which is removed by capture to → streptavidin-coated → magnetic beads). The new strand is eluted at 85 °C, made double-stranded by → primer annealing and extension with → *Taq* and → *Pwo* DNA polymerases (no strand displacement), and the resulting microsatellite-containing M13 DNA again transformed into bacteria. After this second round of amplification, the single-stranded DNA from the plaques can be isolated and sequenced. Then primers complementary to the microsatellite-flanking sequences are designed and used to amplify locus-specific → *s*equence-*t*agged *m*icrosatellite *s*ites (STMS) from genomic DNA(s) in a conventional → polymerase chain reaction.

Microsatellite polymorphism (*s*hort *t*andem *r*epeat *p*olymorphism, STRP): Any difference in the number of → microsatellite repeat units at corresponding genomic loci in two (or more) different → genomes, that can be detected by → sequence-tagged microsatellite polymorphism marker technology.

Microsatellite-*p*rimed *p*olymerase *c*hain *r*eaction (MP-PCR; *i*nter-*s*imple *s*equence *r*epeat amplification, ISSR; ISSR amplification; inter-SSR amplification, [I]SA; *s*ingle *p*rimer *a*mplification reaction, SPAR): A variant of the conventional → polymerase chain reaction that uses → microsatellite sequences as → primers to amplify regions of a genome located between two microsatellites on opposing DNA → strands. MP-PCR detects → polymorphisms in genomic DNA of different organisms of a population. See → anchored microsatellite-primed polymerase chain reaction, → inter-simple sequence repeat amplification, → minisatellite-primed amplification of polymorphic sequences, → simple repetitive DNA.

Microsequencing (protein microsequencing): A technique to increase the sensitivity of conventional → protein sequencing by two to three orders of magnitude into the picomole range. For example, in gas phase microsequencing the reagents of the → Edman degradation are carried by a stream of argon to the protein, that is bound to a polybren film. This reduces the amount of solvents, reagents, and byproducts, which leads to increased sensitivity of the technique.

Microsynteny: The conserved order, sequence, and orientation of genes, conserved gene repertoire and conserved gene spacing (similar length of intergenic regions) in the range of about 100 kb in the genomes of closely related species. See → macrosynteny, → synteny.

microTAS: See → lab-on-a-chip.

Microtiter plate (MTP): A plastic plate with regularly arranged wells. The number of such wells ranges from 96-384 (and more), allowing to run reactions simultaneously, but physically separated and in minute volumes. This miniaturization reduces the costs for the incubation medium (e. g. PCR reaction mixture, culture medium).

Micro total analysis system: See → lab-on-a-chip.

Microtransponder: A light-powered silicon-based miniature radio-frequency (RF) transmitter of minute dimensions (250x250x100 µm) representing an integrated circuit composed of photocells, memory and an antenna, onto which 20–25 nucleotide long target sequences are covalently attached. During attachment, each transponder stores an identification number (ID number) for the attached sequence. Microtransponders allow to detect and identify large numbers of unique DNA sequences in one single assay. The microtransponder is first hybridised to fluorochrome-labeled probes, then the unbound (non-complementary) sequences are removed by washing, and the fluorophors attached to the hybridised probes excited by laser light. The microtransponders are then pumped through the flow chamber of a high-throughput scanner, where the fluorescence signal is detected, transformed to an RF signal, and assigned to a specific oligonucleotide on the transponders surface. The sequence information is stored in the electronic memory of the transponder. The microtransponder technology is used for the determination of DNA or RNA sequences.

Microtube microarray("array tube"): Any → microarray, that is embedded in the bottom of a 1.5 ml microcentrifuge tube. This configuration allows to perform the hybridisation, washing and blocking steps in one and the same tube, avoiding evaporation and contamination. The microarray can be made of either DNAs, oligonucleotides, or proteins, that are either spotted or synthesized *in situ*. Interactions between the spotted molecules and probes can be visualized by either fluorochrome labeling of the probes and laser excitation, or non-fluorescently with gold-induced silver precipitation, in which a → biotinylated target is stained with a gold-strepavidin conjugate. The gold particle catalyses precipitation of silver particles, that in turn can be detected by transmission imaging. Compare → DNA dip stick.

Microwell polymerase chain reaction (microwell-PCR, microPCR): A variant of the conventional → polymerase chain reaction, that uses 96 or 384 2–3 µl microchambers in a supporting glass slide chip, that are loaded by nanodispensors and sealed. Then a thermoelectric heating-cooling device is used for cycling in a flatbed block of a thermocycler. Usually, microwell-PCR is coupled to a gel chip electrophoresis, and the samples separated in ultra-thin gel layers and a combination between slab and capillary gel electrophoresis Moving lasers allow detection of bound fluorochromes during the run ("real time"). Micro-PCR reduces the amounts of chemicals (total reaction volume: 0.5–1.0 µl), and the amplification and electrophoresis times (15–20 minutes and 1.5 to 5 minutes, respectively).

Middle repetitive DNA (moderately repetitive DNA): A fraction of genomic DNA that – after denaturation – forms duplexes fairly late in a → C_0t analysis (i.e. reassociates at medium → C_0t values). It is composed of diverse sequences 100-500 bp in length which each are repeated from 100 to 10000 times (see e.g. → rDNA, → transfer RNA genes and → histone genes). Middle repetitive DNA also encompasses → microsatellite sequences.

MIDGE vector: See → *mi*nimalistic *i*mmunogenically *d*efined *g*ene *e*xpression vector.

Midpoint *d*issociation *t*emperature (Td): A parameter for the characterization of the dissociation dynamics of oligonucleotides and their homologous target sequences. Td is defined as the temperature at which 50 % of the originally bound, short (<50 bp) oligonucleotide → probe molecules dissociate from the membrane-bound target DNA within a specific time period and under specific conditions (e. g. special buffer composition).

Miller spreading: See → Miller spreads.

Miller spreads (Miller spreading, Miller technique): A method to prepare → chromosomes for electron microscopy, starting with the centrifugation of chromosomes isolated from lysed nuclei through 10% formalin in 0.1 M sucrose onto membrane-coated grids. These are then treated with a chemical to reduce the surface tension before being dried. Finally the specimens are stained with phosphotungstic acid and examined with the electron microscope.

Miller technique: See → Miller spreads.

Millicurie (mCi): The amount of a radioactive nuclide in which $3.7 \cdot 10^7$ disintegrations per second (dps) occur.

Millipore filter: The trade name of a series of filters with defined pore sizes ranging from 0.001 to 10 µm. Used to sterilize non-autoclavable solutions, or to trap nucleic acid precipitates.

MIN: See → *mi*crosatellite *in*stability.

Min A min B mutant: An → *E. coli* → double mutant that divides into two cells of different size, a normal wild-type cell and a smaller → mini-cell.

Miniature inverted repeat transposable element (MITE): Any non-autonomous transposable element of about 0.3 kb flanked by 14 bp → terminal inverted repeats, and preferably occurring in → 3' untranslated regions of genes from worms, insects, mammals, and plants. MITEs attain very high copy numbers in most genomes (from 1000 to over 15,000), possess no coding capacity, insert preferentially into non-coding regions of single or low copy sequences and exceptionally function as part of → transcription initiation or → polyadenylation sites. MITEs do not recognize any specific target site, but rather distinct secondary structures of the target DNA. See → heartbraker, → Ping, → Pong.

Miniaturization: The reduction in the size of scientific instruments, and with it, the reduction in reagent volumes, reagent masses, frequently a reduction in time and cost of experimentation.

Miniaturized protein: A synthetic peptide-based model of a naturally occurring protein, which contains a minimum set of constituents necessary for an accurate reconstruction of a defined three-dimensional structure and a reproduction of a defined function. Such miniaturized proteins are model systems for structure-function relationship studies.

Mini-cells: Spherically shaped small cells which are continuously produced during the growth of specific mutant strains of bacteria (e.g. *E. coli*, see → min A min B mutant, or *B. subtilis*) and can be separated easily from normal-sized cells by → density gradient centrifugation. These mini cells do contain plasmids but not chromosomal DNA, are capable of RNA and protein synthesis and therefore serve to detect the expression of plasmid-borne genes and to characterize the proteins encoded by these genes without chromosomal background (in vivo transcription-translation system).

Minichromosome:
a) The circular 5.2 kb duplex DNA genome of → Simian virus 40 after its transfer into the host cell nucleus, where it becomes complexed with host cell → histones H2A, H2B, H3 and H4, and resembles a small chromosome.
b) A synonym for → artificial chromosome (see also → yeast artificial chromosome).

Minichromosome *maintenance* (Mcm): A comprehensive term for a series of nuclear proteins that probably function as replication licensing factors.

Minichromosome *maintenance* protein (MCM protein): Any one of a family of ATP-binding proteins regulating DNA replication such that it occurs only once per cell cycle. Expression of MCM proteins increases during cell growth and reaches a maximum in the transition phase from G1 to S.

Mini-exon: A synthetic → exon flanked by consensus 3'and 5' → splice sites, that contains → open reading frames encoding short (e.g. 40–50 amino acids) peptides. In neither of the three open reading frames any → stop codon exists, and each reading frame encodes a peptide recognized by the same → monoclonal antibody. The 3'splice site includes a consensus branch point sequence, a polypyrimidine tract and the mandatory AG dinucleotide. The 5'splice site in turn carries the mandatory GT dinucleotide. Since exon-intron boundaries tend to map to the surface of the final protein product, the mini-exon peptide will be displayed on the surface of the protein, and be accessible for the antibody. The mini-exon can be inserted into the → introns of genes, so that the small encoded peptide appears in the protein encoded by the host gene. This protein can then be recognized by the antibody. See → mini-exon epitope tagging.

Mini-*exon* *epitope* *tagging* (MEET): A technique for the discovery of genes and their analysis, which is based on the insertion of a synthetic → mini-exon into → introns of a target gene, and permits detection of encoded proteins with the same → monoclonal antibody regardless of the intron class. The resulting protein is altered minimally by the mini-exon and can be analysed by functional assays such as → immunofluorescence or isolated using affinity-purification of the tagged protein.

Mini-gel: See → baby gel.

Minigene: A hypothetical precursor of a present-day → gene, that was formed prebiotically as a small part of a nucleotide sequence with no biological information content. Minigenes were assembled during evolution and became present-day → exons, whereas the intervening "sense-less" sequences still exist as → introns.

Mini hairpin: A small → hairpin-like DNA structure with a stem of only two, and a loop of only 3-4 nucleotides (d[GCGAAAGC]). Mini hairpins are compact structures with a high melting temperature, and are used for → mini-hairpin primed DNA amplification fingerprinting.

Mini-hairpin primed DNA amplification fingerprinting **(mhpDAF):** A variant of the → DNA amplification fingerprinting technique that employs → mini-hairpin primers to amplify genomic sequences in a conventional → polymerase chain reaction. The primers harboring a mini-hairpin at their 5'-termini and an arbitrary core of only 3 nucleotides at the 3'-termini allow to amplify multiple loci in DNAs from → plasmids, PCR-amplified fragments, → bacterial artificial chromosome and → yeast artificial chromosome clones, and small and big → genomes. In contrast to conventional DNA amplification fingerprinting with → arbitrary primers, mhpDAF enhances the detection of polymorphisms in target DNAs.

Mini-hairpin primer: An → oligonucleotide → primer that contains highly stable → hairpin-like structures with a short stem and a 3 nucleotide looped domain at its 5' terminus, and 8 nucleotides long 3' terminal stretches of arbitrary sequences. Such primers select annealing sites during the primer-template screening phase of the amplification reaction much better than normal arbitrary primers, and are effectively anchored at their target sites (e. g. in genomic DNA). Mini-hairpin primers are used in → mini-hairpin primed DNA amplification fingerprinting.

Linear primer: Mini-hairpin primer:

5´-G T A A C G C C-3´

Linear and Mini-hairpin DAF-Primers

Minilibrary (partial gene bank): A laboratory term for a → gene library, that contains preselected and enriched → genomic DNA (genomic minilibrary) or → cDNA sequences (cDNA minilibrary). Such minilibraries contain only part of complex genomes or mRNA populations and are therefore easier to screen for target sequences than complete → genomic or → cDNA libraries. Their establishment, however, requires sequence information(s) and separation and enrichment procedures.

Minimal domain vector: Any → plasmid expression vector, that contains only a truncated version of a cloned transcription → activation domain. For example, the VP16 activation domain from herpes simplex virus as part of the → tet-on/tet-off gene expression system plasmid contains repeats of a 13 amino acids tract, that represents the functional core of the domain. Now, the

→ overexpression of VP16 can be deleterious, because it interacts with specific components of the transcription machinery. Therefore a truncated version of VP16 is less toxic than the full-length activation domain.

Minimal gene set: The minimal number of cellular genes that allows life, estimated for the genome of *Mycoplasma genitalium*. This organism seems to contain a → minimal genome made up of only 517 genes, about 265 – 350 of which are indispensable for life. *Bacillus subtilis* harbors a total of 4071 genes, of which 271 genes are essential for survival (under optimal conditions in the laboratory).

Minimal genome: The smallest set of genes that allows the replication of an organism in a particular environment (i.e. encode proteins for the catalysis of basic metabolic and reproductive functions). See → minimal gene set.

Minimal information about a microarray experiment (MIAME): A guideline for the publication of → microarray data, that sets standards for good performance and reliable evaluation of microarray experiments, and their easy interpretation and independent verification (see the home page of the Microarray Gene Expression Data Society: www.mged.org

Minimalistic immunogenically defined gene expression vector (MIDGE vector): A linear double-stranded DNA vector for the transfer and expression of a gene in target cells (or, more precisely, nuclei), that contains only one single gene with its → promoter and → terminator sequences necessary for its → expression, and single-stranded → loops on either end for its protection from intra- and extracellular → nucleases (especially → exonucleases) and for the covalent introduction of peptide-, glycopeptide- or carbohydrate → ligands (for e. g. binding to cell surface receptor molecules, or directed transport to the nucleus or other cell organelles). MIDGE vectors circumvent the need for → selectable marker genes (e. g. → kanamycin/neomycin resistance genes) whose products may cause undesirable immune reactions in the target organism, and – in case of bacteria-derived sequences – may represent highly immuno-modulatory agents. More over, MIDGE vectors lack → origins of replication, since they are not propagated in bacterial hosts. MIDGEs can be targeted to specific cells by attaching specific peptide sequences to their end, and are vectors of choice for → genetic vaccination and somatic → gene therapy.

Minimal promoter: Any → promoter that consists only of the essential sequences for correct initiation of transcription of the adjacent gene (e.g. the → TATA-box and → cap site).

Minimal protein identifier (MPI): A collection of data that unequivocally identifies a specific protein from thousands of other proteins. Identification is based on a specific peptide map, generated by mass spectrometry, fragment ion spectrum (actually mass fingerprints of proteolytically [mostly tryptically] generated peptide fragments), and peptide fragment sequences. Nuclear magnetic resonance and X-ray crystallographic data can be used as additional identifiers.

Minimal tiling path: Any map or table showing the placement and order of a set of clones (e.g. → bacterial artificial chromosome clones), that completely and contiguously cover a specific segment of DNA.

Minimal vector: Any DNA-based vector that only contains the sequences required for its maintenance.

Mini-me element: Any one of a group of highly abundant Dipteran → retroposons, that contain two internal → proto-microsatellite regions (see also → microsatellite) with the potential to expand. Mini-me elements are flanked by 10 – 20 bp long, → inverted repeats (IRs) with the 3' repeat located sub-terminal (22 – 45 bp from the actual 3' end of the element). A partial duplication of the 5'-IR allows the formation of a → hairpin loop. A highly conserved 33 bp core region is flanked by both proto-microsatellites, the 3' proto-microsatellite consisting of $(TA)_n$ repeats, the 5' one of (GTCY), where Y is either C or T. The elements in different dipteran genomes vary in size from 500 to 1,200 bases, caused by → insertions or → deletions in a socalled variable region 3' of the 3' proto-microsatellite. These elements comprise about 1.2% of the *Drosophila melanogaster* genome and represent sources for new microsatellites. Basically two mechanisms generate these new repeats: (1) preexisting tandem repeats expand by an as yet unknown process, and sequences with high → cryptic simplicity are converted to tandemly repetitive DNA, and (2) the elements move to new genomic loci, where the new environment relaxes constraints on proto-microsatellites such that they expand more rapidly.

Minimum efficient processing segment (MEPS): The minimum length of a DNA sequence that is required for efficient → recombination. MEPSs are 25 – 30 bp in *E. coli*, about 250 bp in yeast, and 250 – 400 bp in plants and cultured mammalian cells. If the size of a previously functional MEPS is reduced below a certain threshold, recombination becomes very inefficient, or ceases.

Minimum tiling set: The smallest number of clones that span the entire length of the DNA molecule from which they were derived for cloning. See → tiling path.

MiniPing: See → Ping.

Mini-prep (mini preparation): A small-scale method to extract and purify DNA and RNA from any source (e.g. phage, bacteria, plant, animal). Mini-preps are specifically adapted to small amounts of material (cells, tissues) and small volumes, and are therefore used to analyze → insert DNA in large numbers of transformants or cloning vectors.

Minipreparation: See → miniprep.

MiniSAGE: A variant of the conventional → serial analysis of gene expression (SAGE) technique for the analysis of global gene expression, that capitalizes on the use of a single tube to perform → messenger RNA (mRNA) isolation, → reverse transcription of mRNA into → cDNA with a → biotin-labeled oligo(dT) → primer, enzymatic digestion of cDNA, binding of digested biotin-labeled 3'-terminal cDNA fragments to → streptavidin-coupled → magnetic beads, → ligation of → linker oligonucleotides containing recognition sits for a tagging enzyme to the bound cDNA fragments, and release of cDNA tags, with only one microgram of starting total RNA. MiniSAGE also reduces the amount of linker oligonucleotides in the ligation reaction, which minimizes their interference with SAGE → ditag amplification and increases the yield of SAGE ditags, and uses a → phase lock gel for the extraction of RNA from the original sample. See → microSAGE, → SAGE-Lite.

Minisatellite: Any one of a series of short (9-64 bp), usually GC-rich, → middle repetitive, tandemly arranged, highly variable (hypervariable) DNA sequences which are dispersed throughout the human genome (but also occur in animal and plant genomes) and which share a common 10-35 bp consensus or core sequence (core repeat unit, tandem repeat unit). The minisatellites show substantial length polymorphism arising from → unequal crossing-over that alters the number of short tandem repeats in a minisatellite, so that arrays about 0.1-20 kb in length are formed. Unequal exchanges may be favored by a recombination signal within the core sequence, especially since this core is similar to the *E. coli* recombination signal (→ chi sequence). A hybridization probe consisting of the core, repeated in tandem, can detect many highly polymorphic minisatellites simultaneously within a genomic digest and may therefore provide genetic markers for → linkage analyses (used in individual-specific → DNA "fingerprinting"). The theoretical probability that the same set of DNA fragments (the fingerprint) is identical in two human beings is so small that every human individual (except identical twins) is expected to have a unique pattern of bands detected with a minisatellite on autoradiograms. The mutation process acts preferentially at the 3' end of a minisatellite, so that most of the sequence variability originates from here. In contrast, the 5' end belongs to a low mutable region, and therefore stabilizes the repeat. The causes for this polarity are unknown.

An extraordinarily high minisatellite variation occurs in African populations (many groups with independent characteristic minisatellite patterns). The term minisatellite overlaps with → hypervariable region and → variable number of tandem repeats. See also → minisatellite-primed amplification of polymorphic sequences and → minisatellite variant repeat. Compare → microsatellite.

```
              JEFFREYS MINISATELLITE
                 CORE SEQUENCES
                (myoglobin gene)


          consensus                   origin

     GGAGGTGGGCAGGAAG           myoglobin
     aagGGTGGGCAGGAAG           clone  33. 1
     GGAGGTGGGCAGGAAX           clone  33. 3
     tGgGGaGGGCAGaAAG           clone  33. 4
     GGAGGYGGGCAGGAGG           clone  33. 5
     GGAGGaGGGCtGGAGG           clone  33. 6
     GGA-GTGGGCAGGcAG           clone  33.10
     GGtGGTGGGCAGGAAG           clone  33.11
     aGAGGTGGGCAGGtGG           clone  33.15
    ────────────────────────────────────────
     GGAGGTGGGCAGGAXG                 core
    ────────────────────────────────────────
     GCTGGTGGGCTGGTGG             chi dimer
    ────────────────────────────────────────

   X = A or G    Y = C or T    - = deleted
```

Minisatellite-primed amplification of polymorphic sequences (MAPS): A technique for the detection of sequence → polymorphisms in → genomic DNA of different organisms, in which single synthetic → minisatellite sequences are used as → primers in a conventional → polymerase chain reaction to amplify regions flanked by them. After amplification, the polymorphic bands can already be detected on → agarose gels with → ethidium bromide. Compare → DNA fingerprinting, → interspersed repetitive sequence polymerase chain reaction, → oligonucleotide fingerprinting.

Minisatellite variant repeat (MVR): A → minisatellite repeat sequence that differs from its neighbouring repeat(s) by only one or few → restriction endonuclease sites. This "interrepeat unit sequence variability" arises by mutations within certain repeats of a minisatellite. See also → variant repeat unit.

Minisequencing: See → pyrosequencing.

Mini-Ti (mini-Ti-plasmid): A small derivative of the → Ti-plasmid of → *Agrobacterium tumefaciens* from which most of the → T-region has been deleted, except the → opine synthase gene and its promoter, a cloning site into which foreign DNA can be inserted, and the left and right → T-DNA borders. This plasmid replicates in *E. coli*, and may be conjugatively transferred into *A. tumefaciens*. The recipient can then transfer the modified T-region into wounded plant cells provided the vir functions are supplied in trans (e.g. by a Ti-plasmid carrying the → *vir* region).

Mini-Ti-plasmid: See → mini-Ti.

Minizyme: A synthetic oligoribonucleotide with hammerhead structure and → ribozyme activity.

Minor base: See → rare base.

Minor gene: Any → gene, whose contribution to the expression of a particular polygenic trait is inferior to the contribution of another → major gene.

Minor groove: The indentation on the surface of a DNA → double helix molecule, formed by the sugar phosphate backbones and the edges of the base pairs (linked by → Watson-Crick base pairing forces), that contain the N3, (in → purines) or O2 atoms (in → pyrimidines). See → double helix, → major groove.

Minor groove binding **probe (MGB probe, MGB ligand, minor groove binder):** Any → oligonucleotide → probe that preferentially hybridizes to target sequences in the → minor groove of the DNA double helix. Such MGB probes can be used for e. g. the detection of → single nucleotide polymorphisms. The term is also used for → fluorochromes, that bind to the minor groove (e.g. → HOECHST 33258 and variants, also → DAPI).

Minor spliceosome: A less abundant variant of the ubiquitous → spliceosome ("major spliceosome"), that assembles on → splice junctions with a consensus sequence deviating from the canonical → GT-AG rule, and occurs in plants, metazoans, and humans, but not in yeast or *Drosophila*. The assembly of the minor spliceosome starts with the recognition of the aberrant splice sequence 5'-exon-AUAUCCUUU–3' of the → pre-messenger RNA by the di-snRNA U11/U12 (as opposed to the conventional U1/U2 binding to normal splice junctions). Therefore this special assembly pathway is called the U12–dependent pathway. The subsequent steps of the spliceosome formation are identical to the conventional assembly (i.e. recruitment of U4/U6 and U5). Minor spliceosomes harbor a set of specific proteins, but also share common proteins with the major spliceosomes (e.g. SF3b, G, F, E, D1, D2, D3, and B) and assemble on pre-mRNAs of e.g. ion channel protein-encoding genes.

Minus: Located → upstream of the → cap site. See → plus.

Minus strand (minus viral strand, – strand):
a) In a single-stranded DNA virus the strand complementary to the → plus strand, which can be transcribed into mRNA.
b) In a single-stranded RNA virus the non-coding strand which is copied by RNA-dependent RNA polymerase into translatable mRNA. Compare → plus strand, definition b.

Minus strand cDNA: See → antisense cDNA.

– 10 sequence: See → Pribnow box.

– 3/– 1 rule: See → von Heijne rule.

Minus viral strand: See → minus strand.

MIP: See → *m*ethylation *i*nduced *p*remeiotically.

MIP: See → molecularly imprinted polymer.

miRNA: See → microRNA.

miRNP: See → micro-ribonucleoprotein.

Misactivated amino acid: Any activated amino acid (i.e. an amino acid condensed with ATP to yield an aminoacyl adenylate) that is erroneously transferred to the 3'end of a → transfer RNA (tRNA), although it is not the cognate amino acid of this tRNA. The individual aminoacyl-tRNA synthetases possess an editing mechanism to detect misactivated noncognate amino acids, which are then hydrolyzed before they can be incorporated into a polypeptide chain during ribosomal protein synthesis.

Mis-cleavage (Mc): A somewhat misleading term for the trypsin-catalyzed cleavage of some, but not all recognition sites within a target peptide or protein.

Miselongation: See → misextension.

Misextension (miselongation): The addition of bases onto the 3' end of a → primer oligonucleotide that have no complementary counterparts in the → template strand. Such errors occur at a rate of 1 per 10,000-30,000 bases in a → DNA-dependent DNA polymerase-catalyzed reaction, but are corrected, if the enzyme possesses a 3' → 5' proofreading activity. Compare → misinsertion.

MISH: See → *m*icrodrop *in situ* *h*ybridization.

Misinsertion: The incorporation of bases into a growing → polynucleotide chain (→ DNA or → RNA) that have no complementary counterparts in the → template strand. Such mismatched bases are normally excised by → mismatch repair systems and replaced by the matching bases. Compare → misextension.

Mismatch: See → base mismatch.

Mismatched primer: Any oligonucleotide → primer used in the → polymerase chain reaction that is not perfectly homologous to its template DNA. Despite such a mismatch, the primer can be used for amplification if its 3' end is well matched to the template.
Figure see page 671.

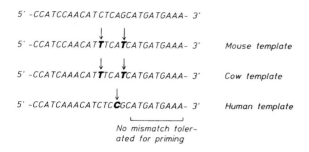

Cytochrome b primer from fish

5´ -CCATCCAACATCTCAGCATGATGAAA- 3´

5´ -CCATCCAACAT**TT**CA**T**CATGATGAAA- 3´   *Mouse template*

5´ -CCATCAAACAT**TT**CA**T**CATGATGAAA- 3´   *Cow template*

5´ -CCATCAAACATCTC**C**GCATGATGAAA- 3´   *Human template*

No mismatch toler-
ated for priming

↓ :Arrows denote mismatches in the primer core
  that does not interfere with amplification.

Mismatch extension: The ability of → DNA polymerases to use even a mismatched base at the 3' end of a → primer to synthesize a new strand complementary to the → template strand. The DNA polymerases extend a → mismatch more slowly than a matched 3' terminus, however, and different mismatches extend at different rates. For example, a G/T mismatch extends readily, whereas C/C, A/G, G/A, A/A and G/G mismatches extend less efficiently.

Mismatch gene synthesis: The *in vitro* synthesis of two single-stranded complementary oligode-oxynucleotides that differ in sequence by only one or a few bases. Upon reannealing, these mismatched bases cannot base-pair with a complementary partner. If such mismatched genes are cloned into a plasmid and transformed into a host cell, DNA repair processes will eliminate the mismatch, using both strands as templates. This leads to a population of two distinct genes, differing by only one or a few bases at a specific site.

Mismatch repair (MMR; postreplication repair): The detection and replacement of incorrectly paired (mismatched) bases in newly synthesized DNA. For example, in *E. coli* a mismatch repair system, encoded by the genes *mutH*, *mutL*, *mutS*, *uvrD* and *uvrE*, screens the newly synthesized DNA strand for mismatched bases. The mispaired bases and a short region surrounding them are excised and then replaced by the → DNA polymerase. This repair mechanism acts before the newly replicated DNA is methylated. Only after its completion the *de novo* synthesized strand is modified e. g. by → dam methylase according to the methylation pattern of the complementary strand (→ maintenance methylation). In eukaryotes, the initial recognition of mismatches is accomplished by a complex of the two proteins MSH2 and MSH6, which binds to mismatched bases. A second complex, consisting of MLH1 and PMS1 proteins then joins the mismatch-MSH2/6 complex and catalyzes excision and repair of the mismatch.

Mispairing: See → base mismatch.

Mispriming: An undesirable artifact generated by the → annealing of → amplimers to non-target sequences and the extension of these amplimers by → *Thermus aquaticus* DNA polymerase in the → polymerase chain reaction. The generation of such artifactual products can be circumvented by the → hot start technique.

Missense mutant: A → mutant carrying one or more → missense mutations.

Missense mutation: Any gene mutation in which one or more → codon triplets are changed so that they direct the incorporation of amino acids into the encoded protein, which differ from the wild type (e.g. UUU, encoding phenylalanine, mutates to UGU, encoding cysteine). The replacement of a wild type amino acid by a missense amino acid in the mutant potentially produces an unstable or inactive protein. Compare → mistranslation, where a "wrong" amino acid is incorporated despite of a correct mRNA.

Missense single nucleotide polymorphism (missense SNP): Any → single nucleotide polymorphism, that occurs in the coding region of a gene, and changes the amino acid sequence of the encoded protein. Such missense SNPs, if responsible for a functional change of e.g. a protein → domain, may cause diseases. See → silent SNP.

Missense SNP: See → missense single nucleotide polymorphism.

Missing contact analysis: See → DNA-protein interference assay.

Mistranslation: The incorporation of an incorrect amino acid into a nascent polypeptide inspite of the presence of an mRNA with the correct sequence. Mistranslation may becaused by the improper function(s) of the → tRNA, the → aminoacyl tRNA synthetases, or the → ribosome. Compare → missense mutation, where a mutated gene causes the transcription of an incorrect mRNA, which consequently directs incorporation of a "wrong" amino acid.

MITE: See → miniature inverted repeat transposable element.

Mithramycin (MIT; also plicamycin, aureolic acid): One of a series of acid oligosaccharide antibiotics produced by different strains of *Streptomyces*. For example, mithramycins A, B and C are synthesized by *S. argillaceus* and *S. plicatus,* mithramycin A being the dominant compound. Mithramycins bind to GC-rich stretches in the minor groove of double-stranded DNA, which can be quantified by the yellow fluorescence of the DNA-mithramycin complex. At the same time, mithramycins prevent RNA synthesis from the complexed DNA *in vitro* and *in vivo*.

Mitochondrial DNA (mtDNA): The circular duplex DNA of mitochondria, which is found in about 5-15 copies per organelle. The mtDNA of most organisms is by far smaller than the nuclear DNA (e. g. 15-20 kb in animals; from 200-2500 kb in flowering plants) and codes for a series of mitochondrial proteins (such as mitochondrial ribosomal proteins, → elongation and termination factors) and RNAs (various → transfer RNAs and → ribosomal RNAs). Most (and especially strategic) mitochondrial proteins are encoded by nuclear genes, and a coordinated interaction between the mtDNA and the nuclear DNA is required for maintenance and division of mitochondria. A series of mutations in mtDNA cause human diseases. Since the proportion of mitochondria with mutations at specific sites in mtDNA varies after repeated cell divisions, mitochondrial diseases are highly variable. For example, *Leber hereditary optic neuropathy* (LHON), *myoclonic epilepsy and ragged red fibers* (MERFF), *Kearns-Sayre syndrome* (KSS) with opthalmoplegia, pigmentary degeneration of the retina, and cardiomyopathy are few examples of such mitochondrial diseases.
Figure see page 673.

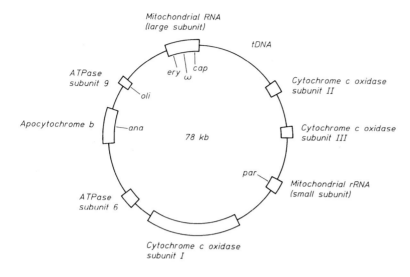

Simplified map of yeast mitochondrial DNA

Ana	:	Antimycin A resistance gene
Cap	:	Chloramphenicol resistance gene
Ery	:	Erythromycin resistance gene
Oli	:	Oligomycin resistance gene
Par	:	Paromomycin resistance gene
ω	:	Locus interfering with recom - bination of mitochondrial genes

***Mi*tochondrial DNA depletion (mtDNA depletion):** The complete removal of → mitochondrial DNA from a → mitochondrium, caused by environmental toxins interfering with mtDNA → replication, as e.g. *azido*thymidine (AZT), that inhibits DNA replication catalysed by the mitochondrial → DNA polymerase γ- mtDNA depletion can be a cause for recessive Mendelian human disorders. Do not confuse with → DNA deletion.

***Mi*tochondrial DNA lineage (mtDNA lineage):** A group of closely related and therefore mostly homologous mitochondrial DNA molecules.

***Mi*tochondrial transcription *termination factor* (mTERF):** A → leucine zipper DNA-binding protein that regulates the transcription termination of mitochondrial genes. For example, a specific mTERF binds to a 28 bp region at the 16S rRNA-leucyl tRNA genes boundary, promotes termination of 16S rRNA gene transcription, and thereby regulates the ratio of rRNA/mRNA in mitochondria.

Mitochondrium (mitochondrion; Greek: mitos-thread, chondrion-grain): Any one of hundreds or thousands of cytoplasmic semiautonomous organelles of eukaryotic cells, that is surrounded by a double membrane and carries a series of → mitochondrial DNA molecules encoding relatively few proteins needed for mitochondrial functions. Most of the mitochondrial proteins are encoded by nuclear genes, whose → messenger RNAs are translated on cytoplasmic → ribosomes. The translated proteins contain socalled signal peptides for their import into the organelle. Among the nucleus-encoded proteins are strategic proteins as e.g. a series of ribosomal proteins, porins, DNA and RNA polymerises, enzymes of the citrate cycle, subunits of the ATPase, cytochrome c oxidase and cytochrome bc_2 complex, to name only few. Main functions of the mitochondria are the electron transport chain with the coupled oxidative phosphorylation, the citric acid cycle, and the oxidative degradation of fatty acids.

Mitogen-*a*ctivated *p*rotein kinase kinase (MAP kinase kinase, MEK kinase): Any one of a family of protein kinases (e.g. MEK–1 to MEK–7), that activate MAP kinases by phosphorylation. For example, the prototype MEK kinase, MEK–1, specifically phosphorylates strategic threonine and tyrosine residues of the sequence H_2N-thr-glu-tyr-COOH in the MAP kinase protein.

Mitomycin C: An aziridine → antibiotic produced by *Streptomyces caespitosus* that cross-links complementary strands of a DNA duplex molecule, and thereby prevents DNA → replication and → transcription.

Mitomycin C

Mixed infection: See → marker rescue.

Mixed oligonucleotide-primed amplification of cDNA (MOPAC): A technique for the isolation of genes of far-reaching homology (e. g. genes of a → gene family) by deducing → primers for → polymerase chain reaction from peptide sequences. Due to the → degenerate genetic code usually a series of primers differing in nucleotide sequence are generated from one peptide sequence and used to amplify the corresponding sequence out of a → cDNA library. Usually two degenerate pools of → oligonucleotides (all sequence variants which may encode the same set of amino acids according to the → degenerate code) to prime → first strand cDNA amplification. In short, mRNA is first reverse-transcribed into first strand cDNA, using → reverse transcriptase. Then two pools of oligonucleotides (pool 1: oligos complementary to all possible sequences encoding a particular tract of amino acids in the target protein; pool 2: oligos complementary to all possible sequences encoding another tract of amino acids in the same protein) are annealed as → amplimers to the first-strand cDNA. The amplified product can then be cloned into appropriate vectors and used as probe to screen → genomic or → cDNA libraries.

Mixed oligonucleotide probe (mixed oligo probe): A mixture of synthetic single-stranded oligo deoxynucleotides about 12-25 bases in length, that differ from each other in one single base only. Mixed oligo probes are used to screen → genomic or → cDNA libraries for a gene whose protein product is known and whose sequence has been inferred from the corresponding amino acid sequence. However, the → codon bias does not allow to deduce the correct nucleotide sequence of the corresponding gene indubitably. The use of mixed oligos increases the probability that at least one perfectly matched oligonucleotide will detect the desired gene. Frequently, inosine is incorporated as the → wobble base, since it base-pairs with most of the other bases.

Mixed primer labeling: See → random priming.

Mixed target *polymerase chain reaction* (mixed target PCR): A conventional → polymerase chain reaction, in which two (or more) target → template DNAs are present and simultaneously amplified.

MLP: See → *multilocus probe*.

MLST: See → *multilocus sequence typing*.

MM: See → *mismatch*.

MM: See → base mismatch.

MMA: See → *multiplex messenger assay*.

MMGT: See → *microcell-mediated gene transfer*.

M-MLVRT: See → *moloney murine leukemia virus reverse transcriptase*.

MMR: See → mismatch repair.

MMS: See → *marker-mediated selection*.

MNP: See → multiple nucleotide polymorphism.

MOB: See → *microsatellite obtained from BAC*.

Mob (mob, mob **functions, *mobilizing functions*):** Two defined regions of a conjugative → plasmid, of which one encodes a mobilizing protein that specifically binds to the *mob* region, and induces a → nick in its so-called → *nic/bom* site (*nic* for *nick*; *bom* for *basis of mobility*). One of the *mob* functions (synthesis of a mob protein) can also be supplied in trans. If for example, one plasmid has lost this property (*mob⁻*), a second plasmid coding for a functional mob protein (*mob⁺*) may complement the *mob* functions so that the deficient plasmid can be transferred from one cell to another. In addition to the *mob* regions functional → *tra* genes are also necessary for plasmid transfer. See also → mobilization.

Mobile domain: Any typically compact, cystein-rich → domain of 30–130 amino acids in a → mosaic protein, that is able to fold independently of other domains and is evolutionarily mobile, i.e. has spread during evolution and now occurs in many functionally unrelated proteins.

Mobile genetic element: Synonym for → transposon.

Mobility-shift DNA-binding assay (band shift, band shift assay, DNA-binding assay, gel electro-phoresis DNA-binding assay, gel mobility shift assay, gel retardation assay): A method to detect specific DNA-protein interactions that is based on an altered mobility of protein-DNA complexes during non-denaturing gel electrophoresis, as compared to free DNA. In short, the target DNA fragment is labeled (e.g. end-labeled with γ-^{32}P-dATP using deoxynucleotidyl transferase) and incubated with a nuclear extract. Specific DNA-protein complexes are detected by low ionic strength → polyacrylamide gel electrophoresis and → autoradiography. The free fragment moves faster than the protein-DNA complex (which is retarded). Usually an excess of hetero-logous competitor DNA is added to saturate the more abundant, non-specific DNA-binding proteins. See also → electrophoretic mobility shift assay.

Mobilization:
a) The directed movement of a non-conjugative plasmid from one bacterium (donor) to another bacterium (acceptor) with the aid of a → conjugative plasmid of the donor.
b) The directed movement of chromosomal genes of one bacterium (donor) to another bacterium (acceptor) with the aid of a → conjugative plasmid of the donor.
c) The movement of → transposons (→ transposition) or → retrosequences.

Mock infection: A laboratory term for a fictive infection of cells with a → bacteriophage or → virus, in which the cells were either not exposed to the infectious agent or treated with killed agents, but otherwise processed as the truly infected cells.

Moderately affected Alzheimer disease DNA (MAD-DNA): The DNA from cells of the hippo-campal region of patients with weak to moderate Alzheimer's disease symptoms. MAD-DNA shows a modified → B-DNA conformation with a small shoulder peak at 290 nm of its spectrum, binds more → ethidium bromide than → severely affected Alzheimer disease DNA, and exhibits an unusual biphasic melting profile with two T_M values of 54°C and 84°C.

Moderately repetitive DNA: See → middle repetitive DNA.

Modification:
a) Any change in a protein or nucleic acid molecule after its synthesis.
b) See → DNA methylation, also → restriction-modification system.

Modification enzyme: Any enzyme that modifies → DNA or → RNA. Typcial examples for modification enzymes are → bacterial alkaline phosphatase, → calf intestinal alkaline phos-phatase, → DNA polymerase I, → DNase I, → exonucleases, → Klenow fragment, → modifi-cation methylase, → mung bean nuclease, → nuclease P1, → nuclease S1, → reverse transcrip-tase, → RNA polymerase, → RNase, → T4 DNA ligase, → T4 DNA polymerase, → T4 RNA ligase, → T7 DNA polymerase, → T7 RNA polymerase, → T3 RNA polymerase, → terminal transferase.

Modification gene: See → modifier gene (b).

Modification methylase (DNA modification methyltransferase, modification enzyme, EC 2.1.1.72): A bacterial enzyme that catalyzes the transfer of methyl groups from → S-adenosyl-L-methio-nine to specific positions of specific bases in DNA. Since the methylation of such bases within the recognition sequence of a → restriction endonuclease prevents the recognition process and thus the cleavage at this sequence, methylation protects bacterial DNA against own and foreign restriction enzymes (→ restriction-modification system). Some modification methylases with high specificity for distinct recognition sequences of particular endonucleases are used to protect internal recognition sequences of a DNA fragment which is to be cloned (e.g. → *Eco* RI me-thylase specifically methylates bases in the recognition sequence of *Eco* RI endonuclease). If for example, such a fragment contains *Eco* RI sites, any cloning into an *Eco* RI site of a → cloning vector would be obsolete, since excision of the insert by *Eco* RI would inevitably destroy it. If, however, the *Eco* RI sites of the fragment are methyl-protected, then the use of *Eco* RI linkers and cloning into *Eco* RI sites of vectors becomes feasible. See also → heteroprostomers and → iso-prostomers, → Dam methylase, → Dcm methylase, and compare → methyltransferase.

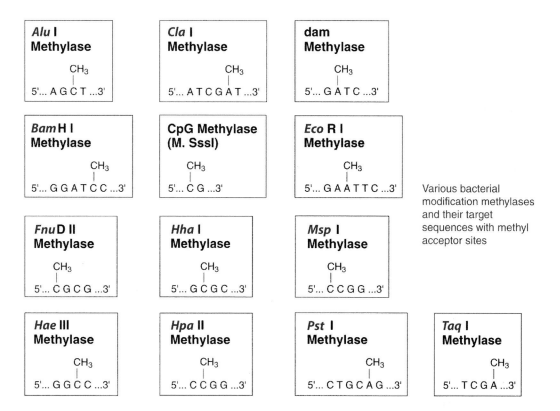

Various bacterial modification methylases and their target sequences with methyl acceptor sites

Modified adenine recognition and restriction system: See → methylated adenine recognition and restriction (Mrr) system.

Modified base:
a) Any nucleic acid base that is altered postsynthetically, e.g. by methylation. See → DNA methylation, → restriction-modification system.
b) See → rare base.

Modified cytosine restriction (Mcr) system: See → methylated cytosine recognition and restriction system.

Modified deoxynucleoside-5'-triphosphate: Any one of a series of synthetic → deoxynucleoside triphosphates, into which an additional group is incorporated, as e.g. an amino, azido, or methyl group (see formulas). Such analogues are used for DNA-protein interactions and inhibition of DNA polymerases. See → modified ribonucleoside–5'-triphosphate.

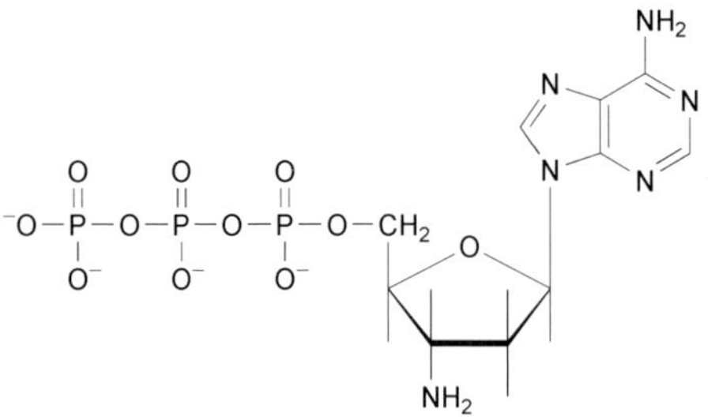

3'-Amino-2', 3'-ddATP

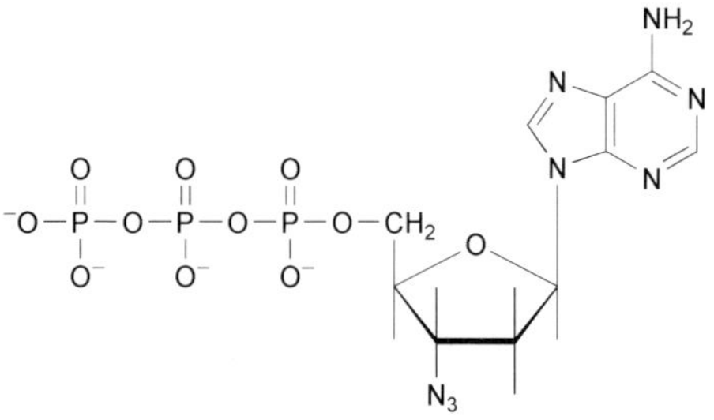

3'-Azido-2', 3'-ddATP

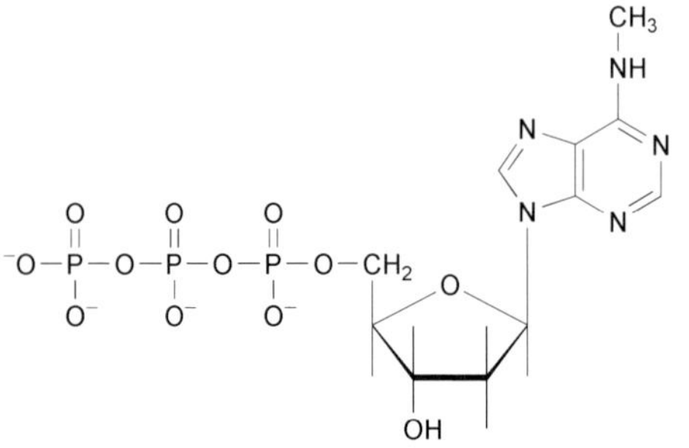

N^6-Methyl-2'-dATP

Modified ribonucleoside–5'-triphosphate: Any one of a series of synthetic → ribonucleoside triphosphates, into which an additional group is incorporated, as e.g. an amino, azido, or methyl group (see formulas). Such 2'-modified analogues prevent nuclease degradation of RNAs, and 3'-modified nucleotides inhibit RNA polymerases. See → modified deoxynucleoside–5'-triphosphate.

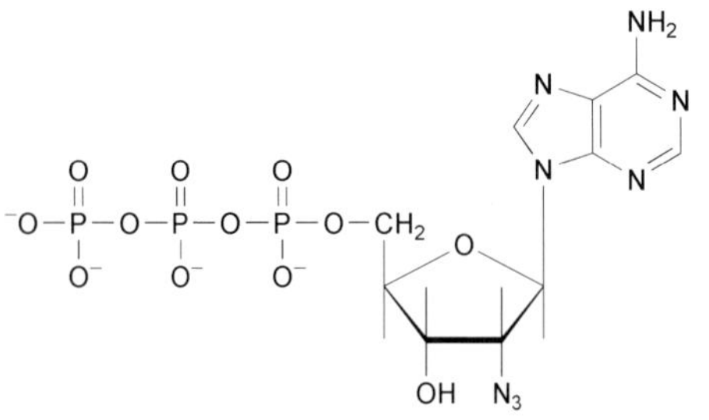

2'-Azido-2'-dATP

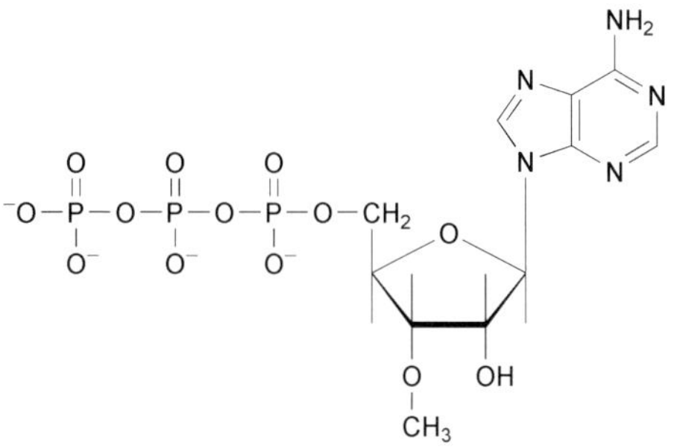

3'-O-Methyl-ATP

NH₂

O O O

$^-O-P-O-P-O-P-O-CH_2$

NH₂ OH

3'-Amino-3'-dATP

Modifier:
a) Any gene that modulates the phenotypic expression of one or more other genes.
b) Any DNA sequence motif that is located 5' upstream of a → promoter and either enhances (→ enhancer) or reduces (→ silencer) the rate of expression of a gene located downstream.

Modifier gene:
a) Any gene that either controls or affects → DNA methylation or → genomic imprinting.
b) Any mammalian gene (or → gene family), that modifies a trait encoded by another gene (or genes), and affects penetrance (i.e. the frequency with which affected individuals occur among carriers of a particular genotype), → dominance, expressivity (i.e. the extent to which specific processes are influenced by a particular → genotype), and → pleiotropy. Modifier genes may be the cause for extreme phenotypes ("enhanced phenotypes"), less extreme ("reduced"), novel ("synthetic") or also wild-type ("normal") phenotypes. For example, a dominant modifier gene on human chromosome 7 reduces penetrance of a non-syndromic deafness gene (linked to *DFNB26*) on chromosome 4q31, that leads to hearing loss in the homozygotic state. However, several individuals in carrier families, that are homozygous for *DFNB26* nevertheless hear normally: an effect of the modifier gene. Or, a modifier gene on chromosome 13q affects genes responsible for familial hypercholesterolaemia, an autosomal dominant trait (affecting one person in 500 and causing elevated cholesterol levels). Familial hypercholesterolaemia homozygotes often die of cardiovascular diseases. However, some individuals in afflicted families have *low density lipoprotein* (LDL) levels 25% lower than expected: again the effect of the modifier gene.
Modifier genes influence various phenotypes, preferentially in mice, rats and humans (e.g. gene *brachyura* [T] modifies tail length. [mouse], *Pax 3* ^{Sp} suppresses spina bifida [mouse], and *Cfm* 1 a meconium ileus [mouse]).
c) Modifier of *mdg4*: A specific *Drosophila melanogaster* gene, of which both DNA strands are transcribed into different pre-messenger RNAs. These in turn are then ligated into one single mRNA, that is translated into a protein. The modifier of *mdg4* corrupts the dogma, that only one strand (see → antisense strand, → coding strand) encodes the mRNA of a gene.

Modular array (modular microarray, modular chip): Any → microarray, that consists of several arrays separated by e.g. microfluidic hybridisation chambers. Each module can be separately used for specific experiments.

Modular microarray: See → modular array.

Modular vector: Any → cloning vector that is composed of a series of easily exchangeable → modules. For example, a modular plant transformation vector could contain the right and left → T-DNA border sequences, unique → restriction recognition sites for the insertion of → foreign DNA, a strong constitutive → promoter (e. g. the CaMV 35S promoter), a → Kozak consensus sequence, a → flag sequence, a → histidine tag, and a → 3' untranslated region as useful modules.

Module: Any DNA sequence that contains one or more conserved sequence → motifs and encodes a specific function (as e. g. the → TATA box as a module of → promoters of regulatable genes) or a specific domain of an RNA or a protein (as e. g. → exons as modules of genes).

```
A G C A C C C GGT   ACACTGTGTC   CT C C C G C T G C   A C C C A G C C C C   T T C A G C G C G A
              GRE

G G C G T C C C C G   A G G C G C A A G T   G G G C G G C C T T   C A G G G A A C T G   ACCGCCCGCG
                                                                                       AP-2

G C C C G T G T G C   A G A G C C G G G T   GCGCCCGGCC   CA G T GCGCGC   GGCCGGGTGT
                                 MRE 4                      MRE 3

T T C G C C T G G A   G C C G C A A G TG   ACTCA G C G C G   G G G C G T G T G C   A G G C AGCGCC
                              AP-1                                                      MRE 2

CGGCCGGGGC   GGG C T T T T G C   A C TCGTCCCG   GCTC T T T C T A   G C TATAA A C A
      Sp-1                              MRE 1

C T G C T T G C C G   C G C T G C A C T C   C A C C A C G C C T   C C T C C A A G T C   C C A G C G A A C C
                                 +1 ————————▶ RNA

C G C G T G C A A C   C T G T C C C G A C   T C T A G C C G C C   T C T T C A G C T C   G C C ATGGATC
```

Modular Architecture of the Metallothionein Gene Promoter

Module-shuffled primer: Any → primer composed of six modules, each consisting of three or four nucleotides. All module-shuffled primers contain the same modules, but in different arrangements. Modules with three nucleotides carry a C at both their 5'- and 3'-termini, modules with four nucleotides a T at both ends. Therefore, the modules of module-shuffled primers are always connected by either C/T or T/C:

    5'-CCC-TTCT-CAC-TGTT-CTC-TCAT-3'
    or 5'-CAC-TCAT-CTC-TTCT-CCC-TGTT-3'
    or 5'-CTC-TGTT-CCC-TCAT-CAC-TTCT-3'

Since the different module-shuffled primers differ only by the order, in which the otherwise identical modules are arranged, their sequences are unique, but their → melting temperatures are identical. Module-shuffled primers hybridise only with → complementary strands and have identical PCR amplification efficiency in conventional → polymerase chain reaction. A mixture of module-shuffled primers labeled with different → fluorochromes (e.g. → FAM, → HEX, → cyanin 5) is used for → module-shuffling primer PCR.

Module-shuffled primer polymerase chain reaction (**MSP-PCR, module-shuffled primer PCR, multiplex PCR with colour-tagged module-shuffling primer):** A variant of the conventional → *polymerase chain reaction* (PCR) for the comparative → gene expression profiling in different cells, tissues, or organs, that employs specifically designed so called → module-shuffled primers to drive the analysis of several to multiple genes in one single reaction tube. In short, total RNA is first isolated, → poly(A)$^{+}$→ messenger RNA (mRNA) extracted and converted to double-strandede → cDNA by any conventional method (e.g. → RNA priming). The resulting cDNAs are then digested with a → four-base cutter → restriction endonuclease, leading to fragments averaging 256 bp. Three different → oligonucleotide → adapters corresponding to three different module-shuffled primers are then prepared and separately ligated to the restriction fragments from three different original cDNA preparations (i.e. from three different sources). The double-stranded adapters are designed such that after → ligation to the cDNAs, they form a Y-shaped end with one recessed strand, which avoids → priming by two module-shuffled primers. The adapter-ligated fragment populations are then mixed in equal quantities and serve as PCR → template. Each target cDNA fragment in the mixture is amplified with a primer pair consisting of one member of the module-shuffled primer mixture and a → gene-specific primer, where the module-shuffled primer discriminates between the sources of each amplified gene. PCR products are then analysed in a → DNA sequencer, and each fragment is identified by its specific electrophoretic mobility and the specific emission light wave length of its → fluorochrome. MSP-PCR circumvents the need for internal standards and a calibration curve, because the same genes from different sources are simultaneously amplified and directly measured and compared in one run. See → adapter-tagged competitive PCR, → enzymatic degrading subtraction, → gene expression fingerprinting, → gene expression screen, → linker capture subtraction, → preferential amplification of coding sequences, → quantitative PCR, → targeted display, → two-dimensional gene expression fingerprinting. Compare → cDNA expression microarray, → massively parallel signature sequencing, → microarray, → serial analysis of gene expression.

Molecular agriculture: See → molecular farming.

Molecular beacon: A single-stranded → oligonucleotide that contains a → fluorochrome (e. g. → fluorescein, → TAMRA, → Cy3, → Cy5, → Texas red) at its 5'-terminus and a non-fluorescent quencher dye (e. g. [4(4–(dimethylamino)phenyl)azo] benzoic acid; DABCYL) at its 3'-terminus. The sequence of such a molecular beacon is designed such that it forms a → hairpin structure intramolecularly, with a 15-30 bp probe region (complementary to the target DNA), and 5 – 7 bp long stem region (self-complementary). In this folded state the fluorochrome is quenched (i. e. any photon emitted by the fluorophore through exciting light is absorbed by the quencher [e. g. TAMRA], and emitted in the non-visible spectrum). After binding to a homologous target sequence, the beacon undergoes a conformational change forcing the stem of the hairpin apart, displacing the fluorochrome from the quencher, and abolishing the quenching (i. e. fluorescence occurs). Such molecular beacons are used for quantitation of the number of → amplicons synthesized during conventional → polymerase chain reaction, for the discrimination of → homozygotes from → heterozygotes, the detection of → single nucleotide polymorphisms, in situ visualization of → messenger RNA within living cells, and the simultaneous detection of different target sequences in one sample, if different fluorochromes with differing emission spectra are used. See → aptamer-beacon, → gene pin.
Figures see page 682 and 683.

Molecular biology: A modern branch of biology that studies biological phenomena and processes at the molecular level, using physical, chemical and physico-chemical methods. Compare → molecular genetics, a specific field of molecular biology.

Molecular breeding: The application of the whole repertoire of → genetic engineering and → molecular marker technologies for the improvement of fungi, plants, and animals.

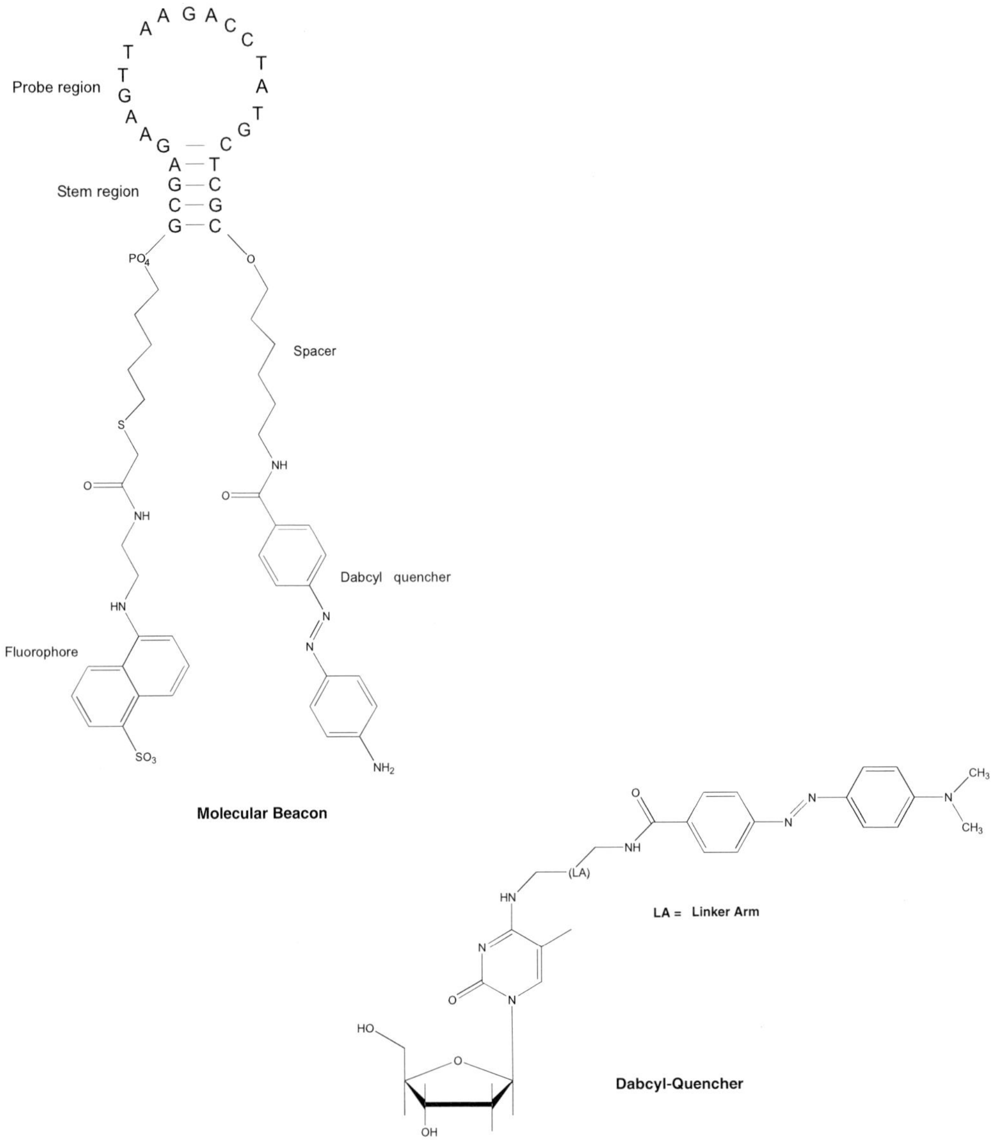

Probe region

Stem region

PO₄

Spacer

S

NH

O

NH

HN

NH

O

Dabcyl quencher

Fluorophore

SO₃

NH₂

Molecular Beacon

(LA)

HN

NH

O

N

N

CH₃

CH₃

LA = Linker Arm

HN

N

O

N

HO

O

OH

Dabcyl-Quencher

Molecular beacon

Molecular brightness (q): The number of photons emitted by a → fluorochrome at a given excitation intensity. A change in molecular brightness as e.g. induced by a specific binding process (of a protein with a second protein, or a protein with its recognition sequence on a DNA) results in a change in total fluorescence intensity.

Molecular chaperone: See → chaperone.

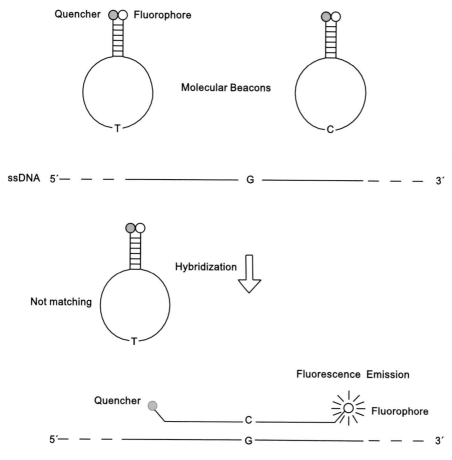

Molecular beacon

Molecular clock: The hypothetical regular rate of nucleotide → substitutions or amino acid replacements over time. Considerations of a molecular clock are part of the study of rates of molecular evolution.

Molecular cloning: See → cloning.

Molecular combing: The stretching of cloned or native DNA molecules on a microscope slide. In short, the termini of solubilized DNA are bound to the silanated surface of a glass slide, the DNA solution covered with an untreated coverslip such that the drop spreads uniformly, and the solution evaporated. This procedure leads to the straightening and stretching of the bound DNA molecules, so that the labeled → probes can more easily be hybridized as compared to condensed chromosomes. See → chromosome stretching, → dynamic molecular combing.

Molecular docking: The prediction of the molecular geometry and binding affinity between two (or more) different molecules (e.g. a protein and a low molecular weight → ligand). It includes the calculation of the relative orientation and the conformational space of both the ligand and the protein, the geometric characterization of all potential molecular interaction(s), energy functions and possible torsion angles, and the major energy contributions (e.g. hydrogen and salt bridges, hydrophobic contacts). The process of molecular docking ("modelling") may start with e.g. the ligand, which is first dissected into smaller fragments. Each of these fragments are then placed into the active center of the protein (in case of enzymes) or a specific domain with a particular function. After finding the best fit, other fragments of the ligand are then successively incorporated in an energetically favourable way, unless the whole molecule is assembled and accommodated in the protein structure.

Molecular epidemiology: An ill-defined multidisciplinary approach to detect causative relationships between the genetic background (genes, mutated genes) and diseases, and its modification by racial and environmental parameters. Molecular epidemiology applies a series of different technologies spanning from the design of family studies for linkage analyses, the definition of factors influencing → linkage, localization of the relevant chromosomal regions or underlying gene(s) to the sequence variants of these genes, their frequencies in different populations, and their diagnosis. For example, the genetic causes of atherosclerosis, asthma, diabetes, schizophrenia or cancerogenesis and association with genetic markers (e.g. a mutation in the chemokine receptor gene as indicator for asthma bronchiale) are research areas of molecular epidemiology. Another research focusses on the distribution of mutant alleles in various populations. For example, the variants IVS10, I65T, E280K, and P281L are frequent in Spanish phenylketonuria patients, whereas R 408W and IVS12 prevail in Scandinavian patients.

Molecular farming (biofarming, molecular agriculture): The exploitation of plants for the production of peptides and proteins of biomedical, therapeutical and/or technical applications (e.g. → antibodies, industrial enzymes). For example, the first commercially available product of molecular forming was a recombinant avidin from chicken egg, produced by transgenic maize (*Zea mays* L.). Other products encircle a secretory IgA antibody against the causal agent of caries, *Streptococcus mutans* (tobacco), an antibody effective against herpes simplex virus (soybean), and a tumor-specific antibody (maize). Compare → gene farming.

Molecular forceps: See → DNA forceps.

Molecular genetics: A branch of genetics that studies the structure and function of genes and genomes at the molecular level, using physical, physico-chemical and chemical techniques. Compare → molecular biology.

Molecular hybridization: See → hybridization.

Molecular imprinting: The generation of a polymeric mold around target molecules. Usually the polymer (e.g. → polyacrylamide) is synthesized from functional monomers, that assemble around the target template. If the target is then removed, it leaves an impression on the polymer's surface, which can e.g. be used to selectively capture molecules that interact with the target molecule. For example, protein capture arrays are based on molecular imprinting. In this case, a glass or quartz slide, onto which target peptides are spotted, is coated with a mixture of monomers and crosslinkers, that polymerises to a hydrogel layer over the peptides. When the film is removed, it carries the imprint of the immobilized peptides. See → molecularly imprinted electrosynthesized polymer, → molecularly imprinted polymer.

***Molecularly imprinted electrosynthesized polymer* (MIEP):** A variant of a → molecularly imprinted polymer, that is produced by growing the polymer around a template molecule on an electrode of a quartz crystal. The circulated charge controls the thickness of the polymer. For example, *poly(o-*phenylene*di*amine, PPD) in the presence of a low molecular weight template molecule (e.g. glucose) can electrochemically be polymerised into thin films (progressively covering the electrode), which can be used as a biomimetic sensor for glucose.

Molecularly *i*mprinted *p*olymer (MIP): Any synthetic support, that mimics the three-dimensional shape of a target molecule and serves as an affinity capture mold largely resistant to biological degradation, acidity, extreme pH values, high temperature and other experimental conditions. Molecular imprinting starts with the polymerisation of functional monomers (e.g.*meth*acrylic *a*cid, MAA, and 2–(*tri*fluoromethyl)acrylic *a*cid, TFMAA) and crosslinkers (e.g. ethylene glycol dimethacrylate) in the presence of a polymerisation catalyser (e.g. 2,2'-azobisisobutyronitrile), a pore former (e.g.chloroform) and an imprinting molecule of choice (e.g. a low molecular weight compound such as a sugar or acid, or a peptide or protein as a template molecule) at 5°C under UV. Then the template molecule is removed from the resultant polymer network to leave a template-fitted cavity with template-selective binding capacity. MIPs are exploited e.g. for ligand-binding experiments, where they mimic the role of → antibodies in immunoassays. See → molecularly imprinted electrosynthesized polymer.

Molecular marker (DNA marker): Any specific DNA segment whose base sequence is different (polymorphic) in different organisms and is therefore diagnostic for each of them. Such markers can be visualized by → hybridization-based techniques (e. g. → DNA fingerprinting, → restriction fragment length polymorphism) or → polymerase chain reaction-based methods (e. g. → DNA amplification fingerprinting, → random amplified polymorphic DNA, → sequence-tagged microsatellite sites). Ideal molecular markers are highly polymorphic between two organisms, inherited codominantly (i. e. allow to discriminate homo- and heterozygotic states in diploid organisms), distributed evenly throughout the genome and easily to be visualized. Moreover, molecular markers should occur frequently in the genomes, should easily be visualized, and be stable over generations. No single marker system fulfills all these criteria. Molecular markers (to which also protein markers, e. g. → isoenzymes belong) are used for the genotyping of single organisms, the detection of genetic variation(s) between organisms, the identification of hybrids, paternity testing, generally genetic diagnostics, and → genetic mapping. See → morphological marker.

Molecular medicine: A branch of medical sciences, that employs techniques of→ molecular genetics and → gene technology to unravel disease processes at the molecular level. Molecular medicine aims at diagnosing, preventing, treating and curing various human and animal diseases and also to develop animal models for human disorders. See → genetic medicine.

Molecular mimicry: See → molecular piracy.

Molecular pathology: A scientific discipline, that focuses on the interaction(s) between a pathogen and its host(s) on the molecular level.

Molecular piracy (molecular mimicry): The capture of host genes by pathogens (e.g. viruses or remainders of viruses as herpes or pox viruses) in host genomes.

Molecular sensor: See → sensor probe.

Molecular sieving: See → gel filtration.

Molecular switch: Any supramolecular assembly that can switch from one to another conformation and back, thereby generating nanomechanical power. For example, a → DNA forceps represents such a molecular switch.

Molecular syringe: A tube-like structure formed by some Gram-negative bacteria pathogenic to plants and animals upon contact with the host cell, that allows to secrete virulence factors directly into the cell. These factors are translocated from the bacterial cytoplasm to the host membrane or even into the host cytoplasm, passing three membranes consecutively. In some cases, the socalled secretion system III is assembled upon host cell contact, and a syringe-like structure formed, that spans both inner and outer bacterial membranes. The system contains a translocator capable of channeling virulence proteins into the host cell. As an example, protein EspD of *enteropathogenic E.coli* (EPEC), as part of a pore, inserts into the host cell membrane (here: small intestinal mucosa cells), and is thought to catalyze such translocation processes.

Molecular tweezers: See → DNA forceps.

Molecular weight: The sum of the weights of all atoms of a molecule.

Molecular weight marker: See → molecular weight standard.

Molecular weight standard (molecular weight marker): A mixture of different peptides, proteins or nucleic acid fragments with known molecular mass that are used for the calibration of the molecular weight of proteins or nucleic acid molecules, after their separation by → gel electrophoresis. Compare → binning marker, see → marker.

Molecular writing: The graphical description of the surface of a material (e.g. a protein, DNA) by → atomic force microscopy.

Molecule: A complex of two or more identical or non-identical atoms that has a specific chemical property or properties different from those of the constituent atoms.

Moloney murine leukemia virus **(M-MLV)** *reverse transcriptase* **(RTase), RNase H minus (EC 2.7.7.49):** A genetically modified → M-MLV reverse transcriptase, from which the → RNase activity has been removed. It is used to synthesize → cDNA, in → RT-PCR, RNA sequencing and → filling in 5' overhangs and → primer extension.

Moloney murine leukemia virus **(M-MLV)** *reverse transcriptase* **(RTase; EC 2.7.7.49):** A single polypeptide enzyme from Moloney murine leukemia virus (M-MLV) that catalyzes the synthesis of a DNA strand from single-stranded RNA or DNA as a template requiring a → primer. The enzyme lacks → endonuclease activity, but has low → RNase H activity. It is used for full-length cDNA synthesis from large mRNAs (up to 10 kb) using oligo(dT) primers annealing to the poly(A) tail of the mRNA, and for → filling-in 5'overhangs. The enzyme is also available as recombinant DNA product completely devoid of RNase H activity, so that it no longer attacks the primer-poly(A) hybrid or the RNA-DNA hybrid arising from the reverse transcription. See → M-MLV RT, H minus.

Monoallelic expression: The transcription of only one of two → alleles in diploid organisms.

Monobromobimane: A → fluorochrome for the labeling of proteins, that reacts with cystein residues. The compound is directly added to protein-containing gels, and the derivatized proteins can be visualized as fluorescent turquoise bands under UV light. Monobromobimane does not allow to label a series of cystein-free proteins (as e. g. myoglobin, concanavalin A, or cytochrome b5).

Monocistronic mRNA (monogenic mRNA): Any messenger RNA that codes for only one single polypeptide chain (in contrast to a polycistronic mRNA that codes for more than one protein).

Monoclonal: See → monoclonal antibody.

Monoclonal antibody (mAb, "monoclonal"): A population of immunoglobulins originating from one single clone of plasma cells, and therefore consisting of structurally and functionally identical antibodies. In the organism, monoclonals are produced by tumorous cells of the immune system (myelomas). However, it is experimentally possible to fuse such myeloma cells with antibody producing plasma cells (B lymphocytes). These hybrid cells proliferate into a clone that can be maintained in tissue culture and be used as a source of monoclonal antibodies. Monoclonal antibodies are important tools for the characterization and identification of proteins. See also → plant antibody.

Monoclonal antibody (mAb, "monoclonal"): Any one of a population of immunoglobulins originating from one single clone of plasma cells, and therefore consisting of structurally and functionally identical antibodies. In the organism, monoclonals are produced by tumorous cells of the immune system (myelomas). However, it is experimentally possible to fuse such myeloma cells with activated antibody-producing plasma cells (B lymphocytes) to socalled hybrid (→ hybridoma) cells, that are immortal, grow permanently *in vitro*, and produce and secrete practically unlimited amounts of identical mAbs. In short, animals (e.g. mice, rats, sheep) are first immunized with the specific → antigen, then the required cells (about 10^8 lymphocytes) are isolated from the spleen, and fused with myeloma cells, that carry a defective gene encoding an enzyme of the nucleotide metabolism such that they can selectively be removed after cell fusion. This fusion can be performed with either *poly*ethyleneglycol (PEG), viruses, electrofusion or also laser fusion. A fusion is a relatively rare event (10^{-4}), and the fused cells have to be selected by HAT (H: hypoxanthine, A: aminopterine, T: thymidine). The fusion products usually are distributed into microtiter plates with 96 wells such that about 5×10^4 cells are contained in one well (half a spleen original material). One week later the supernatants of the fused cells are tested for the secreted mAb by e.g. → enzyme-linked immunosorbent assay. Since several hybridoma clones with different specificities may reside in one well, the cells have to be cloned either by limited dilution or → flow cytometry. For mass production, the mAb-encoding cells are either injected into and grown in a mouse peritoneum ("Ascites technique"), synthesized by → transgenic plants (e.g. tobacco) or → transgenic animals (sheep, cow), or grown in various types of bioreactors (e.g. spinner flasks, stirred tank fermenter, airlift fermenter, hollow fiber reactor). Monoclonal antibodies can also tailored by → genetic engineering. For example, murine antibodies can be "humanized" by coupling the variable region of the murine antibody to the Fc region of a human antibody, creating a chimeric antibody. After integration of the hypervariable regions (*comple*mentarity *d*etermining *r*egions, CDRs) of the murine antibody into a human antibody a fully humanized antibody is regenerated (CDR-grafted or reshaped monoclonal antibody), that still retains its binding specificity. Antibody-encoding DNA can also be tailored by → site-directed mutagenesis. For example, mutations in the hypervariable regions or the Fc region can improve important properties of the resulting antibody (e.g. binding affinity, target specificity, biological half-life time). Monoclonal antibodies can also be expressed in *E. coli* ("coliclonal antibodies"), but cannot be glycosylated by the bacterium. Therefore these antibodies may be immunogenic. Generally, monoclonal antibodies are important tools for the identification and characterization of peptides and proteins, and are also components of → antibody arrays. See → catalytic antibody.

Monocuts: Laboratory slang for two (or even more) fragments of defined size arising through cleavage of → lambda DNA with a → restriction endonuclease that cuts only once.

Monogenic mRNA: See → monocistronic mRNA.

Monomer:
a) Any basic unit from which polymers are made (e.g. amino acid monomers are polymerized by peptide bond formations into the polymer protein).
b) A subunit of a supramolecular, multimeric complex (e.g. a protein consisting of different or identical polypeptide chains).

Mononucleotide editing: A variant of → RNA editing in the mitochondria of the slime mould *Physarum polycephalum* and several other members of the phylum Myxomycota (e.g. *Stemonitis* and *Didymium*), which is characterized by the insertion of mononucleotides in RNAs relative to their mtDNA template. The most commonly inserted mononucleotide is cytidine, although a number of uridine mononucleotides are inserted at specific sites, whereas adenosine and guanosine are not at all inserted. See → dinucleotide editing, → transfer RNA editing.

Monosome:
a) One mRNA-ribosome complex.
b) Any → chromosome that has no homologous counterpart.
c) A single → nucleosome.

MOPAC: See → *m*ixed *o*ligonucleotide-*p*rimed *a*mplification of *c*DNA.

MOPS buffer (*m*orpholino-*p*ropane *s*ulfonic acid buffer): A synthetic zwitterionic buffer with a pK$_a$ of 7.2 widely used in biochemical experiments.

***Morgan* unit (M):** A measure for the relative distance between two genes on a chromosome, or, concomitantly, for the frequency of → recombination between two genetic markers. One Morgan corresponds to the length of a chromosome in which, on average, one recombination event (a → cross-over or a → chiasma) occurs each time a gamete is formed. One Morgan is equivalent to a crossover value of 100 %, a *centi*Morgan (cM) corresponds to 1 % crossover value and to 0.01 Morgan.

Morpholino oligonucleotide ("morpholino"): Any non-ionic oligonucleotide with a backbone different from the phosphodiester backbone of DNA or RNA. Morpholino antisense oligonucleotides are used for e.g. gene silencing. To that end, they are transferred into target cells via → electroporation or → microinjection, bind to complementary target sequences by → Watson-Crick base-pairing and silence the expression of genes. Morpholinos are resistant to exo- and endonucleases and do not interact with e.g. proteins non-specifically, because they are not charged.

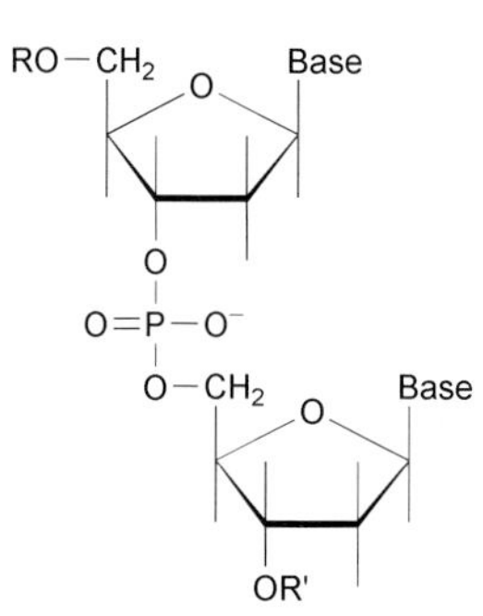

RO—CH₂   Base

O

O=P—O⁻

O—CH₂   Base

OR'

Phosphodiester bond

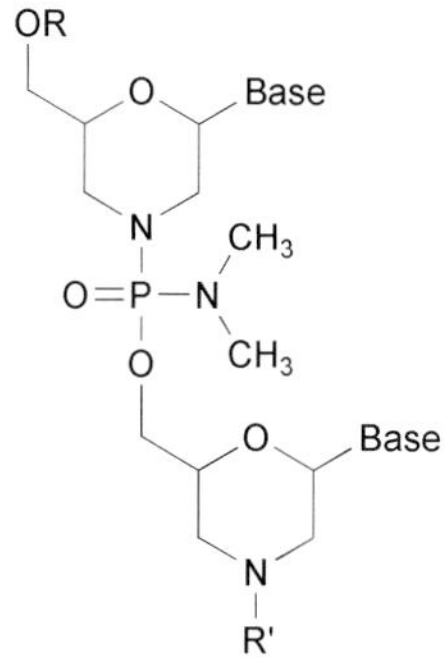

OR

Base

N   CH₃

O=P—N

O   CH₃

Base

N

R'

Morpholino bond

Morphological marker: Any easily identifiable trait (e. g. eye or flower color), that is characteristic for an individual. Morphological markers (slang: "morphos") can be placed on socalled → genetic maps, where they identify → linkage to other markers or traits and help to tag the underlying gene(s). See → molecular marker.

Morphome: The description of all anatomical and histological structures, the organ and body architecture, and their structural relationship in an intact organism.

Mosaic gene: See → split gene.

Mosaic protein: Any protein, that is composed of a series of discrete → domains, where each domain (or a set of different domains) has a specific function for the overall activity of the protein. For example, hemostatic proteases carry large extensions N-terminal to their serine protease domains. These extensions consist of a number of discrete domains with defined functions as e.g. substrate recognition, binding to phospholipid membranes or interaction(s) with other proteins. The majority of metazoan mosaic proteins are extracellular or membrane-bound, and are considered as indicators for the evolution of multicellularity. Compare → mosaic gene.

Motif-primed PCR: A variant of the conventional → polymerase chain reaction that uses → primers complementary to conserved DNA sequence motifs important for gene function and regulation (e.g. parts of → promoters, → consensus sequences for → DNA-binding proteins, or regulatory domains of gene families).

MOUSE: See → *m*icrosatellite *o*btained *u*sing *s*trand *e*xtension.

MP: See → *t*ransport *p*rotein.

MP-PCR: See → *m*icrosatellite-*p*rimed *p*olymerase *c*hain *r*eaction.

M-PVA: See → *m*agnetic *p*oly*v*inyl *a*lcohol microparticle.

Mᵣ: Abbreviation for relative → molecular weight.

MRE: See → *m*etal *r*egulatory *e*lement.

MRI: See → magnetic resonance imaging.

M RNA (*medium* RNA): One of the three linear single-stranded RNAs of the tripartite genome of Tospoviruses (family: Bunyaviridae), that is about 5 kb in length, is associated with the nucleocapsid proteins, and encodes a socalled nonstructural protein and the two viral membrane glycoproteins G1 and G2. The terminal sequences of M RNA carry complementary repeats 65–70 nucleotides long, which allow to form a quasi-circularized (pseudo-circular) molecule. See → L RNA, → S RNA.

mRNA: See → *m*essenger *RNA*.

mRNA display: See → messenger RNA display.

mRNA initiation site: See → cap site.

mRNA profiling: See → messenger RNA profiling.

mRNA-protein fusion: See → messenger RNA display.

mRNP: See → *m*essenger *r*ibonucleoprotein.

MRP: See → *m*apped *r*estriction site *p*olymorphism.

Mrr system: See → *m*ethylated adenine *r*ecognition and *r*estriction system.

MS: See → *m*icrosatellites.

MSAP: See → *m*ethylation-*s*ensitive *a*mplification *p*olymorphism.

Ms DNA: See → multicopy single-stranded DNA.

MSNT (1-*m*esithylene-2-*s*ulfonyl-3-*n*itro-1,2,4-*t*riazole): A → coupling reagent used in → chemical DNA synthesis.

MSP: See → *m*ethylation-*s*pecific polymerase chain reaction.

MS-PCR: See → *m*utagenically *s*eparated *p*olymerase *c*hain *r*eaction.

MSSCP: See → *m*ultiplex *s*ingle-*s*trand *c*onformation *p*olymorphism.

Ms-SNuPE: See → *m*ethylation-*s*ensitive *s*ingle *n*ucleotide *p*rimer *e*xtension.

MTase: See → *m*ethyl*t*ransferase.

mtDNA: See → *m*itochondrial DNA.

mTERF: See → *m*itochondrial transcription *ter*mination *f*actor.

MT gene: See → *metallothionein gene.*

M13: A → filamentous phage of *E. coli* (→ coliphage) containing a circular, single-stranded DNA molecule of 6.407 kb ("plus-strand"). Filamentous phages infect only *E. coli* strains with F pili (containing → F factors), where they adsorb and invade the host cell. The latter is not lysed but grows at a slower rate. Infected cells may thus be recognized as → plaques. As soon as the ssDNA of the phage enters the cell, it becomes converted into a double-stranded → replicative form which multiplies rapidly until an accumulating phage-encoded single-strand-specific DNA binding protein prevents the synthesis of the complementary strand. From then on, only single stranded phage DNA is produced, which is packaged at the host's cell membrane into → capsid proteins which replace the ssDNA-binding proteins. Finally the complete phages are released from the host.

A multitude of M13 derivatives have been developed as → cloning vectors (M13mp series) which contain the *E. coli lac* regulatory region (see → *lac* operon) and the coding sequence for the α-peptide of → β-galactosidase together with either single cloning sites (in M13mpl, for *Ava* II, *Bgl* I, *Pvu* I), symmetrical → polylinkers (e.g. M13 mp7), or asymmetrical polylinker regions (e.g.

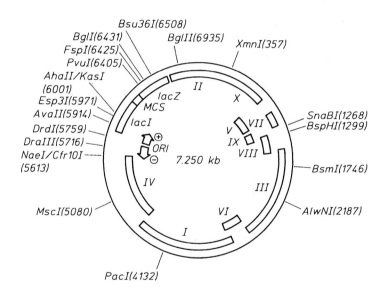

Simplified map of M13mp18
(with unique restriction sites)

I-X   : Viral genes (transcribed clockwise)
ORI   : Origin of DNA replication
⊕ ⊖ : Plus and minus strand replication
MCS : Multiple cloning site

MCS sequence

M13 mp8, mp10, mp11, mp18, mp19). Insertion of foreign DNA into these cloning sites will interrupt the sequence coding for the α-peptide of the β-galactosidase gene. The defective gene does not allow the conversion of the indicator dye → X-gal into its blue derivative upon induction of the *lac* operon by → IPTG so that the phages containing an insert can be easily selected as white plaques. The single-stranded phage DNA can be obtained in bulk quantities, and used directly for large-scale → sequencing of DNA.

M13mp cloning vector: See → M13.

MTN blot: See → *m*ultiple *t*issue *N*orthern blot.

MTP: See → *m*icro*t*iter *p*late.

MTS: See → *m*embrane-*t*ranslocating *s*equence.

Mtx: See → *m*etho*t*re*x*ate.

Mtxʳ: See → *m*etho*t*re*x*ate *r*esistance.

μ:
a) μm, mu: See → micron.
b) See → Mu phage.

Mu: See → *Mu* phage.

***Mu* (mutator):** Any one of a class of transposable elements in the maize (*Zea mays* L.) genome that increases the frequency of mutation of various loci by more than an order of magnitude. *Mu* elements, present in the genome in 10 – 100 copies, comprise maximally 2 kb and are flanked by 200 bp → inverted repeats with adjacent 9 bp → direct repeats. Basically two size classes prevail, of which the shorter ones are derived from longer ones by internal → deletions. *Mu* elements transpose by a replicative mechanism, and can also occur in circular extrachromosomal state (e. g. *Mu*1 [1.4 kb] and *Mu*1.7 [1.7 kb]). Methylation of inserted *Mu* sequences prevents their transposition and stabilizes the mutation, whereas less than complete methylation leads to transpositional activity. See → *Mu* phage, → mutator gene.

***Mu*-AFLP:** See → mutator amplified fragment length polymorphism.

***Mu* array:** A glass chip or a nylon membrane, onto which thousands of → mutator transposon flanking regions are spotted at high density, and which serves to identify specific genes with mutator insertions. The various mutator flanking regions are isolated from individual *Mu* active plants with the → mutator amplified fragment length polymorphism technique, so that each spot on the array represents the *Mu* flanks of an individual plant. Hybridization of these arrays with e.g. → cyanin-labeled or radiolabeled gene probes (e.g. → cDNAs) identifies plants with *Mu* insertions in specific genes.

MUG:
a) 4-*m*ethyl-*u*mbelliferyl-β-D-*g*alactopyranoside; MUGal: A colorless chromogenic substrate for β-galactosidase which is converted into the strongly fluorescent 4-methyl umbelliferone (MU) after cleavage, and used in → enzyme-linked fluorescent assays.

b)   4-*m*ethyl-*u*mbelliferyl-*g*lucuronide: A fluorogenic, synthetic substrate for → β-glucuronidase.

MUGal: See → MUG, entry a.

Multiallergen chip: Any solid support (e.g. glass, quartz, silicon), onto which synthetic peptides or
proteins are spotted, that represent the spectrum of → allergens causing allergic reactions in
sensitive individuals. Though about 20,000 sources for allergens exist, only about 50 allergenic
molecules lead to allergenic responses (e.g. the Betv1, the main allergic protein of the birch tree,
Betula spp.). These 50 allergens can be detected and monitored via the multiallergen chip through
their interaction (binding) with chip-bound molecules.

Multicolor fluorescence *in situ* hybridization (multicolor FISH): A technique to identify several
specific sequences of intact chromosomes simultaneously by → hybridization with different nucleic acid → probes, each of which is labeled with a specific → fluorochrome (e.g. probe A with
→ fluoresceine isothiocyanate, probe B with → rhodamine B isothyocyanate, probe C with a
coumarin derivative), and each of which detects a specific chromosomal site. The simultaneous
use of several differently labeled probes in one single → *in situ* hybridization experiment generates multicolor chromosome pictures.
See color plate 5.

Multicolor spectral karyotyping: A variant of the → spectral karyotyping, which combines visualization of emitted fluorescence light from chromosome-specific fluorescent-labeled probes
(hybridized to a metaphase spread) through a triple band pass filter, sent through an interfero-
meter and imaged with a charge-coupled device (CCD) camera. The interferogram generated for
each pixel is analyzed by Fourier transformation, and the measured spectra converted to display
or classification colors. This technique allows to identify each chromosome in a metaphase
spread after → fluorescence *in situ* hybridization with chromosome-specific probes labeled with
different → fluorochromes in a single experiment.

Multicopy inhibition: The reduction in → transposition frequency of a single-copy → transposon
10, resident in the host chromosome, by the interference of a → multicopy plasmid carrying an IS
10-R sequence (IS 10-R is the *r*ight IS element flanking Tn 10, and encodes a → transposase that
mediates transposition). The mechanism of interference involves pairing of the start region of the
transposase mRNA and a short complementary RNA (→ antisense RNA) transcribed from the
opposite IS 10 strand. As a consequence, translation of the transposase message is impaired.

Multicopy plasmid (high copy number plasmid): A → plasmid that is present in bacterial cells in
copy numbers greater than one per chromosome, because it is under → relaxed control. Its copy
number can therefore spontaneously or artificially be increased (spontaneously to 10-100, artificially to 20-40000 copies; compare → amplification). See also → multicopy inhibition, compare
→ low copy number plasmid.

Multicopy *single*-stranded *DNA* (ms DNA): Any one of a family of small (from 48 to 163 nucleotides) single-stranded DNAs of Gram-negative bacteria, that are present in hundreds of copies per cell ("multicopy"). ms DNAs are apparently always – associated with small (i.e. from 49 to 119 bases), single-stranded RNA molecules, that are joined to the 5'-end of the DNAs by 2', 5' phosphodiester linkages at a specific internal guanosine base in the RNA (sequence context 5'-AGC – 3'), and protein(s) to form an extrachromosomal nucleoprotein complex. The ms DNAs of different bacteria vary considerably in both their DNA and RNA strands, which nevertheless share more or less conserved secondary structures: they fold into stable → stemloop structures, all RNAs contain the G residue (forming the branched, bond with the DNA), and a small RNA-DNA hybrid region forms between the 3' ends of both strands. ms DNA is encided by a → retron in the bacterial chromosome, which contains *msd* (the gene for ms DNA), *msr* and *ret* (the gene for reverse transcriptase) driven by an upstream → promoter. This retron is transcribed into a long → messenger RNA, that is translated to produce → reverse transcriptase, which then uses the upstream *msr-msd* region of the mRNA as the → template as well as the → primer for ms DNA synthesis. An other locus, located → upstream of *msd* (designated *msr*) encodes the RNA strand of the ms DNA complex. Many stream of *msd* (designated *msr*) encodes the RNA strand of the ms DNA complex. Many retrons (in *E. coli*: Ec 48, Ec 67, Ec 73, Ec 83, Ec 86, and so on) are associated with → prophages of the P2 family. The function(s) of ms DNA is still inknown.

Multicopy tag sequence: Any → flag sequence that is reiterated up to 10 copies per tag, so that its detection in the corresponding protein is facilitated, since it produces stronger signals on e. g. → Western blots. The copy number of any multicopy tag has to be optimized, as it should not interfere with the function of the tagged protein. See → epitope tag, → epitope tagging.

Multidomain protein: Any protein that is composed of a set of discrete, structurally and functionally independent modules cooperating to achieve the overall function of the protein.

Multi*d*rug *r*esistance (MDR): The indifference of bacteria against two (or more) different antibiotics. MDR is based on mutations in genes encoding e.g. ABC transporters, and presently represents a serious medical problem.

Multi-epitope imaging: A technique for the simultaneous detection and imaging of many different proteins in a tissue section, that capitalizes on the sequential *in situ* interactions between fluorescently labeled specific → antibodies and their epitopes. Multi-epitope imaging of specific classes of proteins illustrate the localizations of these classes within a cell, and their changes in different phases of the life cycle or after environmental challenges.

Multi-epitope ligand cartography (*m*ulti-*e*pitope *l*igand *k*artographie, MELK): A technique of → topological proteomics, that allows to simultaneously determine both the cellular abundance of about 50 specific proteins and their cellular or subcellular localization in a single cell ("whole cell protein fingerprinting"). MELK produces three-dimensional distribution patterns of proteins by first reacting a specific fluorescent → antibody with its cognate target protein *in situ*, recording the → fluorescence (and with it, the protein localization and quantity), then bleaching out the → fluorochrome, repeating the process with a second fluorescent antibody directed against a second protein, and so on. The signals can be visualized under a microscope, and are collectively compiled into a single panoramic view of the cell with a resolution of about 100 nm. Since only fixed cells can be used, real-time imaging of the target proteins in a living cell is not possible.

Multi-exon gene (multi-exonic gene): Any → gene that contains more than one → exon. See → single-exon gene. Compare → multi-intronic gene.

Multifactorial analysis: The identification of the relative contributions of two (or more) genes and environmental factors to the expression of a distinct → phenotype.

Multi-functional biochip **(MFB):** An integrated chip, that allows the simultaneous detection of DNA-DNA, DNA-RNA, DNA-protein, and protein-protein interactions. The different "chips-on-a-chip" each contain an integrated circuit electro-optical system produced by the socalled *c*omplementary *m*etal *o*xide *s*ilicon (CMOS) technology. The MFB is supplied with photodiode sensor arrays, electronics, amplifiers and all necessary elements for analysis, so that nucleic acid hybridisations can be detected side by side with protein-antibody interactions.

Multi-functional phagemid: Any → phagemid that has been engineered to serve several functions at the same time (e.g. permits DNA cloning, double- or single-stranded sequencing of cloned inserts, *in vitro* mutagenesis, *in vivo* and *in vitro* transcription).

Multi-functional plasmid: Any → plasmid that has been engineered to serve several functions at the same time (e.g. permits → DNA cloning, → Sanger sequencing of cloned inserts, → *in vitro* mutagenesis, and → *in vitro* transcription).

Multigene family (gene family): A set of closely related genes originating from the same ancestral gene by duplication and mutation processes (see → gene amplification). They may either be clustered on the same chromosome (e.g. genes coding for ribosomal RNAs, see → rDNA) or be dispersed throughout the genome (e.g. → heat shock protein genes). Most of the members of such multigene families retain a far-reaching homology in the coding region, but are divergent in the → intron and → promoter regions. See for example → histone genes. Compare also → gene battery, definition b: the genes are related and contiguous in a specific chromosomal region; compare → supergene family: related genes with limited homologies.

Gene Family	Organism	Number of Genes	Clustered (C) Dispersed (D)
Actin	Yeast	1	–
	Slime mold	17	C & D
	Drosophila	6	D
	Chicken	8 – 10	D
	Human	20 – 30	D
Tubulin	Yeast	3	D
	Trypanosome	30	C
	Sea urchin	15	C & D
	Mammals	25	D
α-Amylase	Mouse	3	C
	Rat	9	D
	Barley	7	C
β-Globin	Human	6	C
	Lemur	4	C
	Mouse	7	C
	Chicken	4	C

Multi-gene shuffling (gene-family shuffling, family shuffling, multiple gene shuffling, multiple gene family shuffling): A variant of the → DNA shuffling technique, in which many homologous genes from related organisms are used for creating diversity. The resulting shuffle libraries contain novel chimeras, that differ in many positions. For example, a single cycle of family shuffling of the four cephalosporinase genes from *Citrobacter freundii*, *Klebsiella pneumoniae*, *Enterobacter cloacae* and *Yersinia enterocolitica* resulted in a mutant enzyme, that differs by 102 amino acids from the *Citrobacter*, by 142 amino acids from the *Enterobacter*, by 181 amino acids from the *Klebsiella*, and by 196 amino acids from the *Yersinia* enzyme. In contrast, three rounds of → single gene shuffling yielded only four amino-acid substitutions.

Multigene transformation: The simultaneous transfer of multiple genes into a target organism and their (preferably linked) integration into its genome. For example, up to 14 different genes have been transformed into the genome of rice plants by → biolistic gene transfer techniques, were mostly genetically linked, stable over several generations, and almost all transcribed. Multigene transformation allows to engineer durable traits into target organisms (e.g. resistance to a pathogen will be more effective and lasting, if several resistance genes are involved).

Multigenic trait: See → polygenic trait.

Multi-intronic gene: Any → gene that is composed of more than one → intron. See → multi-exonic gene.

Multi*locus* *probe* (MLP): Any → repetitive DNA sequence that allows to detect two or more, in extreme cases a multitude of loci in a genome, using → labeling and → hybridization techniques. Compare → single locus probe.

Multi*locus* *sequence* *typing* (MLST): A variant of the conventional *multilocus* enzyme *electrophoresis* (MLEE), that involves the sequencing of internal fragments of a series of → house-keeping genes (as e.g. *abcZ* transporter gene, glucose-6-phosphate dehydrogenase [*gdh*], phosphoglucomutase [*pgm*], polyphosphate kinase [*ppk*], 3–phosphoserine aminotransferase [*serC*], adenylate kinase [*adk*], shikimate dehydrogenase [*aroE*], pyruvate dehydrogenase subunit [*pdhC*]) to detect allelic variations between strains of bacterial pathogens. In short, chromosomal DNAs of the test strains are first isolated, and gene fragments of a size between 400 and 600 bp (for convenient sequencing) amplified by conventional → polymerase chain reaction using gene-specific → primers. Both strands of the resulting products are sequenced, the sequences compared, and used to establish socalled *sequence* *types* (STs), i.e. the allelic combination at each locus. Since many loci are involved, the typing identifies multilocus sequence types.

Multimer: A supramolecular complex consisting of two or more identical or non-identical subunits (monomers). For example, a protein molecule, made up of two or more individual polypeptide chains.

Multi-pass protein: Any protein, that spans a plasmamembrane more than once, i.e. contains more than one → transmembrane helix. See → single-pass protein.

Multiple alignment: The → alignment of two (or more) nucleic acid sequences, into which → gaps are introduced such that residues with common features (e.g. pyrimidines versus purines) and/or ancestral residues are ordered in the same vertical line. The most widely used program for multiple alignments is ClustalW.

Multiple allele: Any one of a series of alternative → alleles of a single gene.

Multiple allelism: The occurrence of more than two alleles of a genomic → locus in a population.

Multiple *arbitrary* *amplicon* *profiling* (MAAP): See → arbitrarily amplified DNA.

Multiple *cloning* *site* (MCS): Synonym for → polylinker.

Multiple *displacement* *amplification* (MDA): A variant of the conventional → rolling circle amplification technique, which allows to amplify whole genomes directly from biological samples as e.g. blood or tissue cultured cells. MDA is an isothermal amplification reaction, as such does not require any heating or cooling steps compared to e.g. the → polymerase chain reaction, and differs from rolling circle amplification in that linear → genomes can be amplified. In short, genomic DNA and random, exonuclease-resistant hexamer → primers at high concentrations (50 µM) are first mixed, and the template amplified with the highly processive, strand-displacing φ29 DNA polymerase at 30°C into amplicons more than 10, sometimes 70 kb in length. The polymerase is tightly bound to the template and therefore able to replicate through difficult primary and secondary structures of the DNA. After an initial priming step, branched amplification generates additional single-stranded templates, that in turn serve for primer binding and extension: an exponential cascade of branched amplification ensues ("secondary priming"). Since the polymerase displaces downstream product strands, it creates the templates for multiple concurrent and overlapping rounds of replication. At the end of the reaction (usually after 18 hours), most of the amplified strands are converted to double-stranded DNA and the reaction stopped by heating to 65°C. MDA yields up to 20–30 µg product DNA, starting from only 1–10 copies of genomic DNA. MDA-amplified genomic DNA is used for a whole series of genomic techniques such as → chromosome painting, genotyping of → single nucleotide polymorphisms, → RFLP analysis, → cloning, → subcloning, and DNA sequencing.

Multiple *fluorescence-based* PCR-SSCP (MF-PCR-SSCP): A sensitive mutation detection technique that combines conventional → polymerase chain reaction with two (or more) primers labeled with different → fluorochromes and → single-strand conformation polymorphism analysis. In short, the target sequence is first amplified using forward and reverse primers labeled with two different dyes, respectively (e. g. FAM [blue] and JOE [green]) at their 5'-ends. Amplified products are then heat-denatured, mixed with an internal DNA size marker labeled with a third dye (e. g. ROX [red]), and run in temperature-controlled non-denaturing → polyacrylamide gels in an automated DNA sequencer. Mutations are detected as positional shifts of two-coloured peaks in the electropherogram. The technique allows to diagnose single base exchanges and → loss of heterozygosity, and works without radioactivity.

Multiple gene DNA shuffling: See → DNA shuffling.

Multiple gene family shuffling: See → multi-gene shuffling.

Multiple genes: See → polygene.

Multiple gene shuffling: See → multi-gene shuffling.

Multiple hit: See → superimposed substitution.

Multiple nucleotide polymorphism (MMP): Any polymorphism between two (or more) → genomes that is based on more than one → single nucleotide polymorphism. For example, many human diseases probably are caused by single base exchanges at strategic sites of several genes (e.g. coding for functional → domains of different proteins), that are not present in the wild type genomes and act in concert to cause a disease. These altogether are multiple nucleotide polymorphisms.

Multiple overlapping primer PCR: A variant of the → PCR *in situ* hybridization, using → primers with overlapping sequences to generate large amplification products, that do not diffuse away from their original location in cells or tissue specimens.

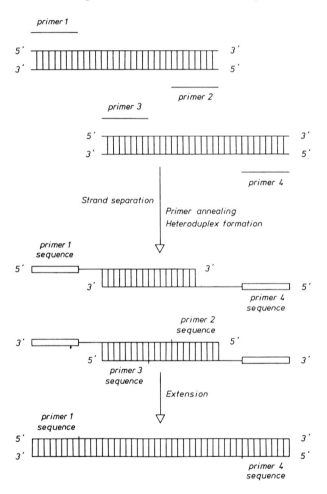

Multiple overlapping primer PCR

Multiple recognition site (multiple recognition sequence): An infelicitous term for the occurrence of more than one → restriction site recognized by a specific → restriction endonuclease. For example, the endonuclease *Acc* I recognizes the four sequences 5'- GTAGAC- 3', 5'-GTATAC- 3', 5'-GTCGAC- 3', and 5'- GTCTAC- 3', and cleaves 3' of the T residue.

Multiplet: Any single band on a → DNA fingerprint gel, that contains two (or more) different DNA molecules. This undesirable comigration of different sequences can be resolved by isolation of the band from the gel, the cloning of the different DNAs, and their sequencing.

Multiple *tissue* *Northern* blot (MTN blot): A ready-to-hybridize → Northern blot that contains poly(A)$^+$-RNAs (→ polyadenylated RNA) from a series of tissues of an organism, separated by denaturing → agarose gel electrophoresis and blotted onto a nylon membrane. Equal amounts of RNA per lane allow to detect tissue-specific expression of genes.

Multiplex DAF: See → multiplex *DNA* *amplification* *fingerprint*.

Multiplex *DNA* *amplification* *fingerprint* (multiplex DAF): A variant of the → DNA amplification fingerprinting technique that uses at least two or multiple primer oligodeoxynucleotides to generate → amplification fragment length polymorphisms.

Multiplex *fluorescent* *in* *situ* *hybridization* (M-FISH): A technique for the simultaneous detection and discrimination of all different chromosomes in a metaphase spread by different colorization. In short, chromosome-specific DNA libraries are first labeled with distinct combinations of → fluorochromes. Then the different, specifically labeled chromosomal DNA libraries are hybridized onto spreads of metaphase chromosomes (or cell nuclei), and the individual fluorochromes detected by epifluorescence microscopy (using all filters to excite all the fluorochromes in the sample) coupled to a *cooled* *charge-coupled* *device* (CCD) camera. The different fluorescence signals allow to unequivocally assign a specific fluorogram (i. e. colour) to a specific chromosome.

Multiplex hybridisation array: A general term for any → microarray, that contains hundreds or thousands of individual genes, cDNAs or oligonucleotides and is used to simultaneously detect DNA-DNA-, DNA-RNA- or DNA-oligonucleotide-interactions by hybridisation (i.e. by → Watson-Crick base pairing).

Multiplex *messenger* *assay* (MMA): A technique for the simultaneous analysis of the expression of many different genes in a cell, tissue, organ, or organism. In short, → cDNA clones corresponding to known genes are picked from → replica plates, spotted onto nylon-based filters in duplicate, and fixed onto the membrane, which can contain up to 50,000 such cDNAs ("high-density filter"). Then complex → probes are prepared separately from cell A and B, respectively, by → reverse transcription of their total RNAs, which are either labeled radioactively or fluorochromated (perferably with two different → fluorochromes with varying emission spectra, so that they can be detected simultaneously). Each of these probes is then hybridized to the high-density array filter, the hybridization signals detected by → autoradiography or → fluorography, quantified, and socalled hybridization signatures established. These signatures reflect the expression profiles of the corresponding genes in the two target tissues. See → cDNA expression array.

Multiplex PCR: See → multiplex *polymerase* *chain* *reaction*.

Multiplex *polymerase* *chain* *reaction* (multiplex PCR): A variant of the → polymerase chain reaction technique, that uses at least two, frequently many → primer oligodeoxynucleotides to either amplify different stretches on a target DNA molecule or different genomic loci simultaneously. Multiplex PCR requires optimization of the → annealing temperature, the concentrations of *Taq* DNA polymerase, nucleotide triphosphates, $MgCl_2$ and the temperature profile of the PCR reaction.

Multiplex ratio: The number of genetic loci that can simultaneously be detected by a specific → molecular marker technique in a single experiment. For example, → amplified fragment length polymorphism techniques display many different loci (their number depending on the → genomic DNA and the amplification → primers used, among others), whereas → sequencetagged microsatellite site methods detect only one single locus.

Multiplex sequencing (Church sequencing): A → DNA sequencing method that allows the determination of base sequences in 10-50 different DNA fragments synchronously. In short, the various fragments are first cloned into → plasmid vectors that differ from each other by sequences flanking the cloning site (so that each cloned fragment has specific unique border sequences). All the different inserts are then excised from the plasmid vector, using a → restriction endonuclease that cuts outside all border sequences. The pooled inserts are then sequenced using the → Sanger sequencing procedure. After electrophoresis of the various resulting fragments, their transfer to and immobilization on membranes the sequences belonging to each of the original fragments are detected by sequential hybridization with synthetic → oligonucleotides that are complementary to the sequences flanking the inserts. After hybridization to one such oligonucleotide and autoradiography, the probe is stripped off and the membrane reprobed with another oligonucleotide to detect sequences from another insert. In this way, some 10-50 different DNA fragments can be sequenced synchronously. (Multiplexing is a term from electronics, describing the mixture of many signals at the begin of an electronic circuit, which are separated from each other later on.) An interesting application of this method is the technique of → multiplex walking.

Multiplex single-strand conformation polymorphism (M-SSCP): A variant of the → single-strand conformation polymorphism detection technique that allows to discover multiple mutations within one gene in one single approach. The technique works with → 5' endlabeled forward → primers and unlabeled reverse primers and → polymerase chain reaction amplification. The amplification products are electrophoresed in thin → polyacrylamide gels and detected by autoradiography. Small changes in base composition (e.g. → deletions, additions, or duplications) appear as a slightly different band position as compared to a reference band. With M-SSCP it is, for example, possible to simultaneously screen for deletions within → exons, and a variety of base substitutions within a multi-intronic gene (e.g. the human dystrophin gene with 79 exons).

Multiplex walking (oligomer walking): A technique for the → sequencing of long DNA stretches. In short, the DNA is first restricted with different → restriction endonucleases, the various fragments are subjected to the reactions for → chemical sequencing and then processed as for → multiplex sequencing. After sequence determination of one fragment → oligonucleotides complementary to its 5' – or 3' -terminus can be synthesized and used as → probes to "walk" to the adjacent fragments. Compare → primer-directed sequencing.

Multiprobe RNase protection assay (multiprobe RPA): A variant of the → RNase protection assay, which allows to simultaneously detect and quantify many → messenger RNAs. In short, a multiprobe template set is first established, consisting of a series of defined → cDNA fragments, each cloned into a plasmid, that encode specific → antisense RNAs. Such a set could e. g. contain cDNAs involved in cytokine expression during T-cell-mediated immune response (IL2, IL4, IL5, IL9, IL10, IL13, IL14, IL15, IFNγ, together with the house-keeping genes L32 and GAPDH as controls). The corresponding antisense RNAs are synthesized by → T7 RNA polymerase, labeled with ^{32}P-UTP, and purified by phenol-chloroform. The antisense probes are then overnight hybridized to total RNA from the target cell under highly stringent conditions. Then → RNase A and → RNase T1 are added, that digest single-stranded RNA. These enzymes are removed by proteinase K treatment, the double-stranded RNAs recovered by phenol-chloroform extraction, and the different RNA hybrids resolved by denaturing polyacrylamide gel electrophoresis. Subsequent → autoradiography or → phosphorimaging allows to analyze the transcripts both qualitatively and quantitatively (band intensity). Multiprobe RPA is about 50-100 fold more sensitive than → Northern blotting analysis.

Multisite mutation: Any mutation that either involves alteration of two or more contiguous nucleotides, or occurs repeatedly at many loci in a given genome.

Mung bean nuclease (EC 3.1.30.1): A single-strand specific, Zn-containing → nuclease from mung bean (*Phaseolus aureus*) sprouts that catalyzes the degradation of single-stranded DNA or RNA molecules into deoxy- or ribonucleoside 5' monophosphates. The enzyme does not attack → double-stranded DNA, or DNA-RNA hybrids, unless very large amounts of enzyme are used. Mung bean nuclease can be used for the trimming of single-stranded overhangs produced by → restriction endonucleases, for the removal of single-stranded regions in DNA hybrids, for the cleavage of → fold-back DNA in → cDNA synthesis, and for transcript mapping.

MUP (4-methyl-umbelliferyl-phosphate): A fluorogenic substrate for → alkaline phosphatase used in → enzyme-linked fluorescent assays.

***Mu* phage (*Mu*, phage *Mu*, m, bacteriophage *Mu*, also → mutator phage):** A 37 kb → temperate bacteriophage with transpositional properties. Its DNA is flanked by 5 bp → direct repeats and contains a transcriptional → enhancer ("internal activation sequence") of about 100 bp in length. The 75 kDa → transposase binds to the two termini and the enhancer, and cuts *Mu* at the 3' end. Transposition itself is mediated by a complex of nucleoproteins, the → transposome. Transesterification at the 3' OH integrates *Mu* into the host genome, but the 5' ends of *Mu* are still attached to the old flanking DNA by a socalled *strand transfer complex* (STC). After nucleolytic cleavage of *Mu* from these old flanks, the gaps are repaired, and transposition is complete. Transposition may occur at many different sites in the host chromosome, whereby inactivation of host genes (→ insertional mutagenesis) or chromosomal rearrangements are caused. Both → mutations manifest themselves as an altered → phenotype.

Muramidase: See → lysozyme.

Mutagen (mutagenic agent): Any physical or chemical agent that increases the frequency of → mutations above the spontaneous background level. Such mutagenic agents include ionizing irradiation, UV irradiation, alkylating compounds and → base analogues. See also → mutagenesis.

Mutagenesis: The induction of → mutations in DNA, either in the test tube (see → in vitro mutagenesis) or in vivo, e.g. by irradiation (irradiation mutagenesis), chemicals (→ chemical mutagenesis) or by the → deletion, → inversion or insertion of DNA sequences (→ insertion mutagenesis). See also → interposon mutagenesis, → transposon mutagenesis. Compare also → mutator gene, → site-specific mutagenesis.

Mutagenic: Capable of inducing → mutations.

Mutagenic agent: See → mutagen.

Mutagenically separated polymerase chain reaction (MS-PCR): A technique for the detection of → point mutations in a known DNA sequence, which relies on conventional → polymerase chain reaction. It allows to amplify normal and mutant → alleles of a gene simultaneously in the same reaction, using allele-specific → primers of different lengths. Additionally, the allele-specific primers differ from each other at several nucleotide positions and therefore introduce new and discriminating mutations into the allelic PCR products (thereby reducing cross-reactions between amplification products during the PCR process). Since both products possess different lengths, MS-PCR "separates" both amplified alleles, that can then be identified by → agarose gel electrophoresis and → ethidium bromide staining.

Mutant: An organism harboring a mutant gene whose expression changes the phenotype of the organism. See → mutation.

Mutant allele-specific amplification (MASA): Any one of a series of → polymerase chain reaction-based techniques, allowing the specific amplification of an → allele, that has undergone a → mutation (e. g. a → deletion, → insertion, → inversion, → transition, → transversion). MASA techniques are presently employed in clinical screening and diagnosis.

Mutant analysis by PCR and restriction enzyme cleavage (MAPREC): A technique for the detection of → point mutations in coding → genomic DNA. In short, RNAs of wild-type and mutant are first isolated, and → reverse transcribed into → cDNAs using a random hexanucleotide → primer. The cDNA serves as template for → asymmetric PCR with primers specific for the gene of interest, more precisely the gene region in which the mutation occurs. The primer is designed such that it creates a → recognition site for the → restriction endonuclease *Mbo*I (5'-GATC-3') in the wild-type target DNA. The mutant sequence will give rise to e.g. a *Hinf*I restriction site. An excess of sense polarity primer ensures that the product is predominantly single-stranded DNA. Now the second strand is synthesized, using a labeled antisense primer (→ biotin labeling), the double-stranded product digested with *Mbo*I, and the resulting restriction fragments separated by → polyacrylamide gel electrophoresis. After → Southern blotting or → vacuum blotting and fixation of the fragments onto the blotting membrane, the fragments are visualized with → streptavidin-conjugated alkaline phosphatase. The wild-type sequence will be fragmented, the mutant sequence will remain uncut.

Mutation: Any structural or compositional change in the DNA of an organism that is not caused by normal segregation or genetic recombination processes. Such mutations may occur spontaneously, or may be induced by → mutagens such as ionizing radiation or alkylating chemicals. The change of a nucleotide base for example, may cause the conversion of one → codon into another one. It is silent, if the codon change does not cause any detectable phenotypic change (if e.g. both codons stand for the same amino acid, see → codon bias). See also → mutagenesis, → mutation breeding, → mutation rate.

Mutational load: See → genetic load.

Mutation analysis: The detection and characterization of a → mutation in DNA, e. g. → deletion, → insertion, → inversion, → mismatch mutation, → point mutation, → translocation. Out of a multitude of techniques for mutation analysis, see → allele-specific hybridization, → amplified restriction fragment length polymorphism, → arbitrarily primed PCR, → arbitrary primer technology, → arbitrary signatures from amplification profiles, → base excision sequence scanning, → capillary electrophoreseis hybridization, → chimeric oligonucleotide-directed gene targeting, → cleavase fragment length polymorphism, → cleaved amplified polymorphic sequence, → digested random amplified microsatellite polymorphism, → direct amplification of minisatellite DNA, → dynamic allele-specific hybridization, → forensically informative nucleotide sequencing, → heteroduplex analysis, → inter-retrotransposon amplified polymorphism, → methylation-sensitive amplification polymorphism, → methylation-specific PCR, → methyl filtration, → microsatellite-primed PCR, → minisatellite-primed amplification of polymorphic sequences, → multiple fluorescence-based PCR-SSCP, → mutagenically separated PCR, → mutant allele-specific amplification, → mutant analysis by PCR- and restriction enzyme cleavage, → mutator amplified fragment length polymorphism, → MutS mismatch detection, → PCR clamping, → PCR-ligation-PCR mutagenesis, → polymerase chain reaction, → restriction fragment length polymorphism, → primer-specific and mispair extension analysis, → random amplified microsatellite polymorphism, → random amplified polymorphic DNA, → retrotransposon-microsatellite amplified polymorphism, → reversed enzyme activity DNA interrogation test, → selective amplification of polymorphic loci, → semi-specific primer technology, → sequence-based amplified polymorphism, → sequence-specific amplification polymorphism, → single nucleotide polymorphism, → single-strand conformation analysis, → temperature modulated heteroduplex analysis.

Mutation breeding: The development of plants with improved characteristics (e.g. resistance against pathogens or environmental stress, increased agricultural productivity) through physically or chemically induced → mutations. Since such mutations are totally at random, no directed genetic change is possible. This method is still used but will be replaced by directed genetic engineering in future. But see → targeting induced local lesions in genomes.

Mutation delay: The time lag between a → mutation event and its phenotypic expression. For example, recessive mutations may only be apparent, if they become homozygous.

Mutation detection electrophoresis (MDE) gel: A gel made of modified → polyacrylamide with slightly hydrophobic properties that selectively alters the electrophoretic mobility of → heteroduplexes such that even single mismatched bases in one kb of duplex DNA can be visualized by a mobility shift.

Mutation rate: The number of → mutations occurring per unit DNA (e.g. → kb or a → gene) per unit time.

Mutator: See → mutator gene, → *Mu*.

Mutator amplified fragment length polymorphism (*Mu*AFLP; amplification of insertion mutagenized sites, AIMS): A variant of the conventional → amplified fragment length polymorphism (AFLP) technique to screen, isolate and characterize → insertions of → *Mu* elements (or → T-DNA in plants) and their flanking sequences in target genomes, and to screen for sequence polymorphisms in the DNA flanking inserted mutator copies. In short, genomic DNA is restricted with *Mlu* I (that cuts within mutator → long terminal repeat sequences) and a four-base cutter (e. g. *Mse* I or *Bfa* I), then biotinylated *Mlu* I and four base cutter → adaptors are ligated to the corresponding ends of the restriction fragments, the biotinylated fragments captured on streptavidine beads and the fragments including insertion sequences amplified by linear → polymerase chain reaction of only 12-15 cycles (to minimize PCR artifacts) using → primers complementary to *Mu* sequences and the four-base cutter adaptor (the latter primer is labeled with ^{32}P). Then the amplified fragments are electrophoretically separated in → polyacrylamide gels, and *Mu* insertions detected by → autoradiography. If a primer labeled with a → fluorochrome is used for *Mu*AFLP-PCR, the number and sizes of the amplified *Mu* flanks can be quantitatively determined by an automated fluorescence reader. Since *Mu* preferentially inserts into → genes, MuAFLP allows to detect insertions that lead to → gene knock-out mutants. See → *Mu* array. *Figure see page 705.*

Mutator gene (mutator): Any gene (*mut* gene) that increases the rate of spontaneous → mutations of one or more other genes. Such mutators may themselves originate from normal genes by mutation. If for example, a gene is mutated whose product normally functions in DNA repair or replication, the mutant protein encoded by the mutated gene may introduce multiple errors (that is mutations) during these processes.

Mutator phage: Any phage that is able to increase the rate of mutation in its host cell (e.g. the → *Mu* phage).

Mutator polymerase: A mutated, nucleus-encoded mitochondrial γ- DNA polymerase, that gives rise to the accumulation of → frame-shifts, → point mutations, and → deletions in the → mitochondrial genome. One of the mutations converting the wild-type DNA polymerase to the mutator polymerase is a point mutation, that changes a highly conserved tyrosine at position 955 (part of the binding pocket responsible for selection of deoxyribonucleotides against ribonucleotides) to cysteine ($Y_{955}C$). This simple base exchange does neither change the catalytic rate nor the intrinsic 3' 5'- exonuclease proofreading activity, but decreases the fidelity of DNA replication, which in turn leads to the mtDNA mutations. These mutations cause a series of diseases. For example, the *progressive external ophthalmoplegia* (PEO) is the consequence of several kb long deletions primarily between short, direct repeats of 10–13 base pairs. These deletions are associated with point mutations caused by T·dTMP mispairing, that occurs 100 times more frequent with mutator as compared to wild-type γ-polymerase. The disease appears in patients at the age of 30–40, and causes a weakness of muscles in general, the eye muscles in particular. As a consequence, the muscles moving the eye (especially the lateral rectus) deteriorate gradually, so that the patients can only follow a moving object by turning their heads. The muscle weakness is a result of impaired electron transport chain activity (depletion of ATP). Another cause of PEO is a mutant mitochondrial → helicase encoded by gene *twinkle*.

Mutator strain: Any *E.coli* strain that carries → mutations in one or several DNA repair pathways, and therefore has a higher random mutation rate as compared to the wild type. Typically, the *mut*D (deficient in 3'-5'-exonuclease activity of DNA polymerase III), *mut*T (unable to hydrolyze 8-oxodGTP), and *mut*S (error-prone mismatch repair) alleles are present in such mutator strains, and consequently the spontaneous mutation rate is increased from 50 to as much as 5000 times in triple mutants. If an → insert (e.g. a → gene) is maintained in a → plasmid of such a mutator strain, it also suffers random mutations at an increased rate (e.g. one base change per 2000 nucleotides per generation).

Muton: The smallest unit of a gene that may undergo → mutation (equivalent to one base pair of DNA).

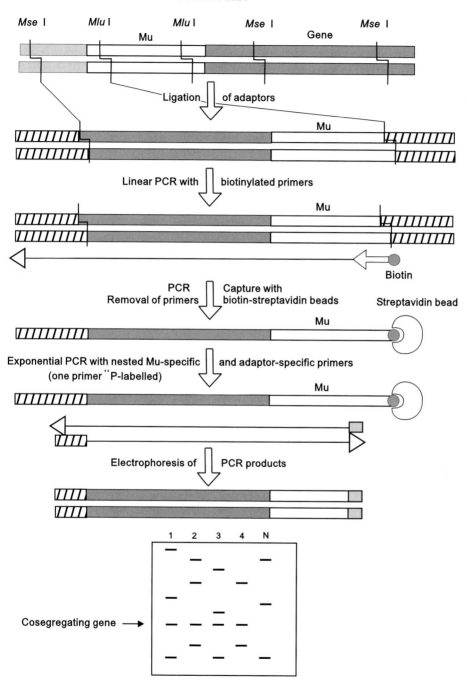

Mutator amplified fragment length polymorphism

Mut S: Any one of a family of *E. coli* methyl-directed → mismatch repair enzymes, that recognize and bind to mismatched bases in target DNA. Mut S is part of a system for the correction of replication errors. See → Mut S mismatch detection.

Mut S mismatch detection: A technique for the detection of single base → mismatches in a target DNA, that exploits the affinity of → Mut S to recognize and bind mismatched bases. In short, the target DNA is first amplified, using appropriate, radioactively endlabeled → primers and conventional → polymerase chain reaction techniques. The PCR products are then heat-denatured and re-annealed, which results in four different DNA duplexes (in case the target DNA is heterozygous at locus A [A/a]: two homoduplexes [AA and aa]), and two heteroduplexes [Aa and aA]). If the heteroduplexes contain e. g. a single base-pair mismatch, the added Mut S protein will bind to this mismatch, and the mutant allele can be detected by → mobility-shift DNA-binding assays in → polyacrylamide gels with subsequent → autoradiography.
Figure see page 707.

MVR: See → *m*inisatellite *v*ariant *r*epeat.

MYB domain: A region of about 52 amino acids in socalled → MYB proteins, that occurs either single or as two or three repeats, respectively, and binds to specific address sequences in target DNA. Each repeat adopts a → helix-turn-helix conformation, which binds in the → major groove of the target. Single MYB domains possess longer C-terminal helices (as e.g. the human telomeric protein hTRF1) than the repeated MYB elements, but all contain regularly spaced tryptophan residues (three per repeat) that contribute to a hydrophobic cluster.

MYB protein: Any one of a large and diverse class of DNA-binding proteins that either function as transcriptional activators, possibly also as repressors, or as structural proteins (as e.g. telomeric G-rich sequence-binding MYB proteins containing a socalled telobox). Common and characteristic feature of all MYB proteins is the occurrence of either one (e.g. StMYB1), two (e.g. ZmMYBC1), or three structurally conserved → MYB domains (e.g. C-MYB), that mediate specific binding to address sequences in target DNA (consensus sequence: 5'-PyAAC (G/T)G–3'. For example, in plants MYB proteins regulate secondary metabolism (e.g. the maize genes ZmMYBC1, ZmMYBPL and ZmMYBP, the snapdragon genes Am-MYBROSEA and Am-MYBVENOSA, and the *Petunia* gene PhMYBAN2), cellular development (e.g. the *Arabidopsis thaliana* gene AtMYBGL1, involved in trichome differentiation), meristem formation (e.g. At-MYB13 and AtMYB103) and the cell cycle (e.g. AtCDC5). The DNA-binding domain(s) of MYB proteins is (are) located close to the amino terminus, and the transcriptional → activation domain lies C-terminal of this binding sequence.

Mycostatin: See → nystatin.

Myeloma cell line: A tumor cell line that originates from a single lymphocyte and produces only one defined immunoglobulin.

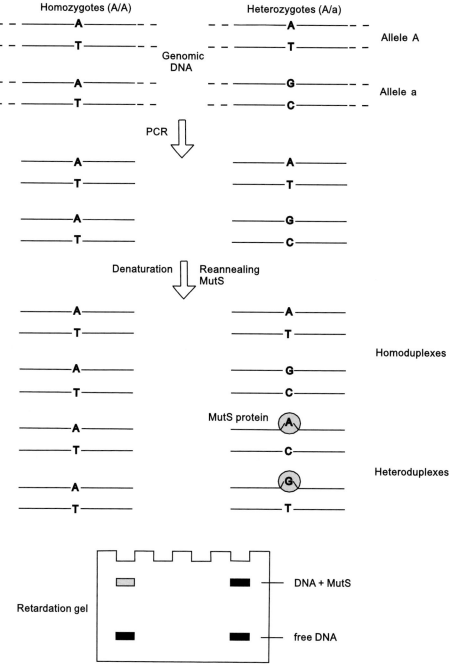

Mut S mismatch detection

N

N:
a) Abbreviation for a*N*y base in DNA (e. g. → A, → T, → G, → C) or RNA (e.g. A, → U, G, C). Synonym for → X.
b) *N*umber of chromosomes in a haploid set.

N-ꟷᶠMet: See → N-formylmethionine.

NAB: See → nucleic acid biotool.

N-*acetyl* *d*eblocking *a*mino*p*eptidase (Ac-DAP): A thermostable, CoCl2–activated exo-amino-peptidase from *Pyrococcus furiosus* catalysing the removal of N-terminal acyl-type blocking groups in proteins and peptides. For example, the tocacco mosaic virus coat protein contains an acetylated N-terminal amino acid residue, which cannot be directly analysed on protein sequencers based on Edman degradation. Therefore, the N-acetylaminoacid has to be released, which is done by Ac-DAP. Since the N-terminus of a commercially available Ac-DAP is acetylated, its amino acid cannot be determined by the conventional Edman degradation. Therefore, the sequence of only the target peptides or proteins is analysed.

NACS™ (*n*ucleic *a*cid *c*hromatography *s*ystem): An ion-exchange resin that is used to separate DNA or RNA from contaminating low-molecular weight substances (e.g. nucleotides, or sulfonated and sulfated polysaccharides which are components of most → agaroses, and inhibit many enzymes used in recombinant DNA experiments, e.g. → restriction endonucleases). During chromatography, nucleic acids are bound to the resin in low-salt conditions. Extensive washings remove all contaminations, and the nucleic acids can then be eluted with a high-salt buffer.

Naked DNA: Any → DNA that is devoid of all proteins, with which it is normally associated in the nucleus. See → chromatin.

***N*aked *e*ye *p*olymorphism (NEP):** Any difference between two closely related organisms that can be detected visually (i.e. by the naked eye). See → visual marker.

Nanoarray: Any solid support (e.g. a gold-coated glass chip), onto which dots of peptides, proteins, oligonucleotides or DNAs are spotted via e.g. → dip-pen nanolithography in arrays of 100 nm (or less) diameter and 100 nm (or less) distance between spots (see → ultra-high density microarray). Interactions between the probes and target molecules on a nanoarray are detected by atomic force microspcopy.

Nanobarcode particle: Any encodeable, machine-readable, sub-micron particle manufactured in a semi-automated process by electroplating inert metals such as gold, nickel, platinum, or silver into metallized templates. These templates define the particle diameter, and are dissolved, thereby releasing the resulting striped nano-rods.

Nanocavity trap: A staphylococcal α-hemolysin transmembrane → ß-barrel, that is engineered to accommodate two different *cyclo*dextrin (CD) adapters (e.g. ßCD, and hepta–6–sulfato-ßCD) within its lumen such that they serve as *cis* and *trans* gates of a nanocavity of several thousand of cubic Å volume. Organic molecules (e.g. → oligonucleotides) can be pulled into this cavity by an electric potential and kept there for hundreds of milliseconds. The trapped molecules shuttle back and forth between the adapters, before leaving the cavity. Nanocavity traps alter the magnitude and selectivity of ion flux (conductivity) in a transmembrane potential, and the build-in adapters bind guest molecules, that block or reduce conduction by the pore.

Nanocircle: See → DNA nanocircle.

Nanocrystal antenna: Any metal (preferentially gold) covalently linked to a biomolecule (DNA, RNA, or protein), that can inductively be heated by an alternating magnetic field and allows to selectively and reversibly control the function of the biomolecule carrier (e.g. the activity of an enzyme).

Nanodroplet: A small volume (100 – 200 nl) droplet, in which cellular reactions can be simulated and analyzed without the problems in liquid cultures (e. g. caused by diffusion of all molecules). For example, the influence of small effectors on protein-protein interactions can be analyzed, if nanodroplets containing defined media, yeast cells and beads carrying photochemically releasable effectors (e. g. galactose) are employed. The yeast cells harbor → two-hybrid system vectors (one carrying a gene fused to the LexA DNA-binding domain, the other one to a transcriptional activation domain proximal to a promoter containing LexA binding sites upstream of a *URA*3 reporter gene. Both constructs are repressed by glucose and activated by a shift to galactose in the medium. The *URA*3 expressing cells are killed in the presence of 5–fluoroorotic acid (5–FOA). If, however, a small molecule disrupts the intracellular protein-protein interaction, proximity of the activation domain to the promoter is diminished, and *URA*3 gene transcription reduced: cells can survive in the presence of 5–FOA.

Nanomechanical transduction: The transformation of forces, generated by DNA-DNA, DNA-RNA or DNA-oligonucleotide hybridizations or protein-protein and protein-ligand interactions into nanomechanical responses of a microfabricated silicon support, on which the interactive processes take place. In short, one side of silicon → cantilevers is first covered with a monolayer of gold. Then synthetic, 5' thio-modified oligonucleotides are covalently immobilized on the gold surface in a monolayer. Cantilever I may carry e.g. 12-mers, cantilever II e.g. 16-mers, and so on. These cantilever arrays are then placed in a liquid cell and equilibrated in hybridization buffer, after which a complementary 12-mer is injected. Hybridization between the 12-mer and the matching oligonucleotides on the cantilever surface leads to a difference in surface stress between the functionalized gold and the non-functionalized silicium surface, which bends the cantilever. The degree of bending is recorded by an optical beam deflection technique. The transduction of Watson-Crick hybridization into surface stress is triggered by electrostatic, steric, and hydrophobic interactions (e.g. the charge density in the sugar-phosphate backbone of the oligonucleotides and their counter-ions during hybridization is increased, as is the packing of the oligonucleotides on the cantilever surface. These changes lead to repulsion and produce compressive surface stress. If cantilever II is now hybridized to a complementary 16-mer oligonucleotide, the same process is repeated and leads to a deflection of cantilever I. This technique allows to detect single base mismatches, which are translated into (minimal) nanomechanical responses. Since the hybridized oligonucleotides can be dissociated from their cantilever-bound counterparts (by e.g. 30% urea solution), the same array can be reused repeatedly.

Protein-protein recognition can also be translated into nanomechanical forces. For example, one cantilever can be loaded with → protein A, the second one with bovine serum albumin as a reference. Then immunoglobulin G (IgG) is bound specifically – through its constant region – to protein A, leading to a deflection of the corresponding cantilever.

Nanomechanical transduction neither requires labeling of probe oligonucleotides or proteins nor optical excitation (by e.g. laser light) or external probes, and can be expanded to a high-throughput format (by e.g. parallel organization of 1000 or more cantilevers).

Nanoparticle-based DNA detection: A technique for the detection of specific DNA sequences and mutations in these sequences (e.g. → single nucleotide polymorphisms), in which → nanoparticle probes are hybridized to single-stranded DNA templates. Upon hybridization with complementary sequences, an extended polymeric network in two and three dimensions is formed. Whereas the non-hybridized particles (in case of gold, Au) have a red color, the hybridization-induced polymerization to aggregates changes the color to purple. This color change is brought about by the shortening of interparticle distances to less than an average gold particle diameter by the hybridization process. This shift is attributed to the → surface plasmon resonance of the Au. Therefore, the hybridization event can be monitored without radioactive label. Nanoparticle-based DNA detection can also be applied to DNA arrays on glass supports (→ DNA chips), where detection of a hybridization event is mediated by the reduction of silver ions to metallic silver through hydroquinone on the surfaces of the gold nanoparticle. Since this process can simply be scanned by a flatbed scanner, the procedure is also called scanometric DNA array detection.

Nanoparticle probe: Any oligonucleotide → probe that is covalently bound onto the surface of gold, silver or platinum particles of a diameter of 5–10 nm. The noble metal particles contain N-propylmaleimide substituents, that can be selectively coupled to sulfhydryl groups (thiol capping with S-trityl-6-mercaptohexylphosphoramidite) at the 3'ends of single-stranded DNAs. Nanoparticle probes are used for → nanoparticle-based DNA detection.

Nanoplex (*nano*meter com*plex*): A complex of nanometer-sized particles (e.g. beads) and DNA (e.g. plasmid DNA), that can be used for → direct gene transfer by e.g. → particle gun techniques.

Nanopore processing: A technique for the detection of single nucleic acid molecules, based on the blockage of ionic current in a single α-hemolysin pore channel caused by the traversing RNA or → ssDNA. Such pores of 1.0-1.2 nm internal diameter are formed by the self-assembly of the 293 amino acid staphylococcal α-hemolysin polypeptide into lipid bilayers. The duration and amplitude of the current blockage is related to both the concentration of the nucleic acid and its chain length. Also, → purine and → pyrimidine nucleotides produce distinct current blockades, and hybridization can be detected directly, since → dsDNA cannot penetrate the nanopore.

Nanopore sequencer: A high-throughput → DNA sequencing machine, based on the movement of the analyte DNA through a channel membrane protein (e.g. bacterial α-hemolysin) anchored in a lipid bilayer membrane (usually Teflon horizontal bilayers), that separates two solutions. When a voltage is applied across the membrane, charged DNA molecules migrate through the 1.5

nanometer pore ("nanopore") pore of the channel protein, and each base identifies itself with a characteristic electrical current, which in turn is measured by a detector. For example, one of the nanopore sequencers is based on α-hemolysin, a 33kDa protein from *Staphylococcus aureus*, that self-assembles in a lipid bilayer to form a nanopore. This pore remains open at neutral pH and high ionic strength, and the diameter at its narrowest point (about 1.5nm) allows single-stranded DNA (but not double-stranded DNA) to pass through the pore. As the ssDNA enters the pore, it blocks the ionic current transiently and it does so specifically for each base. In the future, synthetic nanopores will certainly replace the α-hemolysin central pore for → nanopore sequencing. The socalled ion-beam sculpting technique allows to fabricate 5 nm pores in e.g. silicon nitride membranes, which shares many of the properties with α-hemolysin. The nanopore sequencer concept is still in a developmental stage.

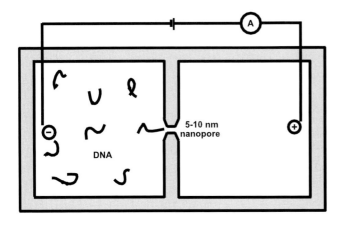

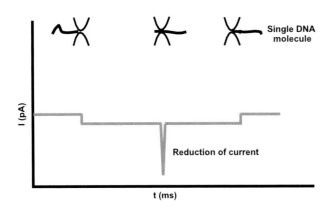

Nanopore sequencer

Nanopore sequencing: The determination of the sequence of bases of a single DNA molecule in a → nanopore sequencer.

Nanotechnology: A series of techniques designed to study molecules at the nanometer level (e.g. scanning tunneling microscopy, STM, scanning the tunneling current, and → atomic force microscopy, AFM, scanning the repulsive atomic forces between a sample and a probe, thereby producing high-resolution surface topographies of proteins and nucleic acids).

NAPPA: See → nucleic acid-programmable protein array.

Naptonuon (*non-aptative nuon*): Any → potonuon that disintegrates over evolutionary times by random nucleotide changes ("nonaptation") and is lost.

Narrow range immobilized pH gradient strip: See → ultra-zoom gel.

NAS:
a) See → network-attached storage.
b) See → nonsense-associated alternative splicing.

NASBA: See → *n*ucleic *a*cid *s*equence-*b*ased *a*mplification.

NASBH: See → *n*ucleic *a*cid *s*canning *by* *h*ybridization.

Nascent polypeptide: A chain of amino acids linked together via peptide bonds that is being formed (in statu nascendi) and still attached to the 50S (bacteria) or 60S (eukaryotes) ribosomal subunit through a tRNA molecule.

Nascent RNA, nascent DNA: A chain of nucleotides linked together via → phosphodiester bonds that is being formed (*in statu nascendi*).

NAT: See → *n*atural *a*ntisense *t*ranscript.

National Center for Biotechnology Information (NCBI): A unit of the National Library of Medicine (which in turn is part of the US National Institutes of Health, Bethesda, Md.), that is organized in (1) an Information Resources Branch (data acquisition, storage and distribution), an Information Engineering Branch (data control) and the Computational Biology Branch (data research). NCBI operation is based on more than 500 CPUs, serves millions of accesses day by day, and distributes a terabyte of data per day. NCBI databases are, for example, Gen Bank and Pub Med. Gen Bank alone will contain about 33 billion bases and 25 million entries by the end of 2002.
URLs:
(1) NCBI Bookshelf
    http://www.ncbi.nlm.nih.gov/entrez/query.fcgi?db=Books
(2) NCBI GeneRif
    http://www.ncbi.nlm.nih.gov/LocusLink/GeneRIFhelp.html
(3) NCBI LocusLink
    http://www.ncbi.nlm.nih.gov/LocusLink/index.html
(4) NCBI Reference Genomes
    http://www.ncbi.nlm.nih.gov/entrez/query.fcgi?db=Genome
(5) NCBI RefSeq
    http://www.ncbi.nlm.nih.gov/LocusLink/refseq.html
(6) US National Center for Biotechnology Information
    http://www.ncbi.nlm.nih.gov/

National Institutes of Health Guidelines: See → NIH Guidelines.

Native DNA: A double-stranded DNA molecule with intact hydrogen bonds between all its base pairs.

NAT pair: Any two → messenger RNAs that form → sense-antisense complexes. See → natural antisense transcript.

Natural *antisense transcript* (NAT): Anyone of a series of naturally occuring → antisense → messenger RNAs in pro- and eukaryotic organisms. NATs function in the regulation of pre-mRNA splicing, control of → translation, the degradation of target RNA ("turnover") or its transport from the nucleus into the cytoplasm. See → cis-NAT, → NAT pair, → trans-NAT.

Natural gene transfer: See → indirect gene transfer.

Natural plasmid: Any plasmid which has not been constructed *in vitro* for cloning purposes. Natural plasmids described in this book are for example the → colicin factor; → F factor, → *Dictyoste lium discoideum*, → *Dictyostelium giganteum*, → *Dictyostelium mucoroides* and → *Dictyostelium purpureum* plasmid; → pSC 101, → pMB 9, → resistance factor, → RP1, → two micron circle.

Natural selection: An evolutionary process, during which individuals carrying a distinct gene possess a greater fitness than those without this gene.

Natural transformation: The modification of the genome of a cell by the active uptake of free DNA from the environment and its integration into the recipient's genome. In nature, DNA is liberated from pro- and eukaryotic cells via autolysis or excretion (in case of bacterial cells, also by bacteriophage-induced lysis), and may accumulate to relatively high concentration in the soil or water (marine ecosystems: 50 mg/l). If this DNA appears in the environment, potential accptor bacteria acquire → competence, interact with the foreign DNA, take it up and integrate it, most frequently via → homologous recombination (divergence of 10–20% is not tolerated). See → transformation (b).

NC: See → nitrocellulose.

NCBI: See → National Center for Biotechnology Information.

ncRNA: See → non-coding RNA.

NcSNP: See → non-coding single nucleotide polymorphism.

NE (*negative element*): See → silencer.

Nearest-neighbor frequency analysis: See → nearest-neighbor sequence analysis.

Nearest-neighbor sequence analysis (nearest-neighbor frequency analysis): A method for the characterization of DNA molecules that is based on the estimation of the relative frequencies with which pairs of each of the four bases lie next to one another. Any deoxyribonucleotide can be covalently bound to any one of the three others or to a nucleotide of the same type by its 3' or 5' hydroxyl group to form a dinucleotide molecule. Since there are 4 different deoxyribonucleotides (dATP, dCTP, dGTP, dTTP), the formation of 16 dinucleotides is possible. The frequency with which each of these combinations occurs is characteristic for a particular DNA. It can be determined by incubating a DNA template with *E. coli* → DNA polymerase and the four deoxyribonucleotides, one of which is labeled with ^{32}P at the α (innermost) phosphate position. The α-^{32}P then links the labeled nucleotide with its nearest-neighbor nucleotide. After synthesis the isolated DNA is digested with → micrococcal nuclease and spleen → phosphodiesterase to yield deoxyribonucleoside 3' monophosphates. The ^{32}P is now attached to the 3' carbon atom of the neighboring nucleoside (see scheme). The four deoxyribonucleoside 3' monophosphates are separated by paper electrophoresis and their radioactivity is measured. This measure gives the frequency with which the originally labeled nucleotide has been bound to the other nucleotides. By using all four α-^{32}P-labeled deoxyribonucleotides in repeated nearest-neighbor analyses the frequency of all 16 dinucleotides can be determined.

Near-*isogenic* *line* (NIL): A genotype, usually derived from repeated backcrossing, which differs from another genotype by only one or a few genes.

Near-*upstream* *element* (NUE): A DNA sequence that is located 4-40 nucleotides upstream the → poly(A) signal site in many plant and some plant virus genes and functions as part of a → termination signal complex in → transcription. NUEs of different genes contain a different core sequence (5'-AAUAAA-3' in the → cauliflower mosaic virus 35S gene complex; 5'-AAUGAA-3' in the zein gene of maize; 5'-AAUGGAAUGGA-3' in the ribulose bisphosphate carboxylase/oxygenase gene of pea). See → far-upstream element.

Nebulization: A simple method to fragment high molecular weight DNA by passing it through a small hole of a device used for inhalation directly onto a plastic hemisphere, where it is broken and dispersed in the surrounding plastic tube. The higher the applied pressure, the smaller the fragments.

Negative control: Any experimental control element, that provides little or no signal or result, irrespective of the results obtained from the actual experimental components. For example, a negative control on an → expression microarray consists of e.g. the → cDNA from a foreign gene, that is therefore not active in the test organism (e.g. a human gene in a plant microarray experiment). If total and fluorescence-labeled cDNAs of a test organism are hybridised to the array with these negative control cDNAs, they will not produce a signal (e. g. a → fluorescence signal) notwithstanding the reaction of the other cDNAs spotted onto the array. Negative controls are necessary for a test of the function of the array. See → positive control.

Negative *element* (NE): See → silencer.

Negative gene control: The termination of gene expression by the binding of a specific → repressor protein to → operator sites upstream of the coding region of many genes which prevents the simultaneous binding of RNA polymerase. See → inducible gene, → inducible operon, for example → *lac* operon.

Negative regulator: A molecule that turns off → transcription or → translation.

Negative selection: A procedure for the isolation of → transformants, in which detection is based on the loss of one or more specific functions. For example, an → insertion of a DNA fragment into the coding sequence of a → selectable marker gene of a vector inactivates this gene (→ insertional inactivation). Transformants can therefore be selected by the absence of marker gene function.

Negative supercoiling: The coiling of a covalently closed circular DNA (→ cccDNA) duplex molecule in a direction opposite to the turns of its double helix (i.e. in a left-handed direction). Compare → positive supercoiling, → supercoil.

Neofunctionalization: The acquisition of a novel, beneficial function by a duplicated gene (see → gene duplication) in evolutionary times, which is preserved by natural selection. The gene copy with the original function is retained. See → nonfunctionalization, → subfunctionalization.

Neoisoschizomer: Any → isoschizomer that cleaves at a position different from its prototype (i. e. the first restriction endonuclease sample of this type isolated).

Neomycin **(Nm):** A broad-spectrum antibacterial aminoglycoside → antibiotic from *Streptomyces fradiae* that binds to the 30S subunit of bacterial ribosomes and causes severe miscoding, inhibits initiation factor-dependent binding of fMet-tRNA and transpeptidation in pro- and eukaryotes, and blocks translocation. It is effective against a wide range of Gram-negative (e. g. *E. coli*) and most Gram-positive bacteria. See → neomycin resistance, → neomycin resistance gene, → neomycin sensitivity.

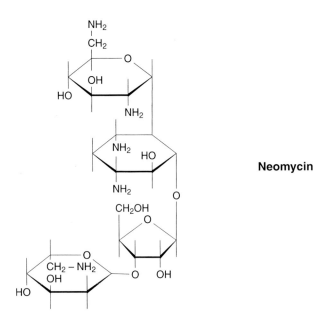

Neomycin

Neomycin phosphotransferase: See → aminoglycoside phosphotransferase.

Neomycin **resistance (Nm^r):** The ability of an organism to grow in the presence of neomycin, an → aminoglycoside antibiotic from *Streptomyces fradiae*. See also → neomycin resistance gene, → neomycin sensitivity.

Neomycin resistance gene (Nm^r gene): A gene (*neo*) from → *transposon 5*, → *transposon 601*, and from *transposon 903*, that encodes an → aminoglycoside phosphotransferase (APH I and II, respectively). These enzymes phosphorylate neomycin and related aminoglycoside compounds, and inactivate them. The neomycin resistance genes can be ligated to eukaryotic → promoters and transferred to eukaryotic cells, where their expression leads to neomycin resistance of the host. The neomycin resistance genes can be used as dominant → selectable markers in bacteria, fungi, animal and plant cells.

Neomycin sensitivity (Nm^s): The inability of an organism to grow in the presence of → neomycin, an aminoglycoside antibiotic from *Streptomyces fradiae*. Compare → neomycin resistance.

NEP: See → naked eye polymorphism.

NES: See → nuclear export signal.

NEST: See → nuclear expressed sequence tag analysis.

Nested oligo procedure: See → nested primer polymerase chain reaction.

Nested on chip (NOC) polymerase chain reaction (NOC-PCR): A variant of the conventional → polymerase chain reaction technique, that combines the advantages of → DNA chip technology (as e.g. parallelism, speed and automation) with the specificity and simplicity of liquid phase PCR in a single carrier system. In short, socalled → nested primers (i.e. oligonucleotides complementary to sequences within an → amplicon) are immobilized on a solid support (glass or plastic chip). Each one of these primers (P3) contains a specific nucleotide sequence characteristic for e.g. a polymorphism within a gene. Now the target sequence (e.g. an → exon of a gene of interest) is amplified in the liquid phase around the chip, using a specific primer pair (P1, P2) targeting conserved regions. The resulting amplification products will bind to the chip-bound primer P3 only if the 3' terminal base of P3 is complementary to the corresponding base in the amplicon. In this case, an amplification of primer P3 on the chip takes place, the amplification products are covalently anchored on the chip, and the non-covalently bound molecules are washed away. If the PCR reaction runs with → biotin-labeled nucleotides, the amplified products on the chip can be detected by → streptavidin-cyanine 5 conjugates and appear as fluorescent spots. NOC-PCR allows e.g. the discrimination between different alleles of socalled *h*uman *l*eucocyte *a*ntigen (HLA) genes on e.g. so called → HLA chips. See → nested primer polymerase chain reaction.

Nested PCR: See → nested primer polymerase chain reaction.

Nested primer: Any → primer whose sequence is complementary to an internal site of a DNA that has been amplified with other primers in a conventional → polymerase chain reaction (PCR). Such nested primers are used to re-amplify the target sequence at sites different from the original primer sites and thereby increase the specificity of the amplification reaction.

Nested primer polymerase chain reaction (nested PCR, nested oligo procedure): A modification of the → polymerase chain reaction which improves the yield of *specific* target sequences. During normal PCR, genomic DNA is denatured and annealed with an excess of two oligonucleotide → amplimers which bind to sequences just up- and downstream of the target DNA. These amplimers are then extended using thermostable → DNA polymerases. The DNA is again denatured, annealed to the same oligonucleotides and extended in a second cycle. This procedure is repeated some 20-30 times. Since the polymerase reads beyond the target DNA, a population of fragments arises, the lengths of which exceed that of the target DNA. In order to reduce the PCR to the target DNA, a second set of amplimers ("nested oligos") is annealed to sequences within the target DNA. After 20-30 cycles of PCR from these new primers, only amplified target DNA accumulates.

Network-attached storage (NAS): A specialized server attached to a local area network and using a streamlined operating and file system, that is employed to extract data from a database ("capture") and serve files to clients. For a better performance, the NAS system can be combined with the → SAN system ("NAS-SAN combo").

Neurogenetics: A branch of → genetics that focusses on the relationship(s) between genes and neuronal function(s) and disfunction(s) on the molecular level. Major research areas of neurogenetics are the development of diagnostic and therapeutic tools for hereditary diseases that afflict the nerve system. For example, a mutation of the *L1* gene (one of a series of genes encoding diverse proteins as e.g. L1, CHL1 [*close homologue of L1*], NrCAM and neurofascin, that represent the socalled L1 family) leads to the socalled CRASH syndrome (symptoms are hydrocephalus and mental retardation). The function of this gene can be deciphered with → knock-out mouse mutants and their anatomical and molecular analysis.

Neurogenome: The total number of genes expressed in both the central and peripheral nervous system at a given time. See → neurogenetics, → neurogenomics.

Neurogenomics: The whole repertoire of techniques for the identification, isolation and characterization of preferably all genes involved in the various functions of the central and peripheral nervous systems (see → neurogenome) and their mutant forms, especially if they cause neuronal disorders or simply changes in behaviour. Neurogenomics still experiments with animal (frequently mouse) models. See → neurogenetics, → neuroproteome, → neuroproteomics.

Neuropharmacogenomics: A branch of → pharmacogenomics, that uses the whole repertoire of → genomics, → transcriptomics, and → proteomics technologies to identify genes and/or mutations in genes involved in neurological disorders and to design and develop new drugs to control such diseases. See → oncopharmacogenomics.

Neuroproteome: The complete set of peptides and proteins expressed in the central and peripheral nervous system at a given time. See → neurogenetics, → neurogenome, → neurogenomics, → neuroproteomics.

Neuroproteomics: The whole repertoire of techniques to characterize the → neuroproteome in molecular detail. See → neurogenetics, → neurogenome, → neurogenomics.

***Neurospora crassa*:** A haploid Ascomycete fungus that grows as a mycelium and exists in two mating types. Fusion of nuclei from two different mating types is followed by meiosis and mitosis with the production of eight ascospores that are arranged linearly in the ascus. This arrangement allows the identification of the various products of meiotic divisions and renders *Neurospora crassa* an ideal organism for genetic studies. Among others, such studies led to the formulation of the "one gene-one enzyme" concept. Transformation of *N. crassa* is possible, and → shuttle vectors have been constructed for the transfer of genes between e.g. *E. coli* and *N. crassa* that increase the → transformation frequency by a factor of 10.

Neutral DNA: A infelicitous laboratory slang term for any DNA, that does not contain genes.

Neutral insertion: The → insertion of a → nucleotide or → oligonucleotide into a coding sequence of a → gene without changing the function of the encoded protein. See → insertion mutation, → neutral mutation.

Neutral mutation: Any → mutation that has no selective advantage or disadvantage for the organism in which it occurs, for example a mutation in a → cryptic gene or other → non-coding DNA.

Neutral substitution: An exchange of one (or more) amino acid(s) in a protein without any change of its function.

NF-I: See → CAAT-box transcription factor.

N-^{f}Met: See → N-formylmethiorine.

NF–1: See → nuclear factor 1.

NF1/CTF: A sequence-specific → DNA-binding protein that recognizes 5'-ATTTTGGCTT GA-AGCCAATATG-3' and represents an initiation factor for → adenovirus DNA replication.

N-*formylmethionine* (N-^{f}Met): A derivative of the amino acid methionine that carries a formyl group at its terminal amino group and functions as starter amino acid in bacterial polypeptide synthesis. Since N-formylmethionine lacks a free amino group it is "blocked", that is, it can only form a peptide bond at its carboxy terminus. Thus it can only be the first amino acid of a polypeptide but cannot be incorporated into the growing chain.

Methionine (Met)   N-formylmethionine (fMet)

NG: See → *n*itrosoguanidine.

NHP: See → non-*h*istone *p*rotein.

***Nic*/*bom* region:** See → *bom* region.

Nick: A break in one 5'-3' phosphodiester bond in one of two strands of a DNA duplex molecule. Compare → cut, → break. See also → nick translation.

Nickase: A general term for an enzyme that introduces nicks (single-stranded breaks) in DNA duplex molecules. See for example → nick translation.

Nick-closing enzyme: A synonym for → DNA topoisomerase I.

Nicked circular DNA: See → open circle.

Nicking: The introduction of → nicks into one strand of a double-stranded DNA molecule.

Nicking-closing enzyme: See → DNA topoisomerase I.

Nick translation: The replacement of nucleotides in double-stranded DNA by radioactively labeled nucleotides using the nicking activity of → DNase I and the polymerizing activity of *E. coli* → DNA polymerase I. In short, → nicks are introduced into the unlabeled ("cold") DNA duplex molecule by a limited digestion with DNase I to generate 3'-OH termini. Then *E. coli* DNA polymerase I is added that starts a DNA replication reaction at the 3' hydroxy terminus of each nick and simultaneously removes nucleotides from the 5' side (5'-3' exonuclease activity of *E. coli* DNA polymerase I), thus extending the nick. Since at least one of the four nucleotides needed for the reaction is labeled (for example α-^{32}P deoxynucleotide triphosphates), the original nucleotides in the duplex DNA molecule are replaced by labeled nucleotides. In the case of radioactive labeling a probe with high specific activity is generated that can be used in hybridization experiments. It is however also possible to use non-radioactively labeled nucleotides (see → non radioactive labeling). Compare → random priming.

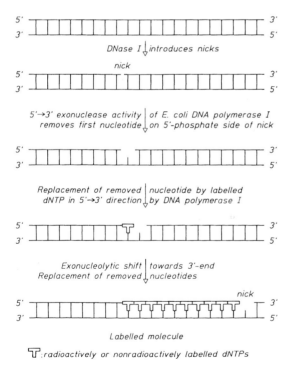

Nick translation

Nif **genes (***ni***trogen-*f*ixation genes, nitrogen-fixing genes):** A set of seventeen genes organized in an → operon in certain soil bacteria, notably *Rhizobium*. The proteins encoded by these genes catalyze the fixation of atmospheric nitrogen (N_2) into ammonia (NH_4^+) and nitrate (NO_3^-).

NIH Guidelines (*N***ational *I*nstitutes of *H*ealth Guidelines):** A compilation of recommmended security measures (see for example → containment) in recombinant DNA experiments, obligatory for laboratories which are funded by NIH grants. These guidelines provided the basis for the establishment of similar guidelines in countries other than the US.

NIL: See → near-*i*sogenic *l*ine.

9-β-D-ribofuranosyladenine: See → adenosine.

*N***itrocellulose (NC; cellulose nitrate):** A nitrated cellulose whose fibers can be used for the production of membrane filters, see → nitrocellulose filter.

*N***itrocellulose filter (NC filter, cellulose nitrate filter):** → Nitrocellulose fibers in the form of membranes with defined pore size (e.g. 0.45 µm). These filters selectively trap dsDNA or DNA RNA hybrids, but no single-stranded molecules. Single-stranded DNA or RNA may, however, be fixed to NC filters by → baking. Such blots can then be used in → Northern or → Southern blotting experiments.

*N***itrogen-*f***ixation genes:** See → *nif* genes.

*N***itrogen-*f***ixing genes:** See → *nif* genes.

*N***itrosoguanidine (NG):** An alkylating mutagenic chemical that adds methyl groups to many positions of all four bases in DNA, notably onto the oxygen at position 6 of guanine (leading to 0^6-alkyl guanine). This causes mispairing with thymine and principally results in GC → AT transitions in subsequent rounds of replication.

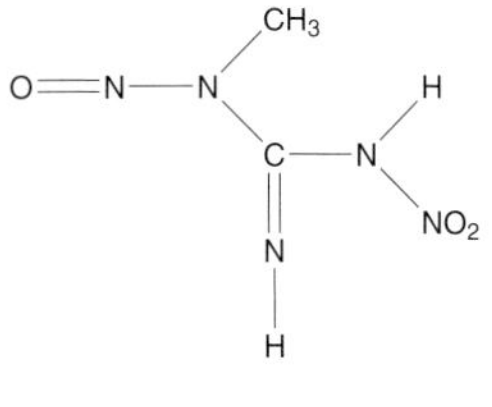

NLS: See → *n*uclear *l*ocalization *s*ignal.

NLS receptor: See → *n*uclear *l*ocalization *s*ignal receptor.

Nm: See → *n*eo*m*ycin.

NMD: See → *n*onsense-*m*ediated mRNA *d*ecay.

NMD: See → nonsense-mediated messenger RNA decay.

N-*Methylanthraniloyl* nucleotide (mant nucleotide): Any ribose-modified → nucleotide, that contains the → fluorophore N-methylanthraniloyl linked to the 2'or 3'carbon of the sugar moiety. Mant nucleotides resemble natural nucleotides in their protein-binding properties and are therefore used for the sensitive detection of conformational changes in a nucleotide-binding protein, protein-protein- and protein-ligand-interactions.

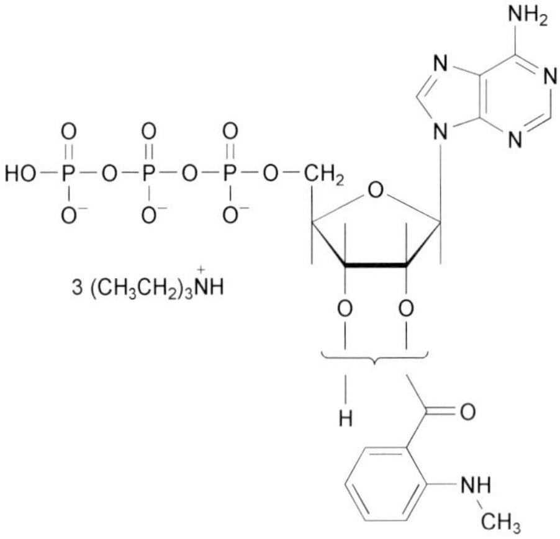

2', 3'-O-(N-Methylanthraniloyl)-adenosine-5'-triphosphate

Nm^r gene: See → *neomycin resistance gene*.

Nm^s: See → *neomycin sensitivity*.

NOC-PCR: See → nested on chip polymerase chain reaction.

NOGD: See → non-orthologous gene displacement.

Noise: A laboratory slang term for → background.

Nomadic gene: See → jumping gene.

Nonautonomous controlling element: A defective → transposon that can transpose only with the aid of a second, autonomous element of the same type.

Non-autonomous sequence element: Any one of a series of genomic sequences, that do not function or move by themselves (autonomously), but require the assistance of other sequences ("helper sequences") for function or movement.

Non-coding DNA: Any DNA that does neither encode a polypeptide nor an RNA. Non-coding DNA is a major constituent of most eukaryotic genomes, and includes → introns, → spacers, → pseudogenes, → centromeres, and most → repetitive DNA.

Non-coding exon (non-coding first exon): Any → exon, that has no coding function (i.e. whose sequence does not contribute to the amino acid sequence of a protein). For example, 16 kb downstream of the → prion protein-encoding *Prnp* gene of mice a second *Prnp*-like gene, called

Prnd and encoding the Dpl protein (German: doppel, for double) is located. These two genes are separated by an intergeneic space containing two intergene exons with no coding function(s).

Non-coding first exon: See → non-coding exon.

Non-coding RNA (ncRNA): Any → ribonucleic acid that does not encode a protein and can therefore not be annotated by a search for → open reading frames. For example, → microR-NAs,→ ribosomal RNAs, → 7SL-RNAs, → small nuclear RNAs, → small nucleolar RNAs, → small interfering RNAs, → small temporal RNAs, → telomerase RNAs, → transfer RNAs, → Xist-RNAs are such ncRNAs.

Non-coding sequence (NCDS): Any → DNA sequence that does not encode an → RNA or a → protein, as opposed to → coding sequences (e. g. a gene). Major NCDSs in eukaryotic genomes are → microsatellites, → minisatellites, → repetitive DNA, → retrotransposons, → satellites.

Non-coding single nucleotide polymorphism (ncSNP): A misleading term for any → single nucleotide polymorphism, that occurs in a non-coding region of the genome (e.g. an → intron). NcSNPs are the most frequent types of SNPs in eukaryotic organisms. See → anonymous SNP, → candidate SNP, → coding SNP, → copy SNP, → exonic SNP, → human SNP, → intronic SNP, → non-synonymous SNP, → reference SNP, → regulatory SNP, → synonymous SNP.

Non-cohesive end: See → blunt end.

Non-conjugative plasmid (non-selftransmissible plasmid): Any → plasmid that does not contain all functions necessary for its own intercellular transmission by → conjugation (e.g. lacks the → *tra* genes).

Non-contact spotting (non-contact printing): The deposition of target oligonucleotides, → cDNAs, DNAs, peptides or proteins on solid supports ("chips") of glass, quartz, silicon or nitrocellulose by an electrically induced discharge of the solution from the pin onto the surface of the chip. The pin does not come into physical contact with the solid support. See → contact spotting.

Non-contiguous translation: The relatively rare → translation of a → messenger RNA, during which part of the message are skipped. For example, during translation of the message derived from bacteriophase T4 gene *60* about 50 nucleotides are skipped.

Non-covalent protein delivery (peptide-mediated non-covalent protein delivery): A technique for the introduction of peptides or proteins into eukaryotic cells, using short synthetic peptides as transient carriers, which dissociate from the cargo protein after crossing the plasma membrane. For example, the 21 amino acid long peptide Pep–1 consists of a hydrophobic, tryptophan-rich motif (targeting the cell membrane and interacting with proteins hydrophobically), a hydrophilic lysine-rich domain derived from the → Simian virus 40 large → T antigen → nuclear localization sequence (that improves intracellular delivery of the peptide vector), and a spacer separating both. The peptides or proteins associate with Pep–1 through non-covalent hydrophobic interactions and form stable complexes, in which each protein is interacting with many Pep–1s (e.g. a 30 kDa → green fluorescent protein is complexed with 12–14 Pep–1 molecules). Once inside the cell, the Pep–1 and cargo rapidly dissociate ("decaging"), and the cargo can then translocate to its proper intracellular compartment. Proteins of up to 500 kDa and whole protein-DNA complexes can be rapidly delivered by this non-covalent protein delivery process.

Non-degenerate code: Any code in which the information is written in one specific sequence of symbols. In molecular biology, the genetic code is non-degenerate, if only one → codon specifies one amino acid.

Non-disjunction: The phenomenon that homologous chromosomes or sister chromatids do not separate at meiosis or mitosis, which leads to the formation of aneuploid cells.

Nonfunctionalization: The prevention of an acquisition of a novel and beneficial function of a duplicated gene (see → gene duplication) by degenerative mutations. See → neofunctionalization, → subfunctionalization.

Non-functional polymorphism: Any sequence → polymorphism, that has no consequences for the function of a protein and is therefore selectively neutral. Compare → functional polymorphism. See → intronic single nucleotide polymorphism, → non-coding single nucleotide polymorphism.

Nongenic DNA: The non-coding part of a → genome, mainly consisting of → microsatellites, → minisatellites, → retrotransposons, → satellite-DNA, → transposons, and in eukaryotes varying from about 3.0×10^6 to 1.0×10^{11} bp.

Non-*histone* *protein* (NHP): Any one of a large group of mostly acidic nuclear proteins of eukaryotes. These proteins serve enzymatic functions (e.g. → DNA and → RNA polymerases, → DNA methylases, RNA → processing enzymes), transport functions (e.g. RNA-binding proteins), regulatory functions (e.g. → transcription factors and → high mobility group proteins) and structural functions (e.g. → nuclear lamins).

Non-*homologous* *end-joining* (NHEJ): A mechanism ("pathway") for the repair of → double-strand breaks (DSBs) in DNA, that rejoins the two ends of this break. NHEJ frequently leads to error-prone repair of DSBs, because the ends are only incompletely processed. Non-homologous end joining is catalysed by the concerted action of ligase IV, Xrcc4, Ku70 and 80, DNA-PKcs, Artemis and Nej1/Lif2 in rodent cells. See → homologous recombination, → single-strand annealing.

Non-homologous recombination: See → illegitimate recombination.

Non-ionic detergent: A → detergent with an uncharged hydrophilic head-group that may be used to solubilize membrane proteins without their denaturation. Non-ionic detergents are for example the Tritons (see → Triton X-100), and octyl glucoside.

Nonliving array (chemical array): A polyethylene support or cellulose membrane, on which peptides or proteins are systematically arranged for high-throughput screening of oligonucleotide-protein, protein-protein, or protein-ligand interactions. The peptides can be synthesized on the polyethylene matrix by a *f*luorenyl*meth*oxy*c*arbonyl (Fmoc) amino acid protection technique in a C- to N-terminus direction, the side chains and → α-amino groups being protected between consecutive cycles. Similarly, peptides can be synthesized on the cellulose membranes ("spot synthesis"), except that the hydroxyl groups of the cellulose can be derivatized by Fmoc-β-alanine groups, and the peptide arrays be synthesized via the cellulose-bound alanine (subsequent to its deprotection). The array size (= number of bound peptides per area unit) can be increased substantially by the combination of solid-phase synthesis with photolithographic techniques. For example, photolabile protective groups such as *nitroveratryloxycarbonyl*, NVOC) on the growing peptide chain are selectively removed by light passing through a mask, similar to masks used for oligonucleotide synthesis (see → photolithography, → DNA chip). Nonliving arrays can be screened for e.g. chemical reactivity with low molecular weight ligands (e.g. pharmaceutically interesting compounds and their derivatives), or interactive peptides, proteins, RNAs, or oligonucleotides. See → living array.

Non-LTR retrotransposon: A → retrotransposon that lacks → long terminal repeats.

Non-Mendelian inheritance: See → cytoplasmic inheritance.

Non-nuclear gene: Any gene, that is localized outside of the nucleus in a eukaryotic cell. For example, chloroplast genes (in plants) and mitochondrial genes (in plants and animals) are such non-nuclear (organellar) genes. See → nuclear gene.

Non-orthologous gene displacement **(NOGD):** The replacement of a gene encoding a protein with a particular function by a non-orthologous (unrelated, or distantly related), but functionally analogous gene during evolution.

Non-overlapping code: A → genetic code that specifies only as many amino acids as are triplets arranged in linear sequence. For example, the sequence UUUCCCUUU encodes only phenylalanine (UUU), proline (CCC) and phenylalanine (UUU). Compare → overlapping genes.

Non-overlapping FRET pair: Any pair of → fluorochromes, whose emission spectra do not overlap, but can nevertheless be used for → *fluorescence resonance energy transfer* (FRET) experiments. Normally, FRET between two fluorophores occurs only, if the emission spectrum of the socalled donor overlaps the excitation spectrum of the acceptor fluorophore. However, also non-overlapping FRET pairs can be employed for such experiments, except that both the donor and acceptor have to come into close vicinity to each other. The excited fluorophore then transfers the energy to the acceptor ("quencher"), and no photons are emitted.

Non-palindromic cloning: The use of recombinant DNA techniques to propagate a DNA sequence inserted into non-complementary (non-palindromic) → cloning sites of a → cloning vector (→ non-palindromic vector). Non-palindromic sites on the vector can be generated by the ligation of non-palindromic → linkers to the termini of a linearized vector molecule. Non-palindromic cloning prevents the self-ligation of the vector molecules and the concatemerization of linkers. Thus dephosphorylation of the vector termini, and additionally any methylation or cutback steps, are superfluous. Compare → linker tailing.
Figure see page 726.

Non-palindromic vector: A → cloning vector that carries non-complementary (non-palindromic) termini at the cloning site. Such vectors allow → non-palindromic cloning.

Nonpenetrance: The absence of → expressivity of an → allele. See → complete penetrance, → incomplete penetrance.

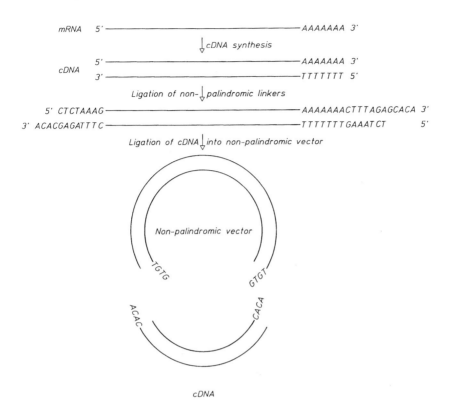

mRNA   5' ———————————————— AAAAAAA 3'

↓ cDNA synthesis

cDNA   5' ———————————————— AAAAAAA 3'
        3' ———————————————— TTTTTTT 5'

Ligation of non-↓palindromic linkers

5' CTCTAAAG ———————————————— AAAAAAACTTTAGAGCACA 3'
3' ACACGAGATTTC ———————————————— TTTTTTTGAAATCT   5'

Ligation of cDNA↓into non-palindromic vector

Non-palindromic vector

TGTG   GTGT

ACAC   CACA

cDNA

Nonpermissive cell: A cell in which a particular virus cannot produce progeny viruses, because it is not capable to complete DNA replication (→ abortive infection).

Nonpermissive condition: A condition that does not allow the survival of a → conditional lethal mutant.

Nonprocessive transcription: Any gene → transcription, whose → initiation occurs normally, but with inefficient → elongation. The transcription complex pauses and is rapidly released from the template, leading to an accumulation of short, non-polyadenylated RNAs, and only rarely full-length → messenger RNAs. See → processive transcription.

Nonproductive base-pairing: The imperfect pairing of bases in DNA, that are not complementary to each other and therefore cannot form hydrogen bonds. For example, A-A, A-G, A-C, G-A, G-G, G-T, C-C, C-A, C-T, T-G, T-C, and T-T are such nonproductive base pairs, which e.g. destabilize hybrids and reduce the melting temperature of a hybrid through reducing the force of interaction between the two strands. See → productive base-pairing.

Non-productive infection: See → abortive infection.

Non-radioactive labeling (chemical labeling): The introduction of nonradioactive groups into a DNA duplex molecule by → nick translation, → random priming, or → endlabeling. The introduced chemical compound (e.g. → biotin, → digoxigenin) can be detected by either colorigenic or luminogenic techniques (see → DNA detection system). See also → sulfonated DNA detection.

Nonribosomal peptide synthesis (NRPS): The synthesis of short peptides between two and 48 amino acid residues, most of them antibiotics (e.g. actinomycin D, bacitracin, bleomycin, cephalosporins, cyclosporins, erythromycin, penicillins or vancomycin) on *nonribosomal peptide synthetases* (NRPSs) of actinomycetes, bacilli, and filamentous fungi, that consist of iterated modules. Each module activates a specific amino acid through a pair of closely coupled domains: an adenylation domain (A) produces an *aminoacyl-O-adenosine monophosphate* (aa-O-AMP), that is then covalently tethered in a thioester linkage to the phosphopantetheinyl prosthetic group of the neighboring thiolation domain (T; also peptidyl carrier protein). The peptidyl chain grows directionally in incremental steps of elongating acyl-S-enzyme intermediates. Peptide bond formation and chain translocation occur each time an upstream donor peptidyl-S-pantetheinyl is attacked by a downstream acceptor aminoacyl-S-pantetheinyl nucleophile. This process is catalysed by the socalled condensation domain (C). Some synthetases are multimeric complexes, others are single massive proteins. For example, the cyclosporin synthetase is composed of 15,281 amino acid residues (1,7 MDa). The resulting peptides can be linear, or adopt nonlinear (e.g. heterocyclic) and iterative configurations (e.g. branched peptide backbones), and include unusual amino acids (e.g. D-amino acids, methylated variants of the standard amino acids, nonproteinogenic, hydroxylated and glycosylated residues, of which more than 300 are known). Most of the modifications are performed during synthesis of the peptide, others are added postsynthetically. In each case, no genetic code and no ribosomes are involved. Probably, functional enzymes can also be synthesized by NRPS. For example, the 60 amino acids enzyme LPXTGase, that cleaves the socalled LPXTG motif of many bacterial cell surface proteins (a prerequisite for an attachment of these proteins to the cell's surface), contains about 30% of unusual amino acids (not accepted by ribosomes). Therefore, this enzyme is most likely synthesized by NRPS. See → polyketide, → ribosome.

Non-selective polymerase chain reaction (NS-PCR): A variant of the conventional → polymerase chain reaction that allows to construct high quality → cDNA libraries employing sequence independent → primers. In short, → polyadenylated RNA is isolated, reverse-transcribed by → reverse transcriptase primed by oligo(dT), and the first strand cDNA is oligo(dC)-tailed at its 3' terminus using → terminal transferase. Then the mRNA template is removed by → RNase H, and the resulting single-stranded cDNA amplified by → *Taq* DNA polymerase using an oligo(dT) primer complementary to the original poly(A) tail (e.g. 5'-GGGGCTCGAG [T_{16}]– 3'), and an oligo(dC) primer complementary to the oligo(dC) tail (e.g. GGGGAATTC[G_{11}]-3'). Both primers contain an *Eco*RI → restriction site at their 5'-termini to facilitate subsequent cloning. The cDNA libraries obtained with NS-PCR are usually representative, i.e. contain each cDNA sequence at least once.

Non-selftransmissible plasmid: See → non-conjugative plasmid.

Nonsense-associated alternative splicing (NAS): A relatively rare intranuclear → splicing process initiated by reading frame-sensitive recognition of → premature termination codons (PTCs) in certain → messenger RNAs (mRNAs) during translation. For example, the exon encoding the hypervariable VDJ region of human *T* cell receptor β (TCR-β) mRNA is a result of gene rearrangements, which probably generates PTCs. NAS leads to the accumulation of the corresponding pre-mRNA, the increased use of potentially alternative, but normally latent → splice sites, and reduced normal splicing of the PTC-containing mRNA. See → nonsense-mediated messenger RNA decay, → nonstop messenger RNA decay.

Nonsense codon: Synonymous with → stop codon, see also → nonsense mutation.

Nonsense-mediated mRNA decay (Nonsense-mediated decay, NMD): The destruction of eukaryotic → messenger RNAs (mRNAs) containing → frameshift or → nonsense mutations, that would otherwise lead to the synthesis of truncated and thus non-functional proteins. All mRNAs are first monitored for errors that would encode potentially deleterious proteins ("RNA surveillance"). During their exit from the nucleus to the cytoplasm, they are recruited for NMD by the shuttle protein Upf3p (in yeast), if they cannot be translated along their full length. In this case they will remain in a transition complex (i.e. associated with mRNP proteins and Upf3p), which triggers their decay. First, Upf3p forms a binary Upf3p-Upf2p complex ("recruitment complex"), then a transient bridge between recruitment and termination complexes (mediated by Upf1p in yeast). Finally, Upf1p-associated ATP-dependent 5' → 3' RNA/DNA helicase activity unwinds the faulty RNA in the 5' → 3' direction and induces a topology change that exposes the 5' → cap, making it accessible to the decapping enzyme Dcp1p. Once decapped, the mRNA is fully degraded by Xrn1p from the 5' end. NMD requires active → translation. Without NMD or similar processes (see → non-stop messenger RNA decay), the eukaryotic cell would produce truncated and most probably non-functional proteins. See → nonstop messenger RNA decay.

Nonsense mutation: Any mutation in a coding sequence that converts a sense codon into a nonsense codon (a → stop codon) or a stop codon into a sense codon. As a consequence, the encoded protein will either be truncated (premature termination) or too long which in turn hampers or abolishes protein function. See also → nonsense suppression, → amber mutation, → ochre mutation and → opal mutation.

Nonsense suppression: A secondary mutation occurring at a chromosomal site separate from the site of a nonsense mutation and correcting the phenotype associated with the latter. See for example → suppressor gene, also → suppressor mutation.

Nonsense suppressor: A → tRNA that is mutated in its → anticodon and recognizes a nonsense (→ stop) codon so that the synthesis of a specific polypeptide can be extended beyond the stop codon. As a consequence, the nonsense codon is ignored (suppressed).

Non-specific transduction: See → transduction.

Nonstop decay: See→ nonstop messenger RNA decay.

Nonstop messenger RNA decay (non-stop decay): A process that eliminates eukaryotic messenger RNAs, that do not possess → termination codons. Such mRNAs ("non-stop mRNAs") are degraded by the → exosome, a highly conserved complex of 3' 5'-exonucleases. Compare → nonsense-mediated messenger RNA decay.

Nonstop transcript ("nonstop messenger RNA"): A laboratory slang term for a → messenger RNA, that does not contain any → stop codon. Such transcripts are usually labile and removed by → nonstop messenger RNA decay.

Non-synonymous sequence change: Any alteration in the nucleotide sequence of a coding region, that changes the amino acid sequence (and possibly the function) of the encoded protein. See → non-synonymous single nucleotide polymorphism, → synonymous sequence change.

Non-synonymous single nucleotide polymorphism (non-synonymous SNP, nsSNP): Any → single nucleotide polymorphism, that occurs in a coding region of a eukaryotic gene and changes the encoded amino acid. NsSNPs may cause the synthesis of a non-functional protein, and therefore be involved in diseases. See → anonymous SNP, → candidate SNP, → coding SNP, copy SNP, → exonic SNP, → gene-based SNP, → human SNP, → intronic SNP, → non-coding SNP, → promoter SNP, → reference SNP, → regulatory SNP, → synonymous SNP.

Non-transcribed spacer: A DNA sequence that separates tandem copies of an expressed gene or an expressed transcription unit, but is not transcribed itself. See for example → rDNA.

Non-viral retroposon: A → transposable element that transposes via an RNA intermediate, but does not contain → long terminal repeat sequences. Usually non-viral retroposons carry sequences with homology to → reverse transcriptase and poly(A) tracts at their 3' end. They probably originate from escaped → messenger RNAs.

Nopaline (N-α-[1,3-dicarboxylpropyl]-L-arginine): An amino acid derivative that is synthesized in plant cells transformed by the soil bacterium → *Agrobacterium tumefaciens* (e.g. strain C58). This bacterium, after contact with wound-exposed plant cell walls, transfers part of a large plasmid (→ Ti-plasmid) into the plant cell where it is integrated into the nuclear DNA. A gene (*nop* gene) close to the right border of the transforming DNA (→ T-DNA) encodes the enzyme → nopaline synthase that synthesizes nopaline from α-ketoglutarate and L-arginine. Nopaline is an → opine. It cannot be used by the host plant cell, but is secreted and serves as a carbon, nitrogen and energy source for agrobacteria possessing *noc* (*nopaline catabolism*) genes on their Ti-plasmid (see → genetic colonization). See also → nopaline synthase gene.

$$HN$$
$$\diagdown$$
$$C - NH- (CH_2)_3 - CH - COOH$$
$$H_2N \diagup \qquad\qquad\qquad | $$
$$\qquad\qquad\qquad\qquad NH$$
$$\qquad\qquad\qquad\qquad |$$
$$HOOC - (CH_2)_2 - CH - COOH$$

Nopaline synthase (Nos, nopaline synthetase): An enzyme present in → crown gall tumor cells and encoded by the *nop* gene of the → T-DNA originating from the → Ti-plasmid of → *Agrobacterium tumefaciens*. Nopaline synthase catalyzes the synthesis of the unusual amino acid → nopaline from L-arginine and α-ketoglutarate.

Nopaline synthase gene (NOP gene, *nop* gene): A gene encoded by the → T-region which is part of the → Ti-plasmid of → *Agrobacterium tumefaciens*. *nop* encodes the enzyme → nopaline synthase and is only expressed in transformed plant cells (see → crown gall). The *nop* gene is frequently used as a → reporter gene in plant transformation experiments, its → promoter (*Pnop*) and → termination sequences (3' t nop) are incorporated in plant transformation vectors. See also → nopaline.

Nopaline synthetase: See → nopaline synthase.

Nopalinic acid (N²-[1,3-D-dicarboxypropyl]-L-ornithine; ornaline): An amino acid derivative that is synthesized in plant cells transformed by the soil bacterium *Agrobacterium tumefaciens*. Nopalinic acid belongs to the so-called → opines. See also → crown gall.

```
H₂N – (CH₂)₃ – CH – COOH
                 |
                 NH
                 |
HOOC – (CH₂)₂ – CH – COOH
```

nop **gene:** See → *nop*aline synthase gene.

NOR: See → *n*ucleolus *o*rganizer *r*egion.

Normalization:
a) The process of dotting → messenger RNAs from → housekeeping genes (e. g. → the ubiquitin gene sequence) onto → hybridization membranes such that a hybridization with labeled → cDNAs from different cells, tissues, or organs will produce consistent hybridization signals for all dots. The strength of these signals – as quantified by → autoradiography or → phosphorimaging – serves as internal standard for quantifying the relative → abundance of other transcripts in e. g. → Northern hybridization.
b) The equalization of the concentrations of various transcripts present in a cell at extremely different levels (e. g. single copy or "rare" or "least abundance" versus abundant or "highly abundant" or "most abundant" RNAs). Since the difference between single copy and highly abundant messages is more than 10^5 in most cells, any cloning of cDNAs will inevitably lead to an overrepresentation of clones from strongly expressed genes, whereas least abundant messages probably escape cloning. Normalization balances the otherwise unequal representation of the various messages in a cDNA library by reducing the proportion of highly expressed mRNAs with concomitant enrichment of rarely expressed messages. An efficient technique for normalization, → phenol emulsion reassociation technique, involves the amplification of cDNAs, its precipitation with ethanol and resuspension in hybridization solution containing 8 % phenol (reduces the aqueous phase and increases the rate of hybridization). Vigorous shaking leads to a mixing of the phases. The resulting emulsion then allows hybridization of abundant cDNAs. Subsequently chloroform-isoamyl alcohol extraction and desalting is performed, and the single-stranded cDNAs (representing single-copy mRNAs) enriched by → restriction of double-stranded cDNAs (representing abundant mRNAs). The efficiency of normalization can be monitored by the loss of distinct bands (overrepresented cDNAs) and an increase of the background smear in → ethidium bromide-stained agarose gels (normalization of previously underrepresented messages).

Northern blot (RNA blot): A nitrocellulose or nylon membrane, onto which RNA molecules are transferred from a gel by e.g. capillary action and fixed by → baking or → cross-linking. Such blots can be hybridised to radioactively labeled → probes, and specific RNAs detected by → autoradiography. A Northern blot is the result of → Northern blotting. Compare → Southern blotting, → South-Western blotting, → Western blotting.

Northern blotting (Northern transfer, RNA blotting): A gel → blotting technique in which RNA molecules, separated according to size by → agarose or → polyacrylamide gel electrophoresis, are transferred directly to a → nitrocellulose filter or other matrices by electric or capillary forces

(Northern transfer). Single-stranded nucleic acids may be fixed to the nitrocellulose filter by → baking and are thus immobilized. Hybridization of specific, radioactively or non-radioactively labeled, single-stranded probes to the immobilized RNA molecules (Northern hybridization) allows the detection of individual RNAs out of complex RNA populations. See also → multiple tissue Northern blot. Compare → Southern blotting, → South-Western blotting, → Western blotting. See → Northern blot.

Northern transfer: See → Northern blotting.

NOS:
a) See → *N*opalin *s*ynthase.
b) *N*itric acid *o*xide *s*ynthase (Nos; L-arginine-NADPH:oxygen oxidoreductase, EC 1.14.23.39): A homodimeric hemeprotein that catalyzes the conversion of L-arginine into nitrogen monoxide (NO) and L-citrulline, consuming molecular oxygen and $NADPH_2$.

Not I library: See → chromosome linking clone library.

Novel gene: Any gene that has not been known before its detection by e. g. → genomic sequencing. The term is misleading, since a novel gene is not really novel (as e. g. a → synthetic gene might be), but normally a component of a genome for millions of years.

NPA: See → *n*uclease *p*rotection *a*ssay.

N-protein: A protein of the → lambda phage (and other → coliphages) that binds to specific sequences of the phage genome (*nut* sites, *N-ut*ilization sites), prevents *rho*-dependent termination of leftward early transcription and induces the expression of adjacent genes. The gene for this antiterminator protein (gene *N*) is transcribed during the early phase of infection (→ early gene).

NPT; *n*eomycin *p*hospho*t*ransferase: See → aminoglycoside phosphotransferase.

NR: See → *n*uclear *r*eceptor.

nRNA: See → *n*uclear RNA.

NS-PCR: See → *n*on-*s*elective *p*olymerase *c*hain *r*eaction.

nt: Abbreviation for *n*ucleo*t*ide(s).

nTaq: Abbreviation for the *n*ative form of → *Thermus aquaticus* → DNA polymerase. Compare → r*Taq*.

N-terminal end (N-terminus; amino terminus, amino terminal end): The terminus of a protein where the amino (NH_2) group does not form part of a peptide bond. Polypeptide synthesis starts at this end. Compare → N-formylmethionine.

N-terminus: See → N-terminal end.

NTP: Abbreviation for any ribo*n*ucleoside-5'-*t*ri*p*hosphate (e.g. ATP, CTP, GTP, TTP, or UTP).

Ntp: See → base pair.

NTT: See → *n*uclear *t*ransportation *t*rap.

N²-di-methylguanosine: A → rare base.

Nu body: The equivalent of a → nucleosome in electron microscopic pictures of negatively stained Miller spreads.

Nuclear cage: See → nuclear lamina.

Nuclear chromosome scaffold: See → nuclear lamina.

Nuclear dimorphism: The presence of two differently sized nuclei in one and the same cell. For example, ciliates possess one or more socalled micronuclei and macronuclei. The smaller micronuclei harbor typical eukaryotic chromosomes with associated histones, divide by mitosis, and are transcriptionally silent during asexual growth of the ciliate. However, they are active during sexual reproduction and responsible for the genetic continuity of the protozoon ("germ-line nucleus"). The macronucleus in turn actively transcribes its genes during asexual growth, replicates during asexual reproduction, but is destroyed and re-formed during sexual reproduction. Therefore, macronuclei do not transmit genetic information to sexual offspring.

Nuclear DNA (nDNA): The DNA that is located within the nucleus of eukaryotic cells, in contrast to the DNA of mitochondria (mtDNA) or chloroplasts (cpDNA). See → mitochondrial DNA and → chloroplast DNA.

Nuclear envelope (NE, nuclear membrane): The double-membrane boundary of nuclei in eukaryotic cells. The outer membrane forms a continuum with some parts of the endoplasmic reticulum (ER) whereas the inner membrane functions in the organization of → chromatin (e.g. by anchoring → looped domains). Both membranes are perforated by complex pores (→ nuclear pore) that consist of a central channel and a peripheral layer of proteins, and mediate im- and export processes.

Nuclear export sequence: See → nuclear export signal.

Nuclear export signal (NES; nuclear export sequence): A glycin- or leucine-isoleucine-rich domain in proteins that are synthesized in the → nucleus and exported into the cytoplasm of a cell. NESs are potential address sites where proteins (e.g. receptor proteins) bind and assist in the nucleo-cytoplasmic exportation process. In the Rev protein from the pathogenic human T-cell leukemia virus type 1 (HTLV-1), the NES consensus sequence is:

$$\text{leu-X}_{2-3}\text{-phe/ile/leu/val/met-X}_{2-3}\text{-leu-X-ile/val.}$$

Also, 5S rRNA is channeled into the cytoplasm after complexing with → transcription factor TF IIIa that contains an NES. See → nuclear localization signal.

Nuclear expressed sequence tag analysis (NEST): A technique for the identification of transcribed (active) genes in the nucleus of eukaryotic organisms. In short, nuclei are first labelled with → fluorochromes (e.g. via direct binding of the fluorophore to nuclear DNA, or indirectly with

→ autofluorescent proteins), isolated by → flow cytometry ("flow sorting"), lysed, and the re-leased → poly(A)$^+$-RNA captured on oligo(dT)-linked → magnetic beads. The captured RNA is then reverse transcribed into → cDNA, restricted with → four base cutters (restriction enzymes with a 4bp restriction recognition site), resulting in 3'-fragments bound to the beads. Then → linkers are ligated to the fragments, the fragments amplified via conventional → polymerase chain reaction techniques, using linker-complementary primers, and the resulting amplicons separated on → sequencing gels, which display characteristic expression profiles of the cells, tissues, organs or organisms of interest.

Nuclear factor: See → transacting factor.

Nuclear factor I: See → CAAT-box transcription factor.

Nuclear *factor* 1 (NF–1): Any one of a large family of eukaryotic → transcription factors, that recognize specific address sites and bind to DNA. The tremendous diversity within the NF–1 family is a consequence of the presence of multiple genes. The diversity of encoded proteins originate from → alternative splicing and heterodimerization.

Nuclear gene: Any gene, that is localized in the nuclear genome of a eukaryotic cell. See → non-nuclear gene.

Nuclear genome: The entire → genetic material of the → nucleus of eukaryotic cells. Synonym of → genomic DNA.

Nuclear halo: An artificial structure generated through the lysis of nuclei and the spread of the DNA as loops. These loops protrude from a central scaffold (→ nuclear lamina) that appears as a halo (gr.-lat.: zone of diffuse light around a light source).

Nuclear import: The process of transporting proteins from the cytoplasmic space into the nuclear space. Nuclear import of such proteins proceeds via several pathways. For example, proteins carrying the classical → nuclear localization signal are bound by an → importin (karyopherin) α/b1 heterodimer that docks at the → nuclear pore complex. The docked protein is then trans-located into the nucleus in an energy-dependent step requiring a set of proteins, including *n*uclear *t*ransport *f*actor 2 (NTF2), the GTPase Ran, and a nuclear pore protein designated nucleoporin p62. Certain RNA-binding proteins are imported by importin β2, some ribosomal proteins by importin β3. See → nuclear transport.

Nuclear lamin: A family of interrelated polypeptides that are the constituents of the → nuclear lamina network and fall into three major types: the neutral A- and C-lamins, and the acidic B-lamins (molecular weight range from 62-69 kDa). Less frequently occurring lamins belong to the D and E categories. The lamins are structurally related to the intermediary filaments, assemble to 10 nm filaments *in vivo*, and possess the typical → coiled coil-configuration of two intertwined α-helices. During nuclear division the lamina disintegrates with concomitant strong phosphorylation of lamins. A single base exchange in the gene encoding lamin A leads to the use of a → cryptic splice site in the → pre-messenger RNA. Consequence: a shorter lamin A is synthesized, that does not function correctly. The underlying mutation therefore is the cause for the *Hutchinson-Gilford Progeria Syndrome* (HGPS), an extremely rare disease leading to severe premature aging.

Nuclear lamina (fibrous lamina, karyoskeleton, nuclear cage, nuclear matrix, nuclear scaffold, nuclear chromosome scaffold): A filamentous meshwork located between the inner nuclear membrane (see → nuclear envelope) and heterochromatin, which provides potential attachment sites for → chromatin and cytoplasmic intermediate filaments.

Nuclear localization sequence: See → nuclear localization signal.

Nuclear localization signal (NLS; nuclear localization sequence): A cluster of basic amino acids (usually containing a proline of glycine, for example the sequence proline-lysine-lysine-lysine-arginine-lysine-valine, PKKKRKV of the SV 40-like NLSs) in proteins larger than 40 kDa, that directs their targeted import into the nucleus. Such NLSs have been identified in a series of yeast, *Drosophila*, amphibian, mammalian and plant proteins, and vary in amino acid sequence (smallest consensus sequence: $K_K{}^RX_R{}^K$). The NLS is not proteolytically removed after translocation of the linked protein, so that the protein retains the capacity to enter the nucleus repeatedly (e. g. after each cell division). Basically, two arrangements of NLS exist in import proteins. Single-cluster NLS consist of one single NLS sequence, bipartite NLS are composed of two interdependent domains with short intervening sequences, that act synergistically, but can also independently, yet less effectively direct proteins into the nucleus. Three main NLS categories can be found:
1. Simian virus 40-like NLSs contain short tandem stretches of 6-8 basic amino acids with either a proline or glycine (PKKKRKV), and occur also in e.g. a transcription activator protein of maize (*Zea mays*).
2. Mating type α2-like NLSs consist of short hydrophobic regions that contain one or more basic amino acids (KIPIK or MNKIPIKDLLNPG).
3. Bipartite NLSs (nucleoplasmin-type NLS) are a combination of two regions of basic amino acids separated by a spacer of approximately ten amino acids, and are ubiquitous.
   Proteins smaller than 40-60 kD may also diffuse through nuclear pores. Larger proteins definitely require ATP and at least one NLS to traverse pores. Compare → nuclear export signal.

Species	Protein	Motif
Xenopus laevis	Nucleoplasmin	KRXXXXXXXXXXKKKK
Homo sapiens	Glucocorticoid Receptor	RKXXXXXXXXXXRKXKK
Homo sapiens	Androgen Receptor	RKXXXXXXXXXXRKXKK
Homo sapiens	p53 Protein	KRXXXXXXXXXXKKK
Simian virus 40	SV40 T-Antigen	PKKKRKV
		K = Lysine R = Arginine
		P = Proline X = Any Amino Acid

Nuclear localization signal receptor (NLS receptor; NLS-binding protein, NLS-BP): A protein that recognizes → nuclear localization signals, interacts with them, and directs the corresponding protein to nuclear pores.

Nuclear matrix: See → nuclear lamina.

*N*uclear *m*atrix *p*rotein *e*nzyme-*l*inked *i*mmunosorbent *a*ssay (NMP-ELISA): A technique for the *in vitro* detection and quantitation of specific nuclear matrix proteins from injured, dying, or dead cells. Specifically, NMP-Elisa detects the socalled *n*uclear *m*itotic *a*pparatus protein (NuMA) or its fragments, that arise after an encounter with a toxic chemical or a pathogenic organism. This 240 kDa protein is restricted to the nucleus during interphase, but redistributed and concentrated at the spindle apparatus during mitosis. If cultured cells are injured or going to die, NuMA is released into the culture medium, where its concentration can be estimated by a detector → antibody. NuMA levels are positively correlated with *in vitro* cell death.

Nuclear membrane: See → nuclear envelope.

Nuclear pore (*n*uclear *p*ore *c*omplex, NPC, "porosome"): A cylindrical channel through the → nuclear envelope that mediates cytoplasmic-nuclear and nuclear-cytoplasmic exchange ("traffic"). A pore complex consists of a ring of eight globular subunits (annular granules) of 100 – 250 Å in size, arranged in a symmetrical, octagonal pattern at each side of the nuclear envelope. These rings border a circular hole of 900 Å in diameter. From the ring at the cytoplasmic side a series of eight, irregularly formed filaments protrude into the cytoplasm. The nuclear ring consists of eight filaments, that unite distally into a ring-like structure ("distal ring") such that a cage-like complex results ("nuclear basket"). A series of 30 → nucleoporin proteins are more or less symmetrically distributed at both the cytoplasmic and nuclear sides. About 100 to more than $5 \cdot 10^7$ pores per nucleus may exist, their number varying with the metabolic state of the nucleus or the cell. Proteins over 40 kDa have to carry a → nuclear localization or → nuclear export signal to be transported through the pore complex, smaller molecules or ions diffuse "passively".
Figure see page 736.

Nuclear pore complex *p*rotein (NUP): Any one of the various proteins tightly bound to the → nuclear pores. Also called nucleoporin.

Nuclear processing of RNA: See → post-transcriptional modification.

Nuclear receptor (NR): A ligand-inducible → transcription factor that binds to cognate *r*esponse *e*lements (REs) and induces the transcription of target genes. The ligand-dependent → transactivation by NRs is mediated by an *a*ctivation *f*unction motif (AF-2) which is present in the *l*igand-*b*inding *d*omain (LBD) of the receptor and functions via *t*ranscriptional *i*ntermediary *f*actors (TIFs). Nuclear receptors (e.g. the estrogen receptor) regulate complex events in early embryogenic development, cell differentiation, and homeostasis.

Nuclear receptor: Any one of a family of ligand-activated → transcription factors that regulate the expression of target genes. All nuclear receptors have at least four different domains, that are differentially conserved between the subfamilies: the DNA-binding C-domain (a → helix-loop-helix or → zinc finger conformation), the ligand-binding/dimerization domain, the A/B transactivation domain, and the socalled hinge (D) domain. The various nuclear receptors bind different, mostly hydrophobic ligands such as dioxin, ecdysone, retinoic acid, steroids, thyroid hormones, and vitamin D, and form distinct complexes with → heat shock protein 90, that assists in domain-folding for ligand binding. The ligand-nuclear receptor complex directly acts upon the DNA, and therefore links extracellular signals to transcriptional response(s).
Based on C-domain sequences and structural data, nuclear receptor genes fall into three subfamilies: subfamily I encodes *ear1* subgroup, retinoic acid and thyroid hormone receptors, subfamily II the orphan receptor genes (orphan: a nuclear receptor, for which no ligand has yet been identified), and subfamily III, harboring the steroid hormone receptor genes.

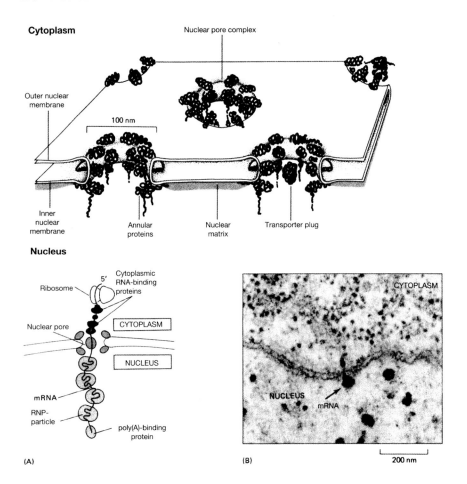

Scheme of nucleo-cytoplasmic transport (A) and an electron microscopic image of this process (B)

Nuclear pore

Nuclear RNA (nRNA): Any RNA that either remains within the nucleus after its synthesis, or is exported into the cytoplasm only after → processing. For example, heterogeneous nuclear RNA (hnRNA), including the primary transcripts of many genes (e.g. pre-mRNA, pre-tRNA, pre rRNA), occurs only in the nucleus. The processed transcripts (e.g. mRNA, tRNA, rRNA) are associated with specific proteins and transported into the cytoplasm.

Nuclear run-off transcription assay: See → run-off transcription.

Nuclear scaffold: See → nuclear lamina.

Nuclear translation: The synthesis of proteins from → messenger RNAs (mRNAs) within the nucleus (*in nucleo*) of a eukaryotic cell. Actually → transfer RNAs, certain translation factors and → ribosomes (most probably not functioning prior to their export into the cytoplasm) are present in the nucleus, nuclear translation sites overlap with RNA polymerase II transcription sites (i.e. transcription and translation of the resulting messenger RNAs are probably coupled), and mRNA translation in the nucleus reportedly reaches 10–15% of the total cellular protein synthesis, nuclear translation is still not unequivocally proven and therefore controversial.

Nuclear transplantation: The → microinjection of nuclei (or pronuclei) from one embryo into a second embryo, or the transfer of an isolated nucleus from one cell into the enucleated cytoplasm of another cell.

Nuclear transport: The import and export of molecules across the → nuclear membrane. The passage may be facilitated by specific proteins, may depend on a guide sequence (e.g. a → nuclear localization sequence), and may preferentially use → nuclear pores. See → nuclear import.

Nuclear transportation trap (NTT): A technique for the identification of → cDNAs (or genes) encoding nuclear transport signals (e.g. → nuclear localization signals, NLSs), that is based on a yeast selection system. The NTT consists of two components: (1) A yeast expression plasmid for → nuclear export signal (NES)-LexAD (activation domain) fusion proteins, that are excluded from the nucleus, because they possess an NES, and (2), a LexAD-responsive leu2 → reporter gene, that is only expressed if the hybrid NES-LexAD protein is actively imported into the nucleus (i.e. contains an NLS).

Nuclease: Any enzyme that catalyzes the hydrolysis of → phosphodiester bonds in nucleic acid molecules and leads to their breakdown. Nucleases can be broadly categorized into → exonucleases (releasing nucleotides from the ends of nucleic acid molecules), and → endonucleases (cleaving the nucleic acid molecule at internal sites). There exist nucleases specific for DNA (deoxyribonucleases, DNases) or RNA (ribonucleases, RNases), and for single-stranded or double-stranded polynucleotides. Nucleases generally present problems during the isolation of nucleic acids from animal and plant tissues and are therefore inhibited by the inclusion of various agents (e.g. EDTA, → RNasin; compare also → nuclease-free reagent) in the extraction buffers. See also → micrococcal nuclease, → mung bean nuclease, → nuclease P1, → Bal 31 nuclease, → repair nuclease.

Nuclease Bal 31: See → Bal 31 nuclease.

Nuclease-free reagent: Any chemical that does not contain even traces of RNases and/or DNases. Such chemicals are used to isolate RNA or DNA from cells, tissues, organs, or organisms that are rich in nucleases.

Nuclease P1 (EC 3.1.30.1): A single-strand specific → nuclease (endo- and exonuclease) from *Penicillium citrinum* that catalyzes the degradation of RNA and single-stranded DNA to 5' phosphomononucleotides. The enzyme also hydrolyzes 3' mononucleotides (ribo- and deoxyribonucleotides) to nucleosides and inorganic phosphate, and is used for the analysis of the 5'-terminal nucleotide of RNA and DNA.

Nuclease protection assay (NPA): A more general term for any technique for the detection, quantitation and characterization of specific → messenger RNA molecules out of complex mixtures of total cellular RNAs. The most frequently used NPAs are → RNase protection assay and → S1 nuclease protection assay.

Nuclease S1: See → S1 nuclease.

Nuclease S1-mapping: See → S1-mapping.

Nuclease S7: See → micrococcal nuclease.

Nucleation: The reannealing of a few complementary bases of two single-stranded DNA or RNA molecules to form a nucleation point for complete renaturation to a duplex molecule.

Nucleic acid: A single- or double-stranded linear polynucleotide containing either deoxyribonucleotides (→ DNA) or ribonucleotides (→ RNA) that are linked by 3'-5'-phophodiester bonds.

Nucleic acid biotool (NAB): A generic name for any synthetic oligonucleotide, that binds specifically to a target protein and interferes with its function(s). NABs are used to interfere with physiological or pathological processes, to tag proteins, or to investigate their function.

Nucleic acid chromatography system: See → NACS™.

Nucleic acid hybridization: See → hybridization.

Nucleic acid microarray: A more general term for any → microarray, onto which DNA, RNA or oligonucleotides have been spotted.

Nucleic acid ordered module assembly with directionality (NOMAD): A cloning strategy for the combinatorial arrangement of different DNA fragments in constructs of predetermined structure. NOMAD works with basically two elements, a socalled "assembly vector" with an insertion site, and individual or combined DNA "modules" which are ligated into this site in a sequential or directional mode. In short, specially designed assembly vectors with insertion sites flanked by convergently oriented → recognition sequences for two different type IIS → restriction endonucleases (cutting at a precise distance outside of these sites and producing → sticky ends) are first digested with the appropriate restriction enzyme (e. g. *Bsa*I [recognition sequence: 5'-GAGACC-3'] and *Bsm* BI [recognition sequence: 5'-CGTCTC-3']), then the desired module(s) with compatible cohesive ends are inserted. The second module can be inserted either 5' or 3' to the first one. When the assembly vector is cut by *Bsm* BI, then the second module is inserted at the *Bsm* BI site, whereas a *Bsa* I cut directs the second module to this site. This procedure leaves the restriction site intact, so that the vector can be cut again with the same endonuclease(s), and other modules can be inserted adjacent to the already inserted ones. It is also possible to ligate previously assembled multimodule blocks ("composite modules"). The modules themselves can be sequentially added in any desired order and can also be released as desired and recloned into another modular construct. NOMAD allows the modular construction of → chimeric genes and therefore composite proteins, and creation of new cloning vehicles (e. g. by recombining modules for → origins of replication, → transcription termination signals, → selectable marker genes, and → reporter genes).

Nucleic acid-programmable protein array (NAPPA): A GST-coated glass slide variant of the conventional protein array, onto which plasmid DNAs, each containing a distinct gene (or genes) of interest, are spotted (immobilized). All the different genes are simultaneously transcribed/translated in a cell-free in vitro transcription/translation system (in which the glas slide is immersed), and the resulting proteins immobilized *in situ*. The slides are the washed to remove unbound protein, and target-probe complexes detected by e.g. fluorescence. NAPPAs allows to detect protein-protein interactions.

Nucleic acid scanning: The search for distinct sequence motifs (e. g. the → TATA box, → start or → stop codons) in a nucleic acid molecule.

Nucleic acid sequence-based amplification (NASBA): An *in vitro* DNA or RNA primer-directed amplification procedure that uses → RNA polymerase and → reverse transcriptase to amplify a template. The amplification of an RNA target for example, starts with the annealing of primer 1 that contains an RNA polymerase → promoter sequence at its 5' end. Then reverse transcriptase synthesizes a complementary DNA strand (→ cDNA) off the primer. → RNase H is added to degrade the template RNA. Next, primer 2 anneals to the cDNA, → second strand synthesis ensues driven by RTase, and the promoter sequence becomes double-stranded and functional. Now RNA polymerase can be used to produce RNA copies of the cDNA that are anti-sense to the original target. In turn, these serve as templates for cDNA synthesis in the cyclic phase of NASBA. Thus the original target is continuously amplified.

Nucleic acid sequence-based amplification (NASBA): An *in vitro* DNA or RNA primer-directed isothermal amplification procedure that uses → RNA polymerase and → reverse transcriptase to amplify a template. The amplification of an RNA target for example, starts with the annealing of primer 1 that contains an RNA polymerase → promoter sequence at its 5' end. Then reverse transcriptase synthesizes a complementary DNA strand (→ cDNA) off the primer, and → RNase H is added to degrade the template RNA. Next, primer 2 anneals to the cDNA, → second strand synthesis ensues driven by RTase, and the promoter sequence becomes double-stranded and functional. Now RNA polymerase can be used to produce RNA copies of the cDNA that are anti-sense to the original target. In turn, these serve as templates for cDNA synthesis in the cyclic phase of NASBA. Thus the original target is continuously amplified without the temperature shifts necessary during a → polymerase chain reaction.

Nuclein: An outdated synonym for DNA, originally coined by Friedrich Miescher, who isolated DNA (probably) for the first time in 1869.

Nucleocapsid: The protein coat (→ capsid) of a → virion or → virus together with the enclosed nucleic acid molecule (DNA or RNA).

Nucleocidin: A → nucleoside antibiotic.

Nucleo-delta peptide (NDP): Any artificial peptide, that forms the basic unit of biopolymers with importance for chip and nanotechnology.

Nucleofection: A technique for the → direct gene transfer into the nucleus of a cell. Basically, current nucleofection methods rely on → electroporation of the foreign DNA into the target cells and its guidance into the nucleus by cell-type specific solutions ("Nucleofector", composition not disclosed), resulting in high transfection efficiencies.

Nucleoid (karyoid, DNA plasm): The region within a prokaryotic cell that contains the DNA. A nucleoid is analogous to the → nucleus of eukaryotic cells, though not engulfed by a nuclear membrane. Nucleoids are also constituents of mitochondria and plastids (e.g. chloroplasts).

Nucleolar organizer: See → nucleolus organizer region.

Nucleolin (C23): A eukaryotic nonribosomal nucleolar phosphoprotein with a tripartite structure. The N-terminal domain interacts with nucleolar → chromatin and is phosphorylated. This phosphorylation, catalyzed by cyclic AMP-independent protein kinase II, modulates chromatin condensation in conjunction with histone H1 and is correlated with nucleolar transcriptional activity. This domain also contains bipartite → nuclear localization sequence motifs. The central domain of nucleolin contains four → RNA recognition motifs. The C-terminal domain consists of *g*lycine- and *a*rginin-*r*ich repeats (socalled GAR repeats). In animals, nucleolin is highly phosphorylated and has a molecular mass of 90-110 kD.

Nucleolin is regulating intranucleolar chromatin organization, → rDNA transcription, and rRNA processing, preribosomal synthesis, ribosomal assembly and maturation. It also is involved in cytoplasmic-nucleolar transport of preribosomal particles from the nucleolus to the cytoplasm.

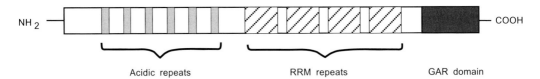

Acidic repeats   RRM repeats   GAR domain

Nucleolus: The spherical or globular subnuclear organelle associated with the so-called → nucleolus organizer region of chromosomes. It consists mostly of primary → rDNA transcripts, attached ribosomal proteins, and a variety of other proteins such as RNA polymerase I (A) and RNA methylases. In electron microscopic pictures the nucleolus is made up of a → fibrillar zone (pars fibrosa; containing rDNA) and a → granular zone (pars granulosa, containing pre-ribosomal particles). An active nucleolus exports large amounts of ribosomal precursors and exhibits special substructures, such as "pulsing vacuoles", less dense regions within the nucleolus that change their volume rhythmically.

Nucleolus organizer: See → nucleolus organizer region.

Nucleolus organizer region (NOR, nucleolus organizing region, nucleolar organizer, NO, nucleolus organizer): A specific chromosome segment containing the ribosomal RNA genes (→ rDNA) and active in the formation of the → nucleolus.

Nucleolus organizing region: See → nucleolus organizer region.

Nucleome: The microscopical and molecular description of all components of a nucleus of a eukaryotic cell. It encircles the DNA with all its constituents (→ genes, → promoters, repetitive sequences as → satellites, → microsatellites, → transposons, → retrotransposons, → telomeres, → centromeres), the RNA (→ ribosomal RNA precursors, all the → small nuclear and → nucleolar RNAs) and the proteins (→ histones, → non-histone proteins, → scaffold proteins, → lamins). See → nucleomics.

Nucleomics: An ill-defined term of the → omics generation for the whole repertoire of technologies applicable to the study of nuclear architecture, → genomes, → transcriptomes, → post-transcriptional modification of transcripts, → post-translational modification of proteins, and nucleo-cytoplasmic interaction(s). See → behavioral gen*omics*, → chemical gen*omics*, → comparative genomics, → environmenral gen*omics*, → epigen*omics*, → functional gen*omics*,

→ gen*omics*, → horizontal gen*omics*, → integrative gen*omics*, → kin*omics*, → medical gen*omics*, → nucleome, → nutritional gen*omics*, → pharmacogen*omics*, phylogen*omics*, → prote*omics*, → recogn*omics*, → structural gen*omics*, → transcript*omics*, → transpos*omics*.

Nucleomorph: A remnant gene-rich nucleus of a formerly free-living eukaryotic alga that has been engulfed by another eukaryotic cell and became an endosymbiont in a process called secondary endosymbiosis (where primary endosymbiosis is the acquisition of mitochondria and plastids by a recipient proto-eukaryotic cell). The process of this secondary endosymbiosis certainly occurred frequently in evolution, but nucleomorphs in only two algae groups, the cryptophytes and chlorarachniophytes have been preserved. These nucleomorphs contain three small linear → genomes (chlorarachniophytes: 380 kb; cryptophytes: 600 kb). Nucleomorph DNA encodes a total of 200–300 genes, among them diverse → housekeeping genes (for e.g. → transcription, mRNA → processing, → translation, protein degradation, and signal transduction) and genes for protein subunits needed in multiprotein complexes partly encoded by nuclear DNA of the alga, partly encoded by chloroplast DNA.

Nucleon: See → nuon.

Nucleoplasm (karyoplasm, karyolymph): The non-chromatin fluid phase of a → nucleus.

Nucleoplasmin: The most abundant nuclear protein in some animals (e.g. *Xenopus* oocytes). It interacts as a → chaperone with histones H2A and H2B during the assembly of → nucleosomes, reducing their positive charges. Nucleoplasmin is a pentameric protein with a molecular weight of about 165 kDa, consisting of a highly charged carboxy-terminal tail and a globular amino-terminal domain.

Nucleoporin (nup): Anyone of a series of about 30 proteins associated with the → nuclear pore complex. For example, the socalled nup 180 (molecular weight: 180 kDa) is located close to the annular pore complex at the cytoplasmic side of the pore, whereas the phenylalanine-glycine (FG)-rich nup 153 localizes to the nuclear side, with its C-terminal domain probably involved in the nucleo-cytoplasmic im- and export of RNA molecules.

Nucleoprotein: A complex of nucleic acid(s) and protein(s). For example, basic → histone proteins together with the associated phosphoric acid backbone of DNA form a nucleoprotein complex, the → nucleosome. Compare → ribonucleoprotein.

Nucleoprotein hybridization: A technique to isolate specific genes of an organism as → chromatin. In short, isolated nuclei are digested with appropriate → restriction endonucleases and lysed with → EDTA. Single-stranded termini of the nuclear chromatin fragments are generated by 5'-exonuclease digestion. Then a synthetic, biotinylated oligonucleotide complementary to the sequence adjacent to the restriction site on the targeted gene is hybridized in solution to the chromatin fragments. The oligonucleotide-chromatin hybrids are then immobilized on an → avidin matrix. They may be eluted by cleavage of the disulfide bond in the linker of the biotinylated probe (compare → biotinylation of nucleic acids). This type of → affinity chromatography allows the isolation of specific genes that retain their original chromatin structure.

Nucleosidase: Any enzyme that catalyzes the hydrolysis of → nucleosides to produce free bases and pentoses.

Nucleoside: A → pyrimidine or → purine base covalently linked to ribose (ribonucleoside) or deoxyribose (deoxyribonucleoside) via N-glycosidic bonds. See also → nucleoside antibiotic.

Nucleoside-α-thiotriphosphate (dNTPα-S): A purine or pyrimidine → nucleotide that contains a phosphorothioate diester bond and blocks the 3' → 5' proof-reading activity of → DNA polymerase I. Such nucleotides are used in → DNA sequencing and in vitro → mutagenesis procedures.

2'-Deoxynucleoside-5'-O-(α-thio)-triphosphate (NTP)

Nucleoside analogue (NA): Any synthetic or naturally occurring substitute for a → nucleoside, that is either incorporated into RNA or DNA and accepted without consequences, or blocks the subsequent synthesis of RNA or DNA. Such nucleoside analogues are used as therapeutic agents to block (or at least interfere) with DNA replication of viruses and tumor cells. The analogue triphosphates (NA-PPPs) are incorporated into the growing DNA chain and lead to an interruption of chain elongation. For example, the triphosphate of the thymidine analogue 3'-*acido*–3'-deoxy*t*hymidine (ACT) is used by the → reverse transcriptase of HIV (AIDS virus) and build into newly synthesized viral DNA. The acido residue at the C3 position of the ribose then blocks chain elongation and interrupts the life cycle of the virus.

Acido-thymidine (ACT) Ganciclovir (GCV)

Nucleoside antibiotic: Any one of a series of → purine or → pyrimidine nucleosides with → antibiotic activity. These compounds are formed in various bacteria and fungi by modification of → nucleosides, either by derivatization of the sugar (epimerization, isomerization, oxidation, reduction or decarboxylation of D-ribose) or the base moiety (methylation). They are antagonists of their naturally occurring nucleosides, and therefore block the metabolism of purines, pyrimidines, and proteins. Examples for such nucleoside antibiotics are amicetin A and B (*Streptomyces fasciculatus, S. plicatus*), 5-azacytidine (*Streptoverticillius lakadamus*), blasticidin S (*Streptomyces griseo-chromogenes*), cordycepin (*Cordyceps militaris*), nucleocidin (*Streptomyces calvus*), puromycin (*Streptomyces albo-niger*) and tubercidin (*Streptomyces tubercidicus*). Some of them are used in molecular biology (see e.g. → azacytidine, → cordycepin, → puromycin).

Nucleoside bisphosphate (nucleoside diphosphate): Any one of a series of ribose-modified nucleotide analogues, that contains phosphate residues at various positions of the ribose moiety, as e.g. the 3'and 5', or the 2'and 5'carbon atoms. Such analogues are used for the mapping of active sites in ribonucleases or other nucleotide-binding enzymes, the inhibition of nucleotide-dependent enzymes, and protein affinity studies.

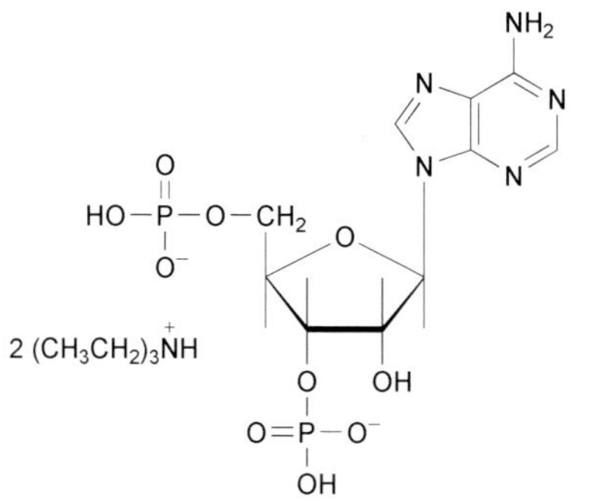

Adenosine-3', 5'-bisphosphate (pAp)

Nucleoside extrusion (base flipping): The opening of base pairs in a DNA double helix, whereby an entire nucleoside is swiveled out of the helix and inserted into the recognition pocket of a DNA-binding protein. Base flipping is induced by the torsional stress imposed onto the double helix by binding the protein.

Nucleoskeleton: An intranuclear network of fibrils (e.g. of actin and myosin) that is thought to coarsely compartment nuclear reactions. The nucleoskeleton contains for example, anchorage or attachment sites for → looped domains of → chromatin.

Nucleosome (nu particle, nu body): A disk-shaped structure of eukaryotic chromosomes consisting of a core of eight → histone molecules (two each of H2A, H2B, H3 and H4) complexed with 146 bp of DNA and spaced at roughly 100Å intervals by "linker" DNA of variable length (8-114 bp; see → nucleosome phasing) to which histone H1 attaches. Nucleosomes mainly serve to package DNA within the nuclei of eukaryotic cells. *In vitro* reconstitution of nucleosomes is possible. See → gradient dialysis. See also → lexosome.
Figure see page 744.

Nucleosome phasing (phasing): The non-random arrangement of → nucleosomes along nuclear DNA of eukaryotic chromosomes. Though the underlying mechanisms of nucleosome phasing are not fully known, it is generally accepted that phasing is a means of gene control. A regulatory sequence within the → promoter of a gene, for example, that is supercoiled around a nucleosome generally is not accessible for trans-acting or regulatory proteins. Once such a nucleosome has been partly dissociated (relaxed; → lexosome), the regulatory sequence becomes fully available for binding proteins.

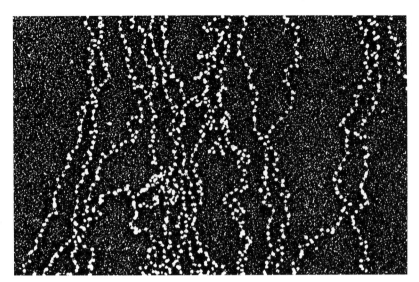

Nucleosomes
of chromatin
of a potato plant

Nucleosome remodeling complex (NuRD): A multiprotein complex capable of changing → chromatin architecture (e.g. the transition from relaxed [active] to tense [inactive] configuration), consisting of at least six different proteins, one of which (called MBD_3) binds to methylated CpG residues and recruits histone deacetylases and histone methylases. Deacetylases convert the acetylated histones such that they are no more inducive for an open chromatin structure. Result: the nearby genes are silenced.

Nucleotide (nt): A → pyrimidine or → purine → nucleoside that is esterified with one, two, or three phosphate groups at the 5' carbon atom of the ribose (ribonucleotide) or deoxyribose (deoxyribonucleotide). Ribose-containing nucleotides include ribo*nucleoside *mono*phosphate (NMP), ribo*nucleoside *di*phosphate (NDP) and ribo*nucleoside *tri*phosphate (NTP), deoxyribose-containing nucleotides include *d*eoxy-ribo*nucleoside *mono*- (dNMP), *d*i- (dNDP) and *t*riphosphates (dNTP). DNA contains deoxyadenylate, thymidylate, deoxyguanylate, and deoxycytidylate nucleotides; RNA adenylate, uridylate, guanylate, and cytidylate.

Nucleotide analogue interference mapping: A technique for the detection of amino acids in a protein that interact with a target RNA. In short, a nucleotide is exchanged in a gene by conventional gene technology, the modified gene re-introduced into the target organism and expressed there. The resulting protein is then tested for its ability to bind to the cognate RNA. By using a series of such mutants, each coding for a protein with a specific amino acid replacement, it is possible to map (localize) such amino acids that are involved in the interaction process (or, the nucleotides in the gene that encode these amino acids). This technique has e. g. been used to determine the amino acids in the → RNase P protein from *E. coli* which interact with → transfer RNA during the binding process.

Nucleotide-binding domain ("nucleotide-binding fold"): Any conserved → domain of a protein, containing specific amino acid sequence motifs, that form a pocket to bind and accommodate a nucleotide (usually as triphosphate, as e.g ATP or GTP). For example, the ATP-binding motif ("Walker motif") of certain proteins consists of a small stretch of hydrophobic amino acids followed by [gly/ala]-X-X-gly-X-gly-lys-thr/ser (where X is any amino acid). The hydrophobic residues form a buried ß-strand, the glycine-rich region a loop ("P-loop"), that interacts with the phosphate of the bound nucleotide. For the isolation of the encoding genes, → primers can be designed against these conserved domains and used to amplify parts of the gene from genomic DNA via conventional → polymerase chain reaction techniques. This strategy is employed for the isolation of → resistance gene *analogues* (RGAs) in plants, where the forward primer could be directed against one of the (three) *nucleotide binding sites* (NBSs), and the reverse primer against the trans-membrane domain or the leucine-rich repeat motifs of the extra-cellular part of such proteins.

Nucleotide diversity: The number of base differences between two (or more) → genomes, divided by the number of base pairs compared.

Nucleotide diversity map: See → diversity map.

Nucleotide diversity per site (Π): The frequency with which any two nucleotide sequences differ at a specific site.

Nucleotide excision repair: A prokaryotic DNA repair system, encoded by genes *uvr*A (encoding an ATPase subunit of endonucleases), *uvr*B and *uvr*C (encoding the endonuclease subunits of *E.coli* excinuclease), and *uvr*D (coding for a helicase removing the excised stretch of DNA), that repairs from few to more than several thousands of nucleotides. First the ABC excinuclease recognizes damaged sites, cuts at two flanking sites and removes the intervening sequences. Then → DNA polymerase I catalyses repair synthesis, gaps are filled by any of the four DNA polymerases, and the ends ligated. See → base excision repair, → mismatch repair.

Nucleotide heterozygosity (η): The average number of nucleotide differences between two nucleic acid sequences selected at random from a particular population of organisms. η depends on the number of polymorphic sites and their frequency in the nucleic acid region in focus.

Nucleotide mapping: A misleading term for the isolation and characterization of nucleotides resulting from an enzymatic digestion or the chemical hydrolysis of a target DNA or RNA.

Nucleotide pair: See → base pair.

Nucleotide replacement site: Any position in a → codon where a → point mutation has occurred.

Nucleotide sequence: See → DNA sequence.

Nucleotide substitution: The exchange of one → nucleotide in a DNA molecule for another one. Such substitutions are neutral, if the → genetic code is not changed, but have massive consequences, if the genetic code is altered (e. g. result in the synthesis of a non-functional protein).

Nucleotide turnover rate: The maximum number of nucleotides polymerized per molecule of → DNA dependent DNA polymerase per minute. Compare → processivity.

Nucleus: An organelle of eukaryotic cells, surrounded by a double-membrane system (→ nuclear envelope) with pores (→ nuclear pore), and containing the → chromosomes in the form of → chromatin (i.e. associated with a multitude of proteins). Compare → nucleoid.

NUE: See → *near-u*pstream *e*lement.

Null allele: Any → allele whose DNA sequence has been changed by one or more → mutations such that (1) it can no longer be detected by → allele-specific probes in → genomic DNA and (2) the encoded protein is no more functional (i. e. can no longer be detected by e. g. → immunoassays).

Nullisomy: The absence of a complete chromosome pair from the → karyotype of a cell (in a diploid organism: 2n–2). See → disomy.

Null mutation: Any mutation that leads to a complete loss of function of the sequence in which it occurs.

Null promoter: A → promoter that does not contain a → TATA box or the → initiator element (TATA⁻Inr⁻), and therefore allows multiple start sites for → transcription initiation. Some of the null promoters share an intragenic sequence motif (*m*ultiple start site *e*lement *d*ownstream, MED-1), that is indispensable for null promoter function and replaces the conventional → consensus motifs (as e.g. TATA box).

Num (*nu*clear *m*igration): Any one of a series of proteins of filamentous fungi, that is composed of three domains (an NH_2-terminal heptad region, a central region with → direct repeats, and a carboxyterminal PH [*p*leck strin *h*omology] region) and can be translocated from cytoplasm to the nucleus.

Nuon (*nu*cleic acid, *nucle*on): Any coding or non-coding DNA or RNA sequence. For example, → genes, → introns, → exons, → retrotransposons, → spacers, → enhancers, → silencers, → microsatellites all are nuons.

n/u orientations: The two orientations possible when a fragment of foreign DNA is inserted into a → cloning vector. N, when both vector and insert have the same orientation; u, when insert and vector are in different orientations.

NUP: See → *nu*clear *p*ore complex protein.

Nu particle: See → nucleosome.

NURD: See → nucleosome remodeling complex.

Nurse cells: See → feeder cells.

NusA tag (Nus tag): A short peptide sequence from the NusA protein of *E. coli*, that can be fused to a target protein and thereby increase the solubility of the fused protein in the bacterial host. The sequence encoding the NusA peptide is cloned into a suitable plasmid vector, fused to the target protein gene, and appropriately flanked by → histidine tag-encoding sequences, a protease cleavage site (for the removal of the tags) and a → T7 RNA polymerase promoter. Expression of the fused protein in the host cell can reach high levels without solubility problems. See → strep tag.

Nutrient _b_roth (NB): A medium rich in mineral salts, vitamins and carbohydrates and otherwise useful compounds that is used for the growth of microorganisms. Contrary to → minimal medium, which contains only basic chemical compounds.

Nutrigenome: The complete set of (still largely unknown) genes, that underlies the nutritional qualities of animals and plants (or parts of them) consumed by humans. See → nutrigenomics, → nutritional genomics.

Nutrigenomics: The whole repertoire of techniques designed to decipher the complex interactions between the genetic predisposition and the uptake, processing and utilization of nutrients, as well as their influence on the immune, digestive and metabolic systems of the consumer. Do not confuse with → nutritional genomics. See → behavioral genomics, → chemical genomics, → comparative genomics, → environmental genomics, → epigenomics, → functional genomics, → horizontal genomics, → integrative genomics, → medical genomics, → nutritional genomics, → omics, → pharmacogenomics, → phylogenomics, → proteomics, → recognomics, → structural genomics, → trans-criptomics, → transposomics.

Nutritional genomics: An infelicitous term for a series of techniques aiming at the improvement of the nutritional quality of plants through the transfer of foreign, novel, or altered genes encoding enzymes that produce nutritional compounds. These genes are used to increase the levels of essential or desirable micronutrients in crop plants. For example, _Arabidopsis thaliana_ converts γ-tocopherol to α-tocopherol (vitamin E) at a very low rate only. The transformation and over-expression of the gene encoding a γ-_t_ocopherol _m_ethyl_t_ransferase (γ-TMT) in this plant allows to increase its α-tocopherol content substantially. See → behavioral genomics, → biological genomics, → cardio-genomics, → chemical genomics, → clinical genomics, → comparative genomics, → deductive genomics, → environmental genomics, → epigenomics, → functional genomics, → horizontal genomics, → integrative genomics, → lipo-proteomics, → medical genomics, → neurogenomics, → neuro-proteomics, → nutritional genomics, → omics, → pathogenomics, → pharmacogenomics, → phylogenomics, → physical genomics, → population genomics, → proteomics, → recognomics, → structural genomics, → transcriptomics, → transposomics.

Nylon macroarray: See → macroarray.

Nystatin (mycostatin; fungicidin): A polyene → antibiotic from _Streptomyces nouresii_ that affects specifically fungal growth through the formation of complexes with membrane-bound cholesterols. These complexes generate "pores" in the membrane and lead to uncontrolled leakage of solutes. Since it is not active against bacteria, it is used to keep bacterial cultures free from fungi.

Nytran™: The trade-mark for a nylon membrane that is used to immobilize nucleic acids and proteins. It is positively charged and therefore electrostatically binds the negatively charged nucleic acids or SDS-protein complexes.

O

OC: See → *o*pen *c*ircle.

O^c: See → *o*perator *c*onstitutive mutation.

OC-DNA: See → *o*pen *c*ircle.

OcDNA: See → open circular DNA.

Ochre codon: The triplet UAA in mRNAs which is not recognized by any → tRNA, but signals the termination of → translation, (→ stop codon). See also → ochre mutation, → ochre suppressor.

Ochre mutant: A bacterial mutant that synthesizes mRNA carrying an → ochre mutation.

Ochre mutation: A base substitution which converts an amino acid specifying → codon into the → stop codon UAA (→ ochre codon). Usually such a mutation leads to premature termination of polypeptide synthesis and the formation of abnormally short polypeptides. Its effect can, however, be neutralized by an → ochre suppressor mutation. See also → nonsense mutation.

Ochre suppressor: A mutant gene coding for a mutant → tRNA, which recognizes the → stop codon UAA and causes the insertion of an amino acid into the growing polypeptide chain at the termination site.

Ocs element: Any one of a family of related, bipartite, *cis*-acting 20 bp sequence elements, usually located in between the → TATA box and nucleotide -200 in the promoters of various bacterial, viral and plant genes. Ocs elements, originally identified in the promoter of the → *Agrobacterium tumefaciens* → *o*ctopine *s*ynthase (ocs) gene, are also present in other promoters of *Agrobacterium tumefaciens* (e. g. → nopaline synthase gene promoter), promoters of viruses (e. g. 35S promoter of → cauliflower mosaic virus, here the element is called as-1; 19S and 35S promoters of the *f*igwort *m*osaic *v*irus, FMV; also in the badnavirus Commelina Yellow Mottle Virus) and plants (e. g. glutathione S-transferase gene promoter). Plant promoters containing ocs elements are activated by the plant hormone auxin and salicylic acid, which is part of a stress response. Ocs elements are target sites for the highly conserved *b*asic domain-leucine *zip*per (bZIP) transcription factors (socalled *o*cs element *b*inding *f*actors, OBFs). The ocs element contains functionally identical, tandemly arranged nuclear protein-binding sites, each site centered around the consensus core sequence 5'-ACGT-3' and harboring a binding site for plant transcription factors ("OTFs"). Occupation of both binding sites is required for ocs element function. The 16 bp palindromic consensus sequence of the ocs element is 5'-TGACGTAAGC-GCTTACGTCA-3' (dashes: variable nucleotides). The ocs element is used for tissue-specific expression of genes in → transgenic plants.

ocs gene: See → *oc*topine *s*ynthase gene.

Octamer-*b*ased *g*enome *s*canning (OBGS): A technique for the detection of sequence length differences between over-represented, strand-biased octamer nucleotide stretches in the *E.coli* genome. The technique exploits the presence of about 150 different over-represented oligomers, whose occurrence is skewed to one strand (the leading strand) of the genome. Of these, 23 octamers are represented from 515–867 times, and probably function as priming sites for discontinuous DNA replication. For OBGS, fluorescently labeled octamer-based primers (octamers from the leading strand) are mixed with unlabeled octamer-primers (octamers biased to the lagging strand) and used to amplify the octamer-octamer regions in a → conventional polymerase chain reaction. The size distribution of the fluorescent products is then measured on automated sequencers followed by establishments of binary files from the absence and presence of bands.

Octamer-binding *t*ranscription *f*actor (OTF): One of several nuclear proteins that bind to the consensus sequence 5'-ATTTGCAT-3' present in promoters of several protein-coding genes (e.g. histone H2B and immunoglobulin light chain genes) and enhancers (e.g. an enhancer of the → RNA polymerase II-dependent U1 and U2 RNA genes). The octamer sequence in → class II genes interacts with two distinct transcription factors (OTF-1, ubiquitous; OTF-2, B-cell specific).

oct gene: See → *oct*opine synthase gene.

Octopine (N-α-[D-1-carboxyethyl]-L-arginine): An amino acid derivative that is synthesized in plant cells transformed by the octopine strain of the soil bacterium → *Agrobacterium tumefaciens*. This bacterium, after contact with wound-exposed plant cell walls transfers part of a large plasmid (→ Ti plasmid) into the plant cell where it is integrated in the nuclear DNA. The expression of a gene (ocs gene) close to the right border of the transforming DNA (→ TL-DNA) leads to the production of the enzyme → octopine synthase that synthesizes octopine from pyruvate and L-arginine. Octopine cannot be used by the host plant cell, but is secreted and serves as a carbon, nitrogen and energy source for agrobacteria possessing *occ* (*oc*topine *c*atabolism) genes on their Ti-plasmid (see → genetic colonization). Octopine is an → opine.

$$\begin{array}{l} HN \\ \diagdown \\ C-NH-(CH_2)_3-CH-COOH \\ H_2N \diagup | \\ NH \\ | \\ H_3C-CH-COOH \end{array}$$

Octopine synthase (octopine synthetase; EC 1.5.1.11): An enzyme present in → crown gall tumor cells and encoded by the *ocs* gene (see → octopine synthase gene) of the → T-DNA originating from the → Ti plasmid of → *Agrobacterium tumefaciens*. Octopine synthase catalyzes the synthesis of the unusual amino acid → octopine from pyruvate and L-arginine.

Octopine synthase gene (ocs gene, oct gene): A gene of the → Ti-plasmid of → *Agrobacterium tumefaciens* that encodes the enzyme → octopine synthase and is only expressed in transformed plant cells (see → crown gall). The *oct* gene is frequently used as a → reporter gene in plant transformation experiments, its → promoter (*Pocs*) and → termination sequences (3' t oct) are incorporated in → plant transformation vectors.

Octopine synthetase: See → octopine synthase.

Oct-protein: Any one of a series of DNA-affine proteins, that bind specifically to *oct*amer sequences. See → homeodomain.

ODN: See → oligodeoxynucleotide.

OFAGE: See → *o*rthogonal-*f*ield-*a*lternation *g*el *e*lectrophoresis.

O'Farrell electrophoresis: See → two-dimensional gel electrophoresis.

O'Farrell gel: See → two-dimensional gel electrophoresis.

O'Farrell gel electrophoresis: See → two-dimensional gel electrophoresis.

Off-gel electrophoresis: A free-flow technique for the separation and isoelectric purification of proteins according to their charge, that works with a flow chamber of minute dimensions (4x0.6x40 mm), where one wall consists of an → *i*mmobilized *p*H *g*radient (IPG) gel such that it buffers a thin layer of solution (without any carrier ampholytes). The protein molecules migrate in this solution rather than in a gel matrix as in conventional electrophoresis methods. An electric field is applied perpendicular to the flow of the solution. Due to the buffering capacity of the IPG gel, proteins with an isoelectric point close to the pH of the gel in contact with the flow chamber stay in solution, because they are neutral. Other proteins are charged, when approaching the IPG gel and are migrating into the gel. The positively charged proteins migrate to the cathode and penetrate the gel, the negatively charged proteins penetrate the gel towards the anoide. The proteins are recovered in free flowing solution. Off-gel electrophoresis separates faster and achieves a higher resolution than gel-based separation techniques. The separated proteins are immediately ready for further analyses such as → two-dimensional gel electrophoresis, crystallization, → protein microarrays or → mass spectrometry.

Off-target effect ("side effect"): The influence of a chemical substance (e.g. an inhibitor) on a process or processes, which are no actual targets for it. For example, → *s*mall *i*nterfering RNA (siRNA) designed to block the expression of a specific gene, also interferes with the expression of other unrelated genes ("off target"). In this specific case, off-target effects can be reduced by decreasing the intracellular, active siRNA concentration (normal range: 5–200 nM).

OH⁸dG: See → 8-hydroxy-2'-deoxyguanosine.

Ohnolog: A region of eukaryotic genomes, in which functionally (and phylogenetically) related genes are clustered. These homologues arose by → gene amplification and are frequently duplicated in a genome. Ohnologs are named after Dr. Ohno, who first postulated, that much of eukaryotic genomes consists of duplicated sequences.

Ohno's rule: The conservation of the → gene content of the X chromosome in mammals. For example, the genes localized on the human X chromosome are also localized on the mouse X chromosome. However, the relative order of the genes has been changed by genomic rearrangements over the past 80-90 millions of years (evolutionary separation of both species).

Okayama-Berg cloning (Okayama-Berg method): An efficient method to construct a → library of full-length → cDNAs, using oligonucleotide-tailed vector fragments that allow cDNA synthesis and cloning in one coordinate experiment. In short, the poly(A)-containing mRNA is first annealed to an oligo(dT)-tailed plasmid primer and → reverse transcriptase used to synthesize a cDNA. The generated vector-mRNA-cDNA hybrid is oligo(dC)-tailed at the 3' OH-terminus of the full-length cDNA using → terminal transferase (non-full-length cDNAs are not efficiently dC-tailed and therefore eliminated). Then the plasmid primer is trimmed with the → restriction endonuclease *Hind* III to remove the unnecessary oligo(dC)-tailed segment and a *Hind* III-linker is annealed to the oligo(dC)-tailed plasmid, and ligated with *E. coli* → DNA ligase. The original mRNA is selectively removed with *E. coli* → RNase H and replaced by *E. coli* → DNA polymerase I. The remaining → nicks are sealed with *E. coli* DNA ligase before transformation of an *E. coli* host. See also → Okayama-Berg cloning vector, → Honjo vector. Compare → Heidecker Messing method.
Figure see page 753.

Okayama-Berg cloning vector (Okayama-Berg vector): Any cloning vector (usually derivatives of → pBR 322) that is specially designed for the → Okayama-Berg cloning of → cDNA. In short, a pBR 322 molecule is first cut with the → restriction endonuclease *Kpn* I and → oligo(dT) tails are attached to the termini using → terminal transferase (see → DNA tailing). A subsequent *Hpa* I digestion removes one oligo(dT) tail, leaving the other for the annealing of the mRNA → poly(A) tail. See for example → Honjo vector.

Okayama-Berg method: See → Okayama-Berg cloning.

Okayama-Berg vector: See → Okayama-Berg cloning vector.

Okazaki fragment: A DNA fragment of several thousands (bacteria) or a few hundred of nucleotides (eukaryotes) that is newly synthesized during DNA replication on the → lagging strand, starting at an RNA → primer (synthesized by an → RNA primase). The Okazaki fragments are covalently linked by ligases to give a continuous strand.

OLA: See → *o*ligonucleotide *l*igation *a*ssay.

Olfactory receptor gene (OR gene): Any one of more than 1000 gene copies in primate genomes, that encodes an olfactory receptor (binding odorant molecules within the nasal epithelium). Whereas in e.g. rodents (with a highly developed sense of smell) almost all *OR* genes are encoding functional receptor molecules, about 70% of all human *OR* genes are non-functional → pseudogenes. This is most probably the cause for a greatly reduced sense of smell relative to other mammals (as e.g. rodents).

Oligo: See → oligonucleotide.

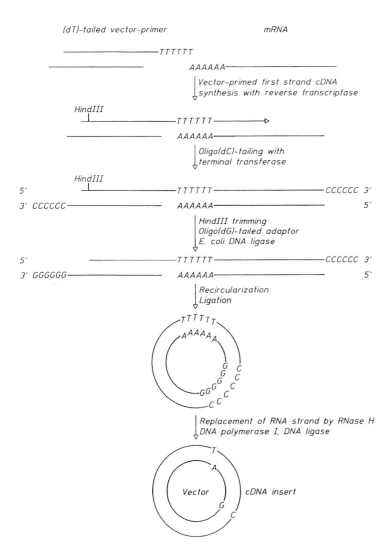

(dT)-tailed vector-primer   *mRNA*

Simplified scheme of Okayama-Berg cloning

Okayama-Berg Cloning

Oligo-capping: A technique for the *in vitro* capping of eukaryotic → *messenger RNA* (mRNA) to define the 5'- cap site accurately. In short, isolated mRNA is first treated with → alkaline phosphatase to remove the 3'-terminal phosphate, and then with → tobacco nucleotide *acid* pyrophosphatase (TAP) to remove the 5'cap of the message. Subsequently, a → T4 RNA ligase is used to ligate a specific 38-mer oligoribonucleotide to the 5'-end of the de-capped message ("re-capping"). The sequence of the 38-mer oligo cap is only rarely represented in mRNA databases. The oligo-capped mRNA is then converted to a stable cDNA by reverse transcriptase employing either a random hexamer or an oligo(dT) primer. The double-stranded cDNA is then purified and used to determine the exact sequence around the original cap site.

Oligodeoxynucleotide: See → oligonucleotide.

Oligo(dT) cellulose: A cellulose matrix, covalently coupled to thymidylic acid oligomers up to 30 nucleotides in length, which is used for the quantitative binding and isolation of poly(A)$^+$-mRNA in oligo(dT) cellulose → affinity chromatography. See also → oligo(U)-sepharose.

Oligo(dT) ladder: A set of single-stranded (dT) oligodeoxynucleotides ranging in size from 4 to 22 nucleotides with 1 bp intervals. This ladder is used as → marker for the precise determination of the size of electrophoretically separated oligodeoxynucleotides (e.g. → linkers, → primers).

Oligo(dT) primer: A synthetic homopolymeric → oligodeoxynucleotide that can be annealed to the → poly(A) tail of polyadenylated mRNA and used as a → primer to drive → first strand → cDNA synthesis by → reverse transcriptase.

Oligo(dT) priming: The use of a 12-20mer oligo(dT) deoxynucleotide for the synthesis of the → first strand in → cDNA cloning procedures. See → oligo(dT) primer.

Oligo(dT) tail: A single-stranded tail of deoxythymidine nucleotides added to the termini of linear DNA molecules by the enzyme → terminal transferase.

Oligolabeling: See → random priming.

Oligomer: A molecule made up of relatively few monomers. See for example → oligonucleotide, → oligopeptide.

Oligomerization: The covalent linkage of identical oligonucleotides to form long DNA molecules (see → concatemer).

Oligomer restriction **(OR):** A rapid → liquid hybridization method for the detection of a specific → restriction endonuclease → recognition site at any location in genomic DNA. To this end a ^{32}P end-labeled oligonucleotide probe is hybridized under stringent conditions to a specific segment of denatured genomic DNA that spans the target restriction site. Any mismatch within the restriction site prevents subsequent endonucleolytic cleavage of the duplex formed between the probe and its genomic target sequence. This allows the detection of allelic variants.

Oligomer skew: The unequal distribution of short → oligonucleotide stretches (e. g. → microsatellites) at specific regions (e. g. → origins of replication) on both strands of → double-stranded DNA in prokaryotes.

Oligomer walking: See → multiplex walking.

Oligo-mismatch mutagenesis (oligonucleotide-primed mutagenesis, oligonucleotide-directed mutagenesis, oligonucleotide-directed double-strand break repair): The introduction of site-specific → mutations into a target DNA molecule by annealing a specifically designed synthetic → oligodeoxynucleotide (7-20 nt long). The oligo is complementary to the region to be mutated except for one or more "wrong" bases which leads to specific mismatches. After hybridization of the oligonucleotide to the denatured target DNA (usually inserted in a → cloning vector), the → Klenow fragment of → DNA polymerase I is used to synthesize a complementary strand, where the double-stranded region serves as → primer. Finally DNA ligase seals the → nick. After introduction of this double-stranded molecule into an *E. coli* host, DNA → mismatch repair processes will lead to the occurrence of a mixed population of molecules that consists of 50% wild-type and 50% site-specifically mutagenized clones (because the repair system uses both the original as well as the mutated strand as template). Oligo-mismatch mutagenesis is a way of → site-specific mutagenesis.

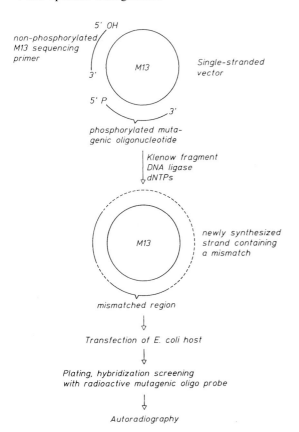

Oligonucleotide (oligo): A short nucleic acid molecule of up to 100 nucleotides in length (oligomer), consisting either of deoxynucleotides (oligodeoxynucleotide; general formula: dN_x; for example dAdAdAdA or $[dA]_4$, a tetramer), or ribonucleotides (oligoribonucleotide; AAAA or $[A]_4$). Oligos may also consist of a mixture of both deoxyribo- and ribonucleotides (in this case deoxyribonucleotides and ribonucleotides are discriminated by the prefix 'd' or 'r' to avoid confusion).

Oligonucleotide adaptor: See → adaptor.

Oligonucleotide array (oligonucleotide chip, oligonucleotide microarray): A two-dimensional arrangement of thousands, hundreds of thousands, or even millions of short → oligonucleotides, immobilized on a membrane, silicon, or glass support, and used to screen for complementary sequences by hybridization. For example, in → *sequencing by hybridization* (SBH), the immobilized oligonucleotides have overlapping sequences and are used to reconstruct the sequence of a target molecule by computer analysis of the resulting hybridization signals. See → cDNA array.

Oligonucleotide capture: See → DNA capture.

Oligonucleotide carrier: Any organic molecule, that is covalently attached to an oligonucleotide (e.g. → antisense oligonucleotide), and functions to introduce the oligonucleotide into a cell. For example, bile acids are such oligonucleotide carriers for e.g. the invasion of liver cells. Constructs of bile acids and an oligonucleotide directed towards the hepatitis C virus are used to inhibit virus replication *in vivo*.

Oligonucleotide cataloguing: An outdated technique for the comparison of sequences (originally → ribosomal RNA sequences) from different organisms. The rRNA molecule was first fragmented by cutting it at every → guanosine residue, then each of the resulting fragments ("oligonucleotides") was again subfragmented by enzymes cutting at different residues, so that a catalogue of fragments characteristic for a specific rRNA molecule was produced.

Oligonucleotide chip: See → oligonucleotide array.

Oligonucleotide-directed double-strand break repair: See → oligo-mismatch mutagenesis.

Oligonucleotide-directed mutagenesis: See → oligo-mismatch mutagenesis.

Oligonucleotide-directed RNase H cleavage: See → RNase H mapping.

Oligonucleotide fingerprinting: A method to screen any genome for the presence of repetitive, highly polymorphic sequences (e.g. short → consensus sequences with varying copy numbers at various loci, such as → variable number of tandem repeats) using synthetic oligonucleotides as probes (see → DNA fingerprint, → DNA fingerprinting).

Oligonucleotide ligation assay **(OLA):** A method to detect single-base substitutions in a target DNA sequence. In brief, the target DNA is denatured and hybridized to two → oligonucleotide probes in a way that the 3' end of one oligo is immediately adjacent to the 5' end of the other ("head-to-tail juxtaposition"). If → DNA ligase is added, it will covalently join the two oligos through the formation of a → phosphodiester bond, provided the nucleotides at the junction are correctly base-paired with the target DNA. This will not be the case if single base substitutions at the junction site have occurred. The ligation of the two oligos may be detected more conveniently if one oligo is labeled with → biotin, and the other one with ^{32}P.

Oligonucleotide microarray: See → oligonucleotide array.

Oligonucleotide mimic: See → DNA mimic.

Oligonucleotide-primed mutagenesis: See → oligo-mismatch mutagenesis.

Oligonucleotide priming: See → oligo(dT) priming.

Oligonucleotide purification-elution cartridge **(OPEC):** A cartridge packed with a resin (e.g. a polystyrene-divinylbenzene copolymer) that allows the purification of → oligonucleotides in a minimum volume of solvent (e.g. acetonitrile) and a minimum of time.

Oligonucleotide screening: A procedure to identify specific clones in a → cDNA library by using specific synthetic → oligonucleotides of 15-30 nucleotides in length. The cDNA library is plated out, the DNA transferred onto appropriate membranes (e.g. → nitrocellulose) and hybridized to either one specific radiolabeled oligonucleotide or a mixture of similar oligonucleotides with slightly differing base composition, taking into consideration the → degenerate code (see → mixed oligonucleotide probe). The hybridized probe can then easily be detected by autoradiography.

Oligonucleotide therapeutic: Any usually synthetic→ oligonucleotide, oligonucleotide analogue or oligoribonucleotide, that exerts a therapeutic influence on the symptoms of a disease. For example, → triplex-forming oligonucleotides, oligonucleotides interfering with → transcription of specific genes or the → translation of the resulting messenger RNAs, and oligoribonucleotides for → ribozymes or → aptamers (see also → protein epitope tagging) are such oligonucleotide therapeutics.

Oligopeptide: A short peptide of up to 40-50 amino acids in length (oligomer). Compare → polypeptide.

Oligoribonuclease: A special → exoribonuclease of *E.coli*, encoded by the chromosomal *orn* gene, that specifically degrades small oligoribonucleotides to nucleoside monophosphates. Such small oligoribonucleotides arise by → RNase II and → polynucleotide phosphorylase-catalyzed fragmentation of messenger RNA. Oligoribonuclease is essential for cell viability.

Oligoribonucleotide: See → oligonucleotide.

Oligo(T)–peptide *n*ucleic acid (oligo(T)-PNA): An artificial, negatively charged → peptide nucleic acid (PNA) with high affinity to polyadenylated → messenger RNA, that is used to isolate and purify poly(A)⁺-mRNA, especially in combination with → trans–4–hydroxy-L-proline PNA (HypNA). Oligo(T)-PNA, like conventional PNAs, lacks polarity (i.e. binds to target RNA in parallel and antiparallel orientation) and cannot be degraded enzymatically.

Oligo(U)-sepharose: A → sepharose dextran matrix to which oligouridylic acid (oligo[U]) chains of more than 10 nucleotides in length are covalently bound. Oligo(U)-sepharose is used in → affinity chromatography to isolate polyadenylated RNA (poly[A]⁺-RNA) from complex RNA mixtures. See also → oligo(dT)-cellulose.

Ome: A frequently used syllable of the genomics era. See → biome, → cellome, → cellular proteome, → chondriome, → complexome, → composite genome, → core proteome, → cybernome, → cytome, → degradome, → enzymome, → epigenome, → expressome, → foldome, → functome, → genome, → glycome, → immunome, → interactome, → kinome, → lipidome, → localisome, → metabolome, → metabolon, → metagenome, → methylome, → methylosome, → microbiome, → morphome, → neurogenome, → neuroproteome, → nucleome, → nutrigenome, → onco-proteome, → operome, → ORFeome, → osmome, → peptidome, → phenome, → phylome, → phosphoproteome, → physiome, → plastome, → protein interactome, → proteome, → pseudo-genome, → ribonome, → secretome, → signalome, → spliceome, → sub-genome, → sublimone, → sub-proteome, → toponome, → transcriptome, → translatome, → transplastome, → unknome.

Omega nuclease: See → meganuclease.

Omega sequence (ω sequence): The sequence H$_2$N-DGRGG-COOH at the → C-terminus of the → virD2-encoded protein of → *Agrobacterium tumefaciens* that seems to be involved in correct folding of the virD2 protein. This folding may be necessary for the targeting of the → T-strand to the nuclear DNA of the recipient plant cell. See → crown gall.

Omega transposase: See → meganuclease.

o-micron DNA: See → two micron circle.

Omics: A funny abbreviation (coined by J. N. Weinstein) for the various newly generated terms of the genomics era (e. g. → array-based proteomics, → behavioral genomics, → biological genomics, → biomics, → cardiogenomics, → cellomics, → cellular genomics, → chemical genomics, → chemical proteomics, → chemogenomics, → clinical proteomics, → comparative genomics, → computational genomics, → crop genomics, → cybernomics, → cytomics, → deductive genomics, → degradomics, → economics, → environmental genomics, → epigenomics, → expression genomics, → expression pharmacogenomics, → functional genomics, → functional proteomics, → genomics, → glycomics, → glycoproteomics, → horizontal genomics, → immunomics, → industrial proteomics, → *in silico* proteomics, → integrative genomics, → interaction proteomics, → interactomics, → kinomics, → lateral genomics, → lingandomics, → lipidomics, → lipoproteomics, → medical genomics, → metabolic phenomics, → metabolomics, → metabonomics, → metagenomics, → methylomics, → microgenomics, → neurogenomics, → neuropharmacogenomics, → neuroproteomics, → nucleomics, → nutrigenomics, → nutritional genomics, → oncopharmacogenomics, → one cell proteomics, → operomics, → pathogenomics, → peptidomics, → pharmacogenomics, → phenomics, → phosphoproteomics, → phylogenomics, → phyloproteomics, → physical genomics, → physiomics, → population genomics, → proteomics, → quantitative proteomics, → recognomics, → reconstructomics, → regulomics, → ribonomics, → riboproteomics, → RNA genomics, → RNomics, → structural genomics, → subcellular proteomics, → telomics, → three-dimensional proteomics, → 3D proteomics, → tissue proteomics, → topological proteomics, → toxicogenomics, → toxicoproteomics, → transcriptomics, → transposomics, → xenogenomics.

On-chip PCR: See → on-chip polymerase chain reaction.

On-chip *polymerase chain reaction* (on-chip PCR, "solid-phase PCR"): A variant of the conventional → polymerase chain reaction technique, which allows to perform all the cycling processes on a glass or silicon chip ("PCR chip"). In short, → primers are first spotted onto silanized glass chips, covalently coupled via their 5'-phosphates by e. g. the EDC-methylimidazol method, blocked with succinic anhydride in dimethylformamide, and then hybridised to a mixture of target DNAs. The target then hybridizes to its complementary sequence on the primer, and an added DNA polymerase extends the primer and synthesizes a chip-bound double-stranded DNA. Then a second PCR cycling denatures this dsDNA, and now a second fluorophore-conjugated primer anneals to the previously synthesized strand and the DNA polymerase completes the synthesis of a fluorescently labelled dsDNA, that can be detected with a laser beam. It is favourable for on-chip PCR to double the amount of primers, dNTPs and DNA polymerase, and to extend the PCR time by 100% (except for the synthesis step). On-chip PCR allows to amplify fragments of up to 7 kb in length.
Figure see page 759.

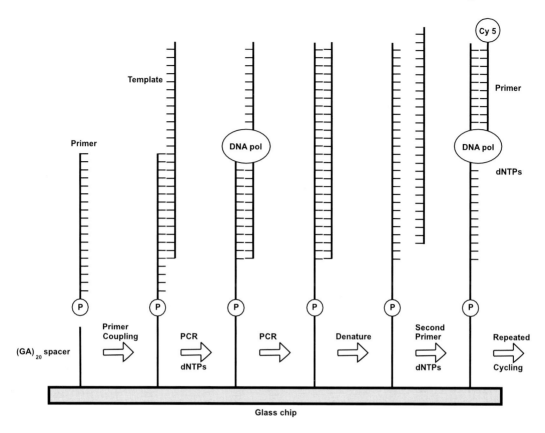

On-chip PCR

On-chip single-base primer extension: A variant of the → primer extension technique, in which the extension reaction occurs on a chip-bound target molecule. This configuration allows to use high → stringency for the hybridization reaction between primer and target.

Oncogene (transforming gene): Any one of a class of mutated and/or overexpressed variants of normal genes (→ cellular oncogenes) in animal and human cells that encodes a protein which transforms the normal cell into a tumor cell. Cellular oncogenes (c-*onc*) may become part of → retroviruses and are then designated as v-*onc* (→ viral oncogene). At present, about 40 different oncogenes isolated from acutely transforming viruses of animals and humans are known (see table). See also → oncogenic virus, → *ras* gene.

Some oncogenes and tumor viruses

Acronym	Virus	Species	Tumor origin
abl	Abelson mouse leukaemia (Ab-MLV)	mouse	Chronic myelogenous leukaemia
erbA	Avian erythroblastosis (AEV)	chicken	
erbB	Avian erythroblastosis (AEV)	chicken	
ets	E26 myeloblastosis	chicken	
fes (fps)	Snyder-Theilen feline sarcoma (SM-FeSV)	cat	Gardner-Arnstein sarcoma
fgr	Gardner-Rasheed sarcoma	cat	
fms	McDonough feline sarcoma (SM-FeSV)	cat	
fps (fes)	Fujinami sarcoma(FuSV)	chicken	
fos	FBJ osteosarcoma	mouse	
hst	Non-retroviral tumor	human	Stomach tumor
int1	Non-retroviral tumor	mouse	MMTV-induced carcinoma
int2	Non-retroviral tumor	mouse	MMTV-induced carcinoma
jun	ASV 17 sarcoma	chicken	
kit	Hardy-Zuckerman 4 sarcoma	cat	
B-lym	Non-retroviral tumor	chicken	Bursal lymphoma
mas	Non-retroviral tumor	human	Epidermoid carcinoma
met	Non-retroviral tumor	mouse	Osteosarcoma
mil (raf)	Mill Hill 2 acute leukaemia	chicken	
mos	Moloney mouse sarcoma (Mo-MSV)	mouse	
myb	Avian myeloblastosis (AMV)	chicken	Leukaemia
myc	MC29 myelocytomatosis	chicken	Lymphomas
N-myc	Non-retroviral tumor	human	Neuroblastomas
neu (ErB2)	Non-retroviral tumor	rat	Neuroblastoma
raf (mil)	3611 sarcoma	mouse	
Ha-ras	Harvey murine sarcoma	rat	Bladder, mammary and skin carcinomas
Ki-ras	Kirsten murine sarcoma (Ki-MSV)	rat	Lung, colon carcinomas
N-ras	Non-retroviral tumor	human	Neuroblastomas, leukaemias
rel	Reticuloendotheliosis (REV-T)	turkey	
ros	UR2	chicken	
sis	Simian sarcoma (SSV)	monkey	
src	Rous sarcoma (RSV)	chicken	
ski	SKV 770	chicken	
trk	Non-retroviral tumor	human	Colon carcinoma
yes	Y73, Esh sarcoma	chicken	

Oncogene amplification: The increase in copy number of one or more → oncogenes in genomes of late-stage cancers of many human organs, probably induced by the inactivation of p53. Oncogene amplification is clearly associated with tumor progression and has prognostic significance. See → gene amplification.

Oncogenesis: The gradual progression of a previously normal cell to a cell with changed genetic, cellular and cytological properties, the most prominent of which are lost contact inhibition and permanent proliferation, altogether leading to the formation of a tumor.

Oncogenic virus (tumor virus): A virus that transforms animal and human cells in culture and induces cancerous growth in animals and humans. Such viruses either contain DNA (e.g. Papovaviridae, Herpetoviridae) or RNA (Retroviridae, see → retrovirus) as genetic material.

Onco Mouse ™: A transgenic mouse carrying an activated *ras* → oncogene in all germline and somatic cells. Developed as a transgenic in vivo model to study oncogenesis, the mouse will predictably undergo carcinogenesis within some months. Compare → knock-out mouse.

Onconase P-30: A 12 kDa basic lectin-like → ribonuclease from oocytes (e. g. *Rana pipiens*), that binds to membrane receptors of a target cell, is channeled actively (i. e. ATP-driven) into the cytoplasm, inhibits ribosomal protein synthesis by the degradation of tRNA, 5S rRNA, 18S and 28S rRNA, and is therefore highly cytotoxic.

Oncopharmacogenomics: A branch of → pharmacogenomics, that uses the whole repertoire of → genomics, → transcriptomics, and → proteomics technologies to identify genes and/or mutations in genes involved in cancerogenesis and to design and develop new drugs to control cancerous proliferation of cells and the dissemination of such cells in an individual. See → neuropharmacogenomics. Compare → pharmacogenetics.

Onco-proteome: A part of the → proteome, that consists of proteins expressed primarily or exclusively in tumor cells. The presently and diagnostically interesting oncoproteins comprise e. g. CA 19–9, CD 44v5, CD 44v6, CEA, c-erb B2, c-myc, kathepsin D, kathepsin L, MDR–1, melanoma, MMP2, MMP9, p53 mutant, PSA, p21 ras, und 3, urokinase, and VEGF. Such oncoproteins are detectable at relatively early stages in tumorigenesis (present level of detection; few cells released from tumors of about 1 cm^3 volume), and serve not only as markers for an early diagnosis, but also for the targeting of the tumor cells and tissues.

One cell proteomics: The whole repertoire of techniques to analyze and characterize the → proteome of a single cell at a given time.

One-chip-for-all: See → universal array.

One gene-one enzyme hypothesis: A hypothesis largely based on the assumption that one single gene codes for a specific enzyme. Since many enzymes are the product of two or more genes, the more precise term would be one gene-one polypeptide chain.

One-hybrid system (yeast one-hybrid system): A technique for the *in vitro* isolation of novel genes encoding proteins that bind to a target DNA (bait) sequence (→ DNA-binding proteins), that is based on the fact, that many eukaryotic transcriptional activators are composed of a target-specific → DNA-*binding domain* (DBD) and a target-independent activation domain (AD). The one-hybrid system uses a cDNA candidate encoding a potential DNA-binding protein fused to a sequence encoding an AD. The complex of DBD and AD drives the expression of a → reporter gene (e. g. *His3* gene, β-galactosidase gene, or others). In short, a cassette containing tandem copies of a DNA target element for DNA-binding proteins is first inserted into a → multiple cloning site immediately → upstream of *His3* or *lac Z* reporter gene → promoters in a socalled yeast → integration vector. This linearized vector is then transformed into competent yeast cells at high frequency, and integrates at specific sites of the genome. Transformants are selected by their URA3 phenotype. These transformants are called reporter strains. Then an activator domain fusion library containing candidate cDNA clones is transformed into the yeast reporter strain. Whereas *His3* reporter gene expression without any induction is very low in the reporter strain, it is greatly enhanced after an AD-DNA-binding protein hybrid interacted with the target DNA element. *HIS3* expression allows yeast colony growth on minimal medium lacking histidine, so that transformants will be selected. His⁺ clones contain cDNAs for putative DNA-binding proteins. The cDNAs can easily be isolated, sequenced, the sequence compared to database entries and used in *in vitro* DNA-binding assays. See → dual-bait two-hybrid system, → interaction trap, → LexA two-hybrid system, → reverse two-hybrid system, → RNA-protein hybrid system, → split-hybrid system, → split ubiquitin membrane two-hybrid system, → three-hybrid system, → two-hybrid system.

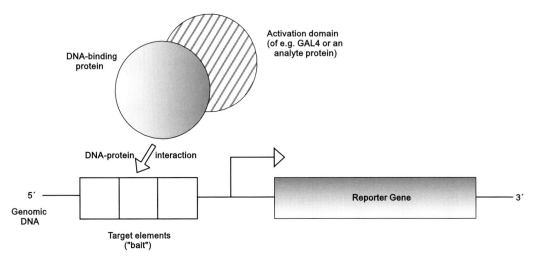

One-hybrid System

1-methylguanosine: See → rare base.

1-methylinosine: See → rare base

One-sided PCR: See → one-sided *polymerase chain reaction*.

One-sided *polymerase chain reaction* (one-sided PCR): A modification of the conventional → polymerase chain reaction for the direct targeting, amplification, and sequencing of uncharacter ized → cDNAs. In short, a specific cDNA sequence ("core region") is selected from a cDNA collection by using two imperfect oligomer → primers that are synthesized in vitro based on sequence information derived from homologous cDNAs (or also proteins) from related organisms. These specific primers can be complementary to any region within the message, can be located adjacent to the region to be amplified or may partially overlap it, and prime the amplification of the core sequence in the polymerase chain reaction. Then, based on this core sequence, specific primers are designed that permit the amplification of regions both upstream and down-stream of the core, if combined with a second nonspecific primer complementary to the 3' → po-ly(A) tail, or to an in vitro enzymatically added d(A)-tail at the 5' end. The pairwise combination of specific and nonspecific primers allows the amplification of the cDNA core with both 3' and 5' flanking regions. The amplified fragments are then inserted into → cloning vectors from which the → insert can be sequenced directly.

One-step gene disruption: The production of a stable, non-reverting gene → mutation in a target genome by (1) transformation of a construct containing the cloned gene interrupted by a → se-lectable marker gene, and (2) its → homologous recombination with the homologous sequence(s) in the target genome via highly recombigenic termini. This process leads to a replacement of the wild-type gene with the disrupted copy.

One-step protocol: An experimental design that combines two normally independent steps of a procedure such that both reactions occur simultaneously in the same reaction tube. For example, the → reverse transcription of eukaryotic → messenger RNA employing an oligo(dT) primer and → reverse transcriptase leads to the synthesis of a double-stranded → cDNA. This cDNA can be used as → template for a subsequent amplification using conventional → polymerase chain reac-tion and → *Thermus aquaticus* DNA polymerase. In a one-step protocol both reactions occur concomitantly in the same reaction tube.

***o*-nitrophenyl-β-D-galactoside (ONPG, o-nitrophenyl-galactoside):** An artificial substrate for β-galactosidase which is cleaved into galactose and the yellowish o-nitrophenol, the concentration of which can be easily measured.

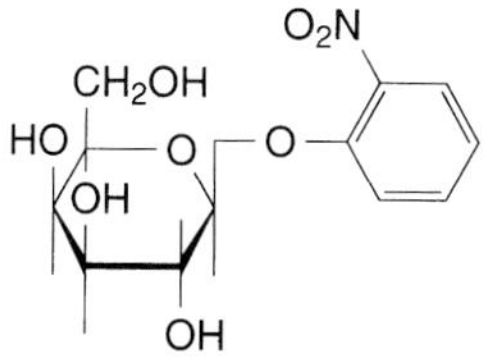

ONPG

***o*-nitrophenyl-galactoside:** See → *o*-nitrophenyl β-D-galactoside.

On-line DNA sequencing: See → automated DNA sequencing.

ONPG: See → *o*-nitrophenyl-β-D-galactoside.

Oocyte translation assay: The translation of foreign mRNA(s) in *Xenopus laevis* oocytes after → microinjection of nanogram amounts of this message.

***Opa*city gene (opa gene):** Any one of a family of constitutively transcribed bacterial genes (e.g. in the veneric disease-causing *Neisseria gonorrhoeae*) that harbors → microsatellite motifs (e.g. [5'-CTCTT-3']$_n$) in the → leader peptide encoding sequence. The encoded proteins are effective antigenic determinants, recognized by defense mechanisms of potential hosts (e.g. the phagocytes of the immune system). Bacteria with such proteins are *opa*que in appearance. However, anti-genic diversity through → frame-shifts as a consequence of → slipped strand mispairing is ex-ploited by pathogenic bacteria to escape immune detection by the host. If the number of the microsatellite motifs in the leader sequence is changed by slipped strand mispairing, then different lengths of the leader peptide result, and → out-of-frame mutations lead to a translational switch in favor of an altered protein that is no longer an antigen for the host organism. For example, if a CTCTT unit is lost during strand mispairing, the resulting shorter protein can no longer bind to the host cells (i.e. the bacterium can no longer penetrate the cell). With a new round of replication, this error can be reversed, and the bacterium regains infectivity ("phase shifting", "phase variation").

Opa gene: See → opacity gene.

Opal codon: The → stop codon UGA. See also → opal mutation, → opal suppressor.

Opal mutation: Any → mutation that converts a → codon into the → stop codon UGA (→ opal codon). See also → nonsense mutation.

Opal suppressor: A gene that encodes a mutated transfer RNA (tRNA), whose → anticodon recognizes the → termination codon UGA (→ opal codon) and allows the continuation of polypeptide synthesis.

Opaque-2: A mutant of *Zea mays* which contains kernels with lysine-rich proteins.

OPEC: See → *o*ligonucleotide *p*urification-*e*lution *c*artridge.

*O*pen *c*ircle (OC, oc-DNA, open circular DNA, relaxed circle, form II-DNA, nicked circular DNA): A non-supercoiled conformation adopted by a circular double-stranded DNA molecule, when one or both polynucleotide strands carry → nicks, so that it cannot form supercoils, but adopts a relaxed conformation.

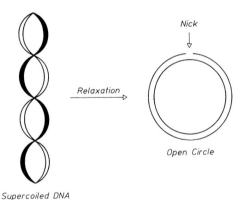

*O*pen *c*ircular DNA: See → open circle.

Open *d*ifferential *g*ene *e*xpression technology (open DGE technology): Any technique for the genome-wide profiling of differentially expressed genes (see → differential gene expression) in two (or more) cells, tissues, organs, or organisms, that does not require any *a priori* knowledge of the → transcriptome, so that the field of discovery is "open". However, most open systems exploit existing expressed genome databases to identify known and novel expressed (or suppressed) genes more efficiently. Such open DGE systems are e.g. → cDNA-AFLP, → differential display reverse transcription PCR, → massively parallel signature sequencing, → serial analysis of gene expression, → total gene expression analysis, and others. Compare → closed differential gene expression technology.

Open promoter complex: A → promoter configuration in which the DNA double helix is locally unwound to facilitate the binding of various → transcription factors and → RNA polymerase to form a → pre-initiation complex.

Open reading frame (ORF):
a) A nucleotide sequence in DNA in between two in-frame → stop codons.
b) A nucleotide sequence in DNA, that potentially can be translated into a protein or be transcribed into an RNA, and begins with an ATG → start codon and terminates with one of the three → stop codons. A good ORF candidate for coding a *bona fide* cellular protein has a set size requirement: it should have the potential to encode a protein of 100 amino acids or more. An ORF is not necessarily equivalent to a → gene or → locus, unless a → phenotype can be associated with a mutation in the ORF, and/or a → messenger RNA or, generally, a gene product generated from the ORF's DNA is detected. The ORFs of e.g *Saccharomyces cerevisiae* are designated by a symbol consisting of three uppercase letters followed by a number and then another letter. For example, Y (for "Yeast"), A to P for the chromosome harbouring the ORF (where "A" is chromosome I, and "P" is chromosome XVI), L or R (for left or right arm), a three-digit number corresponding to the order of the open reading frame on the chromosome arm (starting from the → centromere and counting out to the → telomere), and W or C for the location of the open reading frame on either the → Watson or → Crick strand. Compare → closed reading frame.

Open reading frame (ORF) vector (open reading frame expression vector, fusion vector): A → plasmid → expression vector for the construction of → fusion proteins that carries a bacterial → promoter, a → ribosome binding site and an → initiation codon (ATG) in front of a truncated *lac Z* gene (*'lac Z*). The β-galactosidase encoded by this truncated gene is not active, since it cannot form a tetramer (e.g. because of the deletion of the codons specifying 9 carboxy-terminal amino acids). The first 25 wild-type amino acids at the carboxy terminus may, however, be substituted by up to a few hundred of unrelated amino acids without interfering with tetramerization. Insertion of foreign DNA into cloning sites within this noncritical coding region of *lac Z*, if it is in-frame with a translational start signal, will lead to the formation of a fusion protein with β-galactosidase activity, that can be detected even if it is fused to longer proteins (→ insertional activation).

Open reading frame cloning: The use of → open reading frame vectors for expression cloning.

Open reading frame expressed sequence tag (OREST): Any → expressed sequence tag (EST), that is derived from the central part of a → cDNA rather than its 3' or 5' end.

Open reading frame expressed sequence tags (ORESTES, ORF ESTs): A collection of → expressed sequence tags (ESTs) from → cDNAs of various cells, tissues or organs of an organism, that contains sequences from the central portion of each transcript rather than from its ends (as in most other related techniques). In short, total RNA is first isolated, treated with → DNAse I to remove → genomic DNA, then poly(A)⁺ is extracted, reverse transcribed with an

18–25 nucleotide → primer of random sequence (GC content: 50%), and the single-stranded cDNA again amplified with the same primer. The complexity of the preparation is then checked by → polyacrylamide gel electrophoresis, and the amplification pool with multiple bands cloned into a plasmid vector (e.g. → pUC18). Finally the inserts are sequenced. The ORESTES technique generates a better coverage of the → transcriptome of the cell, and facilitates the construction of → contigs of transcript sequences.

Open reading frame expression vector: See → open reading frame vector.

Operational code: A set of rules by which the aminoacyl-tRNA synthetases recognize their cognate → transfer RNA molecule.

Operational gene: Any gene encoding a protein involved in the catalysis of a step in a normal metabolic pathway. Synonym for → house-keeping gene. Operational genes are horizontally transferred between bacterial (and maybe eukaryotic) genomes, but it is not clear whether this → horizontal gene transfer occurred in few massive ancient transfers before diversification of modern prokaryotes ("early massive horizontal gene transfer hypothesis") or is a continuous process ("continual horizontal gene transfer hypothesis").

Operator: A palindromic nucleotide sequence (→ palindrome) with dyad symmetry, localized at the proximal end of an → operon, that allows the formation of a → cruciform structure. An operator constitutes the recognition site for a specific → repressor protein and controls the expression of the adjacent → cistrons. See also → operator constitutive mutation, → operator zero mutation.

Operator constitutive mutation (O^c): Any mutation of the → operator leading to increased or constitutive expression of the → cistrons in the adjacent → operon. Compare → operator zero mutation.

Operator zero mutation (O^0): Any mutation of the → operator leading to the loss of function of the → operator. The expression of the → cistrons of the adjacent → operon is rendered impossible. Compare → operator constitutive mutation.

Operome: Another term of the excrescent → omics era, describing the part of the → proteome, which contains proteins with as yet unknown functions. Do not confuse with → operon. See also → biome, → cybernome, → genome, → immunome, → interactome, → metagenome, → microbiome, → morphome, → transcriptome.

Operomics: The whole repertoire of technologies for the study of the complete molecular architecture, composition and functions of a cell. Another excrescent term of the → omics era.

Operon: A unit of adjacent prokaryotic cistrons → the expression of which is under the control of a common → operator and leads to the synthesis of a single → polycistronic messenger RNA. See also → inducible operon, for example → *lac* operon; → operon fusion, → operon network.

Operon fusion: The head-to-tail ligation of two → operons by recombinant DNA techniques in a way that the coding sequences of the second operon come under the control of the regulatory sequences of the first operon.

Operon network: A series of → operons with their associated → operators interacting in such a way that the proteins encoded by one operon either activate or suppress another operon.

Operon^{TM} primer: The trade mark for a series of 10mer → primers of arbitrary sequence for the randon amplification of genomic DNA and the generation of → random amplified polymorphic DNA (RAPD) markers.

Opine: One of a series of unusual amino acid or sugar derivatives specifically synthesized by → crown gall tumor cells incited by the soil bacterium → *Agrobacterium tumefaciens*, but not by normal plant tissues. The opine genes reside on the → Ti plasmid close to the right border of → T-DNA. They are not or only weakly transcribed in *Agrobacterium*, but are constitutively expressed once integrated into the plant nuclear genome. Expression results in the appearance of → opine synthases which catalyze the formation of opines. The latter cannot be metabolized by the tumor cells and are therefore secreted. *Agrobacterium* can take up and degrade them because it possesses Ti-plasmid genes for opine catabolism, and thus may use opines as a source of carbon, nitrogen and energy.
Moreover, opines activate the → *tra*-genes of the Ti-plasmid, and thus serve to spread it in a bacterial population. Opines serve as tumor markers in plants. See → agrocinopine, → agropine, → histopine, → leucinopine, → lysopins, → mannopine, → nopaline, → nopalinic acid, → octopine.

Opine synthase (opine synthetase): An enzyme catalyzing the synthesis of → opines (e.g. → octopine, → nopaline).

OPT: See → optical projection tomography.

Optical fingerprinting: A misleading term for the detection of interactions between thousands of → probe molecules (e.g. DNAs, oligonucleotides, RNAs, peptides, or proteins) immobilized on the surface of a chip (e.g. glass, silicon) and target compounds (e.g. DNAs, oligonucleotides, RNAs, peptides, proteins, also low molecular weight compounds such as metabolites) by RAMAN spectroscopy (analyzing the unique structure of the cross-reacting molecules). Compare → DNA fingerprinting.

Optical mapping (visual mapping): The visualization of → genes, generally DNA sequences, along a chromosome, or a chromosome fiber, or along a → BAC or → YAC clone, that are extended (see → DNA combing), by → *in situ* hybridisation of → fluorochrome-labeled → probes (representing e.g. genes), and detection of fluorescence emission. The threshold of direct visual mapping is about 3.0 kb, so that single genes can be detected. Optical mapping is also used for creating e.g. → restriction maps from a series of single DNA molecules. In short, large DNA molecules are first dropped onto specially prepared glass surfaces, linearized in parallel through a fluid flow across the surface, and then affixed onto the glass. Subsequently → restriction endonucleases are added to produce ordered patterns of restriction fragments, which are stained with → fluorochromes and visualized with a fluorescence microscope. The various microscopic images are captured one at a time, processed, and the images of the various restriction fragments aligned to match the restriction sites. Then multiple maps are merged into large → contigs, using map assembly programs. For example, a complete optical restriction map is available for the bacterium *Deinococcus radiodurans*.

Optical noise: An undesirable contribution of reflected light from a → microarray support (e.g. a glass or quartz slide), reflections from any object in the laboratory room, leaking light or even cosmic rays to the readings of the fluorescence detection instrument. See → background subtraction, → dark current, → electronic noise, → microarray noise, → sample noise, → substrate noise.

Optical projection tomography (OPT): A microscopic technique for the production of high-resolution three-dimensional images of fluorescent or also non-fluorescent biological samples of up to 15 mm thickness. The specimen (e.g. a complete mouse embryo) is first stained with a diagnostic fluorescent antibody (e.g. an HNF3β antibody for developing endoderm and the floorplate of the spinal cord, or a neurofilament antibody for developing neurons), then positioned in a cylinder of agarose, and rotated continuously for 360 degrees. Any light emitted by the embryo is focused by lenses onto camera-imaging chips (CICs), and recorded such that a three-dimensional image is generated. OPT allows to map specific messenger RNAs or proteins in intact organs or embryos and can reconstruct gene expression patterns during developmental processes.

Optical trap (optical tweezers): An experimental arrangement, in which the radiation pressure (forces in the picoNewton range) of focussed single-beam infrared lasers trap or move molecules: Optical traps can measure molecular displacements by only few nanometers (e.g. kinesins moving along microtubules and actin-myosin dynamics).

Optimal codon:
a) Any → codon that is utilized very often in a given organism. In → transgenic organisms, → codon optimization is necessary to achieve → overexpression of the → transgene. See → codon bias, → rare codon.
b) Any → codon that is translated more efficiently than its → synonymous codon.

Optimized stringent random amplified polymorphic DNA technique (OS-RAPD): A variant of the conventional → random amplified polymorphic DNA method, that works with optimized amplification reaction mixtures (optimized with regard to concentration of → buffer, Mg^{2+}, dNTPs, → primers, → template DNA, and → Taq DNA polymerase) and DNA amplification at elevated → annealing temperatures, thus increasing → stringency, and avoiding spurious amplification artifacts. OS-RAPD therefore produces reproducible and reliable genomic fingerprint patterns.

OR: See → oligomer restriction.

ORC: See → origin recognition complex.

Ordered array: Any → microarray, onto which regular rows and columns of spots (consisting of oligonucleotides, cDNAs or DNAs) are immobilized.

Ordered clone bank: See → ordered clone library.

Ordered clone library (relational clone library, ordered clone bank): Any → genomic library that contains clones with terminal overlaps which can be arranged so that they represent the complete DNA from which they are derived. See → ordered clone map.

Ordered clone map: A graphical description of the linear arrangement of overlapping DNA fragments, cloned into an appropriate → cloning vector (e. g. → bacterial artificial chromosome, → cosmid, → mammalian artificial chromosome, → yeast artificial chromosome, or even → plasmid). The order of the clones in such a map reflects their original positions on the DNA (or chromosome). See → macro-restriction map.

Ordered fragment ladder far-Western blotting: A technique for the detection of protein-protein interaction(s) and the identification of specific domains involved in such interaction. The method uses a labeled protein → probe, that reacts with fragments of a target protein containing the interacting domain. The interaction is then detected by → autoradiography or → phosphorimaging. In short, the isolated and purified target protein, or a whole cell lysate containing this protein is first cleaved chemically (e. g. with 2-nitro-5-thiocyanobenzoic acid, or hydroxylamine) or enzymatically (e. g. with thermolysin or trypsin), the cleavage fragments separated by → SDS polyacrylamide gel electrophoresis, the separated fragments electrophoretically transferred onto a → nitrocellulose membrane and reacted with a ^{32}P-labeled test protein (e. g. labeled with a kinase). The dried blot is then exposed to X-ray film, or analyzed in a phosphorimager. If a binding of a test protein to one (or more) of the target peptides occurs, the interaction can be visualized by → autoradiography, and the interacting domains of the target protein be identified and mapped ("chemical cleavage mapping").

Oregon Green: The → fluorochrome Oregon Green 488-X, that is used as a marker for → fluorescent primers in e. g. automated sequencing procedures, or for labeling in → DNA chip technology. The molecule can be excited by light of 492 nm wave-length, and emits fluorescence light at 517 nm. Since the wave-length of the excitation and emission maxima is pH-dependent, the exact values vary.

OREST: See → open reading frame expressed sequence tag.

ORESTES: See → open reading frame expressed sequence tags.

ORF: See → *open reading frame.*

ORFan: Any hypothetical gene in an organism, that has no homologues in other organisms. See → orphan gene, → orphon.

ORF clone: Any → clone (e.g. a → plasmid clone) with an inserted → open reading frame (ORF) sequence.

ORFeome:
a) The complete set of → open reading frames (ORFs) in a particular → genome.
b) A laboratory slang term for a set of full-length → cDNA clones transcribed from a particular genome at a particular developmental stage of the carrier, derived from → cDNA chip analysis. Such ORFeomes circumvent → cDNA libary construction, and ideally contain each transcribed gene sequence in equimolar concentrations.

ORF ESTs: See → open reading frame expressed sequence tags.

ORFmer: A set of two → primers that allow the amplification of an → *open reading frame* (ORF) from → genomic DNA using conventional → polymerase chain reaction techniques. One ORFmer (A-primer) contains a 13 bp non-variable sequence (→ adaptamer) including the → start codon 5'-ATG-3' (→ amino-terminus of the encoded protein) and a *Sap* I → restriction site at its 5' terminus, followed by a 20-25 bp gene-specific sequence (ORF sequence I). The adaptamer sequence is 5'-TTGCTCTTCCATG-3'. The other ORFmer (C-primer) also carries a 13 bp adaptamer, containing a *Sap* I site and the → stop codon 5'-TAA-3' (sequence: 5'-TTGCTCTTCGTAA-3') at its 5'-terminus, adjacent to a 20-25 bp gene-specific sequence (ORF sequence II). The stop codon signals the → carboxy terminus of the encoded protein. The length of the gene-specific sequences in each primer is identical to achieve optimal → melting temperature of template-primer duplexes. PCR products generated with both primers contain *Sap* I sites at both termini. These can be cleaved by *Sap* I, producing a product with an ATG start and TAA stop codon as → 5' overhangs, which can be positionally cloned into → cloning vectors containing corresponding *Sap* I sites. Using many different ORFmers it is possible to perform → expression profiling, → expression vector cloning for the characterization of the ORF, and gene → mutagenesis.
Figure see page 771.

ORF vector: See → *open reading frame* vector.

Organellar gene: Any gene that resides on the genome(s) of an → organelle (as e.g. → chloroplast, → mitochondrium). Distinct from → nuclear gene.

Organellar genome: The → genome of a → mitochondrion (in all eukaryotic organisms) and → plastids (in plants), as different from the → nuclear genome. See → chloroplast DNA, → mitochondrial DNA, → nuclear DNA.

Organellar proteome: The complete set of proteins expressed in an organelle (e.g. → chloroplast or its variants, and → mitochondria) at a given time. See → cellular proteome.

Organelle: A membrane-bounded compartment within the cytoplasm of a eukaryotic cell, that contains a specific set of proteins catalyzing reactions in one (or more) specific pathway(s). For example, a mitochondrium is such an organelle, that is specialized on the ß-oxidation of fatty acids, citric acid cycle reactions, electron transport and generation of ATP. Other organelles are the nuclei, plastids, vacuoles, lysosomes, Golgi apparatus.

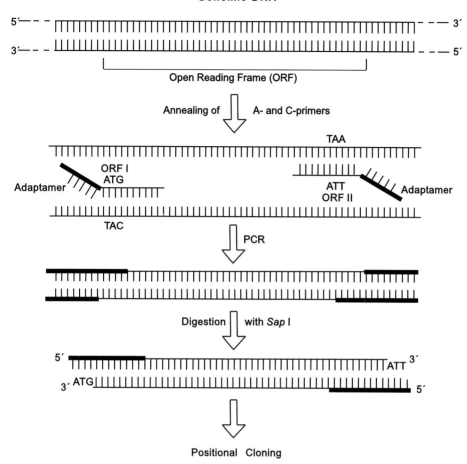

Genomic DNA

Open Reading Frame (ORF)

Annealing of A- and C-primers

Adaptamer

ORF I
ATG
TAC

TAA

ATT
ORF II
Adaptamer

PCR

Digestion with *Sap* I

Positional Cloning

ORFmer

***Organ-specific element* (OSE):** A *cis*-acting DNA sequence motif of 20-100 bp in → promoters of eukaryotic genes that is responsible for their organ-specific expression. If deleted, transcription from the resulting mutant promoter is no longer organ-specifically regulated. The OSEs are target sites for the binding of specific → transcription factors.

OR **gene:** See → olfactory receptor gene.

Ori: *Ori*gin, → see origin of replication, also → oriA, → oriT.

OriA (*ori*gin of *a*ssembly): A specific sequence of *t*obacco *m*osaic *v*irus (TMV) RNA, located within the coding region for P30 (a 30 kDa protein catalyzing the movement of viral RNA from host cell to host cell). The oriA sequence has the potential to form three hairpin loop structures (→ fold-back DNA)and functions in the assembly of coat proteins and viral RNA to new virus particles.

OriC (*origin of chromosome replication*): The sequence of a replicon at which chromosome replication is initiated. For example, the *E. coli* oriC region spans 0.245 kb and contains → consensus sequences for replication initiation proteins. Compare → oriV.

Orientation-specific cDNA cloning: See → forced cloning.

Origin: See → origin of replication.

Origin of assembly: See → oriA.

Origin of chromosome replication: See → oriC.

Origin of replication (origin, ori; replication origin): A specific sequence of a → replicon at which DNA → replication is initiated. See also → oriC, → oriV.

Origin of transfer (oriT): The sequence of a → replicon at which an → endonuclease (in → F factors the products of plasmid genes *tra Y* and *traZ*) introduces a → nick into the → H strand of the replicon, thus generating the substrate for transfer from a donor to an acceptor cell (by e.g. → conjugation).

Origin of vegetative replication: See → oriV.

Origin recognition complex (ORC): A six-subunit → protein complex, that binds to the socalled → *ori*gin of replication (*ori*) and coordinates the assembly of a pre-replication complex (pre-RC) at each *ori* sequence. This pre-RC contains at least six different, but related *m*ini-*c*hromosome *m*aintenance (MCM) proteins.

OriT: See → *ori*gin of *t*ransfer.

OriV (*origin of vegetative replication*): The sequence of a → replicon at which its replication during vegetative growth of the host (vegetative replication) is initiated. Compare → origin of transfer.

Orphan drug: Any drug that has been developed to treat diseases occuring in less than 0.1% of the total population.

Orphan gene (orphan): Any one of a series of → open reading frames discovered in genome sequencing projects, whose function is unknown and whose sequence does not reveal any homology with entries in the sequence databanks. Do not confuse with → orphon. See → fast evolving gene, → pioneer sequence.

Orphan protein: Any protein, for which no substrate or interaction partner is yet found.

Orphan receptor: Any receptor protein, that is known from its encoding genomic sequence, but for which no ligand is yet identified.

Orphon: An isolated → pseudogene that is related to and probably originates from tandemly repeated → multigene families or → gene batteries (for example → histone genes). Orphon genes are not necessarily located close to the gene(s) from which they originate. Compare → orphon gene.

Orthogonal-field-alternation gel electrophoresis (OFAGE): A method to separate DNA molecules in the size range from 50 kb to over 750 kb in → agarose gels by subjecting the molecules alternately to two approximately orthogonal electric fields. Compare → crossed field gel electrophoresis.

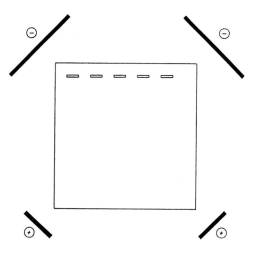

Electrode configuration of OFAGE

Ortholog (orthologous gene): One of two or more genes (generally: DNA sequences) with similar sequence and identical function(s) in two different genomes, that are direct descendants of a sequence in a common ancestor (i. e. withouth having undergone a gene duplication event). Also called "homology by descent". See → paralogs.

Orthologous domain shuffling: The substitution of → exons, encoding specific → domains of a protein in one species, with equivalent exons from the same gene of a different species. See → orthologous exon shuffling, → paralogous domain shuffling, → paralogous exon shuffling.

Orthologous exon shuffling: A variant of → *in vitro* exon shuffling, that allows to replace → exons from a particular gene of one species with the equivalent exons from the same gene of a different species. See → de novo protein assembly, → functional homolog shuffling, → orthologous domain shuffling, → paralogous domain shuffling, → paralogous exon shuffling.

ortho-nitrophenyl-β-D-galactoside: See → o-nitrophenyl-β-D-galactoside.

Osmome (*osmotic response genome*): The entirety of all genes responding to variations in osmotic pressure within a cell, that is sensed as osmotic stress.

Osmotic response genome: See → osmome.

OS-RAPD: See → optimized stringent random amplified polymorphic DNA.

OTF: See → octamer-binding transcription factor.

Ouchterlony technique: See → agarose gel diffusion.

Ouchterlony test: See → agarose gel diffusion.

Outlier: A laboratory slang term for any data point or also cDNA (or, generally, transcript), that differs significantly from the majority of data in repeated sets of → microarray experiments.

Overdigestion: An infelicitous term for the fragmentation of a given DNA substrate by exposing it to an excess of a given → restriction endonuclease and conducting the restriction overnight. Such overdigestion assays allow to test the purity of the enzyme (i. e. to exclude exo- or endo-nuclease contaminations).

Overdrive:
a) The sequence motif 5'-TAARTYNCTGTRTNTGTTTGTTTG–3'adjacent to the right border of the → T-region in → Ti-plasmids, that enhances the efficiency of → T-strand transfer into wounded plant cells. DNA sequences at similar sites with similar → core sequences (5'-TGTTTGTT–3') are present in the → nopaline plasmid pTi T37 and the Ri plasmid pRiA4. The overdrive sequence is also called T-DNA transmission enhancer.
b) A special state of bacterial → DNA-dependent RNA polymerase, in which the enzyme is resistant to → pause, → arrest and → termination signals. Overdrive is e.g. caused by the binding of → bacteriophage λ protein Q, that stabilizes RNA polymerase in the overdrive conformation, thereby optimising RNA-DNA template contacts and inducing RNA polymerase to ignore the signals. Also, → antiterminator proteins prevent → hairpin formation in RNA, that otherwise leads to RNA pausing, or stabilize the elongation complex against disruption by such RNA hairpins. See → antitermination.

Overexpression: The transcription of a gene at an extremely high rate so that its mRNA is more abundant than under normal conditions. Such overexpression usually occurs in host cells that have been transformed with a → cloning vector containing a gene driven by a very strong promoter, allowing the accumulation of its protein product (in some cases this will form up to 40% of the total cellular protein of the host cell). Overexpression may also be due to the presence of a → runaway plasmid in a bacterial cell. In eukaryotes it can be responsible for the transforming activity of → oncogenes.

Overgo hybridisation: A technique to isolate specific sequences from a → bacterial artificial chromosome (BAC) library of the genome of organism A (e.g. human), using genomic sequence information from organism B (e.g. mouse). In short, sequence alignments over the region of interest from both genomes allow to recognize homologous stretches. These in turn are used to design socalled overgo oligonucleotide primer pairs, that are each 24 bases long and overlap by 8 bases (i.e. form → Watson-Crick bonds over 8 base pairs) at the 3'-end. Overgo primers must be non-redundant. They are then exploited as primers to fish BAC clones from the organism with the unknown target region (in the example, human). The overgos are then extended by DNA polymerase in the presence of radiolabeled C or G nucleotides, and the homologous region (e.g. a gene) isolated and characterized.

Overhang: See → protruding terminus. Don't confuse with "hang-over".

Overlap: See → contig.

Overlap hybridization: See → chromosome walking.

Overlapping clone: See → contig.

Overlapping code: See → overlapping genes.

Overlapping genes: Genes with overlapping nucleotide sequences (e.g. gene *E* of phage Φ X 174 which overlaps with gene *D*). Overlapping genes produce two different polypeptides because the corresponding mRNAs are translated in two different → reading frames.

```
   met   gly   gln   tyr   asn   ala   ile   val   thr   gly   phe            gene 1

···A U G G G G C A A U A U A A U G C A A U U G U C A C A G G G U U U···
                         met   gln   leu   ser   gln   gly                    gene 2
```

Overnight culture: Any liquid bacterial culture that has been grown for more than 12 hours (overnight) and has reached its stationary growth phase.

Overproducer: Any → mutant cell or organism producing large quantities of a chemical compound that occurs in the wild type in minute amounts only. See → overexpression.

Overwinding: See → positive supercoiling.

ox: Laboratory slang term for *over-expression* of a gene, or also *overexpressor* (an organism overexpressing a gene).

Oxamycin: See → cycloserine.

Oxford grid: A compilation of probe distribution patterns on homologous chromosomes from different organisms, that allows a direct comparison of their → genome structure.

O⁰: See → *o*perator zero mutation.

P

p:
a) Symbol for a *phosphate* group (e.g. ppCpp).
b) Abbreviation for → *plasmid* (e.g. → pBR 322).

P:
a) Abbreviation for *protein*.
b) Abbreviation for the amino acid *proline*.
c) Abbreviation for *parental generation*, compare → F1, → F2.
d) Abbreviation for → *promoter*.

PAA gel: See → *polyacrylamide* gel.

PAC:
a) See → *protein association cloning*.
b) See → *phage artificial chromosome*.

PACA: See → *polymerase chain reaction assisted cDNA amplification*.

PACE:
a) See → PCR-assisted contig extension.
b) See → *polyacrylamide affinity coelectrophoresis*.
c) See → *programmable autonomously-controlled electrodes* gel electrophoresis.

Packaging (package): The process by which a nucleic acid molecule is encapsulated in a phage (or generally, viral) head particle. This packaging process takes place within the host during normal phage growth. → Lambda phage concatemeric DNA (→ concatemer), produced by a → rolling circle replication mechanism for example, is first cleaved into monomers. One of these monomers is now introduced into the phage head precursor which mainly consists of the major → capsid protein encoded by gene *E*. Then the product of gene *D* is incorporated into the growing capsid, and the products of genes *W* and *FII* (and others) link the capsid to a separately assembled tail to form the complete (mature) phage particle.

"Packed array" hybridization: A method to detect specific sequences in up to 2400 different clones simultaneously by transferring 96 clones from one microtiter plate at one time to an agar plate, and repeating the same process with other microtiter plates containing 96 other clones, except that these are then applied at a slightly different position. This packed array of 2400 clones can be transferred to → nitrocellulose filters (up to 20 times) and hybridized to specific → probes in a → Southern blotting experiment.

Packing ratio (packaging ratio): The ratio of DNA length to the unit length of the → chromatin fiber (e.g. nucleosome or 10 nm, and solenoid or 30 nm fiber) which it forms.

Pacmid: See → P1 cloning vector.

PACS: See → preferential amplification of coding sequences.

Padlock probe: A linear single-stranded oligodeoxynucleotide with target-complementary sequences of 20 bp located at both termini, which are separated by a central spacer element of about 50 bp. Upon hybridization of such a padlock → probe to a target sequence, the two ends of the probe are brought into juxtaposition, in which they can be joined by enzymatic ligation (i. e. by → DNA ligase). This leads to a circularization of the oligonucleotide. This intramolecular reaction is highly specific, and discriminates among very similar sequences from two genomes (that differ by only one or few nucleotides). Padlock probes are used for the detection of gene variants and mutations.

PAGE: Abbreviation for → *polyacrylamide gel electrophoresis*.

pA gene: See → putative alien gene.

Paired-box **gene (pax gene):** Anyone of a series of genes that share a socalled *paired box* (pax) sequence element and encode → transcription factors, which regulate the expression of other genes during ontogenesis in a strict spatial and temporal pattern. Therefore, pax genes themselves are transcribed during the development of e. g. the vertebrate embryo in highly specific patterns. The paired box (from the gene "paired" [prd] of *Drosophila melanogaster* encodes a protein → domain of 128 amino acids with DNA-binding specificity. Many pax genes additionally contain a → homeobox and an octapeptide-encoding sequence in between the paired box and the homeobox. Pax genes are known from echinoderms, molluscs, nematodes, insects, fish, birds, and mammals. See → paired box protein.

Paired box **protein (PAX protein):** Any protein encoded by a → paired box gene, and containing a 128 amino acid "paired box" domain, an octapeptide (consensus sequence: NH_2-HSIDGILG-COOH) and a → homeodomain. Paired box and homeodomains in concert function in sequence-specific binding to DNA. Each paired domain consists of two similar globular subdomains, and each subdomain in turn of three → α-helices, of which helices 1 and 2 run antiparallel, and together almost vertical towards helix 3. The aminoterminal subdomain binds to recognition sequences in the small groove DNA (consensus sequence: 5'-[G/T]T[C/T][A/C][T/C] GC-3') through contacts of terminal amino acids. Helix 3 binds within the large groove. The carboxyterminal subdomain of the paired domain contacts the consensus sequence 5'-(C/G)A-T(G/T)-(C/T)-3' in the next turn of the DNA helix. The complex between protein and DNA leads to a → bend in the DNA. The carboxyterminal regions of most paired box proteins frequently contain *p*roline, *s*erine, and *t*hreonine, the socalled PST region, which functions as → transcription factor. The target genes for the PAX proteins are not known, but they control different programs of organ differentiation (e. g. PAX 6 regulates the development of eyes and frontal lobe of the brain, PAX 4 the inner ear).

Paired-end sequence (mate pair): The → raw sequence of 500–600 bp in length from both termini of a → double-stranded insert of a clone (e. g. a → plasmid, → bacterial artificial chromosome, or → yeast artificial chromosome clone).

Palindrome (Greek: palindromos, running back):
a) Any sequence of letters or words that can be read in either orientation to give the same sence. For example: "Madam, I'm Adam". Or: "A man, a plan, a canal: Panama!"
b) Any sequence in duplex DNA in which identical base sequences run in opposite directions, with the property of rotational (dyad) symmetry, e.g.

Arrows indicate rotational axis.

Palindromes are target sites for various → DNA-binding proteins (e.g. → restriction endonucleases, → RNA polymerases and → transcription factors), and occur in many → promoters, DNA replication → origins and transcription termination sequences. See also → inverted repeat, → perfect palindrome.

Palindromic unit: See → repetitive extragenic palindromic element.

Pancreatic *deoxyribonuclease*: See → pancreatic DNase I.

Pancreatic DNase I (pancreatic *deoxyribonuclease*): An enzyme from bovine pancreas that (in the presence of Mn^{2+}) catalyzes the cleavage of internucleotide bonds in single-stranded and double stranded DNA, preferentially between adjacent purine and pyrimidine residues. The enzyme is used for the limited digestion of DNA, and the removal of DNA from DNA-RNA mixtures. See → DNase I.

Pan-editing: A special type of → RNA editing, in which entire genes are edited, in contrast to partial RNA editing limited to the 5' termini of editing domains (→ 5' editing). Pan-editing is probably the more primitive character (e. g. in ancestral trypanosomatid mitochondria pan-editing is prevalent).

Panning (biopanning): A procedure to screen a → random peptide display library for protein-protein interactions and to enrich interacting clones of this library. It starts with the immobilization of bacteria or phages, displaying a target peptide or protein on a solid phase (e. g. → microtiter plate, glass, agarose, or magnetic beads), which is then incubated with the → random peptide display library and left to react. Then the unbound cells are washed off, and the bound cells released by mechanical shearing. These cells (presenting potential protein-binding proteins) are subsequently grown in suitable media, and plasmid DNA from individual clones is isolated and sequenced.
Figure see page 780.

***par*:** See → *par*titioning functions.

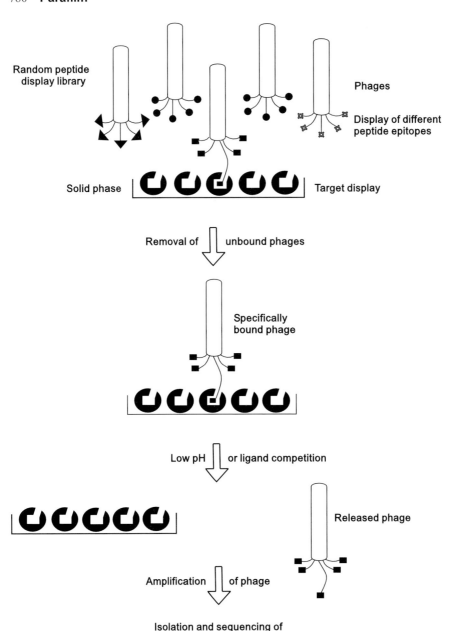

Random peptide
display library

Phages

Display of different
peptide epitopes

Solid phase

Target display

Removal of unbound phages

Specifically
bound phage

Low pH or ligand competition

Released phage

Amplification of phage

Isolation and sequencing of
individual clones

Panning

Parafilm™: A paraffin wax film used to seal laboratory glassware (e.g. tubes, Petri dishes).

Paralogous domain shuffling: The substitution of → exons, encoding specific → domains of a protein, in a gene from one species by homologous exons from different genes of the same species. See → orthologous exon shuffling, → paralogous exon shuffling.

Paralogous exon shuffling: A variant of the → *in vitro* exon shuffling, that allows to replace → exons from a particular gene of one species with the homologous exons from different genes of the same species. See → de novo protein assembly, → functional homolog shuffling, → orthologous domain shuffling, → orthologous exon shuffling.

Paralogs (paralogous genes): Homologous DNA sequences in two organisms A and B, that are descendants of two different copies of a sequence originally created by a gene duplication event in the genome of a common ancestor. See → orthologs.

Paramagnetic particle: See → magnetic bead.

Paramagnetic particle technology: A term encompassing techniques that involve specially pre-pared → magnetic beads (paramagnetic particles) as a solid-phase support for the separation of DNA or RNA molecules from complex mixtures of biomolecules. For example, poly(A)$^+$ mRNA can be affinity-purified using this technique. In short, a biotinylated oligo(dT) probe is hybridized in solution to the 3' → poly(A) tail of eukaryotic mRNA. The biotin-oligo(dT)-mRNA hybrids are then captured with streptavidin covalently coupled to → paramagnetic particles. After removal of nonspecific hybrids by high-stringency washing, the mRNA can be separated from the solid phase by elution with RNase-free water. This procedure yields an enriched poly-(A)$^+$-mRNA fraction after one single step of purification. See also → magnetic crosslinking.

Paramutation: Any change in the activity of an → allele induced by the corresponding allele on the homologous chromosome. Such changes are brought about by → chromatin alterations and do not result in, or depend on changes of the underlying DNA sequence. Compare → mutation.

Paranemic joint: A region of contact between two complementary DNA → strands that do not form a conventional → double helix.

Parent-ion-scan technique: A method for the identification of modified amino acid residues in a protein, that is based on the production and mass analysis of peptide fragments in a triple-quad-ruple mass spectrometer. The generated peptide ions are separated in the first quadrupole according to their mass/charge ratio (m/z), and fragmented in the second quadrupole (collision cell) by collision with inert gas molecules (e. g. argon), a process called *collision induced dissociation* (CID). The third quadrupole again functions as mass analyzer, which permits the transmission of a fragment with a defined m/z only (e. g. m/z = 79, corresponding to the mass of a PO_3^- fragment ion, if phosphopeptides are to be detected). This filter allows to detect peptide ions only, that produce this fragment. The parent-ion-scan therefore filters specific peptides (e. g. the phos-phopeptides) out of a complex mixture of fragment ions.

PARM-PCR: See → *priming authorizing random mismatches polymerase chain reaction.*

PARN: See → poly(A)-specific 3' exoribonuclease.

PARP: See → *poly(ADP-ribose) polymerase.*

***par* region:** See → partitioning function.

Parsing. The use of algorithms to dissect data into components for an extensive component analysis.

par **site:** See → partitioning function.

Partial: See → partial digest.

Partial denaturation: An incomplete unwinding of a duplex DNA. See → denaturation.

Partial digest ("partial", incomplete digest):
a) The fragments arising from endonucleolytic cleavage (see → endonuclease) of a DNA molecule, in which not all the potential cleavage sites have been restricted. Partials with e.g. → four base cutter enzymes, as *Sau*3A, are used as a collection of overlapping DNA fragments for the establishment of an → ordered gene library.
b) An incomplete enzymatic proteolysis. For example, the → Klenow fragment of → DNA polymerase I is obtained by partial proteolysis.

Partial editing: See → 5' editing.

Partial gene bank: See → minilibrary.

Partial intron retention: The inclusion of part of an → intron in a final → messenger RNA. Normally, the introns are spliced out of the pre-mRNA, but in rare cases a → splice site within an intron can be recognized and the corresponding residual intron be left in the message. See → intron retention.

Partially denaturing *h*igh *p*erformance *l*iquid *c*hromatography (partially denaturing HPLC): A variant of the conventional → *h*igh *p*erformance *l*iquid *c*hromatography (HPLC) technique, which allows to discriminate DNA → hetero- from → homoduplexes and is therefore employed for → mutation detection (e.g. the discovery of → single nucleotide polymorphisms). In short, a 200–1000 base-pair target fragment is first amplified from genomic DNA of at least two chromosomes, using → primers flanking the target site. Then the amplified fragments are denatured at 95^0C for some minutes and allowed to reanneal by gradually lower the temperature within the separating column from 95^0C to 65^0C over 30 minutes. In the presence of a mutation in one of the chromosomes, not only the original homoduplexes form upon reannealing, but also the sense and anti-sense strands of either homoduplex form two heteroduplexes. These heteroduplexes are thermally less stable than the homoduplexes. Therefore all these different duplex molecules can be separated from each other by their different retention time in an alkylated non-porous poly(styrene-divinylbenzene) column during their elution with acetonitrile. The more extensive, but still partial denaturation of the heteroduplexes in a temperature range between 50 and 70^0C (which depends on the GC content and size of the fragments, the influence of the nearest neighbour base of both the matched and the mismatched base pairs and column temperature) typically leads to a reduced retention time of the heteroduplexes and their separation from the homoduplexes. As a consequence, one or more additional peaks appear in the chromatogram. See → denaturing high performance liquid chromatography.

Partially *i*nverted *r*epetitive DNA (PIR-DNA): A tandemly repeated sequence family on chicken chromosome 8, whose basic repeat units are 1.43 kb in length and consist of a central core of about 0.6–1.0 kb (in different animals). An 86 bp flanking sequence forms a → palindrome with the core (therefore partially inverted repetitive DNA).

Particle acceleration technique: See → particle gun technique.

Particle bombardment: See → particle gun technique.

Particle gun technique (biolistics, microprojectile bombardment, particle acceleration technique, particle bombardment): A method for → direct gene transfer into cells, tissues, organs or whole plants. Tungsten or gold particles are coated with DNA and shot through target cells. On their way through the cells the DNA on the particle surface is stripped off and may then be inserted into the nuclear genome. The particle gun technique has certain advantages over other direct gene transfer methods, e.g. the transfer does not require → protoplasts, but is possible with intact tissues.

Partition (plasmid partitioning): The → segregation of → plasmids to daughter cells during bacterial cell division. Segregation may depend on random distribution to the daughter cells (as e.g. for → Col E1, or other → multicopy plasmids), or may involve → partitioning functions (as e.g. for → F-factors, or other → low copy number plasmids).

Partitioning **function (partitioning region, *par* site, *par* region, *par* locus, *par*):** A particular nucleotide sequence of → plasmids responsible for their precise → segregation at each cell division. The partitioning activity normally ensures that each daughter cell receives about the same number of plasmids. Not all plasmids, however, have a *par* site. This region has been deleted in e.g. → pBR 322 which is consequently segregated at random during cell division. See → partition.

Partitioning region: See → partitioning function.

PAS: See → primosome.

PASA: See → allele-specific polymerase chain reaction.

PAS gene: See → *p*eroxisome *as*sembly gene.

Passage:
a) The serial infection of different hosts by one and the same parasite.
b) The repeated sub-culture of cells from a cell culture.

Passenger DNA: A synonym for → insert DNA.

Passenger protein (target protein): Any protein of interest expressed in appropriate host cells (e.g. *E.coli, Bacillus* or *Staphylococcus* strains, *Saccharomyces cerevisiae*) as a → fusion with a so called carrier protein, that transports, anchors and exposes the passenger on the cell's surface. See → microbial cell-surface display, → peptide display, → phage display, → ribosome display.

Pasteur pipet: An open-end glass tube with one end pulled out to a capillary. Such Pasteur pipets allow the transfer of small volumes of liquids with the aid of a rubber bulb.

Patched **circle** *polymerase* **chain** *reaction* **(PC-PCR):** A variant of the → polymerase chain reaction technique that can be used for → site-directed mutagenesis. In short, the target DNA is cloned into a specific plasmid → cloning vector between opposing → T3 and → T7 RNA polymerase promoters. Then one amplification primer (→ amplimer) is annealed to sequences within the T7 promoter, and a second amplimer – in opposite direction as compared to the first – is annealed to a sequence flanking the region to be deleted. The supercoiled plasmid then serves as a template for PCR. During the amplification process linear DNA molecules accumulate which lack the region to be deleted. A third oligodeoxynucleotide primer, base-pairing with the two ends of the linear molecules is then used to form patched circles for direct transformation of *E. coli*. Appropriate and rapid screening procedures allow the isolation of clones that lack the deleted fragment. See also → polymerase chain reaction mutagenesis.

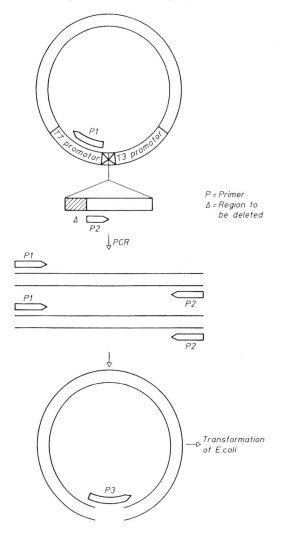

Patched circle PCR

PAT gene: See → *p*hosphinotricin *a*cetyl*t*ransferase gene.

Pathochip: A laboratory slang term for a → DNA chip or → protein chip, onto which either specific genes, gene-specific oligonucleotides, cDNAs, peptides or proteins are spotted, that are diagnostic for a particular pathogen (as e.g. *Staphylococcus aureus*), or even a specific strain or isolate of the pathogen. Such pathochips are used to screen samples from e.g. hospitals for particular pathogens.

Pathogen-*derived* *resistance* (PDR): The resistance of a plant towards a virus, bacterium, or fungus, that is engineered by the stable transformation of the plant with a → transgene derived from the virus, bacterium or fungus. For example, the gene for the coat protein or the movement protein of a virus can be transferred to target plants, stable integrated into their genome, and expressed. The resulting proteins then either coat the viral genome (preventing its replication in the host cell) or inhibit the dispersal of the virus throughout the plant, especially if the movement protein gene is genetically engineered such that its protein product still lines the plasmodesmata (channels between two plant cells, through which the viruses are transported for a systemic infection), but does not support viral transport anymore. At least one variant of PDR works with → RNA interference.

Pathogenesis-*related* (PR) proteins: An operational term encircling a characteristic group of proteins accumulating in pathogen-infected or elicitor-induced plant cells. These proteins have mostly low molecular weight and acidic isoelectric points (e.g. *phenylalanine ammonia lyase*, PAL; 4-*coumarate ligase*, 4CL; ß-1,3-glucanases, chitinases, thaumatin-like inhibitors, proteinase inhibitors, hydroxyproline-rich glycoproteins, peroxidases, and others).

Pathogenicity: See → virulence

Pathogenomics: The whole repertoire of techniques for the sequencing and characterization of genomes of pathogens, for the identification of genes involved in pathogenicity and virulence (e.g. pathogenicity islands) or other functions relevant for the efficiency and fitness of the pathogen, and for the detection of pathogen genes with high homology to host genes (able to mimic host gene function).

Pathway mapping: The estimation of – preferentially all – possible interactions between all proteins of a biochemical pathway (e.g. glycolysis, steroid biosynthesis, protein degradation).

Pathway slide: A laboratory slang term for a → microarray, onto which → cDNAs or → oligonucleotides are spotted, that represent the transcripts of genes encoding enzymes of a particular metabolic pathway (e.g. glyolysis, pentose phosphate shunt, phenyl-propanoid pathway). Pathway slides allow activity profiling of all genes encoding all proteins of such a pathway.

PATTY: See → PCR-*aided transcript titration assay*.

Pauling-like DNA (P-DNA): A specific conformation of DNA, experimentally produced by stretching and overwinding, in which the sugarphosphate backbone is oriented towards the center, and the unpaired bases turned outside. The extreme stretching allows only 2.6 base pairs per turn (→ B-DNA: 10-11 bp/turn). Hypothetically, P-DNA could occur in vivo in front of a moving → RNA polymerase molecule, where a positive torsional stress leads to overwound DNA. A similar configuration was proposed by Linus Pauling before the discovery of B-DNA by Watson and Crick, therefore the somewhat misleading name "Pauling-like DNA". See → A-DNA, → B-DNA, → C-DNA, → D-DNA, → E-DNA, → ε-DNA, → G-DNA, → H-DNA, → M-DNA, → V-DNA, → Z-DNA.

Pause signal: A specific sequence element, at which → DNA-dependent RNA polymerase hesitates to elongate the nascent → messenger RNA. See → arrest signal, → termination signal.

Pax gene: See → paired-box gene.

Pax protein: See → paired box protein.

pBeloBac 11: A 7.507 kb single-copy → bacterial artificial chromosome (BAC) → cloning vector for the cloning of large DNA fragments (up to 1 Mb) in *E. coli*, that contains an oriS replicon of the fertility (F) factor of *E. coli*, Sop AB (Par AB) functions for active → partitioning (acting at Sop C [IncD, Par C] such that each daughter cell receives a plasmid copy during cell division), replication initiation factor Rep E (Rep A) sequences (the encoded protein mediates the assembly of a replication complex at ori 2), a truncated copy of a site-specific recombinase (red F), and a → chloramphenicol acetyltransferase gene from → transposon 9 as a → selectable marker. The cloning region encompasses a → lambda phage → cos site (representing a unique cleavage site and enabling the → packaging into phage particles), a → lox P site, two (in variants of the orginal pBeloBac vector more) cloning sites (e. g. *Bam*HI and *Hind*III sequences), a series of → rare cutter sites (e. g. *Sfi*I), → SP6 RNA polymerase and → T7 RNA polymerase promoters flanking the cloning site (for the generation of RNA → probes from the insert). The large DNA inserts in the single-copy pBeloBac 11 vectors are stable and do not interfere with the viability of the host cell. BAC cloning has superseded the traditional → yeast artificial chromosome (YAC) cloning, because BAC cloning procedures are easier, and → chimerism of inserts a rare event. See → BIBAC, → mammalian artificial chromosome, → P1 cloning vector, → *Schizosaccharomyces pombe* artificial chromosome.

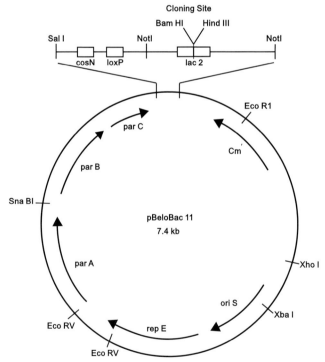

pBeloBac 11

pBR 322 (and derivatives): A series of comparatively small, → multicopy (15-20 copies/cell), → non-conjugative plasmid cloning vectors containing → ampicillin and → tetracycline resistance genes and several unique cloning sites (or, in derivatives, → polylinkers). The latter are located within one or the other resistance gene, so that the insertion of foreign DNA can be detected by → insertional inactivation of the antibiotic resistance function. The notation 'BR' is derived from *Bolivar* and *Rodriguez*, two Mexican molecular biologists who synthesized the plasmid using the tetracycline resistance gene from pSC 101, the origin of replication (ori) and rop gene from the Col E1 derivative pMB1, and the ampicillin resistance gene from → transposon Tn3. The plasmid replicates in *E. coli* under → relaxed control, but is slightly unstable and has a relatively narrow host range (*E. coli, Serratia marcescens*). Therefore, more advanced derivatives have been designed (see → pUC).

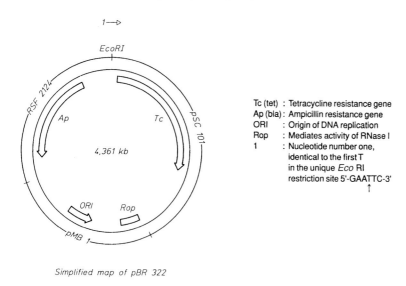

Tc (tet) : Tetracycline resistance gene
Ap (bla) : Ampicillin resistance gene
ORI   : Origin of DNA replication
Rop   : Mediates activity of RNase I
1   : Nucleotide number one,
           identical to the first T
           in the unique *Eco* RI
           restriction site 5'-GAATTC-3'

Simplified map of pBR 322

PBS: See → primer *b*inding *s*ite.

PC-PCR: See → *p*atched *c*ircle *p*olymerase *c*hain *r*eaction.

PCR: See → *p*olymerase *c*hain *r*eaction.

PCR add-on primer (restriction site add-on): A synthetic → oligonucleotide that carries a → recognition site for a → restriction endonuclease and still serves as a → primer for → *Thermus aquaticus* DNA polymerase. Such add-on primers with 5' overhanging termini are annealed to the target DNA and the DNA amplified by the conventional → polymerase chain reaction. The amplified product then contains the desired restriction site(s), and can easily be cloned into appropriately cut → cloning vectors. See also → add-on sequence.

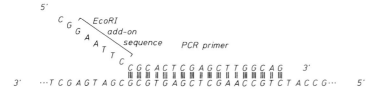

Template DNA

PCR-aided transcript titration assay (PATTY): A technique for the quantification of a specific → messenger RNA, which capitalizes on the co-amplification of a mutated, and therefore different form of the target messenger RNA. In short, first the mutated cDNA is generated by → site-directed mutagenesis such that a single base exchange occurred and a new → restriction recognition site is generated. Then identical amounts of total RNA (containing an unknown amount of wild-type mRNA) are mixed with decreasing, but known amounts of mutated mRNA. After → reverse transcription to cDNA and → polymerase chain reaction amplification with → primers complementary to sites within the target cDNA, the amplified fragments are restricted (only mutated cDNA is cut, and therefore differentiated from the wild-type cDNA). The cDNA (target) or cDNA fragments (mutant) are then separated by → agarose gel electrophoresis and hybridized to a radioactively labeled subfragment of the target cDNA. The hybridization signals then allow to identify one particular sample, which contains equal or nearly equal amounts of both types of cDNAs, reflecting equal starting concentrations of the original mRNAs.

PCR amplification of specific alleles: See → allele-specific polymerase chain reaction.

PCR-assisted contig extension (PACE): A technique for the closure of → gaps remaining in unfinished bacterial genome sequences, that involves the generation of stepwise extensions from the ends of → contigs by a conventional → polymerase chain reaction (PCR), until the closure of the individual gaps is achieved. In short, specific internal and → nested primers are first derived from the sequenced contigs. In a first step, the specific internal forward primer is used in combination with a reverse primer of arbitrary sequence (→ "arbitrary primer") to amplify → genomic DNA outwards of the contig. The nested primers are designed approximately 150 bp from the contig ends and 40 bp apart from each other. In a subsequent second step, the amplification products of the first step are diluted and again amplified with a nested primer and a perfectly matching primer derived from the sequence of the first amplicon under higher → stringency. The products are then electrophoresed in agarose gels, stained with → ethidium bromide, and single bands isolated and sequenced directly with the same specific primer used for amplification. The contig → extensions have to be verified by specific PCR and sequencing.

PCR carry-over prevention: See → polymerase chain reaction carry-over prevention.

PCR clamping: A technique for the detection of → deletions, → insertions, → mutant alleles, or → point mutations in a target DNA, that is based on the increased affinity and specificity of → peptide nucleic acids (PNAs) for their complementary target sequences and the inability of → DNA polymerase to recognize and extend a PNA primer. In short, a 15-18mer peptide nucleic acid complementary to the wild-type sequence is synthesized. The PNA oligomer is then mixed with two DNA primers, one of which is complementary to the mutant allele sequence (forward primer), whereas the other one serves as a → reverse primer to amplify the target sequence. In the subsequent → polymerase chain reaction, the wild-type PNA competes with the mutant DNA primer for the same target priming site. Hybridization of the DNA primer and subsequent amplification will only occur, if the target is a mutant allele (amplification product can be visualized by e. g. → ethidium bromide staining). In the absence of a mutant allele, the PNA will bind to the target and prevent amplification (no amplification product can be visualized). Two PCR clamping configurations are possible. First primer exclusion, where a PNA oligomer competes with a DNA primer for binding at the target site, as described above. The DNA outcompetes the PNA, binds to the target, and allows its extension only when it is fully complementary to the mutant site. Point mutations at various positions in the target can be identified by altering the sequence of the primer. Second, elongation arrest is a result of the stronger binding of PNAs to their targets (PNA/DNA duplexes at physiological ion strength are about 1°C/base more stable than the corresponding DNA/DNA duplexes), which prevents the elongation of a primer, that binds outside the target DNA.

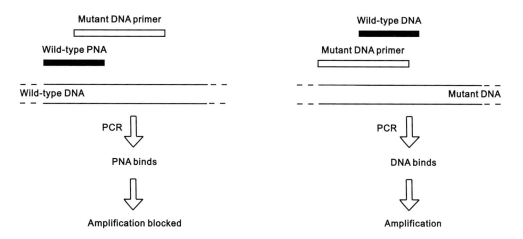

PCR fingerprinting: The amplification of distinct highly polymorphic target DNA sequences (e.g. → simple repetitive DNA sequences), using → polymerase chain reaction techniques to establish a → DNA fingerprint of the target. See for example → arbitrarily primed polymerase chain reaction.

PCR *in situ* hybridization: A variant of the → polymerase chain reaction, in which DNA is amplified and detected within morphologically intact cells or tissues. In short, cells or tissue specimens are fixed, mounted on a silane-coated microscope slide and digested with → protease. All PCR reagents are added and diffuse into the cells. Then the slide, on an aluminum foil, is placed directly on the thermoblock of a → thermocycler, → *Thermus aquaticus* DNA polymerase is added and the slide overlaid with mineral oil. The amplified product can be detected by → *in situ* hybridization or by direct incorporation of → biotin- or → digoxygenin-labeled nucleotides into the PCR product. Since the diffusion of the product away from its original location is a problem, either → multiple overlapping primer PCR or → concatemer PCR are used to generate large PCR products that do not freely diffuse. See → *in situ* hybridization.

***PCR-ligation-PCR* mutagenesis (PLP mutagenesis):** A technique for the generation of → fused genes, site-directed mutagenesis, or introduction of specific → deletions, → insertions or → point mutations into target DNA. For example, the fusion of two (or more) genes starts with the amplification of each gene in a separate → polymerase chain reaction. The amplification products are then phosphorylated using → T4 polynucleotide kinase and ligated with → T4 DNA ligase, creating different combinations of joined fragments. The fused gene is then specifically PCR-amplified out of this heterogeneous mixture with a primer directed to the 5' end of the upstream gene and a primer complementary to the 3' end of the downstream gene. The resulting amplification product is then subcloned into the original target sequence, creating a type of insertion mutation. PLP mutagenesis relies on a DNA polymerase with exonuclease (i. e. proofreading) activity, so that the blunt-ended fragments match exactly with the primer sequence. Compare → splice overlap extension polymerase chain reaction.

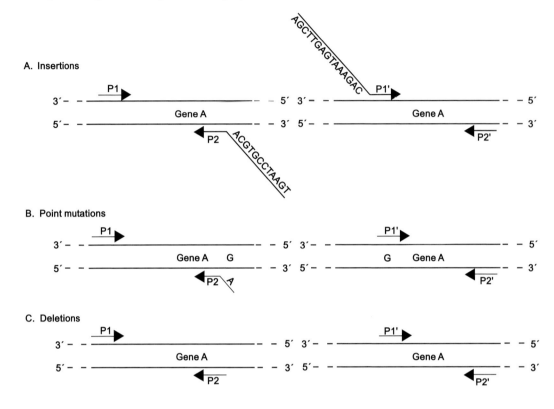

PCR mimic: See → heterologous competitive standard.

PCR mutagenesis: See → polymerase chain reaction mutagenesis.

PCR-RFLP: See → polymerase chain reaction restriction fragment length polymorphism.

PCR-SSCP: See → single-strand conformation analysis.

PCR technology: A myriad of techniques to amplify a specific DNA segment and to modify the amplified sequence simultaneously (e.g. by the application of → PCR add-on primers, the introduction of → mutations in → PCR mutagenesis, or the in-vitro recombination of two specific DNA fragments in → recombinant PCR). See → polymerase chain reaction.

P-DNA: See → Pauling-like DNA.

Pectinase ("driselase"; EC 3.2.1.15): An enzyme that catalyzes the degradation of plant pectin and is therefore used to degrade the cell walls in → protoplast isolation procedures.

PEG: See → *polyethylene glycol.*

Pegylation: The attachment of *polyethylene glycol* (PEG) chains to peptides or proteins to increase their size and to protect them from degradation. For example, pegylation of interferon α-2a increases its half-life time from 9 (unpegylated) to 77 hours and at the same time reduces its renal clearance 100-fold, such that the compound is less susceptible to destruction in the digestive tract and remains in the body for a longer time.

P element: A member of a family of transposable elements (→ transposons) in *Drosophila* species that is randomly distributed in the genomic DNA of so-called P strains (*paternally* contributing) in 30-50 copies. The P element prototype is 2.9 kb in length, and the other members of the family have evolved to 0.5-2.5 kb elements by different internal deletions. Each P element is flanked by perfect → inverted repeats of 31 bp at which excision takes place. The insertion of the P element at another locus is accompanied by a duplication of a short 8 bp target sequence that can be found on either side of the integrated P sequence. The internal fragment of the P element prototype carries four → open reading frames (ORF 0, 1, 2 and 4), one of which encodes an 87 kDa → transposase and another one a 66 kDa transposition repressor protein. → Transposition requires the activity of this transposase, which is active in so-called M (*maternally* contributing) cytotype cytoplasm, but mostly inactive in so-called P-cytotype cytoplasm because of the presence of the transposition repressor protein in the cytoplasm of P strains. Transposition activity of P elements is strictly limited to germ line cells.
P elements may insert into control or coding sequences of genes, which are thereby inactivated. Together with concomitantly occurring chromosome breakages these insertions lead to the disease syndrome of "hybrid dysgenesis" (P-M hybrid dysgenesis, i.e. genetic abnormalities such as chromosomal aberrations, high frequencies of lethal mutations and high rates of sterility).
P elements can be exploited as → gene transfer vectors. Any foreign DNA can be cloned into a P element which in turn can be inserted into a → plasmid. After → microinjection of this plasmid into *Drosophila* embryos the P element together with the foreign DNA can transpose into germ-line chromosomal DNA. P elements can also be used to search for specific genes of *Drosophila* via → transposon tagging. In this case, the P elements function as → mutagens which lead to a loss of gene function through their → insertion (→ insertion mutation).

P element transformation: The integration of specific DNA fragments into germ line chromosomes of *Drosophila* using the transposable → P elements as transposing sequences.

Pellet: Any packed material sedimented by centrifugation.

pEMBL: A family of single-stranded 4 kb → plasmid cloning vectors derived from → pUC, containing the → *bla* (ampicillin resistance) gene as → selectable marker, a short DNA segment coding for the α-peptide of → β-galactosidase that carries a → polylinker, and the intragenic region of the → f1 phage. Upon superinfection with phage f1, these plasmids may be encapsidated as single-stranded DNA, and the virions are excreted into the culture medium. pEMBL vectors can be used for DNA → Sanger sequencing, for → site-directed mutagenesis, → S1-mapping and hybridization to mRNA and cDNA. These vectors are smaller than the → M13 vectors and are relatively stable even with large inserts. Without superinfection, the replication of the double-stranded pEMBL plasmids is initiated at the → col E1 → origin of replication. EMBL stands for *European Molecular Biology Laboratory*. Do not confuse with the lambda phage derived → EMBL vectors.

Penetrance: The frequency (also probability) of the expression of an → allele or → gene. See → complete penetrance, → expressivity, → incomplete penetrance, → nonpenetrance.

Penicillin: Any of a series of → antibiotics synthesized by *Penicillium notatum* and related molds (e.g. *Aspergillus, Trichophyton, Epidermophyton*). Penicillins are derivatives of 6-amino-3,3-dimethyl-7-oxo-4-thia-1-azabicycloheptan-2-carboxylic acid (6-aminopenicillanic acid). Different penicillins differ from each other in the structure and number of the side chains, (one in e.g. penicillin G, penicillin V). Penicillins block the cross-linkage between parallel peptidoglycan chains, and thus prevent the completion of the synthesis of bacterial cell walls.

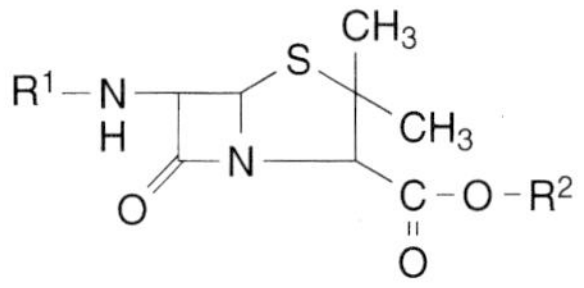

6-Aminopenicillanic acid

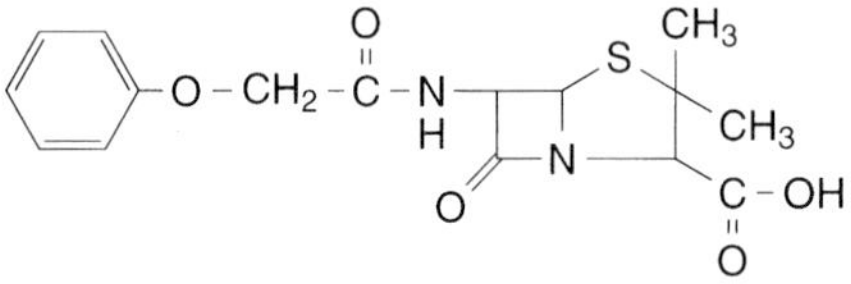

Penicillin V

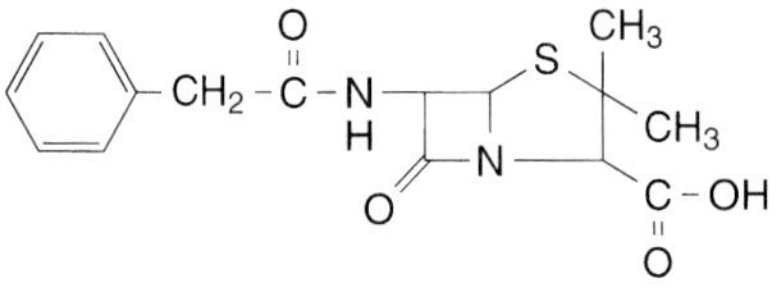

Penicillin G

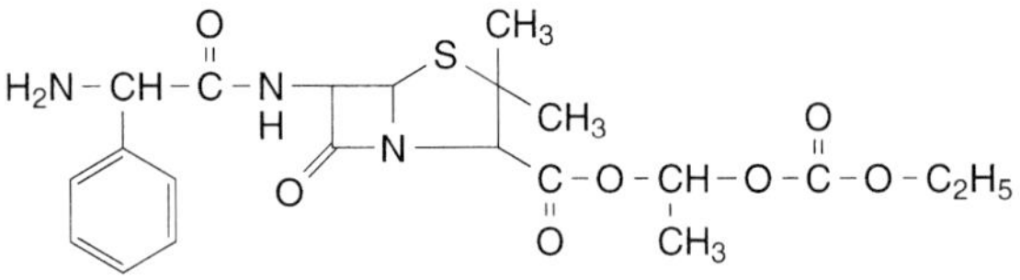

Bacampicillin

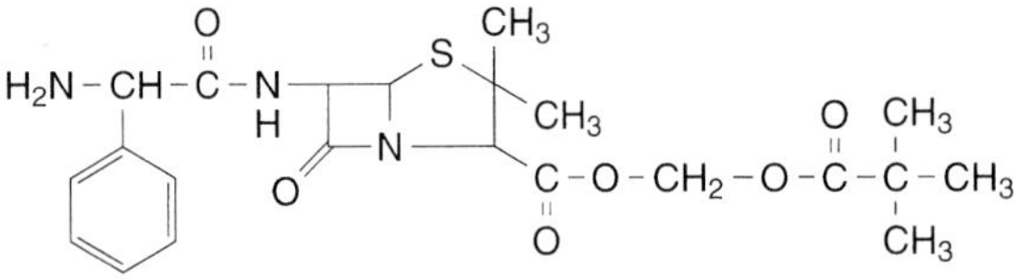

Pivampicillin

Penicillinase: See → (β)-lactamase.

Pentaplex DNA: A self-assembled higher-order structure of DNA, in which the naturally occurring base 2'-deoxy-iso-guanosine (iG) is assembled around a caesium ion in a quintet geometry (iG-quintet/Cs+/iG-quintet), and the caesium ion is positioned between two quintet layers.

Penta-snRNP: A pre-formed complex of all five → small *nuclear* RNAs (snRNAs) U_1, U_2, U_4, U_5 and U_6 and about 13 different proteins, that associates with the → messenger RNA (mRNA) as a single discrete particle. The experimental proof of this particle is somehow conflicting with the socalled stepwise assembly model of the → spliceosome, which predicts that the U_1snRNP first recognizes its substrate mRNA and binds to the 5'splice site. After this interaction, the U_2snRNP contacts the branch point region of the message. And only then a complex of $U_4/U_5/U_6$ j→with the → intron removed and the two adjacent → exons joined.

PEP: See → *primer-extension preamplification.*

PEPSI: See → *polyester plug spin insert.*

Peptide: A molecule consisting of two or more amino acids. See → peptide bond, → peptide map, → polypeptide.

Peptide array (peptide microarray, peptide microchip): An inert membrane or glass slide (or other solid support), onto which thousands of short, 24 amino acids long peptides are spotted in an ordered array to allow the visualization of interactions with labeled ligands (e.g. peptides, proteins, antibodies, low molecular weight effectors). Peptide arrays are used for e.g. → epitope mapping, immunogen selection, vaccine design, and drug discovery. Compare → protein chip.

Peptide bond: Any covalent bond between two amino acids arising from linkage of the α-aminogroup of one to the α-carboxyl group of the second molecule with concomitant elimination of water.

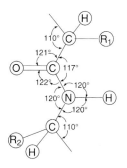

Peptide chip: See → protein chip.

Peptide computer (protein computer): A special variant of a → biocomputer, that performs computational tasks with peptides or proteins. One of the major advantages of peptide computers over → DNA computers is, that every position of a peptide can be occupied by 20 (or more, for example, synthetic or artificial) amino acids as compared to only four bases in DNA). Instead of a hybridisation reaction of two nucleic acid molecules in DNA computers, peptide computers exploit the (usually stereo-specific) interaction(s) of e.g. → antibodies with peptide → antigens.

Peptide display: See → phage display.

Peptide fingerprint: The specific pattern of peptide fragments generated by proteolytic cleavage of a protein and displayed on e.g. stained gels after their electrophoretic separation. Peptide fingerprints are the products of → peptide or → protein fingerprinting. See → peptide fragmentation fingerprint, → peptide mass fingerprint.

Peptide fingerprinting: See → protein fingerprinting.

Peptide fragmentation fingerprint (PFF): The specific pattern of fragments arising from a singl e peptide of a → peptide fingerprint. The target peptide is first isolated from the peptide mixture in the mass spectrometer and subsequently fragmented. The molecular weights of these fragments can be determined precisely and altogether represent a fingerprint of the peptide. See → peptide fingerprint, → peptide mass fingerprint.

Peptide map: Λ characteristic → peptide fragment pattern, generated by → protein fingerprint-ing. The comparison of such peptide maps from two (or more) proteins allows the detection of similarities or dissimilarities between the corresponding proteins on a large scale.

Peptide mapping: A procedure for the establishment of a → peptide map of a protein. In short, peptide mapping starts with the unfolding of the isolated and purified protein, its reduction, and alkylation to prevent re-formation of disulfide bridges. After extensive dialysis to remove excess reagents, the protein is proteolytically digested, and the resulting peptide fragments separated by reversed-phase chromatography. Peptide mapping provides informations about protein struc-ture and reveals substitutions of amino acids and post-translational modifications. See → protein fingerprinting.

Peptide mass fingerprint (PMF): The specific peptide fragment pattern arising from e.g. cleavage of a protein by proteolytic enzymes and analysed by → matrix-assisted laser desorption ioniza-tion mass spectrometry (MALDI-MS). PMFs are specific for the target proteins and can be used as search query against a database of PMF-like entries (e.g. produced by theoretical digests of protein sequences). See → peptide fingerprint, → peptide fragmentation fingerprint.

Peptide-mediated non-covalent protein delivery: See → non-covalent protein delivery.

Peptide mimicry: The synthesis of a biologically active protein or peptide that retains all or most of the structural features in addition to all functional domains. Compare → peptide morphing.

Peptide morphing: The design and synthesis of a derivative of a naturally occuring protein or peptide, that eliminates all non-functional amino acids, but retains its functional domains (for e.g. binding of ligands, protein-protein interactions, catalysis, protein-DNA- or protein-RNA interactions). Peptide morphing aims at increasing the chemical and biological stability of the "morphed" peptide and lowering its polarity by reducing its amide bonds such that it can be used as therapeutically active compound. Compare → peptide mimicry.

Peptide nucleic acid (PNA; polyamide nucleic acid): A relatively simple, synthetic chimeric poly-mer with a neutral achiral polyamide (peptide-like) backbone composed of N-(2-aminoethyl) glycine units, to which nucleic acid bases are covalently bound via carbonyl methylene (-CH$_2$-CO-) linkers. Such PNAs are increasingly used as substitutes of normal DNA. PNAs also form duplexes with complementary PNA strands via Watson-Crick base pairing. The resulting PNA-PNA hybrids are more stable than PNA-DNA duplexes. But also PNA-DNA and PNA-RNA hybrids are more stable than the corresponding DNA-DNA and DNA-RNA complexes, because no repulsion occurs between the charged phosphodiester backbone of DNA (or RNA) and the neutral PNA backbone. In contrast to DNA-DNA hybrids, PNA-DNA duplex stability is little affected by changes in salt concentration. Since single base → mismatches in PNA-DNA duplexes are more unstable than corresponding single base mismatches in DNA-DNA hybrids, a higher specificity results. Therefore, PNAs easily allow discrimination between perfect matches and mismatches of bases. As an artificial molecule, a PNA is no substrate for proteases or nucleases.

Homopyrimidine PNAs invade intact double-stranded DNA. For example, a PNA complementary to CAG repeats binds to its target DNA even in intact chromatin. PNA strand invasion results in stable PNA-DNA complexes, especially within transcriptionally active regions. Following a digestion of chromatin with a mixture of → restriction endonucleases (that do not cleave within the CAG repeats), the CAG-containing fragments are then first bound to mercurated → paramagnetic beads via thiol-reactive → nucleosomes (→ lexosomes), then released from the beads, hybridized to the biotinylated PNA → probe and the resulting PNA-DNA hybrids captured on → streptavidin-coated beads. DNA is then released and tested for CAG triplet content that allows to diagnose → triplet expansion-based diseases.

PNA oligomers serve as hybridization probes in → in situ hybridization, → Northern and → Southern analyses, but cannot be used as → primers in conventional → polymerase chain reaction techniques, since they lack 3' OH groups and therefore cannot be recognised by → DNA polymerases.

PNAs can be labeled by → biotin, → digoxygenin, → Cy 5, → Cy 3, → fluoresceine, → rhodamine, or ^{32}P and ^{125}I. They can also be attached to a solid phase (e. g. a glass or quartz chip, see → DNA chip) and used as probes to screen for complementary → cDNA or DNA sequences. Compare → pyranosyl-RNA.

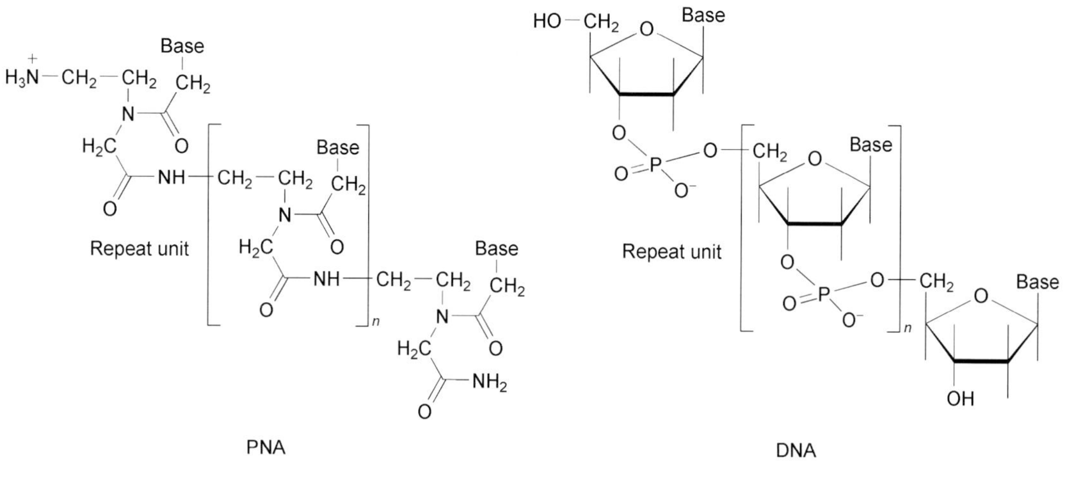

PNA

DNA

Peptide nucleic acid-phosphono peptide nucleic acid cimera (PNA-pPNA chimera): A synthetic hybrid polymer composed of alternating stretches of → peptide nucleic acids and → phosphono peptide nucleic acids. Phosphono peptide nucleic acid monomers of basically two types containing N-(2–hydroxyethyl)phosphono-glycine or N-(2–aminoethyl)phosphono-glycine are linked through amide or phosphonate monoester bonds to PNA derivatives N-(2–hydroxyethyl)glycine or N-(2–aminoethyl)glycine. The resulting chimeric oligomers form stable complexes with complementary single-stranded DNA or RNA molecules, are resistant to nucleases and possess good water solubility. PNA-pPNA chimeras are used for nucleic acid hybridisations.

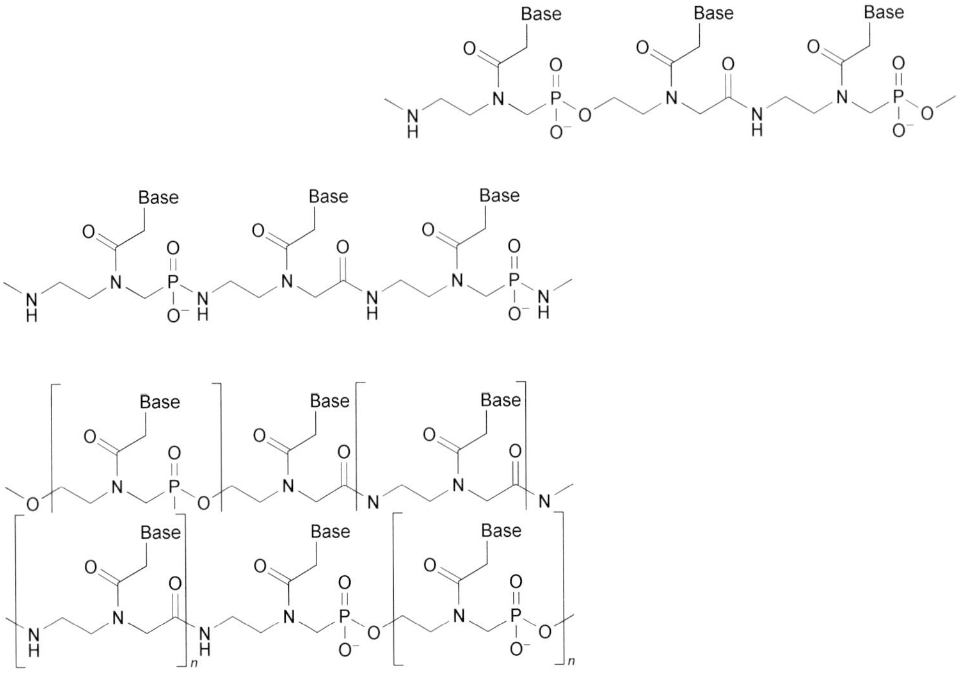

PNA-pPNA chimeras

Peptide topography: The three-dimensional arrangement of the side chains of the amino acids in a peptide.

Peptide vaccine: Any peptide that induces a strong immune response and therefore can be used for vaccination. Such peptides must carry an appropriate allele-specific T-cell epitope for the recipient species, and need to be attached to a carrier protein to enhance immunogenicity (longer peptides elicit a strong humoral response). For example, lipopeptide vaccines trigger humoral and cellular immune responses very effectively. These vaccines are heat-stable, non-toxic, fully biodegradable and are synthesized on the basis of minimized epitopes. They activate the antigen-presenting macrophages and B lymphocytes.

Peptidome: The complete set of (specifically biologically active) peptides in an organelle, a cell or a tissue at a given time. For example, all the peptides secreted by neuroendocrine cells or glands represent the neuropeptidome. The peptidome is very dynamic, i.e. changes during development, cell differentiation, generally with the stage of a cell, and as a consequence of many endogenous and environmental influences. See → peptidomics, → proteome.

Peptidomics (peptide-genomics): The whole repertoire of techniques to detect, analyze and characterize the → peptidome of an organelle or a cell, encircling peptide isolation, chromatographic or electrophoretic fractionation and separation, analysis by → MALDI mass spectrometry, sequencing, including determination of modifications (such as e.g. acetylation, glycosylation, methylation, phosphorylation), and storage and analysis of the resulting informations. See → functional genomics, → genomics, → proteomics, → recognomics.

Peptone: A partially hydrolyzed protein preparation.

Percent *identity* plot (PIP): A graphical depiction of a comparison of two (or more) related nucleotide or amino acid sequences from two (or more) different organisms, that allows to infer the extent of sequence identity. Computer programs such as PipMaker (http://bio.cse.psu.edu/) and VISTA (http://www-gsd.lbl.gov/vista/) assist to establish a PIP.

Percoll: An inert colloidal silica coated with *polyvinylpyrrolidone* (PVP) that is used for generating gradients which allow the separation of subcellular organelles (e.g. nuclei, mitochondria, plastids), viruses, and cells.

Perfect *match* (PM): The complete correspondence of two (or more) bases in two (or more) strands of a DNA molecule. Perfect matches are only possible by → Watson-Crick base pairing of A=T and G≡C pairs, respectively. Any other combination inevitably leads to a → mismatch.

Perfect palindrome: Any sequence in duplex DNA in which completely identical base sequences run in opposite directions (e.g. 5'GAATTC 3'). Such perfect palindromes frequently are recognition sites for → restriction endonucleases. Compare → palindrome.

Perfect repeat: Any stretch of → repeated sequences that consists of elements with identical sequence (e.g. 5'- CATCATCATCAT-3'). See → compound microsatellite, → imperfect repeat.

Perinatal genetics: A branch of → genetics, that focusses on the detection of chromosomal and DNA abnormalities in new-born human beings, using the whole repertoire of classical → cytogenetics and → molecular genetics from chromosome banding to → DNA chip technology.

Periodicity: The number of base pairs per turn of the DNA double helix.

Permanent cell line: Any cell line (→ cell strain) with an unlimited life time.

Permanganate oxidation of DNA: An outdated technique for the detection of methylated cytosines in a DNA molecule, that uses potassium permanganate at pH 4.3 to degrade 5-methylcytosine to barbituric acid derivatives, but does not attack cytosine itself. Since this treatment is not specific for 5-methylcytosine, but also degrades thymine, it was changed to include a combination of hydrazine degradation of C and T (but not 5-methylcytosine) and permanganate oxidation with little further improvement. The hydrazine and permanganate-modified nucleotides can be removed with piperidine and detected by sequencing techniques. See → combined bisulfilte restriction analysis,→ methylation assay, → methylation-sensitive amplification polymorphism, → methylation-sensitive single nucleotide primer extension, → methylation-specific polymerase chain reaction.

Permissive cell (permissive host): Any cell in which a particular virus may cause a production of progeny viruses (productive infection).

Permissive condition: A condition that allows the survival of a → conditional lethal mutant.

Permissive host: See → permissive cell.

Permissive temperature: The temperature at which a → temperature-sensitive mutant is able to grow.

Permissivity: The ability of cells to support the growth of phages (or plasmids).

Peroxidase-conjugated antibody (POD-conjugated antibody; immunoperoxidase): An → antibody to which a horseradish peroxidase (HRP) molecule is covalently attached. Such conjugates are used to detect a specific protein or nucleic acid sequence in e.g. biotinylation- and digoxygenin-based detection systems (see → biotinylation of nucleic acids and → digoxigenin labeling), where the antibody binds to its antigen (e.g. a biotin-avidin complex), and the complex is detected by the H_2O_2-dependent conversion of e.g. luminol (5-amino-2,3-dihydro-1,4-phthalazinedion) with concomitant emission of light. This reaction can be enhanced by the presence of an enhancer, see → enhanced chemiluminescence detection. Compare → immunophosphatase, see also → enzyme-conjugated antibody.

Peroxin: Any one of a series of *peroxi*somal proteins that are synthesized on cytoplasmic → ribosomes and imported into peroxisomes. All peroxins carry one or two targeting signals (*per*oxisomal *t*argeting *s*ignal, PTS) that allow their specific transport to and into peroxisomes. PTS1 is localized at the → carboxy terminus of the peroxins and is composed of the tripeptide serine-lysine-leucine (SKL), or variants of this motif. PTS2 is part of the → amino terminus and contains up to 30 amino acids (consensus sequence: [R/K][L/I/V] X_5 [H/Q][L/A]. In contrast to PTS1, some of the PTS2 signal sequences are processed after the import of the corresponding peroxins. Most peroxins are membrane-bound, some contain → zinc finger domains. See → peroxisome assembly gene.

Peroxisome *a*ssembly gene (PEX gene, PAS gene): Any one of a series of genes encoding proteins (socalled PEX proteins) for the biogenesis of socalled peroxisomes, organelles of eukaryotic cells that contain catalases, peroxidases, a ß-oxidation system, and enzymes of the glyoxylate cycle (plants), or glycolysis (glycosomes of the trypanosomes). PEX genes encode → peroxins.

Perpendicular *d*enaturing *g*radient *g*el *e*lectrophoresis (perpendicular DGGE): A method to determine the → melting behavior of a DNA duplex molecule in an → agarose gel containing a gradient of denaturants perpendicular to the direction of electrophoresis. The DNA is applied to the gel in a single large slot. In the gel region with low denaturant concentration the DNA fragments run far into the gel (i.e. do not melt), in the gel region with high denaturant concentration they do hardly migrate (i.e. melt extensively). In between these extreme positions intermediate mobilities of the DNA fragments may be observed. After → ethidium bromide staining the fragment pattern in the gel resembles a → C_0t curve, and therefore allows the calculation of the number of melting domains in a DNA fragment as well as the estimation of the → T_m for each individual fragment.

Perpendicular DGGE: See → perpendicular *d*enaturing *g*radient *g*el *e*lectrophoresis.

Persistence length (p): The number of → base pairs between two bends in → double-stranded DNA.

PERT: See → *phenol emulsion reassociation technique*.

Petite: A mutant strain of *Saccharomyces cerevisiae*, that suffered mutations in either one or more mitochondrial genes (called vegetative petits), or on nuclear genes (called segregational petites). Petites grow only slowly and as small colonies, a consequence of respiratory deficiency.

Petri dish (Petri plate): A disposable, round and flat plastic culture dish with a lid, that is used for the culture of bacteria or fungi on solid media.

Petrifilm™ plate: Any ready-to-use, water-thin substitute for a conventional agar plate for the culture of bacteria. The Petrifilm plate contains a dehydrated nutrient medium (as e. g. → LB medium), a gelling agent and indicators for → blue-white screening. The bacteria are simply added onto the medium, and the top film used to seal the plate, which can then be incubated. All routine plating procedures (e. g. library screening) can be done with the Petrifilm plates.

PEV: See → *position effect variegation*.

PEX gene: See → *peroxisome assembly gene*.

pEX vector: Any one of a series of → plasmid → expression vectors that is designed for the expression screening of → cDNA libraries in *E. coli*, and for the expression of β-galactosidase → fusion proteins. Each pEX vector contains a *cro – E. coli lacZ* gene fusion driven by the strong P_R promoter. A → polylinker at the 3' end of the *lacZ* gene allows the insertion of a foreign sequence in such a way that it is placed in all three → open reading frames alternatively. In one of these constructs the insert will thus be in frame with the vector, allowing its expression as a hybrid β-galactosidase protein. Downstream of the polylinker site → fd phage transcription terminators and a synthetic translation stop signal are inserted.

PFF: See → peptide fragmentation fingerprint.

PFGE: See → *pulsed-field gel electrophoresis*.

PFM: See → physical functional marker.

PFP: See → *protein fusion and purification technique*.

pfu: See → *plaque forming unit*.

***Pfu* DNA polymerase:** See → *Pyrococcus furiosus* DNA polymerase.

PGRS: See → *polymorphic GC-rich repetitive sequence*.

pGV 3850: A → cointegrate vector for the transfer of foreign genes (generally, DNA) into target plants via → *Agrobacterium*-mediated gene transfer. pGV 3850 is a derivative of the → *Agrobacterium tumefaciens* → Ti-plasmid, in which the T-region has been substituted for a modified → pBR 322. The latter is flanked by the two T-DNA borders in this construct. The pBR 322 portion allows the insertion of foreign DNA into pGV 3850 by homologous recombination with a conventionally constructed recombinant pBR 322 As pGV 3850 does not contain an → ampicillin resistance gene, *Agrobacterium* cells containing such a → cointegrate structure can be selected on ampicillin-containing medium. The foreign sequence can then be transferred to compatible plant cells via *Agrobacterium*-mediated gene transfer because it is flanked by T-DNA border regions. See also → coculture or → leaf disk transformation.

Phage: See → bacteriophage.

Phage bank: See → phage library.

Phage cloning vector (phage vector): A → cloning vector derived from a → bacteriophage. See for example → autocloning vector, → broad host range vector, → expression vector, → lambda phage-derived cloning vector, → P1 cloning vector, → SP6 vector.

Phage conversion (lysogenic conversion; prophage-mediated conversion): The acquisition of new properties by bacterial cells harboring a → prophage (for example the property of immunity against phage superinfection, see → phage exclusion). If the prophage is lost, the new characters disappear.

Phage cross: The exchange of genetic material between phages. Occurs during multiplication of → bacteriophages after their entry into the host cell. If a single bacterium is infected with several phages differing at one (or more) genetic loci, then recombinant progeny phages can be recovered upon → lysis of the host cell. These recombinants carry genes derived from two parental phages.

Phage display (slang: Ph.D., peptide display): A technique for the presentation of distinct peptides or proteins on bacterial surfaces, that uses → bacteriophages (e. g. → M13, fd, f1) as carriers for these display molecules and allows to identify peptides or proteins with desirable binding properties. Genes for the display peptides are integrated in the single-stranded DNA genome of the phage, and the corresponding peptides expressed as → fusion proteins with a viral coat protein. The fusion proteins are then exposed to the surrounding medium. For example, the M13 phage carries a single-stranded circular DNA genome of 6408 bp, that is packaged by various viral DNA-encoded proteins (e. g. g3p, g6p, g7p, g8p, g9p), of which g8p is the major coat protein (about 2700 copies per phage). The phage particle itself is a flexible, 900 nm long filament (diameter: 6 nm), and on its surface the coat proteins (especially g3p) are exposed. If the coat protein is fused to a foreign protein, the latter is also presented. Phage display is used for the establishment of libraries of peptide or protein (e. g. enzyme) variants, or oligopeptide inhibitors for various target molecules, for the isolation of enzyme variants with a better or modified binding affinity for their substrates and changed catalytic properties, or for the detection of enzyme variants with increased stability. See → display library, → panning, → random peptide display. Compare → dual-bait two-hybrid system, → interaction trap, → LexA two-hybrid system, → one-hybrid system, → reverse two-hybrid system, → ribosome display, → RNA-protein hybrid system, → split-hybrid system, → three-hybrid system, → two-hybrid system.
Figure see page 801.

Phage display library: See → display library.

Phage display peptide library: A DNA library, established in → phages, that contains the → insert fused to the gene of the coat protein gene of the phage (→ fused gene), and allows to detect the insert-encoded peptide on the surface of the phage. See → display library, → panning, → phage display, → phagemid display.

Phage Library

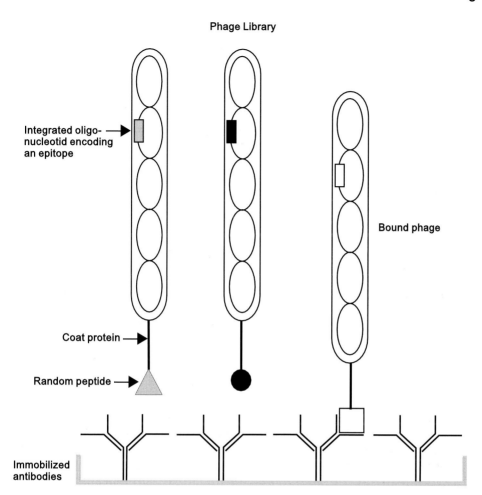

Integrated oligo-
nucleotid encoding
an epitope

Bound phage

Coat protein

Random peptide

Immobilized
antibodies

Phage display

Phage exclusion (phage immunity, prophage interference, superinfection immunity): The immunity of a host bacterium that contains a → prophage in its genome (→ lysogenic bacterium), against a secondary infection by the same or a related phage. The inserted prophage codes for the synthesis of → repressor proteins that bind to the → operator sequences of the superinfecting phage and interfere with its transcription. In the case of the → lambda phage the repressor protein is en-coded by the gene → *cI* and acts in concert with the products of the genes *rex A* and *rex B* (immunity loci, *imm* loci) to suppress the genes necessary for the lytic cycle of the superinfecting phage.

Phage fd: See → fd phage.

Phage f1: See → f1 phage.

Phage immunity: See → phage exclusion.

Phage induction: The stimulation of a → prophage to enter the productive, i.e. → lytic cycle, usually by exposure of lysogenic cells to UV light, X-rays or → mutagens (e.g. nitrogen mustard, hydrogen peroxide). Phage induction allows the initiation of transcription of phage genes, the excision of the prophage from the host chromosome, and the synthesis of phage DNA and capsid proteins.

Phage lambda (λ): See → lambda phage.

Phage library (phage bank): A collection of random DNA fragments, cloned into a phage → cloning vector (e.g. → M13 or → lambda phage-derived vector) and ideally encompassing the entire genome of a given species. See also → gene library.

Phage lifting: See → plaque hybridization.

Phagemid ("*phage*-plas*mid*"): A chimeric → plasmid vector (→ hybrid vector) that contains an → origin of ssDNA replication such as the f1 or M13 intergenic region (IG). Phagemids replicate as normal plasmids in *E. coli*. If the host cells are infected with a helper bacteriophage (→ helper virus, e.g. M13 KO7) that supplies the functions necessary for ssDNA replication and packaging, phage-like particles are synthesized and released through the bacterial cell walls in a non-lytic process. The ssDNA can then easily be recovered from the culture medium. See for example → Bluescript[R], → expression phagemid, → lambda ZAP, → multi-functional phagemid.

Phagemid display: A technique for the presentation of distinct peptides or proteins on bacterial surfaces, that uses → phagemids as carriers for the display molecules. In conventional → phage display, the size of displayed peptides is limited, because the fusion product of target peptide and viral coat protein should not exceed a certain threshold, otherwise the function of the coat protein is inhibited. This size limitation is relaxed in phagemid display.

Phage M13: See → M13.

Phage Mu: See → Mu phage.

Phage φ X 174: See → Φ X 174.

Phage Q-beta: See → Q-beta.

Phage typing: The classification of bacteria on the basis of their susceptibility towards infection by various → bacteriophages.

Phage vector: See → phage cloning vector.

Pharmaceutically *tractable* genome (PTC): A subset of genes from a genome, that represents (preferably) all drug targets (as e.g. genes encoding cell surface proteins such as receptors, circulating proteins, or proteins modulated by small molecules as e.g. drugs). The human PTC probably consists of 6,000 – 8,000 genes.

Pharmacogenetics: The detection, isolation and characterization of → genes and the encoded proteins as potential targets for pharmaceutically active compounds. Moreover, pharmacogenetics aims at establishing individual gene profiles, i.e. to detect sequence polymorphisms at strategic sites of a particular gene between e.g. patients. For example, → *single nucleotide polymorphisms* (SNPs) – the human genome probably contains 3 million SNPs, of which the majority is already mapped – in specific genes may determine the capacity of the encoded proteins such that e.g. a wild-type protein transports a certain drug, the mutated protein does not. The patient with the SNP mutation in the transporter gene does not respond to the drug. Another example is the *multi-drug resistance* (MDR)–1 gene, that encodes the socalled P glycoprotein (a membrane-bound protein, eliminating compounds recognized as xenobiotics). This gene harbors at least 35 polymorphisms, of which the socalled TT variant occurs in about 25% of humans. This mutation leads to a highly reduced production of P glycoprotein in the intestines, so that the uptake of drugs from the intestinal tract to blood proceeds uncontrolled. Therefore the drugs are present in very high concentrations in the blood, increasing the incidence of side effects. Patients with the TT variant can be advised to reduce the drug dosis. Still another important example capitalizes on the genes encoding *cytochrome P450* enzymes (CYP, in this case CYP3A enzymes), which metabolise about 50% of all common therapeutics as well as natural compounds such as estrogene, testosterone, and bile acids. Specific SNPs in the CYP3A genes reduce or abolish the individual's capacity to metabolize a drug (i.e. they determine, how the individual is susceptible to the drug and its side effects). See → functional genomics, → genomics, → medical genomics, → pharmacogenomics, → proteomics, → recognomics. Compare → comparative genetics, → cytogenetics, → developmental genetics, → forward genetics, → interphase genetics, → molecular genetics, → reverse genetics.

Pharmacogenomics: The whole repertoire of techniques to explore the effects of drugs on the structure of → genomes and → genes and the expression of these genes, as well as the implication(s) of → mutations in specific genes and, as a consequence, amino acid replacements in the encoded protein(s) for the effectiveness of pharmaca. For example, the genomes (or particular genes) of socalled responders (individuals responding positively to a specific drug) may be different from the genomes or genes of socalled non-responders (individuals not responding to the drug). By profiling the potential users of such a pharmacon, a prediction can be made about the effectivity of a specific drug application ("right drug for the right patient"). Pharmacogenomics aims at the identification of previously unknown target molecules for the development of fitting drugs ("drug targets"; e. g. the design of cyclooxigenase inhibitors for the treatment of arthritis, based on gene expression analysis), at the recognition of genetic polymorphisms in genes encoding drug-metabolizing enzymes (e. g. the phase I drug metabolizing P-450 superfamily of monooxygenases), or the definition of all genes contributing to a specific disease phenotype coupled to a better, more effective drug application ("personalized drug therapy"; "individualized medicine"). Compare → behavioral genomics, → biological genomics, → cardio-genomics, → chemical genomics, → clinical genomics, → comparative genomics, → deductive genomics, → environmental genomics, → epigenomics, → functional genomics, → horizontal genomics, → integrative genomics, → lipo-proteomics, → medical genomics, → neurogenomics, → neuro-proteomics, → nutritional genomics, → omics, → pathogenomics, → pharmacogenomics, → phylogenomics, → physical genomics, → population genomics, → proteomics, → recognomics, → structural genomics, → transcriptomics, → transposomics.

Pharmacoperone (*pharmaco*logical cha*perone*, chemical chaperone): Any synthetic → chaperone, that binds to proteins and corrects their incorrect folding. For example, the gene encoding the receptor protein for *gonadotropin-releasing hormone* (GnRH) is frequently mutated, and consequently the encoded receptor can be non-functional. This is the cause for e.g. hypogona-dotropic hypogonadism (phenotype: faulty sexual development). Thirteen of the 14 most frequent mutations change a single amino acid in the receptor, each of which leads to an incorrect folding of the receptor protein. If the faulty proteins are exposed to a synthetic GnRH antagonist, the pharmacoperone, then the receptor can be folded normally and regains function. Even unspecific chaperones such as 4–phenyl*butyric a*cid (PBA) can be employed for correct folding of target proteins *in vitro*. Pharmacoperones are potential agents for the causative treatment of many diseases caused by defective folding of proteins, as e.g. → prion diseases, Alzheimer disease, and Chorea Huntington.

Pharmacophore map: A list of descriptors defining the physical, chemical, and structural properties of pharmaceutically potent compounds.

Pharmacoproteomics: The whole repertoire of techniques for the identification and characterization of peptides and proteins in a specific cell, tissue, organ or individual, that are expressed as a consequence of drug administration. Such → protein expression data can be used to predict drug toxicity and efficacy, and to understand the mechanism of action of a drug.

Phase:
a) The arrangement of → codons downstream of the → start codon AUG (or GUG) in → messenger RNA. These codons are *in phase*, if they can be read in → triplets starting from the AUG (i. e. function as codons for amino acids during → translation). They are *out of phase*, if the → reading frame has shifted by one or two nucleotides (i. e. the new arrangement does no longer code for the same amino acids as before the change). See → reading frame shift mutation.
b) The distribution of specific → alleles on → homologous chromosomes. For example, alleles A and a occupy one, alleles B and b another locus. An individual with the phenotype AaBb may have two possible genotypes, the cis and trans phases (in analogy to chemical isomeres). The phase can be determined in a particular individual by a genetic cross with an appropriate (and informative) partner.
Figure see page 805.

Phase *lock* gel (PLG): A gel block that serves to trap the organic phase and interphase material in phenol or phenol/chloroform extraction of nucleic acids. The PLG forms a seal between the organic and aqueous phases after centrifugation and either allows to decant the aqueous phase or to pipet it without contamination.

Phase lock procedure: A technique to separate the organic phase and interphase (containing denatured proteins) from the aqueous phase (containing nucleic acids) during phenol-chloroform extractions. A chemically inert and hydrophobic gel ("phase lock gel") is included that forms a solid barrier between the aqueous and organic phases during centrifugation of the phenol-chloroform mixture, thereby trapping the organic and interphase, leading to complete separation of the phases and easy recovery of nucleic acids from the aqueous epiphase.

Phase shift: See → reading frame shift.

Phase shift mutation: See → reading frame shift mutation.

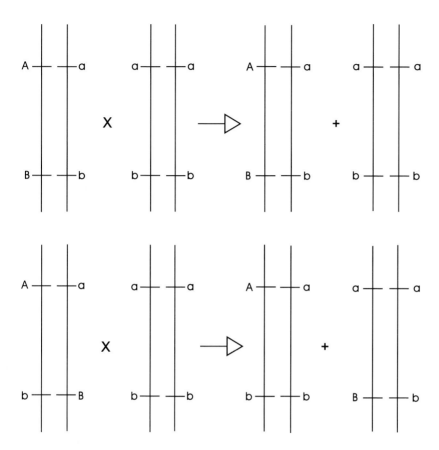

Phase variation: The reversible loss or gain of → intragenic → microsatellite repeats in certain bacteria, leading to a loss or gain of specific function(s). For example, the socalled Opa genes of *Neisseria gonorrhoe* (encoding 12 outer-membrane proteins, that make the bacterial colonies appear *opa*que and allow the bacteria to adhere to and invade epithelial cells, e. g. respiratory tract epithelia) contain a microsatellite composed of multiple copies of 5'-CTCTT-3'. As a con sequence of → slipped strand mispairing during → replication, one such repeat can be lost, which leads to a shorter protein. Cells with such a truncated protein can no longer enter epithelial cells. This deficiency turns into a selective advantage, if the bacterium is unable to invade e. g. pha-gocytotic cells, which would destroy them. See → contingency gene.

Phasing: See → nucleosome phasing.

Phasmid: A → hybrid vector consisting of a → plasmid with a functional → origin of replication and → lambda phage sequences (in particular, the λ origin of replication and one or more → at-tachment site(s). Foreign DNA may be conventionally cloned into the plasmid vector. The recombinant plasmid can then insert into a phage genome ("lifting"), exploiting the function of the λ *att* sites. Phasmids may be propagated in appropriate *E. coli* strains either as a plasmid (non-lytic route), or as a phage (lytic route). Reversal of the lifting process releases the plasmid vector.

***Phenol emulsion reassociation technique* (PERT):** A variant of the genomic subtraction technique that employs phenol to increase the rate of → hybridization. PERT is a form of competitive hybridization between two related, but slightly differing genomes (e.g. genomes of male and female plants) that preserves only the unique sequences of one genome (e.g. the female one) in a clonable form. See → normalization.

Phenol extraction: A procedure for the denaturation and removal of proteins from solutions containing nucleic acids and proteins, using buffer-saturated phenol.

Phenome: See → phenotype.

Phenomic fingerprint ("molecular phenotype"): A vague term for the specific protein expression pattern of a cell, or an organism (e.g. a bacterial cell) at a given time. Such fingerprints can be established by e.g. → protein chips.

Phenomics: The whole repertoire of techniques to decipher all molecular processes leading to a → phenotype (phenome). Phenomics encompasses → transcriptomics, → proteomics, and → metabolomics.

Phenotype (phenome): The observable structural and functional properties of an organism, which results both from its → genotype and the environment.

Phenotype array (*phenotype microarray*, PM): A solid support, onto which single cells or cell colonies are arrayed, and used to detect their interactions with small molecular weight compounds (e.g. metabolites, drugs) in solution by monitoring cell or colony growth (i.e. the phenotype). See → microarray.

Phenotype microarray: See → phenotype array.

Phenotype mixing: The packaging of the genome of one virus into the protein → capsid of a second, unrelated virus.

Phenylmethylsulfonyl fluoride: See → PMSF.

φ X 174: A small icosahedral → bacteriophage infecting *E. coli* (→ coliphage) with a circular single-stranded DNA genome of 5.386 kb. Its replication proceeds through a double-stranded circular → replicative form. Some of its genes have been used for the construction of → cloning vectors.
Figure see page 807.

Phleomycin: One of a series of glycopeptide → antibiotics of *Streptomyces verticillus* that binds and intercalates DNA and destroys the integrity of the double helix by its metal-chelating domain. Phleomycin is an effective selective drug for mammalian cells, but can also be used for prokaryotes, fungi, plants, and generally animal cells.

***pho* A promoter:** See → alkaline *pho*sphatase promoter.

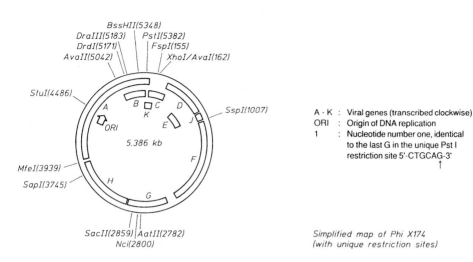

$1 \longrightarrow$

BssHII(5348)
DraIII(5183)   PstI(5382)
DrdI(5171)   FspI(155)
AvaII(5042)   XhoI/AvaI(162)

StuI(4486)

SspI(1007)

5.386 kb

MfeI(3939)

SapI(3745)

SacII(2859) AatII(2782)
NciI(2800)

A - K : Viral genes (transcribed clockwise)
ORI   : Origin of DNA replication
1   : Nucleotide number one, identical
        to the last G in the unique Pst I
        restriction site 5'-CTGCAG-3'

Simplified map of Phi X174
(with unique restriction sites)

pho-box (*pho*sphate uptake box): A regulatory sequence element of the → promoter of bacterial genes involved in phosphate uptake and metabolism. The pho-box is the address site for PhoB, a positive regulator protein inducing or enhancing the transcription of these genes.

Phosphatase: An enzyme catalyzing the removal of phosphate residues from substrates (including polymers such as nucleic acids). See → alkaline phosphatase.

***Phosphinotricin acetyltransferase* gene (PAT gene):** A gene (*bar*) from *Streptomyces hygroscopicus* encoding the enzyme phosphinotricin acetyltransferase that catalyzes the inactivation of the herbicide phosphinotricin (PPT). PPT is an analogue of glutamic acid and inhibits plant glutamine synthase. The PAT gene is used as a → selectable marker gene in plant transformation experiments.

Phosphodiester: An imprecise term for a molecule containing the group depicted below, where R^1 and R^2 are carbon-containing groups. For example, in RNA or DNA the 5' carbon of a pentose (ribose or deoxyribose) and the 3' carbon of an adjacent sugar moiety are linked by a phosphodiester type bond. See → phosphodiester bond.

$$R^1-O-\overset{\overset{O}{\|}}{\underset{\underset{OH}{|}}{P}}-O-R^2$$

Phosphodiesterase: An enzyme that catalyzes the hydrolysis of phosphodiesters into a phosphomonoester and a free hydroxyl group. See for example → phosphodiesterase I.

Phosphodiesterase I (5' exonuclease, snake venom phosphodiesterase; EC 3.1.4.1): An enzyme from *Crotalus adamanteus* that catalyzes the hydrolysis of both DNA and RNA by processive exonucleolytic attack of the free 3'hydroxy terminus to produce 5'-mononucleoside phosphates.

Phosphodiester bond: The covalent linkage between the phosphate group of the → 5' position of one pentose with the hydroxyl group of the → 3' position of the next pentose in a → nucleotide polymer (e.g. DNA, RNA).

Phosphodiester method: See → chemical DNA synthesis.

Phosphono *peptide nucleic acid* (pPNA): A negatively charged → peptide nucleic acid, in which the monomer units are attached to an *N-(2*–aminoethyl)phosphono glycine backbone and connected by phosphonester bonds. These → DNA mimics recognize complementary target DNA or RNA by → Watson-Crick base pairing. If composed of homo-T stretches (e.g. containing a chain of 14 thymine pPNA momomers), pPNA binds strongly to complementary poly(A)$^+$-strands and can be used to isolate polyadenylated → messenger RNAs with e.g. short poly(A)-tails or complex secondary structures (e.g. → stem-loops) around the poly(A)-tail. pPNA oligomers do not possess a → polarity, and therefore bind in both parallel and antiparallel orientation to RNA. They also bind double-stranded RNA by invading the RNA:RNA duplex and displacing one strand, forming a stable → displacement loop (D-loop). Also, pPNAs are excellently water-soluble, but not enzymatically degraded by nucleases and therefore stable *in vivo*. See → peptide nucleic acid-phosphono peptide nucleic acid chimera, → trans–4–hydroxy-L-proline PNA, → oligo(T)-PNA.

Phosphonyl-*methoxypropyl-adenine* (PMPA): An → adenine derivative, that inhibits the activity of → reverse transcriptase of retroviruses and is therefore employed in antiretroviral treatment strategies.

Phosphoproteome: A → sub-proteome, consisting of phosphorylated peptides and proteins of a cell.

Phosphoproteomics: The whole repertoire of techniques to study phosphorylated peptides and proteins, the corresponding phosphokinases and phosphatases and the consequence of one or more phosphorylations of amino acid residues of a protein onto its function(s). See → peptidomics, → phosphoproteome, → proteomics.

Phosphorimaging: A technique for the sensitive detection of radioisotopes, that employs a polyester plate coated with fine crystals of photostimulatable phosphor (BaFBr:Eu^{2+}) as an imaging plate. This plate accumulates and stores the energy emitted by the respective isotope. The sample (e.g. a nylon membrane) is simply covered by e.g. Saran wrap and exposed on the imaging plate (IP) inside a cassette. After exposure, the IP is scanned with a laser beam and emits → luminescence (proportional to the recorded radiation intensity), which is collected into a photomultiplier tube and converted to electrical signals. The IP is reusable, after the image data are erased (e.g. by exposure to light).

Phosphoroamidite technique: See → chemical DNA synthesis.

Phosphoro*dithioate* oligodeoxyribonucleotide (PS$_2$): Any deoxyribonucleotide in which both oxygen atoms of the nucleotides are substituted by sulfur atoms. PS$_2$s are chemically very stable, achiral, resistant towards exonucleases, moderately resistant towards endonucleases, and hybridize with the normal oligodeoxyribonucleotides, though with a decreased stability of the duplex. Also, the → antisense properties of PS$_2$s are inferior to the normal oligodeoxyribonucleotides (as measured by → in vitro translation inhibition of specific → messenger RNAs), and the capacity to bind proteins is reduced. See → phosphorothioate oligonucleotide.

Phosphorolysis: The cleavage of a covalent bond by orthophosphate.

Phosphorothioate group: A modified phosphate group, in which one of the oxygen atoms is replaced by a sulfur atom.

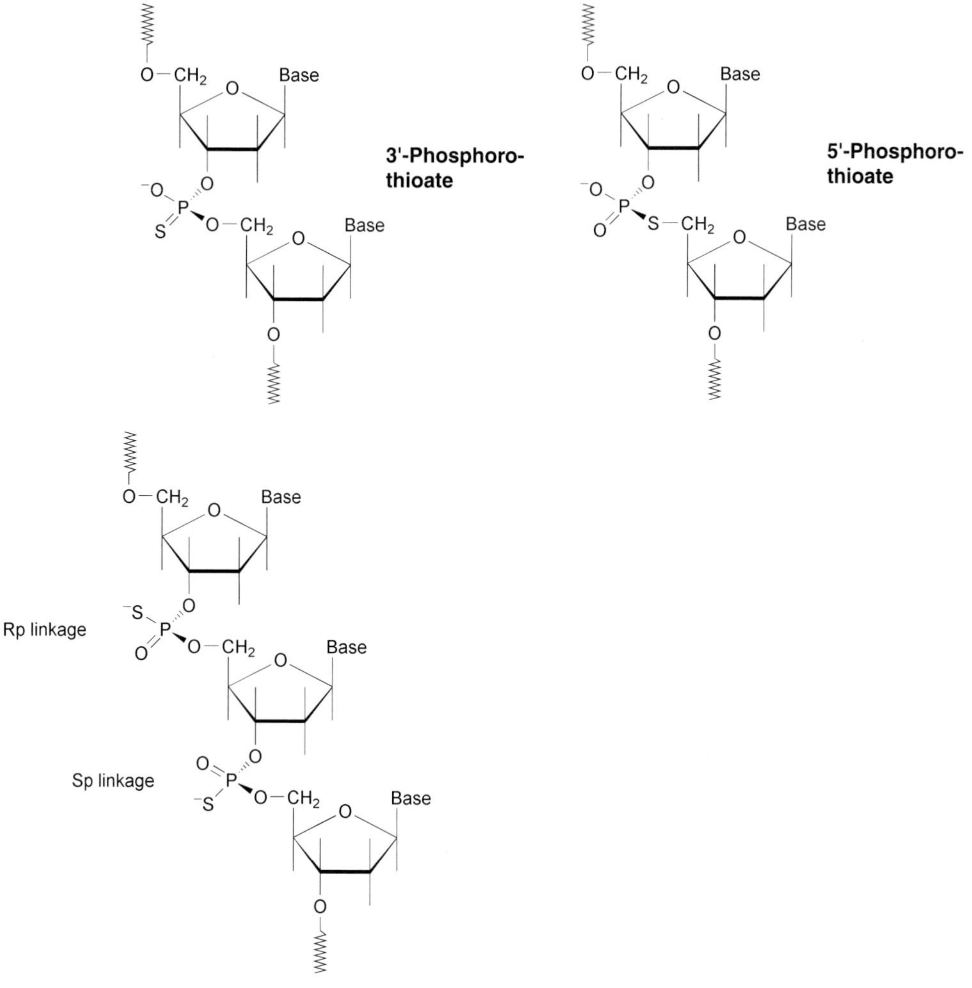

3'-Phosphoro-thioate

5'-Phosphoro-thioate

Rp linkage

Sp linkage

Phosphorothioate group

Phosphorothioate interference: The enzymatic replacement of a nonbridging oxygen atom at a 5'-phosphate group of an oligonucleotide (generally, DNA) molecule with sulfur (see→ phosphorothioate group) and the use of this modification to detect the function of the substituted oxygen or, more precisely, of the specific phosphor atom to which it is covalently linked. For example, binding of a metal ion to a specific phoshate group is changed, if the latter is exchanged for a sulfur atom. This change can be measured, and the interactive phosphate be defined. For interference studies with oligonucleotides, phosphorothioates are generally incorporated by transcription.

Phosphorothioate oligonucleotide ("S-oligo"; phosphorothioate): Any → oligodeoxynucleotide in which some or all of the internucleotide phosphate groups are replaced by → phosphorothioate groups. Such modified oligonucleotides are resistant towards attack of most exo- and endonucleases, and could therefore be useful as intracellular → antisense oligonucleotides. See → phosphorodithioate oligodeoxy-ribonucleotide.

Phosphorothioate sequencing: A method for the → sequencing of DNA that uses Sanger techniques (→ Sanger sequencing) in combination with chemical cleavage reactions (→ chemical sequencing). In short, a synthetic oligodeoxynucleotide (→ primer) is annealed to the single-stranded target DNA. The reaction mixture is then aliquoted into four separate tubes that contain all four deoxynucleoside triphosphates, and additionally a → nucleoside-α-thiotriphosphate (dNTPαS) that also serves as substrate for the polymerization reaction and is incorporated at random. Then 2-iodoethanol or 2,3-epoxy-1-propanol is used to form a phosphorothioate triester with the incorporated dNTPαS. These esters are more easily hydrolyzed than → phosphodiesters. A careful hydrolysis can therefore lead to DNA fragments that can be used directly in the Sanger sequencing procedure.

Phosphorylation: A frequent → post-translational modification of proteins, mediated by specific phosphotransferases.

Phosphorylation *site-specific antibody* (PSSA): An → antibody raised against specific phosphorylated amino acid residues, that is used for the detection and quantitation of the phosphorylation status of these amino acids in target peptides or proteins.

Phosphotriester technique: See → chemical DNA synthesis.

Photoactivated cross-linking: A technique to locate the sites of effective contacts between a nucleic acid sequence (e.g. a → promoter) and its cognate protein (e.g. one or more → transcription factors) by UV irradiation which leads to a complex formation between both partners.

Photoaptamer: Any synthetic single-stranded → oligonucleotide (→ aptamer), into which → BrdU is incorporated instead of thymidine. This BrdU can be covalently cross-linked to a target protein by UV light, if it fits into the three-dimensional structure of its target region on the protein. This extremely specific interaction is exploited with the design of → photoaptamer arrays. Photoaptamers are selected *in vitro* by combinatorial chemistry (see → systematic evolution of ligands by exponential enrichment).

Photoaptamer array: Any glass slide, onto which thousands of → photoaptamers are spotted in an odered array, that allows to detect many proteins of a protein mixture simultaneously on the basis of their specific interactions with the immobilized aptamers. In short, the protein mixture to be analyzed is first incubated with the photoaptamer array, specific interactions take place between the photoaptamers and some cognate proteins, the non-bound proteins are washed away, and the bound proteins cross-linked to their target aptamers by UV irradiation. Then non-bound proteins are removed by washing. After detection of the bound proteins socalled protein profiles can be established.

Photobiotin: A → biotin molecule (vitamin H) attached to a photo-activable azido group via a spacer arm and used for the → non-radioactive labeling of single-stranded RNA and DNA, and double-stranded DNA. The labeling reaction involves exposure of the compound to strong visible light. This converts the azido group into a highly reactive nitrene that forms stable complexes with nucleic acids (single- or double-stranded DNA and RNA). Compare → biotinylation of nucleic acids; similarly used is → photodigoxigenin.

Reaction of photobiotin with DNA (or RNA)

Photobleaching: The irreversible light-catalyzed degradation of any → fluorochrome. Photobleaching determines the half-life time and thereby the utility of a particular fluorochrome for e. g. → fluorescent *in situ* hybridization.

Photodigoxigenin: A → digoxigenin molecule linked to an azido phenyl residue via a hydrophilic spacer, and used to introduce digoxigenin into nucleic acids or proteins by simply exposing the reactants to UV irradiation of 260-300 nm wavelength. Incorporation of digoxigenin by photoactivation is less efficient than enzymatic labeling. See → digoxigenin-labeling; similarly used is → photobiotin.

Photodynamic protein: Any protein, that is capable of transforming light energy into either a change of color ("photochromic protein"), an electromotive force ("photovoltaic protein"), or a change in absorbance ("nonlinear optical protein"), or to use a photon to energize a process.

Photo-footprinting:
a) Psoralene footprinting. A method for the detection of specific contacts between one or several proteins and a DNA duplex molecule, using → psoralene and UV light. In the presence of UV light, psoralene reacts with DNA through the formation of a photo adduct and pyrimidine monoadducts. These complexes influence or prevent the binding of → DNA-binding proteins.
b) Photo-footprinting technique. A method to detect contacts between specific sequences of DNA and regulatory proteins in vivo. After UV irradiation of intact cells, DNA is isolated before cellular → repair of the DNA damage begins. After purification it is subjected to a series of chemical reactions that break its sugar-phosphate → backbone only at the sites of UV damage. The DNA is then denatured, and labeled using direct or indirect methods. After electrophoresis on a polyacrylamide → sequencing gel the resulting fragment pattern is visualized by → autoradiography. Because protein-DNA contacts can inhibit or enhance UV photoproduct formation, differences in the strand-breakage patterns of protein-free and protein-associated DNA can be used to detect protein-DNA contacts at the basepair level.

Photohydrate: Any → pyrimidine base to which a hydroxyl group has been added onto C5 or C6 as a result of ultraviolet light radiation. **Photohydrate:** Any DNA photoproduct, in which hydoxyl groups are attached to carbon 5 or 6 of the pyrimidine bases.

Photolithography: A technique for the light-dependent engraving of a specific pattern on a solid support, used in printing processes. The solid support ("plate") is coated with a light-sensitive emulsion and overlaid by a photographic film. Then the coated plate is illuminated, and the image of the film is reproduced on the plate. Photolithography is employed in → DNA chip technology, where modifications of the usual phosphoramidite reagents are used (i. e. the *di*methoxy*t*rityl [DMT] group, that protects the 5'hydroxyl, is replaced by a photolabile protective group). The synthesis of the oligonucleotides on the chip proceeds by photolithographically deprotecting all the areas that will receive a common nucleoside, and coupling this nucleoside by exposing the entire chip to the appropriate phosphoramidite. This is achieved by socalled masks made from chromium/glass, that contain holes at positions, where deprotection is desired. A more advanced procedure exploits a socalled virtual mask. Up to 480,000 (or even more) digitally controlled micro-mirrors allow the illumination of only defined spots on a DNA chip depending on their precise angular position (→ "mask-less photo-lithography"). After the oxidation and washing steps the procedure has to be repeated for the next nucleoside.
Figure see page 813.

Photolyase: A 54 kDa repair enzyme from *E. coli*, catalyzing the removal of → pyrimidine dimers.

Photon (Greek: phos for "light"): The quantum of light. Its energy is proportional to its frequency: $E = h \cdot \upsilon$ (where E is energy, h is Planck's constant [$6.62 \times 10{-}27$ erg-second], and υ is frequency).

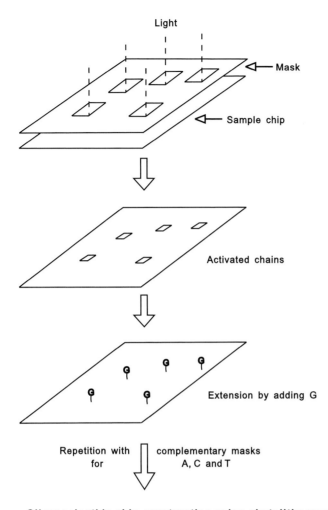

Oligonucleotide chip construction using photolithography

Photolithography

Photoprotective group: Any chemical compound covalently bound to the 5'-deoxyribose of a nucleic acid base, that prevents any reaction of this deoxyribose with the 3'-OH of another deoxyribose. This protective group can, however, be removed by e.g. UV light. For example, for the production of a special kind of → microarray, the oligonucleotides are synthesized directly on the glass support of the array. Each base, that is coupled onto the glass surface, carries a protective group, frequently *methyln*itro*p*iperonyl*oxyc*arbonyl (MeNPOC) on the 5'-OH position. MeNPOC can be removed by half a minute UV irradiation, and a second base can be coupled, that in turn carries a MeNPOC group at its 5'-OH position. See → photolithography.
Figure see page 814.

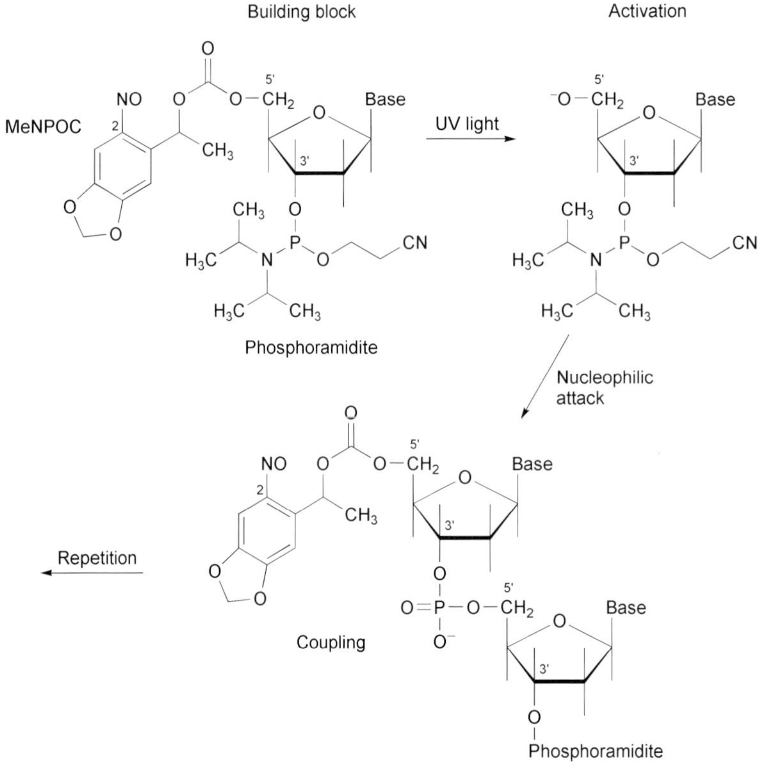

Building block   Activation

NO

MeNPOC

2

CH₃

O

O

CH₃   O

H₃C   N   P   O   CN

H₃C   CH₃

Phosphoramidite

5'

O—CH₂   Base

O

3'

UV light

5'

O—CH₂   Base

O

3'

CH₃   O

H₃C   N   P   O   CN

H₃C   CH₃

Nucleophilic attack

NO

2

CH₃

O

O

5'

O—CH₂   Base

O

3'

O

O=P—O—CH₂   Base

O⁻

3'

O

Phosphoramidite

Repetition

Coupling

Photoprotective group

Photo-reactivation: The breakage of carbon-carbon bonds in the cyclobutane ring of → thymidine dimers generated in DNA by ultraviolet radiation, to restore the normal base sequence. This process is catalyzed by → DNA photolyases, that bind to pyrimidine dimers in the dark, but utilize light energy (365 – 405 nm, 435 – 445 nm) to break the cross-links (photo-reactivation). *Figure see page 815.*

Photoreactive crosslinker: Any → crosslinker, whose chemical reactivity is induced by UV illumination. Photoreactive crosslinkers are used for crosslinking molecules at defined sites (e.g. specific cells or organs). For example, 4,4'- *di*azido*di*phenyl ethane (DADPethane) or 4,4'-*di*azido*di*phenyl ether (DADPether), that react non-specifically with two different biomolecules, sulfo*s*uccinimidyl–6(4'- *azido*–2'-*nitro*phenyl*a*mino)hexanoate (sulfo-SANPAH), sulfo*s*uccinimidyl–4(p-*azido*phenyl)*b*utyrate (sulfo-SAPB), N-hydroxy*s*uccinimidyl–4–*azido* *b*enzoate (HSAB), N-hydroxy*s*ulfosuccinimidyl–4–*azido* *b*enzoate (sulfo-HSAB) or N-hydroxysuccinimidyl–4–*azido* *s*alicylic *a*cid (NHS-ASA), that react with amines, are such photoreactive crosslinkers. See → heterobifunctional crosslinker, → homobifunctional crosslinker.

PHRAP: A software program, that allows to assemble → raw sequence data into sequence → contigs and to assign a specific quality score to each position of the DNA sequence, based on → PHRED scores of the raw sequence reads.

PHRED: A software program, that allows to analyze → raw sequence data, to generate a base call and a linked quality score for each position in the sequence. See → PHRAD.

Photo-reactivation

```
5'  A  C  T  G  T  T  C  A  C  G  3'
       ||  |||  ||  |||  ||  ||  |||  ||  |||  |||
    T  G  A  C  A  A  G  T  G  C
3'                                           5'

              UV   light

5'  A  C  T  G  T==T  C  A  C  G  3'
       ||  |||  ||  |||      |||  ||  ||  |||
    T  G  A  C  A  A  G  T  G  C
3'                                           5'

                        DNA photolyase

5'  A  C  T  G  T==T  C  A  C  G  3'
       ||  |||  ||  |||      |||  ||  |||  |||
    T  G  A  C  A——A  G  T  G  C
3'                                           5'

       light absorption   break of cross-links

5'  A  C  T  G  T  T  C  A  C  G  3'
       ||  |||  ||  |||      |||  ||  |||  |||
    T  G  A  C  A  A  G  T  G  C
3'                                           5'

          restoration of   wild-type DNA

5'  A  C  T  G  T  T  C  A  C  G  3'
       ||  |||  ||  |||  ||  ||  |||  ||  |||  |||
    T  G  A  C  A  A  G  T  G  C
3'                                           5'
```

pHyg: A mammalian → expression vector containing the *E. coli* gene for → hygromycin B phosphotransferase that can be used as a dominant → selectable marker in transfection experiments.

Phylogenetic profiling: A computational screen for proteins, that always occur together in many, if not all organisms. Phylogenetic profiling aims at inferring functional linkage between proteins from their simultaneous presence in a multitude of cells, tissues, organs, or organisms.

Phylogenetic tree: A graphical representation of the genealogical or evolutionary relationship(s) among individuals of a group of molecules or organisms.

Phylogenomics: A branch of → genomics that exploits existing sequence information from various organisms ("evolutionary information") in the databases to assign a specific function to a particular sequence. Functional predictions are improved by concentrating on questions as e. g. *how* genes became similar in sequence during evolution rath than focusing on sequence similarity *itself.* See → behavioral genomics, → biological genomics, → cardio-genomics, → chemical genomics, → clinical genomics, → comparative genomics, → deductive genomics, → environmental genomics, → epigenomics, → functional genomics, → horizontal genomics, → integrative genomics, → lipo-proteomics, → medical genomics, → neurogenomics, → neuro-proteomics, → nutritional genomics, → omics, → pathogenomics, → pharmacogenomics, → physical genomics, → population genomics, → proteomics, → recognomics, → structural genomics, → transcriptomics, → transposomics.

Phylome: Another term of the → ome era, describing the complete set of phylogenetic trees for the genes of a given genome.

Phyloproteomics: A branch of → proteomics, that aims at deciphering phylogenetic relationships between organisms on the basis of peptide and protein sequence and structure.

Physical containment: A package of physical-technical security measures to prevent the escape of living organisms containing → recombinant or otherwise dangerous (e.g. pathogenic) DNA from a laboratory or an industrial production plant. Generally, four levels of biosafety (BL 1-4) of various stringencies are characteristic for the guidelines of most countries:
BL 1: The lowest level does not require a separate laboratory, nor any specific containment equipment or specially trained personnel (that should be familiar with microbiological techniques).
BL 2: This level demands a limited access to the laboratory, biological safety benches and autoclaves in addition to the requirements of BL 1.
BL 3: This more stringent level requires additionally, that the laboratory is only accessible to authorized and specially trained personnel (i.e. personnel familiar with handling of pathogenic or potentially lethal agents) who wears protective clothing. Protected laboratory bench surfaces, biological safety benches, airlocks, and negative pressure within the BL 3 area are obligatory.
BL 4: The most stringent level requires additionally a separate, window-less building, air- and liquid-decontamination, airtight doors and positive pressure protective clothing. See also → biological containment, → containment.

Physical functional marker (PFM): Any → molecular marker generated with → genomic DNA from organism A by the amplification of a gene with a gene-specific primer from organism B (using a conventional → polymerase chain reaction), that can be mapped on the → physical map of organism B.

Physical genomics: The whole repertoire of techniques for the analysis of an organism at the → genome level, encompassing large insert libraries (→ bacterial artificial chromosome libraries, → yeast artificial chromosome libraries), genome sequencing, the establishment of expressed sequence tag databases, the (preferably) complete inventory of the → transcriptome, → proteome and → metabolome, and the relevant → bioinformatics tools. Compare → biological genomics.

Physical map: The linear arrangement of genes or other markers on a chromosome as determined by techniques other than genetic recombination (e.g. → heteroduplex analysis, → DNA sequencing). Usually, map distances are expressed in numbers of nucleotide pairs between identifiable genomic sites (e.g. → contigs, → sequence tagged sites, or → restriction sites). See → contig mapping, → macro-restriction map, → ordered clone map, → restriction mapping. Compare → map, → mapping.

Physiome: An additional term of the → omics era for the description of the complete physiological condition of a cell, a tissue, an organ, or an organism. See → genome, → metabolome, → physiomics, → proteome, → transcriptome.

Physiomics: The whole repertoire of techniques for the quantitative description of the → physiome of a cell, a tissue, an organ, or a complete organism.

Phytochelatin (PC): A member of a class of small, cysteine-rich peptides with high heavy metal ion-binding capacity, which is mediated by thiolate coordination. These plant peptides function as traps for cadmium, copper, lead, mercury and zinc. The synthesis of phytochelatins proceeds without → translation and is catalyzed by phytochelatin synthase.

Π: See → nucleotide diversity per site.

P$_i$: Symbol for *i*norganic *p*hosphate.

pI: Abbreviation for → *i*soelectric *p*oint.

PIC: See → pre-initiation complex.

***Pichia* expression system (*Pichia pastoris* expression system):** An *in vivo* system for the high-level expression of heterologous recombinant proteins, based on the methylotrophic yeast *Pichia pastoris* (or *Pichia methanolica*). This yeast can metabolize methanol as sole carbon source, if the preferred substrate glucose is absent. The first step in methanol utilization is the alcohol oxidase-driven oxidation of methanol to formaldehyde. Expression of this enzyme, which cannot be detected in the absence of methanol and which is encoded by the AOX1 gene, is therefore tightly regulated and induced by methanol to very high levels (e. g. >30 % of the total suluble cellular proteins represent alcohol oxidase). Expression of the AOX1 gene is controlled by the strong AOX1 promoter, which has therefore been cloned into *Pichia* expression vectors. In short, the gene of interest is first cloned into such a *Pichia* vector, designed for intracellular expression, or intracellular expression and secretion, the linearized construct transformed into appropriate competent *Pichia* cells or → spheroplasts, the transformants selected by their resistance phenotype (e. g. if a *HIS*4-selectable marker is used, a histidine-deficient medium is employed; in case of → zeocin-selection, this antibiotic is added to the medium), and analyzed for the integration of the gene of interest at the correct locus and in the correct orientation. Then a small-scale pilot expression by some 10-20 colonies is tested (to verify the presence of the recombinant protein, using → SDS polyacrylamide gel electrophoresis and → Western blot analysis), before an up-scaled production in fermenters is started. Since the AOX1 gene promoter is very strong, expression of the foreign gene leads to extraordinarily high levels of the recombinant protein (e. g. grams per liter on average). For high-level methanol-independent expression, vectors are equipped with the constitutive promoter of the glyceraldehyde-3-phosphate dehydrogenase gene. Also, expression vectors are available, that allow to detect multiple insertion events (occurring spontaneously at a frequency of 1-10 %). These multi-copy *Pichia* expression vectors carry the → kanamycin resistance gene, conferring resistance to → geneticin. Multiple insertions can therefore be identified by increased levels of resistance to this antibiotic.
The *Pichia* expression system combines the advantages of *E. coli* (inexpensive and easy handling, high-level expression) and the eukaryotic *Pichia* (protein folding, post-translational modifications, protein processing, secretion), and allows production of nearly all proteins in high quantity (e. g. enzymes, enzyme inhibitors, membrane proteins, regulatory proteins, antigens and antibodies). Compare → Baculovirus expression system.

Picking robot: An automated machine for the transfer of bacterial colonies onto → microtiter plates. Usually the robot station uses a camera, which generates a digital image of the colonies on the petri dishes. A suitable image analyzing software then transforms the positions of the colonies into robot coordinates. An xyz system moves a picking tool which allows to individually guide picking pins to the colonies. After 96 such pins have taken up different bacterial colonies, they dive into the wells of microtiter plates which are filled with nutrient medium. In between two picking processes, the pins are sterilized in ethanol and dried in a hot air stream. The capacity of picking robots ranges from 5,000-10,000 clones picked per hour.

Picotiter plate: A variant of the → microtiter plate, that contains about 300,000 wells. Such picotiter plates are used for highly multiplexed amplification and sequencing reactions for whole genome analysis.

PICS: See → 7–propynyl *iso*carbostyril.

PID: See → pre-implantation diagnostics.

Pilot protein: Any protein that mediates the transfer of DNA from a donor to a receptor cell during bacterial → conjugation.

Pilus (sex pilus, conjugative pilus): Extracellular filamentous organelle of Gram-negative bacteria containing a → conjugative plasmid. Pili serve to form mating pairs between donor and recipient cells and are the site of adsorption for certain bacteriophages (see e.g. → fd phage, → f1 phage). The F-pilus for example is a hollow cylinder 80 Å in diameter with a 20 Å axial hole and is composed of a single subunit protein (pilin) arranged in four parallel helices with a 128 Å repeat.

PIM: See → protein interaction mapping.

Pin and ring spotter (PARS): An instrument for the → spotting of → probes onto → microarray supports, that works with a circular metal loop ("ring") to load sample liquid by capillary action, and a solid pin moving up and down through the liquid in the loop to deposit the probes onto the microarray by contact printing (i.e. by direct contact with the support).

Ping: Any one of a class of → *mini*ature *i*nverted repeat *t*ransposable *e*lements (MITEs), that spans about 5,500 bp with a central region of 4,900 bp flanked by a 252 bp left and a 178 bp right part, and is present in low copy numbers in the genome of e.g. rice (*Oryza sativa* spp. *japonica*: 60–80; *O. sativa* spp *indica*: 14). A Ping sequence ends in TTA duplications. The central part contains two putative → *o*pen *r*eading *f*rames (ORFs), and can be excised, giving rise to the highly conserved 430 bp socalled *mini*ature Ping element (miniPing or mPing), that in turn is flanked by 15 bp terminal inverted repeats (TIRs), but does not contain any ORF. MPing excision is activated by stress (e.g. culture stress, γ-rays), its reinsertion occurs at new loci, also within exons of genes (e.g. exon 2 of the *Waxy* (*Wx*) gene of maize. Pings seem to be active, giving rise to mutable seed phenotypes. Compare → *pong*.

Pioneer sequence: Any novel DNA sequence, for which no related sequence exists in the databases. Since the sequence databases contain immense amounts of sequences, in particular gene sequence information, pioneer sequences most probably have species-, family- or kingdom-specific functions, evolved to meet special demands of the particular organism or group of organisms, from which they originate. See → orphan gene.

PIP: See → percent identity plot.

PISA: See → protein *in situ* array.

PITC: Phenyl isothiocyanate, a compound used for → protein sequencing.

Pitch: The length of one complete turn of a DNA double helix along its vertical axis (as measured e.g. in Å).

Pixel plot: Any two-dimensional representation of a → microarray spot, in which all pixel intensities are plotted at two fluorescence emission wavelengths (e.g → cyanin 3 versus → cyanin 5). A pixel plot allows the experimentor to verify the quality of the spot (i.e. whether or not the distribution of pixel intensities is uniform and normal, as should be expected from a successfully printed spot).

Plab: See → plant antibody.

PLAC: See → plant artificial chromosome.

Planar array: Any → microarray, whose elements (individual spots) are immobilized on a planar surface, in contrast to any microchip based on microfabricated channels (see → electrophoresis chip, → lab-on-a-chip, → microfluidic chip, → suspension array).

Plant antibody **(plab; plantibody):** Any → monoclonal antibody that is synthesized by → transgenic plants. For example, the genes for the heavy and light chain peptide of → IgG antibodies have been transferred into separate tobacco mesophyll cells by → *Agrobacterium*-mediated gene transfer, which were regenerated to complete plants expressing the foreign gene. Conventional crossing of these two transgenic plants leads to a plant harboring both the gene encoding the κ → light chain and the gene for the γ → heavy chain. This plant is able to synthesize a complete IgG antibody.

Plant artificial chromosome **(PLAC):** A → cloning vector containing plant → centromere DNA and → telomere repeats, that can be introduced and maintained in both yeast and a target plant as a stable autonomous → minichromosome. PLACs additionally are equipped with → selectable marker genes and are designed to optimally function in diverse plant cells and to adopt → genomic DNA in the megabase range. See → bacterial artificial chromosome, → human artificial chromosome, → mammalian artificial chromosome, → pBeloBac11, → P1 cloning vec tor, → transformation-competent articificial chromosome vector, → yeast artificial chromosome.

Plant cloning vector (plant vector; plant cloning vehicle): Any → cloning vector that is designed to introduce foreign DNA into a plant's genome. Such vectors may be based on the → Ti-plasmid of → *Agrobacterium tumefaciens*, or DNA plant viruses. See also → plant expression vector.

Plant expression vector: A → plasmid cloning vehicle, specifically constructed so as to achieve efficient transcription of the cloned DNA fragment(s) and translation of the corresponding transcript(s) within a target plant cell. Such cloning vectors contain either a constitutive and highly active ("strong") → promoter sequence (e.g. CaMV35-promoter, → nopaline synthase promoter) or an inducible (e.g. hormone- or light-inducible) or regulated promoter. Immediately downstream of the promoter appropriate cloning site(s) and a plant transcriptional → termination sequence have been inserted. Any promoter-less foreign gene, cloned into such a vector will be expressed at a high level in the transgenic plant.
Figure see page 821.

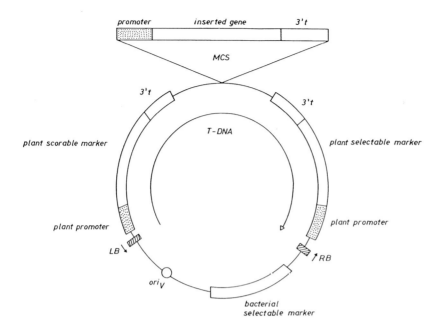

Plant gene therapy: The use of → chimeric oligonucleotide-directed gene targeting to correct or introduce single-nucleotide mutations in plant genomic DNA.

Plantibody: See → plant antibody.

Plant vector: See → plant cloning vector.

Plaque: A clear or turbid area in a bacterial lawn on a culture dish caused by → phage growth and subsequent death and → lysis of the bacteria. Each plaque contains from 10^6-10^7 infectious phage particles. The plaques of virulent phages are generally clear, the plaques of temperate phages are turbid. The term plaque is also used for cell-free areas in cell culture lawns, caused by → viruses. See also → plaque count, → plaque-forming unit, → plaque hybridization, → turbid plaque.

Plaque assay: See → plaque count.

Plaque blotting: See → plaque hybridization.

Plaque count (plaque assay): The determination of the number of complete, infective → bacterio phage particles or infected bacterial cells in a particular suspension. For plaque count, the sus pension is spread onto the surface of an agar plate that is covered with a thin layer of susceptible bacteria ("lawn"). Appropriate dilutions ensure that not more than one phage can infect one host cell, and the number of → plaques are counted that develop on the host cell lawn.

***Plaque forming unit* (pfu):** Usually defined as the number of infectious → virus particles per unit volume, or alternatively, any single infectious particle that generates a single → plaque under defined conditions.

Plaque hybridization (Benton-Davis technique, Benton-Davis procedure, Benton-Davis hybridization; lifting, plaque lifting, plaque blotting, plaque screening, phage lifting): An *in situ* → gene screening procedure for the direct detection of a particular DNA sequence within a population of transformed → bacteriophages harboring vast amounts of different cloned sequences (→ phage libraries). Detection is made possible by hybridization *in situ* with radioactively labeled RNA or DNA → probes with complementarity to the sequence sought. In short, a → nitrocellulose filter is placed on an → agar plate containing bacteriophage → plaques. Unpackaged recombinant phage DNA is bound to the filter (plaque lift, phage lift), denatured and fixed to the filter by baking. The radioactive probe is then hybridized to the filter-bound DNA and the position of the plaque containing the complementary sequence is located by → autoradiography. Interesting plaques can then easily be recovered from the master plate.

Plaque lifting: See → plaque hybridization.

Plaque screening: See → plaque hybridization.

Plasmid: A closed circular, autonomously replicating, extra-chromosomal DNA duplex molecule ranging in size from 1 to more than 200 kb and in copy number from one to several hundred per bacterial cell. The copy number of plasmids may depend upon environmental factors. The average number of e.g. → pBR 322 plasmids per cell on rich (→ LB) medium is 55, immediately before cell divisions it increases to 80. Plasmids generally confer some selective advantage to the host cell (e.g. → antibiotic resistance). → Conjugative plasmids harbor a set of genes capable of transferring the plasmid to other, plasmid-less cells. → Cryptic plasmids are naturally occurring plasmids with unknown genotype and biological function. Different plasmids may interfere with the replication and inheritance of each other, see → plasmid incompatibility. Plasmids are also constituents of mitochondria and plastids in eukaryotic organisms. Bacterial plasmids have been extensively used for the construction of → cloning vectors, see → plasmid cloning vector. See also → chimeric plasmid, → helper plasmid, → multicopy plasmid, → natural plasmid, → non-conjugative plasmid, → plasmid promiscuity, → plasmid rescue, → plasmid sequencing, → plasmid stability.

Plasmid cloning vector: Any → plasmid designed to allow the → cloning of foreign DNA with recombinant DNA techniques. Plasmid vectors are preferentially small in size, replicate under → relaxed control, contain → selectable marker genes (coding for example for → antibiotic resistance), scorable marker genes (coding for enzymes which can easily be monitored), and unique → restriction sites or → polylinkers at locations not necessary for plasmid function (e.g. not in regions needed for replication). Plasmid vectors for a great number of specific experimental needs and different host systems have been developed, see for example → artificial chromosome, → ARS plasmid, → broad host range vector, → expression vector (→ open reading frame vector, → pEX vector), → intermediate vector, → low copy number plasmid vector, → multi-functional plasmid, → mini-Ti, → Okayama-Berg cloning vector, → pBR 322, → pEMBL, → promoter plasmid, → pUC, → restriction site conversion plasmid, → ribozyme auto-cleavage vector, → shuttle vector, → yeast cloning vector, → cosmid vector; also → helper plasmid.

Plasmid conjugation: See → conjugation.

Plasmid curing: See → curing.

Plasmid DNA (pDNA): The covalently closed circular (ccc) double-stranded DNA molecule, that represents a → plasmid.

Plasmid end pair: The sequence reads from both ends of a → plasmid clone. See → BAC end pair.

Plasmid-enhanced PCR-mediated (PEP) mutagenesis: A variant of the → splice overlap extension PCR (SOE-PCR), that allows to introduce mutations (e.g. → deletions or → insertions) into a target DNA. In short, the target DNA is first amplified as two parts using two primer pairs, designed to introduce the mutation and two → restriction sites (which are incorporated into the most distal primers). The internal 5'-phosphorylated primers permit an efficient → blunt-end ligation of the two parts. For a deletion mutation, the targeted sequence is simply omitted, for an insertion it is incorporated into one of the primers. The two parts together with the cloning plasmid are then digested with the two → restriction endonucleases, ligated and used to transform bacterial host cells. The efficiency and orientation of the blunt-end ligation process is controlled by sequence-specific overlapping interactions with the plasmid.

Plasmid incompatibility: The inhibition of replication and thus inheritance of a given → plasmid by the presence of another coresident plasmid in the absence of external selection pressure. Incompatibility is based on several mechanisms. First, in the competition of both plasmids for common membrane binding sites one of them, but not the other may be successful. Usually such binding occurs at the → origin of vegetative replication ($oriV$) of the plasmid and induces → replication. Second, an inc-gene at the origin of → replication of a resident plasmid may encode an RNA (RNA 2) that functions as → primer for DNA replication. If the complementary RNA (RNA 1) is also synthesized, RNA 1 and 2 will anneal and the primer is masked. In this way, the RNA 1 of the resident plasmid may inhibit the replication of the incoming plasmid. See also → plasmid incompatibility group.

Plasmid incompatibility group: A class of closely related → plasmids that are mutually exclusive (i.e. cannot be stably maintained in the progeny of a particular host cell). Since incompatibility is based on the action of inc (incompatibility) genes, incompatibility groups are designated as incA, incB, incC and so on. See → plasmid incompatibility.

Plasmid-like DNA (plDNA): A circular → plasmid of filamentous fungi (e.g. *Podospora anserina*) that is derived from the first → intron of the cytochrome oxidase subunit I gene. The plDNA is involved in age-related rearrangements of mitochondrial DNA of this fungus.

Plasmid maxiprep: The isolation and purification of large amounts (> 100 µg) of → plasmid DNA from comparably large volumes of bacterial cultures (> 10 ml). Compare → plasmid miniprep.

Plasmid miniprep: The isolation and purification of minute amounts (< 20 µg) of → plasmid DNA from comparably small volumes of bacterial cultures (< 1 ml). Compare → plasmid maxiprep.

Plasmid partitioning: See → partition, → partitioning function.

Plasmid promiscuity: The ability of a → plasmid to promote its own transfer to and replication in a wide range of host cells. A promiscuous plasmid is for example → RP4.

Plasmid rescue (homologous assist): The → recombination of a donor → plasmid with a homologous resident plasmid to form a stable → cointegrate preserving the function(s) of the donor

Plate 823

DNA that would otherwise be destroyed by → restriction. This rescue is observed predominantly in *Bacillus subtilis* and several other Gram-positive bacteria. Usually the donor plasmid will be linearized upon entry into the recipient cell and degraded by restriction. It may be rescued, however, by a → homologous recombination with a resident plasmid (homologous helper plasmid), a process requiring the *recE* gene product. Plasmid rescue may be used for → shotgun cloning of heterologous DNA in *B. subtilis*. Monomeric vector DNA is ligated to a foreign sequence and the construct is linearized. After its uptake it pairs with homologous sequences of a resident plasmid, and both the non-homologous vector DNA and insert DNA are rescued.

Plasmid sequencing (double-strand sequencing, "supercoil sequencing"): The → sequencing of linearized → plasmid DNA that has been heat-denatured (i.e. made single-stranded). The single strands can either be separated and each annealed to a synthetic, strand-specific → oligonucleotide → primer, or both strands can remain in the same reaction mixture, if only one strand-specific primer is used. The primer-annealed plasmid strands can then be sequenced following the → Sanger sequencing procedure.

Plasmid shuffling: A technique for the identification of conditional lethal mutations in an essential cloned gene on a → plasmid, preferentially developed for yeast. In short, the chromosomal copy of the essential genes is first deleted (or disrupted), but functionally replaced by an intact copy on a → yeast episomal plasmid, YEp (that additionally carries the *URA3* gene). Then a second plasmid (e. g. → centromere plasmid) with a temperature-sensitive copy of the gene of interest is introduced, relieving the selection pressure on the YEp at permissive temperatures (normally 23-25° C), thereby generating URA⁻ derivatives (loss of the YEp). At non-permissive temperatures (usually 36° C), the YEp is absolutely essential (no URA⁻ segregants), and loss of this plasmid can be monitored by → replica plating single colonies on 5–fluoroorotic *a*cid (5-FOA) plates at permissive and non-permissive temperatures. URA⁻ segregants will then appear as 5-FOA-resistant papillae.

Plasmid stability: The persistence of a → plasmid through many generations of → host cells. Plasmid stability is a function of the number of plasmid copies per cell, and the → partitioning function (*par* sequence) of the plasmid. The *par* sequences direct an equal distribution of plasmid molecules into each daughter cell after cell division.

Plasmid vaccine: See → DNA vaccine.

Plasmid vector: See → plasmid cloning vector.

Plasmone: The entire DNA of both → mitochondria and → plastids of a cell. See → chondriome, → plastome.

Plastome: The genetic information of plastids. See → chloroplast DNA, → mitochondrial DNA. Do not confuse with → plasmone.

Plastome mutation: Any mutation in plastid DNA (→ chloroplast DNA). See → mitochondrial DNA.

Plate: Any Petri dish that contains a solid medium (mostly → agar) for the growth of microorganisms. See for example → gradient plate.

Plateau effect: The decrease of the exponential amplification of a target sequence in the later cycles of a conventional → polymerase chain reaction such that the amplification rate reaches a plateau (the specific product does no more accumulate). In the plateau phase, a preferential amplification of unspecific side products ensues.

Plate lysate: A solution of mature → bacteriophage particles which are released from bacterial → host cells that grow on an → agar plate. Usually a special medium (e.g. SM buffer containing NaCl, $MgSO_4 \cdot 7\,H_2O$, Tris-HCl, pH 7.5, and gelatin) is layered onto the surface of a plate showing confluent lysis, and the phages diffuse into that medium.

Plating: The process of inoculation of a solid growth medium in a Petri dish with microorganisms (in particular bacteria), so that the inoculum is either uniformly distributed or present as stripes.

Plating efficiency (*efficiency of plating*, EOP): The efficiency with which → bacteriophages infect bacteria. A plating efficiency of 1.0 means that each phage particle causes a productive infection (visualized as → plaque). A plating efficiency of $1.0 \cdot 10^{-4}$ indicates that only one out of 10^4 phage particles infects the host cell productively.

Plectonemic winding: The winding of two DNA strands around each other to produce the normal DNA → double helix.

Pleiotropy: The effect of one particular gene on other genes to produce apparently unrelated, multiple phenotypic traits.

Plexisome: Any linear DNA molecule with a known sequence at both ends. Plexisomes can be generated by fragmenting → genomic DNA with a → restriction endonuclease, the → ligation of an → adapter A to one end, subsequent → nick-translation of the fragment such that it stops half way to the 5'end, and by adaptor B ligation to the 5'end. Both ends of this plexisome carry known sequences (adapter A and adapter B).

PLG: See → *phase lock gel*.

PLP mutagenesis: See → PCR-ligation-PCR mutagenesis.

Plus: Located → downstream of the → cap site. See → minus.

Plus-minus hybridization: See → subtractive hybridization.

Plus-minus screening: A nucleic acid → hybridization method for the isolation of tissue- or organ specific or developmentally regulated (generally, inducible) → cDNA sequences. In short, plus minus screening of e.g. hormone-inducible sequences first requires the establishment of a → cDNA library from mRNAs of hormone-treated tissues. Replica filters carrying identical sets of recombinant clones are then prepared. One of these filters is then probed with radiolabeled → messenger RNA (or cDNA) from control cells, and one with radiolabeled mRNA (or cDNA) from hormone-treated cells. Some colonies will give a signal with both probes (i.e. carry cDNA sequences from mRNAs present in both tissues). Some colonies will give a signal with the cDNA from hormone-treated cells only. These represent sequences which are induced by the hormone. The corresponding clones can easily be recovered from the master plate. A similar technique, which allows faster isolation of cell-specific clones is the method of → subtractive hybridization. *Figure see page 825.*

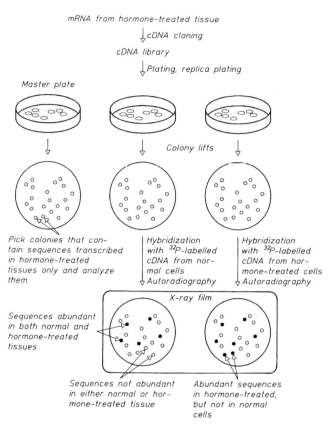

mRNA from hormone-treated tissue
↓ cDNA cloning
cDNA library
↓ Plating, replica plating

Master plate

Colony lifts

Pick colonies that con-
tain sequences transcribed
in hormone-treated
tissues only and analyze
them.

Hybridization
with ^{32}P-labelled
cDNA from nor-
mal cells
↓ Autoradiography

Hybridization
with ^{32}P-labelled
cDNA from hor-
mone-treated cells
↓ Autoradiography

X-ray film

Sequences abundant
in both normal and
hormone-treated
tissues

Sequences not abundant
in either normal or hor-
mone-treated tissue

Abundant sequences
in hormone-treated,
but not in normal
cells

Plus-minus screening

Plus strand (plus viral strand, positive strand, + strand):
a) The DNA strand contained in a single-stranded DNA virus (or any strand with identical sequence).
b) The RNA strand of single-stranded RNA viruses with the same polarity as viral mRNA and encoding viral proteins, compare → minus strand, definition b. See for example → Q-beta, → positive-strand RNA virus.

PM: See → *perfect match*.

PM: See → perfect match.

pMB9: A small multicopy → plasmid carrying a → tetracycline resistance gene and a → Col E1 type → origin of replication, that was used for the construction of the universal → cloning vector → pBR 322. See also → yeast episomal plasmid.

PMF: See → peptide mass fingerprint.

pMON: A series of plant transformation vectors that allow the transfer of foreign genes (gener-ally, DNA) into target plants via a → cointegrate with → Ti plasmid vectors. MON stands for *MON*-SANTO company (USA). See also → *Agrobacterium*-mediated gene transfer.

PMP: *Para*magnetic *p*article, see → magnetic bead.

PMSF (*p*henyl*m*ethyl*s*ulfonyl *f*luoride): An effective inhibitor of trypsin- and chymotrypsin-type proteases that is used to block serine protease activity during protein isolation procedures.

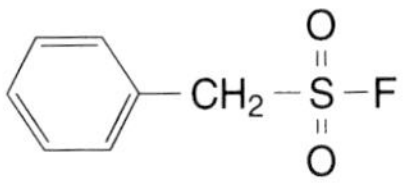

$$\text{C}_6\text{H}_5 - \text{CH}_2 - \overset{\displaystyle O}{\underset{\displaystyle O}{\overset{\|}{\underset{\|}{\text{S}}}}} - \text{F}$$

PNA: See → *p*eptide *n*ucleic *a*cid.

PNA array (PNA microarray, PNA chip): Any → microarray, onto which → peptide *n*ucleic *a*cid*s* (PNAs) instead of conventional nucleic acids with deoxyribose-phosphate backbones are bound via N-terminal groups. PNA arrays do not require any labelling of hybridisation → probes with radioisotopes, stable isotopes or fluorochromes. Another advantage of PNA arrays is the neutral backbone of PNAs and the increased strength of PNA-DNA pairing. The lack of inter-strand charge repulsion improves the hybridisation properties in DNA-PNA duplexes as compared to DNA-DNA duplexes (e.g. the higher binding strength leads to better sequence discrimination in PNA-DNA hybrids than in DNA-DNA duplexes). PNA arrays are used for genome diagnostics, sequencing of DNA or RNA, detection of sequence polymorphisms and identification of expressed genes.

PNA chip: See → PNA array.

PNA opener: A laboratory slang term for a special type of → peptide *n*ucleic *a*cid (PNA), that invades into double-stranded DNA in a sequence-specific way and forms P-loops. The open (looped) region is accessible for hybridization with DNA or RNA → probes and binding of a → sequencing primer or primer for a → polymerase chain reaction. On the other hand, the PNA opener interferes with the binding of enzymes that recognize double-stranded DNA (as e. g. → restriction endonucleases).

PNA-pPNA chimera: See → peptide nucleic acid-phosphono peptide nucleic acid chimera.

PNA2–DNA2 hybrid quadruplex: An artificially produced four-stranded hybrid molecule of two G-rich → peptide nucleic acid (PNA) strands and two homologous G-rich DNA strands, that is extremely stable. Such quadruplexes are used as samples to study the suitability of PNAs as a → probe to target homologous regions in a → genome for e.g. → silencing.

PNK: See → poly*n*ucleotide *k*inase.

PNPase: See → poly*n*ucleotide *p*hosphoryl*ase*.

PNPG (*p-n*itro*p*henyl-*g*lucuronide): A synthetic substrate for → β-glucuronidase.

POD-conjugated antibody: See → *p*er*o*xi*d*ase-conjugated antibody.

Point mutation (microlesion): A → mutation involving a chemical change in only one single → nucleotide. See → transition, → transversion.

pol: The genetic notation for the retroviral gene encoding → reverse transcriptase.

Polarity: The phenomenon, that a → nonsense mutation introduced into a gene transcribed early in an → operon has the secondary effect of repressing expression of genes downstream. See → polar mutation.

Polar mutation (dual effect mutation): Any gene → mutation of the → nonsense or → frame shift type in an → operon producing two effects, namely the repression of nonmutated genes located farther downstream and the failure to express the gene subsequent to the mutation site.

Polished ends (polished termini): The double-stranded termini of duplex DNA molecules generated by the → polishing of → sticky ends with DNA polymerase. Such sticky ends have been generated by restriction with specific type II → restriction endonucleases.

Polished termini: See → polished ends.

Polishing: The synthesis of short sequences complementary to single-stranded protrusions ("tails") generated in DNA molecules by either mechanical shearing or digestion with → restriction endonucleases. Polishing is accomplished by → Klenow polymerase or → T4 DNA polymerase, and leads to complete DNA duplexes.

Pollen electro-transformation: A technique for the introduction of foreign → genes into pollen, which is then used to pollinate flowers. The developping seeds contain the new genes, and can easily be grown to complete plants. Pollen transformation circumvents the tedious transformation of single somatic cells (→ protoplast transformation) and their multi-step regeneration to plants.

Polony: See → polymerase colony technology.

Poly(A): A homopolymer consisting of adenine nucleotide residues. See → poly(A) tail.

Poly(*ADP-*ribose) synthetase (PARS): An enzyme catalyzing the covalent bonding of ADP-ribose to nuclear proteins (e. g. → histones). See → poly(ADP-ribose) polymerase.

Poly(A)⁻-RNA: Any → RNA that does not carry a → poly(A) tail at its 3'-end (e. g. → ribosomal RNA, → transfer RNA, but also → histone messenger RNA). See → polyadenylated RNA.

Poly(A) signal selection: The scanning of an → mRNA molecule for alternative → poly (A) addition signals and the cleavage of one preferred site by an endonuclease recognizing the target sequence 5'-AAUAAA-3'. If several such signals are located on one → primary transcript molecule, then poly(A) signal selection may (1) result in multiple messenger RNAs derived from one transcription unit (if all sites are cut), or (2) the generation of one specific messenger RNA that may be different from cell to cell, or tissue to tissue (if signal selection is cell- or tissue-specific).

Poly(*A*)-specific 3'-exoribonuclease (*poly(A) removing nuclease*, PARN): An enzyme catalyzing the 3'-exonucleolytic removal of adenosyl residues from the → poly(A) tail of eukaryotic → polyadenylated messenger RNAs as a first step towards their complete degradation. The enzyme requires Mg^{2+} and a free 3'-OH group, releases 5'-adenosyl monophosphate (5'-AMP) and thereby initiates the decay of many mRNAs. PARN simultaneously binds to the 3' poly(A) tail and the → cap at the 5'end of → messenger RNA. Absence of the cap reduces PARN activity. This exonuclease belongs to the class of → deadenylating nucleases. See → deadenylation, → decapping enzyme, → destabilizing downstream element.

Poly (A) tail (poly[A] sequence): A sequence of 60-200 adenine nucleotides at the 3' end of most eukaryotic mRNAs, added to the molecule after its transcription by a template-independent → poly(A) polymerase and functioning in the stability of mRNA.

Poly(A) addition signal (poly[A] signal; poly[A] site; poly[A] addition site; poly[A] signal sequence; polyadenylation site):
a) A hexanucleotide → consensus sequence (animals: 5'-AATAAA-3'; plants: 5'-AATAAN-3', generally 5'-AATAA-3' sequence) close to the 3'-end of most eukaryotic genes transcribed by → RNA polymerase II.
b) The consensus sequence 5'-AAUAAA-3' in an → mRNA molecule that directs the cleavage of the message 10-30 bases 3' of the element. The cleaved mRNA then serves as a substrate for processive poly- adenylation. First, the socalled *polyadenylation specificity factor* (CPSF), a tetrameric protein with subunits of 33, 73, 100, and 160 kDa binds to the 5'-AAUAAA-3' signal, then the trimeric *cleavage-stimulating factor* (CstF; 50, 64, and 77 kDA) binds to a GU-rich sequence element further downstream of the RNA. The → poly(A)polymerase (PAP) binds in between the two elements. This complex is joined by two (or more) other proteins, of which the *cleavage factors* CF I and CF II are positioned upstream of the GU-rich box and terminate the mRNA.

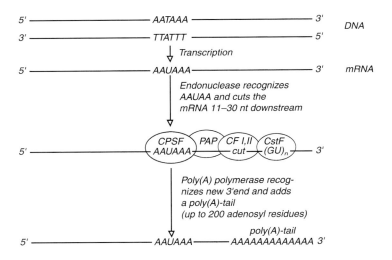

Transportable mRNA

Poly(ADP-ribose) polymerase (PARP): A chromatin-associated 113 kDa nuclear → zinc finger protein, that binds to breaks in → single-stranded and → double-stranded DNA and catalyzes the formation of → homopolymers of poly(ADP-ribose) bound to a series of different nuclear proteins. The enzyme has at least three different domains (a catalytic, an automodification, and the zinc finger DNA-binding domain) and is activated proportionally to the number of strand breaks. Nuclear nicotinamide dinucleotide (NAD) serves as substrate. PARP seems to play a pleiotropic role in various cellular processes, but its major function is most likely DNA repair and maintenance of genome integrity. At the initial processes leading to programmed cell death ($\rightarrow$ apoptosis), a transient activation of PARP with concomitant poly(ADP)-ribosylation of nuclear proteins (e. g. Ca^{2+}/Mg^{2+}-dependent nuclease) is followed by its specific proteolysis by $\rightarrow$ caspase (caspase 3 and 7). The degradation of PARP may be one of the triggers for apoptosis, with the disruption of the nucleus and chromatin (leading to nucleosomal ladders) and the formation of apoptotic bodies. See $\rightarrow$ poly(ADP-ribose) synthetase.

Polyacrylamide *affinity coelectrophoresis* (PACE): A variant of the conventional $\rightarrow$ affinity coelectrophoresis, in which agarose is replaced by *polyacrylamide* (PAA) as electrophoretic matrix, allowing to assay the binding interaction(s) of RNA-peptide or RNA-protein complexes. In short, the gel tray is initially turned by 90° C relative to its orientation during electrophoresis, and a series of PAA plugs with increasing peptide (protein) concentrations sequentially poured such that a discrete step gradient of peptide concentration from the left side of the final gel to the right side is created. Then the gel is turned to its electrophoresis orientation, a PAA mixture without peptide poured onto its top, a $\rightarrow$ comb inserted, and the radioactively labeled RNA (generally: ligand) applied. If any interaction(s) between the peptide and the RNA occurs, then the electrophoretic mobility of the complex is retarded, which can be detected by $\rightarrow$ autoradiography. PACE allows to follow a saturation kinetics, because increasing peptide concentrations will cause increased mobility shifts, unless the RNA is completely bound (i. e. no change in shift occurs anymore). See $\rightarrow$ affinity electrophoresis. Compare $\rightarrow$ mobility-shift DNA-binding assay.

Polyacrylamide gel (PAA gel): An insoluble three-dimensional matrix of acrylamide monomers cross-linked with N,N'-methylene-bisacrylamide in the presence of polymerization catalysts (e. g. TEMED and ammonium *per*sulfate, APS). The crosslinking process starts with the interac-

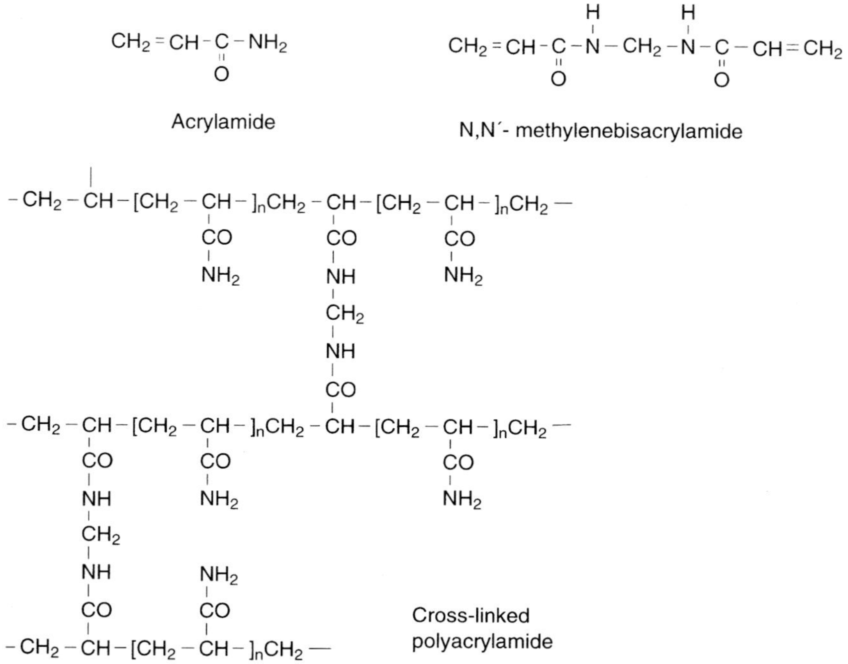

Acrylamide

N,N'- methylenebisacrylamide

Cross-linked
polyacrylamide

Polyacrylamide gel

tion of TEMED and APS, producing a TEMED radical cation, a sulfate radical, and a sulfate anion. The sulfate radical transfers its unpaired electron onto an acrylamide monomer, thereby converting the acrylamide to a free radical, which then reacts with other acrylamide monomers to polymers. These polymeric chains are then statistically crosslinked by N,N'-methylene-bisacrylamide, forming the effective gel matrix, that is used for → polyacrylamide gel electrophoresis. The gel pore sizes can be varied by varying the relative proportions of the ingredients and thus can be adapted to the electrophoretic separation of molecules such as proteins and nucleic acid fragments of various size classes. See also → stacking gel.

Polyacrylamide gel electrophoresis (PAGE): A technique to separate macromolecules (e.g. nucleic acids, proteins) on the basis of their size and structure by electrophoresing them through an inert matrix consisting of cross-linked acrylamide. The separation of proteins is usually carried out in the presence of → sodium dodecyl sulfate (see → SDS polyacrylamide gel electrophoresis). Complete denaturation of proteins before PAGE is achieved by heating, especially in the presence of SDS. Nucleic acids, or the products of DNA-sequencing reactions can also be denatured by heating but furthermore by formamide, urea or methyl mercuric hydroxide before being electrophoresed in strongly → denaturing gels to separate fragments differing by only one single nucleotide. Optimal polyacrylamide concentrations for the separation of e.g. differently sized linear double-stranded (ds) DNA molecules are different. See also → horizontal polyacrylamide gel electrophoresis, → Laemmli gel.

Non-denaturing acrylamide concentrations for resolution of linear dsDNA

Acrylamide (%)	Size of linear dsDNA fragments separated (bp)	Dye migration (bp)	
		Xylene cyanol	Bromophenol blue
3.5	100–1000	460	100
5.0	80–500	260	65
8.0	60–400	160	45
12.0	40–200	70	20
15.0	25–150	60	15
20.0	6–100	45	12

Dye migration in denaturing polyacrylamide gels

Polyacrylamide/urea (%)	Dye migration (bp)	
	Xylene cyanol	Bromophenol blue
5.0	140	35
6.0	106	26
8.0	75	19
10.0	55	12
20.0	28	8

Polyacrylamide gel percentages for resolution of proteins

Gel percentage (%)	Protein size range (kDa)
8	40–200
10	21–100
12	10–40

Polyacrylamide-oligonucleotide conjugate: A copolymer of → polyacrylamide and an → oligo-nucleotide, a → polynucleotide (e.g. DNA) or a → DNA mimic (as e.g. → peptide *n*ucleic *a*cid [acrylamide PNA], → *p*hosphono *p*eptide *n*ucleic *a*cid [acrylamide-pPNA], → trans–4–*h*ydr-oxy-L-*p*roline *p*eptide *n*ucleic *a*cid [acrylamide-HypNA], or chimeras of alternating PNA and pPNA residues or pPNA and HypNA monomers). Such conjugates are produced by (1) deri-vatizing oligonucleotides or DNA mimics with acrylamide groups at their 5'- or 3'- termini, (2) co-polymerizing these acrylamide-modified oligonucleotides (or mimics) with polyacrylamide, and (3) the covalent attachment of the conjugates onto the surface of non-derivatized microscope slides treated with either monoethoxydimethylsilylbutanal, 3-mercaptopropyl-trimethoxysilane or 3-aminopropyltrimethoxysilane, and activated with phenylisothiocyanate. The PAA support owns high chemical and thermal stability and low non-specific adsorption of biological macro-molecules. The polyacrylamide-oligonucleotide conjugates are used for the capture of comple-mentary probe molecules.

Polyadenylated RNA: An RNA molecule, that contains a homopolymeric tail of adenyl residues at its 3'-terminus (e.g. poly[A]$^+$-mRNA). See → polyadenylation.

Polyadenylation: The post-transcriptional addition of → poly(A) tails of up to 200 adenine re-sidues to the 3'-termini of → heterogeneous nuclear RNA and → messenger RNA in eukaryotes. Compare → in vitro polyadenylation, → poly(A) addition signal, see → post-transcriptional modification.

Polyadenylation signal: See → poly(A) addition signal.

Polyadenylation site: See → poly(A) addition signal.

Polyamide dimer: An eight-ring hairpin polyamide molecule consisting of three types of aromatic rings (pyrroles, imidazols, and hydroxypyrroles), that folds back and forms pairs of the three rings. Now the various combinations of each two compounds can discriminate specific → Wats-on-Crick base pairs in the → minor groove of the DNA helix. For example, through direct hydrogen bonding between pairs of pyrroles on the polyamide and the double hydrogen bond acceptor potential of the O2 of thymine and the N3 of adenine, A-T can be distinguished from T-A base pairs. Or an *im*idazole (Im)/*py*rrole (Py) pair discriminates G-C from C-G, and both of these from A-T and T-A base pairs. Likewise, a *h*ydroxy*py*rrole(Hp)/Py pair discriminates T-A from A-T, and both of these from G-C and C-G. The sequence-specific recognition of DNA sequences by polyamide dimers is therefore determined by the side-by-side pairing of the residues in these polyamide dimers.

Poly(A)-PCR: See → poly(A) polymerase chain reaction.

Poly(A)⁺-mRNA: See → polyadenylated RNA.

Poly(*A*) polymerase (PAP; Bollum enzyme): A primer-dependent enzyme that catalyzes the polymerization of AMP from ATP onto free 3' hydroxyl groups of mRNA. The enzyme is used for the addition of → poly(A) tails to RNA and for 3'-endlabeling of RNA.

Poly(A) polymerase *chain reaction* (poly[A]-PCR, *sequence-independent primer reverse transcriptase* PCR, SIP-RT-PCR): A variant of the conventional → reverse transcriptase polymerase chain reaction technique for the global amplification of → messenger RNA (see → global mRNA amplification) from samples containing only little RNA or → low-abundance messenger RNAs at a lower detection limit (e.g. single cells, biopsies, needle aspirates), that capitalizes on limiting the size of the → first-strand cDNA to 300–700 bases, and therefore avoids a bias against long transcripts during the amplification process. In short, whole cells (or isolated total RNA) are first lysed by heating (which also denatures secondary structures in RNA), cooled, and reverse transcribed into first-strand cDNA by → reverse transcriptase using an oligo(dT) primer. After heat inactivation of the enzyme, the first-strand cDNA is poly(A)-tailed using a → terminal transferase to generate a 5'-oligo(dT) and 3'-poly(A)tailed cDNA. Aliquots of the reaction are then directly suspended into a buffer containing sequence-independent 5'-(T)$_{24}$-X–3'(X: A,C, or G, but not T) primer, and the poly(A)-tailed cDNA is amplified by conventional → polymerase chain reaction. For subsequent gene expression profiling, → Northern blot analysis, → quantitative PCR, or cDNA expression arrays (see → microarray) can be employed. SIP-RT-PCR preserves the relative abundances of specific transcripts present in the original mRNA population (i.e. no distortion occurs).

Poly(A) sequence: See → poly(A) tail.

Poly(A) signal (poly[A] signal sequence): See → poly(A) addition signal.

Poly(A) site: See → poly(A) addition signal.

Polybrene transformation: A method for → direct gene transfer into animal cells (e.g. CHO cells), using the polycation polybrene. See → DEAE dextran precipitation.

Poly(C): A homopolymer consisting of cytidylic acid residues.

Polycistronic: See → polycistronic mRNA.

Polycistronic message: See → polycistronic mRNA.

Polycistronic mRNA (polycistronic message): A transcript of an → operon which is transcribed as a single message and codes for all the individual enzymes specified by the operon (e.g. *lac* operon of *E. coli*). Polycistronic messages also occur in eukaryotic organisms, see for example → histone genes. They are occasionally transcribed into → polyproteins.

Polyclonal: The property of molecules or cells to originate from more than one → clone, i.e. from more than one single cell. See for example → polyclonal antibody.

Polyclonal antibody: Any → antibody (immunoglobulin) synthesized by different → clones of B lymphocytes. After challenge with an → antigen, an organism usually produces a heterogeneous mixture of specific antibodies with different binding specificities and affinities. Thus every naturally induced antiserum is polyclonal. Compare → monoclonal antibody.

Polycloning site: See → polylinker.

Polycore probe: Any nucleic acid → probe that consists of a tandem arrangement of several to many so-called → core sequences and serves as a → hybridization probe to detect polymorphic loci in eukaryotic genomes.

Polyester plug spin insert (PEPSI): A simple device for the separation of → agarose and DNA after → agarose gel electrophoresis. The fragment of interest is cut out of the gel, placed on top of a polyester fiber plug in a plastic pipette tip inserted in a microcentrifuge tube, and spun in a small volume of elution buffer. The DNA is recovered in the microcentrifuge tube, and the agarose trapped in the PEPSI. Do not confuse with Pepsi Cola.

Polyethylene glycol (PEG; carbowax): A polymeric hydrophilic chemical compound with the general formula $HOCH_2(CH_2OCH_2)_nCH_2OH$, used to destabilize cellular membranes, for example during → bacteriophage purification, to induce liposome-cell and cell-cell-fusions (see → lipofection and → cell fusion, → protoplast fusion), and to increase the → hybridization efficiency.

Polyethyleneglycol-based brush surface (PEG-based brush surface): A special modification of chip surfaces, that prevents the unspecific binding of non-ligand material (i.e. contaminants) in solution to surface-immobilized → probes (e.g. oligonucleotides, DNA, antibodies, proteins). The charged surfaces of chips are simply coated with densely packed polyethyleneglycol chains, which adsorb spontaneously and form nanometer-thin layers (brushes"). PEG can also be functionalised by the covalent attachment of e.g. → biotin.

Poly(G): A homopolymer consisting of guanidylic acid residues.

Polygene: A misleading term for any gene, that is responsible for only a small effect on a → polygenic trait. The full expression of the trait is then controlled by two (or many) such polygenes.

Polygenic trait (multigenic trait): A (usually phenotypic) feature of an organism, that is controlled by more than one, in extreme cases up to ten different genes. See → polygene.

Polyhistidine tag: See → histidine tag.

Polyketide: Any one of a series of structurally diverse and complex organic polymers, composed of acyl-coenzyme A (CoA) monomers, that are constituents of actinomycetes, bacilli, and filamentous fungi. Their functions for the organism, in which they are synthesized, is not in all cases very clear. However, some compounds own pharmaceutical potential. For example, → rifamycin (inhibitor of RNA polymerase), FK 5506 (immunosuppressant, binds calcineurin) and lovastatin (inhibits hydroxymethylglutaryl-CoA reductase) are such compounds. Synthesis of polyketides circumvents ribosomes. Socalled modular polyketide synthases (PKSs), exceptionally large multi-functional proteins ("megasynthases"), organized as coordinate groups of active sites ("modules"), where each module is responsible for catalysis of one cycle of polyketide chain elongation, replace ribosomal functions. Within each module resides a 75–90 amino acid acyl carrier protein (ACP) domain, to which the growing polyketide chain is covalently tethered. The size and complexity of the synthesized polyketide are controlled by the number of repeated acyl chain extension steps. Four synthetic phases can be discriminated: priming, chain initiation and elongation, and termination, reminiscent of the → non-ribosomal peptide synthesis.

Polylinker (*multiple cloning site, MCS; polycloning site*): Any synthetic → oligonucleotide containing multiple → restriction sites (multiple cloning sites). In a symmetrical arrangement two identical polylinkers flank a DNA fragment (e.g. in → M13 mp 7), in a non-symmetrical arrangement the polylinkers at both ends of a duplex DNA are not identical (e.g. in M13 mp 8, mp 10, mp 11, mp 18 and mp 19). Polylinkers allow the cloning of foreign DNA into any of their restriction sites.

Polymer: Any molecule, that is composed of many identical subunits linked to each other by the process of polymerisation.

Polymerase: An enzyme that catalyzes the assembly of nucleotides into RNA (→ RNA polymerase), or of deoxynucleotides into DNA (→ DNA polymerase).

Polymerase chain reaction (PCR): An → *in vitro* amplification procedure by which DNA fragments of up to 15 kilobases in length can be amplified about 10^8-fold. In brief, two 10-30 nucleotides long → oligonucleotides complementary to nucleotide sequences at the two ends of the target DNA and designed to hybridize to opposite strands, are synthesized. Excessive amounts of these two oligonucleotide → primers (→ amplimers) are mixed with → genomic DNA, and the mixture is heated for → denaturation of the duplexes. During subsequent decrease of temperature the primers will anneal to their genomic homologs and can be extended by → DNA polymerase. This sequence of denaturation, annealing of primers and extension is repeated 20-40 times. During the second cycle, the target DNA fragment bracketed by the two primers is among the reaction products, and serves as template for subsequent reactions. Thus repeated cycles of heat denaturation, annealing, and elongation result in an exponential increase in copy number of the target DNA. The use of thermostable DNA polymerases (e.g. → *Thermus aquaticus* DNA polymerase; → *Pfu* DNA polymerase; → Vent^{TM} DNA polymerase) obviates the necessity of adding new polymerase for each cycle. About 25 amplification cycles increase the amount of the target sequence selectively and exponentially by approximately 10^6-fold. In later phases of the amplification cycle undesirable, incompletely elongated products may accumulate (→ shuffle clones).

Since its introduction the polymerase chain reaction has multiply been modified and its potential been expanded for a great number of sophisticated applications.

See → adapter-tagged competitive PCR, → allele-specific PCR, → *Alu* PCR, → anchored microsatellite-primed PCR, → anchored PCR, → arbitrarily primed PCR, → asymmetric PCR, → balanced PCR, → booster PCR, → bubble PCR, → capture PCR, → cascade PCR, → cDNA PCR, → colony PCR, → colony-direct PCR, → comparative reverse transcription PCR, → competitive oligonucleotide priming PCR, → competitive PCR, → competitve quantitative PCR, → competitive reverse transcriptase PCR, → concatemer PCR, → continuous flow PCR, → degenerate oligonucleotide primed PCR, → differential cDNA PCR, → differential display reverse transcription PCR, → differential PCR, → digital PCR, → direct PCR amplification, → duplex PCR, → electronic PCR, → ePCR, → exclusive PCR, → expression PCR, → extender PCR, → 5' nuclease PCR, → fluorophore-enhanced repetitive sequence-based PCR, → genomic amplification with transcript sequencing, → gradient PCR, → hemicompetitive PCR, → hot PCR, → immuno PCR, → immuno-bead RT-PCR, → in-cell reverse transcriptase PCR, → *in situ* PCR, → *in situ* reverse transcription PCR, → inter-interspersed repetitive element PCR, → inter-retrotransposon amplified polymorphism, → inter-simple sequence repeat amplification, → interspersed repetitive sequence PCR, → interspersed repetitive sequences long range PCR, → inverse PCR, → island rescue PCR, → kinetic PCR, → ligation-anchored PCR, → ligation-independent cloning of PCR products, → limiting dilution PCR, → linker-adaptor PCR, → long and accurate PCR, → long distance PCR, → long-extension PCR, → long fragment PCR, → long range PCR, → long reverse transcriptase PCR, → low copy PCR, → low stringency single specific primer PCR, → message amplification phenotyping, → methylation specific PCR, → micro PCR, → microdissection PCR, → microplate-based PCR, → microsatellite-primed PCR, → microwell PCR, → minisatellite-primed amplification of polymorphic sequences, → mixed oligonucleotide-primed amplification of cDNA, → mixed target PCR, → module-shuffling primer PCR, → motif-primed PCR, → multiple overlapping primer PCR, → multiplex PCR, → multiple fluo rescence-based PCR-SSCP, → mutagenically separated PCR, → nested on chip PCR, → nested primer PCR, → non-selective PCR, → on-chip PCR, → one-sided PCR, → patched circle PCR, → PCR mutagenesis, → PCR RFLP, → PCR single-strand conforma-

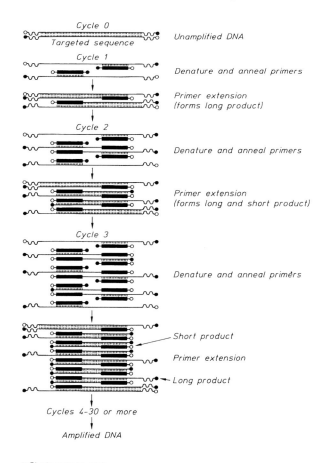

Cycle 0
Targeted sequence — Unamplified DNA

Cycle 1
— Denature and anneal primers

— Primer extension
(forms long product)

Cycle 2
— Denature and anneal primers

— Primer extension
(forms long and short product)

Cycle 3
— Denature and anneal primers

Short product
Primer extension
Long product

Cycles 4–30 or more

Amplified DNA

∘ 5' phosphate end
● 3' hydroxyl end

Polymerase chain reaction (PCR) technique

tion polymorphism analysis, → poly(A)-PCR, → priming authorizing random mismatches PCR, → protein PCR, → quantitative PCR, → quantitative reverse transcriptase PCR, → random PCR, → rapid cycle DNA amplification, → real-time PCR, → recombinant PCR, → relative quantitative reverse transcriptase PCR, → relative reverse transcriptase polymerase chain reaction, → repetitive sequence-based PCR, → restriction endonuclease-mediated selective PCR, → restriction fragment differential display PCR, → restriction site PCR, → reverse transcriptase multiplexed PCR, → reverse transcriptase PCR *in situ* hybridization, → reverse transcription PCR, → RNA amplification with transcript sequencing, → RNA arbitrarily primed PCR, → RNA PCR, → rotary microfluidic PCR, → sequence-independent primer reverse transcriptase PCR, → sexual PCR, → simplex PCR, → single cell PCR, → single molecule PCR, → single-sided PCR, → single specific primer PCR, → single-tube PCR, → solid-phase PCR, → solution PCR, → splice overlap extension PCR, → step-out PCR, → suicide PCR, → suppression PCR, → TAIL-PCR, → tagged random primer PCR, → *Taq*Man PCR, → targeted display PCR, → targeted gene walking PCR, → TD-PCR, → telomerase PCR ELISA, → terminal deoxynucleotidyl transferase-dependent PCR, → thermally asymmetric interlaced PCR, → thermally asymmetric PCR, → touchdown PCR, → transcription chain-reaction, → transcription-mediated amplification, → triplex PCR, → two-stage PCR; → two-step gradient PCR, → two-step PCR, → universally primed PCR, → unpredictably primed PCR, → vectorette PCR, → vector-insert PCR, → whole genome PCR.

An alternative method for the *in vitro* amplification of nucleic acids which is based on continuous cDNA synthesis and transcription is the → self-sustained sequence replication procedure.

Polymerase chain reaction *a*ssisted *c*DNA *a*mplification (PACA): A variant of the conventional → polymerase chain reaction (PCR) that uses two gene-specific degenerate or non-degenerate primers (amplimers) to amplify specific cDNA fragments ("internal PACA"). A modification works with a gene-specific primer and the unspecific oligo(dT)-primer that hybridizes to a stretch of adenyl residues added to the target cDNA by → terminal transferase. See → anchored polymerase chain reaction.

Polymerase chain reaction carry-over prevention: Any measure to avoid the re-amplification of previously amplified → polymerase chain reaction products that are carried over accidentally to new amplification mixtures. In addition to physical procedures excluding such exogenous templates there exists an enzymatic preventive technique. In short, in the PCR mixture dTTP is substituted by dUTP, which does not interfere with most subsequent procedures (e.g. → Sanger sequencing). Before starting a new reaction, the potentially left-over template is destroyed with → uracil-DNA-glycosylase that removes uracil bases from the amplified DNA. Then the enzyme is heat-inactivated and the new reaction started.

Figure see page 837.

Polymerase chain reaction mutagenesis (PCR mutagenesis): A variant of the conventional → oligonucleotide-directed mutagenesis, which allows to introduce → insertions, → deletions, or → point mutations in a target DNA with its concomitant amplification using → polymerase chain reaction techniques. See for example → patched circle polymerase chain reaction.

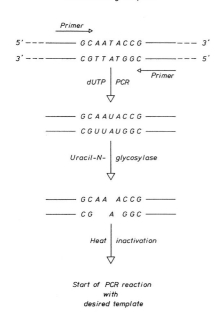

Polymerase chain reaction restriction fragment length polymorphism (PCR-RFLP): Any variation(s) in the length of DNA fragments amplified from → genomic DNA of two or more individuals of a species using conventional → polymerase chain reaction techniques, and subsequently cut with a specific → restriction endonuclease. PCR-RFLP therefore allows to detect restriction endonuclease site variation in two (or more) genomes. The procedure starts with the amplification of a fragment of genomic DNA from organism A and B, respectively, using → primers matching highly conserved → gene sequences (e. g. derived from the chloroplast or mitochondrial genomes). The PCR products are then restricted with a → four-base cutter (e. g. *Rsa* I, or *Dpn* II), and the restriction digests electrophoresed in → agarose gels. The separated fragments are then visualized by staining with → ethidium bromide, and a potential RFLP detected by fluorescence. See → restriction fragment length polymorphism.

Polymerase chain reaction single-strand conformation polymorphism (PCR-SSCP): The difference in the conformation of two slightly differing single strands of a DNA target sequence isolated from two (or more) different individuals. Such polymorphisms are detected by a → polymerase chain reaction single-strand conformation polymorphism analysis.

Polymerase colony (polony) technology: A technique for the → genotyping and → haplotyping of multiple individualized DNA molecules (e.g. single chromosomes) within a thin → polyacrylamide gel attached to a microscope slide. In short, a small amount of DNA (e.g. from a buccal swap) is first mixed with polyacrylamide solution containing all the reagents for a conventional → polymerase chain reaction (PCR), then degassed and poured on a glass slide and polymerised as a 40 μm thick layer. The slide is partially covered with a Teflon coating, that represents a spacer between the glass and a coverslip. Mineral oil is then layered onto the gel (to prevent evaporation of material), the coverslip applied, and the slide as such subjected to a PCR reaction. Two primer pairs (two forward and two reverse primers) flanking the two target SNPs are also included, each amplifying one of the SNPs. This is the simplest arrangement, but several to many different SNPs can be detected in one single experiment by multiplexing. Since the DNA concentration in the gel is low, the chromosome fragments are well separated from each other on the surface of the slide and can be amplified separately side by side. Then the two SNP loci on the chromosomal DNAs are serially amplified in-gel. Since the polyacrylamide matrix restricts the free diffusion of the amplification products, an immobilized colony of double-stranded DNA accumulates around each chromosome ("polony"). Each polony is amplified from a different region of the same chromosome. After amplification, one strand of the amplified DNA remains covalently attached to the acrylamide matrix via the specially designed primers. Therefore it is possible to remove the second strand from all polonies by simple heating to 70⁰C (in 70% formamide) and washing the slide. At the end, single-stranded templates for subsequent → single base extension (SBE) re-actions are produced. For this reaction, the slide is covered with an appropriate chamber, an annealing mixture with the specific primers for SNP 1 added, the unannealed primers washed away, and the templates extended by → Klenow exo-polymerase in the presence of a single-strand binding protein from *E.coli* and → cyanin 3–labeled deoxynucleotides (alternatively, also fluorochrome-labeled → dideoxynucleotides can be employed). The slides are scanned on a scan-ning confocal microscope, and a software used to subtract background. Then the same reaction is run for SNP 2 and → cyanin 5-labeled deoxynucleotides. The images of both reactions are then merged to detect overlapping polonies. The individual polonies can either be genotyped by performing single base extensions with dye-labeled nucleotides, and allow the accurate quanti-tation of two → allelic variants. The polony technology can also determine the → phase or → haplotype, and two (or more) *single- nucleotide polymorphisms* (SNPs) by co-amplifying distally located targets on a single chromosomal fragment. Moreover, the "colonies" of amplified DNA molecules can also be used for fluorescent *in situ* sequencing, a → sequencing-by-synthesis method based on reversibly labeled nucleotides. Almost 10 millions of polonies can be accom-modated on a single microscope slide, which is the basis for a high-throughput sequencing platform. See → nanopore sequencing, → single molecule sequencing.

Polymerase I: See → DNA polymerase I.

Polymer-based biosensor: Any conjugated polyanionic polymer (as e.g. *poly[phenylene vinylene]*, PPV), that normally emits fluorescence light, but can be quenched by a cationic electron acceptor molecule (e.g. methyl viologen, MV^{2+}). If the quencher is covalently linked to e.g. → biotin, and this biotin together with the quencher removed by the addition of → avidin, then the unquenched polymer will emit fluorescence light. Likewise, the quencher can be linked to an → antibody specifically removed by its specific → antigen, or an oligonucleotide, specifically removed by hybridizing to a complementary sequence.

Polymerization fidelity: The accuracy with which a → DNA-dependent DNA polymerase repli-cates a → template, expressed as average number of nucleotides incorporated, before the enzyme introduces an error.

Polymorph: Laboratory slang term for → DNA polymorphism.

Polymorphic *G*C-rich *repetitive sequence* (PGRS): Any one of a series of repetitive sequences in the *Mycobacterium tuberculosis* genome, that contains >80% G+C. PGRS underly the socalled PE (name derived from the motif Pro-Glu at the N-terminus) and PPE (name derived from the motif pro-pro-glu at the N-terminus) multigene families encoding acidic, glycine-rich proteins.

Polymorphism:
a) See → DNA polymorphism.
b) The existence of several forms of a phenotypic or genetic character in a population.
c) A localized change in a specific DNA sequence within a genome, generated by → deletions, → inversions, → insertions, or generally → rearrangements. These mutations lead to the existence of different → alleles for a specific locus in a given population. In the case of → repetitive DNA, variations in the number of repeats may lead to → restriction fragment length polymorphisms, see for example → variable number of tandem repeats. Polymorphisms may be detected by → DNA fingerprinting techniques.

Polynucleotide: A linear sequence of deoxyribonucleotides (in DNA) or ribonucleotides (in RNA) in which the 3' carbon of the pentose sugar of one → nucleotide is linked to the 5' carbon of the pentose sugar of the adjacent nucleotide via a phosphate group (→ phosphodiester bond). Compare → oligonucleotide.

***Polynucleotide kinase* (PNK; T4 polynucleotide kinase):** An enzyme from → T4 phage-infected *E. coli* cells which catalyzes the transfer of the γ-phosphate group of ATP onto the 5' OH termini of RNA or DNA chains. Used to label the 5'-termini of DNA or RNA prior to → sequencing.

***Polynucleotide phosphorylase* (PNPase; EC 2.7.7.8):** An enzyme widely distributed among bacteria that catalyzes the covalent linking of ribonucleotides at random and is used to synthesize artificial RNA (e.g. poly[U], poly[A], or poly[AU] molecules).

Polypeptide: A linear polymer of amino acids that are linked by peptide bonds. Compare → oligopeptide.

Polypeptide tag: Any polypeptide that is conjugated post-translationally (rarely translationally) to target proteins, thereby changing their structure, activity, location, assembly, trafficking or turnover. Conjugation links the C-terminal caboxyl group of the polypeptide tag via a covalent isopeptide bond to ε-lysyl amino group(s) of the target. This process can be reversed by unique proteases (cleaving specifically the isopeptide bond). As examples, → ubiquitin (signal for selective protein degradation by the 26S → proteasome), small *u*biquitin-like *m*odifier, SUMO, also sentrin, UBL1 or PIC1 in animals, SMT3 in yeast (potential role in protein trafficking from nucleus to cytoplasm and vice versa), related to *u*biquitin (RUB), *au*to*p*hagy-defective-12 (APG12), *u*biquitin *c*ross-reacting *p*rotein (UCRP) and *F*inkel-Biskis-Reilly murine sarcoma virus-*a*ssociated *u*biquitously expressed protein (FAU) are such polypeptide tags, whose genes occur in small → gene families. Usually the tags are short (ubiquitin and RUB: 76 amino acids; SUMO: 93 – 115 amino acids; APG12: 96 – 186 amino acids).

Polyphenism: The occurrence of two (or more) different → phenotypes within one species. For example, female social insects exhibit various defined phenotypes such as queens, soldiers, and workers). Polyphenism is not based on a difference in the genomes of the variants, but rather on an epigenetic developmental switch ofdifferential gene expression patterns triggered by worker-controlled nutritional and microenvironmental differences within the nest.

Polyploidy (Greek: polis for "many"; ploid for "fold"): The occurrence of more than two complete sets of chromosomes within a cell, a tissue, an organ, or an organism, resulting from chromosome replication without nuclear division or the recombination of two gametes with differing chromosome sets. The normal set is then diploid (diploidy), a triple set is triploid (triploidy), a quadruple set is tetraploid (tetraploidy), and so on.

Polyprotein: Any protein that is produced by the uninterrupted translation of a → polycistronic mRNA transcribed from two or more adjacent genes.

Polyribosome (polysome): The linear array of → ribosomes attached to a molecule of mRNA. Such polysomes may also contain small → translational control RNA.

Polysome: See → polyribosome.

Polysome display: See → ribosome display.

Polysome selection: See → ribosome display.

Poly(T): A homopolymer consisting of thymidylic acid residues.

Polytene chromosome (giant chromosome): A → chromosome consisting of homologous → chromatids that remain attached to each other (synapsed) after repeated chromosomal → replications without nuclear division. Such polytene chromosomes are characteristic for some Ciliatae, the suspensor cells of some plants, and the salivary gland cells of insect larvae (e.g. *Drosophila*) but also occur in other organisms. The DNA of the original chromosomes in such cells replicates in 10 cycles without separation of the daughter chromosomes so that 2^{10} (1024) chromatids may exist in parallel orientation and in strict register. Polytene chromosomes are visible throughout the interphase and are composed of a series of dark condensed bands that are separated by so-called interbands. The pattern of the bands is specific for each chromosome. For example, the *Drosophila* genome contains about 5000 bands and corresponding interbands, each one consisting of a total of 1024 homologous → looped domains. Each band can be assigned a specific number so that a → chromosome map can be established. If a gene has to be transcribed, the corresponding band is decondensed and forms a so-called → puff. The pattern of puffs is characteristic for the physiological and/or developmental stage of the cell, tissue or organ. Specific genes may be localized on polytene chromosomes by → in situ hybridization.

Poly(U): A homopolymer consisting of uridylic acid residues.

Poly(U) sepharose: A sepharose matrix to which poly-uridylic acid residues (poly[U]) are covalently bound and which is used for the binding, isolation and purification of poly (A)-mRNAs in → affinity chromatography. Compare also → messenger affinity paper.

P1 cloning vector (pacmid): A → cloning vector, derived from the phage P1 of *E. coli*, that allows the packaging of foreign DNA of up to 100 kb without interference with the phage functions and thus has a much higher → cloning capacity than → lambda-phage derived or → cosmid vectors. P1 plasmid vectors contain a P1 packaging site (*pac* site) to initiate packaging of vector and cloned DNA into phage P1 particles, two directly repeated P1 recombination sites (*lox P*) flanking the cloned insert and necessary for the circularization of the packaged DNA after its entry into the host cell, a → selectable marker gene (e.g. → ampicillin or → kanamycin resistance), a P1 plasmid → replicon (stabilizing the vector in the host cell at one copy per chromosome), and a *lac* promoter-regulated P1 lytic replicon (allowing the → IPTG-induced amplification of the DNA, see → *lac* operon). Plasmids are propagated in a special *E. coli* strain containing the P1 *cre* recombinase, which mediates recombination between the two *lox P* sites, and thus the circularization of the infecting linear DNA. Packaging of the DNA is initiated when P1-encoded pacase proteins recognize and cleave the *pac* site in the P1 phage DNA. The DNA on one side of the cleavage point is then packaged into an empty phage pro-head. Once this pro-head is filled, a second, non-specific cleavage occurs that separates packaged from non-packaged DNA. Tails are attached to the filled heads to complete the assembly process.

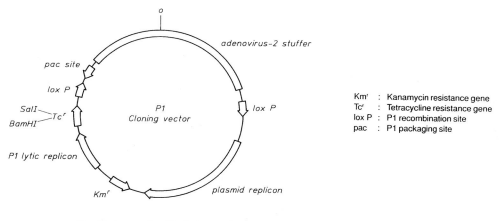

Simplified map of a P1 cloning vector

P1 cloning vector

P1 nuclease: See → nuclease P1

Pong: A variant of the → miniature inverted repeat transposable element *Ping,* characterized by identical 15bp → terminal inverted repeats and a generally similar organization as *Ping* (i.e. two → open reading frames and 5,166 bp length).

Pool: The amount of a defined molecule, that is available for distinct metabolic reactions (e. g. the ATP molecules that can serve as phosphate donors in phosphorylation reactions, as opposed to the AMP molecules fixed in nucleic acids).

Poor man's cloning: See → *in vivo* cloning.

Population genetics: A branch of → genetics, that focuses on the genetic composition (e.g. frequencies of polymorphisms) of whole populations of organisms as influenced by various intrinsic (e.g. mutations, genetic drift, population size and expansion) and environmental factors (as e.g. natural and/or sexual selection, migration, socalled bottle necks) and therefore aims at developing models for evolution. See → population genomics.

Population genomics: The complete repertoire of techniques to develop → molecular markers for genetic variants in whole populations, to associate diagnostic markers with e.g. disease phenotypes by → linkage analysis, and to isolate the underlying genes via → positional cloning. For example, population genomics was applied to different human populations (e.g. Estonians and Icelanders), and markers (i.e. distinct variants of *major histocompatibility complex* [MHC] alleles) were defined for e.g. rheumatoid arthritis.

Population-specific single nucleotide polymorphism (population-specific SNP): Any → single nucleotide polymorphism, that is present in one, and absent in another population. For example, the colonization of Polynesia or the Americas led to the development of single base pair exchanges, that did not occur in ancestral groups of hominids in Asia. These SNPs can therefore be considered as specific for Polynesian or American Indian populations, respectively.

Porosome: See → nuclear pore.

Portable biosensor: Any → biosensor that allows the monitoring of pesticides, chemical warfare agents or, generally, pollutants directly on the spot. For example, *biochemical oxygen demand* (BOD) biosensors can determine the amounts of metabolizable organic material in e.g. waste water by measuring oxygen consumption by immobilized bacteria or yeast cells. The bacteria can also be genetically engineered to express the → lux gene in response to pollutants, which encodes → luciferase generating light during substrate decomposition. See → affinity biosensor, → biomimetic sensor, → electrode biosensor, → enzyme biosensor, → immunosensor, → synthetic receptor.

Portable promoter: Any isolated and fully characterized → promoter or promoter fragment, that contains all regulatory sequence elements for function and can be inserted into any → expression vector and be transformed into any target genome.

Portable SD sequence (portable *Shine-Dalgarno* sequence): A short synthetic oligodeoxynucleotide sequence that contains the → Shine-Dalgarno sequence 5'-AGGAGGU-3', flanked by appropriate → restriction endonuclease → recognition sites, so that it can be easily cloned into prokaryotic → expression vectors.

"Portable sequence":

```
EcoR I
recognition  SD site  poly(A)      start
site                  tract        codon
5'- GAATTCGGAGGAAAAAAATTATG         - 3'
3'-         GCCTCCTTTTTTAATACCTAGG - 5'
                              Bam H 1
                              recognition
                              site
```

Portable *Shine-Dalgarno* sequence: See → portable SD sequence.

Portable terminator: A sequence containing the 3'-terminus of a eukaryotic gene including the transcription → terminator sequence and → poly(A) addition signal(s) flanked by → polylinkers. Such terminators can be ligated to the coding sequence of any gene to construct a → fused gene whose transcript can be correctly terminated and polyadenylated.

Portable *translation initiation site* (PTIS): A double-stranded DNA sequence that contains a five-base → Shine-Dalgarno sequence with an adjacent 3' poly(A) tract flanked by the translation initiation codon ATG eight bases downstream of the SD tract. This configuration is optimal for correct and efficient initiation of translation. Such "portable" sites are usually flanked by specific → restriction endonuclease → recognition sites to allow the → ligation of a PTIS into a → cloning vector.

Positional candidate gene (positional candidate): Any gene linked to a DNA marker co-segregating with a phenotype of interest, and meeting the criteria for a gene, which could be responsible for the trait. It is usually isolated by → positional cloning.

Positional cloning (map-based cloning; *map-assisted cloning, MAC*): The → cloning of a specific gene in the absence of a transcript or a protein product, using → genetic markers tightly linked to the target gene and a directed or random → chromosome walk by linking overlapping clones from a → genomic library.

Position effect: Any change in the expression of one or more genes accompanying a change in its or their position with respect to neighboring (or also distant) genes. Position effects may be brought about by → cross-over or chromosome mutation, and can be seen in → transgenic organisms, where the → chromatin configuration at the integration site determines the expression potential of the foreign gene.

Position effect variegation (PEV): The influence of the chromosomal position on the activity of a → gene. For example, if a normally active gene is introduced into a chromosome at a position adjacent to a → heterochromatin domain, its transcription is repressed. In yeast, genes integrated close to the silent mating-type loci or the → telomeres are silenced. This silencing influence of heterochromatin can spread from 5-10 kb (*Drosophila*) to 20-30 kb (yeast). Position effect variegation is a common observation in → transgenic animals or plants: → reporter genes randomly introduced into the genome have highly variable transcription rates. See → position effect.

Positive control: Any experimental control element, that provides a signal or result, irrespective of the results obtained from the actual experimental components. For example, a positive control on an → expression microarray consists of e.g. the → cDNAs of socalled → house-keeping genes, that are active throughout the life cycle of an organism (e.g. ß-actin- or polyubiquitin genes). If total and fluorescence-labeled cDNAs of a test organism are hybridised to the array with these positive control cDNAs, they will always give a constant signal (e.g. a → fluorescence signal) notwithstanding the reaction of the other cDNAs spotted onto the array. Positive controls are necessary for a test of the function of the array. See → negative control.

Positive cooperation: See → zippering.

Positive feedback activation: The binding of a transcriptional activator protein to its own → promoter and the subsequent activation of the expression of this activator as well as the induction of its target genes. For example, in *Aspergillus nidulans* the activity of the socalled ALCR transcriptional activator is induced by ethanol, and during this induction process the ALCR also activates its own expression by binding to its own promoter. The ALCR promoter contains two ALCR binding sites, and both are essential for its auto-activation and the activation of downstream genes encoding enzymes for ethanol metabolism. See → feedforward loop, → feedforward loop activation.

Positive regulator protein: A protein that activates the transcription of a → gene.

Positive selection: See → direct selection.

Positive selection vector: Any → cloning vector that contains a marker gene or genes, whose mutation can positively and directly be detected. For example, one type of positive selection vectors carries dominant → selectable marker genes. If an → insertion occurs (by e. g. an → insertion sequence element), it may lead to drug resistance or enable the host to grow on a medium containing sucrose. These mutation events generate positively selectable phenotypes in the host cells. Therefore, positive selection vectors are used to isolate mobile genetic elements, as e. g. insertion sequences.

Positive strand RNA virus: A virus containing a single-stranded RNA genome that functions as mRNA template (→ plus strand). The viral RNA is itself infectious. These viruses belong to the Coronaviridae, Flaviviridae, Picornaviridae, Polioviridae, Retroviridae, and Togaviridae.

Positive supercoiling (overwinding): The coiling of a → covalently closed circular DNA duplex molecule in the same direction as that of the turns of its → double helix. Compare → negative supercoiling, → supercoil.

Post-genomic era: The time after the seqencing of the human genome (year: 2001), which is considered to be the time for genome-wide → transcriptomics, the in-depth →.proteomics, detailed → metabolomics, and the deciphering of the molecular mechanisms of development, evolution, and disease, to name only few. Generally, → functional genomics is regarded as the central topic of the post-genomic era. Of course, during this era more and more genomes of both prokaryotes and eukaryotes will be fully sequenced, and the handling of the immense quantities of data will be a challenge for → bioinformatics. The vague starting point for the post-genomic era is arbitrary. It could have been the publication date for the sequence of the genome of the first bacterium (i.e. *Haemophilus influenzae*; year 1995).

Postreplication repair: See → mismatch repair.

Post source decay (PSD): The spontaneous fragmentation ("metastable fragmentation") of ionized molecules during their acceleration in the electrical field or after passage of the accelleration electrode in the field-free drift section of a → mass spectrometer. The analysis of PSD-ions in → reflector time-o-flight mass spectrometers allows to extract informations about the structure of the original ionized molecule.

Posttranscriptional gene silencing (PTGS): A more general term for several protective mechanisms of eukaryotic cells against invading viroids, viruses or moving → retrotransposons (generally RNAs), that can be converted to double-stranded RNAs (dsRNAs) within the cell. These dsRNAs are recognized by → Dicer RNase III and cut into → small RNAs, that incite the silencing of genes encoding homologous RNAs. PTGS was originally described for plants (e.g. *Arabidopsis thaliana, Petunia hybrida*), but its variants are components of defense systems in all eukaryotic organisms. In fungi, → quelling suppresses → transgenes, in invertebrates → RNA interference (RNAi) and → co-suppression are incited by dsRNA, → transgenes and → short hairpin RNA, in vertebrates dsRNA is the prime trigger for RNAi. PTGS can also be transmitted systemically from silenced to non-silenced plant tissues by a → degradation-resistant signal RNA. Invading viruses have evolved more or less effective counter-measures. For example, plant potyviruses encode a protein, HC-Pro, that inhibits maintenance of PTGS, cucumoviruses (e.g. cucumber mosaic virus) encode a 2b protein, that interferes with DNA methylation in the plant nucleus and prevents signal RNA-mediated intercellular spread of PTGS (see → degradation-resistant signal RNA), and the socalled movement protein of potexviruses (e.g. potato virus X) suppresses the release of degradation-resistant signal RNA from infected to non-infected plant cells. See → virus-induced gene silencing.

Post-transcriptional modification (psottranscriptional RNA processing, nuclear processing of RNA, RNA-processing, RNA-maturation): Any one of a series of structural modification(s) of → primary transcripts prior to or during their transport into the cytoplasm. Modifications include → splicing (removal of → introns) → capping of 5'-ends, → polyadenylation of the 3'-end, or → methylation of cytidylic (or adenylic) residues within the RNA molecule, thiolation, isopentenylation, → pseudouridine formation, and association with various proteins. See also → RNA editing, the post-transcriptional modification of mitochondrial RNA.

Posttranscriptional RNA processing: See → post-transcriptional modification.

Post-translational cleavage: The enzymatic cleavage of a large protein molecule or a polyprotein at specific sites to produce smaller functional proteins. A → post-translational modification reaction.

Post-translational modification (protein maturation, protein processing): Any alteration of polypeptide chains after their synthesis (e.g. phosphorylation, acetylation, ADP-ribosylation, glycosylation, oxidation, or also conversion of proenzymes into enzymes by specific proteolytic cleavage). Compare → post-translational cleavage.

Potonuon (*potential nuon*): Any → nuon, that arose by amplification, duplication, recombination or → retroposition and may acquire a new functional role as a new gene (or part of a gene, e.g. an → exon or an→ intron) or new regulatory element (as e.g. an → enhancer, → silencer). If it has acquired the new function, it is called → xaptonuon.

pp: Abbreviation for *phosphoprotein*.

PPD: See AMPPD.

PPi based sequencing: See → pyrosequencing.

pPNA: See → phosphono peptide nucleic acid.

P primer: An → oligonucleotide of arbitrary sequence that serves as a forward → primer in → differential display reverse transcription polymerase chain reaction in combination with a reverse oligo(dT) primer targeting at the poly(A) tail of eukaryotic → messenger RNAs ("T primer"). In advanced differential display techniques, these primers of arbitrary sequence are replaced by primers complementary to common sequence motifs found in a comprehensive collection of messenger RNAs.

Precipitation: The sequestration of an insoluble compound or mixture of compounds in a solution. For example, DNA or RNA can be precipitated from aqueous solutions by extensive dehydration with absolute ethanol. The precipitated material is called precipitate.

Precursor protein: The primary product of the → translation of a → messenger RNA, containing all → exteins and → inteins. See → mature protein.

Precursor RNA: Any → ribonucleic acid synthesized from a gene as a long precursor, that is not yet mature, but still contains many different regions cut out or modified in later processing steps. Such modifications include → capping, → polyadenylation, and → splicing, which altogether lead to its final functional form.

Predicted gene: Any DNA sequence, that has significant homology to → genic sequences deposited in the databanks (e.g. GenBank), and can therefore be considered a gene candidate. Compare → putative gene.

Preferential *a*mplification of *c*oding *s*equences (PACS): A technique for the detection of differentially expressed genes (or better → cDNAs), that specifically targets at the → coding regions of a → messenger RNA (rather than its 5'- or 3'-non-coding parts). In short, total RNA is first isolated, contaminating DNA removed by RNase-free → DNAseI, and single-stranded cDNA produced by → reverse transcriptase polymerase chain reaction with → primers of random sequence. The double-stranded cDNA is synthesized with an ATG-containing → forward primer and a → *double* restricton site primer (DRSP) as a → reverse primer in a conventional → polymerase chain reaction. The ATG- complementary primer specifically selects coding parts of messenger RNAs (mRNAs), since ATG is the → initiation codon of almost all organisms (exceptions: some viral [e.g. human T-cell lymphotropic virus type I], chloroplast [mRNA encoded by the *inf*A gene], plant mitochondrial [atp9–rp116 cotranscript], and bacterial mRNAs [encoding ribosomal proteins]). Moreover, ATG codons occur only rarely → downstream of the proper initiation codon. Therefore, generation of multiple → amplicons from a single mRNA by PACS is unlikely. The ATG primer also contains a → restriction site (e.g. *Bam* H1) at its 5'-end for cloning of the amplification product. These amplification products are then separated electrophoretically in 6% → polyacrylamide/urea sequencing gels, the gels dried and subsequently processed by → autoradiography. The resulting pattern represents a differential mRNA fingerprint of the cell, tissue or organ, from which the RNA was extracted. See → adapter-tagged competitive PCR, → enzymatic degrading subtraction, → gene expression fingerprinting, → gene expression screen, → linker capture subtraction, → module-shuffling primer PCR, → quantitative PCR, → targeted display, → two-dimensional gene expression fingerprinting. Compare → cDNA expression microarray, → massively parallel signature sequencing, → microarray, → serial analysis of gene expression.
Figure see page 847.

Preformed adaptor (ready-made adaptor; conversion adaptor): A short synthetic, single-stranded → oligonucleotide with a → restriction endonuclease → recognition site that allows complete base-pairing with the → cohesive ends of a DNA duplex molecule and a regeneration of a second endonuclease recognition site. For example, the *Eco* RI → *Sma* I preformed adaptor with the sequence 5'-GAATTCCCGGG-3' anneals to cohesive *Eco* RI termini of the target duplex. After → filling-in and ligation, a circular DNA molecule is generated that contains an additional *Sma* I recognition site.

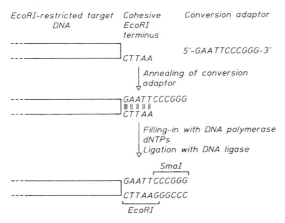

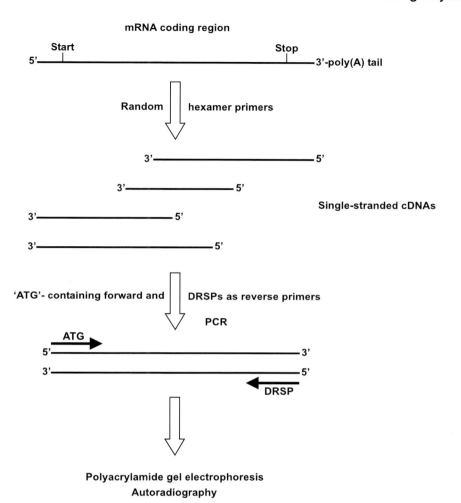

DRSP: Double restriction site primer

Preferential amplification of coding sequences (PACS)

Pre-gel hybridization: A technique for the detection of → mutations (e. g. → deletions, → insertions, → single nucleotide polymorphisms), that circumvents the various steps of → Southern blotting (i. e. pre-hybridization, hybridization, and stringency washes). In short, the double-stranded target DNA is denatured at low salt concentrations and in the presence of a short (i. e. 18mer) labeled → peptide *n*ucleic *a*cid (PNA). At these low salt concentrations, the DNA strands cannot reanneal, and the PNA can bind its complementary DNA target. The resulting PNA/DNA hybrids are then separated by → agarose gel electrophoresis or capillary electrophoresis, and detected after blotting (by e. g. → autoradiography, → luminography, or → fluorescence detection).

Prehybridization: The preparation of → nitrocellulose and nylon-based filters carrying denatured DNA, fixed by → baking or → cross-linking, for hybridization with radioactively labeled → probes. The prehybridization procedure serves to block unspecific binding of the probe to the membrane, and is therefore carried out in solutions containing high concentrations of proteins (→ Denhardt's reagent) and/ or detergents (e.g. SDS). Compare → blocking reagent.

Pre-implantation diagnostics (PID): A series of techniques to detect genetic abnormalities (e.g. chromosomal aberrations) in (usually) a single cell of an e.g. eight-cell embryo, that is a product of an artificial insemination. For that purpose, the target cell is mechanically removed from the morula (by e.g. capillary forces), which in turn will be implanted into the womb after PID. In case of a positive test (i.e. abnormalities found) and no further implantation, a possible abortion is avoided. Moreover, genetically caused repeated miscarriages can be circumvented.

Pre-initiation complex: The assembly of all components of the → basal transcription machinery on the → promoter.

Pre-initiation complex (PIC): An aggregate of general and specific → transcription factors and → DNA-dependent RNA polymerase II subunits, that assembles on → promoter elements following the sequence-specific binding of → transcription factor IID (TFIID) to the → TATA-box. The TFIID-promoter complex is recognized by → transcription factor IIB (TFIIB), the RNA polymerase II-TFIIF complex is recruited (see → transcription factor IIF), and the PIC is completed with the binding of → transcription factor IIE (TFIIE) and → transcription factor IIH (TFIIH). The pre-initiation complex converts to an initiation complex after → promoter melting (local disruption of hydrogen bonds within the DNA helix). Then the first → phosphodiester bond of the nascent → messenger RNA is synthesized, the promoter "cleared" (i.e. the RNA polymerase moves out of the PIC), and the → transcript is elongated. TFIIA (see → transcription factor IIA) interacts and stabilizes the PIC any time after binding of TFIID. Within the PIC, the carboxyterminal domain (CTD) of the large subunit of RNA polymerase II is dephosphorylated, but is phosphorylated by CTD kinases before entering the elongation phase, and leaves the PIC. After termination, the CTD is again dephosphorylated by CTD phosphatases and re-initiates transcription on the → core promoter, if the TFIID complex is still assembled.

Premature polyadenylation: The faulty addition of adenosyl residues within the coding region of a pre-messenger RNA molecule (pre-mRNA), that is recognized by the socalled → non-stop messenger RNA decay system as an aberrant mRNA. Prematurely polyadenylated mRNAs are degraded.

Premature termination codon (PTC): Any → codon in a → messenger RNA, that causes termination of → translation within the message. If such a truncated message would be translated, the resulting protein most likely would be non-functional. However, when a PTC is recognized by a ribosome, → nonsense-mediated mRNA decay (NMD) is activated, reducing the levels of PTC-containing messages to 5–30% of the normal levels.

Pre-messenger RNA: Any complete → primary transcript from a → structural gene before its → post-transcriptional modification. Compare → pre-ribosomal RNA.

Pre-messenger RNA splicing: See → splicing (definition b).

Premutation: Any mutation in a gene, that does not lead to phenotypic consequences, but a predisposition for a disease in the next generation. For example, a normal transmitting male carries a premutation in the FMR1 gene on the distal long arm of the X chromosome. This permutation consists of an increased number of CGG repeats in the 5'-untranslated region of the FMR1 gene (repeat numbers in normal individuals: 6–50; permutation: 55–100). A further expansion of the CGG repeat leads to an inhibition of the transport of the 40S ribosomal subunit from the nucleus and therefore to the suppression of translation, so that repeat numbers beyond 200 inevitably cause the full-blown fragile X syndrome. See → microsatellite expansion.

Prenylation: The covalent addition of either farnesyl (15 carbon atoms) or geranyl-geranyl (20 carbon atoms) isoprenoids to conserved cysteine residues at or near the → C-terminus of proteins via a thioether linkage. Prenylation takes place at the consensus sequence CAAX (C=cysteine; A= any aliphatic amino acid, except alanine; X= carboxyterminal amino acid). First, the three amino acids AAX are removed, and cysteine is activated by methylation (methyl donor: S-a-denosylmethionine). Many membrane-associated proteins are prenylated, and therefore prenylation is probably important for trafficking. Also, prenylation promotes interaction(s) of proteins and cellular membranes, and facilitates protein-protein contacts. Prenylation occurs in e.g. nuclear → lamins, fungal mating proteins, Ras and Ras-related GTP-binding proteins (G proteins), protein kinases and viral proteins.

Preparative comb: A special slot former (comb) for horizontal → agarose gels that allows to apply large volume samples. It contains one tooth spanning most of the length of the comb usually flanked by two small teeth for the electrophoresis of → molecular weight markers.

***Preparative isoelectric membrane electrophoresis* (PrIME):** A technique for the isolation of isoelectrically pure proteins from complex protein mixtures, using → isoelectric focusing in immobilized pH gradient gels on a preparative scale. A variant of this technique works with a series of chambers separated by single-pH → polyacrylamide-immobiline membranes, which act as → isoelectric point-selective barriers. Proteins introduced between the membranes migrate through the membranes to focus in the chamber bounded by one membrane with a pH greater than the pI of the protein, and by another membrane with a pH less than the pI of the protein. PrIME allows to separate proteins differing in pI by only 0.005 pH units.

Pre-replication complex (pre-RC): A → protein machine that is assembled at each → origin of replication during the G1 phase of mitosis in eukaryotic cells. The pre-RC consists of various proteins, of which the ORC proteins (*origin of recognition complex*) bind to the origin in an ATP-requiring reaction, where they remain throughout the cell cycle. The ORC recruits Cdt1 (*cell division target*), Cdc6 (*cell division cycle*), and Mcm (*minichromosome maintenance*) proteins, in higher eukaryotes assisted by geminin, that binds to Cdt1 in the S phase (preventing the formation of a new pre-RC). Cdc6 is rather unstable, and present on the pre-RC only during G1. It becomes phosphorylated at the onset of the S phase and is thereby labeled for degradation. Cdt1 and Cdc6 assemble the Mcm2–7 complex onto the pre-RC. Mcm proteins possess helicase activity. The pre-RC is complete at the end of G1, and converted to the initiation complex (IC) by the action of cyclin-dependent kinases (CDKs) and a helicase, that denatures the RNA at the origin such that the socalled *replication protein A* (RPA) can bind and stabilize the single-stranded DNA. The socalled Cdc45 (*cell division cycle 45*) can then interact with the IC and bind to a subunit of the DNA polymerase α-primase. Thereby the enzyme is activated which leads to the synthesis of a 10 nucleotide RNA primer. This primer is subsequently extended to a 40 nucleotide RNA-DNA primer by a DNA-dependent RNA polymerase, which is displaced by the *replication* *factor* C (RF-C), that loads the socalled *proliferating cell nuclear antigen* (PCNA) onto the RNA-DNA primer (ATP-dependent). PCNA recruits the DNA polymerase δ or ε, which extend the primer by several thousand nucleotides (→ replication).

Pre-ribosomal RNA: The complete → primary transcript from a ribosomal RNA → gene battery (→ rDNA). Its size varies from organism to organism (*Drosophila*: 38S; *Xenopus*: 40S; HeLa cells: 45S). The primary transcript is cleaved in a series of steps to form the → ribosomal RNAs (5.8S, 18S, and 28S rRNA). See → post-translational modification. Compare → pre-messenger RNA.

Pre-spliceosome: A cage-like structure that assembles on the GU splice site of → pre-messenger RNA after ATP-dependent binding of U1 → small nuclear (sn) RNA as a prelude for the formation of a → spliceosome. First a socalled A complex is organized, that harbors U2snRNA (binding to the A of the socalled branch site), U1snRNA, and additionally about 70 different proteins. Then U4/U5/U6 are recruited in an ATP-dependent reaction to form the pre-catalytic socalled B complex.

Press-blot: A simple technique for the detection of nucleic acids or proteins in plant organs (e.g. leaves). The tissues are shock-frozen and then fixed onto hybridization membranes by high pressure. The membrane can then be processed for → Southern (detection of DNA), → Northern (detection of RNA), or Western blotting (detection of proteins).

Prey (P): A part of a hybrid protein component of yeast → two-hybrid systems, encoded by a → hybrid gene consisting of a fusion of a → cDNA or a genomic DNA fragment (the prey *per se,* whose interaction with the socalled → bait has to be tested) and a fused → transcriptional activation domain. If prey and bait interact, the activation domain of the prey construct comes into close proximity with the DNA-binding site, which induces transcription of a → reporter gene.

Prey vector: Any → cloning vector that contains a cDNA-derived sequence encoding a specific protein ("prey") cloned into a → multiple cloning site fused to a sequence encoding a transcription → *activation domain* AD (e. g. B42) upstream. The AD in turn is linked to a → nuclear localization signal and the expression of the prey protein driven by a → promoter. The vector additionally carries replication origins (e. g. the Col E1 origin for replication in *E. coli* and the 2μ origin for replication in yeast) and one (or more) → selectable marker genes. Prey vectors are co-transformed with → bait vectors into socalled yeast reporter strains. Simultaneous expression of the genes on both vectors produces the prey cDNA-derived protein ("prey protein") and the corresponding "bait protein", whose potential interaction can then be detected with the → two-hybrid system.

PRF: See → ribosomal frameshifting.

Pribnow box (Pribnow-Schaller box, -10 box): The 6 bp DNA → consensus sequence 5'-TATAATG-3', located about 10 bp upstream of the → transcription initiation site of prokaryotic → structural genes and functioning as the binding site of the → sigma factor of *E. coli* → RNA polymerase. The Pribnow box facilitates correct initiation and is the equivalent of the eukaryotic → TATA box.

Pribnow-Schaller box: See → Pribnow box.

Primary amplicon: A DNA fragment that is preferentially amplified during → polymerase chain reaction, because the used → primer possesses either complete, or far-reaching → homology to potential target sequences and therefore allows vigorous amplification by → DNA polymerase. Such primary amplicons appear as strong bands in → ethidium bromide-stained → agarose gels, or → silver-stained → polyacrylamide gels, respectively. See → amplicon, → secondary amplicon.

Primary response gene: Any gene that is immediately and directly activated by an external or intrinsic signal. Frequently, such primary response genes encode proteins involved in signal cascades.

Primary transcript: An RNA molecule immediately after its transcription from DNA (i.e. before any → post-transcriptional modifications take place). The primary transcript corresponds to a → transcription unit. See → pre-messenger RNA and → pre-ribosomal RNA.

Primase: See → RNA primase.

PrIME: See → *pr*eparative *i*soelectric *m*embrane *e*lectrophoresis.

Primed *in situ* labeling (PRINS, DNA-PRINS): A sensitive variant of the → *in situ* hybridization technique to detect specific DNA sequences in metaphase chromosomes. In short, metaphase spreads are prepared, synthetic oligodeoxynucleotides or short DNA fragments are hybridized to the chromosomes *in situ*, and used as → primers for → DNA polymerase (e.g. → *Thermus aquaticus* DNA polymerase)-catalyzed extension in the presence of biotinylated or digoxigenin-labeled nucleotides, using the chromosomal DNA as a template. The newly synthesized strand is visualized with fluorescence (e.g. FITC)-labeled → avidin or anti-digoxygenin → Fab fragments. PRINS can be used for the detection of DNA sequences (DNA-PRINS) as well as for the visualization of RNA *in situ* (RNA-PRINS). RNA-PRINS employs oligodeoxynucleotides as primers for → reverse transcriptase (RTase)-catalyzed extension with labeled nucleoside triphosphates, using the RNA (e.g. mRNA) as a template.

Primed synthesis technique: The enzymatically controlled extension of a primer DNA strand in DNA sequencing. See → Sanger sequencing.

Primer: A short RNA or DNA → oligonucleotide which is complementary to a stretch of a larger DNA or RNA molecule and provides the 3'-OH-end of a substrate to which any → DNA polymerase can add the nucleotides of a growing DNA chain in the → 5' to 3' direction. In prokaryotes a specific → RNA polymerase (→ RNA primase) catalyzes the synthesis of such → primer RNAs for DNA → replication (especially of the → Okazaki fragments of the → lagging strand). Primers are also needed by RNA-dependent DNA-polymerases (→ reverse transcriptase).

In vitro, synthetic primers, usually about 10 bp in length, are needed for any DNA polymerization reaction using DNA polymerases or reverse transcriptase. Thus they are necessary for → cDNA synthesis, → Sanger sequencing, the → polymerase chain reaction (see → amplimer), → primer extension and similar techniques. See also → primer adaptor, primer-directed sequencing, primer DNA, → primer hopping, → primed *in situ* labeling, → primer RNA, → primosome; → random priming, → sequencing primer, → unidirectional primer, → universal primer.

Primer-adaptor (adaptor-primer): Any synthetic oligodeoxynucleotide that serves the dual function of a → primer (e.g. for the reverse transcription of a poly(A)$^+$-mRNA by → reverse transcriptase) and an → adaptor (e.g. carrying a → recognition site for a specific → restriction endonuclease). An example for a primer-adaptor is the oligo(dT)-*Xba* I primer-adaptor used in → forced cloning of cDNA.

Primer *binding* site (PBS): A sequence adjacent to the 5' → long terminal repeat of → retroviruses or → retrotransposons, which is complementary to the 3' end of a → transfer RNA. Annealing of the tRNA to the PBS produces a priming site for → reverse transcriptase. In plant retrotransposons the PBS is complementary to the initiator methionine tRNA:

```
LTR                     PBS
5'....CA AGTGGTATCAGAGCCTCGTTT....3'
        || ||| ||| || || || ||| || ||| || ||| ||| |||    ||| || || ||
        ACCAUAGUCUCGGUCCAAA
        3'part of tRNA_i^met
```

Primer dimer: An artifact, representing a non-target amplification product in conventional → polymerase chain reaction techniques, that is caused by sequence homologies within → primers and hence partly double-stranded primer-primer ("primer dimer") adducts. Usually such primer-dimer artefacts also occur in controls (e.g. without any → template) and are of low molecular weight. See → primer-primer artifact.

Primer-directed sequencing (primer-directed walking; primer hopping; primer walking; primer jumping): A technique to sequence DNA fragments of more than 1 kilobases in length. In short, the target fragment is first cloned into an appropriate → cloning vector, and a → forward and → reverse → sequencing → primer complementary to flanking vector sequences used to sequence the → insert from both ends by → Sanger sequencing techniques. This procedure leads to sequence information of about 600-800 bp on both ends of the insert. Now primers are designed from the outermost 100 bp at both ends (walking primers) and used for the second sequencing step, and so on. Primer walking thus allows to sequence long stretches of DNA, that cannot be sequenced by classical sequencing strategies. See → multiplex walking, → uniplex DNA sequencing.

Primer-directed walking: See → primer-directed sequencing.

Primer DNA (DNA primer): A single-stranded DNA fragment required by → DNA polymerase III for DNA → replication.

Primer exclusion: A variant of the conventional → polymerase chain reaction that exploits the competition between a → primer oligonucleotide with a specific sequence and a second oligonucleotide with a slightly different sequence for a common primer binding site. Under conditions of → high stringency the primer with the higher sequence → homology to the target sequence will bind, outcompete (exclude) the competing primer, and gets extended. Primer exclusion allows to detect → point mutations.

Primer extension:
a) A method to precisely map the 5'-terminus of mRNAs and to detect precursors and intermediates of → processing of → messenger RNA. The mRNA is hybridized to a synthetic, 5'
   radiolabeled, complementary oligodeoxynucleotide 30-40 nt in length which is then used by
   retroviral → reverse transcriptase as a → primer. The enzyme completes synthesis of the
   complementary strand (→ cDNA) at the 5'-terminus of the mRNA template. The length of
   the extended primer, and consequently the 5'-terminus of the → transcript, can be precisely
   determined by → polyacrylamide gel electrophoresis and → autoradiography.

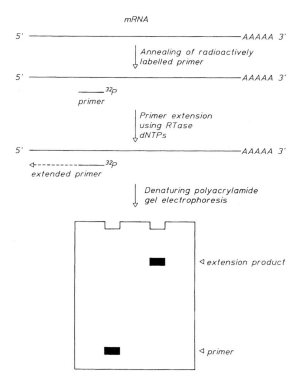

b) Any DNA polymerization reaction using a single-stranded template and starting with an
   → oligonucleotide primer. See for example → Sanger sequencing.
c) A technique to detect socalled → single nucleotide polymorphisms (SNPs) in target DNA. In
   short, the target (e.g. a gene) is first amplified with specific → primers in a conventional
   → polymerase chain reaction and subsequently denatured. Then a single-stranded primer is
   annealed to the single-stranded target DNA such that the primer ends exactly at the SNP site.
   After annealing, the duplex exposes a 3' OH-group for an extension catalysed by DNA polymerase in the presence of all four → dideoynucleoside triphosphates (ddNTPs; instead of
   → deoxyribonucleoside triphosphates, dNTPs), each labeled with a specific → fluorochrome.
   The matching ddNTP will then be incorporated and stops extension. The incorporated
   ddNTP is then identified by the specific fluorescence emission of its fluorochrome. A comparison with the wild-type sequence at the SNP site allows to identify the type of SNP.

***Primer extension preamplification* (PEP):** A technique for the sampling of an entire genome, that uses a mixture of all possible 4^n → primers (excluding primers composed of only one type of nucleotide, e. g. A_n, or G_n). Each primer has a length of about 15 nucleotides. This extremely complex mixture (4^{15} compounds) is then employed in a conventional → polymerase chain reaction to amplify the majority of sequences in a complex genome present in a single haploid cell (e.g. sperm or oocyte). PEP suffers from multiple template-independent primer-primer artifacts. See → tagged random primer PCR.

Primer hopping: See → primer-directed sequencing.

Primer jumping: See → primer-directed sequencing.

Primer-primer artifact: The appearance of amplified products in a → polymerase chain reaction, that arise by primer-primer rather than template-primer interactions. Such artifactual interactions are undesirable, since they give rise to spurious bands on → ethidium bromide-stained gels and withdraw primers from the desired primer-template interactions. Primer-primer artifacts are based on different mechanisms. If the primers partially anneal to each other or to template DNA, the DNA polymerase with its 5'→3' exonuclease activity may remove bases from the 5' end (A). If the primers only partially anneal to the template DNA, the DNA polymerase may add bases onto the 3' end, producing nonspecific amplification products (B). If a primer forms a → hairpin structure with a 3' overhang, the DNA polymerase with its 5'→3' exonuclease may remove bases from the 5' end (C). If a primer forms a hairpin structure with a 5' overhang, the DNA polymerase may add bases to the 3' end (D). If two primers with complementary 3' ends partially anneal, the DNA polymerase may add bases to the 3' ends, resulting in primer-dimer duplexes (E). See → primer dimer.
Figure see page 855.

Primer RNA: See → RNA primer.

***Primer-specific and mispair extension analysis* (PSMEA):** A technique for the detection of single nucleotide variations (e. g. → deletion, → insertion, → transition, → transversion) between two DNA → templates. The method exploits the highly efficient 3' → 5' → exonuclease proofreading activity of → *Pyrococcus furiosus* DNA polymerase, that prevents the → extension of a → primer when (1) an incomplete set of → deoxynucleotide triphosphates is present and (2) a → mismatch occurs at the initiation site of DNA synthesis (i. e. the 3'-end of the primer). For example, in the presence of only dCTP and dGTP, primer 3'-CTCTG····5' can easily be extended on template A (5'-GAGAC····3'), because the crucial nucleotide (in bold face) matches. The same primer cannot be extended on template B (5'-AAGAC····3'), so that genome A can be discriminated from genome B (presence/absence of an extension product). In contrast, the use of dTTP and dGTP allowed the extension of the primer on template B, not on template A. PSMEA therefore allows genotyping of organisms that differ in only one (or few) nucleotide pairs at the 5'end of the primer-binding site.

Primer walking: See → primer-directed sequencing.

Priming: The initiation of the synthesis of a DNA strand by the formation of an → RNA primer or by → self-priming.

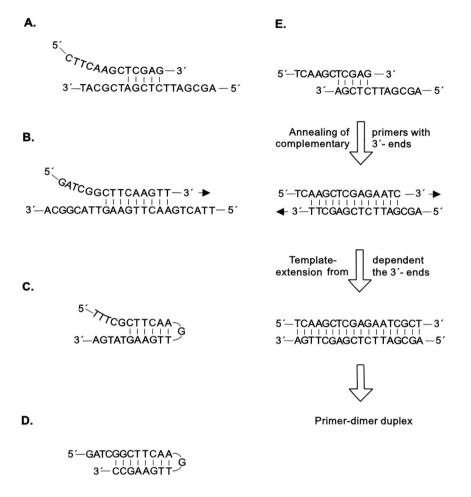

Primer-primer artifact

Priming _authorizing random mismatches polymerase chain reaction_ (PARM-PCR): A rarely used → polymerase chain reaction technique, in which specific → primers are used in combination with highly unspecific annealing conditions (low → stringency) to allow random annealing and therefore universal amplification of target sequences.

Primosome (primosome complex, replisome): A mobile multi-protein DNA replication-priming complex of _E. coli_, consisting of the proteins encoded by genes _dnaB, dnaC, dnaG, dnaT, n, n″_, and the replication factor Y (DNA helicase). The primosome assembles at a specific site on single stranded DNA (primosome assembly site, _PAS_; thought to be located on the → lagging strand at the → replication fork), moves along the DNA in 5' → 3' direction and occasionally synthesizes an → RNA primer. This polymerization reaction is catalyzed by the _dnaG_-encoded → RNA primase.

Primosome assembly site: See → primosome.

Primosome complex: See → primosome.

PRINS: See → primed *in situ* labeling.

Printed microarray: Any glass wafer (chip), onto which small volumes of oligonucleotides are uniformly spotted via a non-contact inkjet process. Alternatively, the term is also used for glass chips, on which oligonucleotides are synthesized base-by-base, using standard phosphoramidite chemistry.

Prion (*proteinaceous infectious particle*): A cellular protein (PrP^c) with as yet unknown function in synaptic transmission. The primary transcript for this protein encodes a pre-protein of 254 amino acids, that is posttranslationally trimmed by 22 amino acids at the amino terminus (the released peptide is a → signal peptide) and 23 amino acids at the carboxy terminus. This trimmed protein contains two N-glycosylation sites and a disulfide bridge, and is anchored in the membrane of neuronal cells via a glycophosphoinositol (GPI) anchor. The prion exists in two isoforms, PrP^c and PRP^SC (sc stands for *sc*rapie, a disease of sheep leading to loss of motoric control and degeneration of the brain). In contrast to PrP^c, the PrP^SC is extremely insoluble, resistant to proteolytic degradation and less sensitive to heat denaturation. PrP^SC is probably a derivative of PrP^c. Once PrP^c is converted to a PrP^SC, then an autocatalytic process progressively increases the concentration of PrP^SC, which is ultimately deposited as socalled amyloid plaques or rods in the brain, leading to the outbreak of the scrapie disease (*spongiform encephalopathy*, SE). The human equivalent of the scrapie disease, the socalled *Creutzfeld-Jacob disease* (CJD), as well as the *bovine spongiform encephalopathy* (BSE) are also associated with prions that are able to "replicate" in the absence of → DNA or → RNA.

PRM: See → protein recognition molecule.

P-RNA: See → pyranosyl RNA.

Probe:
a) A defined and radioactively or non-radioactively labeled nucleic acid sequence used in → molecular cloning to identify specific DNA molecules with complementary sequence(s) by → autoradiography or with non-radioactive → DNA-detection systems. The term is also used for proteins (e.g. a → monoclonal antibody reacting with its target protein).
b) A somewhat misleading, but widely accepted term for any defined nucleic acid sequence (e.g. an → oligonucleotide, a → cDNA), that is covalently bound to a carrier ("chip") made of glass, quartz, or plastic, and hybridized to a → target nucleic acid (mostly cDNAs). Thousands of such probes are assembled on socalled → cDNA expression arrays, → DNA chips, → microarrays, → sequencing arrays and serve for the simultaneous detection of multiple hybridization events ("massively parallel"). Note, that the conventional term "probe" (definition a) has been converted to another meaning in chip technology. Here, probes are also termed "reporters".

Probe complexity: A measure for the number of different nucleic acid sequences in a → probe.

Probe excess: The presence of very high → probe concentrations on the surface of a → microarray such that the array-bound target is saturated. Probe excess is undesirable, because it obscures quantitative differences in hybridisation signals.

Prober sequencing: An outdated laboratory slang term for a → DNA sequencing technique with fluorescent chain-terminating → dideoxynucleotides (ddNTPs), named after J.M. Prober and colleagues. As a variant of the → Sanger sequencing method, it starts with a → polymerase chain reaction driven by a → primer oligonucleotide and using ddNTPs labeled with slightly different succinyl *fluorescein*s (SFs). The different fluoresceins vary in their absorption maxima (e.g. SF–505: 486 nm; SF–512: 493 nm) and emission spectra (e.g. maximum of SF–505: 505 nm; SF–512: 512 nm), and their incorporation leads to chain termination, leaving each molecule labeled at its 3'- end with an SF-ddNTP. During the electrophoretic separation, the various chains are excited with an argon laser at 488 nm and the emitted → fluorescence light monitored by parallel filter/ photomultiplier systems.

Probe set: Any collection of → probes, that altogether represent a target sequence (e.g. a gene). For example, a set of about 20 oligonucleotides of 50 nucleotides each, derived from different regions of a known gene, and thus representing the gene, can be spotted onto a → microarray. Then labeled → cDNAs can be hybridized to the microarray and the expression of the gene of interest be detected. The presence of probes complementary to various regions of one distinct gene allows to discriminate between various gene → homologues or → splicing variants.

Procapsid: The → capsid precursor formed during the assembly of viral capsomers.

Procaryotes: See → prokaryotes.

Processed gene: See → processed pseudogene.

Processed pseudogene (retropseudogene, retrosequence, retrogene, processed gene): An intron-less → pseudogene, that contains a poly(A) tract at its 3' end and is flanked by short → direct repeats of 10-20 nucleotides, which are potential → insertion sequences. Processed pseudogenes most probably arise through reverse transcription of → messenger RNA, after it has been processed (e. g. spliced and polyadenylated), and the integration of the product cDNA into an arbitrary site of the genome. Two classes of processed pseudogenes can be discriminated. The complete re-tropseudogene contains all the → exons of the gene from which it is transcribed, whereas the truncated retropseudogene harbors only a fraction of the exons.

Processing (editing):
a) The → post-transcriptional modification of → primary transcripts.
b) The → post-translational modification of proteins.

Processive DNA polymerase: Any → DNA-dependent DNA polymerase that remains associated with its → template during successive steps of nucleotide incorporations. See → distributive DNA polymerase, → processive enzyme, → processivity.

Processive enzyme: Any enzyme that does not dissociate from its substrate between repetitions of the catalytic event. For example, DNA-dependent DNA polymerase is such a processive enzyme, since it continues to polymerize nucleotides after adding the first nucleotide to e. g. a primer. See → processive DNA polymerase, → processivity.

Processive transcription: Any gene → transcription, where → initiation and elongation are highly efficient, so that high levels of polyadenylated RNAs accumulate. See → nonprocessive transcription.

Processivity (processivity index, processivity value): The extent to which DNA-dependent → DNA polymerases use their → template strand before they dissociate from it (expressed as number of nucleotides incorporated per binding event). The processivity of different, especially purified DNA polymerases in vitro is different. Thus some enzymes allow the synthesis of short DNA strands only, though the template strand is not yet fully copied. See → processive DNA polymerase, → processive enzyme.

Processivity clamp (sliding clamp): A ring-shaped protein (dimer or trimer), that encircles double-stranded DNA and binds to DNA-dependent DNA polymerase, thereby increasing its → processivity.

Producer gene: Synonym for → structural gene (→ Britten-Davidson model).

Productive base pairing: The pairing between two bases in DNA, that are perfectly complementary to each other. For example, A-T, T-A, C-G and G-C are such productive base pairs. See → non-productive basepairing.

Productive infection: See → lytic infection.

Prognostic reporter: A → transcript derived from a gene, that is up-regulated at the onset of tumorigenesis or during the course of tumor establishment, and serves as a → molecular marker ("expression marker") for the clinical outcome (e.g. overall and relapse-free survival). Such prognostic reporters are identified from transcript profiles of a large number of tumor samples, generated by e.g. → expression microarrays, and may outperform currently used clinical parameters. See → reporter gene.

Programmable *autonomously-controlled* *electrodes* (PACE) gel electrophoresis: An improved version of the conventional → pulsed-field gel electrophoresis which utilizes a hexagonal array of 24 computer-controlled electrodes around an → agarose gel, allowing the generation of defined, homogeneous electric fields. The field direction may be alternated during a run in a preprogrammed way, since each electrode or set of electrodes can be individually controlled by a high-voltage operation amplifier driven by a power supply. With this specific arrangement a nearly linear separation of DNA fragments in the range from 500 to more than 10 million bp can be achieved. See → gel electrophoresis.

Programmable chip: Any → DNA chip produced in a completely automated process, that is designed on a computer and customized to given and requested experimental conditions. After hybridisation of target RNA, the binding events are monitored in digital form.

Programmable *melt*ing display *m*icroplate-*array* *d*iagonal g*el* *e*lectrophoresis (melt MADGE): A variant of the → microplate-array diagonal gel electrophoresis (MADGE) technique, that is based on the separation of amplified PCR products (see → polymerase chain reaction) by temporally changing the running temperature during → polyacrylamide gel electrophoresis, which readily distinguishes between the non-mutated (one single band) and the mutated sequences (four bands: two → homo-duplexes, two → heteroduplexes). This type of separation can also be achieved by → denaturing gradient gel electrophoresis. melt MADGE is therefore used for the *de novo* mutation scanning of target DNA, and requires only one hour for separation and little starting material and gel, and additionally is adapted to a → microplate (i.e. 96-well) format.

Programmed ribosomal frameshifting: See → ribosomal frameshifting.

Prohibitin gene: An evolutionary conserved mitochondrial gene, that encodes the protein pro-hibitin functioning as negative regulator of cell proliferation and life span. Prohibitin is associ-ated with senescence and cell death in yeast and mammalian cells.

Prokaryotes (procaryotes): Members of the superkingdom that contains archaebacteria, eubac-teria, and cyanobacteria (formerly, blue-green algae). Prokaryotes did not evolve a membrane bound nucleus with chromosomes, but instead possess a circular DNA genome anchored at the membrane. Neither do they contain mitochondria nor plastids or microtubules.

Promiscuous DNA: See → promiscuous gene.

Promiscuous gene (promiscuous DNA): Any → gene or DNA fragment which has been moved or is still being moved from one organelle to another in the eukaryotic cell (e.g. nuclear genes encoding mitochondrial or plastid functions and believed to originate from the respective orga-nellar genomes).

Promiscuous plasmid: See → plasmid promiscuity.

Promoter (promotor): A → *cis*-acting DNA sequence, 80-120 bp long and located 5' upstream of the initiation site of a gene to which → RNA polymerase may bind and initiate correct → tran-scription (see also → 5' flanking region). Prokaryotic promoters contain the sequences 5'-TAT-AATG-3' (→ Pribnow box) approximately at position -10, and 5'-TTGACA-3' at position -35. Eukaryotic promoters differ for the different DNA-dependent RNA polymerases (see also → core promoter). RNA polymerase I recognizes one single promoter for → rDNA transcrip-tion, RNA polymerase II transcribes a multitude of genes from very different promoters, which have specific sequences in common (e.g. the → TATA box at about position -25 and the → CAAT box at about position -90. The so-called → house-keeping genes contain promoters with multiple GC-rich stretches with a consensus core sequence, 5'-GGGCGG-3'. RNA polymerase III recog-nizes either single elements (e.g. in 5S RNA genes) or two blocks of elements (e.g. in all → transfer tRNA genes) within the gene (→ internal control regions). All these consensus sequences func-tion as address sites for DNA-affine proteins (→ transcription factors) that promote or reduce transcription. Specific promoter sequences can be identified in → genomic DNA using → pro-moter trap vectors, in cloned DNA fragments using → promoter probe vectors. The promoter regions of various → inducible genes are useful tools for the construction of → cloning vectors with specific requirements; see for example → heat-shock promoter, → heavy metal resistance promoter, → light-inducible promoter, → *tac*, *trc* and → *trp*, and → tissue-specific promoters. See also → bidirectional promoter, → cell-specific promoter, → chimeric promoter, → consti-tutive promoter, → core promoter, → cryptic promoter, → decoy promoter, → divergent pro-moter, → downstream promoter, → dual promoter, → Emu promoter, → hybrid promoter, → internal promoter, → low level promoter, → minimal promoter, → portable promoter, → promoter strength, → promoter-up mutant, → pseudo-promoter, →regulated promoter, → split promoter, → strong promoter, → synthetic promoter, → *tac* promoter, → tandem pro-moter, → 35S promoter, → *trc* promoter, → *trp* promoter, → twin promoter, → upstream pro-moter, → weak promoter, → *wun* promoter.

Promoter clearance: The ATP-dependent escape of a transription elongation complex stalled after synthesis of the first 10–17 nucleotides of → messenger RNA, catalyzed by the ERCC3 subunit of transcription factor IIH. See → initiation.

Promoter core: See → core promoter.

Promoter hypermethylation: The methylation of most, if not all, → cytosine residues in a → promoter sequence (see → localized hypermethylation), that leads to the inactivation of the promoter and the adjacent gene. Loss-of-function by promoter hypermethylation is common in several cancer-related genes involved in → DNA repair, cell cycle control, → apoptosis, angiogenesis, cellular differentiation, metatstatic invasion, → transcription and signal transduction.

Promoter insertion: The integration of a → promoter or promoter-containing DNA segment in front of a promoter-less or otherwise inactivated gene with the result of activation of the gene. See also → insertional activation.

Promoter melting: The ATP-dependent and transient breakage of hydrogen bonds ("melting") of about one turn of DNA encompassing the → transcription start site to form the socalled "transcription bubble", catalysed by transcription factor IIH (in humans) or analogous proteins in other organisms.

Promoter module: A structural and functional unit of eukaryotic → promoters, that is composed of two (or more) → transcription factor binding sites (TF sites) in a defined distance from each other. This arrangement allows for synergistic or antagonistic influences of the different transcription factors – that bind to the different TF sites – on each other, which in turn result in stimulation or inhibition of the → transcription of the adjacent gene.

Promoter mutation: Any → mutation, that occurs within the → promoter sequence of a gene. For example, socalled *aphakia* (*ak*) mouse mutants (aphak: without lens), that do not form any lens or pupil in their otherwise normal embryonic development, the underlying gene *Pitx3* on chromosome 19 is absolutely identical to the wild-type gene. However, two deletions of 652 and 1423 bp, respectively, in the promoter and in the transcription initiation region lead to an almost complete silencing of the *Pitx3* gene. As a consequence, the encoded → homeobox → transcription factor is not functional, and the eye development does not occur.

Promoter occupancy: The extent to which all of the → transcription factor-recognition sites of a → promoter are occupied at a given time. Promoter occupancy can be tested with e.g. → chromatin cross linking with immune precipitation.

Promoter opening: The localized formation of a short stretch of melted DNA ("DNA bubble") near the 3'end of a → promoter. The opened structure is accommodated in a channel of the bacterial RNA polymerase, and serves to assemble the preinitiation complex.

Promoter plasmid: A → plasmid cloning vector that contains an → RNA polymerase → promoter that can drive genes cloned into a cloning site 3' downstream of it.

Promoter polymorphism: An imprecise term for any → sequence polymorphism, that occurs in a → promoter of a gene. Usually such polymorphisms are caused by small → deletions, → insertions, or, most frequently, → single nucleotide polymorphisms. See → promoter SNP.

Promoter primer: Any synthetic → oligodeoxyribonucleotide that is complementary to a conserved sequence in either → T7, → T3, or → SP6 RNA polymerase promoters. Such primers are used to sequence inserts in plasmids containing these promoters.

Promoter probe vector: A → cloning vector that contains appropriate cloning site(s) located just 5' (upstream) of a promoter-less → reporter gene (e.g. a bacterial → β-glucuronidase gene). Any foreign DNA cloned into such a vector and possessing promoter elements will drive the expression of the reporter sequence. For example, the vector plasmid pPL 603 designed to clone promoter sequences from *Bacillus subtilis*, contains a → kanamycin resistance gene and a promoter less *Bacillus* → structural gene for → chloramphenicol acetyl transferase (CAT). Immediately upstream of the CAT coding sequence a unique *Eco* RI site is located into which foreign DNA can be inserted. If the foreign DNA contains a promoter element, then the CAT gene will be express ed. Compare → promoter trap vector, which allows the identification of promoter sequences in → genomic DNA after being itself inserted into chromosomal sites.

Promoter proximal pausing: The transient blockage of the → elongation of the nascent → messenger RNA by → DNA-dependent RNA polymerase II about 20–60 bases downstream of the → transcription start site. The release of this blockage relaxes the limiting step in the transcription of e.g. the human *c-myc* gene and the *Drosophila hsp70* gene. The pausing of RNA polymerase II can also be suppressed by elongation factors such as SII, TFIIF, ELL, or elongin.

Promoter scanning: Any procedure to screen → promoter sequences for specific functions. For example, the → methylation interference analysis uses the *in vitro* methylation at → purine residues of the target sequence to prove that this sequence can bind proteins. If so, then this sequence is considered to be an address site for → DNA-binding proteins.

Promoter shuffling: The recombination of individual regulatory elements of different → promoters to create a new promoter with new function(s). Shuffling involves the gain or loss of such elements ("boxes") during evolution. Usually the length, copy number and relative location of the boxes vary in different → orthologous gene promoters, a results of gross changes such as → amplification, contraction, → deletion, duplication, elongation, → fusion, → inversion, → transposition, and the steady accumulation of → single nucleotide polymorphisms. Compare → exon shuffling, → intron shuffling.

Promoter single nucleotide polymorphism (promoter SNP, pSNP): Any → single nucleotide polymorphism, that occurs in the → promoter sequence of a gene. If a pSNP prevents the binding of a → transcription factor to its recognition sequence in the promoter, the promoter becomes partly disfunctional. See → anonymous SNP, → candidate SNP, → coding SNP, → exonic SNP, → gene-based SNP, → human SNP, → intronic SNP, → non-synonymous SNP, → reference SNP, → regulatory SNP, → synonymous SNP.

Promoter strength: The frequency with which an → RNA polymerase molecule can bind to specific → consensus sequences within a → promoter and express the linked gene. It depends on specific sequences (e.g. → TATA box, → CAAT-box) and their exact → spacing within the promoter region. Compare → promoter-up mutant. See → strong promoter, → weak promoter.

Promoter trapping: The use of → promoter trap vectors for the isolation of → promoters.

Promoter trap vector: A → transformation vector that carries a transcriptionally and translationally incompetent (i.e. promoter-less) → reporter gene (e.g. → neomycin phosphotransferase, or → β-glucuronidase gene) and used to select → promoter sequences from a → genomic library. If such a vector integrates at a position in the host genome where promoter sequences are located, the reporter gene will be transcribed, allowing the identification of promoter-containing genomic fragments. See for example → T-DNA mediated gene fusion. Compare → promoter probe vector.

Promoter-up mutant (up-promoter mutant, up mutant): Any → mutant with a → mutation in the → promoter of one of its genes, that leads to a higher rate of expression of this gene. In gene technology such up-promoters are used for the → overexpression of genes encoding useful proteins.

Promotor: See → promoter.

Pronase: A mixture of serine proteases and acid proteases from *Streptomyces griseus* that catalyze the cleavage of → peptide bonds in proteins. Pronase is used to degrade proteins in RNA and DNA isolation procedures, and is especially effective for the isolation of intact RNA molecules through its deleterious effect on → RNases. Compare → proteinase K.

Pronuclear injection (pronuclear microinjection): A variant of the → microinjection technique for the transfer of foreign genes into target cells, which works with a fine glass needle to inject about 200 – 300 linearized copies of a gene of interest (fused to a promoter) into previously fertilized egg cells. After injection, the eggs are briefly cultured and then implanted into surrogate mothers. Only 1–5% of the newborne animals are transgenic (because integration of the → transgene into the host cell chromosome is extremely rare and additionally a totally random process), and only a fraction of these transgenics express the foreign gene strongly. Pronuclear injection suffers from mosaic expression of the transgene (i.e. its expression in some, but not all cells of the → transgenic animal).

Proof-reading ("editing"): The correction of errors in DNA → replication or DNA → transcription. Proof-reading of replicative errors is catalyzed by DNA polymerase, which recognizes and removes incorrectly inserted bases by its 3'-5' exonucleolytic function.

Prophage: A → bacteriophage DNA integrated covalently into the chromosomal DNA of a bacterial host, and replicated as part of the host's genome. Compare → lysogenic bacteriophage, also → temperate phage. See also → phage exclusion.

Prophage interference: See → phage exclusion.

Prophage-mediated conversion: See → phage conversion.

Propidium iodide (PI): A → fluorochrome that binds to DNA. The propidium-iodide-DNA complex can be excited at a wave length of 488 nm, and emits red fluorescent light. PI is the most commonly used dye for → flow cytometry. See → ethidium bromide.

Propyne pyrimidine (C-5 propyne pyrimidine): Any → deoxycytidine or → deoxyuridine, in which a propyne group is introduced at the C-5 position of the base. This modification results in a stronger binding of the base to target RNA sequences.
Figure see page 863.

Propyne-dC

Propyne-dU

Protamine: An arginine-rich, highly basic protein that replaces → histones in sperm heads of various animals to package the DNA into an extremely compact form. The protamines have molecular weights of 3000-5000, contain up to 50% of arginine residues, that are mostly clustered, and can be classified into monoprotamines with one type of basic amino acid (arginine; e.g. clupeine, salmine, iridine, truttine, esocine), diprotamines (arginine and lysine; e.g. barbine, cyprinine, pereine) or triprotamines (arginine, lysine, histidine; e.g. sturgeone).

Protease (proteinase): An enzyme catalyzing the hydrolysis of → peptide bonds in proteins or oligopeptides. Proteinases may be classified according to the chemical nature of the amino acids located in their reactive center (e.g. serine proteases, acid proteases). See for example → pronase, → proteinase K.

Proteasome (26S proteasome): A nuclear and cytoplasmic 2,000 kDa multi-protein complex consisting of 45-50 individual polypeptides organized in several sub-complexes, that altogether function in the proteolytic degradation of abnormal proteins and the bulk turnover of all other proteins. Moreover, sub-sets of proteasomal proteins are involved in → antigen processing, cell cycle control, cell differentiation, stress response, pre-protein cleavage, and programmed cell death.

The degradation of proteins starts with their tagging by multiple enzymatic → ubiquitin (Ubq) binding to the ε-amino group of specific internal reactive lysine residues, catalyzed by a multi-enzyme complex consisting of Ubq-activating, Ubq-conjugating and Ubq-ligating enzymes. The conjugated ubiquitin moieties function as "secondary" → degradation signals (degrons) for proteasomal proteases. The proteasome itself consists of a core complex with multiple peptidase activities (the 20S proteasome with chymotrysin-like sites [cleavage after hydrophobic residues], trypsin-like sites [cleavage after basic residues], and caspases [cleavage after acidic residues]), a barrel-like structure with central catalytic centers. Each degron-tagged protein has to be completely unfolded to be channeled to these centers, which is achieved by an 11S particle (PA 700) consisting of 15-20 different polypeptides, associated with each end of the 20S complex. The whole degradation process requires ATP (at least six AAA-ATPases are part of the 11S particle, that act as reverse → chaperones and unfold the substrate proteins). An average human cell contains about 30,000 such proteasomes. Compare → degradosome.

Proteasome inhibitor: Any one of a series of peptides or peptide conjugates that inhibits the activity of one of the six proteolytic sites on the → rings of the → proteasome. For example, synthetic peptide aldehydes such as Z-leu-leu-leu-al, Z-leu-leu-nval-al, Ac-leu-leu-nle-al, or Z-ile-glu-ala-leu-al reversibly inhibit the proteasomal chymotrypsin-like activity (but also other proteases, e.g. calpains, serine proteases of digestive vacuoles, and lysosomal cysteine proteases). Other proteasome inhibitors are more specific, as e.g. peptide boronates (e.g. Z-leu-leu-leu-boronate), synthetic peptide vinyl sulfones, epoxyketones such as expoxomicin, or the non-peptide lactacystin (that converts to the active form β-lactone in aqueous solution). Proteasome inhibitors are used as antiviral drugs and allow to study the functions of proteasomes.

Protectifer: The direct transfer of nucleosomally organized or otherwise protected genes into eukaryotic recipient cells (see → direct gene transfer). Protection is usually achieved by complexing DNA with proteins (e.g. → histones, → protamine from salmon sperm, protein VII from adenovirus type 2, HMG-1), but low molecular weight compounds such as spermidine have also been used successfully. Protectifer enhances the → transformation rate, the integration of *intact* copies of the transferred gene into the target genome, and generally stimulates expression of the transgene.

Protecting group: Any chemical group that minimizes or prevents undesirable side reactions during *in vitro* DNA or RNA synthesis. It must be resistant to the conditions of synthesis and be easily removed after completion of polymer synthesis.

Protein A: A cell wall protein from *Staphylococcus aureus* that recognizes the Fc portion of most → antibodies only when the antibody is bound to its → antigen. Protein A, either labeled with a fluorescence marker or a radioactive atom (e.g. ^{125}I) is used to detect antibodies in a variety of antibody screening procedures (e.g. → sandwich techniques, → direct immune assays), or functions as ligand in affinity chromatography of immunoglobulins and antigen-antibody complexes. Compare → protein G.

Protein array: See → protein chip.

Proteinase K (EC 3.4.21.14): An enzyme from the fungus *Tritirachium album* that catalyzes the cleavage of peptide bonds in proteins with a slight preference for aliphatic, aromatic or other hydro-phobic amino acids. The enzyme belongs to the subtilisin type serine proteases, and is effectively used to degrade proteins (including → RNases and → DNases) in RNA or DNA isolation procedures. The activity of proteinase K can be stimulated by denaturants (e.g. SDS, urea) and elevated temperature. Autolysis of the enzyme is minimal, and may be prevented by Ca^{2+}. Compare → pronase.

Protein *association cloning* (PAC): A term encompassing several → cloning techniques that use → antibodies to screen → cDNA → expression libraries. Usually the antibody is radioactively labeled and interacts with its target protein expressed in a → bacteriophage or → plasmid expression library. This interaction can be detected by → autoradiography and the corresponding clone carrying the target cDNA be easily isolated.

Protein atlas: A laboratory slang term for an inventory of all peptides and proteins of a cell at a given time, their interactions, and localization. See → protein interaction mapping, → protein linkage map, → protein profiling, → protein-protein interaction map, → protein signature, → proteome, → proteome mapping, → proteomic fingerprint.

Protein biochip: See → protein chip.

Protein biomarker: Any peptide or protein, that is associated with and characteristic for a specific state of a cell, a tissue, an organ, or an organism. Such markers are identified with high-through-put screening techniques such as → protein chips, e.g. using normal and diseased samples.

Protein biosensor: Any protein tagged with a fluorophore, that responds to any environmentally induced change in structure, ligand binding, or catalytic activity of the protein with an increase or decrease of emitted fluorescence. Usually the biosensors are transferred to a target cell by → microinjection. The construction of a protein biosensor may require → protein engineering. For example, to engineer a myosin light chain kinase biosensor, a cysteine residue has to be introduced into the myosin light chain next to the phosphorylated amino acid. This cystein is then specifically labeled with the fluorochrome acrylodan that reacts upon phosphorylation/ dephosphorylation events of the light chain with a change in fluorescence.

Protein blot: See → Western blotting.

Protein blotting: The transfer of electrophoretically separated proteins from a gel (usually a → polyacrylamide gel) to a membrane by diffusion (diffusion blotting), by liquid flow (using capillary forces, or vacuum, → vacuum blotting) or by electrophoretic transfer (→ electroblotting). Most commonly used are → nitrocellulose (binding capacity: 80 µg/cm^2), but also nylon based (binding capacity: 480 µg/cm^2) and polyvinylidene difluoride membranes (binding capacity: about 500 µg/cm^2). After transfer, the filters may be used for → Western blotting or → South-Western blotting experiments. Compare → cDNA expression array, → DNA chip, → expression array, → gene array, → microarray, → sequencing array, → sequencing by hybridization.

Protein chip (protein array, protein biochip, protein microarray, protein microchip, protein-protein interaction chip): A glass slide or other solid support, onto which thousands of proteins are spotted in an ordered array to allow the visualization of protein-protein (e.g. protein-antibody) or protein-ligand interaction(s). In short, the glass slides are first treated with an aldehyde-containing silane, because aldehydes react with primary amines of proteins to establish a Schiff's base linkage (covalent coupling). Alternatively, the glass surfaces can be coated with poly-L-lysine (non-covalent coupling). Then purified proteins, each in a nanodroplet (containing billions of protein copies; the concentration should be 100 µg/ml or less) are spotted onto the glass slide, that is additionally coated with → bovine serum albumin (BSA) to prevent the denaturation of the spotted protein, which is then covalently bound to the glass surface. For peptides or very small proteins, a molecular layer of BSA is first attached to the glass, which is then activated with N,N'-disuccinimidyl carbonate to yield BSA-N-hydroxysuccinimide, and quenched with glycine. The peptides are then displayed on top of the BSA monolayer, which readily exposes them for interaction(s) with target molecules. Then a ligand is fluorescently labeled and exposed to the protein array, and binding of the ligand to a protein (or to many proteins) detected by fluorescence. Using different fluorophores with non-overlapping excitation and emission spectra expands the number of simultaneously detectable interactions. Such protein chips can be used to study the → proteome of a cell. Alternatively, → monoclonal antibodies can be bound onto the surface of microarray chips, to which peptides or proteins from a cell extract are bound. Compare → antibody array, → cDNA expression array, → DNA chip, → expression array, → functional protein array, → microarray, → protein domain array, → recombinant protein array, → sequencing array, → sequencing by hybridization.

Protein complementation assay (CPA): An *in vivo* selection system for peptides or proteins with novel properties, that is based on the functional complementation of the bacterial *di*hydro*f*olate *r*eductase (DHFR) with its murine equivalent (mDHFR). In short, the mDHFR gene is genetically dissected into two fragments, each encoding a part of the enzyme necessary for its function. Each of these fragments is then separately fused to a library of peptides or proteins. *E. coli* cells are then co-transformed with both fusion libraries and plated on selective medium (containing the antibiotic → *tri*metho*p*rim, TMP). TMP in turn specifically inhibits bacterial DHFR, thereby prevents the synthesis of essential purines, also pantothenate and methionine and therefore blocks cell division. The complemented mDHFR is insensitive to low TMP concentrations and allows *E. coli* cells to grow on minimal medium containing TMP. Therefore, the complementation of DHFR as a result of the interaction of members of the different libraries (heterodimerization) restores DHFR enzymatic activity and guarantees survival (growth of colonies). The surviving bacteria are directly accessible for the sequencing of the interacting library partners. See → directed molecular evolution, → ribosome display.

Protein complex: An aggregate of various proteins, whose coordinated activities exert a novel function, that the individual proteins do not possess. For example, the → primosomes, → proteasomes, → repairosomes, → ribosomes, → spliceosomes, and → transcriptosomes are such protein complexes, to name only few.

Protein computer: See → peptide computer.

Protein conformation: The tertiary structure of proteins. See → α-helix, → arginine fork, → β-sheet, → helix-loop-helix, → helix-turn-helix, → random coil; → zinc cluster, → zinc finger and → zinc twist proteins.

Protein cross-linking: The formation of covalent bonds between free amino groups of two individual proteins by the action of bifunctional cross-linking chemicals (as e. g. *N*-hydroxy*s*uccinimide [NHS] esters such as disuccinimidyl tartarate, disulfosuccinimidyl tartarate, dithio bis[succinimidyl propionate], or ethylene glycolbis [succinimidylsuccinate]) that react primarily with the ε-amino groups of lysine side-chains. Protein cross-linking in vivo is possible through the use of water-soluble (or even water-insoluble) agents, that permeate cellular membranes and e. g. preserve intracellular structures (as e. g. actin filaments, actin-binding proteins such as gelsolin, and tubulins) for subsequent analysis.

Protein design: The development of synthetic (i.e. not naturally occurring) proteins with predictable structural and functional properties, supported by special computer programs. The amino acid sequence of such designed proteins may then be translated into a nucleotide sequence, see → gene design. Compare → protein engineering.

Protein domain array: A glass slide or other solid support, onto which native or synthetic peptides representing specific → domains of a protein are spotted at high density. Domain arrays serve to map protein-protein interactions, to detect interactions between ligands and domains and to profile the specificity of enzymes (e.g. kinases or proteases) for their protein substrates. See → functional protein array, → protein chip, → recombinant protein array.

Protein dynamics: The multitude of conformational changes in a protein as a result of its interaction(s) with intrinsic or environmental factors (e.g. ligands).

Protein electroblotting: See → protein blotting, and → electroblotting.

Protein engineering: The modification of the physico-chemical or biological properties (i.e. the reaction kinetics, substrate affinity, substrate specificity, effector sensitivity, thermostability or -lability, intracellular location) of a naturally occurring protein with the ultimate aim of improving the protein's quality for biotechnological processes. One of the techniques used in protein engineering is → in vitro mutagenesis that allows to change the coding capacity of a gene at defined locations (e.g. within the region encoding the active center of the protein). In consequence, the engineered protein adopts different properties that are useful for biotechnology. Compare also → protein design.

Protein expression: An infelicitous and misleading term for → gene expression. The terms "protein expression" and " protein expression profiling" should be avoided, because only genes can be expressed (into → messenger RNAs, which in turn are translated into proteins, the products of gene expression).

Protein expression map: A graphical depiction of the subcellular distribution, abundance and relative concentration of (preferably) all peptides and proteins, also their post-translational modifications, in a cell at a given time. A protein expression map is the result of → protein expression mapping.

Protein expression mapping **(PEM):** The description of the abundance and subcellular distribution of preferably all peptides and proteins of a cell, a tissue, an organ or an organism under defined physiological conditions at a given time. PEM exploits all → proteomics techniques. See → protein interaction mapping.

Protein family expansion: The evolutionary increase in the number of proteins of a protein family, reflecting an increase in the number of corresponding genes. For example, a comparison of protein families in *Anopheles gambiae* versus *Drosophila melanogaster* reveals an expansion of various protein families in the former: many more genes encode proteins for cell adhesion, anabolic and catabolic enzymes involved in protein and lipid metabolism, lysosomal enzymes and salivary gland proteins, reflecting blood feeding and oviposition.

Protein fingerprinting (peptide fingerprinting): A technique to characterize a protein by partial proteolytic cleavage that generates a pattern of → peptide fragments characteristic for this protein. In short, the purified protein is cleaved separately with endoproteinases (e.g. Glu-C, that cleaves at the carboxylic side of glutamic acids, Lys-C that cleaves at the carboxylic side of lysine residues or Arg-C [clostripain], that cleaves at the carboxylic side of arginine residues), and alkaline protease (that cleaves preferentially at aromatic residues). The proteolysis products are then separated according to size by electrophoresis or chromatography and visualized by staining. Protein fingerprints can be used to confirm the identity between the product of a cloned gene and its natural counterpart, or for the analysis of multi-subunit proteins.

Protein fold: The specific three-dimensional structure of a protein adopted in solution, that can be analysed by e.g. X-ray crystallography. See → RNA fold.

Protein fold disease (protein folding-related disease): Any human disease, that is caused by a misfolding of a protein (see → protein fold). For example, so called amyloidoses such as the Alzheimer or Creutzfeld-Jakob diseases (involving deposits of aggregated proteins ["amyloid plaques"] in many tissues of a patient), lung diseases such as cystic fibrosis or hereditary emphysema (involving → mutations leading to the degradation of proteins with vital functions for the respiratory tract), diabetes (misfolded proteins disrupt carbohydrate metabolism or accumulate in the endoplasmic reticulum with toxic consequences), cancer (misfolding of tumor-suppressor proteins abolish their tumor-suppressor function. For example, mutation(s) in the gene encoding → p53, such a tumor-suppressor protein, causes a misfold of the corresponding protein, that in turn fails to recognize its target. About half of all tumors are probably caused by such a p53 misfolding), or infection diseases (pathogens interfere with the host *ER-a*ssociated *d*egradation (ERAD) system, that is responsible for the removal of terminally misfolded proteins) are such folding-related diseases, to name only few.

Proteinformatics (protein informatics): The whole repertoire of software packages for the evaluation of terabytes of data generated by high-throughput → proteomics. Proteinformatics employs software for the identification of peptides and proteins and the prediction of their structure (e.g. Piums), the computation of → peptide fingerprints (e.g. BioAnalyst, biotools, Ettan MALDI-TOF software, Knexus, Mascot, ProtoCall, Radars), → mass spectrometer (MS) peak extraction, including peak picking and reporting (e.g. Pepex, ProID, SNAP), cross-correlation of MS/MS mass spectra of peptides with protein (or nucleic acid) databases (e.g. Mascot, Turbo-SEQUEST), → post-translational modification analysis and de novo sequencing (ProteinLynx Global SERVER). Frequently appropriate software is developed for a specific instrument (e.g. → MALDI) ana sold in combination.

Protein *f*usion and *p*urification technique (PFP): A method for the expression and purification of proteins from cloned genes by fusing them to the *m*altose-*b*inding *p*rotein (MBP). The fusion protein is extracted from the producer cell and purified in a one-step affinity column chromatography. The affinity column retains any MBP (together with the fused protein). The bound proteins can be released in almost pure form. A variant of → protein tagging.

Protein G: A cell wall protein of certain strains of *Streptococcus* that binds to a wide variety of IgG → antibodies by a non-immune mechanism (i.e. does not involve the antigen-binding site of IgG). Protein G is used to purify antibodies, and to detect antibodies in a variety of antibody screening procedures (e.g. → sandwich techniques, → direct immune assays) in which it is either labeled with a fluorescence marker or a radioactive atom (e.g. [125]I). Compare → protein A.

Protein insert: See → intein.

Protein *in* s*it*u *a*rray (PISA): A → protein chip, on which tagged functional proteins are immobilized, that are synthesized by → cell-free expression directly from PCR-amplified genes and simultaneously bound to the chip surface. PISAs can therefore be made in a single step from DNA, avoiding cloning, cell-based expression and protein purification.

Protein interaction domain: Any protein → domain (an independently folding module of 30 – 150 amino acids, whose N- and C-termini are in close proximity to each other, while the ligand-binding site is on the opposite face of the domain) that mediates interaction with another peptide or protein, or other proteins. Protein interaction domains can be characterized by their common sequences or ligands. For example, many cytoplasmic proteins contain one or two socalled SH2 (Src-homology 2) domains recognizing phosphotyrosine moieties and 3 – 6 residues C-terminal to the phosphotyrosine of e.g. activated receptors for growth factors, antigens or cytokines. Other proteins harbor socalled PTB domains, recognizing phosphotyrosine in the context of Asn

– Pro – X – Tyr (NPXY) motifs (which represent ß-turns). These molecular interactions lead to protein networks mediating e.g. signaling from cell surface receptors to intracellular metabolic pathways.

Protein interaction map: See → protein-protein interaction map.

Protein interaction mapping (PIM): The process of establishing a → protein-protein interaction map. See → protein expression mapping.

Protein interactome: See → interactome.

Protein intron: See → intein.

Protein knock-out: The specific inactivation of a particular protein or the inhibition of transcription of its gene, so that the function of the protein is missing and (preferably) leads to a visible or measureable phenotype. Protein knock-outs can be generated by e.g. → chromophore-assisted laser inactivation. See → gene knock-down, → gene knock-in, → gene knock-out.

Protein L: A cell wall-bound receptor protein of 719 amino acids in some strains of the anaerobic bacterium *Peptostreptococcus magnus*, that binds to all → immunoglobulin (Ig) classes (e. g. IgA, IgD, IgE, IgG, IgM), to kappa light chains (without interfering with the binding of the → antigen), and to single chain variable fragments (ScFv). Binding to target proteins is mediated through 5 socalled B domains (each 72-76 amino acids long, sharing 70-80 % homology at both the nucleotide and amino acid level). Protein L can therefore be used to detect and purify ScFv and Fab fragments containing kappa light chains, and IgA, IgD, IgE, IgG, IgM and IgY. See → protein A, → protein LA, → protein L-agarose.

Protein LA: A recombinant protein that combines the immunoglobulin-binding domains of → protein A with those of → protein L, so that it binds strongly to diverse immunoglobulins from many different species with high affinity. Therefore LA is coupled to → agarose, and the resulting LA-agarose used to purify antibodies in a single affinity chromatography step.

Protein L-agarose: A beaded → agarose coupled to → protein L, that is used for the separation of immunoglobulins (e.g. IgA, IgD, IgE, IgG and IgM containing kappa light chains) from a variety of sources, and the purification of → monoclonal antibodies. See → protein A, → protein LA.

Protein library ("proteotheque"): Any collection of artificially created mutant proteins, each differing by a few amino acid residues that change the three-dimensional structure in an interesting → domain. Usually such libraries are established by systematic variation of the corresponding gene sequence, transformation of *E.coli* host cells, selection of the clones with the desirable mutant proteins by → phage display or colony screening techniques for an interaction with an immobilized target protein ("ligand"). Interacting proteins can then be isolated, sequenced and analyzed.

Protein linkage map: A misleading term for all proteins physically interacting with each other in a cell, and identified by chemical crosslinking or → two hybrid or → three-hybrid system techniques.

Protein lipidation: The covalent conjugation of phosphatidylethanolamine to a protein via an amide bond between a C-terminal glycine of the protein and the amino group of phosphatidyl-ethanolamine, which is catalyzed by a ubiquitination system. Protein lipidation plays an essential role in membrane dynamics.

Protein machine: A generic name for any intracellular complex of many proteins, that interact physically and cooperate synergistically. For example, → nucleosomes, → primosomes, → re-pairosomes, → ribosomes, → spliceosomes and → transcriptosomes are such multi-protein machines.

Protein maturation: See → post-translational modification.

Protein microarray: See → protein chip.

Protein microsequencing: See → microsequencing.

Protein mimic: Any peptide or protein possessing special domains ("mimotopes"), that mimic the immunogenic properties of antigens. For example, a peptide mimic of the pneumococcal serotype 4 polysaccharide (pep4) induces much stronger immunoglobulin G (IgG) responses in mice than the polysaccharide itself. Therefore, the mimotopes can structurally mimic carbohydrate anti-gens, induce carbohydrate-reactive B- and T-cell responses after immunization, and enhance boosting responses after priming with carbohydrate. See → DNA mimic, → RNA mimic.

Protein network: The entirety of → protein-protein interactions within a cell. See → protein expression map, → protein expression mapping, → protein interaction map, → protein interac-tion mapping.

Protein PCR: See → protein polymerase chain reaction.

Protein *polymerase chain reaction* (protein PCR): A misleading term for a specific → coupled transcription/translation system, in which an → *open reading frame* (ORF) is first ligated into the → multiple cloning site of a specific vector and efficiently transcribed into an RNA template, that is immediately used in a linked translation system to produce the corresponding protein. The ORF is first amplified in a conventional → polymerase chain reaction, using → *Taq* DNA po-lymerase. The extendase activity of this enzyme adds an adenosyl residue to the 3' end of the amplicon, which allows easy annealing to a T-overhang of the vector. Subsequent ligation of the ORF to the vector obviates the need for cloning. Then a new round of amplification with a vector-specific and an insert-specific → primer (which carries a → T7 RNA polymerase → pro-moter) generates the template for an effective T7 RNA polymerase-driven transcription. Sub-sequent → *in vitro* translation in a → rabbit reticulocyte lysate produces the protein of interest, that can then be analyzed by → denaturing polyacrylamide gel electrophoresis.

Protein polymorphism: Any difference in the amino acid sequence between two (or more) ho-mologous proteins from different cells, tissues, organs, or organisms. These polymorphisms are altogether a consequence of mutations in the → exons of the corresponding genes.

Protein processing: See → post-translational modification.

Protein profiling: The establishment of a – preferably complete – inventory of all proteins (and peptides) of a cell at a given time. Protein profiling encircles the isolation of the various types of proteins (e.g. membrane proteins, "soluble" proteins, glyco- and phospho-proteins), their characterization (e.g. electrophoretic behaviour, estimation of molecular weight), their sequencing (see → microsequencing, → protein finger-printing, → protein sequencing) and the correlation of the sequence data with entries in the protein data banks (e.g. SWISSPROT) in order to functionally categorize the proteins and peptides. In a wider sense, protein profiling may also comprise the identification of specific functions of the proteins (e.g. enzymatic activity, affinity towards ligands, → protein-protein-, protein-DNA-, protein-RNA- and protein-ligand interactions, posttranslational modifications, and others).

Protein-*protein* *interaction* (PPI): The non-covalent and usually transient interaction(s) between two (or more) proteins necessary for the execution of a function or functions, that each protein by itself cannot perform. Protein-protein interactions are the basis of cellular life, and lead e.g. to the aggregation of → protein machines, enable the transfer of signals through signal transduction chains and guarantee high efficiency in metabolic pathways, to name very few. The multitude of PPIs can be represented by a → protein-protein interaction map.

Protein-protein interaction map (protein interaction map, protein linkage map): The establishment of the complex network of interactions between (preferably all) proteins of a given cell at a given time point. Usually → two-hybrid screening procedures are employed to systematically screen for such interactions. See → protein-protein interaction, → proteome.
See color plate 6.

Protein *recognition* *molecule* (PRM): Any synthetic or naturally occurring molecule, that recognizes a target peptide or protein and bind it with high affinity. For example, → antibodies, → aptamers or → molecularly *imprinted polymers* (MIPs) or *molecularly imprinted electrosynthesized polymers* (MIEPs) are such PRMs.

Protein sequencing: The determination of the sequence of amino acids in purified oligo- and poly-peptides. Protein sequencing starts with the identification of both terminal amino acids. The amino-terminus reacts with → dansyl chloride (dansylation) or, alternatively, with dimethyl aminoazobenzoyl-isothiocyanate (DABITC), phenyl-isothiocyanate (PITC), or fluorodinitro-benzol (FDNB), and can be isolated after acid hydrolysis as the corresponding derivative (e.g. dansyl derivative) e.g. by → high-pressure liquid chromatography. The carboxy-terminus is released by carboxypeptidase, using specific reaction conditions. The bulk of the protein is then cleaved into a series of peptide fragments, either by chemical means (e.g. with cyanogen bromide, that cleaves specifically at the carboxy group of a methionine) or enzymatically (e.g. by endo-peptidases). The sequencing of these fragments again starts with the modification of their amino terminal amino acids by phenyl-isothiocyanate, the subsequent cleavage of the neighboring peptide bond, the separation of the modified amino acid from the residual peptide, and its identification. Then the peptide is again modified at its new amino terminus, and the whole cycle (Edman cycle) reinitiated until the peptide fragment is sequenced. The whole process is called Edman degradation. See also → microsequencing.

Protein signature: The complex pattern of all proteins and peptides (see → proteome) of a cell at a given time, as revealed by e. g. → two-dimensional or → three dimensional polyacrylamide gel electrophoresis and subsequent characterization of the individual spots by e. g. → electrospray ionization or → time-of-flight mass spectrometry.

Protein splicing: The posttranslational precise excision of → inteins from a precursor protein, the first product of → messenger RNA → translation. The → excision (definition c) of the intein(s) is catalyzed by intein-extein border-specific endoproteinases, and leads to free exteins, which are subsequently linked to each other by peptide bond formation. Highly conserved amino acid residues mark both splice junctions. A hydroxyl or thiol containing residue (e. g. serine, threonine, or cysteine) is always present at the position that immediately follows both junctions. All inteins own an invariant C-terminal asparagine and, in most cases, a histidine residue in the penultimate position. Protein splicing is a multi-step process involving N-S shift, transesterification, succinimide formation and an S-N shift, is very efficient (the precursor protein rarely accumulates) and leads to a → mature protein. Compare → DNA splicing, → RNA splicing, → splicing.

Protein superfamily: Any cluster of proteins related by sequence homology or structurally and functionally similar or identical conformations. This sequence homology is reflected by the sequence homology of the encoding genes. See → gene superfamily.

Protein tagging: A technique for the detection of a specific protein by fusing it to a second protein that can be easily monitored (by e.g. a specific antibody or by an enzyme assay). Protein tagging is used to screen for → translational fusions. Such fusion proteins are often produced using the α-peptide of → β-galactosidase. See also → protein fusion and purification technique, → expression vector (→ fusion vector).

Protein therapy: The replacement of non-functional peptides or proteins in a cell by the transfer of correctly folded and functional peptides or proteins via e.g. → microinjection, resulting in → complementation. See → gene therapy.

Protein transduction: The direct transfer of peptides and proteins (e.g. → antibodies) into target cells without the interference of a receptor. In nature, transducing proteins (e.g. the HIV–1 TAT protein, or the *Drosophila* Antennapedia transcription factor) contain short peptide sequences rich in basic amino acids (see → protein transduction domain), that catalyze the internalization of the cargo protein. See → non-covalent protein delivery. But see also → cell penetrating peptide.

***Protein transduction domain* (PTD, Trojan horse):** A short peptide sequence motif of about 10–16 amino acids (with a high number of positively charged lysine and arginine residues) in viral or cellular proteins, that transduces across the plasma membrane without the necessity of a receptor. PTDs are present in transducing proteins such as e.g. HIV–1 TAT (consensus sequence: YGRKKRRQRRR), HSV–1 VP22, and the *Drosophila* Antennapedia transcription factor, to name few. The PTD is always covalently bound to the cargo protein, which is rapidly internalised, but excluded from the endocytotic pathway. PTDs can be used to import target proteins almost independently of their size (e.g. proteins in excess of 700 kDa), even → antisense oligonucleotides and → liposomes (with a diameter of more than 200 nm) *in vivo* and *in vitro*. For this purpose, artificial PTDs are synthesized (e.g. with the sequence H_2N-RRRRRRRRR-COOH, or H_2N- KKKKKKKK-COOH, or H_2N-KKKKKKKKK-COOH). A TAT derivative (H_2N-YARAAARQARA-COOH) is 33-times more efficient than the Tat peptide itself. The bonds between a cargo protein and a PTD can also be made reversible, if e.g. disulfide or ester linkages are used, that are reduced or cleaved within the target cell. See → cell-penetrating peptide, → non-covalent protein delivery.

Protein transfection: A technique for the delivery of biologically active peptides or proteins (e.g. antibodies) into living cells. For example, the target protein can be transferred into a cell by → lipofection, i.e. noncovalently complexing cationic lipids with the negatively charged protein, fusing the complex with the cell membrane (if it is not endocytotically taken up), and releasing the protein into the cytoplasm. Ideally, the protein should be non-covalently complexed with a transfective carrier, which protects and stabilizes the protein and preserves its structure and function during the transfection process. Once internalized, the complex should dissociate and the protein be liberated in an active conformation. Usually proteins with a high positive overall charge or minimal hydrophobicity are less efficiently transfected as compared to negatively charged or hydrophilic proteins. Compare → transduction. See → transfection.

Protein translocation: The vectorial movement of a protein from the cytoplasm (where it is synthesized) across a membrane (e.g. into a chloroplast, a mitochondrion, the nucleus, or the intraluminal space of the endoplasmic reticulum).

Protein truncation assay: See → protein truncation test.

Protein *truncation test* (PTT; *protein truncation assay*, PTA): A technique for the detection of one or more → translation termination → mutations occuring in a → gene that lead to early → termination of → transcription and the synthesis of a protein with shorter amino acid chain length (truncated protein) as compared to the wild type protein. This truncated protein is usually inactive ("loss of function"). In short, → genomic DNA or → messenger RNA is first isolated and the target sequence amplified in a conventional → polymerase chain reaction. To that end, the mRNA is reverse transcribed into → cDNA. The amplification uses specially designed primers carrying either a T7, or → SP6 or T3 phage promoter sequence at its 5' end. The resulting PCR products serve as → templates for an *in vitro* RNA synthesis, employing the corresponding phage RNA polymerases. The RNAs in turn are translated into the corresponding proteins, which are then subjected to denaturing → SDS polyacrylamide gel electrophoresis. If a specific mutation in the target gene sequence leads to a premature stop of the translation of the encoded protein, it will be shorter ("truncated") and migrate further in the gel than the wild-type encoded protein. PTTs discover frequent mutations especially in comparatively long coding sequences (e.g. → megagenes; for example the *Duchenne Muscular Dystrophy*, DMD gene).
Figure see page 874.

Proteolytic processing: The highly specific and limited *in vivo* cleavage of substrate peptides and proteins by a variety of proteases. Proteolytic processing controls a series of cellular pathways, as e.g. angiogenesis, → apoptosis, cell cycle progression, cell proliferation, → DNA replication, immunity, or wound healing).

Proteome: The complete set of proteins in a cell or organelle at a given time. The proteome of a particular cell changes dynamically with its developmental stage and changing environment, and comprises many more proteins than → genes are present in the → genome, because (1) → alternative splicing might give rise to → mRNAs that are different from the original mRNA, and (2) proteins are frequently modified by co- or post-translational modification (e. g. acetylation, ADP-ribosylation, C-terminal truncation, → glycosylation, methylation, N-terminal truncation, phosphorylation, sulphation). See → proteome mapping, → proteomics. Compare → proteosome.

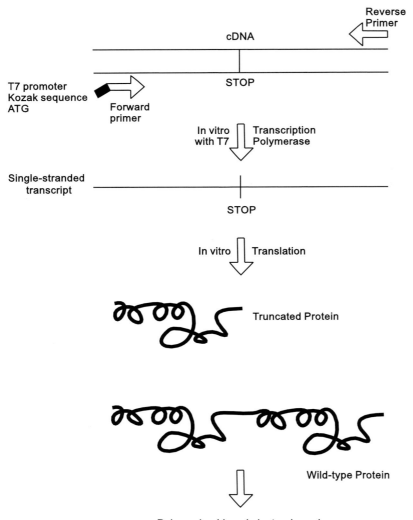

Protein truncation test

Proteome array: A misleading term for a → microarray, onto which full-length → cDNAs or → open reading frame clones, preferably covering the whole → genic space of a genome, are spotted. Compare → proteome chip.

Proteome chip (proteome microarray): Any → microarray, onto which preferably all highly purified proteins of an organism are spotted. The genes encoding these proteins are first cloned, then overexpressed in a bacterium, the overexpressed proteins isolated and purified, and then covalently bound onto the chip's surface. For example, a total of 5,800 *Saccharomyces cerevisiae* genes can be expressed as fusions to sequences encoding glutathione-S-transferase, the corresponding fusion proteins isolated and purified and arrayed on glass slides. Such a chip covers about 94% of all yeast genes (or proteins), and therefore represents a proteome chip. Proteome chips are used for studies of protein-protein interaction(s). Compare → proteome array.

Proteome map: The graphical description of the abundance, quantity and subcellular distribution of (preferably) all peptides and proteins in a cell at a given time. Mostly identical to → protein expression map. See → protein expression mapping, → protein interaction mapping.

Proteome mapping: A misleading term for the categorization of all proteins of a given cell at a given time, using the whole technology of → proteomics, and reflecting the number of → active genes only, without any clue as to where in the → genome these genes map. But see → gene mapping.

Proteome microarray: See → proteome chip.

Proteome signature: The specific patterns of the various → sub-proteomes of a cell at a specific time and a particular developmental stage, as detected by e.g. → two-dimensional polyacrylamide gel electrophoresis of the individual sub-proteomes and subsequent staining of the protein spots. Proteome signatures (or more precisely, characteristic indicator proteins) are characteristic for a specific situation of a cell (e.g. after stress, during mitotic change, after a knock-out of a gene).

Proteomic diversity (proteome diversity): The variations of protein number, protein complexity and protein functions in two (or more) → proteomes. Proteomic diversity is partly a result of → alternative exon usage, → alternative splicing, or → exon shuffling, but is also influenced by various → post-translational modifications.

Proteomic fingerprint: A pattern of proteins that is diagnostic for a specific state of a cell, a tissue, an organ, or organism. First, a global pattern of proteins (comprising potentially all proteins in a given cell type at a given time, usually some 10–30,000 different molecules, but not simply their presence or absence, yet their relative amounts in relation to each other) is established, then compared to other cells of the same type, and characteristic proteins selected. These are then taken as a reference proteomic fingerprint, with which other fingerprints from other cellular states are compared to detect differentially expressed proteins. Compare → peptide fingerprint.

Proteomics (*protein-genomics*, proteinomics): The whole repertoire of techniques to analyze and characterize the → proteome of an organelle or a cell, including protein isolation, fractionation, separation, post-separation analysis (e. g. identification of different proteins by → microsequencing), analysis of post-translational modification by → MALDI mass spectometry (e. g. acetylation, ADP-ribosylation, C-terminal truncation, glycosylation, methylation, N-terminal truncation, phosphorylation, sulphation), bioinformatics (storage and analysis of the resulting informations) and robotics (automation for high-throughput analysis). See → functional genomics, → functional proteomics, → genomics, → industrial proteomics, → *in silico* proteomics, → one cell proteomics, → phosphoproteomics, → recognomics, → subcellar proteomics, → 3D proteomics, → tissue proteomics, → topological proteomics.

Proteosome: The complete set of proteins expressed by a particular → genome in a cell. See → proteome.

Proteozyme: Any protein that possesses an active center and consequently catalytic (enzymatic) properties. Compare → ribozyme.

Protochromosome: An ancestral → chromosome, from which homologous chromosomes of one species or homoeologous chromosomes of related species are derived by various → deletions, → insertions, → translocations, generally genomic → rearrangements.

Protoclone: Any progeny derived from a single → protoplast.

Proto-exon shuffling: The hypothetical combination of precursors of → exons (proto-exons) to functional genes.

Proto-microsatellite: A sequence element of eukaryotes (e.g. diptera), from which → microsatellites are generated by as yet not fully understood processes. For example, the socalled → mini-me elements of *Drosophila melanogaster* contain a $(TA)_n$ repeat 5', and a $(GTCY)_n$ repeat 3' of the highly conserved 33 bp core motif. Frequently the 3' repeat is absent, and instead a cryptically simple sequence (consisting of only C, G and T) can be found (see → cryptic simplicity). Both loci are considered as proto-microsatellites, since they can expand to new $(TA)_n$ and $(GTCY)_n$ repeats, respectively, once inserted at new genomic locations by a possible retroposition of the mini-me element (catalyzed by genes *in trans*).

Proto-oncogene: See → cellular oncogene.

Protoplast: A bacterial, yeast or plant cell from which the cell wall has been removed experimentally (e.g. bacterial walls can be digested with → lysozyme, plant cell walls with a mixture of cellulase, pectinase and polygalacturonase). See → protoplast fusion, → protoplast transformation.

Protoplast fusion: The combination of two related or unrelated → protoplasts, or of an enucleated protoplast and a → karyoplast to form a hybrid cell by either chemical (with → polyethylene glycol or Ca^{2+}) or electrical treatment (electrofusion). Compare → cell fusion.

Protoplast transformation: The integration of foreign DNA into plant DNA using plant cells without cell walls (→ protoplasts). Originally such protoplasts were cocultivated with virulent cells of → *Agrobacterium tumefaciens*. During cocultivation the → T-region from the *Agrobacterium* → Ti-plasmid was transferred to the protoplast genome. After the procedure, the bacteria were killed by an → antibiotic (e.g. carbenicillin, Claforan), and the transformed protoplasts selected by their ability to grow without added growth hormones. Today, the term protoplast transformation is also used when protoplasts are incubated with purified Ti-plasmids or any other DNA. The transformation frequency of protoplasts from especially competent plants such as tobacco is in the range of maximally 10%.

Prototrophy: The ability of a wild-type organism to grow on a → minimal medium.

Protruding end: See → protruding terminus.

Protruding terminus (protruding end, overhanging end, overhang, extension): The end(s) of a DNA duplex molecule where one strand is longer ("protruding") than the other which is referred to as "recessed". See also → cohesive end, → recessed 5'-terminus, → recessed 3'-terminus.

Provirus: Any viral DNA that is an integral part of the host cell chromosome and as such is transmitted from one cell generation to another without → lysis of the host, for example an integrated → retrovirus. See also → prophage.

Proximal: Located close to any fixed point. See → distal.

PR protein: See → *p*athogenesis-*r*elated protein.

Ps: See → pseudouridine.

pSC 101: A small non-conjugative → plasmid carrying a → tetracycline resistance gene, that was used for the construction of the universal → cloning vector → pBR 322. Used as a → low copy number plasmid vector.

Pseudocomplementary *p*eptide *n*ucleic *a*cid (pseudocomplementary PNA): Any → peptide nucleic acid that contains modified nucleobases, as e.g. 2,6–diaminopurine·2–thiothymine or thiouracil, where the former substitutes for adenine, the latter for thymine. Such pseudocomplementary bases recognize their natural A·T and G·C counterparts, but cannot recognize each other. Therefore, pseudocomplementary PNAs are used for → double duplex invasion techniques, where any binding of the two PNA strands to each other would prevent invasion of the DNA target.

Pseudoexon: Any exon-like sequence in eukaryotic → split genes, flanked by → pseudosplice sites, that is ignored by the → spliceosome. Pseudoexons are probably not used, because no functional → splicing enhancer sequences are present in the host gene.

Pseudogene ("silent gene", truncated gene, "dead gene"): A non-functional derivative of a functional eukaryotic gene, that suffered → rearrangements and → mutations preventing normal expression (e. g. lacks → introns and → promoter regions or contains one or more → stop codons). Since pseudogenes are not under selective pressure, they frequently degenerate more rapidly than their functional counterparts. Pseudogenes are thought to represent the DNA copies of mRNA, because they usually carry a poly(dA) sequence at the 3' end. Some pseudogenes may also have arisen from gene duplication and concomitant → deletion of the promoter region or parts of it. Such truncated genes may be present in a particular genome in appreciable numbers (e. g. the human → high mobility group HMG-17 protein → multigene family contains about 30, the actin gene family about 20, and the glyceraldehyde-3-phosphate dehydrogenase gene family about 25 retropseudogenes). See → processed pseudogene.

Pseudogene *m*essenger RNA (pseudogene mRNA): Any → *m*essenger RNA (mRNA), that is transcribed from a → pseudogene under the control of a nearby or adjacent → promoter. Usually such mRNAs represent read-throughs from promoters of adjacent genes.

Pseudogenome: An infelicitous and misleading term for the complete set of → pseudogenes in a genome.

Pseudopromoter: A DNA sequence element that allows → *in vitro* transcription of linked genes, but does not function *in vivo*.

Pseudo *r*esistance *g*ene *a*nalogue (pseudo RGA): Any non-functional (e.g. promoterless) → resistance gene analogue. See → pseudogene.

Pseudosplice site (pseudosplice junction, pseudosite): Any → splice junction, that matches the → consensus sequence of a real splice junction, but is efficiently ignored by the → spliceosome. Pseudosplice sites are abundant in eukaryotic genes. For example, the 42kb human *hprt* gene contains eight real 5'-splice sites, but over 100 5'- and 683 3'-pseudosplice sites.

Pseudouridine (Ps, 5-β-D-ribofuranosyl uracil, ψ): One of the so-called → rare bases, unusual nucleotides found in some → transfer RNAs where the glycosidic bond is associated with position 5 of uracil. See for example → TψC loop.

ψ: See → pseudouridine.

ψ: See → pseudogene.

P-site (*p*eptidyl-tRNA binding *s*ite): The site on the → ribosome to which the growing peptide chain is attached during protein synthesis.

PSMEA: See → *p*rimer-*s*pecific and *m*ispair *e*xtension *a*nalysis.

Psoralen: A photoreactive 6-hydroxy-5-benzofuranacrylic acid-δ-lactone, that intercalates into the two → strands of → double-stranded nucleic acids (→ dsRNA and → dsDNA). Upon UV irradiation, psoralen reacts with the 5-6 position of the → pyrimidine bases (particularly → thymidine) and thereby forms interstrand cross-links. Photoinduced cross-linking can be targeted to a specific genomic sequence by attaching a psoralen to an oligonucleotide complementary to this sequence. Psoralen can be linked to the 5' or/and 3' terminus of an oligonucleotide via a hexamethylene arm. Psoralen-modified oligonucleotides are used in → antisense and triple-helix studies, and in → denaturing gradient gel electrophoresis, → temperature gradient gel electrophoresis, and → temporal temperature gradient gel electrophoresis as an alternative to a → GC clamp. See → triple helix-directed DNA cross-linking.

Psoralen

Psoralene footprinting: See → photo-footprinting, definition a.

Psoralen labeling: The introduction of a psoralen molecule (covalently bound to either → biotin or → fluorescein) into DNA. Psoralen binds to pyrimidine bases (preferentially thymine or uracil, less to cytosine) via hydrophobic interaction(s), which is stabilized by UV crosslinking (i. e. simple irradiation with long wave-length [365 nm] UV light). After removal of unincorporated psoralen-biotin by n-butanol extraction, the probe can be used for DNA-DNA- or DNA-RNA-hybridization. Since psoralen is conjugated to e. g. biotin, the hybridization can be detected by → streptavidin-alkaline phosphatase conjugate (or, in case of fluorescein, with anti-fluorescein-alkaline phosphatase antibody conjugate). Psoralen labeling avoids the problems inherent in → random priming and → nick translation. See → triple helix-directed DNA cross-linking.

pSP 64: A derivative of a → pUC plasmid that contains the → RNA polymerase promoter of phage SP6 (see → SP6 vector).

PSSA: See → phosphorylation site-specific antibody.

PTA: See → protein truncation test.

PTC: See → premature termination codon.

PTGS:
a) See → co-suppression.
b) See → posttranscriptional gene silencing.

pTi: See → Ti-*plasmid*.

PTIS: See → *portable* *translation* *initiation* *site*.

PTS: See → peroxin.

PTT: See → *protein* *truncation* *test*.

Pu: Abbreviation for → *pu*rine.

PU (*palindromic unit*): See → repetitive extragenic palindromic element.

Public sequence databases: A series of databases, that are collected and maintained by public institutions and whose data are fully available to the public (e.g. GenBank, the EMBL data library, and DDBJ. In contrast, the private databases are not available to the public (without licensing fees).

pUC (pUC vector): Any one of a series of relatively small, versatile *E. coli* → plasmid cloning vectors that contain the *Pvu*II/*Eco*RI fragment of → pBR 322 with the β-lactamase gene (coding for → ampicillin resistance), an → origin of replication, and a sequence coding for the α-peptide of the *lac Z* (β-galactosidase) gene with an inserted → polylinker. Insertion of foreign DNA at the polylinker leads to the interruption of the α-peptide gene, so that no functional protein can be synthesized. Consequently, the host cells produce colorless colonies, if grown on media with ampicillin and → X-gal. Strains that are transformed with non-recombinant vectors, develop blue colonies on the same medium. Thus the recombinants can easily be selected. pUC vectors are present in 500-700 copies per host cell.
Figure see page 880.

pUC vector: See → pUC.

Puff (chromosome puff): A local unwinding of → polytene chromosomes where the → chromatin is less condensed and the genes are actively transcribed. Usually a single chromosome band is unwound, though two or more bands can be involved in puffing (Balbiani ring). The pattern of puffing in various animals, especially in the salivary gland cells of *Drosophila* larvae, is cell-specific, organ-specific, and developmentally regulated. It may also be influenced by a series of environmental factors (e.g. ions, or a heat shock).

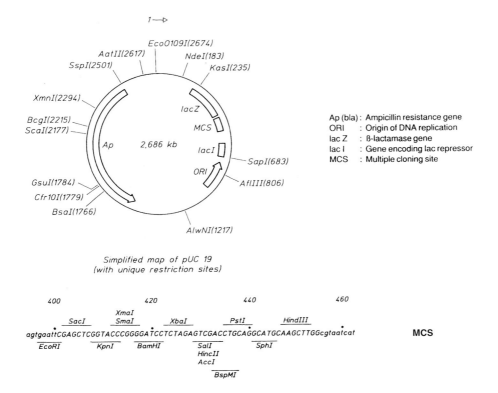

Simplified map of pUC 19
(with unique restriction sites)

Ap (bla) : Ampicillin resistance gene
ORI   : Origin of DNA replication
lac Z   : ß-lactamase gene
lac I   : Gene encoding lac repressor
MCS   : Multiple cloning site

Pull-down: A laboratory slang term for the isolation of specific compounds from complex mixtures by their high-affinity binding to immobilized capture molecules (that are bound to → magnetic beads or stationary phases of affinity columns). See → glutathione-S-transferase pull-down assay.

Pulse-chase analysis: An experiment designed to follow the course of degradation of a molecule within the living cell. First, cells (or cell extracts) are incubated with a radioactively labeled precursor compound *in vitro* for a short period of time ("pulse"), then a large excess of the same, unlabeled compound is added to dilute and to prevent further significant incorporation of radioactivity into potential metabolites. Samples are taken at various time intervals ("chase period") to estimate precursor metabolism.

Pulsed-field gel electrophoresis (PFGE): A technique for the electrophoretic separation of DNA molecules from the size of ordinary → restriction fragments (< 10 kb) to intact chromosomal DNAs of up to 15 million base pairs (15 Mb). The DNA is subjected alternately to two electrical fields at different angles for specific time intervals called the pulse time (τ). Under appropriate conditions reorientation of DNA segments with different size and topology is different, which – among other parameters – leads to the separation of these segments. For example, with each reorientation of the electric field relative to the gel, smaller DNA fragments move in the new direction much faster than the larger ones. Thus, the larger DNA molecules lag behind, resulting in their separation from the smaller DNAs. See → contour-clamped homogeneous electric field gel electrophoresis, → field inversion gel electrophoresis, → gel electrophoresis, → orthogonal-field alternation gel electrophoresis, → programmable autonomously controlled electrodes gel electrophoresis, → pulsed homogeneous orthogonal-field gel electrophoresis, → rotating gel electrophoresis, → secondary pulsed field gel electrophoresis, → transverse alternating field electrophoresis, → zero integrated field electrophoresis.

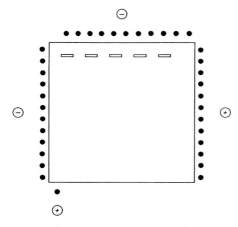

Electrode configuration of PFGE

Pulsed *h*omogeneous *o*rthogonal-field *g*el *e*lectrophoresis (PHOGE): See → orthogonal-field alternation gel electrophoresis.

Pure culture: Any → cell culture that is made up of only one cell type, or one strain of cells.

Pure line: Any organism or population of organisms made homozygous through extended inbreeding.

pUR expression vector: Any one of a series of 5.2 kb → plasmid → expression vectors that is designed for the expression of *lac Z* → fusion genes in *E. coli*. Each pUR vector contains a → polylinker with → recognition sites for *Bam* HI, *Sal* I, *Pst* I, *Xba* I, *Hind* III and *Cla* I in all three → reading frames at the 3'-terminus of a *lac Z* gene. This gene is driven by the *lac* UV 5 promotor. Insertion of a cDNA sequence into the appropriate cloning site allows the expression of a → fusion protein consisting of → β-galactosidase and the peptide encoded by the cDNA.

Purifying selection: Any → selection against DANN sequence changes (→ mutations), that have a deleterious effect on the organism. Compare → adaptive evolution.

***P*urine (Pu):** A heterocyclic molecule consisting of a pyrimidine and an imidazole ring. Purines are constituents of nucleic acids (→ DNA, → RNA). See → adenine, → guanosine; compare → pyrimidine.

Puromycin (6-dimethyl-3'-deoxy-3'-p-methoxy-L-phenylalanylamino adenosine): A → nucleoside antibiotic from *Streptomyces alboniger* (may also be synthesized chemically) that is structurally similar to the 3'-terminal aminoacyl adenosine residue of a → tRNA, binds to the A site on the → ribosome, forms a → peptide bond with the growing peptide chain (see arrow), leaves the ribosome as peptidyl-puromycin and causes termination of elongation.

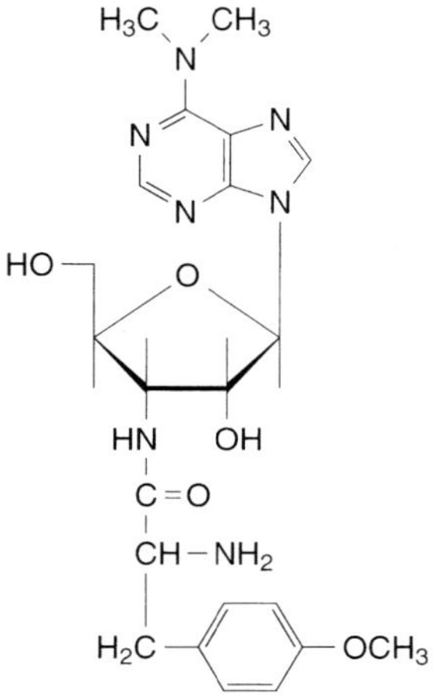

Puromycin

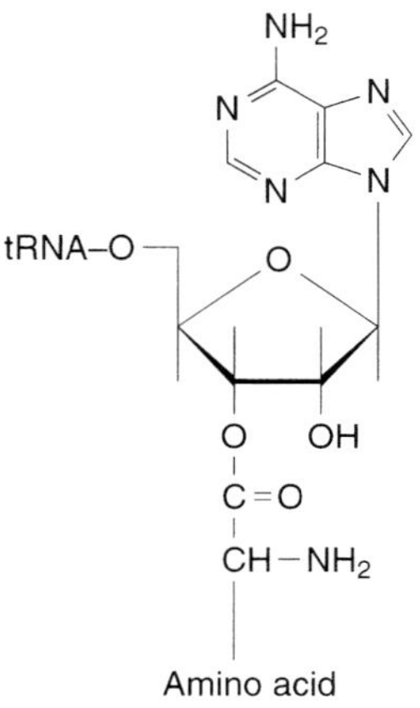

3'-Terminus of a tRNA

Push column: A device for the separation of radioactively labeled DNA or RNA probes from unincorporated nucleotides after → nick-translation or → random priming procedures. The separation is accomplished by the application of pressure to the upper chamber of the push column which in its simplest form is a modified syringe, and is complete after minutes. The high-molecular weight probes elute first from the push column.

***Putative alien* gene (pA gene):** Any gene in a → genome whose → codon usage is different from an average gene in this genome. The difference should (1) exceed a high threshold, i.e. be pronounced, and (2) also be characteristic for genes encoding ribosomal proteins, → chaperones and proteins functional in the protein synthesis machinery. For example, the *cag*A domain of *Helicobacter pylori* strongly differs from the rest of the genome in its → codon bias.

Putative gene: Any → genic sequence, that has homology to proven → genes, whose sequences are deposited in relevant databanks (e.g. GenBank). Compare → predicted gene. See → putative protein.

Putative protein ("probable protein"): Any protein, whose amino acid sequence has only limited similarity to already characterized proteins (as e.g. checked by sequence comparison via SWISS-PROT database).

P-value: The probability in a → BLAST search to obtain, by chance, a pair-wise sequence comparison of the observed similarity, given the length of the query sequence and the size of the searched database. Low P-values symbolize sequence similarities of high significance.

***Pwo* DNA polymerase:** See → *Pyrococcus woesii* DNA polymerase.

Py: See → *py*rimidine.

pYAC: See → yeast *a*rtificial *c*hromosome.

pYC: See → yeast *c*entromere *plasmid.

pYE: See → yeast *e*pisomal *plasmid.

pYH: See → yeast *h*ybrid *plasmid.

pYI: See → yeast *i*ntegrative *plasmid.

pYL: See → yeast *l*inear *plasmid.

pYP: See → yeast *p*romoter *plasmid.

pYR: See → yeast *r*eplicative *plasmid.

Pyranosyl-RNA (p-RNA; ribopyranosyl RNA): A synthetic ribopyranosyl isomer of RNA, a β-D-ribopyranosyl-(4'→2') oligonucleotide, in which the ribose units exist in the pyranose form, and neighbouring ribopyranosyl units are connected by phosphodiester linkages between the positions C (4') and C (2') instead of the 5'→3' phosphodiester bonds in RNA. Double strands of p-RNAs adopt a rigid linear structure and are held together by Watson-Crick purin-pyrimidine and purin-purine base pairing, where e. g. adenine-uracil pairing is stronger than in the corresponding RNA duplexes. Base pairing in such duplexes is also more selective than in RNA, due to the conformational rigidity of the p-RNA pyranosyl ring relative to that of the furanosyl ring of the RNA. p-RNA folds into → hair-pin structures and is able to replicate by non-enzymatic template-directed ligation of 2', 3'-cyclophosphates of short oligomers (e. g. tetramers).

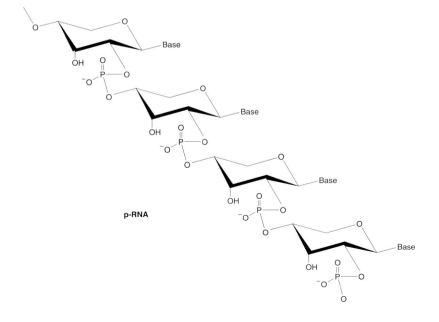

Pyrimidine **(Py):** A heterocyclic 1,3-diazine ring. Pyrimidines are constituents of nucleic acids (→ DNA, → RNA). See → thymine, → cytosine, → uracil; compare → purine. See also → pyrimidine dimer.

Pyrimidine dimer: A structure formed by UV irradiation of DNA in which two → thymidine (or → cytidine) residues or one thymine and one cytosine residue at adjacent positions in the same DNA strand become covalently linked to each other. Such dimers block DNA → transcription and → replication. See → thymine dimer.

Pyrimidine-pyrimidone(6–4)photoproduct (TC pyrimidine-pyrimidone(6–4) photo-product): A DNA photoproduct, resulting from the UV-induced opening of the double bond of the 5' pyrimidine and its reaction across the exocyclic group of the 3' pyrimidine in adjacent pyrimidines. In the (6–4) photoproduct, a rotation of the 3' base by 90° makes it resemble an → abasic site. Pyrimidine-pyrimidone (6–4) photoproducts cause mutations of the C→T or CC→TT type. If these mutations occur in e. g. one of the tumor suppressor genes, for example *Trp53*, they may lead to cancerous proliferation of the afflicted cells.

Pyrithiamine: A thiamine analogue, that is effective against a series of *Aspergillus* species (e.g. *A. fumigatus, A. niger, A. oryzae, A. terris*) and serves as → selectable marker for transformed strains.

Pyrococcus furiosus **DNA polymerase (*Pfu* DNA polymerase; EC 2.7.7.7):** A highly thermostable and hyperactive monomeric → DNA polymerase with a 5'→ 3' polymerase and a 3'→ 5' → proof-reading exonuclease activity, isolated from the hyperthermophilic marine archaebacterium *Pyrococcus furiosus* (*Pfu*), that has originally been detected in geothermal vents off the coast of Vulcano (Italy). This enzyme is used for the → polymerase chain reaction and is considered superior to the convential → *Thermus aquaticus* DNA polymerase, since its 3'→5' proof-reading capacity will excise mismatched 3'-terminal nucleotides from primer-template complexes and incorporate the correct, complementary nucleotides, whereas *Taq* polymerase lacks this activity. Compare → *Pyrococcus* species DNA polymerase, → *Pyrococcus woesii* DNA polymerase.

Pyrococcus species **DNA polymerase (Psp DNA polymerase; EC 2.7.7.7):** A highly processive (see → processivity) and thermostable → DNA polymerase with a 3' → 5' exonuclease (proof-reading) activity, isolated from the thermophilic archaebacterium *Pyrococcus* species. The enzyme does not possess any 5' → 3'exonuclease activity and is used for → polymerase chain reaction experiments. Compare → *Pyrococcus furiosus* DNA polymerase, → *Pyrococcus woesii* DNA polymerase

Pyrococcus woesii **DNA polymerase (*Pwo* DNA polymerase):** A → DNA polymerase of the thermophilic bacterium *Pyrococcus woesii* (*Pwo*) with high thermostability and 3'-5'-exonuclease ("proofreading") activity, but no terminal transferase (→ extendase) activity. Therefore *Pwo* DNA polymerase generates blunt-ended amplification fragments that facilitate their → cloning and → sequencing. In concert with → *Taq* DNA polymerase, the *Pwo* enzyme complements the high → processivity with proofreading, so that in this mixed system ("long template PCR system") fragments of up to 27 kb in size are synthesized on genomic DNA as → template. Compare → *Pyrococcus furiosus* DNA polymerase, → *Pyrococcus* species DNA polymerase.

Pyrogram: The graphical depiction of a → pyrosequencing procedure as a series of spikes, each of which represents a light pulse generated by the oxidative decarboxylation of → luciferin and monitored with a sensitive luminometer or a charge-coupled device (CCD) camera.

PyroMethA: See → pyrosequencing methylation analysis.

***Pyro*phosphat*ase* (PPase; inorganic pyrophosphatase):** An enzyme that catalyzes the hydrolysis of pyrophosphate into two molecules of orthophosphate. The enzyme is used for → DNA sequencing (especially when selective band weakening occurs).

Pyrosequencing (real-time pyrophosphate detection; PP$_i$-based sequencing; minisequencing): A technique to determine the sequence of bases in DNA, that avoids → sequencing gel electrophoresis and radioactivity or fluorescence, needed in other sequencing procedures (e. g. → Sanger sequencing). Instead, pyrosequencing quantitatively measures the pyrophosphate (PP$_i$) released during the DNA polymerase reaction by coupling it to the generation of light by firefly → luciferase. In short, the single-stranded → template DNA is first annealed to a short → sequencing primer. Then DNA polymerase together with an apyrase, ATP sulfurylase and firefly luciferase and only one dNTP (e. g. dGTP) are added. If dG does not form a base pair with the first free base on the template, it is rapidly removed by the apyrase (a mixture of nucleoside 5'-triphosphatase and nucleoside 5'-diphosphatase). Then the next base is added (e. g. dTTP). If a base pair T = A can be formed, the DNA polymerase extends the primer and releases PP$_i$, that is quantitatively converted to ATP by ATP sulfurylase. Now → luciferase utilizes this ATP to oxidatively decarboxylate → luciferin, and the light produced in this reaction is detected by a sensitive luminometer or a *charge-coupled device* (CCD) camera. The light pulse signaling incorporation of dT is shown in real time on a PC. Since apyrase removes the excess of dTTP, and luciferase utilizes the ATP, light emission is transient. Then e. g. dATP is added, which either forms a base pair on the template DNA (initiating a new reaction), or is rapidly removed by apyrase. By repeatedly adding the deoxynucleotides (here in the series G, T, A and C) and monitoring the light pulses, the sequence of the template DNA can be derived. One base is recorded every minute, and 96 samples can be read simultaneously in an automated pyrosequencer. Pyrosequencing is only suitable for the diagnostic sequencing of relatively short DNA fragments (up to 200 bases).

(1) Template DNA + primer ⇌ [template-primer]

(2) [Template-primer] + dNTPs $\xrightarrow[\text{polymerase}]{\text{DNA}}$ template + extended primer + dNMP + PP$_i$

(3) PPi + APS $\xrightarrow[\text{sulfurylase}]{\text{ATP}}$ ATP

(4) ATP + luciferin + O$_2$ $\xrightarrow{\text{luciferase}}$ oxiluciferin + AMP + PP$_i$ + CO$_2$ + h · ν

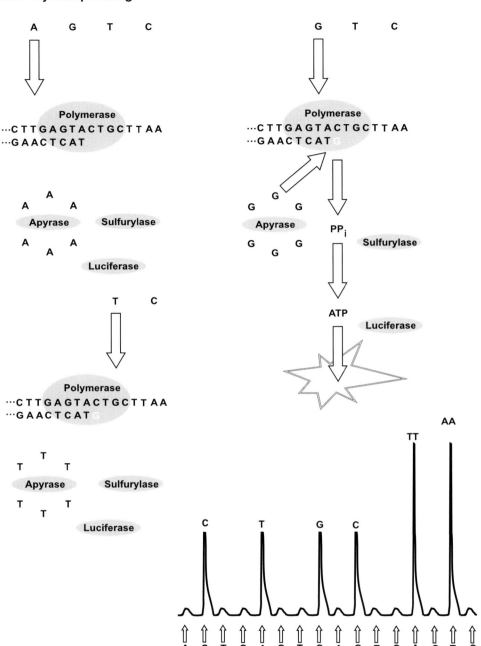

Pyrosequencing

***Pyro*sequencing *meth*ylation *a*nalysis (PyroMethA):** A technique for the quantification of me-
thylated → cytosines at CpG sites in a distinct genomic region, that combines the → *co*mbined
*b*isulfite *r*estriction *a*nalysis (COBRA) and → pyrosequencing. In short, the genomic target fragment is first amplified by conventional → *p*olymerase *c*hain *r*eaction (PCR), using sequence-
specific up-and down-stream flanking → primers. Preferentially PCR fragments of 100–150 bp
are employed for PyroMethA. Then bisulfite treatment converts unmethylated cytosines to uracil, leaving methylated cytosines unaffected. In essence, this process results in a chemically
induced methylation-dependent C→T transition site, that is subsequently detected by pyrosequencing.

pYX: See → *y*east *e*xpression *p*lasmid.

Q

Q: Abbreviation for base Q (queuosine), the nucleoside of queuine. See → rare bases.

Q-beta (Qβ): A small phage of *E. coli* (→ coliphage) with a single-stranded RNA genome (Qβ plus-strand) of 4.2 kb, encoding a coat protein, a maturation protein, and an RNA-dependent → RNA polymerase (Qβ replicase). The → plus strand is used directly as mRNA for the synthesis of these phage proteins. After infection of bacteria that contain an → F-factor, the Qβ replicase begins to synthesize so-called → minus strands, using the plus strand of the phage as → template. The minus strands then serve as templates for the synthesis of Qβ-RNA (the plus strand) which is packaged into phage heads, leading to the generation of fully infectious Qβ phages.

Qβ: See → Q-beta.

Q-beta (Qβ) replicase amplification: The exponential multiplication of an RNA sequence inserted into the template RNA (Qβ plus strand) for Q-beta replicase (Qβ phage RNA-dependent RNA polymerase). Since the inserted RNA does not interfere with the function of Q-beta replicase, the whole recombinant molecule is amplified to about 10^9 copies. The extent of the amplification may be quantified by → ethidium bromide fluorescence. See → Q-beta.

Q-beta plus-strand: See → Q-beta.

QCM-D: See → quartz crystal microbalance with dissipation monitoring.

QDFM: See → quantitative *DNA fiber mapping*.

Q-FISH: See → quantitative fluorescence *in situ* hybridisation.

Qiagen column: The trademark for a small disposable anion exchange column, used for the fast and simple isolation and purification of → nucleotides, → oligonucleotides and polynucleotides (especially → plasmids).

Q-PCR: See → quantitative *polymerase chain reaction*.

QRT-PCR: See → quantitative *reverse transcriptase polymerase chain reaction*.

QTL: See → *quantitative trait locus*.

Quantitative chromosome map (idiogram): The physical arrangement and copy number of → genes, generally DNA sequences, along a → chromosome, as measured by → *in situ* hybridization of fluorochrome-labeled → probes (representing e. g. genes) to chromosome spreads and quantification of the emitted fluorescence. The higher the copy number of the target sequence, the higher is the number of bound probe molecules, and the higher is the fluorescence light emission, which allows a semi-quantitative analysis of the gene content at the detected locus.

Quantitative DNA fiber mapping (QDFM): A technique for the construction of high resolution → physical maps at a resolution of few kilobases, that relies on the hybridization of specific fluorescently labeled probes to target DNA immobilized on a specially prepared glass surface and stretched by hydrodynamic forces to produce linear templates ("fibers") of about 2–2.5 kb per μm. The hybridization events and the distribution of different probes along the stretched fibers can be detected by fluorescence and be imaged for further analysis. See → molecular combing.

Quantitative fluorescence in situ hybridisation (Q-FISH): A variant of the conventional → fluorescent *in situ* hybridisation technique for the quantitative estimation of → telomere lengths, that employs → peptide nucleic acid (PNA) → probes labeled with → fluorochromes (e.g. → cyanin 3) rather than the DNA or RNA probes in the traditional methods. PNA possesses an uncharged backbone and can hybridise to the target DNA (of a chromosome) under extremely low ionic conditions, which inhibit target DNA → renaturation. Under these conditions, fluorescence intensity correlates linearly with the number of bound fluorophores, and therefore allows a quantitative estimate of telomeric repeat numbers.

Quantitative polymerase chain reaction (Q-PCR; kinetic PCR; real-time PCR; *Taq*Man technique): The detection of the accumulation of amplification products during conventional → polymerase chain reactions and their quantification. Basically, the various techniques of Q-PCR fall into two broad categories. First, the intercalator-based methods include intercalating dyes (as e. g. → ethidium bromide) in each amplification reaction, irradiate the sample with UV-light in a specialized → thermocycler, and detect the resulting fluorescence light with a computer-controlled, cooled, charge-coupled device (CCD) camera. By plotting fluorescence increase versus cycle number, amplification plots are generated, allowing to quantify the products. However, this type of Q-PCR suffers from the disadvantage, that both specific and non-specific products generate fluorescence signals, which makes quantitation obsolete. Second, the socalled → 5' nuclease PCR and similar probe-based quantification protocols allow to detect only specific amplification products in real-time. The 5' nuclease PCR assay exploits the 5' nuclease activity of *Taq* DNA polymerase to cleave probe-target hybrids during amplification, when the enzyme extends from an → upstream primer into the region of the probe. This cleavage can be visualized by increased fluorescence, if the oligonucleotide probe contains both a reporter fluorochrome at its 5' end and a quencher dye at its 3' end. The close proximity of both fluorochromes (1.5-6.0 nm) results in a Förster type fluorescence energy transfer, leading to the suppression of the reporter ("quenching") which is relaxed when the probe is hydrolyzed.
Probe-based Q-PCR has been refined to be reproducible. First, the endpoint measurement of the amount of accumulated PCR products is skipped in favour of the more reliable threshold cycle (T_c), which is defined as the fractional cycle number at which the reporter fluorescence generated by cleavage of the probe passes a fixed threshold above base-line. T_c is inversely proportional to the number of target copies in the sample. The quantification is made by calculating the unknown

target concentration relative to an absolute standard (e. g. a known copy number of plasmid DNAs, or a house-keeping gene as internal control). In contrast to the endpoint approach, T_c is measured when PCR amplification is still in the exponential phase (i. e. the amplicons accumulate at a constant rate, the amplification efficiency is not influenced by variations and limitations of the reaction components, and the enzymes and reactants are still stable). Also, primer-primer artifacts are low in number.

In another version of Q-PCR, the → competitive PCR, a synthetic DNA or RNA is used as internal standard (competitor amplicon), that contains the same primer binding sites and (optimally) has the same amplification efficiency as the target, but has a different size to discriminate it from the target. A known amount of this competitor is co-amplified with the target nucleic acid in the same tube. If the amplification efficiency of target and competitor is identical, then the ratio

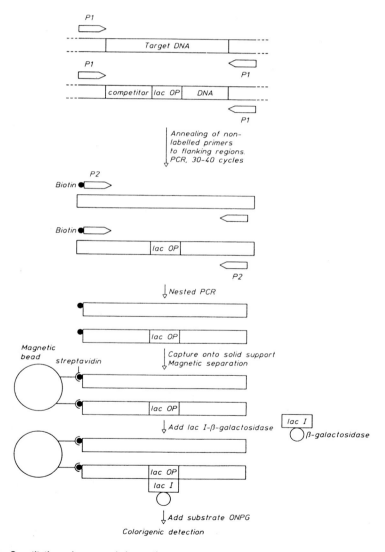

Quantitative polymerase chain reaction

target / competitor will be constant throughout the PCR process. By determining the target / competitor ratio at the end of the process, and accounting for the starting amount of the spiked-in competitor, the initial amount of target can be calculated. As opposed to the superior real-time Q-PCR, competitive PCR is tedious, as it requires to find the most suitable ratio of target to competitor by dilution series, and moreover necessitates construction and characterization of a different competitor for every target to be quantified. Also, a series of experiments have to insure that the amplification efficiencies of target and competitor are in fact identical.

Quantitative proteomics: The whole repertoire of techniques to quantify protein content, masses of the various proteins, numbers of protein isoforms and the stochiometry of protein-protein interaction of a cell. See → proteomics.

Quantitative reverse transcriptase polymerase chain reaction (QRT-PCR): A variant of the → reverse transcription PCR (more precisely, the → relative quantitative reverse transcriptase polymerase chain reaction), that allows accurate quantitation of → messenger RNAs of a cell. This technique is based on the serial dilution of → competitor RNA or → mimic standards, that are then added to a constant amount of sample cDNA. Quantification is achieved by comparison of the relative fluorescence intensities of target and competitor bands after their amplification in conventional → polymerase chain reaction and their electrophoretic separation in an → agarose or → polyacrylamide gel. See → polymerase chain reaction-aided transcript titration assay. *Figure see page*

Quantitative trait locus (QTL): A genomic region with several genes, or two or more separate genetic → loci that contribute cooperatively to the establishment of a specific phenotype.

Quantitative trait locus mapping (QTL mapping): A procedure to localize a → quantitative trait locus (or loci) on a → genetic or also a → physical map.

Quantum dot (QD): A 1–5 nm non-ionic semiconductor nanocrystal (NC) with a cadmium-selenium core and a zinc sulfide cover, that can be excited by UV light and emits fluorescence light over long periods of time (depending on the crystal sizes, over several months). QDs are chemically synthesized at high temperatures, and normally possess a hydrophobic coat, that can be replaced by polyacrylate (which allows e.g. coupling to antibodies) or modified by negatively charged dihydroliponic acid (DHLA). DHLA allows electrostatic binding of specific antibodies, or other proteins. Also, the pores of polystyrene microbeads can be loaded with QDs of different size and emission spectra, by coupling them to short oligonucleotide sequences. These can be hybridised to specific target sequences, and the hybridisation events detected by long-term fluorescence light of different wave-lengths. In QDs, electrons exist at discrete energy levels ("bands"). Any energy input (e.g. as photons) raises an electron from a lower ("valence") to a higher band ("conduction band"). Upon return to the lower band, the excess energy is released as a photon with an energy roughly equalling the gap between the bands (see → photoluminescence). This band gap increases with decreasing size of the dots. Therefore smaller dots release more energy (i.e. emit blue light). QDs fluoresce up to 100times longer than conventional → fluorochromes and are used for the sensitive non-isotopic detection of biomolecules. See → core shell quantum dot, → giant quantum dots.

Quantum efficiency: See → quantum yield.

Quantum yield (QY, quantum efficiency, φ): The number of fluorescence photons emitted (N_{fl}) over the number of photons absorbed (N_{abs}) by a → fluorochrome. QY is a characteristic feature of fluorochromes and is used to describe their quality.

Quartz crystal microbalance with dissipation monitoring (QCM-D): A technique for the real-time detection of interactions of molecules with a surface or with each other, the determination of the mass of extremely thin surface-bound layers and their viscoelastic (structural) properties. A QCM-D sensor consists of a thin plate of crystalline quartz sandwiched between two electrodes. If an AC voltage is applied over these electrodes, an oscillation is induced in the sensor. This oscillation exponentially decays after the AC voltage is switched off. This decay is recorded and the resonance frequency (f) and the dissipation (D) calculated. Since the resonance frequency of the sensor crystal depends on the total oscillating mass, it is used to determine the mass of the molecules deposited on the sensor surface. Also, the thickness can be calculated by dividing the mass by the density of the layer. The dissipation energy (energy dissipated per oscillation over 2π times total energy stored in the system), which represents the frictional losses in the surface film, informs about the structure of this film attached to the sensor surface (which can be composed of metals, polymers such as proteins, but also living bacterial cells). For example, a compact globular protein attached to the surface results in a low dissipation, whereas an elongated protein with many coupled water molecules increases dissipation. An adapted software calculates correct thickness, viscosity and elasticity from the resulting data. QCM-D is used to measure protein adsorption to and their desorption from membranes, and protein-protein interactions.

Quelling: See → co-suppression.

Quencher: Any molecule, that reduces or completely deactivates ("quenches") an excited state of another molecule (e.g. a → fluorochrome), either by energy transfer, electron transfer, or by a chemical reaction. See → quenching.

Quenching: The reduction or complete deactivation of an excited state of a distinct molecule (e.g. a → fluorochrome) by a → quencher.

Query sequence: Any amino acid or nucleotide sequence, that is used in a database search (e.g. for sequence → homology). See → BLAST.

Quick blot: See → quick blotting

Quick blotting (quick blot, fast blot): A technique to immobilize RNA and DNA from cellular extracts without extensive purification on → nitrocellulose filters, in which sodium iodide (NaI) instead of baking is used to fix the nucleic acids onto the support. In short, cellular extracts are deproteinized enzymatically, detergents and NaI are added, and the mixture is filtered through nitrocellulose. NaI promotes dissolution of proteins, causes selective binding of mRNA and DNA (depending on the temperature used), and preserves the biological activity of e.g. → messenger RNA (which still can be reverse-transcribed into → cDNA or translated into protein). Filters are washed and used directly for molecular → hybridization.

Quick-stop mutant: A → mutant of *E. coli*, that rapidly stops its DNA synthesis after a temperature increase to 42°C.

Quinacrine: A → fluorochrome that intercalates into DNA duplex molecules and allows the staining of chromosomes so that a typical pattern of fluorescing bands (Q bands) is generated. Q banding is exploited for the specification of chromosomes and the detection of gross → rearrangements, → deletions, and other abnormalities.

Q vector: A bicistronic → self-inactivating retroviral vector for the transfer of target genes into animal cells, that provides high virus titers, reliable expression levels of the target gene, and reduces → promoter interference. For example, a specialized Q vector contains 3' → long terminal repeats (LTRs) of the self-inactivating class; a 5' LTR, that is a hybrid of the cytomegalovirus (CMV) enhancer and an engineered mouse sarcoma virus (MSV) → promoter; an internal CMV immediate early region promoter to drive the expression of the transduced gene and the → neomycin phosphotransferase (or any other) → selectable marker; an expanded → multiple cloning site (MCS) and a eukaryotic → internal ribosome entry site (IRES). The vector backbone contains an SV40 origin to promote high copy number replication in packaging cell lines, that express the SV40 large T antigen. The target gene is expressed together with the selectable marker, and after integration of the vector into the host cell genome the promoter in the 5'LTR is inactivated. Self-inactivation is a function of a deletion in the U3 region of the 3'LTR. After → reverse transcription of the plus strand of the retroviral genome, it is integrated. During integration, a circular intermediate is formed, that causes the duplication of the deletion in the U3 region of the 3'LTR. This inactivates the CMV-MSV hybrid promoter in the 5'LTR. Therefore, transcription is only driven by the internal CMV promoter immediately upstream of the target gene.

R

R: A grammalogue for → puRine (→ A or → G), used in sequence data banks.

r: A widely used abbreviation for "recombinant" (e.g. rBST means bovine somatotropin that is synthesized by genetically modified bacteria into which a gene encoding somatotropin has been introduced). However, the abbreviation should be replaced by "rec", so that any confusion with "ribosomal" (e. g. → rDNA, → rRNA) is securely avoided.

Rabbit reticulocyte lysate (rabbit reticulocyte system): A complete → *in vitro* translation system from rabbit reticulocytes, including ribosomes, a full complement of → transfer RNAs, → aminoacyl tRNA synthetases, initiation, elongation and termination factors. This system has been treated with → micrococcal nuclease to destroy endogenous mRNAs. Thus the translational activity of the lysate fully depends on exogenously added mRNA(s). In addition to compounds necessary for *in vitro* translation (e.g. ions like K^+ and Mg^{2+}, nucleotides like ATP and GTP and an energy-regenerating system like creatine phosphokinase) hemin can be added to ensure that the *in vitro* translation of exogenous mRNA proceeds to completion. The lysate is usually aliquoted and stored over liquid nitrogen. Compare → wheat germ system.

Rabbit reticulocyte system: See → rabbit reticulocyte lysate.

RACE: See → *r*apid *a*mplification of *c*DNA *e*nds.

Radial *i*mmuno*d*iffusion (RID): A modification of the → agarose gel diffusion technique in which the → antibody is distributed uniformly throughout an agarose gel. The antigenic solution is then placed in a central well of the gel, and in consequence of the diffusion of both reactants a circular immunoprecipitin zone is generated. The precipitate can also be quantified. Alternatively, an antigen-loaded agarose gel can be overlayered with an antibody solution.

Radiation *h*ybrid (RH): Any hybrid cell that contains small fragments of human chromosomes in a *C*hinese *h*amster *o*vary (CHO) genetic background. RH production starts with a CHO cell that harbors a single, but complete human chromosome with a wild-type *h*ypoxanthine *r*ibosyl *t*ransferase (hprt) gene. This cell line is exposed to very high X-ray doses (e. g. 8 krads), that breaks the chromosome (on average, into 5 pieces). After healing, the resulting cells are fused with a CHO cell line (hprt: the inactive hprt gene allele is present). Cell clones showing hprt$^+$ phenotype are selected and maintained on a selective HAT medium (growth on this medium requires an active hprt gene). The selected clones represent a randomly sampled library of fragments of the particular chromosome. See → *r*adiation *h*ybrid (RH) mapping.

Radiation *hybrid* (RH) map: A genome map constructed with *sequence-tagged sites* (STSs) as markers, that are positioned relative to one another on the basis of the frequency, with which they are separated by radiation-induced breaks (expressed as *centirays* [cR], defined as a 1% chance, that a break occurs between two loci). This frequency is determined with a panel of → radiation hybrids. See → radiation hybrid mapping.

Radiation *hybrid* (RH) mapping: A somatic cell hybrid technique for the construction of dense maps of mammalian chromosomes, that involves → radiation hybrids and the statistical analysis of DNA markers to determine their relative order with respect to one another, or the identification of the chromosomal location of a specific gene. For gene localization two → gene-specific primers are used to amplify a single PCR product (usually between 100 and 800 bp). This product can then be labeled and hybridized to a panel of RH clones (consisting of donor DNA fused to hamster or mouse recipient DNA). A hybridization event indicates presence of the probe-complementary sequences in the RH clone. Major advantages of RH mapping: the estimated distances are directly proportional to physical distances and non-polymorphic DNA markers can also be used for mapping (as opposed to → genetic mapping).

Radiative *decay* *engineering* (RDE): The improvement of the physical properties (e.g. stability and increase in quantum yield) of → fluorochromes by increasing their → radiative decay rates. After excitation, the emission rate of a fluorophor depends on the socalled photonic mode density (roughly the spacing) around the fluorophore. This density can be modified by binding fluorophores near or on metallic silver particles (or other conducting metallic surfaces) with the effect of dramatically increased fluorescence emission and fluorophore stability. Also, the number of excitation/emission cycles is consequently elevated.

Radiative decay rate (Γ): The intrinsic emission of fluorescence by → fluorochromes, which depends on the absorption or extinction coefficient of the fluorophore. Fluorochromes with high extinction coefficients possess high emissive rates. Radiative decay rate can be engineered by → radiative decay engineering.

Radioactive half-life (half-life time): The time interval in which half of a given number of unstable radioactive atoms decay into stable non-radioactive atoms. The half-life time is specific for each radionuclide (e.g. 14.3 days for phosphorus-32; 87.1 days for sulfur-35; 12.46 years for tritium).

Radioactive *in situ* *hybridisation* (RISH): A method to identify specific sequences on intact chromosomes (usually metaphase spreads) by hybridisation with complementary nucleic acid → probes (definition a), that are labeled with radioactive isotopes (e.g tritium, also C^{14}). RISH is only rarely used, because radiation-free → fluorochromes can more elegantly be employed in → fluorescent *in situ* hybridization. See → *in situ* hybridization.

Radioactive isotope (radioisotope, radiotope): A form of a chemical element with an unstable nucleus that tends to stabilize by the emission of ionizing radiation (α-particles, corresponding to helium nuclei, or β-particles, representing electrons, or electromagnetic waves). Radioactive isotopes are used as → radioactive label. See also → radioactive half-life.

Radioactive label ("radio-label"): Any radioactive atom introduced into a molecule for its identification. For example, radioactive sulfur atoms (^{35}S) are incorporated into the amino acid methionine which may then be used to identify newly synthesized proteins in → in vitro translation experiments. Or, radioactive phosphorus (^{32}P) is used to label → nucleotides which are incorporated into newly synthesized nucleic acid molecules during the processes of → transcription and → replication. Compare → non-radioactive labeling.

Radioactivity: The spontaneous disintegration of certain nuclides with concomitant emission of one or more types of radiation (e.g. α- or β-particles, or electromagnetic waves).

Radioimmuno assay (RIA): A technique for the quantitative determination of → antigens in which the specific binding of radioactively (e.g. ^{125}I) labeled antigen or → antibody is measured.

Radioisotope: See → radioactive isotope.

Radio-label: See → radioactive label.

Radiotope: See → radioactive isotope.

RAG: See → recombination activation gene.

RAGE:
a) See → rapid analysis of gene expression.
b) See → rotating gel electrophoresis.

RAHM: See → random amplified microsatellite polymorphism.

RAMP:
a) See→ random amplified microsatellite polymorphism.
b) See → ribosomally sythesized antimicrobial peptide.

R-amplicon: The double-stranded DNA molecules arising from the reverse transcription of an RNA template into its → cDNA and its amplification in a conventional → polymerase chain reaction.

RAMPO: See → random amplified microsatellite polymorphism.

RAMS: See → random amplified microsatellite polymorphism.

Random amplified microsatellite polymorphism (RAMP): Any DNA sequence → polymorphism discovered by a combination of → random amplified polymorphic DNA and → microsatellite detection techniques. A chimeric → primer, consisting of a 5' anchor sequence and 3' microsatellite repeats (as e.g. 5'-CCT[GAGAGAGAGA]-3'), is endlabeled and used in a → polymerase chain reaction-driven amplification of genomic DNA in the presence or absence of → RAPD primers. The amplification products are electrophoresed in denaturing → polyacrylamide gels. Only the → amplicons originating from the labeled anchored primer are detected by → autoradiography and reveal microsatellite polymorphisms.

Random amplified microsatellite polymorphism (RAMPO; random amplified hybridization microsatellites, RAHM; randomly amplified microsatellites, RAMS): Any DNA sequence → polymorphism detected by a combination of → random amplified polymorphic DNA and → microsatellite detection techniques. Genomic DNA is first mixed with an arbitrary 10mer → primer, and amplified by conventional → polymerase chain reaction techniques. The amplification products are then separated by → agarose gel electrophoresis and visualized by → ethidium bromide staining (first detection level). After documentation by e.g. photographing, the gel is blotted (→ Southern blotting) and the blot hybridized to radioactively labeled → microsatellite → probes. Hybridization can be detected by → autoradiography of the blot, and reveals a second type of polymorphism (second detection level). Do not confuse with → random amplified microsatellite polymorphism (RAMP).

Random amplified polymorphic DNA (RAPD; *low stringency products*, LSPs): Any DNA segment that is amplified using short oligodeoxynucleotide → primers of arbitrary nucleotide sequence (→ amplimers) and → polymerase chain reaction procedures. In short, → genomic DNA is isolated and single primers, about 10 nt long and of arbitrary sequence are annealed to it. After a conventional PCR amplification the amplified fragments are electrophoretically separated on → agarose gels and visualized by simple staining with → ethidium bromide. If such RAPDs from two related organisms are compared, they may exhibit length polymorphisms. The detected polymorphisms can be used as → genetic markers to construct → gene maps. See → arbitrarily primed polymerase chain reaction, → DNA amplification fingerprinting, → optimized stringent RAPD.

Random amplified transcribed sequence (RAT): A DNA sequence that has been amplified from a genomic → template by conventional → polymerase chain reaction techniques using → primers whose sequences are derived from → cDNAs.

Random clone fingerprinting (random fingerprinting): A technique for the establishment of → physical maps of a genome by using randomly selected clones of a → genomic library. The finger-prints are generated by digesting these clones with one or several → restriction enzymes, → end-labeling the fragments, their separation by denaturing → polyacrylamide gel electrophoresis, and detection by → autoradiography. The number and size of the fragments represent a unique fingerprint of the cloned insert. A comparison of fingerprints from several clones detects over-laps, when the clones share a statistically significant number of common bands. Such over-laps or → contigs are then used for → contig mapping.

Random coil: An irregular secondary structure of macromolecules (nucleic acids or proteins) in comparison to a regular → double helix conformation (nucleic acids) or → α-helix or β-sheet conformation (proteins).

Random elongation mutagenesis: A technique for random mutagenesis of protein-encoding DNA sequences, that capitalizes on the ligation of partially randomized oligonucleotides ahead of the → stop codon of a target sequence such that short chains of about 16 amino acids are introduced into the carboxy ends of the encoded proteins. This leads to the generation of mutant populations, where each mutant carries a different peptide tail, which expands the protein sequence space. This expansion may produce mutants with favorable properties. For example, random elongation mutagenesis generated catalase mutants with increased thermostability of the enzyme. The technique therefore allows in vitro evolution of proteins with new properties (evolutionary molecular engineering). Compare → DNA shuffling.

Random fingerprinting: See → random clone fingerprinting.

Random insertion of antibiotic resistance genes (RIAR): A technique for the introduction of random insertions (see → insertion mutagenesis) of antibiotic resistance genes (e.g. → kanamycin resistance, or → zeocin resistance genes) into plasmids as a prelude to their sequencing. In short, an appropriate plasmid vector (containing an → ampicillin resistance gene, a *Col*E1 → origin of replication, and the target → insert) is first randomly digested with → DNAse I in an Mn^{2+}- containing buffer (which preferably induces double-strand breaks rather than → nicks), the resulting products are then blunt-ended by T4 DNA polymerase, and the linear DNA segments purified by gel electrophoresis. Then the antibiotic gene fragments are ligated into the gaps of the digestion products using → T4 DNA ligase, and the transformants selected with kanamycin or zeocin, respectively. The positions of the insertions are then verified, the inserts amplified with a vector primer and a resistance gene-specific primer, and the amplicon sequenced (preferably in both directions). RIAR is independent of a → restriction map of the plasmid, and does not require high plasmid quality (e.g. nicks and RNA contaminations are not interfering).

Randomly ordered fiber-optic gene array: See → fiber-optic DNA array.

Random mutagenesis: Any non-directed (random) introduction of → mutations into a DNA molecule (e.g. a → plasmid, PCR fragment, → chromosome). Usually, a target DNA is randomly mutagenized by → error-prone PCR with e.g. → *Thermus aquaticus* (*Taq*) DNA polymerase, that introduces a distinct spectrum of mutations into the template. For example, *Taq* polymerase induces the following → transitions or → transversions (in decreasing efficiency): A→T, T→A, A→G, T→C, G→A, C→T, A→C, T→G, G→T, C→A, G→C, C→G. Such mutations can be generated at a low, medium, or high frequency depending on the input template concentration. The mutagenized DNA is then introduced into an *E. coli* strain deficient in several of the primary DNA repair pathways such that the mutation spectrum in the target DNA is maintained.

Random orientation strategy: A technique to clone → cDNA into a specific vector (e.g. lambda gt 10 or gt 11, see → lambda gt vectors), where the orientation of the cDNA insert is totally at random (random orientation). Compare → forced cloning.

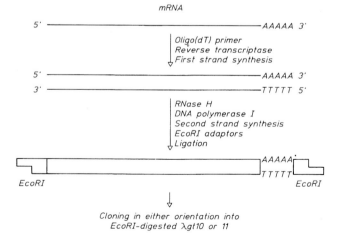

Random orientation strategy

Random peptide display: A technique for the detection of protein-protein interactions, that exploits bacterial flagellin and thioredoxin proteins to display random dodecamer peptides fused to these proteins, on the surface of *E. coli* cells. In short, a library of DNA sequences encoding random dodecapeptides is fused to coding sequences for the active site loop of *thioredoxin* (*trx*A)

inside the dispensable region of the bacterial *flagellin* gene (*fliC*). This fusion is expressed from the bacteriophage major leftward → promoter (P$_l$). The resulting → fusion protein is exported and assembled into partially functional flagella on the bacterial cell surface, where the active site loop of thioredoxin protrudes into the surrounding medium, making the peptides completely available ("maximally represented") for ligand binding ("display of random peptides").
The display of random peptides can easily be screened for interactions with a target protein by → panning, and allows to detect receptor-ligand binding, enzyme-substrate specificity, and anti-body-antigen recognition. Compare → phage display.

Random *polymerase* *chain* *reaction* (random PCR): A variant of the conventional → polymerase chain reaction technique that allows to amplify a whole → cDNA library made from minute amounts of cellular RNA. In short, the RNA from one eukaryotic cell or 100 bacteria is annealed to the oligonucleotide 5'-GCCGGAGCTCTGCAGAATTCNNNNNNN-3' (universal primer) containing a random hexamer at its 3'-end and used to prime → first strand cDNA synthesis with → reverse transcriptase. The → second strand is synthesized by the → Klenow fragment of DNA polymerase I. The double-stranded cDNAs are then amplified in the presence of the universal primer using → *Taq* polymerase to yield high copy numbers of cDNA from low amounts of RNA.

Random primed oligolabeling: See → random priming.

Random primer: Any → oligonucleotide → primer that has a 4-fold → degeneracy at each position, i.e. each of the four → bases of DNA may equally well occupy a position.
Example: 5'-NNNNNNNNN-3' (where N = any of the four bases A,C,G or T).

Random priming (random primed oligolabeling, Feinberg-Vogelstein procedure, mixed primer labeling, oligolabeling, hexamer priming): A method to label single-stranded DNA molecules to high → specific activity (i.e. more than 10^8 cpm/µg DNA) using a mixture of short hexameric → primers which anneal to specific complementary sequences of the denatured target DNA, and serve to prime the synthesis of the second strand by the → Klenow fragment of → DNA polymerase I. Radiolabeled nucleotides are added to the reaction mixture and are incorporated into the growing second strand. Compare → nick translation.

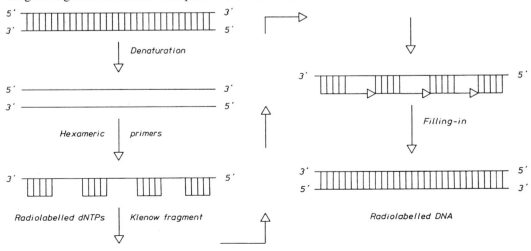

Random priming

Random shear cosmid library: A → cosmid library that is established using mechanically sheared DNA fragments. Mechanical shear (by e.g. → sonication) avoids a potential bias that may be introduced by either non-random distribution of restriction sites or differential kinetics of cleavage, if the genomic DNA is enzymatically fragmented. However, the cloning of randomly sheared DNA fragments is inefficient.

RAP:
b) See → *repeat-associated polymorphism.*
a) See → *RNA polymerase binding protein.*

RAPD: See → *random amplified polymorphic DNA.*

RAP fingerprinting: The use of → RNA arbitrarily primed PCR (RAP-PCR) to detect → messenger RNAs and to isolate the corresponding genes. See → differential display PCR, → RNA fingerprinting. Compare → DNA fingerprinting.

Rapid amplification of cDNA ends (RACE): A variant of the conventional → polymerase chain reaction technique that uses gene-specific oligodeoxynucleotide → primers to amplify → cDNAs reverse-transcribed from → low-abundance messenger RNAs. Basically, the 3' end or the 5' end of a cDNA can be amplified. Accordingly, the somewhat different techniques are called 3'-RACE or 5'-RACE, respectively. In short, **3'-RACE** works with an oligo(dT)-containing → adaptor primer, partly complementary to the → poly(A) tail of mRNAs. This primer allows → first strand synthesis with → reverse transcriptase. After destruction of the mRNA with → RNase H a → gene-specific primer complementary to a region at the 5' end of the original mRNA and a → universal adaptor-primer complementary to its 3' end allow to amplify the cDNA with an intact 3' end (A). The **5'-RACE** technique starts with the annealing of a gene-specific anti-sense primer complementary to the 3' region of the mRNA, first strand synthesis with reverse transcriptase, degradation of the mRNA with RNase H, purification of the cDNA, its → homopolymer tailing with dCTP, the anchoring of an oligo(dG)-sequence, and the amplification of the cDNA using the anchored primer and a nested gene primer, and PCR (B).
Figure see page 902.

Rapid analysis of gene expression (RAGE): A technique for the expression analysis of tens to hundreds of genes in multiple samples. In short, RNA is isolated from the target tissue, converted to → cDNA using a biotinylated oligo(dT) → primer, the cDNAs digested with *Dpn*II, the 3'most *Dpn*II fragment of each cDNA adsorbed to streptavidin-coated magnetic beads and thereby non-biotinylated fragments removed. Then a → linker with a *Dpn*II-generated overhang (B-linker) is annealed to the cDNA fragments on the beads and ligated using → T4 DNA ligase. The preparation is restricted by *Nla*III, the fragments released from the beads recovered and ligated to a linker with an *Nla*III-generated overhang (A-linker). The cDNA fragments containing the gene-specific targets ligated to A- and B-linkers (or B- and A-linkers) are referred to as A/B or B/A → ditags ("bitags"). Then the templates are amplified in a → polymerase chain reaction using linker-complementary RAGE primers, containing 3–4 nucleotide long specificity regions at the 3' end. After electrophoresis on 8% polyacrylamide gels the fragments are simply stained with a fluorescent dye and fluorescent signals digitized in a fluorescence imager. Compare → Long SAGE, → microSAGE, → SAGE-Lite, → serial analysis of gene expression, → Super-SAGE.

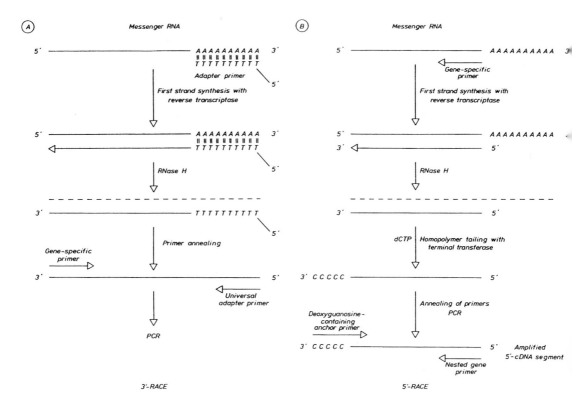

Rapid amplification of cDNA ends

Rapid cycle DNA amplification (rapid cycling, rapid cycle amplification): A technique for the ultra-fast amplification of specific DNA sequences, using microcapillary tubes for the amplification reaction, a hot air temperature control for the cycling process, and extremely reduced → denaturing and → primer annealing times. In short, amplification mixtures are prepared in 96-well microtiter plates, the minute volume samples (5-10 µl) loaded into thin-walled (0.2 mm or less) microcapillary tubes by capillary action, the end of each tube sealed by flame, all tubes placed in the reaction chamber of the special → thermocycler ("rapid cycler") and cycled simultaneously. The thermocycler works with high velocity air streams (1 km/min), resulting in high temperature ramping rates, which guarantee target temperatures within each sample to be reached in less than one second. Therefore a complete 30 cycles amplification program does not require more than 10 minutes, but still produces one specific amplification fragment per specific primer. If compared to normal thermocycling, specificity and yield are improved in rapid cycling, because the annealing time is reduced considerably, and so is the concentration of spurious, undesired amplification products that reflect non-specific amplification. Also, → tracking dyes or → ethidium bromide can be loaded directly onto the capillaries. See → polymerase chain reaction.

Rapid cycling: See → rapid cycle DNA amplification.

Rapid subtraction hybridization (**RaSH**): A variant of the conventional → genomic subtraction technique for the identification of differentially expressed genes. In short, poly(A)⁺-mRNA is isolated from control cells (driver) and treated cells (tester), converted to double-stranded → cDNA, and the cDNA digested with *Eco*RII or *Dpn*II. The resulting small fragments are ligated to → adaptors, using → T4 DNA ligase, and amplified in a conventional → polymerase chain reaction using adaptor-complementary primers. Then only amplification products from the tester population are digested with *Xho* I. The two libraries are then hybridized and the differentially expressed sequences cloned into *Xho*I-digested vectors, resulting in a subtracted cDNA library. The differentially expressed clones are then verified by → Northern analysis.

Rapid translation system: See → continuous exchange cell-free system.

RAP-PCR: See → *R*NA *a*rbitrarily *p*rimed *p*olymerase *c*hain *r*eaction.

RAPping: The procedure for the detection of → *r*epeat-*a*ssociated *p*olymorphism*s* (RAPs). See → RIPping.

Rare base (minor base, modified base): Any one of a series of acetylated, glycosylated, methylated or otherwise modified nucleoside phosphates that are constituents of nucleic acids, especially of → tRNA molecules (e.g. N^1-methyladenosine; N^6,N^6-dimethyladenosine; $N^6(\Delta^2$-isopentenyl) adenosine; 5-hydroxymethylcytidine; 5-methylcytidine; 4-methylcytidine; N^4-acetylcytidine; 1-methylguanosine; N^2-dimethylguanosine; base Y; → inosine; 1-methylinosine; ribothymidine; 5-methyluridine; 5,6-dihydrouridine; 4-thiouridine; pseudouridine, or methylated riboses). *Figure see page 904 and 905.*

Rare codon (low-usage codon): Any → codon that is not utilized very frequently in a given organism. Such underutilized codons, especially if clustered in a gene, slow down ribosome movement along the encoded → messenger RNA. Even if the rare codons cluster at the 3' end of the message, a ribosome jam will transmit back to the 5' end of the mRNA (result: reduction of translation). The same holds for the presence of the cluster at the 5' terminus of the message. See → codon optimization, → optimal codon.

Rare cutter (rare cutting restriction endonuclease): Any → restriction endonuclease that recognizes and cleaves sequences in DNA duplexes occurring only rarely (e.g. *Fse* I, sequence GGCCGGCC; *Not*I, sequence GC/GGCCGC; *Swa*I, sequence ATTT/AAAT; → meganuclease I-*Sce* I, sequence TAGGGATAA/CAGGGTAAT).

Rare cutting restriction endonuclease: See → rare cutter.

ras **gene** (*rat sarcoma gene*): Any one of a series of → oncogenes that encode a transforming protein (p21^{ras}) which strongly binds GTP (GTPase). Such ras genes are for instance *Ha-ras* (*Ha*rvey murine sarcoma), *Ki-ras* (*Ki*rsten murine sarcoma), and *N-ras* (*n*euroblastoma tumor).

RaSH: See → *r*apid *s*ubtraction *h*ybridization.

Ratio PCR: See → relative quantitative reverse transcriptase polymerase chain reaction.

Rat sarcoma gene: See → *ras* gene.

Adenosine derivatives:

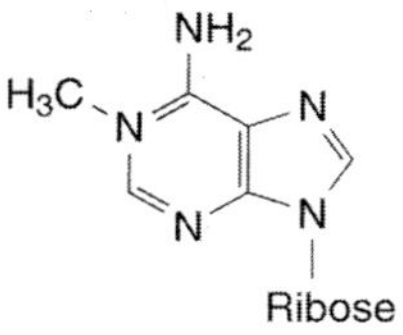

N^1-methyladenosine

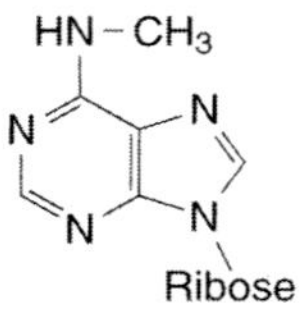

N^6-methyladenosine
(m^6A)

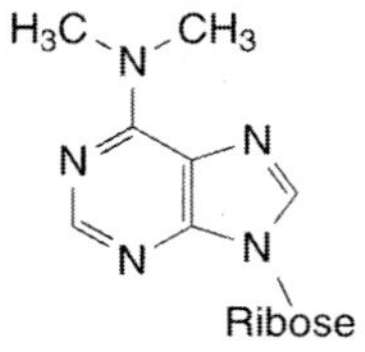

N^6,N^6-dimethyladenosine
(m_2A)

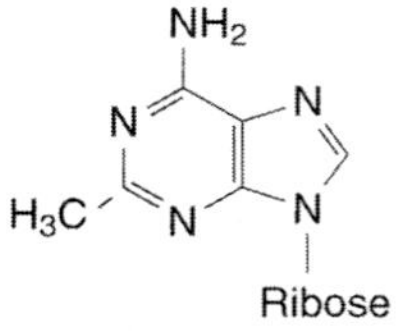

2-methyladenosine

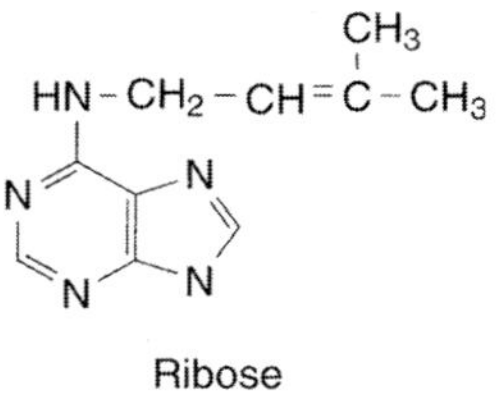

$N^6(\Delta^2$-isopentenyl)-
adenosine (i^6A)

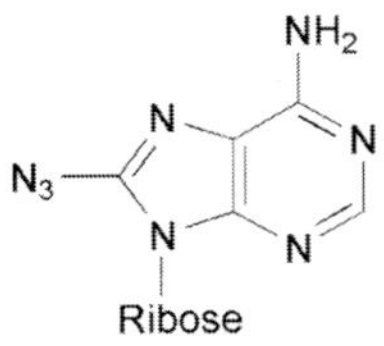

8-azidoadenosine

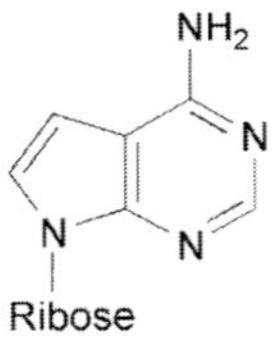

7-deazaadenosine

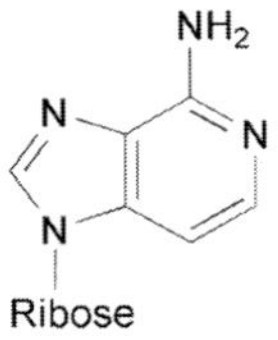

3-deazaadenosine

Cytidine derivatives:

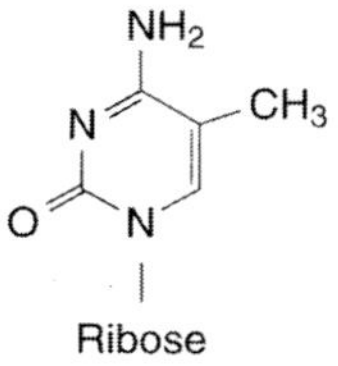

5-methylcytidine
(mC)

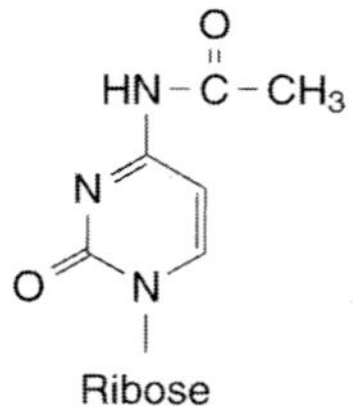

N^4-acetylcytidine

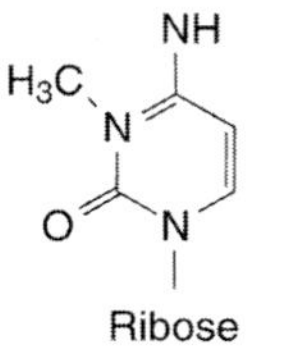

3-methylcytidine
(m^3C)

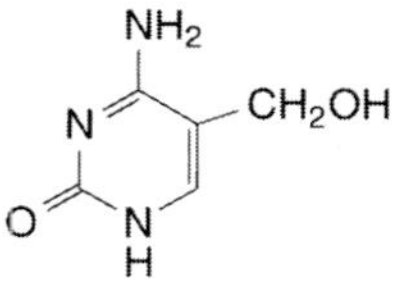

5-hydroxy-
methyl-cytosine
(hmC)

Guanosine derivatives:

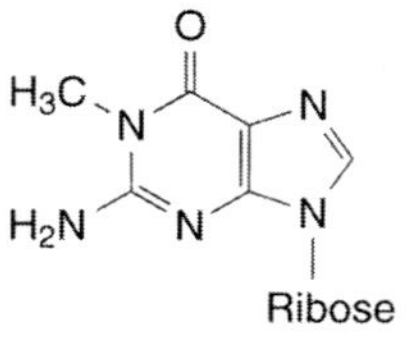

1-methylguanosine
(m^1G)

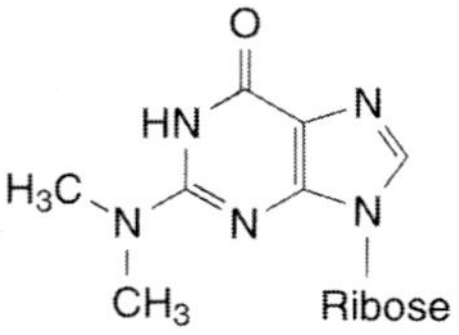

N^2-dimethylguanosine

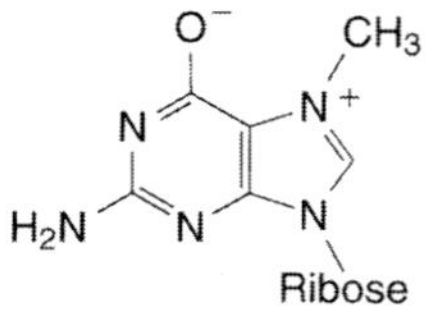

7-methylguanosine
(m^7G)

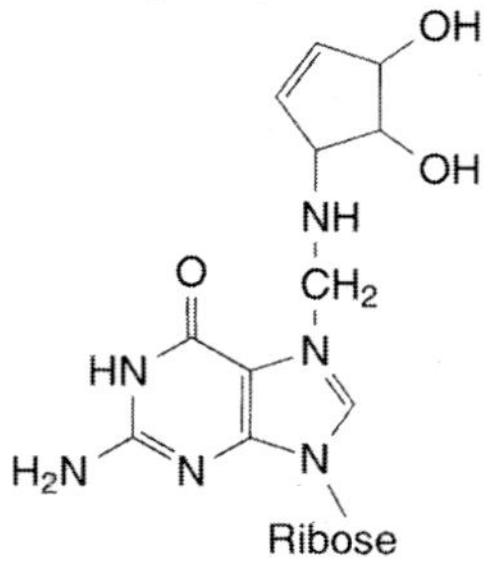

Base Q
(Queuosine)

Guanosine derivatives:

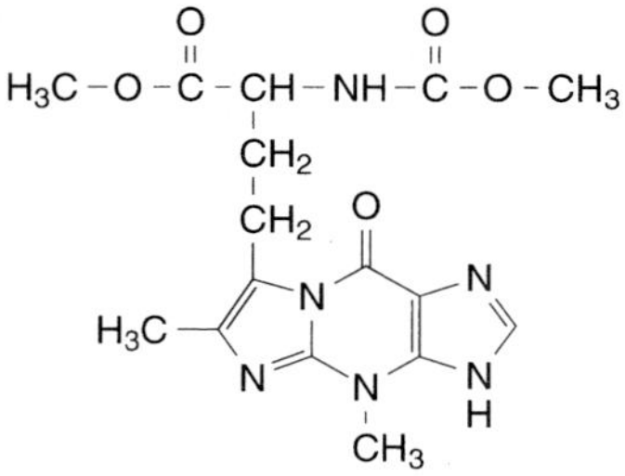

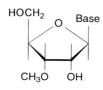

Base Y (Wyosine)

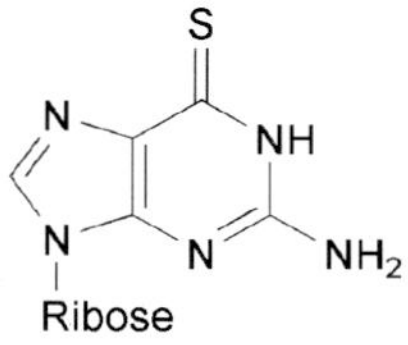

Ribose

6-thioguanosine

Inosine derivative:

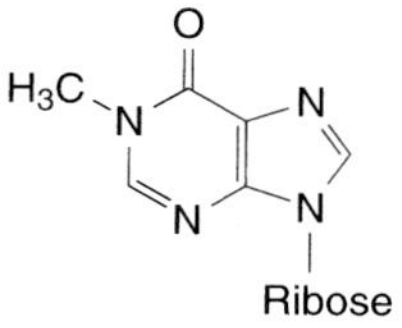

Ribose

1-methylinosine

Ribose derivatives:

3′-O-methylriboside

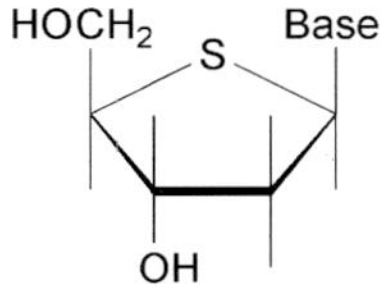

4'-thioanalogue

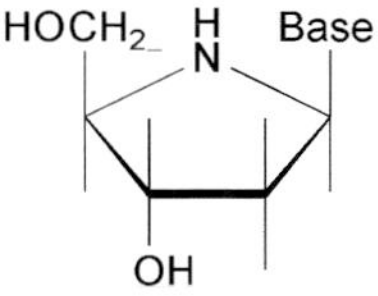

4'-aminoanalogue

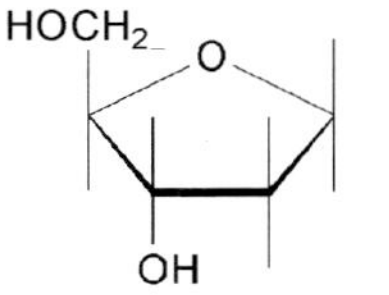

Abasic nucleoside

Thymidine derivatives:

Ribose

Ribothymidine

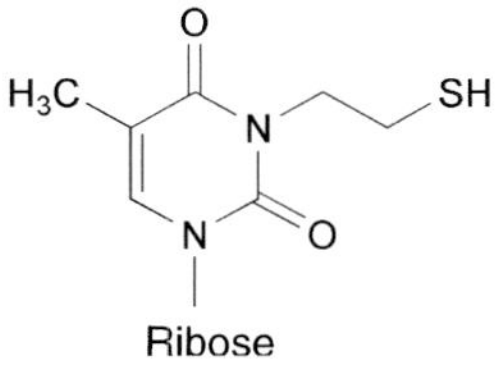

Ribose

N3-thioethylthymidine

Uridine derivatives:

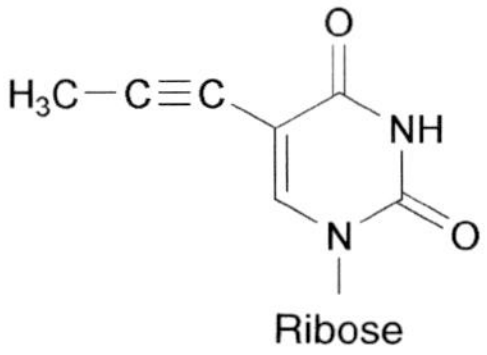

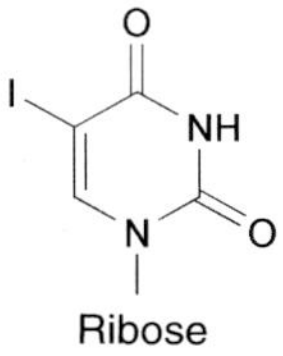

Ribose

Ribose

5-iodouridine

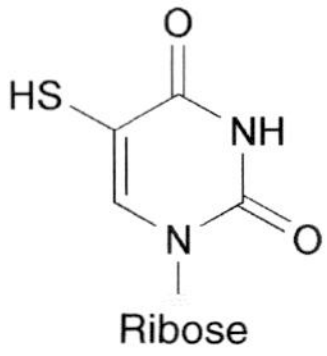

Ribose

5-mercaptouridine

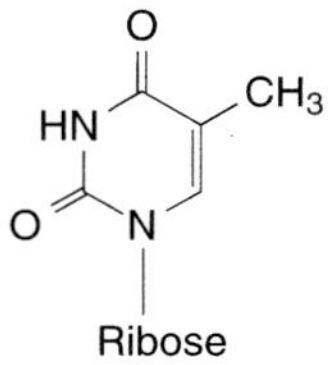

Ribose

5-methyluridine

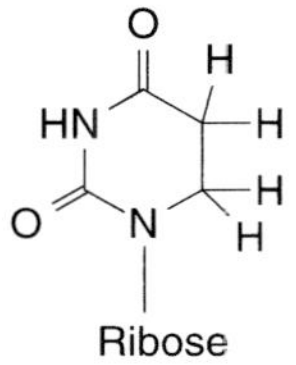

Ribose

5,6-dihydrouridine

(D)

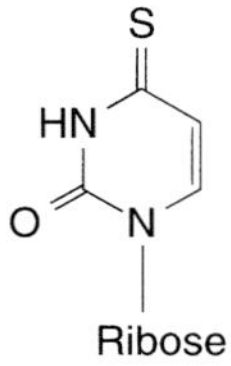

Ribose

4-thiouridine

(S⁴U)

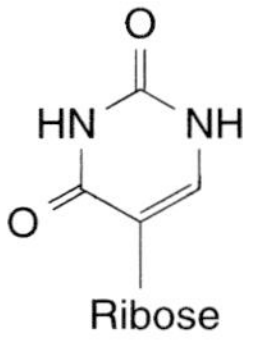

Ribose

Pseudouridine (Ψ)

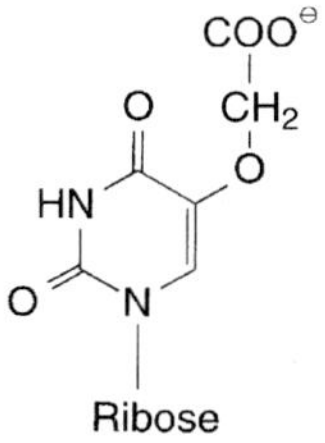

Ribose

Uridine-5-oxyacetate

RAT: See → *r*andom *a*mplified *t*ranscribed sequence.

RAWIT: See → *R*NA *a*mplification *w*ith *i*n vitro *t*ranslation.

Raw sequence: The sequence of the → insert of a clone (e. g. a → bacterial artificial chromosome or → yeast artificial chromosome clone), that is not yet assembled (i. e. linked to other clones in a → tiling path). Compare → paired-end sequence.

RAWTS: See → *R*NA *a*mplification *w*ith *t*ranscript *s*equencing.

RBIP: See → retrotransposon-based insertion polymorphism.

RBS: See → Shine-Dalgarno sequence.

r*Bst* DNA polymerase I: See → r*Bst* polymerase.

r*Bst* polymerase (r*Bst* DNA polymerase I): An enzyme from an overexpressing → recombinant clone of *E. coli*, encoded by the DNA polymerase I gene of the thermophilic bacterium *Bacillus stearothermophilus*, catalyzing the polymerization of deoxyribonucleoside triphosphates into DNA using a single-stranded DNA as template. The enzyme is optimally working at 65° C, and therefore used to sequence DNA in regions of pronounced secondary structure of high GC content. Since r*Bst* polymerase also possesses → reverse transcriptase activity, it is employed for → first strand cDNA synthesis and the generation of templates for → reverse transcriptase polymerase chain reaction techniques.

RCA:
a) See → *R*Nase *c*leavage *a*ssay.
b) See → *r*olling *c*ircle *a*mplification.

RCA polymerase: A lab slang term for → DNA polymerase I capable of replicating circular oligonucleotides and employed in → *r*olling *c*ircle *a*mplification (RCA).

RDA: See → *r*epresentational *d*ifference *a*nalysis.

RdDM: See → RNA-directed DNA methylation.

RDE: See → radiative decay engineering.

RDM: See → RNA-directed DNA methylation.

rDNA (*r*ibosomal DNA):
a) A tandem cluster of genes (see → gene battery) coding for 18S, 5.8S and 28S → ribosomal RNAs of eukaryotes, separated from each other by so-called → internal spacers and transcribed coherently by → RNA polymerase I(A) into a large pre-ribosomal RNA transcript. Two adjacent clusters are interspersed by so-called external (non-transcribed) spacers of varying length. The partially transcribed intergenic sequence contains a series of 240 bp → *Alu* I repeats with potential → enhancer functions. rRNA genes belong to the genomic fraction of → middle repetitive DNA. See also → fibrillar zone, → rDNA amplification.

b) An infelicitous term for → recombinant DNA. See → recDNA.

rDNA amplification: The preferential manifold replication of genes coding for → ribosomal RNA in protozoa (e.g. *Tetrahymena*), insects and *Xenopus laevis*. In *Xenopus laevis*, the rDNA units are amplified extra-chromosomally by a rolling circle mechanism (→ rolling circle replication) which leads to the increase in repeat number from 2000 to more than 2000000.

RDR: See → recombination-dependent replication.

RE: See → *response* element.

Reading: In molecular biology, a one-way linear process by which the information encoded in nucleotide sequences is decoded (e.g. DNA → RNA, or RNA → protein). See → transcription and → translation, → reading frame, → reading mistake, → read-through.

Reading frame: The → codon sequence determined by reading nucleotides in groups of three, beginning at a specific → start codon (AUG). For example, the sequence

5' -AUGAGCACAAAAGGGCUCCACUGA-3'

reads as:

5'-AUG|AGC|ACA|AAA|GGG|CUC|CAC|UGA-3'

See → reading frame shift, → reading frame shift mutation.

Reading frame shift (frame shift, phase shift): An alteration of the → reading frame, caused by the addition or deletion of nucleotides in a number not divisible by three. A reading frame shift leads to the synthesis of a → missense mRNA whose translation usually results in a non-functional protein because the message will be read in triplets specifying wrong amino acids. See also → reading frame shift mutation.

Reading frame shift mutation (frame shift mutation, phase shift mutation): Any mutation (→ insertion or → deletion) resulting in a → reading frame shift. Reading frame shift mutations cause the alteration of the → reading frame of sequences → downstream of the site of mutation. Further, the original reading frame may be restored by a second insertion or deletion which in itself also would be a frame shift mutation.
For example, by insertion of a single nucleotide A in the sequence

5' TTT TGT CAT AAT 3'
3' AAA ACA GTA TTA 5'

after the first T (a+1 frame shift) the code is shifted backwards, and in the resulting mRNA the triplet code -UUU-UGU-CAU-AAU- (encoding the amino acid sequence -Phe-Cys-His-Asn-) will be altered into -UAU-UUG-UCA-UAA-U (encoding -Tyr-Cys-Ser-opal). Thus a stop codon is introduced and the protein product is prematurely ended.
Conversely, deletion of a single nucleotide (a-1 frame shift) shifts the reading frame forward (for example -UUU-UGU-CAU-AAU- → -UUU-GUC-AUA-AU, encoding -Phe-Val-Asn-).

Reading frame shift suppression: An imprecise term for the → reversion of the effects of a → reading frame shift mutation by a second mutation that restores the original reading frame.

Reading frame splicing: The excision of an → intron that interrupts the → reading frame of a gene, and the → splicing of the split exons to restore a functional transcript.

Reading length: The number of nucleotides of a DNA or RNA sequence that can be read on one single → sequencing gel (usually 500-600 bases; maximally 1-1.5 kb).

Reading mistake: The incorrect incorporation of an amino acid into the growing polypeptide chain during protein synthesis. See → translation.

READIT: See → *reversed enzyme activity DNA interrogation test.*

Read through:
a) The → translation of an mRNA beyond its → stop codon (by e.g. a → nonsense suppressor tRNA), leading to the production of a so-called read-through protein.
b) The → transcription of a gene beyond a termination signal sequence through an occasional failure of → RNA polymerase to recognize the → stop codon. This leads to the synthesis of a so-called → read-through messenger RNA.

Read-through messenger RNA (read-through messenger): See → read-through, definition b.

Read-through protein: See → read-through, definition a.

Ready-made adaptor: See → preformed adaptor.

Real-time PCR: See → quantitative PCR.

Real-time pyrophosphate detection: See → pyrosequencing.

Reannealing: Spontaneous realignment (→ renaturation) of two single DNA strands to a DNA double helix.

Rearrangement: Any structural change in a nucleotide sequence, a gene, or a chromosome.

Reassociation: See → renaturation.

Reassociation kinetics: See → C_0t analysis.

Reassortment: The rearrangement of genes between two distinct viral strains to produce a novel virus.

REBASE (restriction enzyme database): A comprehensive database on → restriction endonucleases, their → recognition sequences and cleavage sites, → isoschizomers, → methylation sensitivities, and physical characters (e. g. amino acid sequences) as well as informations on → methyltransferases, → homing endonucleases, → nicking enzymes and commercial availability. See → databases (in Appendix).

rec:
a) See → *rec (rec genes).*
b) Abbreviation for "recombinant", better than the widely used "r" (which is easily confused with e. g. "ribosomal"; see → rDNA, → rRNA).

rec (*rec* genes): A collective term for all genes of the → *rec*ombination repair system of *E. coli*. See for example → recA protein, → recBC protein.

recA protein: A multifunctional 40 kDa protein of *E. coli* encoded by the *recA* (*rec*ombination A) gene, whose expression is induced by the damage of DNA (e. g. by UV light). recA catalyzes the → homologous recombination of single-stranded DNA with a superhelical DNA duplex, during which a loop of single stranded DNA (→ displacement loop) is formed. The same protein also functions as a protease that specifically degrades the → cI gene product of the → lambda phage. *E. coli* mutants with a defective *recA* gene (*recA⁻*) are frequently used as → hosts in recombinant DNA experiments, since they do not allow any recombination of introduced foreign DNA with the host's chromosome. For *in vitro* experiments, recA protein is used to catalyze homologous recombination. The protein first binds preferentially to single-stranded DNA, the resulting nucleoprotein filament complex in turn binds to duplex DNA and scans the DNA for regions of homology. If such homologous sequences are found, → strand displacement and exchange starts. Therefore, recA protein is used for site-directed mutagenesis through displacement loops, targeted site-specific cleavage of DNA, and visualization of specific DNA sequences by electron microscopy. See also → *rec*, → recBC protein.

recBC protein: A protein of *E. coli*, consisting of two subunits, subunit recB (140 kDa) that unwinds double-stranded DNA, and subunit recC (128 kDa) that functions as ATP-dependent nuclease specifically cleaving linear double-stranded DNA. See also → *rec*, → recA protein.

recDNA: Abbreviation for → *rec*ombinant DNA.

Receptor gene: A gene in a set of genes that hypothetically controls the expression of the → producer genes (→ structural genes) of the same gene set (→ Britten-Davidson model).

Receptor mapping: An infelicitous term for the description of the three-dimensional geometric features of a ligand binding site of a receptor protein.

RecE/RecT recombination: See → ET recombination.

Recessed 5'-terminus (5' recessed terminus): The end of a DNA duplex molecule in which the 3'-terminus protrudes, the corresponding strand being longer than the 5'-terminus strand. Compare → recessed 3'-terminus, see → protruding terminus.

Recessed 3'-terminus (3' recessed terminus): The end of a DNA duplex molecule in which the 5'-terminus protrudes, the corresponding strand being longer than the 3'-terminus strand. Compare → recessed 5'-terminus, see → protruding terminus.

Recessive allele: Any → allele in a heterozygote, that has no effect on the phenotype of the carrier. The phenotypically silent → allele. See → dominance, → co-dominance.

rec gene: See → *rec*.

Recipient (acceptor): Any cell or organism that receives genetic information in the form of DNA or RNA. Compare → donor.

Recircularization: The joining of two complementary ends (→ cohesive end) of the same DNA molecule to produce a circular DNA duplex.

rec⁻ mutant (*recombination-deficient mutant*): Any mutant that is defective in → *rec*ombination (and thus also highly susceptible to → mutagens).

Recoding: The reprogramming of the information content contained in a → messenger RNA molecule. Recoding can change the meaning of → codons (e. g. stop codon UAG is reprogrammed to encode glutamine, or UGA to encode tryptophan or selenocysteine), or reset the → reading frame, or skip internal nucleotides in mRNAs (see → translational bypassing).

Recognition helix: See → helix-turn-helix.

Recognition sequence: See → recognition site.

Recognition site (recognition sequence, address site, binding site, cognate sequence):
a) Generally a defined sequence in a DNA molecule to which a specific protein can be bound. See also → recognition site affinity chromatography.
b) More specifically the DNA sequence to which a certain → restriction endonuclease binds. For class II restriction enzymes this recognition site is identical to the → target site (i.e. where the enzyme cleaves the DNA duplex).

Recognition site affinity chromatography: A technique for the isolation of specific → DNA-binding proteins that uses an inert matrix with a covalently bound DNA sequence containing their → recognition site(s). Cellular (or, more appropriately, nuclear) proteins are applied to columns packed with the matrix and its bound affinity ligand. Only cognate protein(s) are bound. Frequently synthetic → oligonucleotides with a sequence identical to the recognition sites are used as affinity ligands.

Recognition site matrix: Any matrix that lists the frequencies with which each of the four nucleotides occur at every position in the binding sites of → transcription factors.

Recognomics: The whole repertoire of techniques to analyze the interaction(s) between two molecules, specifically the conformation- or sequence-dependent binding of → ligands (e. g. substrates or hormones) to their cognate acceptor molecules (e. g. enzymes or receptor proteins, respectively). Recognomics encircles nucleic acid-nucleic acid-, nucleic acid-protein-, and protein-protein interactions, and the multitude of transient reactions between small molecular weight compounds ("effectors") and macromolecules (DNA, RNA, proteins). The methodology of recognomics is mostly identical to the technology used in → proteomics. See → behavioral genomics, → comparative genomics, → environmental genomics, → epigenomics, → functional genomics, → genomics, → horizontal genomics, → integrative genomics, → medical genomics, → nutritional genomics, → omics, → pharmacogenomics, → phylogenomics, → proteomics, → structural genomics, → transcriptomics, → transposomics.

Recombinant: A term describing a molecule, a cell or an organism composed of different DNA sequences originating from different sources, that are combined by → recombinant DNA technologies (see → recombinant DNA) or → genetic recombination (recombinant offspring of a cross between two parents).

Recombinant antibody: Any → antibody, that is composed of chains of different (heterologous) origin, i.e. whose antigen-binding fragments (Fab or Fv) are produced in one, the Fc fragments in another organism.

Recombinant clone: See → recombinant DNA.

Recombinant Dicer: An → RNaseIII or a complete → Dicer ribonucleoprotein complex produced by → recombinant DNA methodology. Recombinant Dicer is used to digest a synthetic 72 bp long double-stranded substrate RNA homologous to a specific gene (or its exons) into a series ("pool") of 19–21 nucleotide → small interfering RNAs (→ siRNA library). Such pools are employed in gene silencing experiments.

Recombinant DNA (recDNA, also not correctly rDNA): A novel DNA sequence formed *in vitro* through the ligation of two or more nonhomologous DNA molecules, for example a → recombinant plasmid containing one or more → inserts of foreign DNA cloned into its → cloning site or its → polylinker. Organisms containing such *in vitro* constructed DNA are also referred to as recombinant, for example a recombinant phage or a recombinant clone of bacteria.

Recombinant DNA technology (recDNA technology, also not correctly rDNA technology): The multitude of techniques for the *in vitro* recombination of two or more DNA molecules from different organisms. See → recombinant DNA. This comprises the different methods of → cloning, starting with the isolation and purification of the DNA of interest, also for example the synthesis of → cDNA from an mRNA molecule. Furthermore the choice and preparation of a suitable → cloning vector, depending among other things on the size of the → insert that is to be cloned and on the → host organism to be used. Of great importance is the preparation of the insert DNA, e.g. digestion with → restriction endonucleases and e.g. → DNA tailing, then the method of *in vitro* ligation chosen and finally the diverse techniques of direct or indirect → gene transfer.
This is, of course, only a short and fragmentary description of the methods involved. Recombinant DNA technology is the central experimental field of gene technology and the fundamental techniques are explained in this book. *In vitro* recombination of RNA is also possible. See → recombinant RNA technology.

Recombinant drug: Any drug that is produced by → recombinant DNA technology. For example, blood clotting factor VIII and IX, a DNase, erythropoietin, erythropoietin-ß, growth hormone, human insulin ("humulin"), interferons (IFN-á2a, –2b and 2c, IFN-ß1b, IFN-ã and IFN-ã1b), interleukin–2, tissue plasminogen activator), but also recombinant vaccines (e.g. *Haemophilus influenzae*, *Hepatitis*-B, *Vibrio cholerae*) are few of a plethora of such already marketed recombinant drugs.

Recombinant inbred: See → recombinant inbred line.

Recombinant inbred line (RIL; recombinant inbred, RI): One of a number of lines derived from the same cross by repeated selfing of the F2 individuals, until each line is near-homozygous. RILs which differ by the presence or absence, respectively, of a gene of interest, can be used to link these genes with → markers. See → gene tagging. Compare → near-isogenic line.

Recombinant organism: An organism containing → recombinant DNA.

Recombinant PCR: See → recombinant *polymerase chain reaction*.

Recombinant phage: See → recombinant DNA.

Recombinant plasmid: See → recombinant DNA, → hybrid plasmid.

Recombinant *polymerase chain reaction* (recombinant PCR): A variant of the conventional → polymerase chain reaction that allows the *in vitro* recombination of two different DNA fragments (e.g. a → promoter of gene *A* and a coding sequence of gene *B*), using two anti-parallel → PCR add-on primers that carry complementary 5'-termini. These primers are used to amplify the different sequences separately. After amplification they can reanneal by their complementary ends, thus producing a fusion product of promoter *A* and gene *B*.

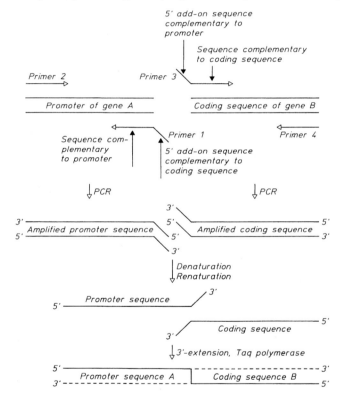

Recombinant PCR product

Recombinant PCR

Recombinant protein: Any protein encoded by a → gene fusion and synthesized as a compound polypeptide, with one part originating from one, the other part from the other gene. See → fusion protein, a specialized recombinant protein.

Recombinant protein array: A glass slide or other solid support, onto which thousands of recombinant proteins, each one produced by the expression of → chimeric genes in e.g. bacterial hosts, are spotted at high density. Such arrays are used to detect interactions between the recombinant proteins and ligands, peptides, proteins, RNAs or oligonucleotides. See → functional protein array, → protein chip, → protein domain array.

Recombinant protein production: An *in vitro* technique for the synthesis of larger amounts of a recombinant protein, that can also easily be recovered. In short, a gene of interest is first amplified by conventional → polymerase chain reaction, the amplified fragment cloned into an → expression vector, and the vector used to synthesize the encoded protein in an *in vitro* transcription-translation system (for example, → continuous exchange cell-free system). After its synthesis, the protein is usually run in 12% SDS-polyacrylamide gels, and afterwards purified by → affinity chromatography or other techniques.

Recombinant restriction endonuclease: Any → restriction endonuclease that is produced by bacteria transformed with the corresponding restriction endonuclease-encoding gene. Usually the cloned gene is overexpressed. Appropriate hosts allow to reduce or abolish contaminating non-specific → endo- or → exonucleases, and do not produce similar enzymes (e.g. isolation of *Hae* III from *Haemophilus aegypticus* is difficult, because the closely related *Hae* II endonuclease is copurifying).

Recombinant RNA: Any → RNA molecule composed of two or more heterologous RNAs ligated *in vitro* by → T4 RNA ligase.

Recombinant RNA technology: The techniques to ligate two or more heterologous RNA molecules *in vitro* (by e.g. → T4 RNA ligase), and to replicate the ligated RNA (by e.g. → Q-beta replicase). Compare → recombinant DNA technology.

Recombinant vaccine: Any vaccine, that is a product of an engineered gene.

Recombinant vector: See → hybrid vector.

Recombination (genetic recombination): The combination of genes in a progeny molecule, cell or organism in a pattern different from that in the parent molecules, cells or organisms. This may be due to the exchange of DNA sequences between two chromosomes, or the new association of genes in a recombinant, arising from independent assortment of unlinked genes, from → cross over between → linked genes, or from intra-cistronic cross-over. See → homologous recombination, → site-specific recombination, → unequal crossing-over, → somatic recombination. The natural process of recombination is exploited *in vitro* for → cloning of DNA using → recombinant DNA technology. See also → recombination frequency, → recombination repair.

Recombination activation gene (RAG, recombination activator gene): Any one of a series of genes encoding recombinase proteins, catalyzing the site-specific V(D)J (*v*ariable, *d*iversity, *j*oining) recombination, that assembles immunoglobulin and T cell receptor genes.

Recombinational hot spot: See → hot spot.

Recombinational map: See → genetic map.

Recombination-deficient: See → *rec⁻* mutant.

Recombination-*d*ependent replication (RDR): Any DNA → replication that is initiated on a → recombination intermediate. RDR repairs → breaks in double-stranded DNA, re-establishes replication at broken → replication forks, initiates de novo replication and even rescues → telomeres in cells lacking → telomerase.

Recombination fraction (θ): The number of recombinant progeny as a function of the total number of individuals among offspring of a particular mating. The construction of → linkage maps is based on the estimates of the recombination fraction between all pair-wise combinations of loci.

Recombination *f*requency (RF): The number of recombinants divided by the number of progeny individuals. The recombination frequency serves to estimate the relative distances between two (or more) loci on a → linkage map (e.g. an RF of 50% is indicative for an independent segregation of two loci; the lower the frequency, the smaller the distance between them). See → recombination, → cross-over, → linkage analysis.

Recombination frequency map: See → RN map.

Recombination load: A potentially disadvantageous consequence of genetic → recombination, caused by the disruption of favourable sets of genes that accumulated by selection in the course of evolution.

Recombination modulator: A major gene in *Petunia hybrida* (gene Rm_1) that regulates the frequency of → chiasma formation. The presence of Rm_1 in the heterozygous state enhances the recombination rates along all seven chromosomes, and the insertion of Rm_1 in a chromosome region that normally recombines at a very low rate, induces an increased recombination frequency.

Recombination repair: A → DNA repair system that recognizes defective sites in DNA during → recombination processes (e.g. → cross-over). In *E. coli* this system involves the action of the products from genes *recA*, *recBC*, *recF*, *recJ*, *recK*, *sbcB*, and *recE*, where *sbcB* and *recE* code for the subunits of an → exonuclease. See also → recA protein, → recBC protein.

Recombinator: Any nucleotide sequence in a DNA duplex molecule that enhances general → recombination (recombinational → hot spot; e.g. the → chi-sequence in *E. coli*).

Recon: The shortest distance between two adjacent markers on a → genetic map, that cannot be separated by → recombination. See → muton.

Reconstructomics: Still another term of the → omics era, that encircles the *in silico* assembly and annotation of sequences from a part of a genome of a single species. The reconstructed sequence context is coined reconstructome.

RED: See → *r*epeat *e*xpansion *d*etection.

Red biotechnology: A laboratory slang term for the application of the methodological repertoire of → biotechnology to animal (including human) cells. Compare → blue biotechnology, → green biotechnology, → grey biotechnology, → white biotechnology.

Red blood cell fusion: A method for the transfer of small macromolecules (e.g. proteins below 300 kD) into living animal cells, that uses red blood cells loaded with these molecules during hypotonic hemolysis. After their resealing these loaded cells are then fused to cultured cells with the aid of the Sendai virus. This technique does not work well with nucleic acids.

Red fluorescence protein (RFP): Any → fluorescent protein, that emits fluorescent light of a wavelength around 600 nm. For example, an RFP from the tentacles of the sea anemone *Anemonia sulcata* (called AsFP) is naturally non-fluorescent, but emits an extremely low level of red fluorescence at λmax = 595 nm (therefore AsFP595) under certain conditions. This level is, however, increased by green light, yet quenched by blue light. Compare → blue fluorescent protein, → destabilized enhanced green fluorescent protein, → enhanced blue fluorescent protein, → enhanced cyan fluorescent protein, → enhanced green fluorescent protein, → enhanced yellow fluorescent protein, → farnesylated enhanced green fluorescent protein, → red-shifted green fluorescent protein.

Red fluorescent protein (RFP; Ds Red): An autofluorescent protein from the Indo-Pacific sea anemone *Discosoma striata* (Ds) that absorbs light of a wave-length of 558 nm and emits red fluorescent light at 583 nm. The gene encoding this protein can be used as → reporter gene. RFP fluorescence does not require any cofactors (e. g. a substrate). See → blue fluorescent protein, → enhanced blue fluorescent protein, → enhanced green fluorescent protein, → enhanced yellow fluorescent protein, → farnesylated enhanced green fluorescent protein, → green fluorescent protein, → red-shifted green fluorescent protein.

Redox RNA: Any synthetic RNA molecule, that catalyses a redox (*reduction-oxydation*) reaction with an artificial substrate (e.g. benzyl alcohol) in the presence of *nicotinamide adenine dinucleotide* (NAD$^+$). The protein-independent reaction depends on the presence of Zn^{2+} ions. See → ribozyme.

Red-shifted green fluorescent protein (rsGFP): A → green fluorescent protein from the sea pansy *Renilla reniformis*, that can be excited by blue light at 498 nm (as opposed to the GFP from the jellyfish *Aequora victoria*, that is excited by light at 395 nm), and emits fluorescent light at 510 nm (as does the *Aequorea* GFP). Red shifted GFP genes can be used in combination with GFP genes as → reporter genes, since their protein products can be detected by simply exciting the host cell at different wave lengths.

Reduced genome: Any → genome streamlined in evolutionary times such that it lost a series of genes present in an ancestor. Usually adaptation to extreme environments favor such genome reductions. For example, the genome of the primary endosymbiont of carpenter ants, *Blochmannia floridanus*, lost genes involved in → DNA replication initiation (as e.g. *dna*A, *pri*A, *rec*A) and is therefore a reduced genome.

Reduced representation shotgun sequencing (RRS sequencing): A technique for the establishment of → single nucleotide polymorphism (SNP) maps, that uses reproducibly prepared subsets of a genome (reduced representation), of which each subset contains a manageable number of SNP loci. Subsets are generated by restriction of the DNA with e.g. *Bgl* II, and purifying restriction fragments of a distinct size (e.g. 500–600 bp). For RRS sequencing, DNAs from many individuals are mixed, restricted with *Bgl* II, the fragments separated by → agarose gel electrophoresis, fragments between 500 and 600 bp excised from the gel, libraries from the appropriately sized fragments prepared, and the clones randomly sequenced. The sequences are then analyzed by a SNP detection algorithm (as e.g. PHRED, allowing to estimate sequence accuracy).

Reduction-of-function mutation: Any → mutation within a → gene that does not abolish the function of the encoded protein, but only reduces its catalytic or regulatory properties. See → gain-of-function mutation, → loss-of-function mutation.

Reductive evolution: The decrease in number of genes or genomic complexity in certain organisms (e.g. parasites) in evolutionary times. For example, the tuberculosis bacillus *Mycobacterium tuberculosis* has nearly 4,000 protein-encoding genes and only 6 pseudogenes. In contrast, its close relative *Mycobacterium leprae*, the causative agent of leprosy, owns a mere 2,720 genes, of which only 1,604 are active protein-coding genes. Therefore, since its divergence from the last common ancestor about 2,000 *M. leprae* genes were lost in reductive evolution. See → gene elimination, → gene decay.

Redundancy: The repetitive occurrence of the same sequence motif within the eukaryotic genome. See → repetitive DNA.

REF: See → restriction endonuclease fingerprinting.

Reference DNA: Any DNA sequence that is present in constant copy numbers in all samples (e.g. cells, tissues, DNA or RNA preparations). Usually a → single-copy gene is used for gene dosis quantification, or the transcripts of a → house-keeping gene serve to normalize mRNAs on e.g. → Northern blots.

Reference hybridization: The measurement of the concentration of → cDNAs or → oligonucleotides on a membrane or → DNA chip, before complex hybridization with appropriate target DNAs is carried out. Reference hybridization allows to compensate differing amounts of cDNAs, so that the hybridization signal intensity reflects the amount of target sequence more correctly. See → normalization.

Reference single nucleotide polymorphism **(refSNP, rsSNP, rsID, SNP ID):** Any → single nucleotide polymorphism at a specific site of a genome (or part of a genome, e.g. a → BAC clone), that serves as reference point for the definition of other SNPs in its neighborhood. The rsID number ("tag") is assigned to each refSNP at the time of its submission to the databanks. See → anonymous SNP, → candidate SNP, → coding SNP, → exonic SNP, → gene-based SNP, → human SNP, → intronic SNP, → non-synonymous SNP, → promoter SNP, → regulatory SNP, → synonymous SNP.

Reflectometric interference spectroscopy **(RIfS):** An optical technique for the direct observation of interactions between analytes and chip-bound target molecules. RIfS is based on the multiple reflection of white light at thin transparent layers, that allows to determine the optical thickness of these layers by interference (as a measure for interaction). First, a functionalized target (e.g. a peptide or protein, an antibody, oligo- or polynucleotide, a lipid, carbohydrate or low molecular weight pharmaceutical agent) is covalently immobilized on e.g. a glass chip surface via amino-, carboxy- or thiol-groups. Then a very dilute solution of the analyte (e.g. an antigen for antibody-antigen reactions, a peptide or protein for protein-protein- or protein-DNA interactions, an oligonucleotide for DNA-DNA-hybridizations, a drug or substrate for drug-receptor- or substrate-enzyme binding) is reacted with the "sensor" chip, and the increase in optical thickness continuously monitored. The resulting interferogram ("sensogram") describes the surface layer thickness (in nm) versus time (sec), and allows to quantitatively characterize the binding event. Therefore, RifS does not require labeling of reactants.

Reflector time-of-flight mass spectrometer (reTOF-MS): A variant of a → MALDI post source decay mass spectrometer, in which the ionized peptide fragments are reflected at least once during their flight to the detector. The reflector leads to a focussing of the ions, to a higher resolution of the mass spectra and a more precise mass determination. See → linear time-of-flight mass spectrometer.

Reflectron: A reflector in → time-of-flight mass spectrometers, first retarding and then reversing fragment ion velocities, thereby correcting for the flight times of these ions, that have different kinetic energies.

Refractive index: The ratio of light velocity in vacuum to its velocity in a given medium (e.g. CsCl solution), used to monitor e.g. → CsCl gradients for their density profile.

RefSNP: See → reference single nucleotide polymorphism.

Regenerated restriction site: A specific → restriction endonuclease → recognition site newly generated by the ligation of two DNA duplex molecules whose ends are appropriately trimmed. For example, two DNA fragments alternatively cut with *Eco* RI and *Bam* HI can be ligated at low efficiency, if the *Eco* RI → overhang is filled in with → T4 DNA polymerase, and the *Bam* HI 5'-overhang removed with the 5' → 3' exonuclease activity of *E. coli* → DNA polymerase I. Subsequent → blunt-end ligation of both termini will regenerate a new *Eco* RI site.

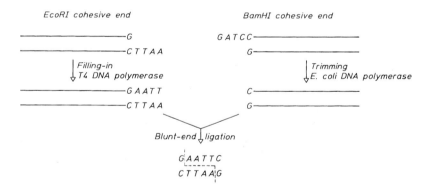

Regenerated restriction site

Regional clone: See → regional sequence.

Regional sequence (regional clone, "internal clone"): A laboratory slang term for DNA sequences located within large subgenomic DNA fragments, isolated e.g. by → pulsed-field gel electrophoresis.

Region of increased gene expression (RIDGE): Any region of a → genome, where the transcription of genes per unit DNA length is at least 5 times higher than the average and about 20 to 200 fold higher than in the weakly expressed regions. For example, the *major histo*compatibility locus (MHC) on human chromosome 6 represents such a RIDGE. Most of the RIDGEs are characterized by high → gene densities and locate to subtelomeric regions.

Region-specific mutagenesis: Any change in the base sequence of a DNA duplex molecule caused by the → insertion or → deletion of larger fragments.

Regulated gene: Any → gene whose expression is conditional, i. e. both the → transcription of this gene and the → translation of the corresponding → messenger RNA is tightly controlled by intra- and extracellular parameters. See → house-keeping gene.

Regulated promoter: Any → promoter whose activity is limited to times when an inducer (a → transcription factor recognizing a consensus sequence in the promoter region) is present. Compare → constitutive promoter.

Regulator gene: A gene coding for a protein → repressor that blocks the activity of an → operator gene and in this way prevents → transcription of the adjacent → operon. The induction of transcription can be effected by low molecular weight compounds (effectors) which are recognized by the repressor protein, change its three-dimensional structure after binding, and thus prevent repressor-operator interactions.

Regulatory sequence: A DNA sequence regulating the expression of genes (i.e. → attenuator, → enhancer, → promoter).

***Regulatory single nucleotide polymorphism* (regulatory SNP, rSNP):** A relatively rare → single nucleotide polymorphism, that affects the expression of a gene (or several genes). Usually this SNP is located in the promoter of the gene. See → anonymous SNP, → candidate SNP, → coding SNP, → exonic SNP, → gene-based SNP, → human SNP, → intronic SNP, → non-synonymous SNP, → promoter SNP, → reference SNP, → synonymous SNP.

Regulomics (*regul*ation-gen*omics*): Another term of the → omics era, that encompasses all molecular mechanisms underlying the regulation of genes, proteins and protein networks, sometimes even metabolic pathways, and the whole repertoire of techniques to detect and characterize such mechanisms.

Regulon: A set of → structural genes or → operons, located in different regions within the same genome but having a common mechanism for the regulation of their expression (e.g. the 8 genes for arginine synthesis in *E. coli*).

Rehybridization: The repeated use of one and the same DNA → blot for → hybridization with different radioactively labeled → probes. After the first hybridization the hybridized probe is washed off (with e.g. water of > 60°C), the blot monitored for the absence of radioactivity and hybridized to the second probe. Nitrocellulose blots usually allow 5-8 rehybridizations, but then become brittle and break. Nylon-based membranes stand some 10-12 rehybridizations, and are thus comparable to dried → agarose gels.

R 1822 plasmid: See → RP1 plasmid.

R 1822 plasmid: See → RP1 plasmid.

Reiterated sequence: See → repetitive DNA.

Related to empty sites (RESites): An *in silico* approach to search for a recent mobility of → transposable elements. Sequences immediately flanking transposon insertions are used as queries to identify potential genomic target sites without inserts. These in turn are taken as evidence for recent transposition of the corresponding element.

Relational clone library: See → ordered clone library.

Relative quantitative reverse transcriptase polymerase chain reaction (relative quantitative RT-PCR): A variant of the → reverse transcription PCR, that allows to quantitate any → messenger RNA. It involves the amplification of an internal control mRNA simultaneously with the → transcript of interest. The internal control is used to normalize the samples, so that after → normalization the quantity of a specific mRNA can be estimated by direct comparison of the amplified products (e. g. by → ethidium bromide fluorescence), analyzed in the linear range of amplification. Moreover, the control should be expressed at a constant level in all experimental samples (i. e. not be affected by external parameters). Constitutively expressed genes (β-actin, cyclophilin, ubiquitin, glyceraldehyde phosphate dehydrogenase genes) are best for such internal controls, provided they do not vary in their expression (which is often the case). See → competitive reverse transcription polymerase chain reaction, → PCR-aided transcript titration assay, → quantitative RT-PCR.

Relative reverse transcriptase polymerase chain reaction (relative RT-PCR): A variant of the conventional → reverse transcriptase polymerase chain reaction, that allows a comparison of the transcript abundances between multiple samples by amplifying an internal control in addition to the → cDNA under study, thereby normalizing for different amounts of RNA in the different samples Relative RT-PCR requires, that the → primers for the gene of interest (see → gene-specific primer) and the internal control are compatible, and that the amplification rate of both sequences are within the linear range of amplification. See → competitive reverse transcriptase polymerase chain reaction. Compare → relative quantitative reverse transcriptase polymerase chain reaction.

Relative synonymous codon usage (RSCU): The ratio between the observed → codon frequency and the codon frequency inferred from statistical distribution of codons.

Relaxation protein: See → DNA topoisomerase I, also → helix-destabilizing protein.

Relaxed circle: See → open circle.

Relaxed control: A mechanism which allows resident → plasmid(s) to escape tight coupling of their → replication to host chromosomal DNA replication. Such plasmids are therefore present in copy numbers of 10-250. See → multicopy plasmid. Compare → tight control, → runaway plasmid.

Relaxed DNA: Any DNA that is not supercoiled (→ supercoil).

Relaxed plasmid: A → plasmid which is replicated under → relaxed control.

Relaxing enzyme: See → DNA topoisomerase I.

Relaxing protein: See → DNA topoisomerase I, also → helix-destabilizing protein.

Release factor: See → termination factor.

REMAP: See → *re*trotransposon-*m*icrosatellite *a*mplified *p*olymorphism.

REMI: See → *r*estriction *e*nzyme-*m*ediated *i*ntegration.

Remodeling and spacing factor (RSF): A 400–500 kDa complex of basically two proteins, that alters the structure of → nucleosomes in an ATP-dependent process and thereby disrupts their ordered array, leading to a rearrangement of chromatin architecture at → promoter regions. After this remodelling, another protein complex called FACT (for *fa*cilitates *c*hromatin *tra*n-scription) catalyzes elongation through downstream nucleosomes.

REM-PCR: See → *r*estriction *e*ndonuclease-*m*ediated selective *p*olymerase *c*hain *r*eaction.

Renaturation (reannealing, reassociation): The transition of a nucleic acid or protein from a denatured state (e.g. single-stranded in case of DNA, random coil in case of protein) to a fully active, native three-dimensional configuration. The renaturation behaviour of DNA depends on its sequence: → repetitive DNA will reanneal faster than → unique DNA. This provides the basis for the characterization and comparison of specific genomic DNAs in the so-called → C_0t ana-lysis. See also → in-gel renaturation.

REP: See → *r*epetitive *e*xtragenic *p*alindromic element.

Repair: See → DNA repair.

Repair nuclease: Any one of a series of → nucleases that functions in → DNA repair, either by recognizing and removing an incorrect base or an otherwise damaged site. Repair nucleases may act as an → endonuclease, or as → exonuclease, removing nucleotides on one strand from the end of the duplex.

Repairosome: The protein complex catalyzing → DNA repair.

Repeat: See → repeating unit.

***Repeat-associated polymorphism* (RAP):** Any nucleotide → polymorphism that is generated by an elevated → mutation frequency around repeated sequences (e. g. → variable number of tandem repeats). For example, in the socalled control region of → mitochondrial DNA (with the → origin of replication and → transcription) a series of repeated sequences are located in tandem, which undergo expansion or contraction as a result of → insertion or → deletion of repeat units (by e. g. → slipped strand mispairing during mtDNA replication). These processes actually lead to the generation of mtDNA length variants. Now, the nucleotides surrounding the original site of the repeat are also mutated. Since genome expansion and contraction events occur indepen-dently in germ cells of different individuals, unique repeat-associated polymorphisms are created that serve to identify individuals, demes, and whole populations. Compare → repeat-induced point mutation, → RIPping. See → RAPping.

Repeat expansion detection **(RED):** A method to detect trinucleotide repeats and their multimers in genomic DNA, which is heat-denatured, and oligonucleotide(s) complementary to either strand of the repeat target is (are) hybridized at a temperature close to the melting point. Then thermostable → DNA ligase is added and will covalently join only those adjacent oligonucleotides that are not separated by gaps. Then the ligation products are separated from their template sequences by denaturation. As a consequence, a mixed population of single-stranded DNA molecules is generated. The ligation step is then repeated from 180 to 400 times. Then the single stranded multimers are size-separated in denaturing → polyacrylamide gels, subsequently blotted onto hybridization membranes and hybridized to radiolabeled complementary oligonucleotides.
The RED technique can also be modified to include several different oligonucleotide repeats simultaneously (multiplex RED). In order to differentiate between the (RED) products formed by each type of repeat, oligonucleotides of different length have to be used (e. g. $[CCG]_{11}$, $[TAG]_{12}$, $[CTG]_{13}$, $[CAT]_{14}$, and so on).
This technique can be used to e. g. detect the expansion of trinucleotide repeats in human diseases (e.g. fragile X, myotonic dystrophy, Huntington's disease, spinobulbar muscular atrophy, and others).

Repeat-induced gene silencing: See → co-suppression.

Repeat-induced point mutation **(RIP; originally: *rearrangement induced premeiotically*):** A mechanism to inactivate repetitive sequences, to reduce their → copy numbers, or both, by introducing → transition mutation(s), predominantly affecting $G \equiv C$ pairs (and some $G \equiv C$ pairs more than others), that are replaced by $A = T$ pairs in *Neurospora crassa* and some other fungi. RIP involves enzymatic deamination of cytosines or 5-methylcytosines to uracil or thymine, respectively, and occurs after fertilization, specifically in the subsequent mitoses prior to nuclear fusion.

Repeating unit (repeat): Any, usually short nucleotide sequence motif that is repeated in a tandem cluster (→ tandem repeat). A number of specific repeats are described in some detail, see → *Alu* I sequence, → delta sequence, → enterobacterial repetitive intergenic consensus sequence, → LINES, → minisatellite variant repeat, → repetitive DNA, → satellite DNA, → simple repetitive DNA, → SINES, → variable number of tandem repeats. See also → direct repeat, → indirect repeat, → inverted repeat, → tandem repeat.

Repeat unstable nucleotides **(RUN):** A general term for unstable repeats (i. e. repeated sequences whose repeat unit numbers increase or decrease during the life cycle of an organism or during meioses). See → loss of heterozygosity, → microsatellite expansion, → microsatellite instability.

REP element: See → *repetitive extragenic palindromic* element.

Repetition frequency: The frequency with which a given DNA sequence occurs in the haploid genome. Compare → repetitive DNA.

Repetitive DNA (repetitious DNA, reiterated sequence): Any nucleotide sequence that is present in multiple copies per genome. Repetitive sequences often occur clustered in specific chromosomal regions. Repetitive DNA can be broadly categorized into several classes by → C_0t analysis. A highly repetitive fraction (→ highly repetitive DNA) contains short sequences (5 – 100 bp) repeated up to a million times, a middle repetitive fraction (→ middle repetitive DNA) is made up of sequences 100-500 bp in length which are repeated from 100 to 10000 times each (e.g. genes for → ribosomal RNA, → transfer RNA and → histones). Compare → unique DNA.

Repetitive extragenic palindromic (REP) element (palindromic unit, PU): The highly conserved 38 bp consensus sequence

$$\text{5'-GCC}^{G}_{T}\text{GATGNCG}^{G}_{A}\text{CG}^{C}_{T}\text{NNNNN}^{G}_{A}\text{CG}^{C}_{T}\text{CTTATC}^{C}_{A}\text{GGCCTAC-3'}$$

of *E. coli*, *Salmonella typhimurium* and related bacteria, that is located within untranslated regions of → operons and is palindromic (i.e. forms a stable → stem-and-loop structure). REP elements are dispersed throughout the bacterial genome, bind → DNA topoisomerase II and → DNA polymerase I, and are potential sites for transcription termination, → mRNA stability, or chromosomal → domain organization.

Repetitive extragenic palindromic polymerase chain reaction: See → repetitive sequence-based polymerase chain reaction.

Repetitive sequence-based polymerase chain reaction (rep-PCR; repetitive extragenic palindromic polymerase chain reaction, REP-PCR): A variant of the conventional → polymerase chain reaction, in which oligonucleotide → primers complementary to dispersed → repetitive DNA sequences (e. g. the → repetitive extragenic palindromic [REP] elements) in bacterial and fungal genomes are used to amplify unique sequences between two opposing primer binding sites. Since repetitive DNA is spread throughout genomes, multiple amplification products arise, creating a → DNA fingerprint. Due to various mutations both in the → single-copy and repetitive sequences, the amplified products vary in size from strain to strain, and therefore allow unequivocal discrimination between closely related organisms. See → fluorophore-enhanced repetitive sequence-based polymerase chain reaction.

Repetitive simple sequence: See → microsatellite.

Replacement: The substitution of a naturally occurring gene by a cloned gene that has been mutated *in vitro*.

Replacement synthesis (replacement synthesis labeling): A method for the rapid radioactive labeling of DNA fragments directly from → restriction endonuclease digestion reactions, using both the 3' → 5' exonuclease and 5' → 3' polymerase activities of → T4 DNA polymerase. In short, in the absence of deoxynucleoside-triphosphates the 3' → 5' exonuclease activity removes nucleotides from the 3' ends of the DNA fragments. Then deoxynucleoside triphosphates, one of them carrying an α-^{32}P label, are added, which block the 3' → 5' exonuclease activity and activates the 5' → 3'-polymerase activity of → T4 DNA polymerase. This fills in the recessive ends of the DNA fragment, thereby introducing ^{32}P label. As opposed to 5' → endlabeling with → polynucleotide kinase, replacement synthesis labeling incorporates ^{32}P into internal phosphodiester bonds so that any contaminating phosphatase will not interfere with the labeling process.

Replacement synthesis labeling: See → replacement synthesis.

Replacement vector (substitution vector): A derivative of a wild-type → cloning vector in which a pair of → restriction sites span a DNA segment that can be foreign DNA (e.g. the → stuffer fragment of the → lambda phage; see → lambda-derived cloning vectors). Compare → insertion vector.

Replica-plating (Lederberg technique): A method to reproduce identical patterns of bacterial colonies that uses a cylindrical block covered with velveteen which is pressed onto a Petri dish containing the bacterial colonies (master plate). About 20% of the bacteria will attach to the block and may then be transferred to secondary plates containing appropriate growth media. A single pad usually gives 5 to 8 replicas.

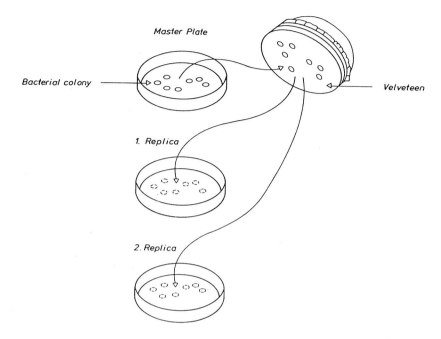

Replicase: An enzyme catalyzing the faithful reproduction of a copy from a template (either RNA or DNA). The term replicase is almost exclusively used for an enzyme that replicates certain virus RNA genomes (RNA dependent → RNA polymerase). Compare → DNA polymerase, → reverse transcriptase, → RNA polymerase. See also → Q-beta.

Replicatable reporter: See → replicatable RNA reporter, → reporter amplification.

Replicatable RNA reporter: A recombinant RNA molecule which consists of a probe sequence embedded within the sequence of a replicatable RNA (e.g. the template RNA for → Q-beta → replicase), which may be used for the amplification and thus detection of rare target sequences complementary to the probe. The probe together with the amplifyable reporter are hybridized to this target sequence. Since the inserted probe sequence does not interfere with Q-beta replicase, the whole recombinant molecule is amplified exponentially by this enzyme to about 10^9 copies which may be quantified by → ethidium bromide fluorescence detection.

The detection limit for rare sequences can be improved by the selective removal of nonhybridized probes. This is performed by → RNase III digestion of nonbound probe RNA. RNase III splits the reporter RNA at a sequence motif which forms a double-stranded → recognition site. However, when the probe sequence is hybridized to its target, the complementary sequences are forced apart, eliminating the RNase III recognition site ("molecular switch"). Compare → reporter amplification.

Figure see page 924.

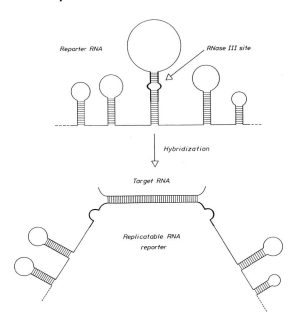

Replication: The process of faithful reproduction of a copy from a template. The term "replication" is mainly used for the synthesis of a new DNA strand using a DNA template strand (DNA-dependent DNA polymerization), that is for the faithful copying of the information content of the genome. DNA replication is semiconservative (→ semiconservative replication). The basic steps of this complex process involve the separation of the two strands of the DNA duplex molecule to generate the → replication fork, the binding of → DNA polymerase and the addition of → complementary nucleotides starting at the 3' end. The strand which is continuously replicated (→ leading strand) has to be discriminated from the other (→ lagging) strand which is replicated discontinuously in short pieces (→ Okazaki fragments → bimodal DNA replication, compare → semidiscontinuous operation). Following their synthesis the Okazaki fragments are ligated to form the complete lagging strand. See also → replicative form, → replicon, → rolling circle replication. → saltatory replication, The synthesis of an RNA strand along an RNA template is, however, also a replication process. See → replicase.

Replication bubble: See → replication.

Replication-defective virus: Any virus with → mutations in one or more genes coding for functions needed to complete the infective cycle.

Replication er*rror (RER):
a) Any erroneously introduced base during → replication of DNA.
b) The appearance of multiple alleles at → microsatellite loci in genomes of tumor cells as a consequence of a mutation in a → DNA repair gene. These alleles are either not or only weakly represented in the microsatellite profile of the normal tissue, but possess diagnostic value. For example, RER can be detected in almost all *hereditary nonpolyposis colon cancers* (HNPCC). RER has also prognostic value, since in sporadic colon cancer cells, RER$^+$ genotypes are associated with a better prognosis than RER$^-$ genotypes. Compare → loss of heterozygosity.

c) The consequence of → replication slippage at → microsatellite loci, leading to differences in → allele lengths that can be visualized by band shifts after → polyacrylamide gel electrophoresis and → autoradiography (if ^{32}P or ^{33}P-labeled primers are used) or → silver staining. RERs are more frequent during tumor development (e. g. colorectal cancer) as compared to normal tissue, and possess diagnostic value.

Replication fork: A structure of DNA duplex molecules at which the → double helix is locally destabilized by the interaction of DNA and the proteins of the so-called → primosome, forming a Y-shaped opening. This structure can also be visualized in the electron microscope (as "replication bubble") and represents the location of active → replication of DNA.

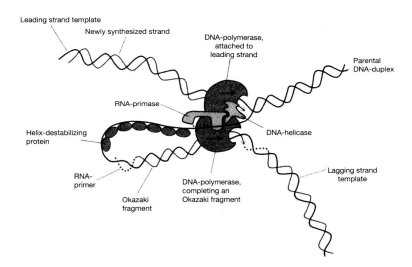

Replication fork barrier (RFB): A DNA-protein complex limiting or arresting the movement of a → replication fork such that it does not collide with e.g. an approaching → transcription complex. For example, replication of mammalian → ribosomal RNA genes occurs concomitantly with their transcription. Now a replication complex proceeding through an rRNA gene against the direction of its transcription would collide with the RNA polymerases repeatedly. Collision is avoided by a sequence-specific arrest of the replication fork. In mice, an RFB is located in the non-transcribed spacer of the rDNA and composed of an 18 bp socalled Sal box 2 (repeated 10 times downstream of the 3' end of the pre-rRNA coding region) and its flanking sequences. The upstream flank contains a cluster of 20 contiguous cytosine residues followed by a 19 bp GC- and a 26 thymidine residues stretch. The Sal box is recognized and bound by *t*ranscription *t*ermination *f*actor TTF–1, which is prerequisite for replication fork arrest 122 bp downstream of RNA polymerase I transcription termination within the GC stretch. TTF–1 interacts with proteins of the replication machinery, cooperatively assisted by the nuclear → Ku protein (a complex of 70 and 86 kDa proteins, originally detected as autoimmunoantigen reacting with antibodies from patients with rheumatic disorders), that binds to the cis-acting termination region at Sal box 2.

Replication fork collapse: The detachment of one arm of the → replication fork, resulting in a double-strand break of the template DNA, when the → primosome encounters a pre-existing → nich or gap.

Replication origin: See → origin of replication.

Replication protein A (RPA): A single-strand-specific eukaryotic DNA-binding protein consisting of three subunits (RPA14, RPA32, and RPA70), that stabilizes single-stranded regions during DNA metabolism (e. g. → DNA replication) and functions to unwind long stretches of double-stranded DNA.

Replication protein A affinity chromatography (RPA affinity chromatography): A technique for the isolation of proteins that interact with → replication protein A, a eukaryotic single-stranded DNA-binding protein required for DNA → replication, DNA→ excision repair and → homologous recombination. The protein consists of a stable complex of three subunits of 14, 32, and 70 kDa, which can be expressed in transgenic *E. coli*, and be linked to a gel matrix (e. g. Affi-gel 10). RPA affinity chromatography allows to isolate and purify RPA-affine proteins as e. g. → DNA helicases.

Replicative form (RF): A DNA structure necessary for the → replication of phages or virus particles, which is different from the DNA of these phages or viruses; e.g. the double-stranded circular DNA intermediate of → filamentous phages (e.g. → M13, → fl phage, → fd phage) which arises during intracellular replication of the single-stranded genomes.

Replicative vector: Any → cloning vector that promotes its own → replication (by e.g. the possession of an → autonomously replicating sequence).

Replicon: A segment of DNA under the control of one adjacent replication-initiation locus and behaving as an autonomous unit during DNA → replication (e. g. plasmids, or bacterial chromosomes). The number of replicons per organism varies, but roughly correlates with evolution in eukaryotes (e. g. yeast 500; *Drosophila*: 3500; *Xenopus*: 15000; *Mus*: 25000; *Vicia faba*: 35000). The size of a replicon varies from 4.7 Mbp (*E. coli*), 40 kb (yeast, *Drosophila*). 150 kb (mouse), 200 kb (*Xenopus laevis*) to 300 kb (*Vicia faba*).

Replicon rescue: A technique for the → cloning of genes disrupted by → plasmid → insertions, that is based on the double selection for antibiotic resistance and for the presence of a plasmid → origin of replication (ori). In short, DNA from a plasmid insertion mutant is restricted with two → restriction endonucleases generating (1) heterologous → sticky ends (e.g. the eight-base rare cutter *Sse* 8387I and *Apa* I) and (2) a plasmid ori sequence at the junction between the plasmid and the disrupted gene. This plasmid replicon is then rescued by its → ligation to an → antibiotic resistance gene (e.g. → tetracycline) cassette with matching sticky ends, transformed into *E.coli*, and selected by its tetracycline resistance and its ability to autonomously replicate in the host cell. Unwanted clonal species in the background (either with no resistance gene, or with only one, or with only an origin, or with neither sequences) are not stably maintained and therefore eliminated.

Replisome: See → primosome.

Reporter:
a) See → probe (definition b).
b) Any protein, whose presence or activity is easily detectable.

Reporter amplification: A method to detect rare target sequences with a → probe embedded within a sequence of replicatable and thus amplifyable reporter sequences (replicatable reporter). An example for this strategy is the use of → replicatable RNA reporter sequences which flank a probe sequence and after hybridization may be amplified to some 10^9 copies per target using → Q-beta replicase.

Reporter gene: Any gene that is well characterized both genetically and biochemically, may easily be fused to regulatory regions of other genes, and whose activity is normally not detectable in the target organism, into which it is transferred. Most reporter gene activities can be easily tested by simple assays (for example the enzymatic activity of the protein product, as for → β-galactosidase, → β-glucuronidase, → chloramphenicol acetyl transferase, → luciferase, → neomycin phosphotransferase II, → nopaline synthase, or → octopine synthase). Also, a series of reporters are → autofluorescent proteins, as e.g. → green fluorescent protein and the various analogues. See also → genetic marker, → selectable marker. Compare → replicatable RNA reporter.

Reporter plasmid (*cis*-reporting plasmid): Any → plasmid vector that contains a → reporter gene (e. g. → β-glucuronidase gene, → green fluorescent protein gene, → luciferase gene) driven by a → minimal promoter (e. g. a → TATA box in its simplest version), fused to multiple → enhancer elements (e. g. the *interferon-stimulated response element*, ISRE, or the interferon *gamma-activating sequence*, GAS, or others). If a plasmid expressing a gene of interest is cotransfected into host cells together with a reporter plasmid, an increased reporter gene expression demonstrates presence of → transcription factors binding to and activating both promoters.

Reporter vector: Any → cloning or → expression vector that contains one or more → reporter gene(s) as e. g. genes encoding → β-galactosidase, → β-glucuronidase, → chloramphenicol acetyltransferase, → green fluorescent protein and its derivatives, human growth hormone, → luciferase, → secreted alkaline phosphatase.

REP-PCR: See → *repetitive extragenic palindromic element polymerase chain reaction*.

REP-PCR fingerprinting: The establishment of a → DNA fingerprint using → repetitive sequence-based polymerase chain reaction.

Representation (also called amplicon): Any population of DNA fragments, representing a subset of the genome, generated by → restriction endonuclease digestion of genomic DNA, and having less complexity than the genome from which it is derived. See → cDNA representational difference analysis, → representational differential analysis.

Representational difference analysis **(RDA; genomic RDA):** A technique to detect and clone small sequence differences between two (or more) complex genomes (or DNA populations). In principle, RDA eliminates fragments present in both populations, leaving only the differences. This elimination requires a vast excess of the socalled driver DNA to compete out cell sequences present in both → representations. In short, first the complexity of both the target ("tester") and driver genomes is reduced by digestion of genomic DNA with → restriction endonucleases recognizing 6 bp → recognition sequences. A high proportion of the resulting fragments do not fall into the amplifiable range of 0.1-10 kb, and therefore the complexity of the representation is reduced to about 2-10 % of the complete genome. Then oligonucleotide → adaptors are ligated to the ends of the restriction fragments, and adaptor-complementary → primers used to amplify the fragments in a conventional → polymerase chain reaction. Only low molecular size fragments

(< 1 kb) are effectively amplified, and represent a representation. After producing both representations, their adaptors are removed by cleavage, and only tester fragments are now ligated to new adaptors at their 5' ends. Then the target representation is mixed with excess driver, melted, and reannealed. The termini of the formed → duplexes are filled-in with DNA polymerase, and PCR used to amplify the entire reaction. Self-annealed target (tester) duplexes are exponentially amplified, because they contain two 5' adaptors, and thus could be filled-in at both 3' ends. Heteroduplexes are only linearly amplified, and driver DNA is not amplified at all. Single-stranded molecules are destroyed by → mung bean nuclease, and the cycles of hybridization and amplification repeated. Reiteration of this process (i. e. cleavage of PCR products and ligation of new 5' adaptors) leads to exponentially increasing enrichment of the target. The enriched targets can then be cloned and sequenced.
Figure see page 929.

Figure see page 929.

Repression: The inhibition of the expression of a → gene or an → operon (e.g. through binding of a → repressor protein to an → operator sequence). See also → long term repression, → repressor.

Repressor (*trans*-silencing protein): A multimeric protein, encoded by a → regulator gene, that binds with high affinity to an associated → operator gene. The bound repressor blocks the movement of the → RNA polymerase complex along the → promoter to initiate RNA synthesis. The affinity of repressors for the operator is modulated by small molecules (effectors). So-called → inducers for example, bind to repressor proteins, and change their conformation so that they no longer bind the operator target sequence with high affinity. So-called co-repressors do the reverse. Many of the well-known repressors (e.g. the → cI protein or the cro protein of the → lambda phage) adopt a specific three-dimensional structure (the so-called → helix-turn-helix conformation) that allows them to bind to DNA duplexes with the protein helices fitting into consecutive major grooves.

Reprobing: The repeated use of the same membrane filter (for e.g. the hybridization of nucleic acids to different probes, or for the immunological detection of proteins with different antibodies). See → rehybridization.

Reproductive cloning: The removal of the nucleus from an egg cell of a female donor, its replacement with the nucleus from a somatic cell of a donor (which resembles fertilization), and the induction of the egg to divide and develop into a multicellular embryo, which then is implanted into a surrogate mother's womb ("gestation"), where it develops into a fetus, that is carried to term. The resulting baby is (supposedly) a virtual copy of the person, who donated the genetic material. Though suggestive, reproductive cloning has nothing to do with → molecular cloning or → genetic engineering. Compare → therapeutic cloning.

Reptation: The end-on, *rept*ile-like mig*ration* of DNA molecules larger than 50 kb in conventional → agarose gel electrophoresis, responsible for their nearly size-independent mobility.

RER: See → *replication error*.

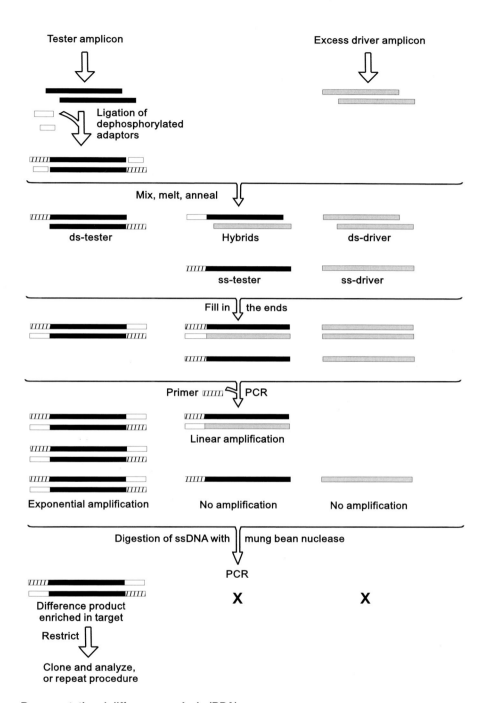

Representational difference analysis (RDA)

Re sequencing:
a) The repeated → sequencing of a distinct RNA or DNA (e.g. an amplification product, a → clone, a → BAC clone, or a synthetic → oligonucleotide) to reconfirm the already known base sequence.
b) The sequencing of a genomic, subgenomic or organellar DNA region from individual A and its comparison to the same, already known region from individual B ("reference"). Re-sequencing is necessary to detect e.g. → single nucleotide polymorphisms in many individuals of a population, which is a prerequisite for → SNP profiling (here mainly the detection of sequence variations related to disease phenotypes). Re-sequencing in this sense is also called comparative sequencing. In essence, the human genome is sequenced only once, but will be re-sequenced many times to detect variations of the target region in different individuals. See → comparative genomics.

Residual DNA: A DNA component of pharmaca, that is present in low quantity in a drug produced by → transgenic organisms and originating from host cell and/or vector DNA. Residual DNA is regarded as undesirable contaminant and as potentially hazardous (e. g. containing → oncogene or tumor suppressor gene sequences, that may recombine with the patient's → genomic DNA). The tolerance level of residual DNA in a drug is arbitrarily set to 100 pg per therapeutical dose. However, the US *F*ood and *D*rug *A*dministration (FDA) recommends a threshold value of 10 pg per dose.

Residue: Any component of macromolecules. For example, amino acids are residues of proteins, nucleotides residues of RNA or DNA.

Resistance:
a) The ability of an organism to grow in the presence of otherwise toxic or lethal compounds (e.g. → antibiotic resistance, → heavy metal resistance). See → resistance gene.
b) The resistance of an organism against infection by a parasite (e.g. → virus or pathogenic bacterium) conferred by → immunity of this organism.

Resistance-associated gene (RAG): An infelicitous term for a gene, that is induced in a plant cell after pathogen attack and encodes a protein involved in a defense reaction of the host. RAGs act in concert with other genes to restrict a pathogen from invading more cells, and ultimately to destroy it.

Resistance expressed sequence tag (R-EST): An → expressed sequence tag that possesses complete or partial → homology to genes encoding resistance against pathogens (e. g. viruses, bacteria, fungi).

Resistance factor (R factor, resistance transfer factor, R plasmid): Any naturally occurring → plasmid that carries one or more genes conferring → antibiotic resistance(s) to the host cell.

Resistance gene (R gene):
a) A prokaryotic → gene that encodes a protein catalyzing the destruction of a toxin (e.g. an → antibiotic). Such resistance genes are frequently used in → cloning vectors to facilitate selection of → transformants containing the vector (e.g. → antibiotic resistance genes for → ampicillin resistance, → chloramphenicol resistance, → kanamycin resistance, → tetracycline resistance; also → heavy metal resistance genes). See also → resistance factor. Compare → resistance gene candidate.

b) A eukaryotic gene, that encodes a protein involved in resistance to a pathogen. The gene may be dominant (R-gene) or recessive (r-gene). See → resistance gene candidate.

Resistance gene analogue polymorphism (RGAP): Any sequence difference between → resistance gene candidates from different organisms, detected by → polymerase chain reaction-mediated amplification of → genomic DNA with → primers directed to conserved regions of these genes (e. g. the leucine-rich repeat, nucleotide-binding site, or kinase domains). The amplified fragments are eletrophoretically separated in high-resolution denaturing → polyacrylamide gels and visualized by silver-staining.

Resistance gene candidate (RGC; resistance gene analogue, RGA; resistance gene-like sequence, RGL; candidate disease resistance gene, CDR gene): Any genomic sequence amplified with → primers directed against conserved regions of already identified → resistance genes, using conventional → polymerase chain reaction techniques. For example, a specific class of plant resistance genes encoding → receptor kinases carry conserved nucleotide-binding and transmembrane motifs. Therefore → degenerate primers based on these motifs can be used to amplify resistance gene analogues of the receptor kinase type from any plant genome. See → candidate gene, → resistance gene homologue.

Resistance gene homologue (RGH): Any gene whose sequence is identical or mostly identical to another gene encoding a protein with a proven role in the resistance of an organism towards a pathogen.

Resistance gene-like sequence: See → resistance gene candidate.

Resistance transfer factor: See → resistance factor.

RESites: See → related to empty sites.

Resolution: The degree of density of → molecular markers mapped on a stretch of DNA (e.g. a clone, a → BAC clone, a → gene, a → genome). The higher the marker density, the higher the resolution of the map. High resolution is a prerequisite for → map-based clong of a distinct gene.

Resolvase: An → endonuclease catalyzing the site-specific recombination between two DNA molecules.

Resonance light scattering (RLS): The generation of intense monochromatic light by oscillating electrons in gold or silver atoms on the surface of gold or silver nanoparticles (diameter: 40–120 nm), that are exposed to a white light source. The emission wavelength depends on the diameter and the composition of the particle, so that particles can be manufactured that emit at wavelengths spanning the entire spectrum. Such particles can be covalently bound to oligonucleotides, cDNAs, peptides, antibodies, generally proteins, and therefore be used as label in e.g. → microarray hybridisation experiments.

Response element (RE): Any one of a series of short → consensus sequences in DNA occurring in the → promoters or → enhancers of a number of genes that are controlled by the same external stimulus (e.g. temperature: → heat-shock element; hormones: glucocorticoid response element, GRE; heavy metals: → metal regulatory element).

Response regulation map: A compilation of all → genes of an organism or an organell that respond the same way to the same exogenous (environmental) or endogenous signal, i.e. are either up- or down-regulated, or are not affected at all. A response regulation map does not simply list all expressed (or silenced) genes, but also informs of the time course of expression (or silencing), and the intensity with which each single gene responds. Originally, such maps were derived from → two-dimensional gel electrophoretic protein patterns, more recently complemented by → cDNA expression array techniques. See → serial analysis of gene expression.

R-EST: See → *resistance expressed sequence tag*.

Restless: A → transposon of the filamentous fungus *Tolypocladium inflatum*, that is flanked by 20 bp terminal inverted repeats (TIRs), and contains a transposase gene which is transcribed. The primary transcript is alternatively spliced and the two resulting messenger RNAs translated into two polypeptides tnp 803 (803 amino acids) and tnp 157 (157 amino acids). The latter protein functions as DNA-binding protein. The genome of *T. inflatum* harbors about 15 copies of restless, which are flanked by 8 bp target site duplication (TSDs) as a result of the insertion process.

Restorer gene: Any one of a small gene family that reverts ("restores") a socalled cytoplasmic male sterility (CMS) gene defect in the mitochondria of flowering plants. The multiple phenotypic consequences of this mutation include meiotic irregularities, disintegration of the tapetum, prevention of tetrade formation or pollen maturation, production of sterile pollen, inability to discharge pollen, and others. The mechanism of restoration is presently not known.

Restricted template random amplified polymorphic DNA (RT-RAPD): A variant of the conventional → random amplified polymorphic DNA technique, in which the → template DNA is restricted by a → restriction endonuclease prior to its amplification in the → polymerase chain reaction. RT-RAPD is usually used to detect DNA polymorphisms which cannot be discovered without restriction.

Restricted transduction: See → transduction.

Restriction: Exclusion of foreign DNA through degradation in bacteria. The restriction system uses → restriction endonucleases which recognize specific target sequences for cutting (→ recognition site). Bacterial DNA is protected from restriction by modification, see → restriction-modification system. See also → restriction fragment length polymorphism, → restriction mapping.

Restriction analysis: An infelicitous term for → restriction mapping.

Restriction endonuclease (restriction enzyme, restriction nuclease, ENase; EC 3.1.21, 3-5): Any bacterial enzyme recognizing specific target nucleotide sequences (→ recognition site) in double stranded DNA and catalyzing the breakage of internal bonds between specific nucleotides within these targets or in a specific distance thereof (at the → target site) to generate double-stranded breaks with either → cohesive or → blunt ends (endodeoxyribonuclease, endo-DNase). They are part of a bacterial → restriction-modification system, protecting the cell against foreign DNA, which is cleaved, while the cell's own DNA is protected by methylation of the recognition sites. For the nomenclature of restriction endonucleases the following system is used. First an abbreviation stands for the bacterial species from which the restriction enzyme originates (*E*scherichia coli, *Eco*; *B*acillus *amy*loliquefaciens, *Bam*; *S*erratia *ma*rcescens, *Sma*), followed by a subscript for the strain or type identification (e.g. *Eco* R; *Bam* H). If a particular host strain has several restriction systems, these are identified by Roman numerals (e.g. *Eco* RI, *Eco* RII, *Eco* RIII; *Hind* I, *Hind* II, *Hind* III, *Hind* IV). Recognition sequences are written → 5' → 3', usually one strand only. If cleavage occurs within the recognition sequence, this is indicated by an arrow or stroke at the site of cleavage. See also → star activity.

A^+C^+:   Superscript + indicates the *inhibition of cleavage* by N^6-methyladenine or 5-methylcyto-sine.

A^mC^m: Superscript m indicates that an N^6-methyladenine or 5 methylcytosine is *required for cleavage.*

A^0C^0:   cleavage not influenced by N^6-methyladenine or 5-methyladenine.

Restriction enzymes can be classified in three main categories:

a) Type I (class I) restriction endonucleases are composed of three subunits (R, restriction; M, methylation; S, specificity, i.e. sequence recognition) and show both modification and restriction functions. They require Mg^{2+}, ATP and → S-adenosyl methionine, restrict only one strand of a duplex DNA at a distance of 1-5 kb from their recognition site and produce a → gap of about 75 nucleotides in length. A double-strand break is completed by a second enzyme. For example, *Eco* B recognizes the sequence TGA(N)$_8$TGCT (N stands for any nucleotide), binds there, and cleaves the DNA non-specifically in an ATP-requiring reaction at a 3' site about 1 kb away. If the recognition sequence is unmethylated in one strand, the enzyme first methylates that strand and then cleaves. Because the site of the cut is non-specific, type I enzymes are not used in → recombinant DNA technology. There exist more than 30 different type I enzymes (predominantly in members of the Enterobacteriaceae).

b) Most type II (class II) restriction endonucleases recognize and split dsDNA within particular sequences consisting of tetra-, penta-, hexa-, hepta- or octanucleotides with an axis of rotational symmetry (→ palindrome), e.g. for *Eco* RI:

$$\downarrow$$
$$\text{5'-GAA TTC-3'}$$
$$\text{3'-CTT AAG-5'}$$
$$\uparrow$$

The recognition site may, however, also be asymmetric.

These class II enzymes are differentiated into various subclasses:

- Subclass-IIP ENases recognize tetra-, hexa- and octanucleotide palindromes with internal deoxyadenosine, deoxyguanosine, deoxycytidine, or thymidine nucleotides, but also include palindromic recognition sequences with ambiguous nucleotides (e.g. *Eco* RI, *Hae* II and *Hae* III).

- Subclass-IIW ENases recognize penta-and heptanucleotide palindromes with internal deoxyadenosine, deoxyguanosine, deoxycytidine or thymidine nucleotides, but include sequences with ambiguous nucleotides within the flanking palindromic di- or trinucleotides (e.g. *Hinf* I, *Sau* I, *Dra* II).

- Subclass-IIN ENases recognize interrupted palindromes with internal deoxyadenosine, deoxyguanosine, deoxycytidine or thymidine nucleotides (e.g. *Asp* I, *Sfi* I).

- Subclass-IIS ENases recognize largely asymmetric, 4-6 bp long recognition sequences, where the cut positions are located 1-20 nucleotides distal (e.g. *Fok* I, *Mme* I).

- Subclass-IIT ENases recognize asymmetric recognition sequences, and at least one cut position is located within the recognition sequence (e.g. *Bsm* I, *Gdi* II).

- Subclass-IIU ENases recognize asymmetric recognition sequences for which the cut positions are unknown. Nearly all these numerous type II enzymes are used in → recombinant DNA technology, because the sites of the cut are specific. Furthermore, enzymes can be selected which generate specific → cohesive ends or → blunt ends.

c) Type III (class III, class IIS) restriction endonucleases are composed of two subunits, recognize a short asymmetric sequence, bind there, and catalyze a staggered cut about 24-26 bp away in an ATP-requiring reaction (e.g. *Eco* P1). They require Mg^{2+} and ATP, but use S-adenosyl methionine only for stimulation. Type III enzymes are rare (less than 10 presently known). See also → four base cutter, → six base cutter, → rare cutter, → isohypekomer, → isoschizomer, → heterohypekomer, → heteroschizomer, → meganuclease, → site preference.

Restriction endonuclease fingerprinting (REF): A variant of the → single strand conformation polymorphism (SSCP) technique, that allows to detect → mutations in a target DNA. In short, genomic DNA is first amplified in a conventional → polymerase chain reaction and then cleaved in separate reactions by several groups of → restriction endonucleases (e. g. I: *Alu*I; II: *Hinf*I and *Hae*III; III: *Dde*I and *Sty*I; IV: *Mbo*I and *Ase*I; V: *Mnl*I). The endonucleases are selected such that the average restriction fragment size is about 150 bp. After heat-inactivation of the restriction enzymes, the restriction fragments of all five reactions are mixed and 5' endlabeled with → T4 polynucleotide kinase and γ-^{32}PATP. The denatured products are then electrophoresed in a nondenaturing → polyacrylamide gel to produce a pattern of single-stranded segments. Point mutations are identified by e. g. the loss of a → restriction site in the mutant (in this case, the mutant shows one large fragment, the wild type two smaller ones) or abnormal migration in the SSCP gel (mobility shift). The mutated regions can be isolated from the gel and the mutated site characterized by → sequencing.

Restriction endonuclease-mediated selective polymerase chain reaction (REM-PCR): A variant of the conventional → polymerase chain reaction that allows the detection of a single base exchange in a target DNA. REM-PCR is based on the action of a thermostable → restriction endonuclease (e. g. *Bst* NI) that is added to the PCR mixture and cleaves its restriction recognition site only in the wild-type DNA, but not in the mutant DNA (since the base exchange occured within the restriction site).

Restriction enzyme: See → restriction endonuclease.

Restriction enzyme-mediated integration (REMI): A special type of non-homologous integration of DNA fragments, cut with → restriction endonucleases, into genomic → restriction sites. In short, a → REMI vector, designed for the organism to be transformed, is cut with *one* restriction enzyme and the complete digest mixture transformed into protoplasts (or → spheroplasts) with → polyethylene glycol. Then the protoplasts are plated out and the → transformants selected (e.g. with an → antibiotic, if the REMI vector contains the corresponding antibiotic → resistance marker). These transformants contain the integrated REMI vector, and integration occurs preferably at genomic sites cut with the still functioning restriction endonuclease. If the REMI vector is integrated into a gene encoding a phenotype, this gene can be isolated by cutting with a restriction enzyme other than the cloning enzyme. REMI represents an alternative to chemical or UV → mutagenesis.

Restriction fragment: Any oligo- or polynucleotide arising from non-random cleavage of a larger DNA molecule by a → restriction endonuclease. See → restriction fragment length polymorphism.

Restriction fragment differential display polymerase chain reaction (RFDD-PCR): A variant of the → differential display reverse transcription polymerase chain reaction, that allows to estimate the number of expressed genes and to detect differences in gene expression in different cell types. In short, the complete set of → messenger RNAs of a particular cell type is used as → template for → reverse transcriptase to synthesize → cDNAs, employing an oligo(dT)$_{12\text{-}18}$ primer. The double-stranded cDNAs are then restricted with the → four base cutter *Taq*I, which generates a → 5' overhang. Two different, but specific → adaptors are now ligated to the cDNA ends. One of the adaptors, the EP ("extension protection") adaptor possesses a 5' overhang and a 3' extension protection group, preventing 3'→5' fill-in reactions. The other adaptor represents a standard oligonucleotide. Now, primers complementary to both adaptors are employed to amplify the cDNA population. The socalled 0-extension 5' primer (annealing to the EP adaptor) is radioactively labeled, and used in combination with the socalled display 3' primer, which extends into the cDNA sequence by three arbitrary nucleotides. The extension protection group on the EP adaptor prevents any amplification of cDNA fragments with EP adaptors on both termini. Also, cDNAs with standard adaptors on both termini will not, or only rarely be amplified (since both cDNA ends would need to carry identical three bases for successful amplification). In contrast, amplification of cDNAs with a standard adaptor on one, and an EP adaptor on the other end will proceed preferentially. By using 64 different display primers (differing in their selective nucleotides) in separate PCR reactions, about 17,000 distinct cDNA fragments can be visualized (~ 200 amplification products per reaction; "expression window") by → sequencing gel electrophoresis and → autoradiography.

Restriction *fragment* *length* *polymorphism* (RFLP; DNA polymorphism): The variation(s) in the length of DNA fragments produced by a specific → restriction endonuclease from → genomic DNAs of two or more individuals of a species. RFLPs are generated by → rearrangements or other → mutations that either create or delete → recognition sites for the specific endonuclease. They may also be due to the presence of → repetitive DNA in different copy numbers on a specific chromosomal region (see for example → variable number of tandem repeats). RFLPs are now widely used to localize specific genes in complex genomes (→ linkage analysis), to discriminate between closely related individuals (→ fingerprint tailoring) and to establish → genome maps. *Figure see page 936.*

Restriction fragment tagging: The ligation of → adaptor sequences to → restriction fragments using → DNA ligase. Tagged fragments can be processed in e.g. → selective restriction fragment amplification.

Restriction *landmark* *genomic* *scanning* (RLGS): A technique to type a genome. In short, genomic DNA is first restricted with a → rare-cutter (e.g. *Not*I) to produce several thousand fragments (the exact number depending on the → complexity of the genome). All these restriction fragments are labeled by → filling in the termini using → terminal transferase and α^{32} P-deoxynucleotide triphosphates. Then the size of the labeled fragments is reduced by a second restriction with a six-base cutter (e.g. *Eco*RV). The resulting secondary restriction fragments are separated by → agarose gel electrophoresis. Then a gel strip is excised and the fragments contained within it are again restricted with a four base cutter (e.g. *Hinf* I). The resulting tertiary fragments are then subjected to a two-dimensional → polyacrylamide gel electrophoresis and the radioactive spots visualized by → autoradiography. RLGS resolves up to several thousand restriction fragments originating from all over the genome (about one fragment per Megabase of an average eukaryotic genome). This technique detects subtle differences between genomes.

ORIGIN OF RESTRICTION FRAGMENT LENGTH POLYMORPHISM (RFLP)

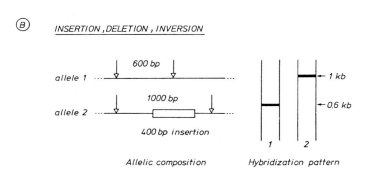

Ⓐ   LOSS OR GAIN OF RESTRICTION SITE

allele 1   1000 bp

allele 2   600 bp   400 bp

Allelic composition

1 kb
0.6 kb
0.4 kb

1   2

Hybridization pattern

Ⓑ   INSERTION, DELETION, INVERSION

allele 1   600 bp

allele 2   1000 bp

400 bp insertion

Allelic composition

1 kb
0.6 kb

1   2

Hybridization pattern

Restriction fragment length polymorphism (RFLP)

Restriction map (restriction site map): A linear array of → restriction sites on a particular DNA molecule. See → restriction mapping. → circular restriction map, → long-range restriction map.

Restriction mapping (restriction site mapping, restriction analysis): A technique to construct a → restriction map of a particular segment of DNA that allows to arrange different type II → restriction endonuclease → recognition sites on a linear scale. See also → bottom-up, → top down-, → Smith-Birnstiel mapping, → fine mapping.

Restriction-mediated differential display: See → cDNA-AFLP.

Restriction-*m*odification (R-M) system (*h*ost-*c*ontrolled *r*estriction-*m*odification, hcR-M): A bacterial defense system that is designed to restrict infectious → phage or → plasmid DNA molecules. This is achieved by sequence-specific endonucleolytic cleavage of the infecting DNA (see → restriction endonuclease). The foreign DNA may become resistant towards the restriction enzymes by → modification, i.e. its sequence-specific methylation, as is also all the cellular DNA. This modification is caused by DNA methyltransferases or → modification methylases that catalyze the transfer of a methyl group from → S-adenosyl methionine onto one of the nucleotides within the recognition sequence on each strand of DNA. The recipient nucleotide may be adenine (converted to N^6-methyladenine, m^6A), or cytosine (converted to 5-methylcytosine, m^5C or N^4-methylcytosine, m^4C). Restriction and methylation may be catalyzed by subunits of one specific enzyme.

R-M systems fall into three broad categories, according to the requirements of the corresponding enzymes, see → restriction endonuclease. Compare → DNA methylation, → methylated adenine recognition and restriction system, → methylation assay, → methylation protection, → modified cytosine restriction system.

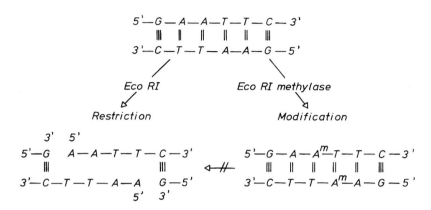

Eco RI Restriction Site

Restriction nuclease: See → restriction endonuclease.

Restriction site: A → restriction endonuclease → recognition site.

Restriction site add-on: See → PCR add-on primer.

Restriction site conversion plasmid (*restriction site mobilizing element*, RSM element): A → plasmid cloning vector that contains two identical → polylinkers either in tandem arrangement, or separated by a short spacer sequence. Any foreign DNA, cloned into one of the cloning sites of the polylinkers, can be isolated as a molecule with two different termini by cutting with two different → restriction endonucleases for which → recognition sites are present in the polylinkers.

Restriction site mobilizing element (RSM): See → restriction site conversion plasmid.

Restriction-site oligonucleotide (RSO): A synthetic → oligonucleotide carrying one (or more) → restriction endonuclease → recognition sequence(s), a → T7 phage promoter sequence and additionally an anchor sequence for priming. This RSO is used as an → amplimer in e.g. → restriction-site polymerase chain reaction. In combination with primers complementary to known sequences (e.g. a gene) RSOs allow the amplification of DNA regions of unknown sequence and the → T7 RNA polymerase-driven expression of the PCR products.

Restriction-site PCR: See – restriction-site polymerase chain reaction.

Restriction-site polymerase chain reaction (RS-PCR, restriction-site PCR): A variant of the conventional → polymerase chain reaction that allows to retrieve unknown DNA sequences adjacent to known sequences. Specially designed → primers, so-called → restriction site oligonucleo-tides (RSOs) specific for a given → restriction site → recognition sequence, are first annealed to target DNA. Then a first round of PCR is performed in different tubes with a set of RSOs, and a primer specific for the known region. The resulting amplification product is then reamplified using the same RSOs, but another specific primer internal to the first one. The amplified products are finally transcribed with T7 RNA polymerase, that uses its own → promoter linked to each of the RSOs, and can be sequenced directly with the → genomic amplification with transcript sequencing (GAWTS) technique, employing a third specific endlabeled primer internal to the second one. See → gene walking polymerase chain reaction, → inverse polymerase chain reaction, → nested polymerase chain reaction.

Restrictive transduction: See → transduction.

Retentate mapping (SELDI retentate mapping): An infelicitous term for the detection of interaction(s) between a → probe molecule (e.g. a low molecular weight chemical compound, an → antibody, a → receptor protein, a peptide or protein, or an → oligonucleotide) bound onto a glass or silicon chip, and an analyte molecule (e.g. a peptide or protein as component of a biological fluid as serum or urine). The chip is first incubated with the analyte and then washed to remove buffers and contaminants. This procedure retains the analyte ("retentate"), which then can be characterized by e.g. → surface-enhanced laser desorption/ionization (SELDI). This method therefore "maps" (i.e. identifies) the retentate.

Retroarray (retrovirus chip, retrochip): A laboratory slang term for a → microarray, onto which various → retrotransposon sequences are spotted at high density. Retroarrays are used to detect expressed retrotransposons and to monitor their differential expression in various tissues or organs in response to different environmental influences. See → human endogenous retrovirus chip.

Retrochip: See → retroarray.

Retroelement: See → retroposon.

Retrogene: See → processed pseudogene.

Retron: A prokaryotic → retroposon that transposes via an RNA intermediate (→ reverse transcription), has no extracellular phase and does not possess retroviral-like → long terminal repeat sequences (e.g. the msDNAs of mycobacteria and *E. coli*). Compare → retrotransposon.

Retronuon: An abbreviation for various mobile elements in a genome (e.g. → retrotransposons), that are generated by → reverse transcription of → messenger RNA and integration of the corresponding → cDNA into random positions in the genome. Do not confuse with → retron. Together with the terms → naptonuon, → nuon, → potonuon, and → xaptonuon, retronuon belongs to the vocabulary of evolutionary genomics (see → evolutionary developmental genetics).

Retroposition (retrotransposition): The process of integration of → retrotransposons into target DNA.

Retroposon (retrosequence, retroelement): A → transposable element of pro- and eukaryotic organisms that transposes via an RNA intermediate and involves → reverse transcriptase activity. Retroposons may be of:
- *viral origin* (viral-like retroposon, viral-like element, retroviral-like element, retrotransposon that share structural similarities with integrated → retroviruses), or
- *non-viral origin* (non-viral retroposon).

Retropseudogene: See → processed pseudogene.

Retroregulation: The regulation of the → translation of a → messenger RNA by a sequence downstream of the translation → initiation codon.

Retroselection: A somewhat misleading term for the isolation of a gene with the help of an → antibody raised against the protein encoded by the gene. In short, → monoclonal *antibodies* (Mabs) are generated from surface proteins of target (e. g. tumor) cells. Then a → cDNA library from RNA of the target cells containing the surface protein marker is established in an → expression vector. The surface protein-specific Mab is now used to screen the proteins expressed from the cDNA library. Positive clones are isolated, and their cDNAs isolated and sequenced. The sequence information can now be used to surf in data banks for identical or similar sequences, or to construct → probes for the isolation of the corresponding gene from e. g. → genomic libraries.

Retrosequence:
a) See → processed pseudogene.
b) See → retroposon.

Retrotherm™ RT: The abbreviation for a thermostable DNA polymerase, that also contains → reverse transcriptase (RTase) activity and is isolated from a thermophilic bacterium. The enzyme has little or no → RNase H or → exonuclease activity, and can be used for → cDNA synthesis from an RNA template or DNA synthesis from a DNA template.

Retrotransposition: See → retroposition.

Retrotransposon (viral-like retroposon, viral-like element, retroviral-like element, RL): A → transposable element of eukaryotic organisms that has a gene organization, replication cycle and integration mechanism similar to those of → retroviruses (i.e. contains → reverse transcriptase, protease, → RNase H and → integrase coding regions). RL elements are flanked by → long terminal repeats including terminal TG......CA repeats, a putative → promoter, and transcription termination signals. Retrotransposons move by producing a copy-RNA that is reverse transcribed by → reverse transcriptase into DNA which is then inserted at a new target site. The reverse transcriptase is not necessarily encoded by the element itself, but may also be supplemented in trans. Such retrotransposons are present in yeast (→ Ty element), echinoids, dipterans (→ copia element of *Drosophila melanogaster*), lepidopterans, gymnosperms, angiosperms and vertebrates, and form families of up to 400 members that differ slightly in sequence.

Retrotransposon-based insertion polymorphism (RBIP): Any sequence difference between two (or more) genomes, that is based on either the presence or absence of a → retrotransposon at a specific chromosomal → locus, which can be detected by its site-specific amplification in a conventional → polymerase chain reaction (PCR). This PCR employs genomic DNA as template, and three different → primers (see → triplex PCR), two targeting at the genomic DNA flanking a retrotransposon at its 5'- and 3' side, respectively, and a third primer complementary to sequences within the retrotransposon. The flanking primers define a specific genomic locus, the internal primer in combination with one of the flanking primers allows to amplify part of the retrotransposon sequence. The polymorphism is based on the presence (amplification product detectable) or absence (no amplification product) of the retrotransposon. Therefore, RBIP markers are → codominant (i.e. different allelic states at a locus can be revealed). Since this type of polymorphism is based on either presence or absence of a PCR product, there is no need for gel-electrophoretic separation of bands. Instead, a simple → dot blot hybridization assay is sufficient, and amenable to high-throughput automation. Since retrotransposon → insertions are frequent events in eukaryotic genomes, many different loci can be targeted each with locus-specific primers, and therefore many different codominant markers be generated. See → expression transposon insertion display, → inter-retrotransposon amplified polymorphism, → retrotransposon-microsatellite amplified polymorphism, → transposon insertion display.

Retrotransposon-microsatellite amplified polymorphism (REMAP): Any difference in DNA sequence between two genomes, detected by → polymerase chain reaction-mediated amplification of the region between a → long terminal repeat of a → retrotransposon and a nearby → microsatellite. Any observable → polymorphism is the consequence of mutation(s) (i. e. → insertions, → deletions) primarily in the region between the retrotransposon and the microsatellite. Compare → inter-retrotransposon amplified polymorphism.

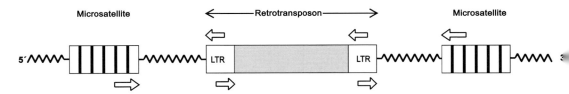

Retroviral-like element: See → retrotransposon.

Retroviral vector: One of a series of specially designed → plasmid cloning vectors that contains elements of a → retrovirus. In its simplest version, the viral genes for group-specific antigens (*gag*), the → reverse transcriptase (*pol*) and the viral envelope proteins (*env*) have been deleted and replaced by a → selectable marker gene (e.g. → neomycin phosphotransferase gene). Additionally a → polylinker has been inserted, that allows the cloning of foreign genes. The plasmid background (e.g. → pBR 322) allows the replication of this vector in *E. coli*. Since the vector still contains the viral sequences necessary for integration (e.g. → long terminal repeat sequences), the cloned foreign gene may be introduced into the host cell's genome and transcribed from the → promoter of the 5'-LTR region (retroviral expression vector, see for example → exon trap).

Retrovirus: Any one of a class of single-stranded RNA → viruses that infects eukaryotic cells where its RNA genome is reverse-transcribed into a DNA copy (→ cDNA) with the aid of a virus-encoded RNA-dependent DNA polymerase (→ reverse transcriptase). This double-stranded cDNA is then circularized in the cell nucleus and finally covalently integrated into the genome of its host cell at border sequences containing → long terminal repeats (LTRs) via a mechanism similar to → transposition. The integrated viral genome (endogenous retrovirus) is called → provirus. The LTRs contain → enhancer and → promoter elements, and flank the coding region, that harbors the *gag*-gene (encoding the viral → capsid proteins), the *pol*-gene (encoding → reverse transcriptase), and the *env*-gene (encoding glycoproteins of the viral lipid coat). The termini of the viral genome carry characteristic R segments, direct repeat sequences of about 10-80 nucleotides in length, that are flanked by specific regions (U5 at the 5' end, about 80-100 nucleotides long; U3 at the 3' end, about 170 to more than 1000 nucleotides long).

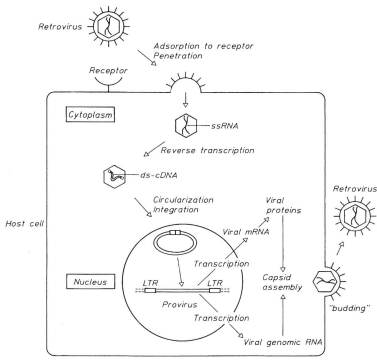

Life cycle of a retrovirus

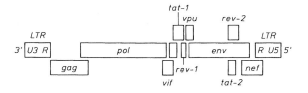

Schematic structure of the HIV-1 genome

Retrovirus

The infection cycle starts with the release of the viral genomic RNA from its capsid, continues with its reverse transcription and results in the covalent integration of one or more such DNA copies into the recipient genome, as described above. The transcription of the endogenous retroviral genome leads to the accumulation of transcripts that are translated into e.g. capsid or reverse transcriptase proteins. Reassociation of the viral RNA and the protein components generates new virus particles which leave the host cell through a process called "budding". Some retroviruses have been engineered to function as vectors for cloning of mammalian genes (→ retroviral vector).

Retrovirus chip: See → retroarray.

Reverse chromosome painting: A derivative of → chromosome painting which uses → probes from aberrant chromosomes for → *in situ* hybridization to metaphase spreads from specimen cells and detection of hybridization events by → fluorescent *in situ* hybridization. Abnormal chromosomes are characterized by highlighting normal chromosomes. In short, aberrant chromosomes have first to be isolated by → flow cytometry. Then they are amplified and labeled by → degenerate oligonucleotide-primed PCR and used for *in situ* hybridization under suppression conditions (i.e. the avoidance of undesirable background signals). Reverse chromosome painting is used to identify marker chromosomes, to analyze chromosome breakpoints, and to characterize → deletions and → insertions.

Reversed base: Any nucleic acid → base whose deoxyribose is linked to a second deoxyribose via a 3'→ 3' bond (rather than a 3' → 5' bond in e.g. DNA). This unusual configuration is called inverted linkage, and the base a reversed base, which can be incorporated into an oligonucleotide and effectively blocks elongation.

***Reversed enzyme activity DNA interrogation test* (READIT):** A technique for the detection of specific DNA sequences (e. g. → foreign genes in → genetically modified organisms, mutations as e. g. → single nucleotide polymorphisms, or chromosome → translocations) that is based on an enzymatic luminogenic reaction. In short, a complementary DNA → probe is hybridized to the target sequence, and a pyrophosphorylase added that uses pyrophosphate to remove deoxynucleotide triphosphates from the probe-target hybrid, which serve as substrate for a kinase to transfer the γ-phosphate onto ADP to generate ATP. This ATP in turn is used by → luciferase to produce light that can be easily quantified by a standard luminometer. The emitted light intensity is directly proportional to the concentration of the target DNA. Compare → pyrosequencing.

Reversed field gel electrophoresis: See → field-inversion gel electrophoresis.

Reverse DNA gyrase: An enzyme from the hyperthermophilic lithotroph *Pyrolobus fumarii* (optimal growth at 113° C), that catalyzes the introduction of positive ("reverse") rather than the usual negative → supercoils into circular DNA. See → DNA topoisomerase II.

Reverse dot blot: A method to detect specific amplified DNA sequences using an → oligonucleotide probe that is immobilized on a membrane and to which a labeled DNA fragment generated by the → polymerase chain reaction (PCR fragment) is hybridized. In short, the oligonucleotide probe is tailed with poly(dT) using → terminal transferase and UV-cross-linked to a nylon membrane. The amplified DNA fragment, labeled during the amplification process by using biotinylated primers (→ amplimers, see → biotinylation of nucleic acids), is then hybridized to the immobilized oligonucleotide probe. The presence of the specifically bound PCR fragment is detected using a → streptavidin horseradish → peroxidase conjugate. The reverse dot blot allows to screen different genomic fragments synchronously for mutations.

Reverse format array: A specific variant of the conventional → microarray technique, that capitalizes on complex → cDNA populations (representing the entire → messenger RNAs of a cell in their original complexity and abundance) spotted onto nylon membranes fixed to glass slides. A radiolabeled probe for a gene of interest is then hybridised to the array, and hybridisation events and intensities detected by → phosphorimaging. Rather than using a single cell or tissue sample to determine the expression of thousands of genes, the reverse format array allows to profile the expression of one single gene in hundreds, if not thousands of cells or tissues.

Reverse genetics ("surrogate genetics"): The isolation of a gene without reference to a specific protein or without any functional assays which could be exploited for its detection. This approach has been successfully employed in research of hereditary disorders (e.g. Duchenne type of muscular dystrophy, chronic granulomatous disease) and involves
a) the establishment of the map position of the gene (using cytogenetic methods, supported by studies of → restriction fragment length polymorphisms, with a resolution of roughly several million bp).
b) the identification of a specific gene within this region of interest in which → mutations are strictly correlated with the disease (using → cloning and → chromosome walking procedures).
c) the search for an RNA transcript and its disruption or altered expression in a disorder.

Reverse mutation (back mutation, reversion): The restoration of the original nucleotide sequence of a gene previously mutated by a "forward" → mutation. See also → unstable mutation, → suppressor mutation.

Reverse Northern hybridization ("cDNA Southern"): A variant of the conventional → Northern blotting technique for the quantification of the → abundance of specific → messenger RNAs of a cell, tissue, or organ. In short, genomic or → cDNA sequences of specific → genes are dotted onto → hybridization membranes at high concentrations, and hybridized to labeled cDNAs from tissue A. The strength of the resulting hybridization signal reveals the abundance of the corresponding messenger RNAs. Then the cDNAs are quantitatively removed from the membrane by → stripping, and the membrane re-hybridized with cDNAs from tissue B. This procedure allows to detect differentially expressed messenger RNAs as well. Depending on the durability of the membrane, this process can be repeated 6-10 times.

Reverse sequencing primer (reverse primer): An → oligonucleotide that is complementary to the minus-strand of an → M13 sequencing vector (→ Sanger sequencing) and allows to sequence this strand from its 5' terminus (5' → 3'). Usually, the plus-strand is sequenced first starting at its 3' end (3' → 5'), using a normal → sequencing primer. Minus-strand sequencing serves to confirm the base sequence determined by plus-strand sequencing.

Reverse Southern: See → reverse Southern hybridization.

Reverse Southern hybridization (reverse Southern): A technique to detect specific DNA sequences by hybridizing them in solution to a suitable radioactively labeled → probe. The thymidine residues of this probe are chemically modified by a psoralen derivative that allows the UV-light induced → cross-linking of the probe-target DNA hybrids. A subsequent electrophoresis in a denaturing gel allows the separation of cross-linked hybrid molecules from single strands. The hybrids can then be visualized by → autoradiography. Compare the conventional → Southern blotting procedure, where DNA fragments are separated by gel electrophoresis and transferred to a membrane prior to hybridization.

Reverse transcriptase **(RTase, RNA-dependent DNA polymerase; EC 2.7.7.49):** A retroviral multifunctional enzyme that synthesizes a double-stranded DNA using a single-stranded RNA template and a → primer (in vivo a → transfer RNA). In some → retroviruses, reverse transcriptase is a monomer, in others a dimer. For example, the larger β-subunit of → avian myeloblastosis virus reverse transcriptase catalyzes the unwinding of double-stranded DNA. Its limited proteolysis generates the α-subunit with polymerase and → RNase H activities. Reverse transcriptases are probably also involved in the → mobilization of → retrotransposons (e.g. → Ty elements in yeast) and the generation of → pseudogenes. They are used in → recombinant DNA technology for the synthesis of → cDNA from → messenger RNA → cDNA cloning). See also → Moloney murine leukemia virus reverse transcriptase, → reverse transcription polymerase chain reaction. Compare → *Thermus thermophilus* DNA polymerase.

Reverse transcriptase multiplexed polymerase chain reaction **(RT-MPCR):** A variant of the conventional → polymerase chain reaction in which multiple target → messenger RNAs are simultaneously amplified in the same reaction tube using multiple → primer pairs. The presence of many primers in a single reaction may lead to e. g. increased formation of misprimed PCR products, → primer dimers, or preferential amplification of shorter mRNAs. Therefore the RT-MPCR conditions have to be optimized (e. g. primer pairs with similar → melting temperature allow to use → stringent conditions; the reaction buffer components decrease competition among → amplicons and normalize the amplification of longer mRNAs). Following RT-MPCR, the resulting fragments are separated by → agarose gel electrophoresis and visualized by → ethidium bromide staining.

Reverse transcriptase polymerase chain reaction in situ **hybridization:** See → *in situ* polymerase chain reaction.

Reverse transcription: The synthesis of a DNA on an RNA template by → reverse transcriptase.

Reverse transcription polymerase chain reaction **(RT-PCR, cDNA-PCR, RNA-PCR):** An *in vitro* RNA amplification procedure that uses retroviral → reverse transcriptase or thermostable → *Thermus thermophilus* (*Tth*) DNA polymerase to produce a → cDNA on the RNA → template. This cDNA is then amplified using conventional → polymerase chain reaction techniques. Tth reverse transcriptase catalyzes both the reverse transcription in the presence of $MnCl_2$, and the amplification of the resulting cDNA in the presence of $MgCl_2$. Furthermore its catalytic activity is unimpaired by elevation of the reaction temperature to destabilize complex secondary structures of the RNA for stringent → primer annealing. Thus all reactions can be done in one test tube.

RNA-PCR allows the amplification of cDNA derived from small amounts of purified mRNA, tRNA, rRNA and viral RNAs and the detection of specific RNAs at a very low copy number, and is therefore used in the analysis of gene expression at the RNA level (for example the study of → posttranscriptional modifications like → alternative splicing; also for the analysis of mRNA populations of very small cell populations, ideally even of single cells). RNA-PCR may also be combined with → *in vitro* translation, see → RNA amplification with *in vitro* translation. *Figure see page 945.*

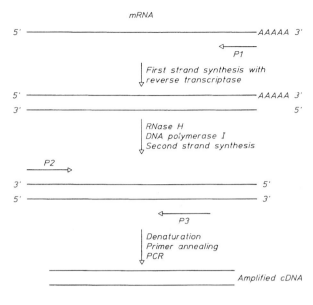

mRNA

First strand synthesis with reverse transcriptase

RNase H
DNA polymerase I
Second strand synthesis

Denaturation
Primer annealing
PCR

Amplified cDNA

P1	:	Internal primer for first strand synthesis
P2	:	Internal 3' cDNA primer
P3	:	Internal 5' cDNA primer

Reverse transcription PCR

Reverse transfection: A technique for the high-throughput analysis of the function of foreign genes in mammalian cells on a → cell chip. In short, a glass or plastic support is first printed with different → cDNAs cloned in → expression vectors. Then mammalian cells are grown on top of the printed cDNAs, where each printed spot (diameter: 120–150 μm) contains from 50–80 cells. These cells take up the DNA ("reverse transfection") and are transformed. Reverse transfection can be used to identify the effect of the transfected cDNA on the cellular phenotype, or the interaction between cellular receptor proteins and effectors (as e.g. an immunosuppressor). Since the experiments are performed with living cells, they also detect transient changes in cellular performance and their kinetics. A scale-up of slides with 10,000–20,000 cell clusters per array provides a means to rapidly screen thousands of cDNAs. See → live cell microarray.

Reverse translation: A technique for the isolation of → genes or their → messenger RNAs that uses short → oligodeoxynucleotides which are synthesized according to known protein sequences. These oligonucleotides are exploited as probes to fish the corresponding genes (in → genomic libraries) or mRNAs (in → cDNA libraries) by hybridization.

Reverse two-hybrid system: A variant of the conventional → two-hybrid system, designed for the detection of protein-protein interaction(s) in *Saccharomyces cerevisiae*, that is based on negative selection, i. e. expression of interacting hybrid proteins increases the expression of a counterselectable marker, toxic under particular conditions. Under these conditions, any failure of the proteins to interact provides a selective advantage (survival). Therefore, mutants, in which protein-protein interaction is **not** possible, can be identified from non-growing yeast colonies in which interaction occurs. Similarly, compounds preventing the interaction can be selected from large collections the same way. The most widely used toxic reporter gene is *URA*3, encoding orotidine-5'-phosphate decarboxylase (part of the uracil biosynthetic pathway). Yeast cells with wild-type *URA*3 grow on media without uracil (ura$^+$ phenotype). Yet the enzyme also catalyzes the conversion of the non-toxic 5-*fluoroorotic acid* (FOA) to the toxic 5-*fluorouracil* (FU). Now a *URA*3 allele was designed, whose expression responds tightly to a reconstituted transcription factor, conferring a Ura FOA^r phenotype, if no transcription factor is present (means, the absence of protein-protein interactions). In its presence, the cell exhibits a Ura$^+$ FOA^s phenotype. Therefore, reverse two-hybrid systems allow to select a few FOA^r cells among the majority of cells with an FOA^s phenotpye (expressing wild-type interacting hybrid proteins). See → dual-bait two-hybrid system, → interaction trap, → LexA two-hybrid system, → one-hybrid system, → RNA-protein hybrid system, → split-hybrid system, → split-ubiquitin membrane two-hybrid system, → three-hybrid system.

Reversible gel: A gel matrix consisting of gellan gum, a bacterial carbohydrate polymer of a linear repeating tetrasaccharide monomer composed of ß-D-glucose, ß-D-glucuronic acid, ß-D-glucose and α-L-rhamnose. In the presence of divalent metal ions (e.g. $CaCl_2$, $MgCl_2$, or diamines as e.g. 1,3–*dia*mino–2–y*h*xdroxy*p*ropane, DAHP), gellan gum forms gels composed of intertwined helix-like molecules. Reversible gels are used for → gellan electrophoresis.

Reversion: See → reverse mutation.

Revertant: An organism with an allele that underwent → reverse mutation.

RF:
a) See → *r*ecombination *f*requency.
b) See → *r*eplicative *f*orm.

R factor: See → *r*esistance factor.

RFB: See → replication fork barrier.

RFDD-PCR: See → *r*estriction *f*ragment *d*ifferential *d*isplay *p*olymerase *c*hain *r*eaction.

RFE: See → rotating gel electrophoresis.

RFLP: See → *r*estriction *f*ragment *l*ength *p*olymorphism.

RFP: See → *r*ed *f*luorescent *p*rotein.

RFP: See → red fluorescence protein.

RGA: See → resistance gene candidate.

RGAP: See → *r*esistance *g*ene *a*nalogue *p*olymorphism.

RGE: See → *r*otating *g*el *e*lectrophoresis.

R gene: See → *r*esistance gene.

RGH: See → resistance *gene ho*mologue.

RH: See → radiation hybrid.

***Rhizobium*:** A Gram-negative soil bacterium that establishes a symbiotic relationship with the roots of some higher plants (e.g. legumes), and forms so-called root nodules in which it is converted into the nitrogen-fixing bacteroid form. Some, but not all *Rhizobia* contain one or more → plasmids, one of which is needed for the colonization of a particular plant (e.g. legumes such as beans, cowpeas, chickpeas).

Rhodamin B isothiocyanate: A fluorescent dye (→ fluorochrome) that can be used as a marker for → fluorescent primers in e.g. → automated DNA sequencing procedures.

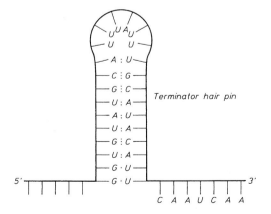

Rho factor (ρ): An oligomeric protein of *E. coli* that binds to specific sequences at the 3' end of a gene and signals stop of → transcription. See → rho factor-dependent termination, and → rho factor-independent termination.

Rho factor-dependent termination: The stop of → transcription of a specific gene in *E. coli*, when the → RNA polymerase molecule recognizes the complex between the → rho factor and specific → terminator sequences at the 3' end of the gene. Termination may also be rho-factor-independent, see below.

Rho factor-independent termination: The stop of → transcription of specific genes in *E. coli*, when → RNA polymerase recognizes specific sequences (→ terminator sequence) as termination signals in the absence of → rho factors. Compare → rho factor-dependent termination.

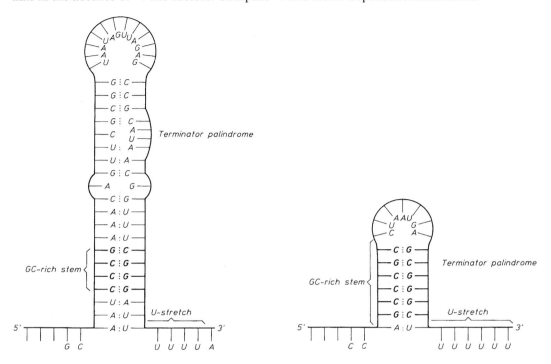

Rho factor-independent termination

RIA: See → radio*immunoassay*.

RIAR: See → random insertion of antibiotic resistance genes.

Ribocyte (RNA cell): A hypothetical primordial cell that functions exclusively on RNA, not DNA. Ribocytes are considered as the predecessors of the present-day prokaryotic and eukaryotic cells. According to the ribocyte theory, some molecular machines of ribocytes are still functioning in cells, as e.g. the peptidyltransferase center of the → ribosome, or the ribonucleotide- containing coenzymes (as e.g. NAD⁺). See → ribo-organism.

Ribonome: The complete set of → ribonucleic acid molecules in a cell at a given time.

Ribonomics (RNomics, RNA genomics): The whole repertoire of techniques to analyze and characterize the entirety of ribonucleic acids (RNAs) in a cell, comprising the identification of novel RNA-encoding genes, the development of techniques for high-throughput determination of RNA structures and RNA-protein complexes, the creation and maintenance of a centralized database of RNA sequences, structures and functions, the design of modelling tools, and the application of *nuclear magnetic resonance* (NMR), cryo-electron microscopy, chemical probing, and modification interference mapping (where modifications are introduced with alkylating agents or phosphorothioates or cleavage with free radicals). See → ribonome.

Ribonuclease: See → RNase.

Ribonuclease **A**: See → RNase A.

Ribonuclease **A technique**: See → RNase A mismatch detection.

Ribonuclease **B**: See → RNase B.

Ribonuclease **CL3**: See → RNase CL3.

Ribonuclease **D**: See → RNase D.

Ribonuclease **E**: See → RNase E.

Ribonuclease **from** *B. cereus*: See → RNase from *B. cereus*.

Ribonuclease **H**: See → RNase H.

Ribonuclease **I**: See → RNase I.

Ribonuclease **P**: See → RNase P.

Ribonuclease **Phy M**: See → RNase Phy M.

Ribonuclease **Phy 1**: See → RNase Phy 1.

Ribonuclease *protection* **assay (RPA)**: See → RNase protection assay.

Ribonuclease **S**: See → RNase S.

Ribonuclease **III**: See → RNase III.

Ribonuclease **T$_1$**: See → RNase T$_1$.

Ribonuclease **T$_2$**: See → RNase T$_2$.

Ribonuclease **II**: See → RNase II.

Ribonuclease **U$_2$**: See → RNase U$_2$.

Ribonucleic acid: See → RNA.

Ribonucleoprotein **particle (RNP):** Any complex consisting of both RNA and protein(s). For example, → ribosomes, → and heterogeneous nuclear ribonucleoproteins are such complexes. See also → informosome, → messenger ribonucleoprotein, → RNase P, → small nuclear and → small nucleolar ribonucleoprotein, → signal recognition particle, → small cytoplasmic ribonucleoprotein, → spliceosome, → telomerase.

Ribonucleoside: A → purine or → pyrimidine base covalently linked to a → ribose molecule. See also → ribonucleotide.

Ribonucleotide: An organic molecule consisting of a → purine or → pyrimidine base covalently linked to → ribose which is esterified with a phosphate group at its 5' position (ribonucleoside-5'-triphosphate, → NTP). Constituent of → RNA.

Ribonucleotide reductase **(RNR):** An enzyme catalyzing the reduction of ribonucleotides to de-oxyribonucleotides, i. e. the replacement of the 2'-OH group in ribonucleotides by hydrogen, involving free-radical action. RNRs can be divided into three classes based on primary structure and radical generation mechanism. Class I RNRs consist of two dimeric subunits, R1 and R2. The R2 subunit generates a stable free tyrosyl radical through the activation of O_2 by a dinuclear iron center. The radical is then transferred to the active site on R1. Class II RNRs are monomers or homodimers and use adenosylcobalamin to generate transient 5'-deoxyadenosine and cysteine radicals. Class III RNRs are expressed by some strictly or facultatively anaerobic bacteria under anaerobic conditions, have a quaternary structure ($\alpha_2\beta_2$), where the active site and allosteric regulatory sites are located on the larger α_2 subunit. The smaller β_2 subunit ("activase") catalyzes cleavage of S-adenosylmethionine (AdoMet) with the aid of an iron-sulfur cluster to produce a stable glycyl radical.

Ribo-organism: A hypothetical vesicle, consisting of long-chain fatty acids (e.g. oleic acid) form-ing an archaic membrane-like structure, and two → ribozymes. One ribozyme would synthesize the membrane component, the other one would replicate itself and synthesize the first ribozyme. Such vesicles, if supplied with → nucleotide triphosphates and membrane precursors, could grow and divide (i.e. show primitive characteristics of life). Such a ribo-organism may have been a precursor of a true unicellular organism. See → RNA world.

Riboprobe (RNA probe): Any labeled single-stranded RNA → probe used as → antisense RNA to detect specific → messenger RNAs in e.g. the → RNase protection assay. For the generation of a riboprobe, a DNA fragment spanning the region of the interesting gene is cloned into an → ex-pression vector (with e.g. → T7, → T3, or → SP6 bacteriophage promoters) in an orientation such that transcription produces antisense RNA. Then the → plasmid vector is linearized with a → blunt end cutter, and used as → template for the synthesis of a radiolabeled riboprobe between 50 and 250 nucleotides in length. This probe can then be employed for RNA : RNA → hybri-dization experiments. Riboprobes can also be used for → *in situ* hybridisation to RNA targets. RNA:RNA hybrids are more stable than DNA:RNA hybrids. Usually excess of unhybridized probe is reduced by treating the sample with → ribonuclease H. Riboprobes are only rarely used for hybridisation to DNA targets.

Riboproteomics: The molecular characterization of RNA-protein interaction(s) in a cell, inclu-ding the chemistry of RNA-protein recognition, protein binding, structure and function(s) of RNA-protein complexes and their localization, dynamic changes and half-life time (i.e. decom-position or degradation).

Riboregulation: The regulation of the expression of a → gene by an RNA molecule, acting either in → *cis* or in → *trans*. Various small RNAs riboregulate gene activity in e. g. *E. coli*. For ex-ample, the 10Sa RNA (= tm RNA) binds and inactivates → repressor proteins, or the small regulatory RNA, DsrA, functions both as gene anti-silencer and → translation modifier. It over-comes nucleoid-associated H-NS-protein mediated transcriptional silencing of genes, and stimulates the translation of the stationary phase stress σ factor, RpoS. Riboregulation is one of the gene-controlling processes in bacteria, plants, and mammals.

Riboregulator: Any one of a family of diverse unstable regulatory RNAs of bacteria, that controls gene expression at the post-transcriptional level. For example, the 87 nucleotide untranslated riboregulator DsrA affects several target genes simultaneously. It acts in trans, and activates and represses, respectively, the synthesis of two transcriptional regulators, HN-S (a histone-like protein suppressing the activity of a number of genes) and RpoS, the stationary phase σ-factor of DNA-dependent RNA polymerase. DsrA-RNA affects the stability of *rpoS* and *hns* mRNAs: *hns*-mRNA is turned over more rapidly, *rpoS*-mRNA is stabilized. These effects are brought about by RNA-RNA interactions, mediated by one of three stem-loop structures of the riboregulator. Some of the riboregulators are induced by e.g. oxidative stress.

***Riborgis eigensis* (*ribosomal organism*):** A hypothetical organism of the prehistoric RNA world, that had linear chromosomes and in which RNA played the roles of present-days DNA and proteins. The species name derives from the German scientist Manfred Eigen.

Ribose (β-D-ribose): A five-carbon sugar (pentose) characteristic for RNA. Compare → deoxy-ribose.

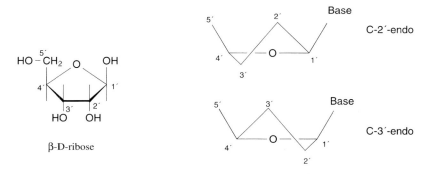

Potential conformations of β-D-ribose

Ribose zipper: A hydrogen bond network formed between two RNA strands that come into close proximity and run antiparallel, in which the 2'-OH groups of two consecutive riboses on both strands interact. The 2'-OH of the 5'-nucleotides in both strands hydrogen bond to the oxygens of the 2'-OHs of the nucleotides on the other strand (using their protons as donors). Compare → leucine zipper.

Ribosomal DNA: See → rDNA.

Ribosomal frameshifting (*programmed ribosomal frameshifting*, PRF): The directed change in the translational → reading frame of → messenger RNAs (mRNAs) in viruses (especially retroviruses, e.g. HIV–1), → retrotransposons, and bacterial → insertion elements (but also eukaryotic mRNAs). Ribosomal frameshifting is caused by a slip of the translating ribosome by one base in either the 5'(–1) or the 3'(+1) direction, and allows the production of one single protein from two or more overlapping genes. It is controlled by the nucleotide sequence of the corresponding → messenger RNA and by its secondary and/or tertiary structure. For example, the –1 PRF (used by many viruses) is directed by a bipartite heptameric *cis*-acting mRNA signal sequence X XXY YYZ ("slippery site"), that is followed by a 3'secondary structure (typically a pseudoknot, consisting of two nested stems, the loop of one stem forming the base pairs of the second). Such structures cause elongating ribosomes to pause over the slippery site such that the transfer RNAs (tRNAs) slip one base in the –1 (5') direction and the non-wobble bases of both the A- and P-site tRNAs re-pair with the new –1 frame codons. The pseudoknot is melted out, translocation takes place, and translation resumes in the –1 frame. The +1 shift occurs, when the A-site is empty.

Ribosomally synthesized **a**nti**m**icrobial **p**eptide (**RAMP**): Any one of a series of short peptides between two and 50 amino acid residues in size, synthesized on cytoplasmic → ribosomes (in contrast to peptides synthesized by → non-ribosomal peptide synthesis) and functioning as → antibiotics against various (pathogenic) bacteria and fungi. For example, certain RAM peptides are secreted by silkworm moth pupae and destroy invading bacteria. Similar RAMPs or their derivatives show therapeutic potential, as e.g. the derivative P–113 of the human saliva peptide histanin against oral candidiasis in immuno-compromised patients, MBI–594, an indolicidin analogue of a peptide from bovine neutrophils, against acne infections, pexiganan, derived from the African clawed frog skin peptide pexiganin, against infected foot ulcers in diabetics, rBPI–21, derived from the human neutrophil peptide BPI, against pediatric meningococcemia and Morbus Crohn, and iseganan, derived from pig leukocyte protegrin–1, against pneumonia. Do not confuse with → *r*andom *a*mplified *m*icrosatellite *p*olymorphism (RAMP). Do not confuse with → random amplified microsatellite polymorphism (RAMP).

Ribosomal protein: Any one of a series of proteins that are structural and functional parts of ribosomal subunits (→ ribosome). Ribosomal proteins fall into two broad categories, the proteins of the large subunit (L-proteins), and those of the small subunit (S-proteins). Prokaryotes possess some 31-34 L- and 21 S-proteins, eukaryotes 45-50 L- and 33 S-proteins. They are complexed with → ribosomal RNA.

Ribosomal RNA (rRNA): A structural component of the → ribosome. Ribosomal RNAs can be broadly classified into the RNAs of the large and those of the small subunit. Prokaryotic organisms possess a 5S and a 23S rRNA in the large subunit and a 16S rRNA in the small subunit. Eukaryotes have 5S, 5.8S and 28S rRNAs in the large, and 18S rRNA in the small subunit (25S in the large and 16S rRNA in the small subunit for plants). Ribosomal RNA is transcribed from → rDNA as → pre-ribosomal RNA (→ primary transcript). In the ribosome it is complexed with → ribosomal proteins. Assembly occurs already in the nucleus, concomitantly with the → processing of the transcripts.
Figure see page 953.

Simplified scheme of yeast ribosomal RNA processing

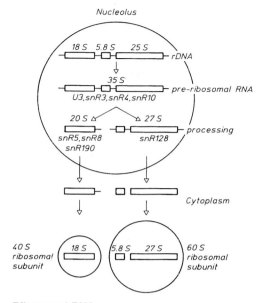

Ribosomal RNA

Ribosome: An asymmetric cytoplasmic → ribonucleoprotein particle of 10–20 μm in diameter, composed of two subunits complexed with Mg^{2+}, which is the site of → messenger RNA binding, its decoding, and polypeptide synthesis (→ translation). The attached, mostly globular ribosomal proteins possess mainly structural functions. The small subunit (in bacteria: 30S; in eukaryotes: 40S) contains a single → ribosomal RNA chain (prokaryotes, plants, and chloroplasts: 16SrRNA together with 21 proteins; eukaryotes including man: 18SrRNA with some 33 proteins) and binds the mRNA (see → Shine-Dalgarno sequence). The proteins of the small subunit are called L (for "*l*ight", coming from the lighter subunit, as inferred from the sedimentation behaviour of this subunit in sucrose gradient centrifugation). The large subunit (prokaryotes: 50S; eukaryotes: 60S) is a complex of a single large rRNA chain (prokaryotes: 23SrRNA; plants and mitochondria: 25SrRNA; animals: 28SrRNA), one or two small rRNAs (prokaryotes: 5S; eukaryotes: 5S and 5.8 S), and more than 30 (prokaryotes) and 49 (eukaryotes) ribosomal proteins (l-proteins, for "*l*arge"). The rRNAs form a scaffold structure, onto which the proteins are bound. Ribosomal RNAs possess a complex structure. In addition to many conventional tertiary structures, they contain novel motifs such as the socalled A-minor motif, in which an adenosine forms hydrogen bonds with base pairs in the → minor groove of adjacent DNA helices. This interaction stabilizes the binding of → transfer RNA (tRNA) in the decoding center of the 16SrRNA as well as the binding of the 3'- terminal adenosine of the tRNA to the 23SrRNA within the *peptidyl transferase* (PT) center. This PT center is located in the large subunit, and the 23SrRNA catalyses the peptide bonding. The ribosomal PT activity is an example for a naturally occurring → ribozyme with amino acid polymerizing function.
Figure see page 954.

Ribosome attachment site: See → Shine-Dalgarno sequence.

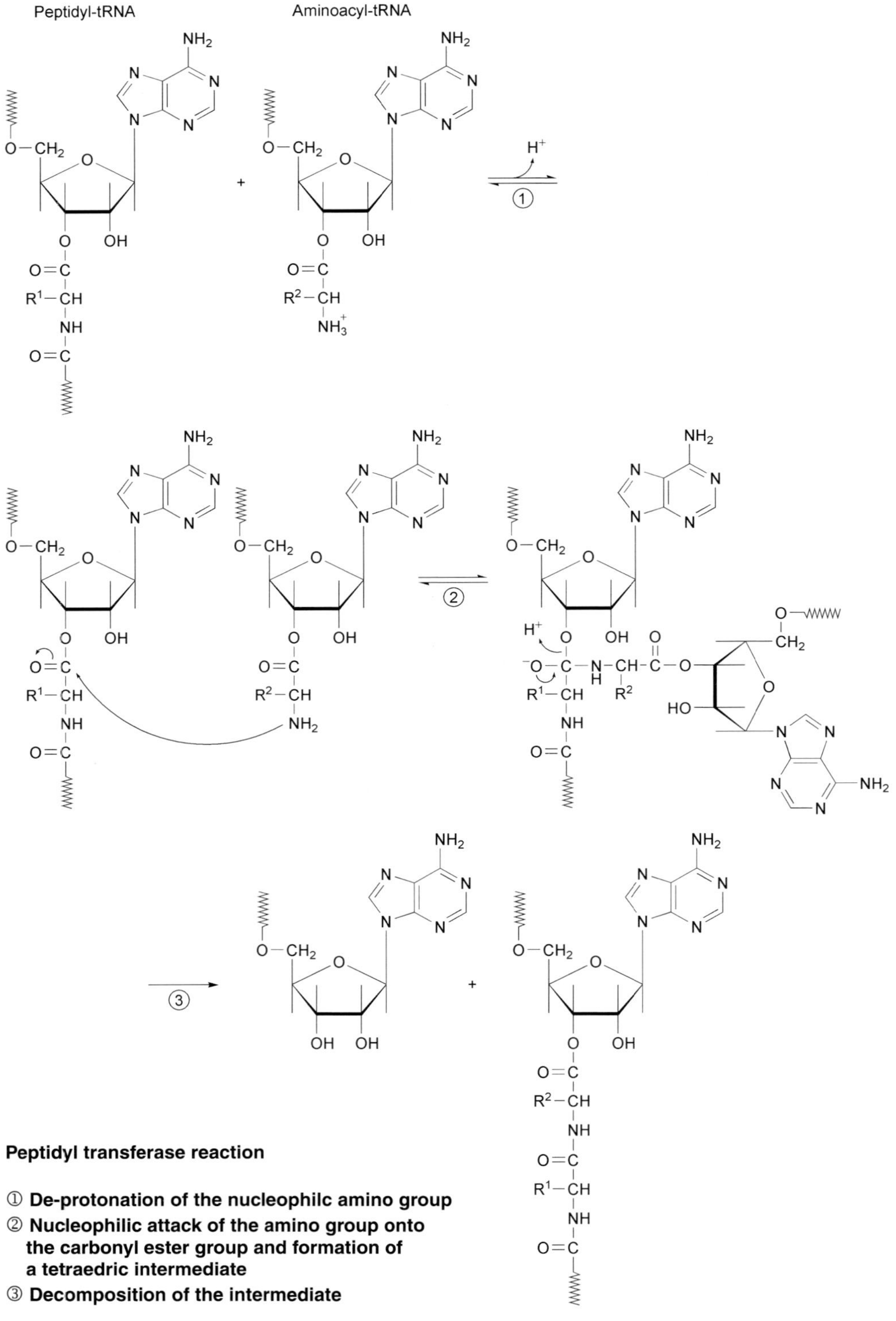

Peptidyl-tRNA   Aminoacyl-tRNA

Peptidyl transferase reaction

① **De-protonation of the nucleophilc amino group**
② **Nucleophilic attack of the amino group onto
the carbonyl ester group and formation of
a tetraedric intermediate**
③ **Decomposition of the intermediate**

Ribosome-binding site: See → Shine-Dalgarno sequence.

Ribosome display (RD, *in vitro* ribosome display, polysome display, polysome selection): An *in vitro* technique for the exposure of a nascent, correctly folded protein on the surfaces of → ribosomes in cell-free lysates, in which the exposed protein is physically coupled to its → messenger RNA (i.e. the protein does not leave the ribosome at the end of translation, and the mRNA does not dissociate from the ribosome-protein complex). In short, an → expression vector is first constructed, that contains a → promoter (e.g. a → T7 RNA polymerase promoter), 3'downstream of this promoter a → translation initiation sequence (for the 30S-*E.coli* translation system: a → Shine-Dalgarno box; for the rabbit reticulocyte or wheat germ system: an optimized → Kozak sequence), then the gene of interest (usually a PCR product) cloned into the cloning site, followed by a socalled tether or spacer sequence without any → stop codon. A stop codon would be recognized by termination factors (that are also present in every *in vitro* translation system), which then would lead to the hydrolysis of the nascent polypeptide and the dissociation of the ribosomal complex with subsequent release of the messenger RNA and the failure of the experiment. The function of the tether sequence is to keep the important part of the nascent protein sufficiently far away from the ribosomal synthesis tunnel, so that it can fold correctly and be exposed for the binding of e.g. ligands. During its synthesis, the nascent protein is first guided through the synthesis tunnel of the ribosome and then released from it at the end of the tunnel opposite to the peptide synthesis centre. The folded protein remains still attached to its peptidyl-tRNA, which in turn is linked to the messenger RNA and hence the ribosome via → codon-anticodon interactions. Each protein is therefore coupled to its messenger RNA. Translation is then stopped by cooling on ice, and the ribosome complex stabilized by increasing the magnesium concentration.

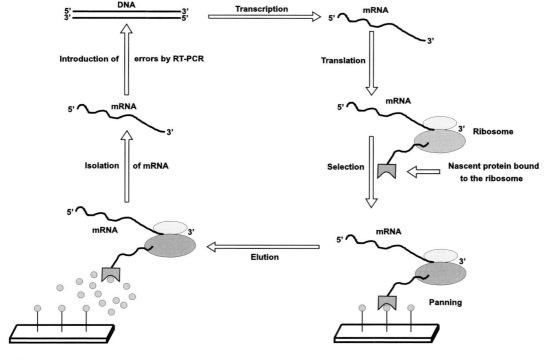

Ribosome display

The exposed protein can now be used for molecular selection (e.g. binding to surface-immobilized ligands), and the corresponding mRNA of potential candidates can directly be used for sequencing. To that end, isolated mRNA is reverse-transcribed to cDNA, and the cDNA further amplified by conventional → polymerase chain reaction techniques. Ribosome display is experimentally relatively simple, avoids → transformation protocols, and can be coupled to e.g. → DNA shuffling or → staggered extension process techniques for more efficient molecular selection. See → differential display,→ phage display, → protein complementation assay.

Ribosome-*i*nactivating *p*rotein (RIP): One of a series of animal or plant N-glycosidases that inactivate the 60S subunit of eukaryotic → ribosomes by cleaving the N-glycosidic bond of specific adenines of 28S rRNA (e.g. adenine-4324 in rat liver ribosomes as an *in vitro* model). The 60S subunit can then no longer bind → elongation factor EF-2 which leads to a stop of protein synthesis. Some of these proteins are active against fungal, but not plant ribosomes.
RIPs can be grouped into two classes. Type 1 RIPs are unusually stable, single-chain proteins, frequently glycoproteins that are synthesized as a prepro-protein and activated by endopeptidase cleavage. Examples are dianthin (from *Dianthus caryophyllus*, carnation), tritin (from *Triticum aestivum*, wheat) and asparin (*Asparagus officinalis*, asparagus). Type 2 RIPs are proteins with a galactose-binding lectin domain that allows them to enter cells through an interaction with galactosyl-terminated receptors of the cell membrane. Among the type 2 RIPs are ricin (from *Ricinus communis*, castor bean), abrin (from *Abrus precatorus*, jequirity bean), and viscumin (from *Viscum album*, mistletoe). RIPs possess anti-viral, cytotoxic and abortifacient activities. The corresponding genes are transferred to target plants to specifically increase their resistance towards pathogenic fungi.

Ribosome-*i*nactivating *p*rotein assay (RIP assay): A technique for the detection of substances that interfere with the → translation of distinct → messenger RNAs on isolated ribosomes (see → *in vitro* translation). A ribosome-inactivating protein (e. g. ricin or lectin of *Viscum album*, mistletoe), that hydrolyzes an N-glycosidic bond in eukaryotic 28SrRNA and thereby interrupts the elongation step of translation, is added to a → coupled transcription-translation system containing a → luciferase → reporter gene. Whereas the control ribosomes without RIP unabatedly produce luciferase and light (given all necessary compounds are present), the photon emission is drastically reduced in the presence of the toxin, whose effectivity can therefore be quantified.

Ribosome loading: The number of ribosomes associated with a single→ messenger RNA molecule. If ribosome loading is weak, then only one or few ribosomes are loaded onto the message, which indicates low translational efficiency ("under-translation"). In contrast, an mRNA loaded with many ribosomes (→ polysomes) is efficiently translated into protein. Both states can be separated by → sucrose density gradient centrifugation. See → translation state array analysis.

Ribosome *n*ascent-chain *c*omplex (RNC): The multiprotein machine consisting of the → ribosome and a newly synthesized protein, that remains associated transiently with the ribosome and (at least) two → chaperone subunits which in turn easily bind to the ribosome ("ribosome-associated complex, RAC"). The chaperones assist in the correct folding of the nascent protein.

Ribosome recognition sequence: See → Shine-Dalgarno sequence.

Ribosome *r*ecycling *f*actor (RRF): A conserved protein (probably present in all organisms except Archaea), that catalyses the last step in protein synthesis, the disassembly of the posttermination complex (containing the 70S ribosome, bound → messenger RNA, an empty A-site, and a deacetylated → *t*ransfer RNA (tRNA) in the P-site. This process is also called ribosome recycling. RRF consists of basically two domains, the three-helix bundle domain I and the three-layer ß- α-ß sandwich domain II. The overall three-dimensional structure of RRF almost completely mimics a tRNA molecule.

Riboswitch: A special configuration of a specific region 5'-upstream of the coding parts of a → messenger RNA (mRNA), that is able to recognize and bind a small molecular weight metabolite (e.g. a vitamin). If bound, the mRNA-metabolite complex prevents → translation of the mRNA. For example, the thiamine derivative TPP binds to one of several possible conformations of the riboswitch in mRNAs from genes controlling synthesis of B1 vitamins, and the complex TPP-mRNA arrests translation of the message. This represents an example of a negative feed-back regulation on the level of the transcript: if too much thiamine is present in the cell, the mRNA for its synthesis is inactivated.

Ribothymidine: See → rare base.

Ribotype: A formal intermediate between → genotype and → phenotype, comprising all ribonucleic acids of a cell at a given time (i. e. thousands of → messenger RNAs, → transfer RNAs, → ribosomal RNAs, → small nuclear RNAs).

Ribotyping: A variant of the conventional → restriction fragment length polymorphism analysis. In short, genomic DNA is isolated from the organisms to be compared (e.g. A and B), and → primers specific for conserved regions at the 3' and 5' end of e.g. 16S → rDNA are used to amplify the partial or complete 16S rRNA gene sequence. Then the → amplicon is cut with appropriate → restriction endonuclease(s), and the products are electrophoretically separated and stained with → ethidium bromide. The resulting patterns reflect the distribution of restriction sites on the 16S rRNA gene, and differences between genome A and B are a consequence of gain or loss of such sites. Ribotyping is used to detect inter- and intraspecific → polymorphisms. See → amplified ribosomal DNA restriction analysis.

Ribozyme: An RNA molecule with the enzymatic properties of a sequence-specific endoribonuclease (→ RNase), that catalyzes the cleavage of single-stranded RNA substrates (e.g. → small nuclear RNAs that are associated with → small nuclear ribonucleoproteins and function in → pre-messenger RNA splicing, or the → intron of the → ribosomal RNA precursor of *Tetrahymena*, that excises itself from the large precursor in a process called self-splicing, compare → splicing). The ribozymes also catalyze the joining of the trimmed RNA fragments and peptide bond formation (as e.g. the ribosomal peptidyl transferase center) and additionally possess esterase activity. The *Tetrahymena* ribozyme can also ligate multiple oligonucleotides aligned on a template strand to yield a fully complementary daughter strand (RNA-catalyzed RNA replication). Ribozymes are tools for the physical → mapping of RNA, the study of the secondary structure of RNAs and their in vitro metabolism, and may be used for → RNA sequencing. Ribozymes can also be engineered to bind to target RNA sequences, that are cleaved and thereby inactivated. See for example → CUCU ribozyme, → hairpin ribozyme, → hammer head ribozyme, → RNase P, also → ribozyme auto-cleavage vector.
Figure see page 958.

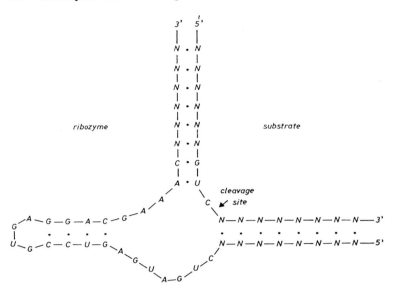

Ribozyme auto-cleavage vector: A → plasmid cloning vector that produces large quantities of specific RNA molecules by → ribozyme auto-cleavage. The vector contains a → cloning site located between two sequences encoding two ribozymes. RNA transcribed *in vitro* (→ *in vitro* transcription) using the circular supercoiled plasmid vector initially contains both sequences from the cloned foreign DNA and the ribozymes. Large amounts of RNA containing exclusively sequences transcribed from the insert DNA are subsequently generated by self-cleavage catalyzed by the ribozymes.

Ribozyme-PEI complex: The non-covalent and reversible association of a → ribozyme molecule with synthetic low molecular weight *poly*ethylene*i*mines (PEIs). Such complexes are effective → transfection carriers for ribozymes (and DNA), because the ribozyme is (1) compacted by PEI such that the resulting small colloidal particles can easily be taken up by endocytosis, and (2) the ribozyme is protected from degradation by intracellular → ribonucleases, since PEIs are acting as an efficient buffer system ("proton sponge effect"). Ribozyme-PEI complexes are therefore used to protect exogenous ribozymes from intracellular degradation, so that they can be used to silence specific genes. See → ribozyme targeting.

Ribozyme targeting: The development of specially designed synthetic → ribozymes for the silencing of specific genes, their introduction into target cells (see → ribozyme-PEI complex) and their action on the → messenger RNA encoded by the target gene. Compare → gene targeting. See → antisense RNA, → co-suppression, → quelling, → RNA interference.

Ricin: A heterodimeric, extremely toxic protein of castor bean (*Ricinus communis*), that consists of a receptor-binding lectin-like 32 kDa B subunit or B-chain (with two galactose-binding sites interacting with cell surface receptors) and a ribosome-inactivating subunit A or A-chain, both linked together by a disulfide bond. The 30 kDa A-chain is partly organized into α-helices and partly into ß-sheet structures, and exerts its ribosome-inactivating effect by removing an adenosine residue in an exposed loop of the 28S RNA of the large ribosomal subunit (without cleavage of the RNA backbone). This adenosine is involved in interaction of the large ribosomal subunit with elongation factors, and therefore its removal interrupts protein synthesis. Ricin bound to cell surface receptors is taken up by endocytosis (partly from coated pits), and only a small fraction transported to the *trans*-Golgi network and backwards through the Golgi apparatus and the intermediary compartment to the endoplasmic reticulum. Only relatively few molecules are then translocated to the cytoplasm via the socalled sec61 pathway, and escape the → proteasomes (probably by their low content of lysine; see → ubiquitin). The receptor-binding part of ricin can be replaced by other binding sequences (as e.g. → antibodies directed towards cell-surface antigens). If the latter are directed towards tumor surface antigens, the uptake of the genetically engineered ricin could lead to death of the cancer cell. Unfortunately, ricin has shown potential for → bioterrorism, fortunately at a limited scale.

RID: See → radial *i*mmuno*d*iffusion.

RIDGE: See → region of increased gene expression.

Rifampicin (3-(4-methyl-1-piperazinyl-iminomethyl)-rifamycin SV): A synthetic derivative of a naturally occurring → rifamycin → antibiotic that binds noncovalently to the β-subunit of bacterial DNA-dependent → RNA polymerase and prevents the formation of a → transcription initiation complex.

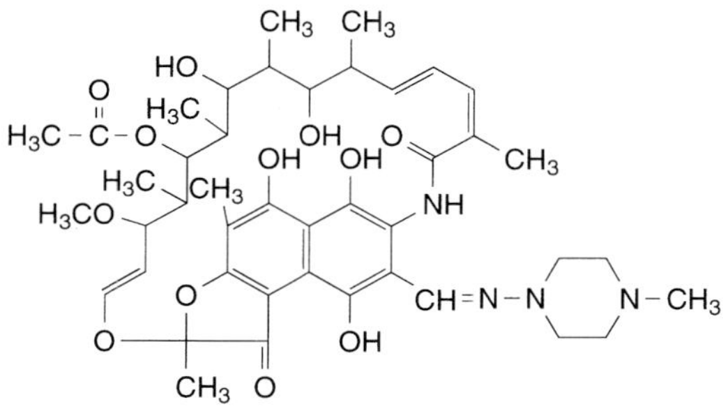

Rifamycin: Any one of a series of macrocyclic naphthalene → antibiotics produced by *Streptomyces mediterranei* that binds to the β-subunit of bacterial DNA-dependent → RNA polymerase and prevents the formation of a → transcription initiation complex. Rifamycin-resistance may occur in mutants with a modified β-subunit of RNA polymerase that is no longer able to bind the antibiotic. See also → rifampicin.

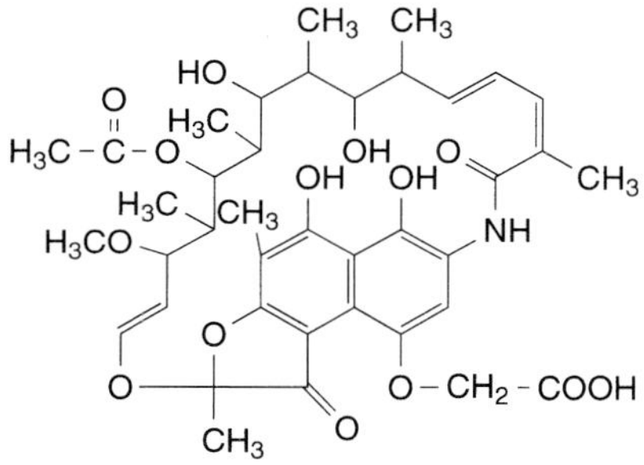

RIfS: See → reflectometric interference spectroscopy.

Right splice junction: See → acceptor splice junction.

Right splicing junction: See → acceptor splice junction.

RING finger: An amino acid motif of a series of proteins (e.g. involved in → ubiquitination, more specifically mediating ubiquitin ligase activity) with the consensus sequence H_2N-CX2CX(9–39)CX(1–3)HX(2–3)C/HX2CX(4–48)CX2C-COOH, where the cysteins and histidines represent zinc-binding domains. RING fingers fall into two classes depending on whether a cystein or a histidine occupies the fifth coordination site.

RIP:
a) See → *r*epeat *i*nduced *p*oint mutation.
b) See → *r*ibosome-*i*nactivating *p*rotein.

RIP assay: See → *r*ibosome-*i*nactivating *p*rotein assay.

Ri-plasmid (*root-i*nducing plasmid): A large → conjugative plasmid of the Gram-negative soil bacterium → *Agrobacterium rhizogenes* (Rhizobiaceae) that is the causative agent of the so-called → hairy root disease in dicotyledonous plants. After contact with susceptible plants part of this plasmid (→ T-DNA) is transferred to the host cells nuclear genome. Its expression leads to permanent proliferation of the transformed cells into hairy roots, and the accumulation of so called → opines (e.g. → agropine, cucumopine, mannopine). The transferred DNA from *A. rhizogenes* strains that produces agropine, is organized in two sections, the T_L (left) and T_R (right)-DNA. The T_R-DNA carries the genes encoding the opine and the plant hormone auxin. The T-DNA of some Ri-plasmids does not code for auxins but still produces the hairy root phenotype.
Tissue transformed with *Agrobacterium rhizogenes* readily regenerates into plants that synthesize opines. Compare → Ti-plasmid. See → crown gall.

RIPping: The procedure for the detection of → *r*epeat-*i*nduced *p*oint mutations (RIPs). See → RAPping.

RISC: See → RNA-induced silencing complex.

RISH:
a) See → radioactive *in situ* hybridization.
b) See → *R*NA *in situ* hybridization.

RL element: See → *r*etroviral-*l*ike element.

RLGS: See → *r*estriction *l*andmark *g*enomic *s*canning.

RLM-RACE: See → RNA ligase-mediated rapid amplification of cDNA ends.

R-loop: See → D-loop.

R-loop mapping: See → D-loop mapping.

***Rluc* gene:** A gene from the anthozoan coelenterate sea pansy (*Renilla reniformis*), encoding the 31 kDa monomeric enzyme → luciferase, which catalyzes the oxidative conversion of its substrate coelenterazine to coelenteramide and CO_2 with concomitant emission of blue light at 480 nm. The *Rluc* gene is used as a → reporter gene and → co-reporter gene in → cotransfection experiments. Compare → *luc* gene, → *lux* gene.

RMDD: See → cDNA-AFLP.

R-M system: See → restriction-*m*odification system.

RNA (*ribonucleic acid*): A mostly single-stranded → polynucleotide characterized by its sugar component (→ ribose) and by the presence of the pyrimidine → uracil (instead of → thymine in → DNA). Single-stranded or also double-stranded RNA are constituents of many viral genomes. In pro- and eukaryotic organisms, RNA serves very different functions. It mediates information flow (e.g. → transfer RNA, → messenger RNA), has enzymatic functions (e.g. → ribozyme), and serves as structural backbone for subcellular particles (e.g. → ribosomal RNA). The cellular RNAs can be classified into three main groups: ribosomal RNA (rRNA; about 80-90% of the total cellular RNA), transfer RNA (tRNA; about 6-8%) and messenger RNA (mRNA; usually less than 2%). In cells RNA probably never occurs free, but is complexed with proteins forming → ribonucleoprotein particles. Such RNPs also contain a great number of specific small RNAs which do not belong to any of the main groups described above (see for example → adjacent hairpin RNA, → ambisense RNA, → amplified RNA, → antisense RNA, → catalytic RNA, → chromosomal RNA, → complementary RNA, → degradation-resistant signal RNA, → guide RNA, → hairpin RNA, → heterogenous nuclear RNA, → intron-containing hairpin RNA, → micro RNA, → non-coding RNA, → precursor RNA, → pre-messenger RNA, → ribosomal RNA, → sense RNA, → 7SL RNA, → short hairpin RNA, → short interfering RNA, → short stop RNA, → SLRNA, → small auxin up RNA, → small cytoplasmic RNA, → small endogenous RNA, → small interfering RNA, → small non-mRNA, → small nuclear RNA, → small nucleolar RNA, → small regulatory RNA, → small RNA, → small temporal RNA, → spatial development RNA, → spliced leader RNA, → stress-response RNA, → subgenomic RNA, → TAR RNA, → telomerase RNA, → tiny RNA, → tiny expressed RNA → transacting RNA, → transfer RNA, → U-RNA, → Xist RNA). Some more important applications of RNAs include: the synthesis of RNA for → *in vitro* translation, chimeric DNA/RNA for targeted gene repair in → gene therapy, the use of → ribozymes for modulation of gene → expression, for diagnostics and for research on ribozyme/substrate interaction(s), → aptamers for specific inhibition of protein function(s), and specific double-stranded RNA for → RNAi. See also → activator RNA, → adaptor RNA, → aRNA, → cell-cycle RNA, → circulating RNA, → cRNA, → cognate RNA, → competitor RNA, → copy RNA, → decoy RNA, → degradation-resistant signal RNA, → double-stranded RNA, → fatal RNA, → intronic snoRNA, → isoacceptor tRNA, → mature RNA, → nuclear RNA, → primer RNA, → pRNA, → pyranosyl-RNA, → redox RNA, → SRNA, → tmRNA, → transitive interference RNA, → translational control RNA, → universal reference RNA, → untranslated RNA.
Figure see page 962.

RNA

RNA amplification with in vitro translation (RAWIT): A method to generate multiple copies of a specific protein encoded by a specific → messenger RNA using → in vitro translation, after the message has been amplified by a → reverse transcription polymerase chain reaction using a → primer containing a phage → promoter (e.g. from phage T7). Thus the resulting amplified cDNA may be transcribed, and its message translated in a conventional in vitro translation system (e.g. → wheat germ system).

RNA amplification with transcript sequencing (RAWTS): A rapid and sensitive method for → direct sequencing that combines the advantages of amplification of a certain → cDNA by the → polymerase chain reaction and its amplification by phage promoter-driven transcription. The method involves cDNA synthesis with oligo(dT) or an mRNA-specific oligonucleotide → primer, subsequent PCR with a primer or primer(s) containing a phage → promoter (e.g. from phage T7), transcription from the phage promoter using → T7 RNA polymerase, and → reverse transcriptase-mediated → Sanger sequencing of the transcript, which is primed with a nested (internal) oligonucleotide. RAWTS allows to amplify a specific RNA more than one billion times and may be used to detect very low-abundance mRNAs. → RNA aptamer selection, or → systematic evolution of ligands by exponential enrichment.

RNA aptamer: Any synthetic, single-stranded, 30–50 nucleotides long RNA oligonucleotide, that folds into a distinct three-dimensional configuration, thereby recognizing target molecules (e.g. peptides, proteins, or low molecular weight compounds) with affinities and specificities comparable to monoclonal antibodies, and binds to the best-fit targets. Aptamers are isolated from combinatorial oligonucleotide libraries, consisting of random sequences, by iterative in vitro selection (see → systematic evolution of ligands by exponential enrichment, SELEX). The libraries are incubated with the target molecule (e.g. a protein). After removal of non-binding compounds, the rare, but specific RNA oligonucleotide ligands are reverse transcribed and/or amplified by conventional → polymerase chain reaction techniques. RNA aptamers are susceptible to nucleases. For example, secondary structures of a particular RNA aptamer allow the specific binding of the → amino acid L-citrullin with high affinity. Point → mutations within the binding center change binding specificity (e.g. a mutant aptamer does no longer bind L-citrullin, but instead complexes L-arginine). The specificity of an aptamer can therefore be modified by targeted mutation. Aptamers can be engineered to bind antibiotics, amino acids, coenzymes and proteins, and to discriminate between different conformations of one and the same protein. See → aptamer-beacon, → aptamer chip, → DNA aptamer, → intramer, → RNA aptamer selection, → RNA modulator, → signalling aptamer.

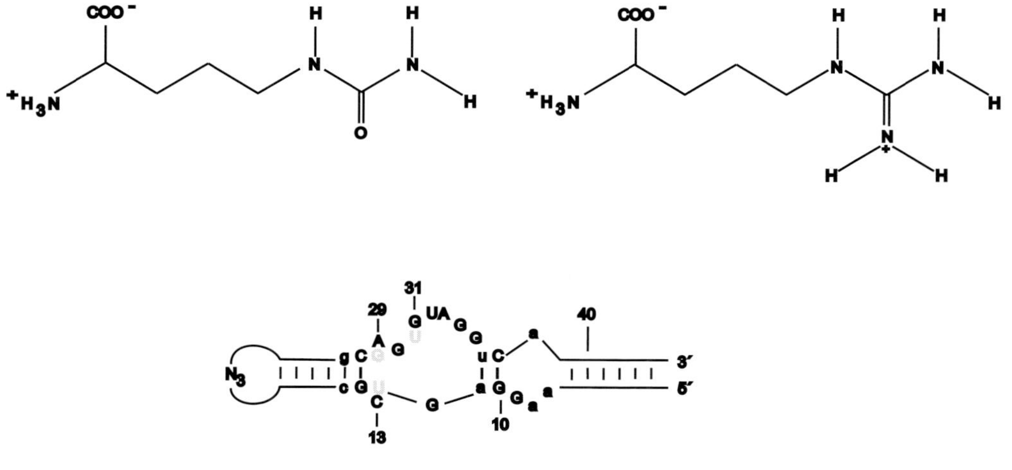

Cit44 / Arg44

RNA aptamer selection: A procedure to generate → RNA aptamers. In short, an initial DNA pool of approximately 10^{15} different random sequences of equal length is reverse-transcribed into RNAs and incubated with the protein of interest at a 1:1 molar ratio. The RNA-protein mixture is then filtered, so that RNA-protein complexes are retained, and unbound RNA molecules pass through the filter. The complexes are eluted, the protein destroyed by phenol-chloroform treatment and the pre-selected RNA again reverse-transcribed into → cDNA. Then the cDNA pool is transcribed into RNA, and the procedure repeated, i.e. the RNA incubated with the target protein, except that the protein: RNA ratio is increased to selectively amplify the RNA aptamers with highest binding affinity. RNA aptamer selection is used to e.g. find RNA inhibitors of proteins. See → aptamer, → systematic evolution of ligands by exponential enrichment.

RNA arbitrarily primed polymerase chain reaction (RAP-PCR): A method to detect differential gene expression, using a → primer of arbitrary sequence to prime both → first and → second strand cDNA synthesis from a → messenger RNA population. The mixture of cDNAs is then amplified by conventional → polymerase chain reaction, the amplification products electrophoresed and visualized by → ethidium bromide fluorescence. If RNAs from two tissues are used for comparison, tissue-specific fingerprints are produced. Differences between such fingerprints arise from differently expressed genes. Fingerprint bands of interest can be isolated from the agarose gel, cloned, and used as → expression-tagged sites. See → differential display reverse transcription polymerase chain reaction.

RNA-*b*inding *p*rotein (RNA-BP): Any protein with an → RNA recognition motif, that allows the protein to bind to specific sites of a ribonucleic acid.

RNA biochip: See → expression array.

RNA blot, RNA blotting: See → Northern blotting.

RNA caging: A process leading to the partial or complete inactivation of an RNA molecule (e.g. → messenger RNA) by the covalent linkage between the phosphate moiety of the RNA and a socalled caging group, usually photosensitive synthetic compounds as e.g. 6–bromo–4–*diazo*methyl–7–*h*ydroxy*c*oumarin (Bhc-diazo), 1–(4,5–*di*methoxy–2–*ni*trophenyl)*e*thyl (DMNPE), or O-(2–nitrobenzyl). In short, both the caging chemical and the RNA are mixed in *di*methyl*s*ulf*o*xyde (DMSO), the reaction run at room temperature for 1–2 hours, the free caging group removed by Sephadex G–50 column chromatography in DMSO, and the caged RNA precipitated by isopropanol. The dried caged RNA can then be solubilized in an appropriate buffer and injected into target cells or tissues. Caged mRNA is extremely stable and translationally inactive. However, the caging group can be removed from the RNA ("uncaging") by illumination with UV light (usually 350–365 nm wave-length) of low energy (100mJ/cm2, depending on the caging group), which leads to a complete recovery of translational activity. RNA caging allows the precise control of the translation of a specific mRNA in living cells or tissues without crtical damage. See → DNA caging.

RNA-catalyzed RNA replication: See → ribozyme.

RNA cell: See → ribocyte.

RNA chip: See → expression array.

RNA-chromophore-assisted laser inactivation: The laser-induced destruction of a specific RNA, into which an → aptamer sequence is engineered that binds *m*alachite green (MG). The substrate RNA is designed such that it contains a well-defined asymmetric internal → bulge within its duplex part (bulge sequence: 5'-ACGUAAGCAAUGAGC-3'). This bulge binds *m*alachite green (MG aptamer bulge), that produces free radicals after laser irradiation at 630 nm. Its vicinity to the aptamer allows MG to destroy the tagged RNA. If the RNA is a → messenger RNA, laser irradiation separates the 5'-cap or the poly(A)-tail from the body of the mRNA (depending on whether the aptamer is inserted into the 5'- or 3'-untranslated region). This makes the message untranslatable and unstable, and targets it to complete nucleolytic degradation (leading to a block in gene expression).

RNA decoy: Any synthetic RNA, that is designed to bind to a target protein and modulate its activity or movement in the organism.

RNA degradosome: See → degradosome.

RNA-dependent DNA polymerase: See → reverse transcriptase.

RNA-directed DNA methylation (RdDM, RNA-mediated DNA methylation, RDM): The transcriptional silencing of → transgenes by the sequence-specific hypermethylation of cytosyl residues in → promoter sequences of homologous genes (endogenous genes with high or complete sequence similarity to the transgene) induced by double-stranded RNA. For example, viral or viroid RNAs (from e.g. potato spindle tuber viroids, poty- or potexviruses, or viral satellite RNA) in the replicating double-stranded form is processed to 21–25 nucleotide small dsRNAs by the invaded cell, and these → small RNAs guide a methyltransferase complex to target gene promoters. The methyltransferase then methylates cytosyl residues at strategic positions in the promoter and induce silencing of the adjacent gene. This phenomenon is exploited in plant gene silencing. If, for example, dsRNA with homology to promoter sequences is expressed in plants, it results in methylation of the corresponding promoter and the transcriptional silencing of the adjacent gene. RdDM has been evolved to protect plants from invading viroids or viruses and acts in concert with → RNA interference. See → virus-induced gene silencing.

RNA display (RNA fingerprint): The visualization of all, or a subset of all → messenger RNA molecules of a given cell at a given time by techniques such as e. g. → differential display reverse transcription polymerase chain reaction.

RNA-driven hybridization: See → saturation hybridization.

RNA editing: A transesterification process whereby mitochondrial and chloroplast, but also nuclear mRNA is post-transcriptionally modified ("edited"), and which leads to the creation of new codons or alteration of reading frames. Editing follows a 3' to 5' polarity and either inserts or deletes uracil residues (e. g. in socalled kinetoplasts, the mitochondria of trypanosomes), replaces cytosine by uracil (e. g. in mitochondria of mosses and higher plants), or transforms an adenine into inosine (is read out as guanine), guided by base-pairing between a → guide RNA (gRNA) and the edited RNA, at least in trypanosomes. RNA editing takes place predominantly within coding sequences and may create AUG initiation codons from e.g. ACG codons (as in maize chloroplast transcripts rpl2, tobacco psbL, and mitochondrial nad1 mRNA of wheat) as well as new stop codons. Yet editing sites are also present in → introns (e.g. _nad1_, _nad2_ and _nad5_ introns of _Oenothera_ mitochondria). Only 25 editing sites occur in the entire maize chloroplast genome, whereas the transcripts of nearly all mitochondrial protein-coding genes harbor at least one editing site. An editing site sequence does not show specific features (only the immediately flanking bases are defining editing specificity). The biological role of RNA editing is not yet clear, but thought to represent an additional level at which gene expression can be regulated. Some of the editing events improve base pairing in secondary structures of the RNA and may be necessary for the correct → processing of introns. Moreover, editing may change protein function. For example, in the glutamate neuroreceptor GluR-B gene transcript an adenosine is exchanged for an inosine, which leads to the replacement of an arginine by a glutamine in the ion channel domain of the receptor. Consequence: reduced calcium penetrance. Because of RNA editing, amino acid sequences of mitochondrial or chloroplast proteins cannot be deduced from the corresponding gene sequences with certainty. Lack of editing at a single position may result in a mutant phenotype, for which reason RNA editing is considered an essential step in chloroplast RNA processing. However, differential editing (unedited and edited transcripts from the same gene) of the apolipo-protein-B mRNA produces two proteins with distinct characteristics in lipid transfer

and metabolism. The edited and unedited transcripts of the glutamate-receptor subunit gene give rise to proteins differing in calcium permeability. Since the ratio of edited to unedited transcripts varies developmentally in mammalians and trypanosomes, it may be important for the life of the organism. Compare → post-transcriptional modification, i. e. the processing of nuclear RNA. See → deletion editing, → differential editing, → insertion editing, → silent editing site, → substitution editing, → transfer RNA editing.
RNA editing website: www.rna.ucla.edu/index.html

DNA

CGACGATAGCGCCATCTGATTCGCTTAGACCTCGCCGACC
GCTGCTATCGCGGTAGACTAAGCGAATCTGGAGCGGCTGG

RNA

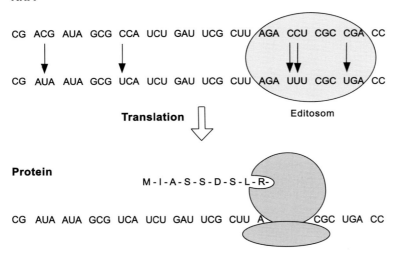

RNA - Editing

RNA fingerprint: See → RNA display.

RNA fingerprinting: See → differential display reverse transcription polymerase chain reaction.

RNA fold: The specific three-dimensional structure of a → ribonucleic acid molecule adopted in solution, that can be deciphered by e.g. X-ray crystallography. See → protein fold.

RNA gene: A sloppy laboratory slang term for any gene encoding an RNA, that is not translated into a protein. For example, genes encoding → microRNA, → non-coding RNA, → ribosomal RNA, → short hairpin RNA, → short interfering RNA, → small RNA, → small endogenous RNA, → small non-messenger RNA, → small regulatory RNA, → small temporal RNA, → tiny RNA and → tRNA are such RNA genes. However, → siRNAs are not encoded by discrete genes.

RNA genomics: See → ribonomics.

RNAi: See → RNA interference.

RNA immobilization: The attachment of → ribonucleic acid molecules onto supports. For example, RNA can be bound to gold-coated silicium wafers, onto which *di*thiobis*s*uccinimidyl propionate (DTSP) is immobilized, by interaction with the thiol endgroup of DTSP.

RNA-induced silencing complex **(RISC):** A ~ 500 kDa protein complex, consisting of an endo- and exonuclease, an RNA helicase, a homology-searching enzyme, and possibly other proteins (e.g. Argonaute proteins, a single-stranded RNase), that serves as an anchor for → *small interfering* RNAs (siRNAs). Within this complex, these siRNAs recognize homologous → messenger RNAs and guide their cleavage by the constituent nuclease. See → RNA interference.

RNA in situ hybridization **(RISH):** A variant of the conventional → *in situ hybridization* (ISH) technique for the detection of → messenger RNAs in tissue sections. First, a → probe cDNA is cloned into a → cloning site of an appropriate vector, which is flanked by → SP6 and → T7 RNA polymerase promoters. These allow to transcribe the cDNA both as → sense and → antisense RNA in the presence of the corresponding RNA polymerases and radiolabeled or biotin- or digoxygenin-labeled ribonucleotides. The probe is then hybridized to appropriately prepared target sequences in e. g. tissue sections, and hybridization detected by *in situ* hybridization detection techniques. Do not confuse with → radioactive *in situ* hybridization.

RNA *interference* (RNAi, RNA-mediated interference, double-stranded RNA-mediated messenger RNA degradation): A process of sequence-specific, post-translational gene silencing in all eukaryotic organisms, that is initiated by double-stranded (ds) RNA homologous to the silenced gene ("RNAi pathway"). In short, RNAi can be divided into two phases. In the socalled initiation phase, dsRNA is processed by the → RNase III family nuclease → Dicer to produce 21–23 nucleotide long double-stranded socalled → *small interfering* RNAs (siRNAs) with symmetric two-nucleotide 3'-overhangs for local interference (and 24–26 nucleotide long siRNAs for systemic interference). In the subsequent socalled effector phase, these siRNAs are incorporated into the multiprotein complex → RNA-induced silencing complex (RISC), that targets transcripts by base-pairing between one of the siRNA strands and the endogenous RNA (generally → messenger RNA). A nuclease associated with the RISC complex then cleaves the mRNA-siRNA duplex and thus targets cognate mRNA for destruction. Therefore, the RNAi pathway silences specific genes and interferes with gene expression. In *Caenorhabditis elegans*, the dsRNA is amplified and moves from cell to cell, causing a systemic response and ensuring a robust RNAi. RNAi represents a protection mechanism against viruses, → retrotransposons, → transposons, also → transgenes and → aberrant single-stranded RNAs. It is also involved in → heterochromatin stability (e.g. regulation of → histone H3 lysine–9 methylation) of fission yeast, or genome rearrangements in *Tetrahymena*.

RNAi has potential to engineer the specific control of gene expression and to serve as potent tool for → functional genomics. For these purposes, 21 nucleotide siRNAs with 2 nucleotides 3'-overhangs are designed for the inhibition of specific genes (i. e. for the degradation of the messenger RNAs encoded by these genes). These siRNAs can either be prepared by chemical synthesis, → *in vitro* transcription by e.g. → SP6 *in vitro* transcription system, or the digestion of long double-stranded RNA by RNase III or Dicer. The synthetic siRNAs are then introduced into target cells by → electroporation, → lentiviral vectors, → microinjection, → retroviral vectors, → transfection, or other techniques, without inducing → antiviral response. Also, animals can be fed with

bacteria, that contain plasmids with cloned siRNA-expressing genes. The siRNAs are then liberated in the digestive tract and taken up. Or siRNA-producing cassettes can be stably integrated in embryonic stem cells and transmitted in the germ-line. The design of a distinct siRNA includes selection of a region located 50–100 nucleotides downstream of the AUG → start codon of the corresponding mRNA. In this region, the sequences AA(N19)TT or AA(N21) is searched, and its G/C percentage calculated (should be 50%, but must be less than 70 and more than 30%). Then a → BLAST (using e.g. the NCBI EST database) for the nucleotide sequence fitting the above criteria is performed to ensure that only one single gene is silenced. More than one siRNA for any given target mRNA can be designed to be more effective. Also, siRNAs consisting of negatively charged → peptide nucleic acids ("gripNA") can be employed for gene silencing, since they are more resistant to nucleases and display better sequence specificity than conventional siRNAs. See → antisense RNA, → co-suppression, → quelling.
Figure see page 969.

RNA ladder: See → ladder.

RNA library: A collection of → ribonucleic acid molecules, generated by → in vitro transcription of e.g. synthetic or natural DNA cloned into an appropriate cloning → vector (as e.g. → T7–RNA expression cassette) and driven by a → T7 promoter. Such RNA libraries are prerequisites for the selection of → aptamers that specifically bind to e.g. a peptide or protein of choice. usually, the T7 promoter is located upstream of the aptamer-encoding DNA inserted between 5' and 3' stem-loop structures, that function as RNA-stabilizing motifs and additionally are necessary for correct termination of the T7 transcripts.

RNA ligand: Any RNA, that has the potential to bind to a specific protein and interfere with its function. Synthetic RNA ligands are used to modulate or block the action of target proteins *in vitro and in vivo.*

RNA ligase: See → T4 RNA ligase.

RNA *ligase-mediated rapid amplification of cDNA ends* (RLM-RACE): A technique for the generation of → cDNA libraries and the analysis of the 5'and 3'ends of the generated cDNAs. In short, RNA is first isolated from the target organism, and treated with → alkaline phosphatase to remove free 5'-phosphate groups from abundant RNAs (e.g. → ribosomal RNA or → transfer RNAs) or contaminating DNA, such that subsequent ligation of RNA → adaptors to these RNAs is impossible. This treatment leaves the → cap structure of intact 5'-ends of mRNA. Then the RNA is exposed to → *tobacco acid pyrophosphatase* (TAP) to remove the → cap structure from → messenger RNA, generating mRNAs with a 5'-monophosphate end. These ends are then ligated to 45 base long RNA adaptor oligonucleotides using → T4 RNA ligase. Then thermostable → *Thermus thermophilus* DNA polymerase, that possesses an Mn^{2+}-dependent → reverse transcriptase function with an optimal activity between 70 and 74°C, and is therefore able to melt out secondary structures in the RNA (especially frequent at higher G+C contents) is employed to produce first strand cDNA, using a → gene-specific primer and a RACE primer (see → rapid amplification of cDNA ends). This cDNA is then amplified by a → nested PCR using a → forward primer complementary to the RNA adaptor and a → reverse primers from within the target gene. This procedure results in the generation of longer products and prevents the skipping of RNA secondary structures (e.g. stable stem-loops) and premature termination by normal reverse transcriptase.

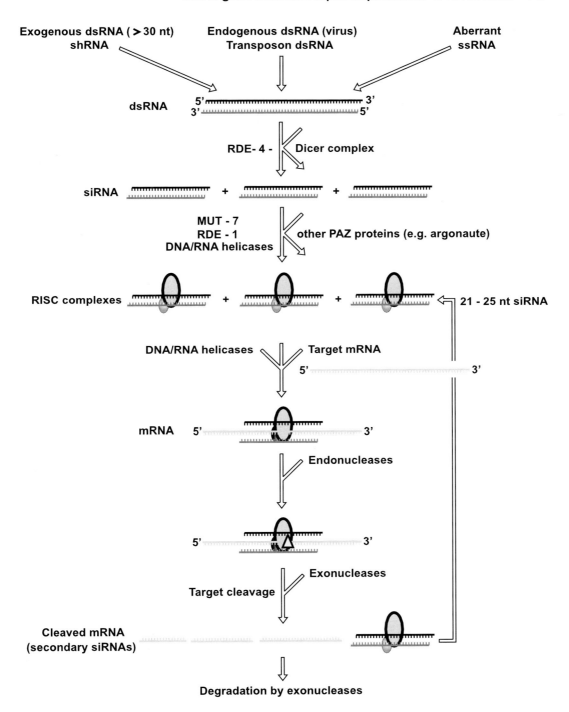

RNA interference

RNA localization: A process by which → messenger RNA (mostly in a complex with protein, as → heterogeneous nuclear ribonucleoprotein) is concentrated in specific compartments of a cell, tissues of an organ, and organs of an organism. The active process of RNA localization is more effective than protein trafficking, since a single mRNA can give rise to many protein molecules, if it is associated with the → translation machinery. For example, specific mRNAs are localized at the anterior pole of *Drosophila* embryos, maternal mRNAs at the anterior and posterior poles of the developing *Drosophila* oocyte, where the encoded proteins are essential for early development, and β-actin mRNA at the growing pole of motile chicken embryo fibroblasts, where it is required for fibroblast morphology. Localized RNAs contain cis-acting signals ("RNA zipcodes") targeting them to specific domains within the cell. These localization signals usually lie in the → 3'-UTRs of mRNAs, that are rich in secondary structures, and function as address sites for RNA-binding proteins ("*z*ipcode-*b*inding *p*roteins, ZBPs"). The actual transport of the RNA from its site of synthesis to the site of localization is mediated by both microtubular and actin cytoskeleton elements.

RNA machine: A generic name for any intracellular complex of several to many RNAs, that interact physically and cooperate synergistically. For example, → ribosomes are such RNA machines, in which different → ribosomal RNAs act in concert with → transfer RNAs and the RNA core of peptidyltransferase to synthesize proteins. See also → protein machine, → spliceosome.

RNA maturation: See → post-transcriptional modification.

RNA-mediated DNA methylation: See → RNA-directed DNA methylation.

RNA-mediated interference: See → RNA interference.

RNA-mediated silencing: See → homology-dependent resistance.

RNA melting: See → melting, also → denaturation.

RNA mimic: Any oligoribonucleotide, that carries a methyl group at the 2'position of its ribose and resembles RNA in its binding affinity to complementary RNA. RNA mimics are more resistant towards nucleases than their unmodified counterparts. See → DNA mimic.

RNA-modifying enzyme: Any one of a group of enzymes that introduce modifications in RNA (e.g. → reverse transcriptase, → RNases, → RNA polymerases).

RNA modulator: Any synthetic ligand-binding → RNA aptamer isolated from a combinatorial RNA library, that is synthesized from a PCR-amplified synthetic DNA pool by →*in vitro* transcription (using e.g. a T7 promoter) and selected by a functional assay (e.g. binding to a specific domain of the target protein). The RNA modulator-encoding sequences are integrated into an expression cassette and transfected into target cells, where they are highly expressed. The resulting *intra*cellular apta*mers* (→ "intramers") bind to specific domains of target proteins (e.g. receptor proteins) and block their function. RNA modulators have potential for therapeutic regulators. See → aptamer-beacon, → aptamer chip, → DNA aptamer, → RNA aptamer, → signalling aptamer.

RNAP: See → RNA *p*olymerase.

RNA patterns method: A technique for the visualization of differential gene expression between different cells, tissues, or organs, that circumvents amplification of rare → *m*essenger RNAs (mRNAs) by conventional → *p*olymerase *c*hain *r*eaction (PCR), and therefore most probably presents realistic relative abundances of mRNAs in the sample. In short, total RNA from two (or more) samples is first isolated, → poly(A)+-mRNA extracted and reverse transcribed with Moloney murine leukaemia virus (MMLV) → reverse transcriptase, short and 5'-^{32}P-labeled oligonucleotide → primers (usually 9mers, that are variable in the three bases at their 3'-ends), three dNTPs (e.g. dATP, dCTP, and dGTP) and a chain terminating nucleoside triphosphate (e.g. ddUTP, see → chain terminator). Thereby several short, labeled → cDNA products are generated, that are characteristic for a specific mRNA population, and separated by → denaturing polyacrylamide gel electrophoresis and visualized by → autoradiography. As a result, detailed RNA fingerprints ("RNA patterns") with 10–60 individual bands of the two (or more) samples are produced, that can be compared. Differentially synthesized cDNAs can be excised from the gel and either amplified by PCR, cloned and sequenced, or directly sequenced (see → chemical sequencing). See → adapter-tagged competitive PCR, → enzymatic degrading subtraction, → gene expression screen, → linker capture subtraction, → module-shuffling primer PCR, → preferential amplification of coding sequences. Compare → massively parallel signature sequencing, → microarray, → serial analysis of gene expression.
Figure see page 972.

RNA-PCR: See → reverse transcription polymerase chain reaction.

RNA plasmid: Anyone of a series of double-stranded RNAs of yeast and plant mitochondria, that are replicating autonomously (i. e. independently of the → mitochondrial DNA). In maize, RNA plasmids preferentially occur in mitochondria of socalled S-type cytoplasms (i. e. mitochondrial types conferring male sterility, when present in specific nuclear genetic backgrounds).

RNA polymerase:
a) (DNA-dependent RNA polymerase, nucleoside triphosphate: RNA nucleotidyltransferase, transcriptase; EC 2.7.7.6; RNAP): An enzyme catalyzing the formation of RNA using the → antisense strand of a DNA duplex as template. In prokaryotes two types of RNA polymerases exist, one synthesizing the RNA primer (→ primer RNA, see → RNA primase) necessary for DNA → replication, the other transcribing structural, ribosomal and transfer RNA genes (see also → sigma factor). In eukaryotes, three distinct nuclear RNA polymerases with different template specificities transcribe → rDNA (polymerase I,A), → tDNA, 7S-DNA, snDNA and 5S-DNA (polymerase III,C) and the → structural genes (polymerase II,B), and can be discriminated by their different sensitivity towards → α-amanitin. See → T7 RNA polymerase, → T3 RNA polymerase.
b) RNA-dependent RNA polymerase: An enzyme which also catalyzes the formation of RNA, but uses RNA as template, for example Q-beta → replicase (see → Q-beta).

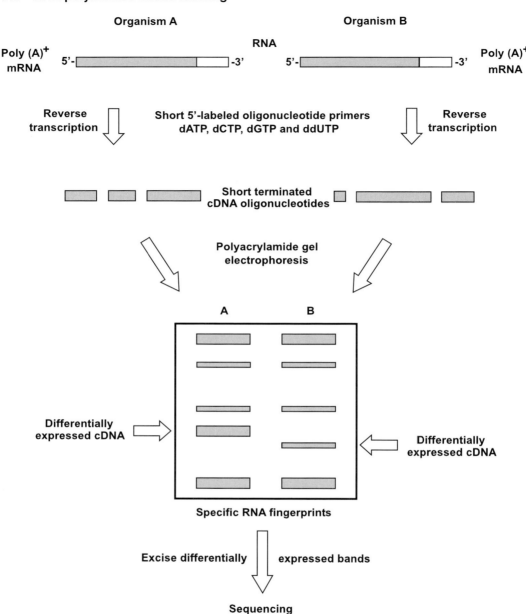

RNA patterns method

RNA polymerase-based labeling: A procedure to generate large amounts of specific, homoge-neous, biologically active and radiolabeled RNA *in vitro*. In short, a template DNA is first cloned into the → polylinker of a plasmid → expression vector. This vector contains a phage → promot-er (e.g. → SP6 RNA polymerase promoter) located 5-8 bp upstream of this polylinker, allowing the transcription of the cloned insert. Transcription is mediated by the addition of specific DNA-dependent → RNA polymerases (e.g. from *Salmonella typhimurium* phage SP6, → coliphages

T3, T5 or T7). If radiolabeled nucleotides are included in the reaction mixture, the produced RNA will be radioactive and can be used in nucleic acid → hybridization experiments. A strand-specific RNA → probe is generated by this procedure, which does not need to be denatured prior to hybridization.

RNA polymerase binding *protein (RAP):* A nuclear protein that binds strongly to DNA-dependent → RNA polymerase II(B) of mammals and stimulates the activity of the enzyme.

RNA polymerase chain reaction: See → reverse transcription polymerase chain reaction.

RNA polymerase holoenzyme: The complex between → RNA polymerase II (B) and its → mediator.

RNA polymerase I trap (RNA polymerase I trapping center): See → tandem promoter.

RNA polymerase II complex (RPC): A relatively stable multi-protein complex consisting of → RNA polymerase II, → general transcription factors (e.g. TFIID, TFIIA, TFIIB, TFIIF, TFIIE, and TFIIH, in the order of their binding to the complex) and and associated factors (e.g. → activators), required to initiate → transcription from RNA polymerase II → promoters.

RNA precursor (precursor RNA): Any primary transcript, that contains sequences not necessary for its final function. These sequences are post-transcriptionally removed or modified to yield the mature and functional RNA. For example, the → pre-messenger RNA is such an RNA precursor, that is trimmed to the exportable messenger RNA by excision of the → introns, ligation of the → exons, → capping at its 5'-end and → polyadenylation at its 3'-end. See → post-transcriptional modification.

RNA primase (primase): A DNA-dependent → RNA polymerase from *E. coli* catalyzing the polymerization of so-called → RNA primers needed in DNA replication for the synthesis of the → lagging strand. Among other proteins, RNA primase is a part of the so-called → primosome (primosome complex).

RNA primer (primer RNA): A short oligonucleotide, about 10-15 nucleotides in length, that is synthesized by → RNA primase, anneals to the → lagging strand (see → replication) in intervals of 200 bp, and serves as a starter molecule (→ primer) in the → DNA polymerase-catalyzed synthesis of an → Okazaki fragment. After the synthesis of adjacent Okazaki fragments the primer is removed by the repair function of DNA polymerase (5' → 3' exonuclease) that replaces it by DNA. The → gaps between the DNA segments are then covalently closed by → DNA ligase. See also → primosome.

RNA priming (Gubler-Hoffmann procedure): A modification of the conventional → second strand synthesis in → cDNA cloning procedures. For RNA priming the mRNA template is partially removed from the → first strand-mRNA hybrid by controlled action of → RNase H or alkali which leaves short RNA stretches with free OH groups. These RNAs serve as → primers for → second strand synthesis, catalyzed by → reverse transcriptase.
Figure see page 974.

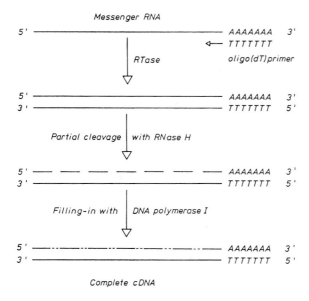

Messenger RNA

5' ———————————————— AAAAAAA   3'
                          ← TTTTTTT
                      oligo(dT)primer

RTase

5' ———————————————— AAAAAAA   3'
3' ———————————————— TTTTTTT   5'

Partial cleavage │ with RNase H

5' — — — — — — — AAAAAAA   3'
3' ———————————————— TTTTTTT   5'

Filling-in with │ DNA polymerase I

5' —-·———————-·—-·— AAAAAAA   3'
3' ———————————————— TTTTTTT   5'

Complete cDNA

RNA-PRINS: See → primed *in s*itu labeling.

RNA probe: Any RNA, that is used as a → probe in hybridization experiments. Usually such RNA probes are generated by *in vitro* transcription of cloned genes by → T7 RNA polymerase and labeled by either radioisotopes (e.g. ^{32}P or ^{125}I), → fluorochromes, or → biotin. The RNA probes exist in plus-sense, minus-sense and anti-sense versions.

RNA processing: See → post-transcriptional modification.

RNA profiling: The isolation, separation and visualisation of (preferably) all → ribonucleic acids (RNAs) of a cell, tissue, organ or organism.

RNA-protein hybrid system: A variant of the conventional → three-hybrid system designed for the detection of RNA-protein interaction(s) *in vivo*, in which a hybrid RNA molecule (RNA bait) links two hybrid proteins ("three component system"). Hybrid protein A contains an RNA-binding domain 1 fused to a → DNA-*binding domain* (DBD), hybrid protein B carries a different RNA-binding domain 2 fused to a → transcriptional → *activation domain* (AD). Since the hybrid RNA contains recognition sequences for both RNA-binding proteins, it combines ("bridges") both, and thereby activates a → reporter gene, whose activity can easily be assayed. In a specific case, hybrid protein A is composed of a LexA DNA-binding protein fused to the coat protein of → bacteriophage MS2. Hybrid protein B consists of the *GAL*4 protein transcription activation domain, fused to the RNA-binding domain of a protein, whose RNA-binding properties are to be detected (prey). The hybrid RNA in turn contains two MS2 RNA-binding sites, one of which is recognized by the MS2 coat protein of the protein A complex. The interaction between RNA and RNA-binding domain of the prey brings the AD into close proximity to the promoter of the reporter gene, thereby inducing its transcription in the *Saccharomyces cerevisiae* host. The DNA sequences of all these components are cloned into separate → expression vectors and cotransformed into a suitable yeast strain, that carries two integrated reporter genes (*HIS*3 and *lac*Z) downstream of LexA-binding sites. These sites may be present as multimers (e. g. four in the *HIS*3 and eight in the *lac*Z promoter). It also owns the gene encoding a LexA DNA-binding domain – MS2 coat protein fusion, which specifically recognizes and binds to the LexA binding sites upstream of the promoters driving the reporter genes. If the prey protein and bait RNA interact, they assemble at the LexA binding site: the reporter gene is activated.

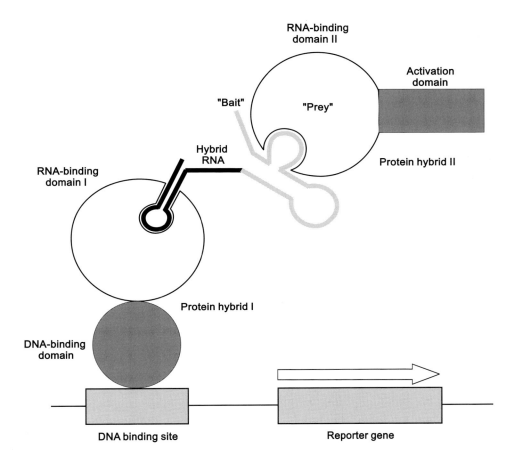

This hybrid system allows to detect even transient or weak interaction(s) between RNA and protein(s), as they occur during e. g. RNA processing, → translation and RNA virus assembly, and permits the identification of new RNA-binding proteins, the isolation of different RNAs that bind to the same protein, the design of synthetic RNA ligands with selective affinity for specific proteins (see → pharmacogenomics) and the synthesis of inhibitors of RNA-protein interactions. See → dual-bait two-hybrid system, → interaction trap, → LexA two-hybrid system, → one-hybrid system, → reverse two-hybrid system, → split-hybrid system, → split-ubiqutin membrane two-hybrid system, → three-hybrid system, → two-hybrid system.

RNA recognition motif (RRM): A more or less conserved amino acid sequence motif of proteins binding to specific address sites in ribonucleic acids. For example, the human proteins AUF1 (= RRM 2), 2UP1, PABP and related proteins contain phenylalanine at specific positions, that – in concert with flanking amino acid residues – contact specific adenine moieties on the target RNA.

Most of the interactions between RNA and protein, however, are non-base-specific stacking and van der Waals interactions rather than base- or sequence-selective contacts. Or, as an example from plants, in a protein with an RRM, a glycine-rich stretch binds to the address site 5'-AAAATATCT – 3' in the promoter of the *At* (*Arabidopsis thaliana*) *grp7* gene, that specifies circadian rhythms.

RNA replication: See → replicase, → replication.

RNA-responsive element binding protein: Any protein capable of binding metal ions, that as a protein-metal complex recognizes a → stem-loop structure ("regulatory element") in the 5'-untranslated regions of specific → messenger RNAs, binds there, and influences → translation of the message.

RNA-RNA interaction: Any interaction between two (or more) identical or different RNA molecules, usually initiated by loops and/or single-stranded stretches of the RNAs. For example, the base-pairing between the → codon of a → messenger RNA and the → anticodon of a → transfer RNA represents such an RNA-RNA interaction.

RNase (*ribonuclease; EC 3.1.4.22*): Any enzyme that catalyzes the cleavage of → phosphodiester bonds (phosphodiesterase) in RNA. See → RNase A and the following entries, compare → ribozyme.

RNase (*ribonuclease*) from *Bacillus cereus*: An enzyme catalyzing the cleavage of 3' → phosphodiester bonds adjacent to → pyrimidine residues and used in → RNA sequencing analysis.

RNase A (*ribonuclease A, pancreatic ribonuclease; EC 3.1.27.5*): An enzyme from bovine pancreas which catalyzes the cleavage of → phosphodiester bonds between → pyrimidines and adjacent nucleotides in RNA. It is the major active component of bovine pancreatic ribonuclease preparations and is predominantly active on heteropolymers; pure purine polymers (e.g. poly[A]) are attacked only by high enzyme concentrations. See also → RNase A mismatch detection. Compare → RNase S.

RNase A mismatch detection (ribonuclease A technique, RNase A technique): A method for the detection of base mismatches in RNA-DNA- or RNA-RNA hybrid molecules, using the ability of → RNase A to recognize such mismatches and to catalyze the cleavage of → phosphodiester bonds in the RNA at the mismatch site. First, the RNA is radioactively labeled (e.g. → antisense RNA resulting from → in vitro transcription) and hybridized to the denatured DNA or RNA of interest. Alternatively, labeled DNA probes may be hybridized to cellular RNA. Then the DNA RNA or RNA-RNA hybrids are exposed to RNase A, that cleaves the RNA at the mismatch, and subsequently electrophoresed through denaturing → sequencing gels. After autoradiography, a fully base-paired complex will show up as a single band, whereas mismatch-induced cleavage(s) will lead to the appearance of bands of lower molecular weight.
Figure see page 977.

RNase A technique: See → RNase A mismatch detection.

RNase B (*ribonuclease B*): An enzyme from bovine pancreas which catalyzes the cleavage of → phosphodiester bonds in RNA. This enzyme is only a minor component of bovine pancreatic ribonuclease preparations.

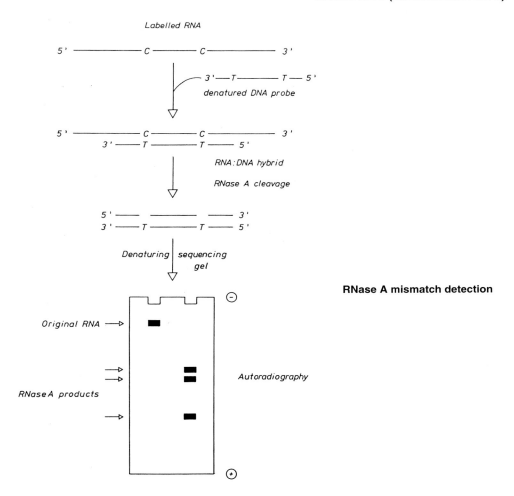

Labelled RNA

RNase A mismatch detection

RNase cleavage: See → RNase cleavage assay.

RNase cleavage assay (RCA, RNase cleavage): A technique for the detection of single base pair exchanges and → mismatches in relatively long regions of DNA (> 1 kb). In short, the target DNA is first amplified in a conventional → polymerase chain reaction, using target-specific → primers that carry 20 bp → SP6 and → T7 RNA polymerase → promoters at their 5' ends. The amplified target is then transcribed *in vitro* by adding the corresponding phage RNA polymerase, and hybridized to form double-stranded RNA. If wild type and mutant transcripts are hybridized, mismatches occur, that can be recognized by added → RNase I and → RNase T₁. Both enzymes cleave the basepair mismatches on both strands. The double-stranded cleavage products are then separated on a native → agarose gel and visualized by → ethidium bromide fluorescence under UV light. Fragments occuring only in the mutant are proof for mismatches, which can be mapped on the target DNA. See → RNase A mismatch detection.

RNase CL 3 (ribonuclease CL 3): An enzyme from chicken liver that catalyzes the cleavage of → phosphodiester bonds in RNA predominantly at Cp/N bonds to produce RNA fragments with 3'-terminal cytidine phosphate. Since Ap/N, Gp/N and Up/N bonds are cleaved to a minor extent only, RNase CL3 is used as a "minus U"-specific RNase in → RNA sequencing procedures.

RNase D (*ribonuclease* D): An enzyme from *E. coli* catalyzing the 3' exonucleolytic removal of nucleotides from an RNA chain, possibly participating in → transfer RNA → processing.

RNase E (*ribonuclease* E): An enzyme from *E. coli* catalyzing the cleavage of RNA 1 (that is involved in the replication of the → Col E1 plasmid) and the processing of → pre-ribosomal RNA to 5S rRNA. RNase E represents a core part of the → degradosome complex.

RNase-free DNase: A → DNase preparation that does not contain any detectable → RNase activity, and is used to remove any residual DNA in RNA extracts. See → DNase-free RNase.

RNase H (*ribonuclease* H; EC 3.1.4.34): An enzyme (e.g. from *E. coli*) that catalyzes the cleavage of an RNA strand in an RNA-DNA hybrid producing oligoribonucleotides with 5' phosphate termini, and single-stranded DNA, but does not attack single-stranded or double-stranded RNA. The enzyme is used to remove e.g. → poly(A) tails from mRNA after hybridization with poly(dT), to remove the mRNA from the mRNA-cDNA complex after → first strand synthesis in → cDNA cloning, and for oligodeoxyribonucleotide-directed cleavage of RNA. RNase H activity is also found in retroviral → reverse transcriptases. See also → RNase H mapping.

RNase H mapping (H-mapping, oligonucleotide-directed RNase H cleavage): A technique for the determination of the length of the → poly(A) tail of specific → messenger RNAs in a population of RNAs. In short, the RNA is annealed to an oligonucleotide 20-25 nucleotides long and complementary to a region about 300-400 nucleotides upstream of the 3' end of the mRNA. The RNA:DNA hybrid is then cleaved with → RNase H, and the resulting fragments electrophoretically separated on a denaturing → agarose or → polyacrylamide gel, transferred to a membrane (see → Northern blot) and hybridized to a radioactively labeled → probe complementary to the 3' fragment that contains the poly(A) tract. The mobility of the 3' fragment is a measure for the length of the poly(A) tail. See also → RNase T1 protection.
Figure see page 979.

RNase I (*ribonuclease* I): An enzyme from *E. coli* catalyzing the degradation of most RNA molecules to oligonucleotides with 3'-phosphate and 5'-hydroxy termini.

RNase P (*ribonuclease* P; EC 3.1.26.5): A processing enzyme from *E. coli* and *B. subtilis* that catalyzes the removal of 5' proximal nucleotides of → transfer RNA precursor molecules in *E. coli*, generating the 5'-termini of mature tRNAs. The enzyme is composed of an RNA (400 nucleotides) and a 14 kDa protein (it is a → ribonucleoprotein particle), where the RNA alone is capable of binding and precise cleaving of tRNA precursors. RNase P functions as a → ribozyme.

RNase Phy M (*ribonuclease* Phy M): An enzyme from *Physarum polycephalum*, catalyzing the cleavage of → phosphodiester bonds between adenine and uracil, respectively, and the adjacent nucleotides. The enzyme is used for → RNA sequencing analysis.

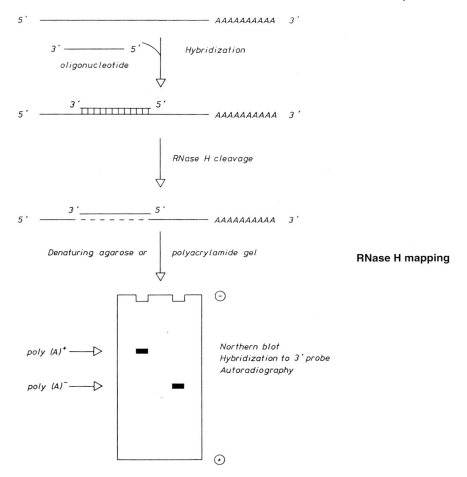

RNase H mapping

Northern blot
Hybridization to 3' probe
Autoradiography

RNase Phy 1 (*ribonuclease* Phy 1; EC 3.1.26.1): An enzyme from *Physarum polycephalum*, catalyzing the cleavage of → phosphodiester bonds between guanine, adenine, and uracil, respectively, and the adjacent nucleotide in RNA, producing 3' phosphate and 5' hydroxy ends. The enzyme is used for the discrimination between cytosine and uracil in → RNA sequencing analysis (by comparing its cleavage pattern with that of the → pyrimidine specific → RNase from *B. cereus*).

RNase *p*rotection *a*ssay (RPA; ribonuclease protection assay):
a) A method for the detection and quantitation of specific → messenger RNA molecules out of total cellular RNA, for the mapping of → transcription start and termination sites, the determination of intron-exon structures and the discovery of mutations (e. g. → deletions and → insertions). In short, total RNA is isolated and hybridized in solution to an excess of specific, labeled → anti-sense RNA probe. This probe is generated by the cloning of a → cDNA fragment of interest into a plasmid in between → SP6, → T7 and T3 DNA-dependent RNA polymerase → promoters. Addition of the respective RNA polymerases allows to synthesize high-specific activity radiolabeled anti-sense RNA probes. After hybridization of

these probes, → RNase A and → RNase T1 are added, which digest single-stranded RNAs and free anti-sense probes, but do not attack RNA-RNA hybrids. After inactivation of the RNases, the RNase-protected hybrids are electrophoresed in denaturing → polyacrylamide gels, which are subsequently dried and exposed to X-ray films. The quantity and the lengths of these hybrids can then be determined by → autoradiography or → phosphorimaging. Absolute quantitation of an mRNA can be achieved by comparing the hybridization signal intensity with the signal strengths of a calibration curve produced with synthetic sense strand target RNA. See → nuclease protection assay.
Figure see page 981.

b) A technique to localize regions of contact between an RNA chain (e.g. a → transfer RNA) and a cognate protein (e.g. a cognate → aminoacyl-tRNA synthetase). Protein-free sequences are digested by a series of → RNases, whereas those regions which are in close contact with the specific binding protein will be protected. The resulting difference in the splicing pattern can be visualized by analysis of the reaction products on → sequencing gels. Thus the binding site can be mapped precisely. Compare → footprinting.

RNA sequence space: A laboratory slang term for all RNA sequences that are able to bind specifically to any given protein. See → RNA aptamer.

RNA sequencing: The determination of the sequence of bases in RNA, using either chemical or enzymatic methods. The chemical method is similar to the → chemical DNA sequencing procedure in which endlabeled RNA is chemically cleaved by base-specific compounds. The resulting RNA fragments are separated in a → sequencing gel, and the sequence is read directly from the resulting autoradiogram. The enzymatic method is based on the action of different → RNases that cut more or less precisely at specific bases (e.g. → RNase T1 cuts at G residues, → RNase U_2 preferentially at A residues, → RNase Phy 1 discriminates C from U, and cuts U > G and A > C, and → RNase CL3 cuts preferentially at C). See also → RNase Phy M, → RNase from *B. cereus*, → ribozyme.

RNase S (*ribonuclease* S; ribonuclease modified by *subtilisin*): A 119 amino acid enzyme derived from pancreatic → RNase A through digestion of the main peptide chain with subtilisin protease from *B. subtilis*, that catalyzes the cleavage of → ribonucleic acids down to single → ribonucleotides. The enzyme consists of two components, a peptide of 15 amino acids (S-peptide, S-tag), and a protein of 104 amino acids (S-protein), both of which are enzymatically inactive. After the binding of the S-tag to the S-protein (Kd = 10^{-9}M), the active ribonuclease S is generated. The S-protein: S-tag peptide system allows the detection and purification of → fusion proteins in socalled → S-tagging techniques.

RNase III (ribonuclease III; endoribonuclease III, EC. 3.1.26.3): An enzyme from *E. coli* catalyzing the endonucleolytic cleavage of double-stranded regions (e. g. → stem-and-loop or hairpin structures) in RNA molecules into single-stranded RNA stretches of about 15 nucleotides in length. The enzyme is a homodimer of about 50 kDa, each subunit a polypeptide of 226 amino acids and 25.5 kDa, and encoded by the *rnc* gene. RNase III is involved in pre-ribosomal RNA processing (e.g. the cleavage of 30S rRNA precursor during maturation of 16S and 23S rRNA; → post-transcriptional modification), the release of monocistronic → messenger RNAs from the polycistronic → early gene transcript (→ polycistronic mRNA) in bacteriophage T7, and the cleavage of *double-stranded* (ds)RNA into 21 nucleotide long dsRNA fragments with 2–3 nucleotides overhangs, and 5' phosphate and 3' hydroxyl termini as a key step in eukaryotic → RNA interference. See → Dicer.

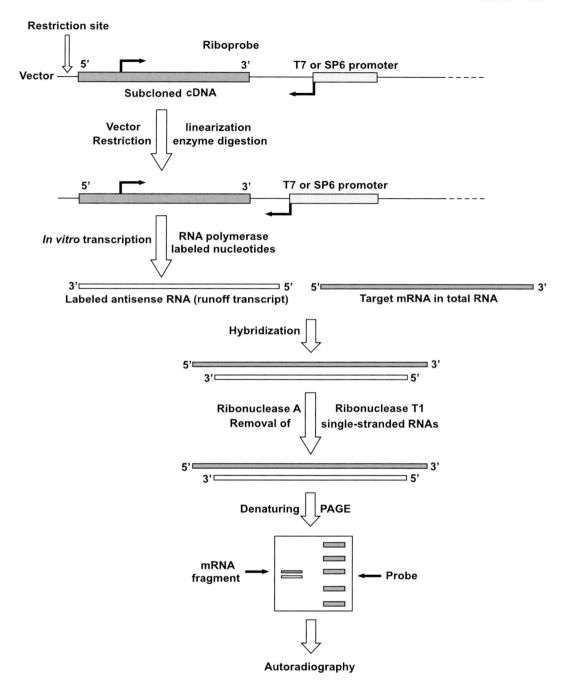

RNase protection assay (RPA)

RNase T1 (*ribonuclease* T₁; EC 3.1.27.3): An enzyme from *Aspergillus oryzae* catalyzing the cleavage of → phosphodiester bonds between 3' phosphate groups of guanosine nucleotides and the 5' hydroxyl group of adjacent nucleotides in single-stranded RNA. The enzyme is used for → RNA sequencing analysis.

RNase T1 protection: A technique for the determination of the length of the → poly(A) tail of specific → messenger RNAs in a population of RNAs. In short, the RNA is annealed to a labeled RNA → probe that spans the entire 3' end of the mRNA. The RNA:RNA hybrids are cleaved with → RNase T1 that specifically attacks non-hybridized RNA 3' to → guanine residues. The T1-resistant RNA including the poly(A)-tract is electrophoresed on a native gel that preserves the hybrid, and detected by → autoradiography. See also → RNase H mapping.

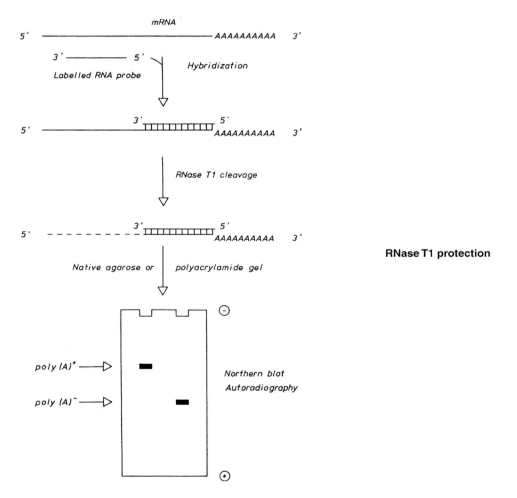

RNase T1 protection

RNase T₂ (*ribonuclease* T₂; EC 3.1.27.1): An enzyme from *Aspergillus oryzae* catalyzing the cleavage of → phosphodiester bonds in RNA, with a preference for adenylic acid bonds. The enzyme is used for 3' terminal analysis of RNA.

RNase II (*ribonuclease* II; EC 3.1.13.1): An enzyme from *E. coli* catalyzing the removal of nucleoside-5'-phosphates from the 3' end of an RNA molecule with no or only little secondary structure.

RNase U$_2$, (*ribonuclease* U$_2$; EC 3.1.27.4): An enzyme from *Ustilago sphaerogena* catalyzing the cleavage of phosphodiester bonds between → purine nucleotides in RNA to yield 3' nucleotides with an intermediate formation of purine nucleoside 2'-3' cyclic phosphates. The enzyme is used for → RNA sequencing analysis.

RNase V1: An enzyme from cobra (*Naja naja oxiana*) venom catalyzing the cleavage of phosphodiester bonds between nucleotides in double-stranded RNA (preferentially at 5'-GAUGA-CUG–3', 5'-CCGCAU–3', 5'-GUGCAG–3' or 5'-CAGUCGUC–3' sequences). The enzyme is used for RNA sequencing, protein footprinting and structural analysis of RNA.

RNase Z: An endonuclease from archaea and a majority of eukaryotes catalysing the cleavage of → *transfer RNA* (tRNA) precursors close to or directly 3'to the socalled discriminator, which is located 5'to the terminal CCA motif and is an identity element in many tRNAs. After RNase Z cleavage, the tRNA nucleotidyl transferase adds the CCA sequence to the 3'-terminus, thereby generating the mature tRNA.

RNasin: A 51 kDa acidic protein originally isolated from human placenta, that inhibits most → RNases noncompetitively (exception: → RNase H, → RNase T1). It binds to RNases noncovalently in a 1 : 1 molar ratio, probably via leucine-rich repeats (LRRs), and forms with them an enzymatically inactive complex (binding constant K = 10^{-14}). RNasin does not inhibit → S1 nuclease, → *Taq* polymerase, → reverse transcriptase, or → SP6, → T7 or → T3 RNA polymerases, and is used in RNA extraction procedures to ensure high yield of undegraded RNA, to improve the size and yield of → cDNAs, to increase the integrity of RNA synthesized in → in vitro transcription, to improve the yield of high-molecular weight polypeptides in → in vitro translation and to prevent degradation of mRNAs in isolated → polysomes. RNasin can also be obtained by expression of a recombinant → plasmid in *E. coli* (rRNasin).

RNA *single strand conformation polymorphism* technique (RNA-SSCP, rSSCP): A variant of the conventional → single strand conformation polymorphism method (designed to detect single-base mutations in a target DNA) that allows to identify single base differences in RNAs of 200-400 bases in length. In short, the genomic DNA of interest is first amplified in a conventional → polymerase chain reaction with → primers, of which one or both contain a 23-29 base → T3, → T7, or → SP6 RNA polymerase promoter core sequence. The amplified products are then transcribed with the corresponding RNA polymerase, the abundant single-stranded RNA transcripts denatured (to dissolve secondary structures), and electrophoresed in a non-denaturing → polyacrylamide gel, and the band visualized by → ethidium bromide staining and fluorescence. Alternatively, α-^{32}P-UTP can be used to label the RNA transcripts, and → autoradiography employed to detect the bands on a dried gel. RNA-SSCP is superior to DNA-SSCP, i. e. it detects almost 100 % of single-base mutations in short (200-400 bases) target DNAs.

RNA splice site: See → splice junction.

RNA splicing: See → splicing and → RNA editing.

RNA-SSCP: See → RNA *s*ingle *s*trand *c*onformation *p*olymorphism technique.

RNA synthesis:
a) RNA-dependent RNA polymerization, see → replicase.
b) DNA-dependent RNA polymerization, see → transcription.

RNA therapeutic: Any RNA, that is used for the treatment of a disease. For example, → antisense RNAs, → catalytic RNAs such as diverse → ribozymes, and → small interfering RNAs are such RNA therapeutics.

RNA topology: The three-dimensional arrangement of a single-stranded RNA chain by the formation of internal → fold-backs (hairpin loops) and → stem- and -loop structures, its folding into tertiary structures and the changes of these structures in response to physical (e.g. temperature) or chemical parameters (e.g. → intercalating agents, or proteins).

RNA track (transcript track): Any trace of a newly synthesized RNA molecule (e.g. a → messenger RNA) from the location of its gene (usually within the interior of the eukaryotic nucleus) to the periphery of the nucleus and the → nuclear pores. The RNA movement can be traced through the socalled *i*nter-*c*hromatin *d*omains (ICDs), regions between the individual → chromosome territories, to the cytoplasm.

RNA transfection: See → lipofection-mediated RNA transfection.

*R*NA *t*ranslation *e*nhancer (RTE): A sequence from an untranslated region of the myelin basic protein messenger RNA, that enhances protein expression in mammalian cells by increasing the rate of mRNA transport from the nucleus to the cytoplasm and activating ribosomes translating the mRNAs containing an RTE.

RNA world: A pre-biotic era in which → RNA was the genetic → template (not DNA), and able to replicate itself (autocatalytically) and to modify other RNAs (heterocatalytically; analogous to → ribozymes).

RNA zipcode: See → RNA localization.

RNC: See → ribosome nascent-chain complex.

RN map (recombination frequency map): A → genetic map or a → chromosome map in which the locations of no, little, or high → recombination frequencies are depicted.

RNomics: See → ribonomics.

RNP: See → *ribo*nucleo*p*rotein particle.

RNR: See → *ribo*nucleotide *r*eductase.

rNTP: Any → ribonucleotide 5'-triphosphate.

Robustness: A (desired) property of a chemical reaction or a scientific instrument, characterized by a relative insensitiveness toward exogenous (but also intrinsic) perturbations and irritations.

Rocket immunoelectrophoresis: See → electroimmunoassay.

Roller bottle hybridization: A modified → filter hybridization procedure that avoids sealable plastic bags but instead uses roller bottles as containers for blot(s) and radioactive → probe. The bottles can be easily loaded, sealed, and handled, protect from β-particles emitted by the ^{32}P labeled probes, and accomodate several blots in a minimal volume of hybridization solution. Roller bottle hybridization produces better signal-to-noise (→ background) ratios than the conventional plastic bag hybridizations.

Rolling circle amplification (RCA): A highly sensitive technique for the amplification of single and small DNA or RNA molecules (e. g. 20-50 bp), that uses DNA polymerase I capable of replicating circular oligonucleotides and amplifying the target DNA via → strand displacement. A 50 nucleotides → probe is first coupled to a → primer sequence, then hybridized *in situ* to a target DNA (or chromosome). A preformed circle, that is partly complementary to the primer sequence, is added. Then DNA polymerase I extends the primer and synthesizes a new strand complementary to the template circle, producing a duplex DNA. However, the circle becomes constrained and forces polymerization to stop. This constraint is relaxed by unwinding of the lagging end of the duplex. If this process is combined with continued polymerization, a long single-stranded DNA containing multiple tandem repeats complementary to the original circle is produced. This RCA product can then easily be detected with fluorescently labeled oligonucleotides. RCA allows to detect → nucleotide substitutions, → deletions, and → insertions in a DNA target sequence. RCA is different from → rolling circle replication in that it starts the rolling circle type synthesis reaction on a very short circular template.

Rolling circle replication: A mechanism for the → replication of DNA molecules generating → concatemers of duplex molecules (e.g. of → lambda phage). The first step in the rolling circle replication is a single-strand → nick which exposes a free 3' OH end for strand extension by → DNA polymerase. The newly generated strand displaces the original parental strand. After several cycles a long DNA strand, containing several unit genomes, is synthesized. The displaced strand then may serve as → template for the synthesis of a complementary strand.
Figure see page 986.

Root hair disease: See → hairy root disease.

Root-inducing plasmid: See → Ri-plasmid.

ROS: Abbreviation for *reactive oxygen species*. Not to be confused with → ros box, → ros gene → ros repressor.

ros box: An → upstream sequence of 40 bp containing the 9 bp → inverted repeat 5'-TATATTT-CA-N$_7$-TGTAATATA-3' in the → promoter region of the → vir-region operons *vir* C and *vir* D. The ros box serves as address site for the → ros repressor protein encoded by the → *ros* gene of the → *Agrobacterium tumefaciens* chromosome. After binding, the ros repressor silences transcription from *vir* C and *vir* D promoters, probably by competing out the vir G protein, a transcriptional enhancer. During *vir* gene induction by plant wound substances, the phosphorylated vir G protein overcomes the effect of the ros repressor.

Circular DNA

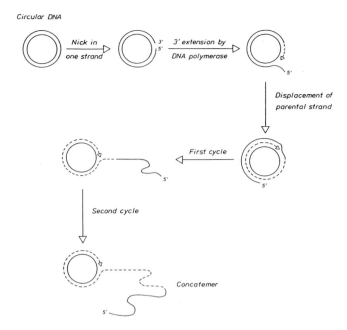

Rolling circle replication

Rosetta stone protein: Any protein that is composed of two (or more) different → domains, which became separated in evolution such that the different domains now represent separate functional proteins. Usually the prokaryotic Rosetta stone proteins are fusion proteins, whereas the eukaryotic analogues are separated into two (or more) different individualized genes.

ros **gene:** A 726 bp chromosomal → gene of → *Agrobacterium tumefaciens* encoding the → ros repressor protein.

ros repressor: A 15.5 kDa → repressor protein encoded by the → *ros* gene of the → *Agrobacterium tumefaciens* and → *A. rhizogenes* chromosome, that interacts with the socalled → ros boxes in the → promoters of → virulence genes *vir* C and *vir* D on the → Ti plasmid via a single → Zink finger motif (type Cys-N$_2$-Cys-N$_{12}$-His-N$_3$-His; N = any amino acid) at the → carboxy terminus. The ros repressor downregulates the activity of *vir* C and *vir* D genes and acts antagonistically to the phosphorylated → vir G protein. See → vir-region.

Rotary microfluidic polymerase chain reaction: A variant of the → continuous-flow polymerase chain reaction technique, that allows to amplify a target DNA in a circular flow channel of a silicon chip with integrated valves and pumps. The different temperatures for → denaturation, → primer annealing and → extension are controlled by microfabricated tungsten heaters. The amplification takes place in 12 nl volumes and is finished in 30 minutes.

Rotating agarose gel electrophoresis: See → rotating gel electrophoresis.

Rotating field electrophoresis: See → rotating gel electrophoresis.

Rotating gel electrophoresis (RGE; rotating agarose gel elelectrophoresis, RAGE; rotating field electrophoresis, RFE): A method to separate large DNA molecules in the size range from 50 kb to more than 7000 kb using a single homogeneous electric field, but changing the orientation of the field in relation to the gel by periodically rotating the gel. See → gel electrophoresis for a list of related techniques.

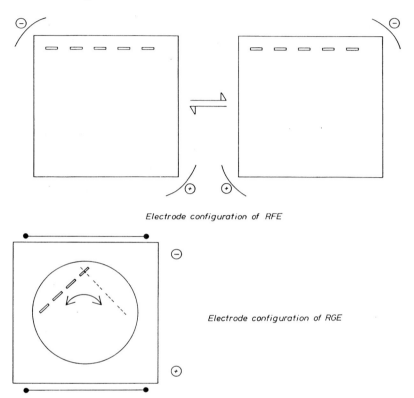

Electrode configuration of RFE

Electrode configuration of RGE

Rotating gel electrophoresis

Rotational base substitution: A specific → transversion mutation induced by radiation, which causes a → break in both complementary strands of a DNA duplex molecule. In this reaction, the bond between base and sugar moiety of a nucleotide molecule is broken, and as a consequence the two hydrogen-bonded complementary bases are detached from their backbones. The transversional mutation is then generated by rotation of the base pair before its reinsertion into the duplex.

Rough draft: An incomplete sequence of any → genome, interrupted by so called → gaps, most of them containing highly → repetitive DNA. The presence of too many gaps makes it impossible to order and orientate the relatively small runs of bases, that are the raw products of → genome sequencing. Therefore, a genome sequence is not finished completely, if even one single gap exists, but rather is represented as a rough draft.

Rox: The → fluorochrome 6-carboxy-X-rhodamine, that is used as a marker for → fluorescent primers in e. g. automated sequencing procedures or for labeling in → DNA chip technology. The molecule can be excited by light of 587 nm wave-length, and emits fluorescence light at 607 nm. Since the wave-length of the excitation and emission maxima is pH-dependent, the exact values vary.

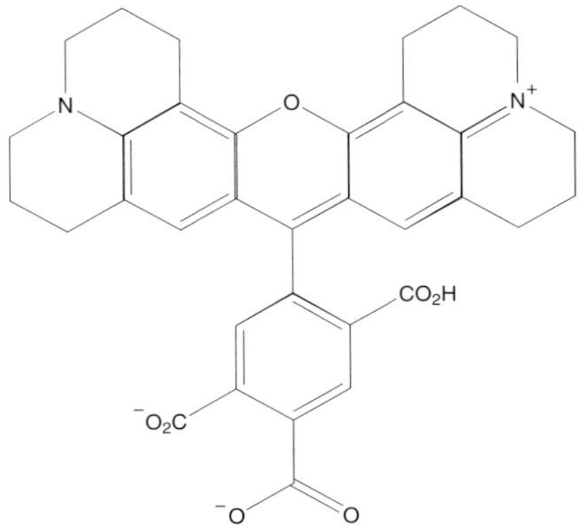

ROX

RPA:
a) See → RNase (ribonuclease) protection assay.
b) See → random priming.

RPA affinity chromatography: See → replication protein A affinity chromatography.

RPC: See → RNA polymerase II complex.

RP4: A broad host range → plasmid from *Pseudomonas aeruginosa*, used as → cloning vector for DNA cloning in a broad spectrum of Gram-negative bacteria (→ broad host range vector).

R plasmid: See → R factor.

Rpm (rpm): *Revolutions per minute*, a term describing the rotor speed in centrifuges.

RP1 plasmid (R 1822 plasmid): A → conjugative plasmid that occurs in 1-2 copies in enterobacteria and pseudomonads, and codes for → carbenicillin, → kanamycin and → tetracycline resistance.

RQ 1 DNase: An → endonuclease from bovine pancreas that catalyzes the degradation of single- and double-stranded DNA molecules to produce 3' hydroxyl oligonucleotides. RQ 1 DNase is used to produce random DNA fragments for → shotgun cloning, to remove DNA from RNA transcribed in vitro by → SP6, → T3, or → T7 RNA polymerase systems (→ *in vitro* transcription), to prepare DNA fragments for → nick translation, and to analyze DNA-protein complexes.

RRF: See → ribosome recycling factor.

RRM: See → RNA recognition motif.

rRNA: See → ribosomal RNA.

rRNasin: See → RNasin.

RRS sequencing: See → reduced representation shotgun sequencing.

RSCU: See → relative synonymous codon usage.

R segment: The → direct repeat sequences 10-80 nucleotides in length, localized at the termini of the genomic RNA of → retroviruses.

rsGFP: See → red-shifted green fluorescent protein.

RsID: See → reference single nucleotide polymorphism.

RSM element: See → restriction site conversion plasmid.

RSO: See → restriction site oligonucleotide.

RS-PCR: See → restriction-site polymerase chain reaction.

RSS: See → repetitive simple sequence.

rSSCP: See → RNA single strand conformation polymorphism technique.

RsSNP: See → reference single nucleotide polymorphism.

r*Taq*: Abbreviation for the recombinant form of → *Thermus aquaticus* → DNA polymerase. Compare → n*Taq*.

RTase: See → reverse transcriptase.

RTE: See → RNA translation enhancer.

RT-MPCR: See → reverse transcriptase multiplexed polymerase chain reaction.

RT-PCR: See → reverse transcription polymerase chain reaction.

RT-PCR *in situ* hybridization: See → *in situ* polymerase chain reaction.

RT-RAPD: See → restricted template random amplified polymorphic DNA.

RUN: See → repeat unstable nucleotides.

Runaway plasmid: A → plasmid whose → replication is tightly controlled (→ stringent control) below, but uncontrolled (→ relaxed control) above a certain temperature threshold. For example, pBEU1 is present in moderate → copy numbers per cell at 30°C. Above 35°C, its copy number increases continuously, which finally affects the growth and viability of the host cell. Runaway replication of → plasmid cloning vectors is a simple and rapid method to multiply the vector plasmid, and to overproduce a specific plasmid-encoded gene product (→ overexpression).

Runaway plasmid vector: See – runaway plasmid.

Runaway replication: Uncontrolled multiplication of a → plasmid to several thousand copies per host cell, usually induced by elevated temperature (i.e. above 35-37°C). See → runaway plasmid.

Runaway replication vector: See → runaway plasmid.

Run-off transcription (nuclear run-off transcription assay): A method to measure the relative → transcription rate of genes in isolated nuclei, after the inducing signal (e. g. light, hormone, or elicitor) has been removed. In short, nuclei are isolated from the material of interest, RNA is elongated in vitro from pre-initiated complexes in the presence of radioactively labeled nucleotide precursors, and the labeled RNA is hybridized to cloned DNA sequences dotted onto a → nitrocellulose or nylon-based filter, or to DNA that has been electrophoretically separated in an → agarose gel and blotted onto such filters. Quantitation of the signal seen in a resulting → autoradiograph allows the estimation of the amount of message obtained in this way.

Run-on transcription: A technique to quantify the transcription rate of specific nuclear or organellar genes. In short, intact nuclei or organelles (e. g. plastids or mitochondria) are first isolated, a reaction mixture containing a radiolabeled ribonucleotide (e. g. α-^{32}P-UTP) added, the organelles hypotonically lysed, and the ribonucleotide incorporated into already initiated transcipts ("run-on"). After the completed run-on transcription the transcripts are quantified by hybridization to gene-specific DNA or RNA probes and subsequent → autoradiography. Since only few transcripts are initiated de novo under these conditions, run-on transcription reflects the true intra-organelle (*in organello*) state of transcription of the particular genes. The procedure is identical to the → run-off transcription assay.

R_0t (R_0t value): In RNA-driven hybridization (→ saturation hybridization) the analog for → C_0t values of DNA-driven hybridization reactions, defined as the product of an original RNA con centration times time of incubation.

S

S:

a) See → sedimentation coefficient.

b) Abbreviation for "strong", i.e. the relatively strong interaction(s) between hydrogen-bonded G ≡ C base pairs in DNA. Used to symbolize either C or G.

c) Incorrect abbreviation for → switching site.

SAAT: See → *s*onication-*a*ssisted *A*grobacterium *t*ransformation.

***Saccharomyces cerevisiae*:** See → yeast.

***Saccharomyces cerevisiae* plasmid:** See → two micron circle.

SAD-DNA: See → severely affected Alzheimer disease DNA.

SADE: See → SAGE adaptation for downsized extracts.

S-*adenosyl*-L-*methionine* (SAM, AdoMet): A carrier and donor of activated methyl groups for e.g. RNA and → DNA methylation. Also required for the binding of class I → restriction endo-nucleases, see → restriction-modification system.

SAFE: See → specificity assessment from fractionation experiments.

SAGA complex (*Spt-Ada-Gcn5*–*A*cetyltransferase complex): A large nuclear protein complex, originally isolated from *Saccharomyces cerevisiae*, that modulates the packaging ratio of → chromatin ("chromatin remodelling") and thereby leads to an assembly of the → transcription machinery on → promoter sequences and the activation of transcription of the adjacent gene. The SAGA complex falls into three main protein categories: the Ada2p/Ada3p/Gcn5p class (interacts with activators and is essential for the *h*istone *a*cetyl*t*ransferase [HAT] activity of the whole complex, where Gcn5p functions as the catalytic subunit stabilized by Ada2p/Ada3p), the Spt3p/Spt8p class (interacts with the → TATA-box-binding transcription factor), and Spt20p/Ada5p/Spt7p class (functions in stabilizing the structural integrity of the complex). The HATs catalyse the transfer of acetyl residues onto lysine residues at the N-terminal domain of histones, that are responsible for DNA packaging. The action of the SAGA complex then results in chromatin decondensation and the derepression of the underlying genes.

SAGE: See → *s*erial *a*nalysis of *g*ene *e*xpression.

SAGE *a*daptation for *d*ownsized *e*xtracts (SADE): A variant of the conventional → serial analysis of gene expression (SAGE) technique for minute amounts of starting tissue (e.g. 0.5 mg tissue, or 50,000 cells, or less from e.g. microdissected specimens), that is run in a single tube from tissue lysis to cDNA tag recovery. See → microSAGE, → SAGE-Lite.

SAGE class: Any one of four groups of → SAGE tags that differ in their location relative to known → ORFs and is assigned a color in graphic displays.
Class 1: within an ORF (orange).
Class 2: within 500 bp 3' of an ORF (violet).
Class 3: none of the above (bright pink).
Class 4: on the strand opposite an ORF (yellow).

SAGE-Lite: A variant of the → serial analysis of gene expression (SAGE) technique for global analysis of gene expression patterns that allows to reduce the starting material tremendously and thus the amount of total RNA (to less than 50 ng). SAGE-Lite is therefore used for expression analysis in rare specimen, bioptic probes and micodissection material. See → microSAGE.

SAGE map: A computer program that allows to compare gene expression profiles between selected libraries generated by the → serial analysis of gene expression technique. Web site: http://www.ncbi.nlm.nih.gov/SAGE//

Salivary gland chromosome: Any one of the polytenic large-sized chromosomes of salivary glands of *Drosophila* and related Dipteran species, that is clearly banded and can easily be isolated and observed under a microscope. In short, third instar larvae with a still soft cuticle are placed into aceto-orcein or any other DNA stain, decapitated, and the salivary gland pulled out of the head part. After about 10–20 minutes in aceto-orcein, the chromosomes are stained such that they can be spread by gentle pressure. Changes of the salivary gland chromosome structure after stresses and various other environmental impacts (see → Balbiani ring, → puff) were investigated in the early days of → cytogenetics.

Salmon sperm DNA: The → genomic DNA of salmon sperms. In → Southern and → Northern blotting procedures the fragmented and denatured DNA is used to block unspecific binding sites on membrane filters to reduce undesirable → background on autoradiographs (→ blocking re-agent).

Saltatory replication: The → amplification of a specific DNA sequence either in the absence or presence of general chromosomal → replication, that leads to the tandem arrangement of multiple copies of that sequence. For example, → satellite DNA and → minisatellite DNA possibly originate from saltatory replication steps.

Salvage pathway: A metabolic pathway that reuses → purines and → pyrimidines, originating from the intracellular degradation of nucleic acids, for the re-synthesis of nucleic acids.

SAM:
a) The genetic notation for an → amber mutation in the S gene of → lambda phages. Gene *S* (together with gene *R*) is involved in the → lysis of the host cell.

b) See → *S*-*a*denosyl-L-*m*ethionine.
c) See → self-assembled monolayer.

Samesense mutation: See → silent mutation.

SAMPL: See → *s*elective *a*mplification of *m*icrosatellite *p*olymorphic *l*oci.

Sample loading buffer: See → loading buffer.

Sample noise: The undesirable contribution of sample parameters such as → probe reflection, salt crystals, non-specific attachment of probe molecules and buffer-specific influences to the reading of fluorescence signals of a fluorescence detector instrument in → microarray experiments. See → background subtraction, → dark current, → electronic noise, → microarray noise, → optical noise, → sample noise, → substrate noise.

SAN: See → storage area network.

Sandwich blotting: See → bidirectional transfer.

Sandwiched gene: A coding sequence that has been cloned between two regulatory sequences (e.g. between a → start codon or → promoter, and a → stop codon or → trailer sequence) from different source(s). For example, → reporter genes for plant transformation are frequently driven by a promoter from → cauliflower mosaic virus (CaMV 35S promoter) and terminated by a sequence from a → nopaline synthase gene of → *Agrobacterium tumefaciens* → Ti-plasmid. See also → fused gene, → transcriptional fusion.

Sandwich fusion: A chimeric gene produced by the fusion of sequences encoding a so called carrier protein (e.g. *E. coli* outer *m*embrane *p*rotein A [OmpA]) and a so called → passenger protein, such that the passenger is inserted close to the center of the the carrier ("sandwiched"). The term also describes the → chimeric protein expressed from such a → fused gene. Sandwich fusions allow to expose the passenger protein on the bacterial or yeast cell surface. For example, the *E. coli* OmpC can be used as a sandwich fusion partner, which displays peptides of more than 160 amino acids.

Sandwich hybridization: A method to identify a specific DNA sequence with two → probes which are homologous to different parts of the target DNA. In short, one probe ("catcher") serves to anchor the target DNA at a solid phase (e.g. a microtiter plate, plastic-coated support, or chemically activated paper). Anchoring may be achieved by baking or by specific interactions (e.g. the affinity between solid phase-bound → streptavidin and → biotin that is incorporated in the catcher probe). Then a second probe ("detector", reporter) is added that recognizes other regions of the target DNA, and serves to detect the sandwich complex, e.g. via an enzyme reaction that leads to a colored product at the sites of hybridization (for example, the reporter probe can be attached to a → digoxygenin moiety via an allylamine spacer, an antibody against digoxygenin will bind to the complex, and an antibody-bound → alkaline phosphatase converts the colorless substrate → X-phos into a colored indigo product). Sandwich hybridization can also be performed in solution.

Sandwich technique: A method to identify specific proteins encoded by recombinant DNA sequences and expressed in bacteria. In short, one → antibody serves to bind the desired protein and to anchor it at a solid support (e.g. a plastic-coated support or chemically activated paper).

Then a second antibody is added that recognizes the same antigen and is either labeled with ^{125}I, or is complexed with ^{125}I-labeled → protein A from *Staphylococcus aureus* (protein A binds to the Fc structure of many antibodies). The antigen thus is fixed between two antibody molecules ("sandwich"). Similar techniques can also be used for the detection of non-radioactively labeled DNA, if the first antibody is directed against the chemical group used for → labeling. See for example → sulfonated DNA detection.

Sanger method: See → Sanger sequencing.

Sanger sequencing (Sanger method, enzymatic method, dideoxy method, dideoxy sequencing, chain terminator method, chain terminator sequencing procedure, enzymatic DNA sequencing, dideoxy mediated chain termination technique): A technique for the sequencing of single-stranded DNA. Basically, in this method a → sequencing primer or → reverse sequencing primer (usually a synthetic oligonucleotide) is annealed to the single-stranded target DNA. The reaction mixture is then aliquoted into four separate tubes, all four deoxynucleoside-triphosphates – one of them labeled with ^{32}P, e.g. ^{32}P-αdCTP – and a 2', 3' → dideoxynucleoside-triphosphate (ddNTP) are added. Each tube contains a different ddNTP (either ddATP, ddCTP, ddGTP or ddTTP). Now the → Klenow fragment of → DNA polymerase I is used to synthesize a complementary copy of the single-stranded target sequence in a → primer extension reaction. If a ddNTP is incorporated into the growing chain instead of the corresponding dNTP, the 3' end of the growing chain lacks a hydroxyl group and cannot be used for elongation: the chain is terminated. In this way, each reaction mixture contains a population of radioactively labeled DNA fragments with a common 5' end (the primer), but a variant 3' end. After the reactions are completed the DNA is denatured, electrophoresed in adjacent lanes of a thin polyacrylamide gel (→ sequencing gel) and the radioactive bands are detected via → autoradiography. The sequence of the original target DNA can then be read directly from the autoradiogram. See also → DNA sequencing, → exometh sequencing, → exonuclease III technique, → multiplex sequencing.
See color plate 7, and figure page 995.

SAP:
a) See → *s*hrimp *a*lkaline *p*hosphatase.
b) See → *s*pecific *a*mplicon *p*olymorphism.

SAR:
a) See → *s*caffold-*a*ssociated *r*egion.
b) See → *s*ystemic *a*cquired *r*esistance.

Sarcoma-inducing gene: See → *src* gene.

SAR-SAGE: See → small amplified RNA-SAGE.

SAS: See → systemic acquired silencing.

SAS-DNA marker: See → *s*imultaneously *a*mplified *s*ingleton DNA marker.

SatDNA (sat-DNA): See → *sat*ellite DNA.

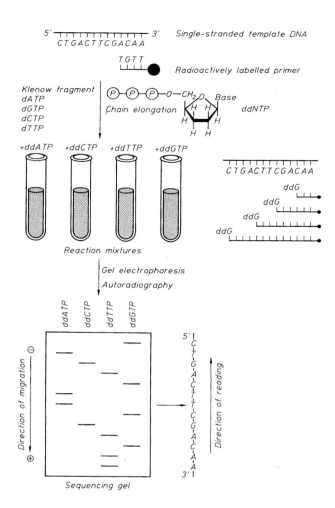

Sanger sequencing

Satellite:
a) Any distal segment of a → chromosome separated by a secondary restriction from the rest of the chromosome. See → satellite DNA.
b) See → satellite RNA.
c) A normally quiescent cell type, present under skeletal muscle basal lamina in small numbers. Satellite cells re-enter the cell cycle after traumatization of the skeletal muscle, and produce myoblasts, and ultimately new muscle fibers.

*Sat*ellite DNA (satDNA, sat-DNA): A specific type of ubiquitous, abundant → tandem repeats of simple sequences in → genomic DNA of eukaryotes. SatDNA may be separated from bulk chromosomal DNA by → isopycnic centrifugation in CsCl gradients. SatDNA bands either at higher or lower density apart from the main DNA, depending on its GC/AT content. The function(s) of satDNA are not known. It may be merely → "junk DNA", but it may also be necessary for maintaining the fitness of an organism (e.g. the Rsp [*Responder*] satellite of *Drosophila* *melanogaster*). A specific class of satDNA, the → minisatellites and → microsatellites, plays an important role in genome research (see → DNA fingerprinting).

Satellite DNA-based artificial chromosome (SATAC): A large capacity → artificial chromosome vector formed by the *de novo* amplification of pericentric heterochromatin. Exogenous plasmid DNA, introduced into the vicinity of mammalian pericentric heterochromatin, will form dicentric chromosomes as a result of the amplification of endogenous pericentric heterochromatin. Subsequent breakage creates 10–360 Mbp chromosomes, that are largely composed of repetitive sequences (→ satellite). These SATACs can accomodate up to 1.0 Mbp of foreign DNA and be injected into target cells by → electroporation or → microinjection. Compare → bacterial artificial chromosome, → mammalian artificial chromosome, → plant artificial chromosome, → yeast artificial chromosome.

Satellite RNA (satRNA): One of a series of single-stranded small, self-splicing RNAs from 194-393 bases in length, associated with a variety of plant viruses (e.g. tobacco ringspot virus), and encapsidated within the viral coat. These satRNAs are able to modulate the disease symptoms caused by their virus. Compare → viroid.

SatRNA: See → *sat*ellite RNA.

Saturation hybridization: A specific form of → hybridization reaction, in which one polynucleotide is added in great excess which drives all complementary sequences to form duplex molecules (e.g. excess of RNA and constant amount of DNA in RNA:DNA hybridization, RNA-driven hybridization; similarly in DNA: DNA hybridization, DNA-driven hybridization).

Saturation mapping: A procedure for the enrichment of specific regions of an already established → genetic map with → molecular markers such that marker density is extremely high. A saturated map is a prerequisite for e. g. the → map-based cloning of a gene of interest, which is only possible, if closely linked markers are available that were mapped in large segregating populations. Saturation mapping generates a socalled high-resolution map, and usually incorporates markers generated with different marker systems (e. g. multi-locus systems like → amplified fragment length polymorphism and → random amplified polymorphic DNA, but also single-locus systems like → restriction fragment length polymorphism and → sequence-tagged microsatellite sites). See → chromosome mapping, → clinical mapping, → comparative gene mapping, → comparative mapping, → contig mapping, → cross-mapping, → deletion mapping, → expressed sequence tag mapping, → function mapping, → gene mapping, → genome mapping, → heteroduplex mapping, → integrative mapping, → interphase mapping, → intron-exon mapping, → long-range restriction mapping, → map, → nucleotide analogue interference mapping, → optical mapping, → peptide mapping, → proteome mapping, → radiation hybrid mapping, → saturation mapping, → visual mapping.

Saturation mutagenesis: A technique for the introduction of single base-pair → mutations at all potential sites in a specific DNA inserted into an appropriate → cloning vector. Saturation mutagenesis leads to a collection of mutated DNA molecules which can then be used to detect the importance of specific nucleotides or pairs of nucleotides for the biological function of the DNA segment.

SAUR: See → *s*mall *a*uxin *u*p *R*NA.

SBAP: See → *s*equence-*b*ased *a*mplified *p*olymorphism.

SBE: See → single base extension.

SBE-TAGS: See → *s*ingle *b*ase *e*xtension-tag array on glass slides.

SBH: See → *s*equencing *b*y *h*ybridization.

SBS: See → simultaneous bidirectional sequencing.

Scaffold (sequence-contig scaffold): A collection of ordered → contigs separated by gaps of known lengths. Such gaps are spanned by clones with end-sequences flanking the gap. Any gap within a scaffold is called a sequence gap, a gap between scaffolds is called a physical gap (no clones available that span the gap). See → sequenced clone contig scaffold.

Scaffold-associated region (*scaffold attachment region, SAR; matrix attachment region, MAR*): Specific, AT-rich DNA sequences of about 200-500 bp in length that are part of eukaryotic chromosomes and interact with → nuclear lamins to anchor large chromatin → looped domains to the → nuclear matrix. These DNA segments closely flank active or potentially active genes or their 5' regulatory and also 3' untranslated regions. SARs carry a consensus sequence for their interaction with → DNA topoisomerase II (consisting of a 10 bp oligo(dA) stretch [A-box; 5'-AATAAATCAA-3'] and a 10 bp oligo(dT) stretch [T-box; 5'-TTATAATTTATTT-3']). In yeast, SARs co-map both with → autonomously replicating sequences and → centromere domain elements. SAR elements generally enhance → promoter functions (e.g. in MMTV, SV 40).

Scaffold attachment factor (*SAF; SAF-A; scaffold attachment protein; also hn RNP-U*): An abundant protein of the nuclear scaffold or → chromatin, that also occurs in → heterogenous nuclear RNA complexes, and probably functions to stabilize nuclear architecture and to regulate RNA metabolism. SAF has a → scaffold-associated region (SAR)-specific bipartite DNA binding domain at the aminoterminus and an RNA binding site (RGG-box) at its carboxy terminus. The DNA-binding domain is cleaved by a protease of the → caspase family during → apoptosis. Cleavage results in the detachment of SAF from the nuclear scaffold.

Scaffold attachment region: See → scaffold-associated region.

Scaffolding protein: See → scaffold protein.

Scaffold protein (scaffolding protein): One or several proteins that associate to form a temporary structural frame-work ("scaffold") to facilitate the assembly of a → bacteriophage head. The scaffold is eliminated after pro-head formation, or degraded, and the degradation products are re-cycled.

SCAGE: See → single cell analysis of gene expression.

Scale-up: The conversion of a reaction or process from a small to a larger scale.

Scanning model (entry site model): The hypothetical description of transcription initiation, where → enhancer and/or → upstream regulatory sequences are recognized by → RNA polymerase II (or one of its subunits, or a → transcription factor). After binding, the corresponding protein slides along the DNA, until it reaches proximal → promoter elements where it stimulates the formation of a → transcription initiation complex.

Scanning probe microscopy (SPM): A general term for advanced microscopy techniques, that allow to scan probes (e.g. DNA molecules, complexes of DNA and a protein) at the atomic level. *Scanning tunnelling microscopy* (STM), *near-field scanning optical microscopy* (NSOM) and → atomic force microscopy are such scanning probe microscopy techniques

SCAR: See → *sequence characterized amplified region*.

Scatter plot: A graphical representation of e.g. → microarray data, in which the ratio of fluorescence signal intensities (e.g. cy3/cy5) of two samples is plotted along the x and y axes. The result is a diagonal crowd of values, that allows to draw a regression line. Any deviation from this line is indicative for either gene activation (usually plotted above the line) or repression (usually plotted below the line).

SCGE: See → *single cell gel electrophoresis assay*.

sc gene: See *single-copy gene*.

Schizosaccharomyces pombe artificial chromosome (SPARC): A high-capacity → cloning vector that contains → telomeres from*Schizosaccharomyces pombe* at each terminus, and appropriate cloning sites (e. g. a *Not* I restriction recognition sequence), into which DNA fragments of an average size of 120 kb (maximum: 400 kb) can be inserted. Since the size of the three chromosomes of *S. pombe* are well above 3.5 Mb, SPARCs migrate faster during → pulsed field gel electrophoresis, and therefore can be separated from the chromosomes easily. Compare → bacterial artificial chromosome, → human artificial chromosome, → mammalian artificial chromosome, → pBeloBac 11,→ plant artificial chromosome, → P1 cloning vector, → yeast artificial chromosome.

Scintillation proximity assay (SPA): A method to detect radioactively labeled ligands bound to acceptor molecules on the surface of fluomicrobeads (SPA beads). The ligands (e.g. biotinylated nucleic acids) are labeled with ^{3}H, that emits low-energy radiation. The ligand is then bound in solution to → streptavidin-coated beads, and thus comes into close contact with the fluorine which is activated by the radiation. The produced light is detectable in a → liquid scintillation counter. The free ligands cannot activate the fluorine so that they need not be removed from the solution.

SCOMP: See → single cell comparative genomic hybridization.

Scorpion primer: A specially designed single-stranded→ primer oligonucleotide that allows the detection of allelic variants and → single nucleotide polymorphisms (SNPs). A scorpion primer consists of the actual → probe sequence that is forced into a → hairpin loop structure by complementary stem sequences on the 5' and 3' sides of the probe (as e.g. is the case for → molecular beacons). A fluorophore is attached to the 5' end, but quenched by a molecule (e.g. methyl red) joined to the 3' end of the loop. The hairpin loop is linked to the 5' end of the primer via a socalled PCR stopper (blocker; cannot be amplified). In short, the scorpion stem and the double-stranded target DNA are first denatured, then cooled down, the scorpion primer annealed and extended in a conventional → polymerase chain reaction. After extension, the specific probe sequence is bound to its complement within the same strand of DNA. This hybridization event relaxes the hairpin loop (the resulting structure resembles a scorpion's tail) such that fluorescence is no longer quenched: an increase in fluorescence signal intensity ensues. Therefore, the scorpions, as opposed to → *Taq*Man primers or molecular beacons, work with a unimolecular mechanism, and avoid enzymatic cleavage of the probe (as is necessary in e.g. *Taq*Man procedures). See → duplex scorpion primer.

scp: See → two micron circle.

Scrambling: A laboratory slang term for the massive rearrangements, that followed ancient large → segmental duplications of genomes.

Scrape-loading: A simple and rapid method for the introduction of macromolecules (e.g. foreign DNA or protein) into cells adherent to glass or plastic tissue culture dishes through mechanical perturbation of the plasma membrane (by scraping the cells off their support). See also → direct gene transfer.

Screening: See → gene screening.

scRNA: See → *s*mall *c*ytoplasmic RNA.

scRNP: See → *s*mall *c*ytoplasmic *r*ibo*n*ucleo*p*rotein.

SCS: See → *s*pecialized *c*hromosome *s*tructure.

scyrp: See → *s*mall *c*ytoplasmic *r*ibo*n*ucleo*p*rotein.

SDA: See → *s*trand *d*isplacement *a*mplification.

SD-box: See → *S*hine-*D*algarno sequence.

SDI test: See → *s*uccinate *d*ehydrogenase *i*nhibition test.

SDS: See → *s*odium *d*odecyl *s*ulfate.

SD sequence: See → *S*hine-*D*algarno sequence.

SDS-PAGE: See → SDS-*p*oly*a*crylamide *g*el *e*lectrophoresis.

SDS-*p*oly*a*crylamide *g*el *e*le*c*lectrophoresis (SDS-PAGE): A method to separate protein molecules on the basis of their molecular weight. The proteins are first loaded with sodium dodecyl sulfate that binds and denatures them, concomitantly conferring to them a net negative charge. Thus migration of the SDS-protein complexes through polyacrylamide gels will largely depend on their size. Complete denaturation of the protein molecules and thus also complete separation of protein subunits is ensured by heating the sample before it is separated on a → Laemmli gel.

SEAP: See → *s*ecreted *a*lkaline *p*hosphatase.

SeaPlaque™ agarose: A trade-mark for a → low melting point agarose.

SEC: See → siRNA expression cassette.

SECIS: See → *se*leno*c*ysteine *i*nsertion *s*equence.

Secondary amplicon: A DNA fragment, that is only weakly amplified during → polymerase chain reaction, because the used → primer possesses no complete → homology to potential target sequences and therefore allows only spurious amplification by → DNA polymerase. Such secondary amplicons appear as weak bands in → ethidium bromide-stained → agarose gels, or → silver-stained → polyacrylamide gels, respectively. See → primary amplicon.

Secondary gene transfer: The transfer of genes (or generally, DNA sequences) from an endosymbiont genome (e. g. mitochondrial or plastid genome) to the nuclear genome of the host cell after the endocytotic event.

Secondary *p*ulsed *f*ield *g*el electrophoresis (SPFG): A variant of the conventional → *p*ulsed *f*ield *g*el *e*lectrophoresis (PFGE), that uses periodic, short and intense electrical pulses along the direction of net DNA migration. This procedure dramatically increases the overall rate of DNA motion, which in some cases results in improved resolution and allows to separate larger DNA molecules (as compared to PFGE). The mechanism of SPFG is not clear, but may be a consequence of the accelerated release of DNA molecules from obstacles in the gel by secondary intense pulses.

Second filial generation: See → F2.

Second strand: The strand complementary to the → first strand in conventional → cDNA cloning. The second strand is synthesized by → DNA polymerase after the RNA template (on which the first strand has been made by → reverse transcriptase) is removed by either alkaline hydrolysis or treatment with → RNase H. See also → RNA priming.

Second strand synthesis: The formation of a → second strand in conventional → cDNA cloning.

Secreted *a*lkaline *p*hosphatase (SEAP): An → alkaline phosphatase protein with sequences that direct its secretion. Genes encoding SEAP are usually cloned into → promoter trap vectors and used as → reporter genes which are only active in the presence of a promotor. This promotor is usually originating from anonymous genomic sequences cloned into the cloning site of the promotor trap vector, which is located → upstream of the SEAP gene. Transcription of the SEAP sequences leads to the accumulation and excretion of alkaline phosphatase that can be easily measured in the surrounding medium.

Secretion vector: Any → cloning vector (e. g. a → plasmid), containing a → *m*ultiple *c*loning *s*ite (MCS) for the insertion of a → gene or → cDNA and a sequence encoding an NH_2-terminal → signal peptide upstream of the MCS, that directs the protein transcribed from the inserted gene to the secretary pathway (via the endoplasmic reticulum). The signal peptide is cleaved off, and the protein secreted into the medium, from which it can easily be isolated.

Secretome: An additional term of the ome mania for the complete subset of proteins, that are actively secreted (either into the intraluminal cisternae of the endoplasmic reticulum, the lysosomes or vacuoles, or into the extracellular space).

Sedimentation: The migration of molecules in natural gravitational fields (e.g. towards the gravitational center of the earth) or artificial gravitational fields (e.g. generated by centrifugal forces, see → ultracentrifugation) that is driven by mass attraction. See → sedimentation coefficient, → sedimentation rate.

Sedimentation coefficient (sedimentation constant, Svedberg constant, Svedberg unit, S value, S): The sedimentation rate of a macromolecule per unit of applied gravitational force, as defined by

$$S = \frac{dr}{\omega^2 r \cdot dt}$$

r,   radius: i.e. the distance in cm between the molecule and the rotation center;
ω,   the angular velocity of the rotor (in radians/s);
$\frac{dr}{dt}$, the rate of particle movement (in cm/s).

The unit of S is reciprocal seconds. For convenience, the basic unit is taken as 10^{-13} s and coined one Svedberg unit (S).

Sedimentation constant: See → sedimentation coefficient.

Sedimentation rate: The velocity in cm/s of a spheric particle sedimenting under specific conditions. It is proportional to the square of the particle diameter and to the difference between the particle density and the density of the surrounding medium, as defined by

$$\text{Sedimentation rate} = \frac{d^2(\rho_p - \rho L) \cdot g}{18\,\mu}$$

d, diameter of the particle;
ρ, the density of the particle;   μ, the viscosity of the surrounding medium;
L, the density of the surrounding medium;   g, gravitational force.

Seed: Any short base sequence in a sequence database, that serves as reference sequence for the analysis of an unknown sequence. Many sequence search methods start by finding such a seed sequence, then extend this seed into longer, non-matching sequences. The size of the initial seed ("word length" in BLAST) is important for a fast sequence analysis.

Segmental duplication: An amplification of a distinct region of a genome, such that two versions of the same sequence are present. Usually these duplicated regions share far-reaching → homology and → synteny, but became different by rearrangements (→ translocations, → deletions, → inversions) in evolutionary times.

Segmentation gene: A gene in segmented animals (e.g. insects) that regulates the location, number and polarity of body segments. In a hypothetical hierarchy of developmentally important genes, they range between maternal effect genes which determine the polarity of zygotic cytoplasm, and → homeotic genes, which are essential for the ontogenesis of the adult animal (see → homeotic mutation). In *Drosophila*, for instance, they can be classified in three major groups according to their mutant phenotype:
a) mutant "gap genes" (e.g. *Krüppel, Knirps*) lead to larvae which lack several adjacent segments.
b) mutant "pair-rule genes" (e.g. *fushi tarazu*) lead to a regular loss of parts of body segments over the whole larva (for example, in fushi tarazu mutants the second half of a segment together with the first half of the following segment is deleted).

c) mutant "segment polarity genes" (e.g. engrailed) lead to a disturbance of the dorso-ventral axis of the segments.

Segmented genome: Any genome which consists of two or more nucleic acid molecules (e.g. the alfalfa mosaic virus possesses four RNA molecules differing in size and coding capacities).

Segregant: A hybrid organism produced by crossing two genetically different parental organisms. See → segregation.

Segregation:
a) The separation of homologous chromosomes (or chromatids, genes, alleles) at meiosis, or homologous or non-homologous plasmids into different daughter cells during cell division. See also → partition, → partitioning function. Compare → co-segregation.
b) The existence of different → phenotypes among offspring, which is a consequence of chromosome or allele separation in their heterozygous parents.

Segregation distorter (SD): A → meiotic drive system of *Drosophila melanogaster*, that transmits the *SD* chromosome from *SD/SD$^+$* male animals in vast excess over its homologue, whereas transmission from females is normal. The gene underlying this phenotype (i.e. required for distortion) is *Sd*, that suffered a → gain-of-function mutation, which deleted 234 amino acids at the C-terminus. The functional *Sd* product is a truncated version of the nuclear transport protein RanGAP (the *Drosophila* homolog of mammalian RanGAP1, the guanosine triphosphatase (GTPase) activator for the Ras-related nuclear regulatory protein Ran. Strong distortion also requires several linked modifier loci, including *Enhancer* (*E/SD*), *Modifier* (*M/SD*), and *Stabilizer* (*St/SD*). The target of distortion is the socalled *Responder* locus (*Rsp*), consisting of an array of repeated → satellite sequences, whose copy number is directly correlated with sensitivity. Chromosomes carrying *Rsp^s* (sensitive) or *Rsp^{ss}* (supersensitive) loci distort, whereas *SD* chromosomes carrying *Rsp^i* (insensitive) are resistant. Distortion involves sperm dysfunction (initially visible as failed chromatin condensation in half of the developing spermatids).

Segregation map: See → genetic map.

Segregation of partly melted molecules (SPM): A variant of the → denaturing gradient gel electrophoresis technique, that allows the separation and enrichment of DNA fragments with both a GC-rich → domain and a non-GC-rich domain (as is the case with regions from the periphery of → CpG-rich islands). The strand dissociation of such fragments is low in a denaturing gradient gel, and such molecules are retained in the gel as a partly melted structure (helical at the GC-rich domain, dissociated at the non-GC-rich domain) after prolonged exposure to an electric field. Other molecules disappaear through strand dissociation.

Segregative instability: Any loss of → plasmids due to defective → partitioning functions.

SELDI: See → surface-enhanced laser desorption/ionization.

SELDI chip: See → surface-enhanced laser desorption/ionisation.

Selectable marker (selectable marker gene): Any gene that encodes resistance against an → antibiotic or toxic compound (e.g. heavy metal) and allows to select for its presence in an organism. Such selectable marker genes are commonly localized on or cloned into → plasmids which are then transformed into bacterial host cells. Since the transformation frequency is rather low, only few transformants will arise that can be identified by their new resistant phenotype. Most commonly used selectable markers are genes conferring resistance against → ampicillin, → chloramphenicol, → kanamycin, → neomycin and → tetracycline. Compare → reporter gene, → genetic marker. See also → yeast chromosomal marker.

Selection:
a) Generally, a procedure to exclude organisms with an undesired → genotype from a group of organisms with a desirable genotype, as for instance in animal or plant breeding. Selection in this case leads to the establishment of specific → strains.
b) In genetic engineering, selection is often based on the presence of a functional → antibiotic resistance gene (→ selectable marker) in the desirable genotype (which can therefore grow in the presence of the corresponding antibiotic), and its absence or non-functionality in the non-desired genotype (which cannot grow under the same conditions). See for example → selective plating.

*Selective **a**mplification of **m**icrosatellite **p**olymorphic **l**oci (SAMPL):* A technique for the detection of genetic→ polymorphisms that combines the advantages of → amplified fragment length polymorphism (AFLP) and → microsatellite marker technologies in one single assay. SAMPL allows to amplify random genomic regions by using both an AFLP → primer and a → compound microsatellite primer in conventional → polymerase chain reaction techniques, and detects length polymorphisms in the microsatellite repeats. This technique does not require previous sequence information of the target DNA. See → DNA fingerprinting, → sequence-tagged microsatellite sites, → simple sequence repeats.
Figure see page 1004.

Selective plating: A technique for the → selection of recombinant cells (e.g. bacteria). Two different auxotrophic mutants are plated together on a → minimal medium, which only allows the growth of recombinants carrying the normal → allele from each mutant.

*Selective **r**estriction **f**ragment **a**mplification (SRFA):* A technique for the amplification of selected restriction fragments of a genomic DNA. Genomic DNA is first restricted with a suitable → restriction endonuclease, then → oligonucleotide → linkers (or → adaptors) are ligated to the ends of the restriction fragments ("tagged restriction fragments"). Subsequently → primers complementary to the adaptor sequences are annealed and conventional → polymerase chain reaction techniques used to amplify the restriction fragments simultaneously (→ multiplex PCR). Two adaptors with different sequences can also be employed. The complexity of the amplifiable restriction fragments arising by digestion of large genomes can be reduced, if the adaptor contains additional nucleotides at its 3'-end. If the annealing conditions are stringent, only those fragments that perfectly match with the adaptor sequences, are amplified. Mismatches are excluded. Therefore only a fraction of all genomic restriction fragments is selected for amplification.

Selective variant: Any mutant that exists under conditions which kill the wild type (e.g. a mutant that acquired, by mutation, resistance against an → antibiotic).

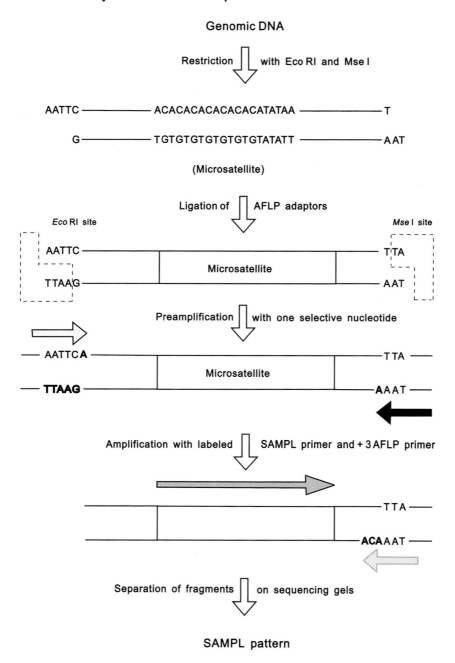

Genomic DNA

Restriction with Eco RI and Mse I

AATTC ——————— ACACACACACACACATATAA ————————— T

G—————— TGTGTGTGTGTGTGTGTATATT ——————— AAT

(Microsatellite)

Ligation of AFLP adaptors

Eco RI site   Mse I site

AATTC ——————————— TTA

Microsatellite

TTAAG ——————————— AAT

Preamplification with one selective nucleotide

—— AATTCA ——————————— TTA ——

Microsatellite

—— TTAAG ——————————— AAAT ——

Amplification with labeled SAMPL primer and + 3 AFLP primer

——————————— TTA ——

——————————— ACAAAT ——

Separation of fragments on sequencing gels

SAMPL pattern

Selenocysteine insertion sequence (SECIS): A → hairpin structure in the 3'-UTR of selenoprotein → messenger RNAs, that recognizes UGA as selenocysteine codon. Effective selenoprotein translation is dependent on socalled SECIS-binding proteins, that facilitate recoding (i.e. recognition of UGA as selenocysteine codon, and not as → stop codon).

SELEX: See → *s*ystematic *e*volution of *l*igands by *e*xponential enrichment.

*Self-a*ssembled *m*onolayer (SAM): Any two-dimensional, regularly ordered molecular aggregate, that forms spontaneously and in which the molecules are hold together by non-covalent interactions. For example, SAMs consisting of carbohydrates with a terminal thiol group can be assembled on a gold substrate, and represent mechanically stable structures, that can be modified by coupling biological molecules (e.g. receptor proteins). Such SAMs can be used as biological sensors.

Self-assembly: The spontaneous aggregation of monomeric macromolecules into multimeric complexes which are stabilized by a series of bonds (e.g. hydrogen bonds, ionic and hydrophobic interactions). In vivo, the assembly of such monomers may be catalyzed by specific proteins (see → chaperones, → chaperonins).

*Self-in*activating expression vector (SIN; SIN vector; SI expression vector): An → expression vector designed for retroviral → gene transfer and → expression of transferred genes from host cell → promoters in mammalian target cells. For example, a derivative of a vector from *Mo*loney *mu*rine *l*eukemia *v*irus (MoMuLV) contains a 299 bp deletion within the 3'-LTR, engulfing the 72 bp repeat → enhancer and → promoter functions, but leaving the → TATA box intact. Upon → transfection of viruses derived from such a vector into a packaging cell line and → reverse transcription, the defective 3'-LTR is copied and replaces the 5'-LTR of the vector, resulting in the inactivation of the proviral 5'-LTR promoter in the infected cell (i. e. no transcription of viral sequences occurs). Since the loss of enhancers and promoter sequences during → replication of the retrovirus leads to the inactivation of the provirus in the target cell, this type of vector is called *self-in*activating expression vector SIN. SIN vectors allow the cloning of a host cell promoter and the gene it regulates, are used to study tissue-specific gene expression, and are potentially safe vectors for human → gene therapy.

Selfish DNA: See → junk DNA.

Selfish genetic elements: See → junk DNA.

Self-ligation: The (undesired) intramolecular recircularization of either → cloning vector or → insert ends in contrast to the formation of recombinant molecules (see also → circularization).

*Self-o*rganizing *m*ap (SOM): A computer program that allows the comparison of two (or more) gene expression profiles from two (or more) tissues, organs, or organisms and the identification of distinct expression patterns common to the compared profiles.

Self-priming: The utilization of hairpin structures (→ fold-back DNA) at the 3'-terminus of the first → cDNA strand for → DNA polymerase I-catalyzed synthesis of the → second strand. After the synthesis of the first cDNA strand the corresponding → mRNA is removed from the cDNA-mRNA hybrid by either alkali or heat treatment. The liberated single-stranded cDNA forms a hairpin loop at its 3' end (corresponding to the 5' end of the mRNA template) which can be used as primer for the synthesis of a double-stranded cDNA molecule. The remaining hairpin can be eliminated with → S 1 nuclease (which frequently removes additional sequences from the 3' end, so that some information of the original mRNA may be lost).

Self-probing amplicon: A specially designed amplification → primer (→ Scorpion primer) for the detection of the products of conventional → polymerase chain reactions. This primer contains an integral extension (tail) consisting of a → probe sequence element as a loop, and a pair of self-complementary stem sequences (forming a → hairpin structure), and a fluorophore/fluorescence quencher pair (e.g. tetrachloro-6-carboxyfluorescein or 6–carboxyfluorescein and a methyl red monomer, respectively). This tail is attached to the PCR primer by a *hexa*ethylene glycol (HEG) monomer linker, that prevents the copying of this extension towards the 5'end. After a round of PCR extension from the primer, a newly synthesized target region is now attached to the same strand as the probe. Following the next round of denaturation and annealing, the probe (i.e. the single-stranded loop) and the target sequence hybridize, leading to a displacement of the fluorophore from the quencher and to emission of fluorescence. The probe element is thus designed such that it hybridizes to its target only, if this target is incorporated into the same molecule by extension of the tailed primer (unimolecular probing).

Self-quenching: The loss of measurable fluorescence emission energy by energy transfer between adjacent fluorescence molecules on e.g. a → microarray, which obscures the actually emitted fluorescence light as a direct measure for fluorescence signal strength (and hence the extent of hybridisation between → probe and → target). Self-quenching is therefore undesirable. See → signal compression.

Self-replicating elements: A generic name for all extrachromosomal elements of pro- and euka-ryotes, that carry → origins of replication (allowing them to initiate their own DNA synthesis).

Self-splicing: See → ribozyme.

Self-sustained sequence replication **(SSSR, 3SR):** A technique for the isothermal *in vitro* → am-plification of a target nucleic acid sequence, that allows to start with the RNA transcribed from the target. In short, the RNA is first reverse-transcribed (by → reverse transcriptase) into the → first strand of a → cDNA using a → primer complementary to the 3' end of the template and carrying a → T7 RNA polymerase → promoter sequence. The RNA template is destroyed by *E. coli* → RNase H, and the → second strand synthesized. Transcription-competent cDNAs are then used to produce multiple (50-1000) copies of → anti-sense RNAs of the original target with T7 RNA polymerase. The amplified anti-sense transcripts are immediately transcribed into dou-ble-stranded cDNA copies, using a second T7 promoter-containing primer. These cDNAs in turn can be used as transcription templates. This process continues in a self-sustained mode under isothermal conditions (e.g. 42°C), until enzymes are inactivated or compounds in the reaction mixture become limiting.
These continuous cycles of reverse transcription and RNA transcription lead to the production of up to 10^8 copies of the target molecule in only half an hour. It is thus an interesting alternative to the widely used → polymerase chain reaction.

Self-transmissible plasmid: See → conjugative plasmid.

Semiconservative replication: An expression describing the fact that during DNA duplication, the two strands of a DNA duplex molecule separate and a complementary strand is synthesized using each of the two parent strands as template. As a result of semiconservative replication, two duplex molecules are generated, half derived from the parent template, half being newly syn-thesized DNA. The term has been developed, before the exact mode of DNA → replication was known and it was still discussed whether dsDNA might not be synthesized completely de novo ("conservative replication").

Semidiscontinuous replication: The type of DNA duplication (→ replication) in which one new strand is polymerized continuously (→ leading strand), and the other is polymerized discontinuously with so-called → Okazaki fragments (→ lagging strand).

Semi-dry blot: See → semi-dry blotting.

Semi-dry blotting **(semi-dry blot, SDB, semi-dry transfer):** A modification of the conventional → Southern, → Northern or → protein blotting procedure that uses an electrical potential to drive the transfer of DNA, RNA or protein molecules from → agarose or → polyacrylamide gels onto nylon-based membranes. The gel with the separated molecules and the filter are sandwiched between layers of buffer-saturated filter paper that are positioned directly against a platinum coated titanium anode plate and a stainless steel cathode plate in order to optimize field strength for efficient transfer. The transfer process occurs in a minimum of liquid ("semidry").

Semi-dry transfer: See → semi-dry blotting.

Semi-random amplified polymorphic DNA (semi-random RAPD): Any DNA fragment that has been amplified using one specific → primer (e.g. complementary to a dispersed repeat or → splicing → consensus sequence) and a second primer of arbitrary nucleotide sequence (→ amplimer) and → polymerase chain reaction procedures. Semi-random PCR combines the advantages of → *random amplified polymorphic DNA* (RAPD) and → *sequence-tagged site* (STS) technologies. Compare → random amplified microsatellite polymorphism.

Semi-random RAPD: See → semi-random amplified polymorphic DNA.

Semi-specific primer technology **(SSPT):** A comprehensive term for a series of → genome scanning techniques that employ either → microsatellites, → anchored microsatellites, or anchored microsatellites in combination with → arbitrary → primers to amplify genomic regions containing microsatellites by conventional → polymerase chain reaction methodology. If one and the same primer amplifies regions of different size from different genomes, it allows to detect → polymorphisms, which can be used as → molecular marker. See → anchored microsatellite-primed PCR, → microsatellite, → microsatellite-primed PCR, → random amplified microsatellite polymorphism. Compare → arbitrary primer technology, → locus-specific primer technology.

Senescence-associated gene **(SAG):** Any one of a series of plant genes, that are activated in the course of senescence in plants, characterized by e.g. loss of chlorophyll and appearance of carotenoids in leaves, degradation of proteins and lipids, finally programmed cell death and leaf shed. For example, genes encoding asparagine and glutamine synthetases, catalase, cysteine endopeptidase, cysteine protease, glutathione-S-transferase, MAP kinase, metallothionein, PEP carboxykinase, phospholipase, polyubiquitin, RNAses, ubiquitin carrier protein and WRKY transcription factor belong to the SAG class.

Sense codon: Any of the 61 triplet → codons specifying an amino acid. Compare → stop codon.

Sense gene: See → structural gene.

Sense RNA (sense mRNA): The transcript of a → structural gene. Sense RNAs possess the same sequence as the → coding strand (sense strand), but a complementary sequence to the → template strand (→ antisense strand).

Sense strand: See → coding strand.

Sensor gene: A hypothetical gene that controls the activity of one or several → integrator genes (→ Britten-Davidson model).

Sensorgram: The graphical depiction of interaction(s) between two molecules. For example, → surface plasmon resonance (SPR) allows to measure the binding of a molecule in a test solution to a target molecule (e.g. a protein) on the surface of a chip. Binding results in an increase, dissociation in a decrease of the refractive index. The continuous, real-time monitoring of the association and dissociation of interacting molecules are recorded by a sensorgram.

Sensor probe (molecular sensor): Any synthetic oligodeoxynucleotide sequence (→ probe) that allows to detect a → point mutation in a target DNA in concert with a second synthetic probe sequence. In short, one of the sensors with a perfect match to the target sequence (i. e. a complete complementarity to its target) is labeled with fluorochrome A (e. g. → Cy 5) and hybridized to the target DNA ("anchor probe"). Then the second sensor is labeled with another fluorochrome B (e. g. → FITC), and also hybridized to the target DNA side by side to the anchor probe, such that both fluorochromes are directed towards each other. Energy exchange can only occur between the fluorophores if they are in direct contact, and therefore the complex emits fluorescence light. Now the complex between target DNA and both probes is slowly heated. If the second probe ("sensor probe") does not fully match its target sequence (e. g. by a single base exchange), then the sensor probe dissociates from the complex and the fluorescence is (drastically) reduced.

Sentrin: A cytosolic protein, that shares a high homology with → ubiquitin and also functions to tag other proteins destined for degradation. Sentrin consists of 18 kDa monomers with a C-terminal glycyl-glycine, which is adenylated using an ATP-Mg^{2+} complex. The target protein will be lysed by free proteases or the → proteasome after sentrinylation.

Separation chip: A variant of a → lab-on-a-chip, that allows the electrophoretic separation of fluorescent nucleic acids or proteins in a very short time and the visualization of the separated molecules in real-time. The glass or silicon chip with microfabricated microchannels (50x10 μm) and micro-reservoirs is first loaded with the sample (in one chamber), and the buffer (in another chamber), and the separation channels filled with dilute separation matrix (e.g. → polyacrylamide). Then the reactants are mixed by electric force (programmable voltages) and the samples separated in the separation channel. Moving lasers allow to detect the migration of the analytes in real time.

Sephacryl: A trade-mark for a covalently cross-linked, beaded allyl dextrose, that can be hydrated and used as a gel matrix to separate molecules in the size range of 5000 to 1.5 million Da in → gel filtration.

Sephadex: A trade-mark for a cross-linked beaded dextran that can be hydrated and used as a gel to separate molecules according to their size in → gel filtration.

Sepharose: A trade-mark for a beaded → agarose that can be hydrated and used as a gel to separate molecules in → gel filtration.

Sequenase: See → T7 DNA polymerase.

Sequenase™: A chemically modified → T7 DNA polymerase that still catalyzes the 5' → 3' polymerization of deoxyribonucleoside-triphosphates into duplex DNA, but lacks the 3' → 5' exonuclease activity. The enzyme possesses a high processive activity (dissociates from the primer only after incorporation of more than 50 nucleotides) and high polymerization speed. Sequenase™ is used in DNA → Sanger sequencing reactions.

Sequence alignment: The computational juxtaposition of two (or more) linear sequences of nucleotides (in DNA or RNA) or amino acids (in proteins or peptides) for the identification of the extent of → homology, sequence variants (e.g. → single nucleotide polymorphisms) and stretches of unique and conserved target sequences for e.g. the design of → primers. See → global alignment, → local alignment.

Sequence analysis: See → sequencing.

Sequence-based amplified polymorphism (SBAP) detection: A technique to discover sequence → polymorphisms between two (or more) → genomes, generally, DNA sequences, that combines the multiplexing capacity of → amplified fragment length polymorphism and the simplicity of → random amplified polymorphic DNA methods. In short, → genomic DNA is first amplified with 17 nucleotides → primers, each consisting of a fixed GC-rich sequence of 14 nucleotides at its → 5'-terminus and three selective bases at its → 3'-terminus, using conventional → polymerase chain reaction techniques. The second amplification is primed by 19 nucleotides primers, each consisting of 16 nucleotides at its 5'-terminus complementary to an AT-rich template sequence and three selective bases at its 3' end. One of the primers is end-labeled using → polynucleotide kinase and γ^{32}P-ATP. After PCR, the amplification products are separated in → denaturing polyacrylamide gels, and the bands detected by → autoradiography.

Sequence-based typing: The direct sequencing of HLA alleles amplified from genomic DNA by conventional → polymerase chain reaction (PCR) with allele-specific → primers. For example, HLA-A and HLA-C genes are amplified in a single-step PCR with locus-specific intron-derived primers, that are generic for all alleles of each locus. For the B locus, that is highly polymorphic (large number of alleles; e. g. 19 alleles at each of the B13, B14, B16, B17 and B18 loci), two amplification reactions are employed. The first one uses two group-specific 5' primers (that define the group), and a second one with a B-locus-specific 3' primer. The resulting amplification products are directly sequenced, and the sequences compared to all possible combination of alleles deposited in an appropriate database. Such sequence-based typing of the main HLA loci increases the chances of success in e. g. bone marrow transplantations.

Sequence characterized amplified region (SCAR; sequence-tagged amplified region, STAR): A genomic DNA fragment originating from a single, genetically defined locus that is identified by → polymerase chain reaction amplification using a pair of specific oligonucleotide → primers. In short, the amplification product is isolated, cloned and sequenced. A SCAR is then constructed using the original primer sequence and adding 10-20 adjacent bases of the genomic DNA. SCARs are superior to → random amplified polymorphic DNA markers, because they detect single loci only. SCARs serve as genetic markers in → physical mapping procedures and in marker-assisted selection and plant cultivar identification. See → sequence-tagged site.

Sequence-contig scaffold: See → scaffold.

Sequence coordinates: The start and stop coordinates of a specific → ORF on a chromosome with informations on exons and introns.

Sequence coverage: The average number of times, with which each base in a defined DNA sequence is independently read with a → base quality score of at least 99% accuracy.

Sequenced clone: Any usually large insert clone (e.g. a → BAC clone), that is partly of fully sequenced, deposited in a public database, and characterized by an → accession number. See → finished clone.

Sequenced clone contig (SCC, "barge"): Any → contig identified by merging overlapping fully sequenced clones (e. g. → plasmid, → bacterial artificial chromosome, or → yeast artificial chromosome clones). See → sequenced clone contig scaffold.

Sequenced clone contig scaffold: Any scaffold produced by joining → sequenced clone contigs on the basis of overlapping sequence information.

Sequenced clone layout: The assignment of a fully sequenced clone (e. g. → plasmid, → bacterial artificial chromosome, or → yeast artificial chromosome clone) to the → physical map of → fingerprint clone contigs.

Sequence gap ("fragment gap"): Any gap between adjacent → sequenced clone contigs in a → draft genome sequence, that does not represent a → fingerprint clone contig gap. See → FCC gap.

Sequence homology: See → homology.

Sequence-independent **DNA** *amplification* **(SIA):** A technique for the amplification of multiple genomic regions, using → degenerate primers in a → polymerase chain reaction. In short, primer A, consisting of a 5 nucleotide random 3' segment and a specific 16 nucleotide 5' segment (containing a → restriction recognition site) with the sequence 5'-TGGTAGCTCTTGAT-CANNNNN–3' is annealed to denatured DNA at low temperature (4° C). Annealing therefore occurs at multiple random sites in the target genome. Then → T7 DNA polymerase is employed to extend the primers. In a second PCR primer B (sequence: 5'-AGAGTTGGTAGCTCTT-GATC–3'), identical to primer A in its 5' most nucleotides, is annealed at intermediate temperature (42° C) and extended. Finally, additional cycles are carried out at a specific annealing temperature (56° C). SIA produces amplicons of 0.2 – 0.8 kb, that can be isolated from a separation gel and be used as chromosome-specific hybridization probes in e. g. → fluorescent *in situ* hybridization, if the DNA is originating from individual or microdissected chromosomes.

Sequence-independent primer reverse transcription polymerase chain reaction: See → poly(A)-PCR.

Sequence insertion vector: A → cloning vector that is able to insert *in toto* into an endogenous → locus thus disrupting original sequences and leading to → mutations (e.g. the inactivation of target genes). See also → integrative vector; compare → sequence replacement vector.

Sequence map: The linear projection of nucleotides (in RNA or ssDNA) or nucleotide pairs (in dsDNA or a chromosome). See → chromosome expression map, → chromosome map, → cytogenetic map, → denaturation map, → diversity map, → expression map, → frequency distance map, → gene map, → genetic map, → integrated map, → landmark map, → linkage map, → map, → marker map,→ nucleotide diversity map, → quantitative chromosome map, → recombination frequency map, → response regulation map, → restriction map, → RN map.

Sequence-related amplified polymorphism (SRAP): A technique for the preferential amplification of → genic sequences (i.e. → open reading frames). In short, → genomic DNA is first amplified with two specially designed 17 or 18mer primers. Of these, the forward primer consists of a core sequence of 14 bases, where the first 10–11 bases from the 5'-end are so called "filler" sequences of no specific constitution (i.e.arbitrary), followed by 5'-CCGG–3' (targeting at → exons that are embedded in GC-rich regions) and three selective nucleotides at the 3'-terminus. The reverse primer shares identical components except that the "filler" is followed by 5'-AATT–3'(targeting at AT-rich regions, normally found in → promoters and → introns). Both primers contain 40–50% GC, do not form → hairpins or other secondary structures, and are labeled with γ^{33}P-ATP. After amplification of genic sequences in a conventional → polymerase chain reaction, the amplification products are separated by → denaturing polyacrylamide gel electrophoresis and the polymorphic bands detected by → autoradiography. Polymorphisms are caused by → insertions or → deletions, rarely by single base exchanges in introns, promoters and spacers. SRAP can be used to generate → co-dominant markers for → genetic mapping, → gene tagging, genomic and cDNA fingerprinting and → map-based cloning.

Sequence replacement vector: A → cloning vector that replaces endogenous DNA sequences with exogenous sequences, thus disrupting original sequences and leading to → mutations (e.g. the inactivation of target genes). See also → integrative vector, → integration vector; compare → sequence insertion vector.

Sequence-specific amplification polymorphism (S-SAP): A technique that combines → amplification fragment length polymorphism (AFLP) with sequence-specific → polymerase chain reaction to identify DNA → polymorphisms between related organisms. As a sequence-specific → primer an → oligonucleotide complementary to the conserved 5'terminus of → long terminal repeats (LTRs) of → retrotransposons (e.g. → Ty1 copia, and others) is used. In short, genomic DNA is first restricted with a rarely and a frequently cutting → restriction endonuclease (e.g. *Eco*RI and *Mse*I, respectively), then *Eco*RI and *Mse*I → adaptors are ligated to the fragments, which are then preamplified with adaptor-homologous primers in conventional PCR. The preamplified products are then selectively amplified with a ^{33}P- or ^{32}P-labeled oligonucleotide complementary to the end of the LTR of a suitable retrotransposon, and either a rare or frequent site adapter-homologous oligonucleotide. The resulting fragments are denatured and separated on → sequencing gels. The detected polymorphisms may result from sequence variation at flanking or internal restriction sites of the genomic DNA or from variation(s) at the 5'terminus of the LTR of the targeted retrotransposon, or also from → INDEL events within the retrotransposable element. In e.g. barley, S-SAP reveals more polymorphisms than AFLP alone.

Sequence-specific oligonucleotide (SSO) typing: A technique for the unequivocal discrimination between sequence variants (e.g. the different alleles of a specific human leucocyte antigen [HLA] locus), that starts with the specific amplification of a target gene with locus-specific → primers ("sequence-specific oligonucleotides") in a conventional → polymerase chain reaction with e.g. → biotin-labeled nucleotides. The biotin-labeled amplification products are then hybridized to discriminatory sequence-specific oligonucleotides (e.g. bound to a chip surface, see → nested-on-chip polymerase chain reaction), and the hybridization event detected with → streptavidin-cyanine 5 conjugates. As a result, the specific detection of the alleles in a sample is possible. See → sequence-specific priming.

Sequence-specific primer (SP): Any → primer that allows to amplify a specific sequence out of a complex genome. For example, gene-specific primers belong to the SP group. Compare → arbitrary primer.

Sequence-specific priming (SSP): A variant of the conventional → polymerase chain reaction, that allows to detect and discriminate sequence variants (e.g. the different alleles of a specific *human leucocyte antigen* [HLA] locus). SSP employs DNA → primer combinations that specifically amplify distinct targets (e.g. alleles), but only, if the 3'-termini of both primers are complementary to the corresponding target sequences. Compare → sequence-specific oligonucleotide typing. See → *nested on chip* (NOC) polymerase chain reaction.

Sequence-tagged amplified region: See → sequence-characterized amplified region.

Sequence-tagged connector (STC): A short DNA sequence of about 500 bases, generated by → sequencing the end of a → bacterial artificial chromosome (BAC) clone. Since this sequence allows to detect other BAC clones carrying inserts derived from the same original DNA (e.g. a specific region of a genome), if functions as "connector" for these clones. Moreover, STCs can be used for the design of → *sequence-tagged sites* (STSs), that represent DNA stretches for the → polymerase chain reaction-based amplification of specific loci without amplifying unwanted genomic regions. Also, STCs may serve as → molecular markers (human genome: one STC every 3-4 kb). Compare → sequence-tagged site.

Sequence-tagged microsatellite site (STMS): Any → genomic DNA sequence that flanks → microsatellite clusters or preferentially a specific microsatellite sequence in eukaryotic genomes.

Sequence-tagged restriction site (STAR): A short DNA sequence, that identifies the sequences flanking the → recognition site of a → restriction endonuclease.

Sequence-tagged site (STS): Any short track of about 200-500 base pairs that is unique to a given → genomic DNA fragment and serves to identify that fragment among thousands of other fragments used to construct a → genetic and → physical map of a eukaryotic genome. STSs, if known for all genomic DNA fragments used for the mapping procedure, eliminate the need to store and exchange clones. If a clone from a specific part of the genome is needed, a database search for an STS mapped to the region of interest will allow the design of primers for the → polymerase chain reaction amplification of the STS. The amplification product is then labeled and used as a probe to fish the corresponding DNA fragment from a → gene library. About 30000 to 50000 STSs distributed throughout e.g. the human genome at about 100 kb-intervals are sufficient to construct a physical map of the entire human genome (STS map).

Sequence-tagged site database (dbSTS): A database containing sequences, mapping data and other relevant information on → sequence-tagged sites. Compare → single nucleotide polymorphism database.

Sequence-tagged sites map (STS map): A → physical map of a genome that is based on the alignment of → sequence-tagged sites.

Sequencing (sequence analysis): The determination of the sequence of bases in DNA or RNA, or of amino acids in proteins. Sequence analyses also comprise the study of secondary and tertiary structures, and of the occurrence of → consensus sequences and other molecular characteristics via computer analysis. See → automated DNA sequencing, → chemical sequencing, → Church Gilbert sequencing, → cycle sequencing, → direct sequencing, → megasequencing, → multiplex sequencing, → phosphorothioate sequencing, → plasmid sequencing, → pyrosequencing, → single colony sequencing, → single molecule sequencing, → shotgun sequencing, → tetra-octa sequencing, → thermal cycle sequencing, → trash DNA sequencing, → uniplex DNA sequencing, → whole chromosome shotgun sequencing, → whole genome shotgun sequencing.

Sequencing array: A high-density → DNA chip onto which multiple short oligonucleotides are fixed, that allows to determine the sequence of a DNA fragment which is hybridized to it. Since the sequence of each oligonucleotide is known, hybridization patterns can directly be converted into base sequence information. See → cDNA expression array, → expression array, → gene array, → microarray, → sequencing by hybridization. Compare → protein chip.

Sequencing by hybridization (SBH; hybridization sequencing; *fragmentation sequencing, FS*): A technique for the rapid sequence determination of DNA fragments, which is based on the hybridization of short oligodeoxynucleotides (6-10 bases) to a target sequence under extremely stringent conditions and the detection of hybrids. SBH techniques fall into two broad categories. Category I works with an array of synthetic hexa- to decameric oligodeoxynucleotides fixed onto a solid support (e. g. a nylon membrane, glass or quartz slides, polypropylene chip). A labeled DNA fragment with unknown sequence of some hundred bases in length is then exposed to such a chip, where all base sequences complementary to the set of oligonucleotides (which may be 65,536 in number) will hybridize.

DNA sample:	5'-A C G T A C A A T G G C G A T-3'
octamer 1	T G C A T G T T
octamer 2	G C A T G T T A
octamer 3	C A T G T T A C
octamer 4	A T G T T A C C
octamer 5	T G T T A C C G
octamer 6	G T T A C C G C
octamer 7	T T A C C G C T
octamer 8	T A C C G C T A
Deduced sequence:	5'-A C G T A C A A T G G C G A T-3'
	3'-T G C A T G T T A C C G C T A-5'

Since the octamers produce a nested array, i.e. the first is complementary to bases 1-8 of the unknown DNA, the second to bases 2-9, the third to bases 3-10, and so on, the overlaps can be used to put the octamers in order, so that a computer can determine the full sequence of the unknown DNA.

Category II is based on the hybridization of labeled oligodeoxynucleotides to the immobilized target sequence. Potential drawbacks of SBH involve (1) mismatched oligonucleotides (for an oligo of length n, only one single perfect match, yet six terminal and 3 (n-2) possible internal mismatches exist, leading to an increase in background signal intensity) and (2) repetitive sequences (especially sequences of a length of n-1, where n = length of the complete oligonucleotide).

Sequencing by *h*ybridization to *o*ligonucleotide *m*icrochips (SHOM): A variant of the → sequencing by hybridization technique, that works with → oligodeoxynucleotides immobilized on a gel-based microchip element. In short, a matrix of → polyacrylamide gel is first polymerized on a glass plate previously treated with binding silane. Then strips of the gel are removed in x-y directions by mechanical scribing or by laser evaporation, so that an array of small-sized gel elements (size: 40×40 μm), separated by gel-free spacers, is formed. After repel silane treatment of the spacers (to prohibit oligonucleotide binding), the gel is activated by substitution of some amide groups with hydrazide groups (hydrazinehydrate treatment), and activated pre-synthesized oligodeoxynucleotides are spotted onto the gel elements with a robot, and immobilized there. Theoretically, all 65, 536 possible 8mers can be accomodated on the chip. Probe DNA is then labeled with a → fluorochrome (e. g. → FAM, → HEX,), hybridized to the microchip, and the fluorescence signals monitored with a two-wavelength epifluorescence microscope. The gel support increases the capacity for oligonucleotide immobilization and facilitates hybridization and → mismatch detection. However, longer oligodeoxynucleotides cannot diffuse into the gel and have to be fragmented. SHOM allows to detect mismatch mutations and gene → polymorphisms and is therefore used for the diagnosis of genetic diseases.

	Point mutation	Solution hybridization	Chip hybridization
I	GACGCTAC**T**AAGATC		I
II	GACGCTAC**G**AAGATC	I II	

Octamer oligonucleotides:

		Solution I	Solution II
1	*G**ATTCTAG	+	+
2	*TG**ATTCTA	+	+
3	*ATG**ATTC	+	+
4	*GATG**ATTC	+	+
5	*CGATG**ATT	+	+
6	*GCGATG**AT	+	+
7	*TGCGATG**A	+	−
8	*CTGCGATG	+	+
9	*G**CAAGATG	−	+

Chip hybridization:
I: 1■ 2■ 3■ / 4■ 5■ 6■ / 7■ 8■ 9□
II: 1■ 2■ 3■ / 4■ 5■ 6■ / 7□ 8■ 9■

Sequencing by hybridization

Sequencing error: Any base introduced into a newly synthesized DNA strand, that does not match the complementary base on the corresponding template strand. Sequencing errors are a source for misinterpretations of sequence data (e.g. → can suggest the presence of a → single nucleotide polymorphism at a particular locus, that does not exist in reality) and can only be corrected by → re-sequencing.

Sequencing gel: An ultrathin high resolution gel for the fractionation of single-stranded (denatured) DNA fragments according to their length, containing from 6-20% polyacrylamide and 7 M urea (to minimize effects of DNA secondary structures on the electrophoretic mobility). Sequencing gels allow the separation of fragments that differ in length by only one single nucleotide. See → DNA sequencing.

Sequencing primer: A short synthetic → oligonucleotide that is complementary to a sequence at one end of a single-stranded DNA target molecule, hybridizes to it, and allows the → Klenow fragment of → DNA polymerase I to synthesize a complementary copy of the DNA target starting at the 3' hydroxyl terminus of the primer. Such oligonucleotide primers are used for → Sanger sequencing procedures. See also → reverse sequencing primer, → primer, → splinker, → SP6 sequencing primer, → T7 sequencing primer.

Sequencing *primer linker*: See → splinker.

Sequential *insertion site* (SIS): A DNA sequence inserted into → NOMAD vectors, that consists (from 5'→3') of a *Bsm* BI → restriction recognition site (sequence: 5'-CGTCTC-3'), a spacer (sequence: 5'-TCTTGA-3'), and a *Bsa*I restriction site in opposite orientation (sequence: 5'-GAGACC-3'). If such NOMAD vectors are cut within the SIS with either *Bsa*I, *Bsm* AI or *Bsm* BI, cohesive ends are produced (3'-GAAC-5' and 5'-CTTG-3') that are rotationally non-equivalent and therefore allow directional insertion of a module with complementary sticky ends into the vector. A second module can be inserted 5' or 3' to the first one by cutting the vector with *Bsm* BI and *Bsa*I, respectively. This procedure is repeated until a recombinant molecule of the desired modular arrangement is assembled. See → excisable sequential insertion site.

Sequential ligation: A method for the *in vitro* synthesis of a gene. Single-stranded, complementary and overlapping → oligonucleotides are first associated to form duplexes that are covalently linked to each other by → DNA ligase. The constructs are cloned into an appropriate → cloning vector. Each clone contains part of the desired gene. Then the different synthetic inserts are isolated from the vector and sequentially ligated to form the complete gene. See → gene synthesis.

Sequential *structure alignment program* (SSAP): A computer program for the assessment of similarities between proteins, that compares the structural environments of the constituent amino acids. It returns a score of 100 for identical proteins, and >80 for homologous proteins. More distantly related proteins with scores of >70 are grouped in the same protein family, if they perform similar biological functions.

Sequential transformation: A process whereby foreign DNA is transferred into target cells in two consecutive steps, leading to the formation of a → double transformant.

Serial *analysis* of *gene* *expression* (SAGE): A high-throughput technique for the simultaneous detection, identification and quantitation of virtually all genes expressed in a given cell at a given time, that additionally allows to identify unknown genes, → novel genes, up or down-regulated genes, to monitor patterns of gene expression at various developmental stages and define disease marker transcripts. SAGE is based on the isolation of a short, 9-14 bp socalled SAGE tag from a defined location within a transcript, that contains unique and sufficient information to identify specifically this transcript ("diagnostic tag"). Such tags from various transcripts are then concatenated serially into a single long DNA molecule for efficient sequencing and for identification of the multiple tags simultaneously. The expression pattern of any transcript population can be quantitated by determining the abundance of individual tags and identifying the gene corresponding to each tag. The sequence data is analyzed by special software to identify each gene expressed in the cell, and to determine its expression level.

In short, total polyadenylated → messenger RNAs is first prepared from the target cell or tissue, reverse transcribed into → cDNAs in the presence of a biotinylated oligo (dT) → primer (biotin-5'T$_{18}$-3') such that they all carry biotin at their 3'-termini. Then the cDNAs are cleaved with

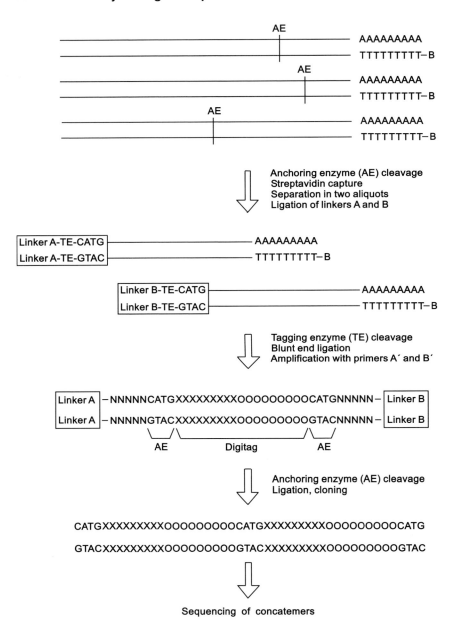

Serial Analysis of Gene Expression

the → restriction endonuclease *Nla*III ("anchoring enzyme"), and the 3'-terminal cDNA fragment captured with → streptavidin-coated magnetic beads. After ligation of an oligodeoxynucleotide linker containing the → recognition site for *Bsm* FI ("tagging enzyme", that cleaves 14-20 bp away from its asymmetric recognition site), the linkered cDNA is released from the beads by digestion with *Bsm* FI. The resulting → overhang of the released tag is filled-in with the → Klenow fragment (or → DNA polymerase I), the tags are ligated to one another, concatemerized, and amplified in a conventional → polymerase chain reaction to create hundreds of copies of each tag. From 30-50 such tags are serially ligated in a single DNA molecule, which is cloned and sequenced.

The number of times each tag is represented correlates with the number of mRNAs originally present in the cell or tissue (i. e. is an index for the expression of the corresponding gene). However, SAGE neither detects transcripts that lack an *Nla*III site, nor very → low abundance messenger RNAs. See → LongSAGE, → SuperSAGE.

Serial *i*nvasive *s*ignal *a*mplification *r*eaction (SISAR): An isothermal technique for the detection and high-throughput analysis of → single nucleotide polymorphisms and → point mutations in a genome. In short, two oligonucleotides (a socalled probe and an upstream oligonucleotide) are first annealed to the target DNA. The probe contains two regions, a target-specific region that forms a duplex with the target, and a non-complementary 5' arm region, that serves as reporter molecule precursor. The 3' terminal nucleotide (a T) of the upstream oligonucleotide overlaps with the terminal AT base pair of the probe-target duplex. This overlap or "invasion" represents the substrate for the structure-specific 5' nuclease *Afu* FEN from *Archaeoglobus fulgidus,* which cleaves the probe on the 3' side of the first base-paired nucleotide at the position dictated by the 3' end of the upstream oligonucleotide. Cleavage releases the noncomplementary 5' arm together with one nucleotide of the complementary region of the probe.

In a secondary reaction, the target strand and the probe adopt a → hairpin structure ("secondary probe"), whose 5' end is labeled with one fluorochrome (e. g. → fluorescein), whereas its interior carries another fluorochrome (e. g. → cyanine 3). Both fluorophores function as a → *f*luorescence-*r*esonance *e*nergy *t*ransfer (FRET) pair, in which the fluorescence energy of the fluorescein is quenched by cyanine 3. If this secondary probe anneals to the 5' arm released in the primary reaction, then the *Afu* nuclease again cleaves this structure and relaxes the distance between both fluorophores: increased fluorescence is a quantitative measure for the concentration of the target DNA. Since the 5' nuclease owns an extremely high cleavage specificity and discriminates substrates differing by only one nucleotide, SISAR allows to screen for single-nucleotide polymorphisms in a high-throughput format.

Ser*i*ne *p*roteinase *i*nhibitor (Serpin): Any member of a protein superfamily of serine proteinase inhibitors that are involved in inflammation, blood coagulation and fibrinolysis.

Seromycin: See → cycloserine.

Serpin: See → serine proteinase inhibitor.

Serum profiling: The determination and quantitation of the (preferably) complete set of peptides, proteins and their variants (e.g. glycoproteins) of the serum. Serum profiling serves to detect differences between the sera of e.g. healthy and diseased individuals.

SEV: See → *s*plit *e*nd *v*ector.

7–*H*ydroxy-*co*umarin–3–carboxylic acid (HYCO): A → fluorochrome, that is used as marker for → fluorescent primers in e.g. → polymerase chain reactions, labeling of probes for hybridisation experiments, and in multicolour labelling. The molecule can be excited by light of 419 nm wave-length, and emits fluorescence light at 447 nm. Since the wave-length of the excitation and emission maxima is pH-dependent, the exact values vary.

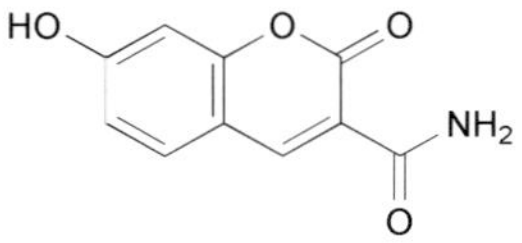

HO⟋⤴⟍⟋O⟍⟋O
       NH₂
       ‖
       O

7–Propynyl isocarbostyril (PICS): A synthetic (i.e. nonnatural) nucleic acid base that can pair with its itself. If incorporated into a target DNA, PICS – as opposed to other nonnatural bases – does not prompt → mispairing of the conventional bases A,C,G, and T.

7SL-RNA: The RNA component of the → signal recognition particle.

Severely affected *Alzheimer* *d*isease DNA (SAD-DNA): The DNA from cells of the hippocampal region of patients with advanced and severe Alzheimer's disease symptoms. SAD-DNA has a substantially higher proportion of left-handed → Z-DNA than normal control DNA, binds less → ethidium bromide than B-DNA, and melts at relatively high temperature (T_M = 79°C; B-DNA: 59 – 65°C). See → moderately affected Alzheimer disease DNA.

SEV system: See → split end vector.

Sex-duction: The transfer of chromosomal genes from one bacterium (donor) to another bacterium (acceptor) via → conjugation.

Sex factor: See → conjugative plasmid.

Sex pilus: See → pilus

Sexual polymerase chain reaction: See → DNA shuffling.

S-gal (3,4–cyclohexeneoesculetin-ß-D-galactopyranoside): A colourless substrate for → ß-galactosidase, which is hydrolyzed at the O-linkage between the galactoside and the cyclohexenoesculetin sidegroup. Two of the resulting sidegroups chelate Fe^{3+} (which is added to the reaction as ferric ammonium citrate) and stain black. Any recombinant colony (i.e. a bacterium containing a plasmid with an → insert, which insertionally inactivates ß-galactosidase) can then easily be discriminated as white from the black wild-type (→ "black-white screening"). As compared to → X-gal, S-gal possesses improved solubility, is less light-sensitive, but fully autoclavable, and provides darkly stained colonies or → plaques (indicating absence of an → insert). Therefore, the → black-white screening with S-gal is advantageous as compared to → blue-white screening.

SGE:
a) See → junk DNA.
b) See → transposon.

Sg mRNA: See → *s*ubgenomic *m*essenger RNA.

sgRNA: See → *s*ubgenomic RNA.

Shared enhancer: Any → enhancer element that functions bidirectionally to control the expression of both a specific gene upstream and another gene downstream of its location. For example, in chicken erythroid cells such an enhancer was localized about 1.6 kb 3' to the promoter of the β-globin gene, and 1.5 kb 5' to the promoter of the adjacent ε-globin gene. The shared enhancer stimulates the expression of both genes in a developmentally regulated way.

Sharkstooth comb: A special slot former (comb) for → agarose or → polyacrylamide gels whose individual teeth are arranged and formed like teeth of a shark. The distance between slots formed by such a sharkstooth comb is small (e.g. 3 mm) so that lane-to-lane comparisons on e.g. → sequencing gels can more accurately be made.

Shatter: The fragmentation of a genome or a DNA molecule by → restriction endonucleases which yield → blunt ends so that the fragments cannot reassemble in the presence of ligase, except when → T4 DNA ligase and very high substrate concentrations are used.

Shearing:
a) A technique for breaking up cells or molecules by high-speed stirring in a blendor or by the use of ultrasound (ultrasonication).
b) A method to break down high molecular weight DNA into small fragments of approximately 250-450 nucleotides (e.g. by passing it through a French press or forcing it through a hypodermic needle).

Sheath flow cuvette technique: A method to determine the length of DNA fragments that are loaded with the intercalating dye → ethidium bromide. The fragments are passed through a laser beam in a very thin stream of water in a sheath flow cuvette. The laser excites the ethidium bromide in the fragments, that emits a short burst of light, which can be detected and recorded. Since there is a linear relationship between the number of dye molecules bound and the length of a DNA fragment, the intensity of the emitted light is proportional to the fragment length.

Shift:
a) Band shift, see → band shifting.
b) Intrachromosomal translocation, see → translocation.

Shine-Dalgarno box: See → Shine-Dalgarno sequence.

Shine-Dalgarno sequence (SD sequence, Shine-Dalgarno box, SD-box; ribosome recognition sequence, ribosome-binding site, RBS; ribosome attachment site): The ribosome binding site on prokaryotic → messenger RNA, which in *E. coli* is located 3-11 nucleotides upstream of the AUG initiation codon in the → leader sequence. It is complementary to a highly conserved sequence at the 3' end of 16S → ribosomal RNA. Hybridization of the SD sequence to that part of 16S rRNA anchors the mRNA tightly on the ribosome:

E. coli **SD-core**

5'....AAGGAGGU....3' mRNA
3'....UUCCUCCA....5' 16S rRNA

E. coli **16S ribosomal RNA**

Together with the → start codon AUG the Shine-Dalgarno box forms the "ribosome binding site" of prokaryotic mRNA. See also → portable SD sequence.

SHOM: See → *s*equencing by *h*ybridization to *o*ligonucleotide *m*icrochips.

Shooter mutant: A mutant of → *Agrobacterium tumefaciens*, in which the → auxin genes have been partially or completely deleted. Therefore, shooter mutants induce an enhanced cytokinin level in transformed plant tissue. In some plant species, transformation with a shooter mutant leads to a "shooty" phenotype, i.e. the appearance of many shoots, so that these mutants are preferably used to induce regeneration of species with a low regeneration capacity.

Short *d*ispersed *r*epeat (SDR): Any → direct or → inverted repeat sequence of 50-1000 bp of organelles (chloroplasts or mitochondria). SDRs may comprise more than 20 % of the organellar genome (e. g. in chloroplast genomes of *Chlamydomonas reinhardii* and higher plants).

Short *h*airpin *RNA* (shRNA, small hairpin RNA): Any one of a series of artificial→ small RNAs, either synthesized exogenously by a → T7 RNA polymerase system and transfected into a target cell, or endogenously transcribed from corresponding genes incorporated into the target cell genome and controlled by → RNA polymerase III promoters. The shRNAs consist of short, usually 20–30 bp stems and a loop of unpaired bases and suppress the expression of target genes through a mechanism resembling → RNA interference. If expressed constitutively in target cells, hsRNAs can silence specific genes permanently and therefore allow to establish continuous cell lines or transgenic organisms. The presence of a spliceable → intron in the shRNA transgene enhances its silencing efficiency. See → non-coding RNA, → short interfering RNA, → small endogenous RNA, → small non-messenger RNA, → small regulatory RNA, → tiny RNA.

Short *i*nternal *e*liminated *s*equence (short IES): Any unique, usually AT-rich, and → ORF-less low-copy number → internal eliminated DNA sequence, that is deleted from the → micronucleus of hypotrich ciliate protozoa (e.g. *Oxytricha nova*, *O. fallax*, *O. trifallax*, *Stylonychia lemnae*, *S. pustulata*, and *Euplotes crassus*), ranges from 10 to 539 base pairs in length, and represents the precursor for the newly arranged genome of the → macronucleus. More than 60,000 short IESs are excised during macronuclear development in O. nova, and about 40,000 IESs exist in E. crassus and O.fallax. IESs are flanked by short direct repeat sequences of 2–7 bp (e.g. in E. *crassus*: 5'-TA-3'), and one copy of such a direct repeat is maintained in the genome of the macronucleus after IES excision. Some of the short internal eliminated sequences resemble te-lomeric repeats (Euplotes: "telIESs"). See → chromosome fragmentation,, → nuclear dimor-phism, → transposon internal eliminated sequences.

Short *i*nterspersed *e*lements: See → SINES.

Short *o*ligonucleotide *m*ass *a*nalysis (SOMA): A technique for the generation and analysis of small DNA fragments from → polymerase chain reaction-produced amplification products for mass spectrometry. In short, the genomic target region is first amplified with 35–38 nucleotide long → primers containing an artificial recognition site for the type IIS → restriction endonuclease *Bpm*I (5'- CTGGAG–3') embedded in sequences perfectly complementary to the targeted region. The PCR product is then digested with *Bpm*I, which cuts away from its recognition site. As a result, an internal sequence of 7–20 bases in length is generated. The DNA is purified by standard organic extraction and ethanol precipitation, and the fragments analysed by → electrospray ionization mass spectrometry. If the target DNA from two (or more) individuals contains single-base polymorphisms (or larger → deletions or → insertions), then SOMA will detect these. The-refore SOMA is used for the analysis of genetic variants, that differ at given position(s) in an e.g. disease-associated gene, common in the population under study.

Short patch repair: The excision of about 20 nucleotides around and including a site of DNA damage (e.g. a missing base or a → thymine dimer) and the repair of the resulting → gap by → DNA polymerase that uses the undamaged strand as → template. The genes involved in this type of repair (*uvrA, uvrB, uvrC, uvrD*) are constitutively expressed in *E. coli*, but their expression is enhanced by the induction of the → SOS repair system. Compare → long patch repair.

Short-period interspersion: The separation of non-repetitive sequences of about 1 kb length by sequences of about 0.3 kb in length (moderately repetitive DNA) in the genome of eukaryotes.

Short *response* gene (*sre* gene): Any one of a family of plant genes encoding peptides of 20–150 amino acids in length, that are induced by bacterial or fungal pathogens very early after pathogen attack. The encoded peptides accumulate in high concentrations at the infection site and represent one of the first responses of the attacked plant. The transient induction of the *sre* genes culminates about 12 hours after the first encounter with the pathogen, and afterwards declines rapidly ("short response").

Short *single copy* sequence (SSC; *small single copy* sequence): A 12-20 kb region of the → chloroplast DNA, that is flanked by the two → inverted repeat regions and carries unique chloroplast genes as single copies (e. g. genes *ndh*A, D, E, G, H, I, *psa*C, and *rps*15). Together with the → long single copy sequence it represents the unique part of the → plastome.

Short-stop RNA: Any → messenger RNA that is shorter than the normally occurring full-length primary transcript of a gene. Short-stop RNAs arise from early → transcription termination by e. g. the presence of a → poly(A) addition signal in one of the → introns of the gene.

Short tandem repeat: See → microsatellite.

Short *tandem repeat* polymorphism: See → microsatellite polymorphism.

Short template: Any DNA strand, synthesized during the → polymerase chain reaction, that carries a → primer sequence at one terminus and a sequence complementary to the second primer at the other terminus. Such short → templates are amplified exponentially to high copy numbers, whereas the long templates (only one primer sequence at the terminus, no sequence complementary to the second primer) are amplified linearly, so that their concentration is comparably low at the end of the PCR.

Shotgun cloning ("shotgun experiment"): A technique to establish a cloned → gene library containing the entire → genomic DNA of an organism in the form of randomly generated fragments (i.e. → restriction fragments).

Shotgun collection: See → gene library.

Shotgun gene synthesis: The procedure to synthesize a set of overlapping oligodeoxynucleotides with single-stranded → overhangs that are complementary to → cloning sites of a suitable → cloning vector. Both the oligonucleotides and the vector are transferred into appropriate bacterial host cells, which then combine oligonucleotide and vector molecules to a circularized and stable entity.

Shotgun sequencing: The determination of the sequence of bases in a complete genome, that involves the fragmentation of the target genome (either by physical or enzymatic means), the cloning and sequencing of the resulting fragments, and the reconstruction of the whole sequence from the individual sequences. See → BAC shotgun sequencing, → clone-based shotgun sequencing, → whole genome shotgun sequencing.

Shrimp alkaline phosphatase (SAP): An enzyme from arctic shrimp (*Pandalus borealis*), catalyzing the removal of the 5'-terminal phosphate group from linear DNA and RNA molecules. SAP is completely and irreversibly inactivated by 15 min heating to 65°C. See → alkaline phosphatase, → calf intestinal alkaline phosphatase.

ShRNA: See → short hairpin RNA.

Shuffle clones: Polymerase chain reaction products whose synthesis has been terminated prematurely, but which may serve as allelic → primers on the templates present in the reaction mixture, thus giving rise to mosaic PCR products. Such artificial alleles most likely accumulate in later cycles of the polymerase chain reaction, because in this stage an insufficient number of polymerase molecules is present to extend all available templates.

Shufflon: A region of → plasmid R64, in which multiple → inversion systems (consisting of six or seven → recombination sites and a shufflon-specific recombinase gene, *rci*) are clustered. The invertible DNA segments are flanked and separated by 19 bp repeat sequences functioning as DNA recombination sites. Random recombinations between any inverted repeat result in inversion of DNA segments, independently or in groups, and the generation of different plasmid isomers. The R64 shufflon determines the recipient specificity in matings. Variations in the combinations of the C-terminal segments of PilV proteins in donor cells broaden the spectrum of accepted recipient bacterial cells in liquid mating.

Shuttle vector (bifunctional vector): A → plasmid cloning vector containing two different → origins of replication, which allow its selection and autonomous replication in two different organisms (e.g. *S. cerevisiae* and *E. coli*, or *A. tumefaciens* and *E. coli*). For example, → yeast episomal plasmid variant pJDB219 replicates in both *E. coli* (via PMB9 origin) and *Saccharomyces cerevisiae* (via the → two micron circle origin). See for example → artificial chromosome, → exon trap vector.
Figure see page 1023.

SIA: See → *sequence-independent DNA amplification*.

Sib selection: The division of a large → cDNA library into smaller sub-libraries consisting of 10-100 clones, and the search for a sequence of interest in such → minilibraries.

SID: See → specific interacting domain.

SI expression vector: See → *self-inactivating expression vector*.

σ^A: A protein subunit of → DNA-dependent RNA polymerase of *E. coli*, whose gene is constitutively expressed. σ^A binds to the core complex of RNA polymerase, and induces it to transcribe → house-keeping genes.

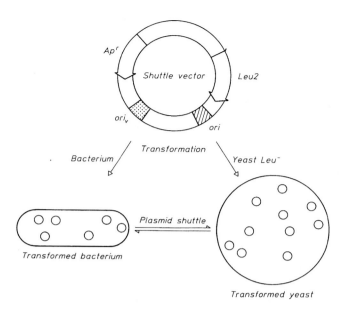

ApᴿＡ : Ampicillin resistance gene
Leu2 : ß-Isopropylmalate dehydrogenase gene
Leu⁻ : Yeast strain with leucine mutation
ori : Yeast origin of replication
oriᵥ : *E.coli* origin of replication

Schuttle vector

σ^{B}: A protein subunit of → DNA-dependent RNA polymerase of *E. coli*, whose gene is induced by environmental stress (e. g. by ethanol, glucose deficiency, heat, salt). Therefore, σ^{B} is also known as "stress σ factor". σ^{B} is completing the RNA polymerase complex, which then transcribes about 250 different genes, whose protein products are involved in stress responses.

Sigma factor (σ factor): A subunit of DNA-dependent → RNA polymerase of *E. coli* which recognizes the → Pribnow box in → promoter regions of genes and thereby directs the core enzyme (consisting of α, β, β', and ω subunits) to initiate → transcription correctly. Once transcription has started, the σ subunit leaves the holoenzyme complex and may serve another core enzyme to initiate transcription of the same (or any other) gene, provided it contains a recognition sequence. A series of σ factors are known, for example in *B. subtilis*, where different factors recognize different sequences (e.g. σ^{28}: CCGATAT; σ^{29}: CATATT; σ^{32}: TANTGTTNTA; σ^{37}: GGNATTGNT; σ^{43}: TATAAT).

Signal: Any measurable output of a hybridisation experiment. For example, the intensity of a spot or a band on an → autoradiogram (a consequence of silver deposition induced by irradiation of the underlying radioisotope) or the fluorescence emission from a labeled target or probe on a → microarray represent such signals. See → signal-to-noise ratio.

Signal compression: The loss of fluorescence emission energy through → self-quenching of adjacent fluorochrome molecules on a → microarray, that is hybridized to a fluorescently labeled → probe. Signal compression leads to an underestimation of the number of probe molecules hybridised to their targets bound to the chip, and therefore obscures any quantification of the microarray data.

Signalling aptamer: Any → aptamer, that undergoes conformational changes during binding of its cognate analyte, which in turn "signals" the presence of the analyte. For example, a → fluorochrome appropriately bound to an aptamer is fully excitable, if not restricted conformationally or quenched by nearby molecules. If the conformation of the aptamer is changed as a consequence of its interaction with a ligand, then the molecular environment for the fluorochrome changes, and its fluorescence is quenched. Thereby it can report the presence of a ligand by sensing the change in aptamer conformation. Signalling aptamers are exploited for the quantitation of metabolites in solution. So, a fluorochrome is bound near the binding site of an ATP aptamer (optimised to bind ATP with high affinity), and exposed to the analyte solution. ATP can unequivocally be quantified via fluorescence light intensity, without any cross-reactivity with GTP. See → aptamer-beacon.

Signalome: The complete set of signalling molecules in a cell, a tissue, an organ, or an organism. For example, ions like Ca^{2+}, nitric oxide, oxygen radicals, also H_2O_2, cyclic AMP, cyclic GMP, hormones, peptides, proteins such as G-coupled receptor proteins or receptors generally, enzymes such as mitogen-activated protein kinases, to name only very few, are constituents of the signalome.

Signal patches: Any amino acid residues, that are distant to one another in the primary sequence of a protein, but close to each other in the three dimensional space, when the protein is properly folded.

Signal peptide (transit peptide, leader peptide; also leader or signal sequence; leader sequence peptide): About 15-30, mainly hydrophobic amino acid residues at the extreme N-terminus of proteins which are transported across membranes. The signal peptide is essential for the efficient and selective targeting of nascent protein chains to the cytoplasmic membrane (prokaryotes) or to the endoplasmic reticulum (eukaryotes). In eukaryotes, interaction of the signal peptide with membrane components is mediated by the → signal recognition particle. This particle in turn interacts with its receptor protein in the ER membrane (docking protein) which ensures correct targeting of the nascent polypeptide into the ER. The hydrophobic amino acids then guide the protein through the lipid bilayer of the membrane. After passage across the membrane the signal peptide is recognized by a signal peptidase ("clippase") on the lumenal side of the ER membrane, and is proteolytically removed. The protein destined for export is finally released into the intralumenal cisternae of the ER.

Signal recognition particle (SRP, signal sequence recognition particle): A universally conserved cytoplasmic 11S → ribonucleoprotein complex, that mediates cotranslational targeting of secretory and membrane proteins to cellular membranes. All proteins, that are inserted into or transported through membranes contain an NH_2-terminal → signal peptide. This peptide is bound by the SRP as it emerges from the ribosome. The resulting signal peptide-SRP complex then targets the nascent peptide chain-ribosome complex to the endoplasmic reticulum (ER) in eukaryotes, and to the plasma membrane in prokaryotes. In higher eukaryotes, the SRP contains six proteins and one molecule of 7SL RNA of about 300 nucleotides with a pronounced secondary and tertiary structure (three helices, helix 6, 7 and 8, where 6 and 8 are closed by → tetraloops). The SRP is composed of two domains, the *Alu* domain and the S domain. Whereas the *Alu* domain contains the two proteins SRP9 and SRP14 and is responsible for retarding the ribosomal elongation of the nascent peptide chain, the S domain binds to the signal peptide and mediates the binding of the SRP to the SRP receptor ("docking protein") on the ER membrane in a guanosine 5'- triphosphate-dependent process. The S domain consists of the four proteins SRP19, the GTPase SRP54 and the SRP68/72 heterodimer, as well as the central part of the SRP RNA. SRP 19 is essential for the assembly of the SRP, and binds free SRP RNA already in the nucleolus, which is a prerequisite for the subsequent binding of SRP54 to helix 8 of the SRP RNA, adjacent to the GAAA tetraloop, and the tetraloop itself. The primary binding site of

SRP19 is helix 6, that is also closed by a highly conserved tetranucleotide hairpin loop of the unusual GNAR type (GGAG). The adenine at the third position is strictly conserved and essential for SRP19 binding.

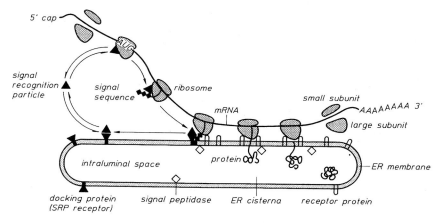

Signal recognition particle

Signal sequence:
a) A nucleotide sequence coding for a → signal peptide.
b) Synonym for → signal peptide.

Signal sequence vector: See → expression-secretion vector.

Signal-to-noise ratio: The ratio between the contribution of a fluorescence signal on a → microarray (hybridized to a → probe) and the contribution of the background noise (see → dark current, → electronic noise, → microarray noise, → optical noise, → sample noise) to the readings of the fluorescence detector instrument. See → background subtraction.

Signal transducer and activator of transcription (STAT): Any member of a comparably small family of → DNA-binding proteins that mediates specific cellular responses upon ligand-receptor binding. After activation by cell-surface receptors or their associated kinases, STAT proteins dimerize, translocate to the nucleus and bind to specific promoter sequences on their target genes. For example, in response to interferon IFNα/IFNβ, a DNA-binding complex is rapidly formed that consists of two STATs (STAT1 and 2) and an additional DNA-binding protein (p48), which binds to an *IFN*-*stimulated response element* (ISRE). Different STATs possess a similar overall structure, with a highly conserved central DNA-binding domain, a highly divergent carboxy-terminal domain influencing transcriptional activation potential, a conserved amino-terminal domain, and two conserved domains (SH2, able to bind phosphotyrosine-containing peptides; SH3, unclear function). DNA binding of STATS is totally dependent on the phosphorylation of a single tyrosine site (e.g. Tyr-701 in STAT1), which results in heterodimerization (e.g. STAT1 and 2).
The 9bp consensus binding site for the different STAT dimers is highly conserved: 5'-TTCC (X) GGAA-3', where X may be G > C (STAT1), G = C (STAT3), G > C (STAT4), A > T (STAT5), A > T, N (STAT6), and C = G (dSTAT).

Though limited in number, STATs affect a series of important cellular functions. STAT3, for example, induces the expression of various genes dramatically responding to tissue injury and inflammation ("acute phase response genes"), STAT5 regulates the expression of milk proteins in mammary tissue responding to prolactin.

Signal transduction: The transmission of a molecular signal from the surface of a cell to cyto-plasmic, nuclear, or organellar targets, leading to a response. For example, if membrane-bound socalled receptor kineases interact with their ligand (e. g. a hormone), then they phosphorylate proteins which in turn are transferred to the nucleus where they bind to target sequences, and turn genes off or on (response).

***Signature-tagged mutagenesis* (STM; signature-tagged transposon mutagenesis):** A → transposon mutagenesis technique, in which transposons carrying unique sequence tags are used for the isolation of bacterial → virulence genes. In short, each transposon is first tagged with a unique DNA sequence tag, composed of a variable 40 bp central region flanked by arms of 10–20 bp constant sequences (common to all tags). The central region can be varied to yield a huge diversity of transposon tags. These double-stranded tags are then ligated into a vector, transformed into *E.coli* (producing a pool of *E.coli* mutants, each of which carries another tag), and transferred into the bacterium under investigation (the pathogen) by → conjugation. Transposition leads to the stable and random single-copy integration of the transposon together with its specific tag into the target bacterial genome. Then transposon-tagged mutant → exconjugants are grown in 96-well → microtiter plates. These arrays are → replica-plated onto two membranes. The 96 mutants of each microtiter plate are then pooled (input pool), an aliquot retained for DNA extraction, and the rest injected into an animal (usually mouse) susceptible to the disease caused by the test bacterium. In case of mice, the spleens are removed after three days (symptoms of disease visible), and mutants that multiplied within this organ recovered by plating of spleen homogenates onto growth medium. Some 10,000 colonies are combined (recovered pool), and DNA is isolated. Then both the tags of the original pool and the recovered pool are amplified with → primers complementary to the invariable arm regions of the tag, and PCR radiolabeled. After release of the arms by *HindIII* digestion, the tags are used to probe the replica colony blots from the microtiter plate. Colonies hybridizing to the probe from the injected pool but not to the probe from the recovered pool are mutants with attenuated virulence (carrying mutated virulence genes).

Silanized nucleic acid: Any nucleic acid (e.g. cDNA or oligonucleotide), that is covalently con-jugated to an active silyl moiety (derived from [3–mercaptopropyl]-trimethoxysilane), which allows the DNA to be immobilized onto and be covalently attached to a glass surface by manual or automated → spotting. See → DNA chip.

***Silence information regulator* (SIR, silent information regulator):** A group of eukaryotic → nu-clear proteins, that interact with the tails of → histones H3 and H4 protruding from → nucleo-somes and transform a previously euchromatic region of the genome into a heterochromatic conformation. Therefore, SIR proteins function as → silencers of → transcription of genes in the converted region. See → euchromatin, → heterochromatin.

Silencer:
a) Transcriptional silencer (*negative element*, NE): A *cis*-acting DNA sequence motif of 20-100 bp in → promoters of eukaryotic genes that reduces or abolishes (silences) their expression. If this motif is deleted, transcription from the resulting mutant promoter is possible. Silencers

are probably target sites for the binding of specific → *trans*-acting factors. See also → up-stream regulatory sequence, compare → enhancer.
b) A *cis*-acting sequence element in eukaryotic genomes, that regulates transcriptional silencing at specific loci and functions during S-phase passage of the cell cycle to establish the silent state. Silencers are required for the maintenance of silent chromatin. See → telomere silencing.

Silent base change: Any exchange of a particular base (e.g. A) for another base (e.g. G) in a protein-coding DNA sequence, that does not change the function of the encoded protein. See → silent mutation.

Silent editing: An → RNA editing of certain nucleotides in mitochondrial → messenger RNA that does not change the coding information of the message. See → degenerate code.

Silent editing site: Any → RNA editing site in mitochondrial, chloroplast, kinetosome or also nuclear transcripts, whose editing does not change the encoded amino acid. See → deletion editing, → insertion editing, → substitution editing.

Silent gene: See → cryptic gene; also used for → pseudogene.

Silent mutation (samesense mutation, silent site mutation): Any gene → mutation that is of no consequence for the → phenotype of the organism (e.g. → a point mutation in the third position of a → codon which does not change the amino acid specificity of the mutated codon; e.g. change of UAU to UAC which both code for tyrosine).

Silent single nucleotide polymorphism (silent SNP): Any → single nucleotide polymorphism, that occurs in the coding region of a gene, but does not change the amino acid sequence of the encoded protein. See → missense SNP.

Silent site mutation: See → silent mutation.

Silent SNP: See → silent single nucleotide polymorphism.

Silicon fiber-mediated gene transfer (fiber-mediated DNA transfer): A method for → direct gene transfer into cells and tissues, that uses silicon fibers for the perforation of cellular membranes. These fibers are softly ground with the cells (or tissues), and the DNA enters the cells through fiber-induced pores in the membranes. Since the silicon fibers are potentially hazardous, the survival of target cells is not satisfactory and the transformation frequency not better than with other techniques of direct gene transfer. Fiber-mediated DNA transfer will probably remain exotic.

Silver-stained random amplified polymorphic DNA (ssRAPD): A variant of the → random amplified polymorphic DNA (RAPD) technique in which the fragments generated by → amplification from → primers with random sequence (e.g. → OPERON primers) in a → polymerase chain reaction are separated in thin → polyacrylamide gels and stained with silver salts. Compare → DNA amplification fingerprinting, → arbitrarily primed PCR.

Silver staining: A technique for the visualization of proteins and nucleic acids in → polyacrylamide gels, in which silver ions react with the macromolecules at pH > 10, form complexes, and are subsequently reduced to elementary silver. The silver is deposited at the sites of reduction and can be visualized easily.

Simian virus: Any virus of a group of papova viruses that infect non-human primates. See → simian virus 40.

Simian virus 40 (SV 40): A small icosahedral papova virus with a 5.2 kb circular duplex DNA genome, originally detected in the African green monkey, *Cercopithecus aethiops*, and infecting cultured cells of non-human primates. The viral DNA is encapsidated in a protein coat consisting of viral proteins VP1, VP2, and VP3. After infection, it becomes associated with host cell → histones H2A, H2B, H3 and H4 in a so-called → minichromosome within the nucleus, where only the ori-sequences are not organized in → nucleosomes. This *ori* region represents a control region for viral gene expression, and separates an early region from a late region of the viral genome. The early region (→ early genes) encodes a → polycistronic mRNA, which is spliced into two messages. One encodes the 95 kDa → T antigen, the other one the 18 kDa t antigen. The latter protein is a → transcription factor for RNA polymerase II and III promoters.

After infecting the host cell, SV 40 can enter two life cycles depending on the type of host. In *permissive cells* (e.g. monkey cells), the virus is first uncoated (removal of the 420 subunits of VP1), the viral DNA is transferred to the cell nucleus, the early genes are transcribed, a virus induced stimulation of host cell DNA synthesis occurs, the → late genes (coding for VP1, VP2 and VP3) are transcribed, and the viral DNA is produced. After virus assembly the cell lyses. In *nonpermissive cells* (e.g. mouse cells) the virus cannot complete DNA replication and no lysis occurs. In some cases, the SV 40 sequences are integrated into the host cell genome, but are often amplified and/or rearranged.

A series of potent → cloning vectors for use in animal cells (SV 40 vectors) have been constructed from SV 40 which contain the foreign DNA either instead of the deleted early region (early region → replacement vector) or the deleted late region (late region replacement vector). In each case the missing functions have to be complemented (e.g. by propagation in the → COS cell line or by a → helper virus). See also → late gene.

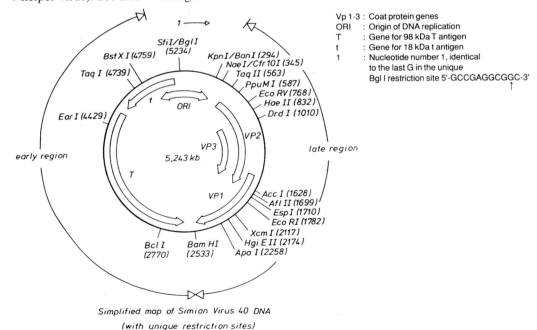

Simplified map of Simian Virus 40 DNA (with unique restriction sites)

Simian virus 40

Similarity: The extent, to which any two nucleic acid or amino acid sequences are related, usually expressed as percent sequence identity.

Simple *long tract repeat* (SLTR): An extended tract of random combinations of → simple repetitive DNA sequences, for example $(TC)_n$, $(CAG)_n$, $(CGTG)_n$, $(GGTGT)_n$, $(GGGTCT)_n$, or $(GGGTGGA)_n$, where n > 100. SLTRs are used as probes for → DNA fingerprinting to screen for abnormally expanded simple repeat alleles in human diseases (e.g. Alzheimer'-s disease, Machado-Josef disease, schizophrenia), in routine paternity testing, for the detection of germline mutations in populations and somatic mutations in human tumors, and phylogenetic studies in microorganisms.

Simple *quadruplet repeat* sequence (sqrs): An oligonucleotide sequence of four base pairs, originally found in reptiles, that is repeated at least four times and is present in most eukaryotic genomes. The biological function(s) of sqrs are unknown, but synthetic sqrs (e.g. $[GATA]_4$ or $[GACA]_4$) serve as probes for → DNA fingerprinting. See → simple repetitive DNA.

Simple repeat sequence: See → simple repetitive sequence.

Simple *repetitive* DNA (SR-DNA; simple-sequence repeats): Mostly short (up to 10 bases), usually tandemly arranged, rather uniform sequence motifs in eukaryotic genomes. Frequently shorter repeat units are located within longer repeats. Subclasses of SR-DNA are e.g. → simple repetitive sequences and → simple quadruplet repeat sequences. See also → simple-sequence length polymorphism. Compare → microsatellite.

Simple *repetitive sequence* (simple repeat sequence, SRS, repetitive simple sequence, RSS): A series of short (up to 10 bases), tandemly arranged sequence motifs (e.g. GAGAGAGAGAGA) in eukaryotic genomes, whose function is unknown. Due to → mutation, → slipped-strand mispairing, or → recombination the number of such repeats can be either reduced or expanded so that SRS probes can be used to detect polymorphisms in human, animal and plant genomes. See → microsatellite, → simple-sequence length polymorphism. SRS are a subclass of → simple repetitive DNA.

Simple-sequence *length polymorphism* (SSLP): The variation(s) in the length of DNA fragments containing → simple repetitive sequences in → genomic DNA from two or more individuals of a species. SSLPs are used to discriminate between closely related individuals and to detect relationships in population genetics. See → simple-sequence length polymorphism DNA fingerprinting.

Simple-sequence *length polymorphism* DNA fingerprinting (SSLP fingerprinting): A variant of the → DNA amplification fingerprinting technique that aims at detecting → amplification fragment length polymorphisms based on the presence of a variable number of → simple repetitive sequences at specific loci (→ simple-sequence length polymorphism). The synthetic oligodeoxy-nucleotide → primers (→ amplimers) used in this procedure are complementary to genomic DNA flanking → simple repetitive sequences. DNA from two or more individuals of a species is compared.

Simple-sequence repeat: See → microsatellite.

Simple transposon: See → insertion sequence.

Simplex *polymerase chain reaction* (simplex PCR): A variant of the conventional → polymerase chain reaction technique, that uses only one short oligodeoxynucleotide → primer of arbitrary sequence (→ amplimer) to amplify anonymous DNA segments of a genome. This technique produces banding patterns (→ RAPD profiles) that are different from the profiles generated with two primers. See → duplex polymerase chain reaction.

Simultaneous bidirectional sequencing: A variant of the conventional → Sanger sequencing technique, in which a forward → primer labeled with fluorophore A and a reverse primer labeled with fluorophore B are used to simultaneously sequence both strands of a duplex DNA in the same sequencing reaction.

Sindbis virus expression system (Sindbis expression system): A → transient expression system based on elements of the Sindbis alpha virus, that infects a wide range of eukaryotic cell types. The gene of interest is first cloned into a Sindbis → expression vector, containing viral genes for in vivo replication of recombinant RNA molecules. Then the vector is electroporated into competent cells (e. g. *b*aby *h*amster *k*idney [BHK] cells) together with a defective Sindbis helper virus, where recombinant virus particles are synthesized (that can be used to infect a variety of eukaryotic cell lines), and the gene of interest is expressed. Expression can be monitored by e. g. the incorporation of ^{35}S-methionine into proteins, their separation in high-resolution → polyacrylamide gels and the detection of newly synthesized proteins by → autoradiography. The protein of interest can be identified by a specific antibody coupled to e. g. a → fluorochrome.

SINEmorph: Any polymorphic single-copy sequence that flanks mammalian → SINES and is amplified by a → polymerase chain reaction, using specific primers (→ amplimers, SINE-PCR). Such polymorphisms are generated by the appearance or disappearance of → restriction endonuclease → recognition sites (caused by mutations), or by a → variable number of tandem repeats at the 3' end or flanking regions of SINES.

SINES (*short interspersed elements*, SINE elements; short interspersed repeat elements): A fraction of → repetitive DNA that is made up of short repetitive sequence elements of less than 0.5 kb in length, alternating with → single-copy sequences and present in about 10^5 copies in the genome of some mammals (human genome: $5 \cdot 10^5$. The most prominent SINES belong to the so-called → *Alu* I family. Compare → LINES.

PROTOTYPE

Single *amino acid polymorphism* (SAP): Any → polymorphism, that is based on an exchange of a distinct amino acid in the sequence of the wild type protein by another amino acid. SAPs are usually detected by → minisequencing.

Single base extension (SBE): A technique for the detection of → *single nucleotide polymorphisms* (SNPs), which uses → primers ending directly adjacent to the SNP mismatch and used to incorporate the complementary, fluorescently labeled ddNTP (see → dideoxynucleotide) in a conventional → polymerase chain reaction with → *Thermus aquaticus* DNA polymerase. After incorporation of the ddNTP, the reaction is terminated, and the incorporated nucleotide can be detected by laser-induced fluorescence. In one single reaction, several SNPs can be discovered, if primers with different 5'- tails and different → fluorochromes are used (e.g. ddATP: green fluorescence; ddCTP: blue; ddGTP: yellow; ddTTP: orange).

Single base extension tag-array on glass slides (SBE-TAGS): A technique for the highly parallel (simultaneous) genotyping of hundreds or thousands of → *single nucleotide polymorphisms* (SNPs), that combines *single base extension* (SBE) with the hybridization of the SBE products to a generic → microarray. First, the various SNPs are amplified with 16–25 bases long SBE hybrid → primers, containing a generic sequence tag at the 5'end and a locus-specific sequence terminating at the base 5' to the individual SNP. The extension reactions are catalyzed by → DNA polymerase in the presence of fluorescently labeled → dideoxynucleoside triphosphates. Three different fluorochromes (e.g. → TAMRA, → cyanine 5, and ROX) are sufficient for discrimination of the SNPs. Subsequently, the fluorescently labeled SBE products are hybridized to a microarray containing the full-length reverse complements of the tags (generic tag array). The spotted oligonucleotides are synthesized with 15 dT residues at their 5'ends to facilitate their quadruplicate attachment to the poly-L-lysine-coated (or silane-treated) microscope slides. The hybridization is performed in a sealed hybridization chamber. The arrays are then washed at high → stringency, dried by centrifugation, and scanned with an argon laser and appropriate filter sets to discriminate between the different → fluorochromes. The four replicate spots on the array have to show identical signals to be accepted. SBE-TAGS with bifunctional primers (function 1: unique sequence tag; function 2: locus-specific sequence) allows single base extension at multiple loci. Because each locus is identified by a distinct tag sequence, the SNP screening procedure is highly multiplexed. The resulting product mixture can then be demultiplexed through hybridization to the reverse complements of the sequence tags on the glass slide array.
Figure see page 1032.

Single base pair replacement: A technique for the substitution of a correct base pair in a target DNA for an incorrect one. In short, the double-stranded target DNA region is first nicked and digested by → exonuclease such that a single-stranded → protrusion (i.e. single-stranded section) is generated. Then this single-stranded overhang is repaired with → DNA polymerase in the presence of only three bases (e.g. dATP, dGTP, dCTP). This forces the polymerase to insert a mismatching base, whenever a thymidine occurs in the template strand. In the next replication round this mismatched base is complemented, resulting in a replacement of the original base pair for a new one.
Figure see page 1033.

Single base pair replacement

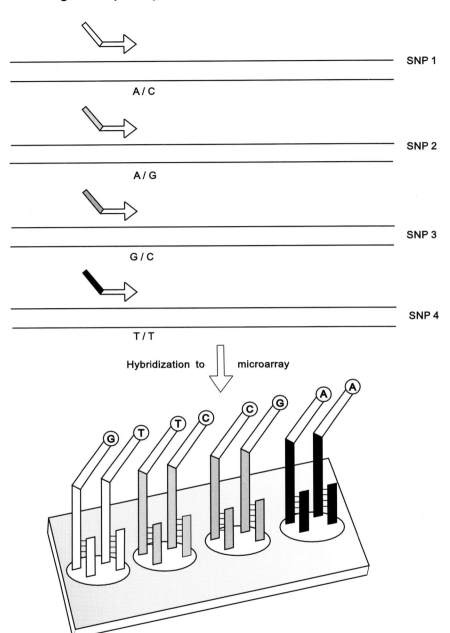

Single base extension tag-array on glass slides

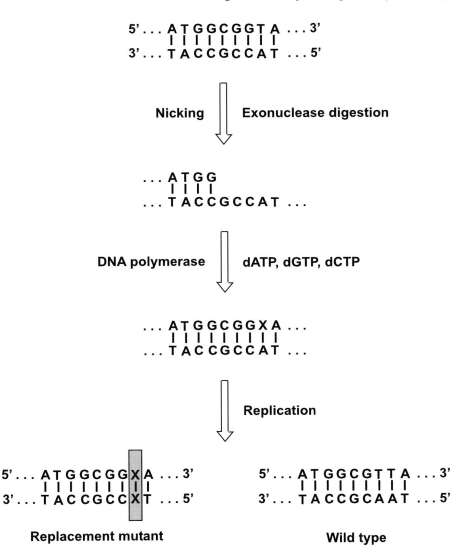

Single base pair replacement

Single cell analysis of gene expression (**SCAGE**): A technique for the analysis of (preferably all) transcripts of a single cell, that starts with the lysis of this cell, isolation of its → messenger RNA through binding to oligo(dT)-coated solid supports (chips) and the chip-based → reverse transcription of the bound RNA into → cDNA, using random primers with a 5'-oligo(dC)-tail and a 3'-tailing reaction with dGTP, employing terminal deoxynucleotidyltransferase, generating a 3'-oligo(dG) flanking region (see → DNA tailing). Subsequently the cDNAs are amplified with a single poly(dC) primer (e.g. 5'-TCA GAA TTC ATG CCC CCC CCC CCC CCC–3') in a conventional → polymerase chain reaction. An aliquot is then reamplified in the presence of labeled nucleotides. Specific transcripts can then be detected by PCR using gene-specific primers, or by hybridisation of the labeled transcripts to specific → microarrays (e.g. a → medium-density array loaded with cancer-specific gene sequences). The mRNA profiling of a single cell requires about 60 pg of mRNA for a representative expression analysis (amount in a single cell: 3–6 pg). SCAGE allows to detect and characterize rare cells, as e.g. occult systemically spread tumor cells, or stem cells.

Single cell *comparative genomic hybridisation* (SCOMP): A technique for the complete and representative amplification of → genomic DNA from single cells or tiny tissue samples embedded in paraffin or formalin. In short, a single cell is first isolated, its proteins digested with proteinase K, its genomic DNA extracted, and completely digested with the → restriction endonuclease *Mse* I, which leads to fragments of relatively homogeneous length (200–300 bp) and TA → overhangs. Then → adaptors are ligated to the ends, and the adapted fragments amplified by conventional → polymerase chain reaction using adaptor-complementary → primers. An aliquot of the resulting fragment mixture is then again amplified in the presence of labeled nucleotides and primers complementary to the → restriction recognition sequence. The amplicons are finally hybridised to chromosomal spreads or flow-sorted chromosomes. SCOMP allows to visualize any chromosomal aberration in a single cell (e.g. a trisomic configuration of chromosome 21 in a single lymphocyte from a Down syndrome patient). Compare → *Alu*-PCR, → DOP-PCR, → primer extension preamplification.

Single cell *gel electrophoresis assay* (SCGE; COMET assay; *microgel electrophoresis*, MGE): A technique for the detection and (semi)quantitation of the genotoxic potential of environmental hazards (e. g. ionizing irradiation, heavy metals, toxic chemicals, and gases as ozone and CO), that is based on the capability of these noxious agents to introduce single- or double-stranded breaks in nuclear DNA. In short, single cells are first embedded in an → agarose layer and then exposed to the corresponding toxic agent, which induces strand breaks in the nuclear DNA, leading to a relaxation of chromatin. Then the cell is lysed in a strongly alkaline electrophoresis buffer (denaturation of DNA), and the relaxed DNA forced out of the (partially damaged) nucleus by an electrical field. After electrophoresis and staining with → ethidium bromide, the released DNA appears as a characteristic "tail of comet" (therefore COMET assay), whose size and fluorescence intensity correlates with the genotoxicity of the tested agent.

Single-cell PCR: See → single-cell *polymerase chain reaction*.

Single-cell *polymerase chain reaction* (single-cell PCR; SC-PCR): A variant of the conventional → polymerase chain reaction that allows to amplify → RNA, or also → messenger RNA from one single cell. In short, the target cell is first isolated from a tissue, then lysed, and the released RNA primed with an oligo(dT)-primer annealing to → poly(A)-tails of messenger RNAs, using → reverse transcriptase. The resulting → cDNA is then further amplified to amounts detectable on → ethidium bromide-stained → agarose gels.

Single-cell *simple tandem repeat* profiling (single-cell STR profiling): The generation of a → DNA fingerprint of genomic DNA isolated from a single cell (e. g. single sperm cell, single dandruff flake cell, small samples left on e. g. weapons or keys), which is therefore employed for forensic identification. See → DNA fingerprinting, → microsatellite, → simple-sequence length polymorphism, → single cell polymerase chain reaction.

Single-cell STR profiling: See → single-cell simple tandem repeat profiling.

Single colony lysate: The preparation of bacterial cells from a single colony for the isolation of DNA.

Single colony sequencing: The sequencing of double-stranded (ds) plasmid DNA obtained from the osmotic lysate of one single bacterial colony, using → linear amplification DNA sequencing procedures.

Single-copy DNA: See → unique DNA.

Single-copy gene (sc gene): Any → gene that is present in only one single copy per haploid genome (e.g. the fibroin gene of the silk moth *Bombyx mori*). Compare → middle repetitive DNA.

Single-copy plasmid: See low copy number plasmid.

Single domain antibody (sdAb): Any → antibody, that consists of only one chain. In a series of antibodies, such a chain is sufficient to bind to the target → antigen.

Single exon gene: Any → gene that contains only one single → exon. See → multi-exon gene. Compare→ multi-intronic gene.

Single-gene cassette: Any → cassette that contains only one → gene. Compare → double-gene cassette.

Single-gene shuffling: A variant of the → DNA shuffling technique, in which one single gene is used for creating diversity. The resulting shuffle library contains novel chimeras, that differ in only relatively few positions from the wild type ancestral gene. For example, three rounds of a ß-lactamase single-gene shuffling yields only four amino-acid substitutions in the resulting protein, whereas a single cycle of → multi-gene shuffling ("family shuffling") of the four cephalosporinase genes from *Citrobacter freundii*, *Klebsiella pneumoniae*, *Enterobacter cloacae* and *Yersinia enterocolitica* creates a mutant enzyme, that differs by 102 amino acids from the *Citrobacter*, by 142 amino acids from the *Enterobacter*, by 181 amino acids from the *Klebsiella*, and by 196 amino acids from the *Yersinia* enzyme.

Single locus minisatellite (SLM): Any → minisatellite repeat sequence that is present at one single locus in a genome. Such SLMs are detected by using → primers, directed towards the minisatellite-flanking (and mostly unique) sequences, in a conventional → polymerase chain reaction (SLM-PCR). Therefore not all minisatellites of a genome with the same core sequence are amplified. The detected locus will be visualized as one (homozygous condition) or, at most, two bands (heterozygous condition). SLMs own a high degree of variability, with 3-50 (or more) alleles existing in most cases. Therefore, only few highly polymorphic SLM probes are sufficient for a detailed analysis of e. g. kinship. See → sequence-tagged microsatellite sites.

Single locus probe (SLP): Any DNA sequence that allows to detect one single → locus in a genome, using → labeling and → hybridization techniques. Compare → multilocus probe.

Single molecule array: A high-density → gene array, onto which a series of different molecules are spotted such that each spot contains only one single molecule (e.g. a DNA fragment). Since only one molecule is present at each site, it is possible to create arrays of extreme density (for example, 10^8 sites per cm^2, or more). Single molecule arrays are used for high-throughput sequencing of genomic DNA (that is spotted on the array as 100–200 bases long single-stranded fragments) and rapid genotyping of two (or more) individuals.

Single molecule dilution PCR: See → single molecule *polymerase chain reaction*.

Single-molecule DNA sizing (SMS): A technique to accurately and rapidly measure the sizes of DNA fragments ranging from 2 – 200 kb, that requires only femtograms of DNA ("single molecule") and minutes to e.g. separate restriction digest products, that is based on microfabri cated microfluidic channels in a silicon elastomer (produced by socalled soft lithography). The DNA sample is applied to small aluminum cylinders in the silicon wafer, from which they are moved by capillary action and electro-osmotic flow. If the DNA molecules are labeled with a → flurochrome, then its movement and also concentration can be measured with appropriate laser fluorography.

Single molecule fluorescence spectroscopy (SMFS): A technique for the visualization of single molecules by → fluorescence spectroscopy. In short, the target molecule (e.g. a virus, an antibody) is first labeled with few → fluorochrome molecules (e.g. → cyanin 3, → cyanin 5, → rhodamine). Then the fluorophore is excited by laser pulses, and the emitted fluorescence light signal monitored in a CCD camera cooled with liquid nitrogen. The signals are computed, and allow to trace the movement of single molecules or molecular complexes in a cell. For example, adenoviruses can be endlabeled with 1–2 molecules cyanin 5, then allowed to infect HeLa cells. A single cell is then laser-pulsed, and the emitted light monitored in 40 msec intervals, which allows to visualize the movement of the adenovirus (into the host cell nucleus).

Single molecule polymerase chain reaction (single molecule PCR, single molecule dilution PCR, SMD-PCR): A variant of the conventional → polymerase chain reaction in which the target DNA is present as a single molecule that can be amplified with appropriate → amplimers. Such SMD-PCR allows e.g. the separation of → alleles of the same locus that comigrate on → agarose gels and therefore cannot be resolved by electrophoretic techniques.

Single-molecule sequencing: A future technology for the determination of the sequence of bases in a single molecule of DNA. The DNA is synthesized or labeled with base-specific fluorochromes, immobilized by light pressure from two laser beams suspended in a flow system and processively digested with an → exonuclease, that sequentially removes base by base from one end of the immobilized DNA. The released bases are pushed through a single-molecule fluorescence detector and identified by their specific fluorescence. The base sequence of the fragment is then computed from the sequence of fluorescence signals. Or, alternatively, a → DNA polymerase carrying a distinct donor → fluorochrome, that interacts with the acceptor fluorochromes on incoming nucleotides. As each tagged nucleotide is incorporated into the growing strand, the donor fluorophor transfers energy to the acceptor, releasing light energy, whose color and intensity reflects the base identity. See → nanopore sequencing, → polymerase colony sequencing.

Single molecule spectroscopy (SMS): A series of spectroscopic techniques for the visualization of single molecules. See → single molecule fluorescence spectroscopy.

Single molecule tracking: The visualization of a single molecule and the pursuit of its movement within a cell by e.g. → single molecule fluorescence spectroscopy.

Single *n*ucleotide *p*olymorphism (SNP; pronounce "snip"): Any polymorphism between two → genomes that is based on a single → nucleotide exchange, small → deletion or → insertion. Statistically, an SNP will occur every kilobase in the human genome (soybean: 3-4 SNPs/kb). The genomic distribution of SNPs is biased towards genic DNA (1 SNP/2000 bp) as compared to extragenic DNA (1 SNP/500 bp). The number of SNPs in the human genome is estimated to 3-30 millions. This relatively good genome coverage and the distribution of SNPs throughout the genome in both coding and noncoding regions make SNPs highly informative markers for → mapping procedures. Since specific SNPs correlate with increased risk for a particular disease, these polymorphisms are diagnostic ("disease-associated SNPs"). The basic difference between → point mutations and SNPs lies in their different frequencies (point mutation: $\leq 1\%$; SNP: $\geq 1\%$ of a population). See → anonymous SNP, → candidate SNP, → causative SNP, → clone overlap SNP, → coding SNP, → common SNP, → copy SNP, → electronic SNP, → exonic SNP, → expressed SNP, → gene-based SNP, → haplotype SNP, → human SNP, → interacting SNP, → intergenic SNP, → intronic SNP, → missense SNP, → multiple nucleotide polymorphism, → noncoding SNP, → non-synonymous SNP, → point mutation, → promoter SNP, → reference SNP, → regulatory SNP, → silent SNP, → synonymous SNP, → tetra-allelic SNP, → tri-allelic SNP, → transition, → transversion. Compare → singleton, → singleton polymorphism, → SNP map.

Single *n*ucleotide *p*olymorphism (SNP) typing: The scanning of specific regions (e. g. genes) in two (or more) genomes for → single nucleotide polymorphisms to detect an SNP composition typical for each of the genomes. SNP typing generates a specific SNP profile of each genome, in which associations of a specific SNP (or SNPs) with a particular phenotype (e. g. a disease) have diagnostic value.

Single *n*ucleotide *p*olymorphism chip (SNP chip): Any → microarray, onto which specific PCR-amplified regions of genes known to contain one (or more) → single nucleotide polymorphisms are spotted. Using → primer extension or → quantitative polymerase chain reaction techniques, single nucleotide polymorphisms can be detected on the chip.

Single *n*ucleotide *p*olymorphism *d*ata*b*ase (dbSNP): A database for → single nucleotide polymorphisms (i.e. single base exchanges), short → deletions and → insertion polymorphisms. Web site: http://www.ncbi.nlm.nih.gov/SNP
HGVbase maintains an extensive list of other SNP databases at:
    http://hgvbase.cgb.ki.se/databases.htm

Single *n*ucleotide *p*olymorphism density (SNP density): The frequency of → single nucleotide polymorphims per unit length of → genomic DNA (usually SNPs/100 kb). SNP density is different for different regions of genome. For example, regional heterogeneity of SNP density is characteristic for different chromosomes of *Anopheles gambiae*, possibly caused by the introgression of divergent Mopti and Savanna cytotypes (chromosomal forms). SNPs are therefore distributed along the chromosomes in a bimodal way: one mode contains about one SNP/10 kb, the other one about one SNP/200 bp. SNP density is high in intergenic and intronic regions, as compared with genic SNPs. Since introgression is excluded from the X chromososme, its SNP density is lower and not bimodal. In the human genome, SNP density is about 12.13 SNPs/10 kb, in the mouse genome SNP density is only 0.821 SNPs/10 kb. See → SNP frequency.

Single *n*ucleotide *p*olymorphism map (SNP map): The linear arrangement of → single nucleotide polymorphisms along a specific region in two homologous genomes.

Single nucleotide polymorphism mass spectrometry (SNP-MS): A technique for the detection of → *single nucleotide polymorphisms* (SNPs) between two (or more) genomes. In short, a SNP (e.g. an AC mismatch) is first localized in a specific region of a genome (e.g. within a gene), and flanking → primers used to amplify this region in a conventional → polymerase chain reaction. Then the amplification product is single-stranded, a → probe oligonucleotide annealed immediately 5'- adjacent to the SNP, and a primer → extension reaction started with one → dideoxynucleotide complementary to the matching nucleotide in the wild-type strand (here: ddTTP). The matching nucleotide can be incorporated and the probe be extended by one nucleotide, whereas the extension is not possible on the other strand carrying the mismatch (here: C). Therefore both products differ by one base, and this difference in mass can be detected by → mass spectrometry.

Single nucleotide polymorphism scanning (SNP scanning): The *in silico* search for → single nucleotide polymorphisms in a sequenced stretch of DNA (e.g. a → BAC clone, a genomic segment, in extreme cases, whole → genomes).

Single nucleotide polymorphism scoring (SNP scoring): The search for → single nucleotide polymorphisms in two (or more) DNA sequences (in extreme cases, → genomes), their characterization and use for genome analysis (e.g. establishment of a → single nucleotide polymorphism map). See → single nucleotide polymorphism typing.

Single nucleotide primer extension (SNuPE): A technique for the detection of a → *single nucleotide polymorphism* (SNP) in a genome. In short, the locus of interest is first amplified with → primers flanking the SNP in a conventional → polymerase chain reaction, the unincorporated primers and dNTPs are removed, and a → single base extension started: a socalled detection primer, that anneals to the target sequence immediately 3' of the SNP, is extended by the matching → dideoxynucleotide, that is fluorescently labeled. Unincorporated ddNTPs are removed, the extension product(s) electrophoresed and the fluorescence (i.e. the incorporated base) detected by laser scanning. Since each ddNTP differs in its mass, and the extended primer varies correspondingly, incorporation of the various ddNTPs can also be detected by → mass spectrometry.

Single-pass protein: Any protein, that spans the plasmamembrane only once, i.e. contain only one → transmembrane helix. See → multi-pass protein.

Single-pass sequencing: The determination of the sequence of bases in one strand of a double-stranded DNA, that is not confirmed by a simultaneous or subsequent sequencing of the complementary strand. Single-pass sequences may there contain sequencing errors introduced by erroneous synthesis or reading.

Single primer amplification reaction: See → microsatellite-primed polymerase chain reaction.

Single-primer DNA amplification: Any → polymerase chain technique that employs one → primer only. See → arbitrarily primed PCR, → DNA amplification fingerprinting, → random amplified polymorphic DNA.

Single primer mutagenesis: An *in vitro* technique to introduce → mutations into a target DNA. An oligodeoxynucleotide containing a single base → mismatch ("mutagenic oligonucleotide") is first annealed to the single-stranded template DNA and then extended by the → Klenow fragment of

E. coli DNA polymerase I in the presence of → T4 DNA ligase. As a result, a double-stranded *c*ovalently *c*losed *c*ircular DNA molecule (cccDNA) is generated, that contains a single base pair mismatch at the mutated site. Subsequently, this heteroduplex is transformed into a suitable host cell. Segregation of the two strands of the heteroduplex leads to a mixed progeny containing either wild-type or mutant DNA.

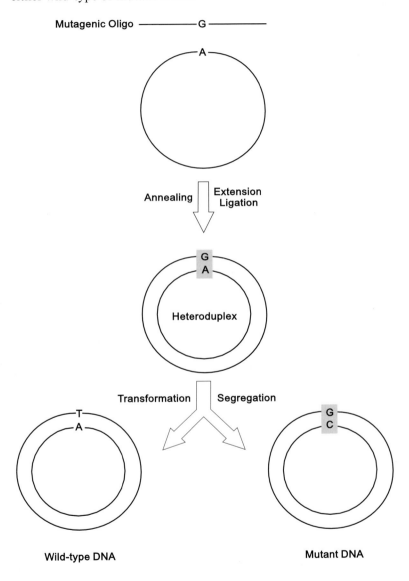

Single primer mutagenesis

Single-sided PCR: See → single-sided *p*olymerase *c*hain *r*eaction.

Single-sided *polymerase* *chain* *reaction* **(single-sided PCR):** A variant of the conventional → polymerase chain reaction, that allows to amplify genomic sequences of which only a few bases are known. In short, genomic DNA is first fragmented by → restriction endonucleases, and then → adaptors of known sequence ligated to the termini of the fragments. Then a → primer complementary to the adaptor sequence and a second primer complementary to the known sequence are used to selectively amplify one side of the target. A serious problem in single-sided PCR is the accumulation of undesirable products (arising from ligation of the adaptors to all other restriction fragments), that obscure the target amplicon. Therefore, single-sided PCR is usually combined with an additional selective method (as e. g. → capture PCR).

Single-side sequencing: A laboratory slang term for the → sequencing of a target DNA molecule from only one terminus. For example, the generation of → expressed *sequence* *tags* (ESTs) requires the sequencing of 550 bases from either the 5' ("5'-ESTs") or 3' terminus ("3'-ESTs"). See → dual-side sequencing.

Single *specific* *primer* *polymerase* *chain* *reaction* **(SSP-PCR):** A somewhat infelicitous term for a technique to amplify DNA with only one sequence-specific → primer using conventional → polymerase chain reaction technology. The procedure is based on the restriction of the target DNA and the ligation of the resulting fragments into a suitable, linearized → cloning vector (e. g. a → plasmid). Then two primers are employed for the → insert amplification, one specific primer complementary to the cloned DNA (e. g. a → gene or → transposon), the other one complementary to the vector DNA. A first round amplification (10-15 cycles) with both primers provides a → template for a second round of PCR with the insert-specific primer and a nested vector-primer (i. e. located closer to the → multiple cloning site into which the inserts are cloned) with the aim to increase the specificity of amplification.

Single-strand *annealing* **(SSA):** A mechanism ("pathway") for the repair of → double-strand breaks in DNA, that requires the presence of repeated sequences on both sides of the break. Repair occurs by → annealing of these complementary repeated sequences after the specific resection of both 5'-ends and leads to the loss of the intervening DNA. See → homologous recombination, → non-homologous end-joining.

Single-strand break: See → nick.

Single-strand *conformation* *analysis* **(SSCA;** *single-strand* *conformation* *polymorphism* *analysis,* **SSCPA;** *polymerase* *chain* *reaction* *single-strand* *conformation* *polymorphism* [PCR-SSCP] **analysis):** The detection and characterization of → single-strand conformation polymorphisms in genomic DNAs from different organisms, that is based on subtle differences of electrophoretic mobility between a non-mutated single-stranded sequence and its mutated counterpart. Such differences arise from mutation-induced changes in the three-dimensional structure of the target DNA. In short, the PCR → primers (→ amplimers) are first radioactively labeled and annealed to the target sequence. Then → polymerase chain reaction is started and the amplified sequences are denatured and separated in thin → non-denaturing → polyacrylamide gels (see → denaturing gradient gel electrophoresis). The single-stranded PCR products and the mutation-induced shift in mobility of one such product can then be detected by autoradiography.

Single-strand conformation polymorphism: See → polymerase chain reaction single-strand conformation polymorphism.

Single-stranded: The property of a nucleic acid molecule to consist of only one polynucleotide chain ("strand"), in contrast to double-stranded molecules containing two intertwined chains (see → DNA). For example, all → ribonucleic acids (→ catalytic RNA, → messenger RNA, → ribosomal RNA, → small cytoplasmic RNA, → small interfering RNA, → small non-mRNA, → small nuclear RNA, → small nucleolar RNA, → small temporal RNA, → tmRNA, → transfer RNA, → translational control RNA, and others) and the genomes of certain phages (e.g. → M13) are such single-stranded nucleic acids, but all may partly form double-stranded regions by intra-strand base pairing.

Single-stranded DNA (ssDNA): A deoxyribonucleotide polymer that consists of only one chain of nucleotides (as opposed to the two base-paired deoxyribonucleotide strands in the DNA → double helix). Single-stranded DNA probably is a transient structure found in all DNA genomes as a consequence of "breathing", and can be detected with → S1 nuclease. Many viruses (e.g. Parvoviridae) possess ssDNA genomes. *In vitro*, ssDNA can be produced from DNA duplexes by heat-denaturation (separation of the two strands), and rapid cooling (prevention of renaturation).

Single-stranded DNA agarose: An → agarose matrix that contains covalently attached denatured (i.e. single-stranded) DNA (usually calf thymus DNA). This type of agarose is used for the purification of enzymes with a high affinity towards ssDNA (e.g. → DNA and → RNA polymerases, type II → restriction endonucleases).

Single-stranded DNA-binding protein (ssDNA-binding protein, SSB): Any protein which binds to specific sequences in single-stranded DNA only (e.g. → RNA polymerase, → recA protein, the T4 gene 32 protein, the vir E2 protein encoded by the → Ti-plasmid from → *Agrobacterium tumefaciens*). See also → DNA-binding protein.

Single-strand linker ligation method (SSLLM): A technique for the cloning of full-length → cDNAs with preserved → caps ("cap-trapped cDNAs"), that uses the ligation of double-stranded → linkers with random 6 bp → protruding ends to full-length, single-stranded cDNA by → DNA ligase. In short, full-length → first-strand, single-stranded cDNA (generated by e.g. → cap trapping) is ligated to double-stranded linkers with 3' overhangs (e.g. dN$_6$ or dGN$_5$) by DNA ligase. These linkers can ligate to any cDNA sequence, thereby facilitating the production of cDNA libraries with high titers. Then second strand synthesis is performed, the resulting double-stranded cDNA restricted with e.g. *Bam*H1 and *Xho I*, and the fragment cloned into a → phage or → plasmid vector. Any → mutation that occurs only once in a genome. See → singleton polymorphism. Compare → single nucleotide polymorphism.

Singleton:
a) A laboratory slang term for a gene or a protein, that has no → paralogs.
b) Any sequence (e.g. a gene), that is present only once in a library.
c) Any → mutation that occurs only once in a genome. See → singleton polymorphism. Compare → single nucleotide polymorphism.

Singleton clone: Any → bacterial artificial chromosome clone, that does not overlap with neighboring clones. Many singleton clones are simply incorrectly placed in the → draft genome sequence → assembly.

Singleton polymorphism: A single base difference between otherwise completely monomorphic (i. e. identical) alleles, genes, genomes, generally DNA sequences. See → singleton.

Single transformant: Any organism into which foreign DNA has been transferred in a single step. Compare → double transformant.

Single-tube *polymerase chain reaction* (single-tube PCR): A variant of the conventional → polymerase chain reaction, which allows to perform all steps of a PCR procedure (from DNA extraction to amplification of the target sequence) in one single tube.

Single *virus tracing* (SVT): The visualization of a single virus (e.g. HIV) during a complete infection cycle, starting from the contact with the host cell membrane, its interaction with the receptor, its penetration of the membrane, its diffusion when packaged in the endosome, its release into the cytoplasm, its transport to the nucleus and integration into nuclear DNA, all in a living cell. SVT starts with the labelling of the viral capsid or its DNA(in viruses with a capsid covered by a lipid double layer) with one single → fluorochrome molecule (to avoid interference with virus-cell interactions). Then the virus infection path is monitored with a highly sensitive single-molecule microscope (spatial resolution: 40 nm; temporal resolution: 10 msec). The tracing is performed in real time and documented with a high-resolution CCD camera.

SIN vector: See → self-*in*activating expression vector.

Sipperchip: Any solid support (e.g. quartz or glass), into which a network of nanochannels (typical width: 50 μm; depth: 10 μm) and reservoirs for reagents for microfluidics applications are etched via nanofabrication, and which additionally contains a 380 μm long fused silica-quartz capillary with an inner diameter of 50 μm ("sipper"), through which the analyte sample (usually a few nanoliters only) is sucked directly into the chip. After the sipper has imported the sample (e.g. a nucleic acid) into the chip, an internal standard is added, and the sample stained with an intercalating dye (e.g. → ethidium bromide). Then the sample is fractionated by e.g. electrophoresis in an electrophoresis channel and the DNA fragments detected with *l*aser-*in*duced *f*luorescence (LIF). Sipperchips allow high-throughput analyses of analytes (e.g. from 96– or 384-well → microtiter plates), with each analysis taking about 30 seconds.

SIP-RT-PCR: See → poly(A)-PCR.

SIR:
a) See → silence information regulator.
b) See → *s*ubterminal *in*verted *r*epeat.

SiRNA: See → small interfering RNA.

SiRNA cocktail: A population of several → *s*mall *i*nterfering RNAs (siRNAs) of only 12–15 bp, produced by complete digestion of long *d*ouble-*s*tranded RNA (dsRNA) with RNase III, that can be used for an effective silencing of genes in mammalian (and, maybe, other) cells. In short, a target gene template is cloned into a vector containing opposing → T7 RNA polymerase → promoters, and transcribed into dsRNA, which in turn is digested with RNase III (from e.g. *E. coli*) into 12–30 bp dsRNA fragments with 2–3 nucleotide 3'overhangs and 5'-phospate and 3'- OH termini. This cocktail targets at different sites in a message and is highly effective for → gene silencing.

SiRNA expression cassette (SEC): A DNA construct containing an → RNA polymerase III → promoter adjacent to a sequence encoding a → hairpin siRNA and an RNA polymerase III transcription termination site, that is used for the synthesis of → small interfering RNA (siRNA) in transfected cells. This hairpin RNA is processed by the double-stranded RNA-specific protein complex → Dicer, that generates a functional siRNA. This RNA then silences its target gene. Since the cassettes contain appropriate → restriction endonuclease recognition sites at their ends, any SEC with a highly effective siRNA sequence can be isolated and cloned into an → siRNA expression vector for stable expression and long-term silencing experiments. SECs are effective against various genes, and are functional in a whole variety of cell types (e.g. HeLa cells, HepG2, and NIH3T3 cells).

SiRNA expression vector (siRNA vector): A vector for the → expression of → small interfering *RNA* (siRNA) directly within target cells for long-term → knockdown of genes (see → gene knockdown). In short, a suitable → plasmid vector is first linearized with two different → restriction endonucleases, and an oligonucleotide complementary to the target → messenger RNA is directionally cloned to the vector (see → forced cloning). Preferentially, → short hairpin RNA-encoding inserts (mimicking the structure of an siRNA) are used, that are composed of two 19 nucleotides long inverted repeats separated by a specific 9 base loop structure (e.g. 5'-UUCA-AGAGA–3'). The vector additionally contains either a mouse U6 or a human H1 → promoter, recognized by RNA polymerase III, which expresses relatively large amounts of siRNAs in mammalian cells. Since a string of thymidines at the 3'end of the insert acts as a polymerase III transcription termination site, the enzyme terminates transcription by incorporating 3–6 uridines. Usually a → G418, → hygromycin or → puromycin gene allows for selection of transformants. See → hairpin RNA vector, → siRNA expression cassette.

SiRNA library("siRNA pool"): An imprecise term for a series of → small interfering RNAs (siRNAs) produced by → Dicer from various regions of a longer (e.g. 72 bp) double-stranded substrate RNA. Either the complete pool is used in gene silencing experiments, or a few functionally tested and highly efficient siRNAs out of this pool.

SiRNP: See → small interfering RNA-protein complex.

SIS: See → sequential insertion site.

SISAR: See → serial invasive signal amplification reaction.

Site-directed gene targeting: The insertion of a gene into a specific chromosomal environment, using → homologous recombination. For example, a foreign gene (e.g. a → reporter gene) can be cloned into another gene (e.g. an alcohol dehydrogenase gene) in vitro, so that it is flanked by sequences derived from the latter. The foreign sequence will then be transferred into the chromosomal region, containing the cellular counterpart of the targeting gene (e.g. a resident *Adh* gene) by homologous recombination. See also → gene targeting, compare → transposon tagging.

Site-directed mutagenesis: See → site-specific mutagenesis.

Site preference: The property of almost all → restriction endonucleases to cleave multiple copies of their → recognition sequence within the same DNA duplex molecule at different rates and in a non-random fashion. Complete digestion of a DNA substrate molecule may therefore need different times with different enzymes depending on their specific site preference (e.g. cleavage of → lambda phage DNA with *Hind*III is complete in only 1 h; it takes 18 h with *Rsr*II).

Site-specific mutagenesis (site-directed mutagenesis, directed mutagenesis, targeted mutagenesis, localized mutagenesis): The introduction of single base-pair mutations (→ point mutations) at specific sites in a target DNA. See for example → oligo-mismatch mutagenesis, → site-directed gene targeting.

Site-specific recombination: Any → recombination of two DNA molecules that occurs only at specific sites (which may or may not share homologies), for example, the integration of a → pro-phage into the host cell's genome. Site-specific recombinations are mediated by specific proteins (recombinases) and are conservative (i.e. occur without any synthesis or degradation of the DNA). Compare → homologous recombination.

Site-specific replacement: The incorporation of an unnatural amino acid at a specific site in a protein. For example, O-methyl L-tyrosine is inserted into proteins of the thermophilic bacterium *Methanococcus jannaschii* instead of L-tyrosine, catalysed by mutant variants of both tyrosyl-tRNA synthetase and tRNA^{Tyr}. Also, in an *E.coli* mutant defective in the editing function of valyl-tRNA synthetase, every fifth valine in a protein is replaced by the unnatural amino acid *amino*butyrate (Abu). Site-specific replacement can be used to generate proteins with improved or altered properties, and is employed for the study of protein function. See → general replacement.

Six base cutter (six base pair cutter): A type II → restriction endonuclease that recognizes a hexanucleotide sequence and introduces a double-strand → cut within it. Theoretically, such hexanucleotide target sites occur once in every 4^6 or 4096 nucleotide pairs of a random DNA sequence. An example for a six base cutter is *Bam*H1 with the following recognition sequence:

```
5'.....GGATC C......3'          5'.....G              GATCC.....3'
                         →                      +
3'.....CC TAGG.....5'           3'.....CCTAG          G.....5'
```

6–FAM: A 6-carboxy → fluorescein derivative, that is used as a marker for → fluorescent primers in e.g. → automated sequencing procedures or for labeling in → DNA chip technology. The molecule can be excited by light of 494 nm wave-length, and emits fluorescence light at 518 nm. Since the wave-length of the excitation and emission maxima is pH-dependent, the exact values vary.

6OS subunit: See → ribosome.

Size: A laboratory slang term for the number of basepairs in a specific DNA fragment, that are not interrupted by a → sequence gap.

SKY: See → *spectral karyotyping.*

Slab gel: An agarose, starch or polyacrylamide gel, that is cast in the form of a slab.

Slab gel electrophoresis: The separation of biological molecules (e.g. DNA, RNA, or proteins) in a matrix (e.g. agarose, starch, or polyacrylamide), that is cast as a slab. Slab gels can be run horizontally (as agarose) or vertically (as polyacrylamide).

Slalom chromatography: A method for the separation of DNA fragments larger than 10 kb, using silica columns for high-performance gel chromatography. During this procedure smaller fragments are eluted faster than larger fragments.

Slide: A special surface-activated glass support, similar to a microscope slide, that possesses a low intrinsic fluorescence and a high DNA-binding capacity. Slides serve as carriers for the non-covalent or covalent → spotting of → probes (i.e. oligonucleotides or → cDNAs). These probes can be applied onto the slides by various techniques (e.g. socalled split or solid pin, pin and ring, or piezoelectric procedures) and are immobilized via e.g. amino modifications. Loaded slides are called → DNA chip, → gene array, → microarray, or similar. See → arrayer, → chip spotting.

Slide cycler: Any → thermocycler that allows to amplify DNA target sequences *in situ*, i.e. in tissue, cell, or chromosome preparations fixed on a microscope (glass) slide. See → *in situ* polymerase chain reaction.

Sliding: The movement of → DNA-binding proteins along the contour length of the DNA → double helix with concomitant displacement of bound positive counterions. Sliding is brought about by thermal fluctuation and unidirectional association-dissociation processes. It is a means for specific target localization by DNA-binding proteins (e.g. → RNA polymerase binding to → promoter sites, *Eco* RI → restriction endonuclease → recognition site localization, and *lac* → repressor-operator interaction).

Sliding clamp (processivity factor): A ring-shaped protein, that encircles template DNA and tethers the catalytic subunit of replicative DNA polymerases to the DNA template for highly processive chain elongation. In *E. coli*, the ß-subunit clamp is active as a dimer, whereas the gp45 clamp of bacteriophage T4 and the eukaryotic PCNA clamp (for " *p*roliferating *c*ell *n*uclear *a*ntigen") acts as a trimer. The clamp cannot assemble itself on the DNA template, but needs a socalled clamp loader complex, that recognizes a primed template junction and hydrolyzes ATP to assemble the clamp around the DNA. The clamp loader is also composed of different subunits. The *E. coli* clamp loader consists of five g-subunits, the T4 clamp loader of five subunits from genes 44/62 (gp 44/62), and the eukaryotic clamp loader is composed of five RF-C (also called activator–1) proteins. After replication is complete, the DNA polymerase dissociates from the clamp, and the clamp loader also unloads the clamp from the DNA, consuming ATP.

Slightly deleterious allele: A variant → allele, that is under negative selection pressure, but survives, since the selection coefficient is relatively low. Such slightly deleterious alleles can be preserved in a population for extended periods of time, but at a low frequency.

*Slipped-strand m*ispairing (SSM): An intrahelical process that leads to the → mispairing of → complementary bases at the site of short, tandemly repeated sequence motifs in viral, bacterial, and eukaryotic DNA. In short, local → denaturation (unwinding) of the DNA duplex, displace ment of the two strands in the region of → tandem repeats and a slip during → renaturation may lead to non-paired single-stranded loops that are target sites for → excision and repair. Both processes can lead to one or more → duplications, → deletions or → insertions of tandem repeat units. Slipped-strand mispairing is regarded as the driving force in the expansion of → simple repetitive sequences in eukaryotic genomes.

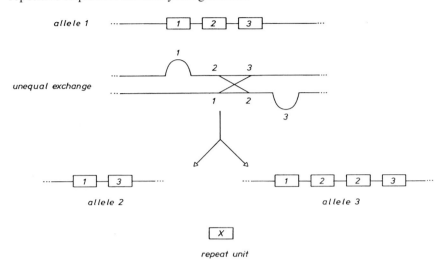

Slipped-strand mispairing

Slippery sequence: See → homopolymer.

SLM: See → single *locus m*inisatellite.

Slot blot: A variant of the conventional → dot blot procedure in which the denatured, non-radioactive DNA is applied to a → nitrocellulose filter or other matrices via slots in a plexiglass mold, usually under slight vacuum.

Slow component: A laboratory term for the DNA that reanneals slowly in a → C_0t analysis and usually consists of non-repetitive DNA.

Slow-stop mutant: A → mutant of *E. coli*, that completes any commenced replication of DNA, but does not start a new replication cycle after a temperature increase to 42°C.

SLP: See→ single *locus p*robe.

SL RNA: Any one of a series of small nuclear RNAs (*less* than 150 nucleotides in length) sharing only low sequence similarity. SL RNAs may function in socalled Sm → sn RNPs, a special class of RNA-protein complexes that bind seven common, highly conserved Sm proteins sharing an unusual cap structure (*tr*imethyl *g*uanosine, TMG, cap). SL RNAs also bind these proteins. See → *s*pliced *l*eader (SL) trans-splicing.

SLTR: See → *s*imple *l*ong *t*ract *r*epeat.

SMC: See → structural maintenance of chromosomes.

Small *amplified* *R*NA-SAGE (SAR-SAGE): A variant of the conventional → serial analysis of gene expression (SAGE) technique, that is adapted for small amounts of starting tissue (e.g. → laser captured microdissected tissue or needle aspirates), and capitalizes on a linear amplification step of small → messenger RNA (mRNA) fragments containing the SAGE tags. In short, a low amount of → total RNA (50 ng or less) is first converted to → cDNA, and the resulting double-stranded cDNAs restricted with *Nla* III (see the procedure of SAGE or → microSAGE). Then an → adapter containing the → T7 RNA polymerase → promoter → upstream of a 16bp spacer and a CATG-terminus is ligated to the → cohesive ends of the *Nla* III-cleaved cDNAs attached to the oligo(dT)$_n$ paramagnetic beads. Transcription is then catalysed by → T7 RNA polymerase (or a mix of enzymes from different biological sources), and extends from the last *Nla* III site of the transcripts to the poly(A) tail. Several rounds of transcription are performed successively with the same cDNA preparation. The resulting mRNA fragments are then separated from the beads, and further processed as in the conventional microSAGE protocol. SAR-SAGE starts with minimal amounts of target tissue (or RNA) and uses a cDNA amplification loop to increase the amount of cDNA and with it the mass of template molecules. See → SAGE-LITE.

Small *auxin* *up* *R*NA (SAUR): Any one of a series of → messenger RNAs that are strongly expressed in elongating regions of plant hypocotyls and epicotyls. They are symmetrically distributed in epidermal and cortical cells of hypocotyls of seedlings grown under normal vertical orientation but become asymmetrically distributed in seedlings exposed to gravitational fields.

Small *cytoplasmic* *ribonucleo*protein (scyrp, scRNP): A member of a family of RNA-protein complexes involved in the → splicing of nuclear precursor RNA, that remains attached to the mature messenger RNA after its transport into the cytoplasm. Scyrps are released from the mRNA before its translation into a protein starts. Compare → small nuclear ribonucleoprotein particle.

Small *cytoplasmic* RNA (ScRNA, scRNA): The RNA component of → small ribonucleoproteins that occur in the cytoplasm of eukaryotic cells.

Small endogenous RNA: See → small RNA.

Small hairpin RNA: See → short hairpin RNA.

Small *interfering* RNA (siRNA, also called *small* *inhibitory* RNA): Any 21–22 nucleotide long double-stranded RNA (dsRNA) molecule with a 3' overhang of two nucleotides, that is generated by → ribonuclease III (see → Dicer) from longer double-stranded RNAs, and mediates sequence-specific → messenger RNA degradation in eukaryotic cells (see → RNA interference). SiRNAs efficiently recruit cellular proteins to form an endonuclease complex (see → siRNA-protein complex), that specifically recognizes the homologous target RNA and destroys it. Since siRNAs are stable over several cell generations, they owe potential for gene-specific therapeutics, especially since they are effective at concentrations that are several orders of magnitude below those of conventional → antisense or → ribozyme gene-targeting approaches.

Small *interfering* *RNA–protein* (SiRNP) complex (RISC): An aggregate between → small interfering RNA and cellular proteins, that guides → messenger RNA recognition and targeted endonucleolytic cleavage, and thereby mediates sequence-specific messenger RNA degradation in eukaryotic cells.

Small molecule microarray: See → chemical microarray.

Small non-mRNA (SnmRNA): A subgroup of → small nucleolar RNAs (snoRNAs) that either belong to the C/D box or H/ACA box snoRNAs, but lack the characteristic 10–21 nucleotide long region with complementarity to → ribosomal RNA (and therefore are not able to methylate the 2'-O-ribose of rRNA, as do the complete snoRNAs). Most of the snmRNAs are predominantly expressed in the brain of vertebrate animals, though low levels are also present in muscle, kidney, and lung tissues.

Small nuclear ribonucleoprotein particle (snRNP, U-snRNP, snurp): A member of a family of nuclear RNA-protein complexes involved in the → splicing of nuclear precursor RNA (pre RNA). See → spliceosome. Compare → small nucleolar ribonucleoprotein (a class of RNP-complexes needed for the processing of pre-ribosomal RNA) and → small ribonucleoprotein.

Small nuclear RNA (snRNA): An abundant class of relatively small (100-300 nucleotides) uridine-rich RNAs (U1-U10) associated with → small nuclear ribonucleoprotein particles, which are found in the nucleus and needed for RNA → splicing. See for example → U-RNA.

Small nucleolar ribonucleoprotein (snoRNP): A member of a family of small nucleolar RNA protein complexes involved in the → post-transcriptional modification of → pre-ribosomal RNA (pre-rRNA) and ribosome assembly. In yeast, more than eight snoRNPs are associated with pre-ribosomal particles, and serve in all pre-rRNA processing reactions, e.g. modification(s) of pre-rRNA and nucleo-cytoplasmic transport of the newly synthesized ribosomes. One of the snoRNP proteins, nucleolar protein 1 (NOP 1), is functionally conserved between yeast and man. Compare → small nuclear ribonucleoprotein particle, a class of RNP complexes needed for messenger RNA processing.

Small nucleolar RNA (snoRNA): Any one of a large family of small, metabolically stable, nucleolar RNAs, that fall into two major categories: the socalled C/D box and H/ACA box snoRNAs, designated after common sequence motifs involved in the assembly of → small nucleolar ribonucleoprotein (snoRNP) complexes. The vast majority of snoRNAs function in the post-transcriptional modifications of nucleotides in → ribosomal RNA, and only few are required for pre-rRNA cleavage. Each 2'-O-ribose methylation or pseudouridylation (see → rare bases) in eukaryotic rRNA is selected by a cognate snoRNA of either the C/D box or H/ACA box family, by forming specific base pairs that span the rRNA modification site. C/D box snoRNAs contain two short, conserved sequence motifs, box C (5'-RUGAUGA–3'; R = any purine) and box D (5'-CUGA–3'), constantly located a few nucleotides away from the 3' or 5' ends, respectively. Immediately upstream of the boxD (or from a second CUGA motif, "box D' ") 10–21 nucleotide long regions complementary to rRNA are localized, that span the 2'-O-methylation sites. In the corresponding snoRNA-rRNA duplex, the 2'-O-methylation is directed to the rRNA target nucleotide paired with the fifth snoRNA nucleotide upstream from box D (or box D'). C/D and H/ACA box snoRNAs are encoded by introns of → house-keeping genes belonging to the → 5' terminal oligopyrimidine (TOP) class of genes, and produced by → processing of the pre-mRNA intron. Compare → small nuclear RNA. See → host gene, → intronic snoRNA, → small non-mRNA.

Small regulatory RNA: Any RNA of comparably small size (i.e. 18–80 nucleotides long), that regulates a nuclear process in eukaryotic cells. For example, → short hairpin RNAs, → small interfering RNAs and → small temporal RNAs are such regulatory RNAs.

Small RNA (small endogenous RNA): A more general term for a series of small (18–80 nucleotides long) RNAs of eukaryotic cells. There exists a confusing multitude of terms for (probably) one and the same molecular species (for example, → tiny RNA, → microRNA). Some of the small RNA families harbor → short hairpin RNAs, → small interfering RNAs, or → small temporal RNAs, and some of the RNAs possess known regulatory functions (e.g. → small regulatory RNA), or have no obvious functions yet (e.g. → small non-messenger RNA). The group of small RNAs in eukaryotic cells is complex and contains probably more than 500 different sequences.

Small single copy sequence: See → short single copy sequence.

Small t (t): As opposed to large T (→ T antigen), see → simian virus 40.

Small temporal RNA (stRNA): Any one of a class of more than 100 non-coding, highly conserved single-stranded, 20–25 nucleotides long → small RNAs, transcribed as a ~75 nucleotide long precursor predicted to form a loop-stem structure, that is processed into mature 22 nucleotide stRNA sequences by → Dicer RNase III and control major developmental transitions. One of the strands of the double-stranded stems of the precursor then represents the stRNA. The → antisense strand of stRNAs recognizes complementary sequences in the 3'-untranslated region of target → messenger RNAs, and bind there with bulges and mismatches, such that it forms a characteristic interrupted hybrid, which cannot be translated into a protein. This process in turn triggers transition to the next developmental stage. For example, the 21 nt long stRNAs encoded by genes *lin–4* and *let–7* of *Caenorhabditis elegans* regulate developmental timing by blocking translation of target mRNA. This translational block relaxes other genes, which incites a gene cascade responsible for the next developmental transition (e.g. transition from the first to the second larval stage of *C. elegans* requires the action of the 22 nucleotide *lin-4* stRNA, transition from late larval stage to adult cell necessitates the 21 nucleotide *let-7* stRNA). Therefore stRNAs do not encode proteins, but regulate cognate mRNA stability. See → RNA interference. Compare → non-coding RNA, → short hairpin RNA, → short interfering RNA, → small RNA, → small endogenous RNA, → small non-messenger RNA, → small regulatory RNA, → tiny RNA.

Small upstream reading frame (SMURF): Any → open reading frame within the → 5'-untranslated region (5'-UTR) in about 5–10% of eukaryotic → messenger RNAs. SMURFs are probably controlling → translation or tissue-specific expression of the encoded protein.

SMART:
See → databases (as appendix).
See → switching mechanism at 5'-end of RNA template.

SMC: See → structural maintenance of chromosomes.

SMD-PCR: See → single molecule polymerase chain reaction.

Smear: The digestion of high molecular weight → genomic DNA with type II → restriction endonucleases gives rise to millions of fragments with a broad range of molecular weights. After gel electrophoresis and → ethidium bromide staining they cannot be distinguished as single bands, but appear as "smear".

SMFS: See → single molecule fluorescence spectroscopy.

Smiling effect: The appearance of curved bands of proteins or nucleic acids on one or both sides of a gel (→ gel electrophoresis) due to slower migration of the most marginal molecules. Frequently a consequence of better heat dissipation at these locations, which leads to lower mobility of the molecules.

Smith-Birnstiel mapping (Smith-Birnstiel restriction mapping): A technique to construct a → restriction map of a particular segment of DNA. This segment is first labeled at one end (see → end-labeling), then one aliquot completely, another aliquot only partially digested with an appropriate → restriction enzyme. The fragments of both → digests are separated by → agarose gel electrophoresis, and the labeled fragments detected by → autoradiography of the gel (or a → blot). By comparison of both autoradiograms and an analysis of the length differences the correct order of the various restriction fragments can be arranged in their original order and a restriction map of the DNA segment can be established.

Smith-Birnstiel restriction mapping: See → Smith-Birnstiel mapping.

Sm^r, sm^s: See → streptomycin resistance, → streptomycin sensitivity.

SMS: See → single-molecule DNA sizing.

SMS: See → single molecule spectroscopy.

Sm site: The consensus sequence 5'-AAUUUUGA–3' in U1 → small nuclear RNA, that represents the binding site for specific spliceosomal proteins.

SNA: See → synthetic nucleic acid.

Snake venom phosphodiesterase: See → phosphodiesterase I.

Snap-back DNA: See → fold-back DNA.

SnmRNA: See → small non-mRNA.

Snorbozyme (small nucleolar RNA-ribozyme): A fusion of a → hammerhead ribozyme to a → small nucleolar RNA as carrier, that localizes the snoRNA-ribozyme complex to the → nucleolus of a (yeast) cell, where the ribozyme in trans completely cleaves target RNA (the carrier-type small nucleolar RNA). Snorbozymes possess a conserved structure motif, the box C/D motif, that stabilizes the molecules in vivo. Snorbozyme activity practically leads to a knock-out of the snoRNA gene(s).

Sno RNA: See → small nucleolar RNA.

snoRNP: See → small nucleolar ribonucleoprotein.

Snorthern analysis: A technique for the estimation of gene expression in restricted tissue sources (e.g. yielding less than 100 ng of cDNAs), that starts with the isolation of → messenger RNAs from the target tissue, cDNA preparation, its digestion with three → restriction endonucleases that recognize four base → restriction sites (as e.g. *Dde*I, *Dpn*II, or *Mbo*I and *Nla*III) and the ligation of specific → adaptors to the ends of the restriction fragments. The adaptor-ligated cDNA digests are separately amplified using conventional → polymerase chain reaction, then combined, the amplified products electrophoresed in an → agarose gel together with → ethidium bromide, and finally blotted onto an appropriate membrane (see → Southern blot). The resulting Snorthern blot (i.e. *S*outhern blot providing expression information, otherwise proven by → *N*orthern analysis) is then successively hybridized to a series of radiolabeled gene probes and hybridization events (i.e. expression) visualized by → autoradiography.

SNP: See → *s*ingle *n*ucleotide *p*olymorphism.

SNP cluster: The accumulation of → single nucleotide polymorphisms in relatively small genomic regions. Such clusters can extend over 1 kb (or more), are relatively rare, but very old, and originate from ancestral chromosome fragments inherited in the extant species.

SNP density: See → single nucleotide polymorphism density.

SNP discovery: The detection of new → *s*ingle *n*ucleotide *p*olymorphisms (SNPs) in a genome.

SNP frequency: The frequency, with which → single nucleotide polymorphisms occur along a defined stretch of DNA or a chromosome (usually expressed as SNP/kb). SNP frequency varies between genomic regions in the same individual and between related individuals. For example, in some regions of the maize genome, SNP frequency is about 1/65 bp, in other regions 1/85 bp. Generally, SNP frequency is much higher than the frequency of → insertions or → deletions (1/250 bp).

SNP ID: See → reference SNP.

SNP image map: A graphical depiction of the distribution and number of confirmed and candidate → single nucleotide polymorphisms along a stretch of DNA, a → bacterial artificial chromosome clone, or a chromosome. Web site: http://lpg.nci.nih.gov/html-cgap/validated.html

SNPing: A laboratory slang term for the detection of → single nucleotide polymorphisms.

SNP LD (*single nucleotide polymorphism linkage disequilibrium*): Any → single nucleotide polymorphism allele, that is closely linked to another one (see → linkage disequilibrium) such that an → association can be inferred.

SNP map: A linear array of → single nucleotide polymorphism sites along a particular DNA molecule (e. g. a chromosome). Since most genomes contain many such sites (e. g. human genome: 3-30 millions of SNPs), SNP maps are highly saturated, i. e. the distance between two neighbouring markers very short (i. e. 1-10 cM). Clustering of SNPs on such a map indicates genic DNA (1 SNP/500 bp), random distribution of SNPs is characteristic for extragenic DNA (1 SNP/2000 bp).

SNP minisequencing: A technique for the detection of → single nucleotide polymorphisms (SNPs). In short, the genomic target region is first amplified with appropriate forward and reverse → primers and conventional → polymerase chain reaction, and the amplification product purified (by e. g. separation of the amplicon from primers and dNTPs). Then a primer is hybridized to the denatured target DNA such that it ends exactly one base 3' upstream of the SNP ("SNP primer"), and extended in the presence of → dideoxynucleoside triphosphates labeled with different fluorochromes (e. g. ddATP-TAMRA, ddCTP-Cy5, ddGTP-Cy3, ddTTP-fluorescein). The → primer extension reaction allows the incorporation of only the nucleotide matching the template, but is then blocked (hence the name SNP minisequencing). The product is then denatured, electrophoresed in capillaries (see → capillary electrophoresis), and the type of fluorescence analyzed. SNP minisequencing allows to identify the target locus as e. g. homozygous wild-type, homozygous mutant, or diallelic heterozygote.

SNP-MS: See → single nucleotide polymorphism mass spectrometry.

SNPper (pronounce snipper): A web-based tool package (http://snpper.chip.org/), that allows to screen public databases for → single nucleotide polymorphisms (SNPs). Users can search for SNPs in defined positions of a chromosome or in genes, and SNPper records all SNPs of the corresponding region, with informations about each SNP, the number of SNPs, alleles, the SNP position, the genomic sequences surrounding a SNP, and links to the relevant pages in the public dbases. The "Save SNPset" command allows to store the SNPset on the server, so that it can be reloaded later on. Web site: bio.chip.org/biotools

SNP primer: Any → primer oligonucleotide, that matches a → single nucleotide polymorphism site at its 3'- or 5'-flank, hybridises to this site and can be extended by one fluorescently labelled dideoxynucleotide (see → primer extension).

SNP profiling: The establishment of a profile of a particular → single nucleotide polymorphism (SNP) in a whole population.

SNP-rich segment: Any one of a series of genomic regions, in which → single nucleotide polymorphisms (SNP) occur at a considerably higher frequency than in the rest of the genome. For example, in the mouse genome such SNP-rich segments contain about 40 SNPs per 10 kb, whereas other regions contain much less SNPs (e.g. the intermitting sequences harbor about 0.5 SNPs per 10 kb).

SNP scanning: See → single nucleotide scanning.

SNP scoring: See → single nucleotide polymorphism scoring.

SNP typing: See → single nucleotide polymorphism typing.

snRNA: See → small nuclear RNA.

snRNP: See → small nuclear ribonucleoprotein particle.

SNuPE: See → single nucleotide primer extension.

Snurp: See → *s*mall *nu*clear *r*ibonucleo*p*rotein.

Sodium *d*odecyl *s*ulfate (SDS, sodium lauryl sulfate): An anionic detergent with the formula $CH_3(CH_2)_{11}OSO_3Na$, used for protein solubilization and electrophoresis. Proteins may bind 1.4 g SDS per gram of protein which confers to them a net negative charge. Therefore their electrophoretic separation after SDS-loading will depend largely on their size (→ SDS polyacrylamide gel electrophoresis).

Sodium lauryl sulfate: See → sodium dodecyl sulfate.

SOE-PCR: See → *s*plice *o*verlap *e*xtension *p*olymerase *c*hain *r*eaction.

Soft agar: A semi-solid → agar that is added to culture media for animal cells.

Solenoid: The 30 nm chromosome fiber formed by a → supercoil arrangement of → nucleosomes, possibly fixed by → histone H1.

Solid-phase cDNA synthesis: A variant of the conventional → cDNA synthesis technique in which the → primer extension reactions proceed on a solid support e.g. an oligo(dT) bead. In short, polyadenylated → messenger RNA is first annealed to an oligo(dT) stretch covalently linked to a solid bead support, and reverse-transcribed. After removal of the mRNA, a primer specific for the 5' end of the first strand allows to synthesize the second strand. A modification of this technique includes a single-sided → polymerase chain reaction with an oligo(dT) primer, and a further amplification of the product using a → nested primer polymerase chain reaction. See → magnetic bead, → solid-phase oligonucleotide synthesis.

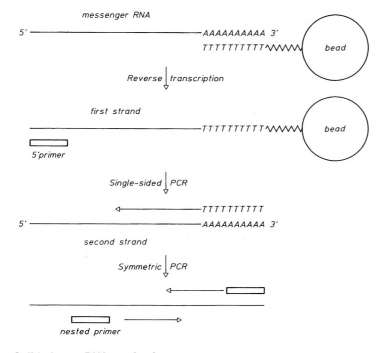

Solid-phase cDNA synthesis

Solid-phase oligonucleotide synthesis: A complex multistep phosphoramidite-based procedure for the production of pure oligonucleotides, that proceeds on a solid *c*ontrolled *p*ore *g*lass (CPG) or polystyrene-copolymer support. The cycle starts with the linkage of the 3'-hydroxyl group of the first nucleoside to the support, the detritylation of the 5'-*di*methoxy*t*rityl (DMT, trityl) protecting group of this (or generally terminal) nucleoside, the coupling of the next 5'-DMT protected deoxyribonucleoside via a nucleophilic attack of a free 5'-hydroxyl group on the activated 3'-phosphoramidite function, the capping of the remaining 5'-OH ends of unreacted terminal monomers via acetylation, the oxidation of unstable phosphite triester to stable phosphate triester by iodine, and the cleavage from the support and simultaneous base and phosphate deprotection by treatment with concentrated ammonia solution. After oxidation, another cycle of nucleotide addition starts to produce the oligonucleotide of the desired length. The coupling reactions generally are more than 98% efficient (i.e. in every cycle, only 98% of the growing chains actually add the next base), so that 2% undesirable components of shorter length ("capped failures") accumulate progressively as a function of chain length with each cycle. Therefore the crude oligonucleotide reaction mixture, containing full-length oligonucleotrides, the shorter capped failures, low level side reaction products and ammonia have subsequently to be purified. Compare → solid-phase cDNA synthesis.

Solid-phase PCR: See → on-chip polymerase chain reaction.

Solid pin: A metal device for the → spotting ("printing") of target molecules onto → microarrays, consisting of a solid (usually metal) shaft and a flat tip, that takes up solution by adsorption and delivers the adsorbed sample onto the microarray support (e.g. glass or quartz) by direct contact.

S-oligo: See → phosphorothioate oligonucleotide.

Solitary gene: See → single-copy gene.

Solo *l*ong *t*erminal *r*epeat (solo LTR): Any → long terminal repeat that has been individualized from a previously intact LTR → retroposon by intramolecular → recombination(s) leading to the total → deletion of the interior part of the retroposon and one LTR. Solo LTRs are flanked by the → target site duplications generated during the → transposition of the originally complete element. Solo LTRs are frequently occurring elements in genomes of higher organisms.

Solo LTR: See → solo *l*ong *t*erminal *r*epeat.

Soluble RNA: See → transfer RNA.

Solution hybridization: See → liquid hybridization.

Solution PCR: See → solution *p*olymerase *c*hain *r*eaction.

Solution *p*olymerase *c*hain *r*eaction (solution PCR; also solution phase PCR): A somewhat infelicitous term for a variant of the conventional → polymerase chain reaction that allows to detect specific → DNA or → RNA sequences and to quantify these sequences in a cell → lysate. Cells are lysed to release the templates, which are then amplified in solution. The amplification products can be separated by → agarose gel electrophoresis and visualized by → ethidium bromide staining.

SOMA: See → short oligonucleotide mass analysis.

Somaclonal variation: The increase in → mutation rate of plant cells kept in culture for a long time, so that various mutant cells can be isolated from a previously uniform → cell culture, and the mutant cells (somaclonal variants) be selected for desirable new traits. The mutations encompass DNA rearrangements, chromosome mutations, and also activation and/or inactivation of genes, and are probably induced by the stress of *in vitro* culture conditions.

Somatic embryo (adventive embryo, embryoid): An asexually generated embryo that originates from a somatic cell (or cells) of a plant. In favorite cases, such embryos can be induced to regenerate into complete plants. One way to introduce foreign DNA into target plants is the direct → transformation of normal plant cells (by → direct gene transfer methods e.g. the → particle gun technique) and the regeneration of transformants via somatic embryogenesis.

Somatic gene therapy (somatic cell genetic engineering): The use of → recombinant DNA technology to introduce new, foreign, or modified genes into the genome(s) of somatic cells, with the aim to correct defects in somatic cells, especially human cells. Basically two approaches are available. First, in vivo gene therapy works with highly selective *non-viral vectors*, that are injected and find their target organs or cells within the organisms, where they regulate the synthesis of the correct protein. Second, for the socalled ex vivo gene therapy, cells from the corresponding target tissue are first isolated from the organism, maintained in cell culture, and transformed by *viral vectors* with high transfection rate (but low selectivity) carrying the correct copy of the gene of interest. The transformants are then transferred into the target organism, where they produce the correct protein.

Somatic gene therapy: See → somatic cell genetic engineering.

Somatic hybrid (somatic cell hybrid): Any cell arising from the → fusion of two different somatic cells. See → cell fusion.

Somatic hybridization: The production of → somatic hybrids through the → fusion of two different somatic cells.

Somatic *hypermutation* (SHM): The creation of point mutations in the *v*ariable region of immunoglobin (IgV) genes, that occurs in the lymph node germinal centers, as B cells compete for limiting numbers of → antigen molecules. Only those cells, that possess highest affinity for the antigen, will survive. SHM underlies → affinity maturation, i.e. the increase in affinity of an antibody for its antigen in the course of immune responses.

Somatic mutation: Any → mutation occurring in somatic, but not germline cells. Somatic mutations lead to the formation of chimeric tissues, if the mutant cell undergoes mitoses.

Somatic recombination:
a) The rearrangement of DNA sequences (by excision and splicing) in B lymphocytes during their differentiation, that leads to the diversity of immunoglobulin genes, and to diversity in → antibody production.
b) The low-frequency rearrangement of DNA sequences via homologous → cross-overs in *Aspergillus, Saccharomyces, Drosophila*, various cultured mammalian cells, and probably other organisms as well.

***S1-h*ypersensitive *site* (SHS):** A short region in → chromatin that is more sensitive to the single strand-specific → S1 nuclease by a factor of 50-100 as compared to neighboring regions. SHSs mark positions within → promoter regions of active, but not inactive genes and are address sites for → transcription factors. They are probably generated by certain conformations of DNA (e.g. close to the borders of homopurine-homopyrimidine stretches that favor → Z-DNA formation).

S1-mapping (S1 nuclease mapping, nuclease S1-mapping; Berk-Sharp mapping, Berk-Sharp technique): A technique to precisely determine the coding region of a → gene and the number of its exons and introns, to map the transcription start (cap site) and the transcription termination site of a gene, and the direction of its transcription, by forming mRNA-DNA hybrids and removing unpaired DNA with → S1 nuclease. In short, the mRNA or its → cDNA is hybridized to the corresponding radioactively labeled, single-stranded genomic DNA, and the hybrid digested with → S1 nuclease that removes all single-stranded 5'- or 3'-overhangs and the intron loops. During the subsequent electrophoretic separation of the hybrid molecules in denaturing → po lyacrylamide gels the RNA is hydrolyzed. The determination of the length of the resulting DNA fragments then allows to map the transcription start site. The length of the intron(s) can be calculated after digestion of the mRNA-DNA hybrids with → exonuclease VII that removes specifically single-stranded termini, but leaves the intron loops intact.

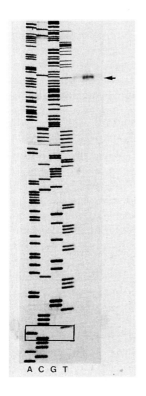

A C G T   **S1-mapping**

S1 nuclease (nuclease S1): An enzyme from *Aspergillus oryzae* that catalyzes the endonucleolytic cleavage of single-stranded RNA or DNA to yield 5' mono- or oligonucleotides. Double-stranded DNA or RNA, or DNA-RNA hybrids are resistant to S1 nuclease unless very large amounts of the enzyme are employed. Used to trim single-stranded termini (e. g. → sticky ends of → restriction fragments), to remove the loops from → stem-and-loop structures during the synthesis of double-stranded → cDNA, to analyze mRNA-DNA hybrids (→ S1-mapping), to produce → deletions in DNA molecules together with → exonuclease III, and to detect S1-hypersensitive configurations (→ S1 hypersensitive sites) in → chromatin.

S1 *nuclease protection assay* (S1-NPA): A method for the detection, quantitation and characterization of specific → messenger RNA molecules out of complex mixtures of total cellular RNAs. In short, total RNA is isolated and hybridized in solution to an excess of specific, labeled → antisense RNA probe. This probe is generated by the cloning of a → cDNA fragment of interest into a plasmid vector in between → SP6, → T7 and T3 DNA-dependent RNA polymerase → promoters. Addition of the respective RNA polymerases allows to synthesize high-specific activity radiolabeled anti-sense RNA probes. After hybridization of these probes to total RNA, → S1 nuclease is added, which digests single-stranded RNAs and free anti-sense probes, but only marginally attacks RNA-RNA hybrids. After S1 nuclease inactivation, the RNase-protected hybrids are electrophoresed in denaturing → polyacrylamide gels, which are subsequently dried and exposed to X-ray films. The quantity and the length of these hybrids can then be determined by → autoradiography or → phosphorimaging. See → lysate RNase protection assay, → nuclease protection assay, → RNase protection assay, → S1-mapping.

Sonication: The use of audible sound waves, produced by a sonifier (sonicator), to disintegrate cells in liquid medium or to remove particles from items to be cleaned. See also → sonication loading.

Sonication-*assisted Agrobacterium transformation* (SAAT): A variant of the → *Agrobacterium*-mediated gene transfer to plant cells, which employs a brief sonication step during infection of cells by → *Agrobacterium tumefaciens*. This sonication probably enhances interaction of the bacteria with slightly disintegrated cell walls of the host cell, leading to higher → transformation frequencies.

Sonication loading: A rapid method for the introduction of macromolecules (e.g. DNA, see → direct gene transfer) into cells in suspension (e.g. the amoebae of *Dictyostelium discoideum*) using brief sonication.

SOS repair (error-prone repair): The repair of damaged DNA in *E. coli*, catalyzed by enzymes that are coordinately induced by a complicated mechanism. First, the damaged DNA somehow activates the so-called → recA protease which cleaves the so-called lexA repressor protein. This event induces the activity of SOS boxes in the → operator of at least 11 SOS repair genes of the → long patch repair and the → recombination repair systems, which are normally repressed by lexA. SOS repair is a cellular reaction to extensive DNA damage (e.g. due to irradiation) and base mutations which inhibit replication. Thus for instance → crosslinking of DNA by alkylating compounds or → intercalating agents induces this repair system. Since the damage has often completely destroyed the correct template, its repair frequently leads to mutations ("errorprone"). Therefore it is regarded as a last resource of the cell to regain a certain degree of viability.

SOS response: The induction of a set of DNA repair, recombination and replication genes and subsequent expression of the corresponding proteins in bacteria as a consequence of DNA damage by e.g. UV light, oxygen radicals or mutagens. See → SOS repair.

Source DNA: Any DNA that serves as starting material for the isolation of a specific sequence, as e.g. a → gene, a → satellite or → transposon.

Source gene (master gene, founder gene): Any ancestral gene that gave rise to gene families.

Southern blot: A nitrocellulose or nylon membrane, onto which DNAfragments (generated by e.g. → restriction of → genomic DNA), separated by → agarose gel electrophoresis, are transferred by → Southern blotting. After transfer the fragments are immobilized by → baking or → cross-linking (definition b). Southern blots are used for → hybridization of radioactively or fluorescently labeled → probes and the detection of hybridization events by → autoradiography or laser excitation and fluorography, respectively. First developed by Ed Southern.

Southern blotting (Southern hybridization, Southern transfer, capillary blotting): A gel blotting technique in which DNA fragments, separated according to size by → agarose gel electrophoresis are denatured *in situ* and transferred to a → nitrocellulose filter or other matrices by electric or capillary forces (Southern transfer). Single-stranded nucleic acids may be fixed to nitrocellulose by baking and are thus immobilized. Hybridization with specific radioactively or non-radioactively labeled probes helps to identify individual fragments out of complex mixtures of millions of other fragments (Southern hybridization). Variations of the original technique have been developed in great numbers, see for instance → alkali blotting, → bidirectional transfer, → dry blotting, → genomic blot/blotting, → packed array hybridization, → reverse Southern hybridization, → Southern blot, → thermoblotting, → Zoo blot.
Figure see page 1059.

Southern hybridization: A technique for the visualization of a hybridization event between a → restriction fragment of target DNA and a specific, radioactively or fluorescently labeled → probe on a → Southern blot. See → Southern blotting.

Southern transfer: See → Southern blotting.

South-Western blotting (South-Western blot mapping): A method for the rapid characterization of both a DNA-binding protein and its specific binding site on genomic DNA. In short, proteins are separated on an → SDS polyacrylamide gel, renatured by removal of SDS in the presence of urea, blotted onto → nitrocellulose by diffusion, and immobilized in their native configuration. The genomic target DNA is restricted, fragments are endlabeled and allowed to bind to the separated proteins in the presence of unspecific competitor DNA [e.g. poly(I)-poly(C)]. The specifically bound DNA can be eluted from each individual protein-DNA complex and both the DNA as well as the protein can be analyzed. A special application of South-Western blotting is the → DNA ligand screening technique. Compare → Northern blotting, → Southern blotting, → Western blotting.

SP: See → *sequence-specific primer*.

SPA: See → *scintillation proximity assay*.

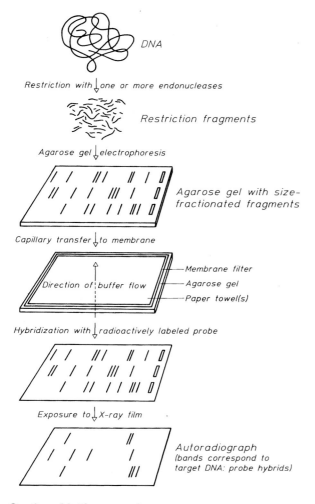

Southern blotting procedure

Spacer (spacer DNA):
a) Untranscribed repetitive sequences (→ repetitive DNA) of eukaryotic and some viral genomes flanking functional genes.
b) DNA sequences in between functional address sites (boxes) for → DNA-binding proteins in eukaryotic → promoters. See → split promoter.
c) Generally, DNA sequences separating specific repeated gene units (e.g. in → histone genes or → rDNA). See → spacing.
d) A DNA region which separates the two parts of a → fusion protein in frame and provides a cleavage point for the post-translational separation of these parts, see → cleavage fusion.

Spacing: The separation of two specific sequence elements in double-stranded DNA molecules by a → spacer sequence. Spacing is one of the structural prerequisites for correct function of e.g. protein target sites. If for example, two target sites for different regulatory proteins are too close to one another, the binding of one protein and interaction between the two proteins may be impaired or impossible. Thus in genes, a specific distance between the → start codon and the → TATA-box, or the → stop codon and the → terminator sequence is required for correct function. Correct spacing is also important for the construction of → expression vectors.

SPAR: See → microsatellite-primed polymerase chain reaction.

SPARC: See → *Schizosaccharomyces pombe artificial* chromosome.

Spatial development RNA (sdRNA): Any → microRNA, that is involved in the regulation of pattern formation during development of an organism.

Spatial expression pattern: The differential expression of multiple genes of an organism in different organs, tissues, or cells at a given time. This spatial expression pattern changes during cell, tissue and organ development. See → temporal expression pattern.

SPE: See → enhancer.

Specialized chromosome structure (SCS): A somewhat vague term for DNA sequences located at the junctions between decondensed → chromatin of a transcribed gene and the adjacent condensed chromatin (→ heterochromatin). SCS elements have neither an inhibitory nor activating effect on the genes within the chromosomal domain they flank, but prevent the spread of the repressive influence of heterochromatin, or the activating influence of an → enhancer on these genes. SCSs are → locus control regions.

Specialized transduction: See → transduction.

Special transduction: See → transduction.

Specific activity:
a) The activity of an enzyme expressed as the amount of substrate converted (in μ-mol)/minute × mg protein or one mole of substrate converted/second under defined conditions (kat, katal).
b) The ratio of radioactive to non-radioactive atoms or molecules of the same kind, defined as Ci/mol.

Specific amplicon polymorphism (SAP): Any → polymorphism detected by the amplification of a specific genomic sequence, using sequence-specific primers and conventional → polymerase chain reaction techniques. Compare → random amplified polymorphic DNA.

Specific interacting domain (SID): Any → domain of a protein A, that specifically interacts with a domain of another protein B. Such SIDs are the basis for protein-protein interactions in a cell.

Specificity assessment from fractionation experiments (SAFE): A technique for the evaluation of the specificity of hybridisations on → DNA chips. SAFE exploits the fact, that a complete hybridisation between immobilized → probe and target sequence (i.e. complete complementarity) leads to a very stable DNA duplex in comparison to a cross-hybridization of the probe with a less complementary target. The specificity of hybridisation on a chip is tested with increasing → stringency. The chip is exposed to increasing concentrations of e.g. formamide (from zero to 94.5%), and the → transition stringency of selected probes determined. The dissociation of probe and target at lower transition stringencies is indicative for (non-specific) cross-hybridization. In addition to confirming techniques such as SAFE, reproduction of the chip hybridisations and the use of different probes covering different regions of the gene in question are recommended (or mandatory) for acceptable chip experiments.

Specificity protein: See → SP 1.

Spectinomycin: An aminocyclitol → antibiotic produced by *Streptomyces* strains, that interferes with peptidyl tRNA translocation from the A site to the P site on the → ribosome and therefore inhibits bacterial protein synthesis. See → spectinomycin amplification.

Spectinomycin amplification (spectinomycin enrichment): A method to increase the → copy number of → relaxed control plasmids in *E. coli*. The antibiotic → spectinomycin inhibits the peptidyl tRNA translocation step and therefore bacterial protein synthesis, but leaves plasmid replication unaffected. See also → chloramphenicol amplification.

Spectinomycin enrichment: See → spectinomycin amplification.

Spectral genotyping: A technique for the identification of the allelic composition of an individual. In short, the target region is amplified via conventional → polymerase chain reaction in the presence of two → molecular beacons as probes, one labeled with e.g. → fluorescein (for the detection of the wild-type allele), the other with → tetramethylrhodamin (for the detection of the → mutant allele). The amplification occurs in a real-time PCR machine. If only green fluorescence appears during amplification, then a homozygous wild-type is present. Also, if red fluorescence can only be detected, a homozygous mutant can be inferred. The detection of both green and red fluorescence indicates a heterozygote. Compare → spectral karyotyping.

Spectral karyotyping (SKY): A technique for the discrimination of all the different chromosomes of a cell (→ karyotype) by a sophisticated combination of fluorescence microscopy, spectroscopy, and high-resolution digital CCD camera imaging. In short, chromosome-specific DNA → probes are labeled with 5 (or more) different → fluorochromes (e. g. → cyanine 3, → cyanine 5, → Texas Red, and Spec Green), and combinations between them, and in situ hybridized to metaphase chromosome spreads (see → fluorescence *in situ* hybridization, FISH). Each chromosome is then characterized by a specific fluorescence spectrum, which can be detected simultaneously with the spectra of the other chromosomes using a socalled "triple-pass filter". A total of 5 different fluorochromes allow 31 different combinations ($2^5 - 1 = 31$), more than are needed to image e. g. all 24 human chromosomes. SKY detects chromosomal aberrations (e. g. → translocations) that can be characterized by spectral analysis. See → multicolor spectral karyotyping.

Supergene: A somewhat misleading term for a gene complex, that is conserved in genomes of different genera of the same family. For example, the order of genes responsible for flowering is almost identical in different grasses, e.g. rice, wheat, oat, barley, maize and millet, though these species have evolved differently. See → synteny.

SPFG: See → *s*econdary *p*ulsed *f*ield *g*el electrophoresis.

SPH: See → *S*-*p*rotein *h*omologue.

Sphaeroplast: See → spheroplast.

S-Phase: See → cell cycle.

Spheroplast (sphaeroplast): A bacterial, fungal or plant cell from which most of the cell wall has been enzymatically removed so that this cell adopts a spherical shape.

Sphingosine-based liposome: A cationic phosphatidylcholine-containing variant of a conventional liposome, that can be used to form a colloidally stable complex with DNA. Sphingosine-DNA lipoplexes are used for direct gene transfer.

Spiegelmer: Any one of a series of extremely stable synthetic mirror-image DNA or RNA molecules, that are designed via an evolutionary selection process. Spiegelmers cannot be degraded by nucleases or proteases, do not bind non-specifically to targets, but extremely selectively to e.g. protein targets for which they were selected.

***Spi* phenotype (*s*ensitive to *P2* *i*nterference):** A characteristic of certain *E. coli* → lysogens carrying the → prophage P2 that does not allow the growth of wild-type → lambda phages (*Spi*⁺ phenotype). This restriction is due to the products of genes *red* and *gam* (they are involved in → recombination), and can be overcome if the λ phage lacks these two genes. In this case, it displays the *Spi*-phenotype and is able to grow in P2 lysogens, but needs a functional → chi sequence and a *rec*⁺ host strain. In gene technology, the → stuffer fragment of λ phages containing *red* and *gam* genes is replaced by foreign DNA, which leads to a *Spi* phenotype whose growth is not inhibited by P2-lysogenic *E. coli* hosts, whereas the *Spi*⁺ phenotype is. This phenomenon is exploited for the selection of recombinant λ-phages.

Splice acceptor site: See → acceptor splice junction.

Spliced *leader* (SL) trans-splicing: A probably ancient nuclear → RNA processing step in a series of lower eukaryotes (as e.g. protozoans [and here in the kinetoplastid], nematodes, some flat worms, euglenoid protists, and possibly chordates and cnidaria). Trans-splicing in these organisms starts with the attachment of a small common terminal 5' exon ("spliced leader", → spliced leader RNA), derived from an → SL RNA, to the 5' most exon of any pre-mRNA.

Spliced leader RNA ("spliced leader"): Any one of small RNAs, that provide the short leader sequences (termed SL1 to SL5) trans-spliced to the 5'end of pre-messenger RNAs by → spliced leader trans-splicing.

Splice donor site: See → donor splice junction.

Spliced protein: See → mature protein.

Splice junction (splicing junction, splice junction signal, RNA splice site): → Consensus sequences at the ends of → introns which are involved in excision and → splicing reactions during the → post-transcriptional modification of → primary transcripts from eukaryotic → split genes. The junction signal at the 5' end of an intron transcript is the → donor splice junction, the signal at the 3' end the → acceptor splice junction.

```
    Exon                        Intron                      Exon
          C       A                                    C
5' ────── AG GT  AGT ───────────────────────── N  AG G ────── 3'
        A       G                              T
    Donor  splice  junction           Acceptor  splice  junction
```

Splice junction signal: See → splice junction.

Splice oligonucleotide array: Any → microarray, onto which oligonucleotides complementary to sequences on both sides of → splice junctions are spotted in an ordered array. Hybridization of → messenger RNAs of a cell to this microarray reveals the number of → splice variants present in a cell at a given time.

Spliceome: The complete set of → splice variants (i.e. all diverse → messenger RNA isoforms) of a cell at a given time. See → alternative exons, → alternative splicing.

Spliceosome (splicing complex): A multicomponent complex of about 200 *splicing factors* (SF) and → small nuclear RNAs catalyzing the → splicing of nuclear precursor RNA (pre-RNA) to form translatable mRNA in eukaryotes. The assembly of the spliceosome is based on the sequential multiple interactions of → small nuclear ribonucleoprotein particles (in particular, uracil-rich U2, U4, U5 and U6 snRNP particles), pre-mRNA and at least six → DEAD-box proteins. In a first step, U2 snRNP binds to a region upstream of the 3' → splice junction in pre-mRNA, a process which depends on ATP. In a second step the complex associates with a multi-snRNP particle containing U4, U5 and U6 snRNPs, which involves at least one → DEAH-box protein (Prp 22 of yeast) whose function is modulation of RNA structure (as RNA → helicase). During the splicing process the U4 snRNP particle is released from the spliceosome. The final step of splicing results in the dissociation of the ligated exons and the → lariat form of the → intron bound to U2, U5 and U6 snRNPs. See → pre-spliceosome.
Figure see page 1064.

Splice overlap extension polymerase chain reaction (SOE-PCR): A technique for the joining of two (or more) different genes, site-directed mutagenesis, or generation of → deletions and → insertions in a target DNA. For example, the fusion of two genes starts with the amplification of each gene in a separate → polymerase chain reaction. The ends of the internal → primers contain overlaps of complementary sequences. Then both PCR products are mixed, denatured, and allowed to anneal. The complementary ends of the products hybridize and serve as primers for the extension of the complementary strand: the two genes are spliced together. Compare → PCR-ligation – PCR mutagenesis.

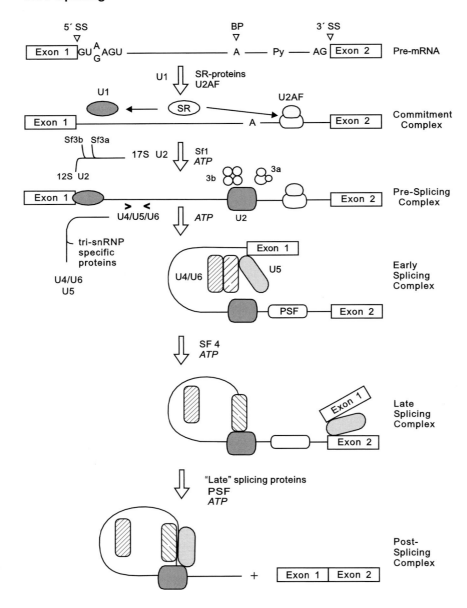

Spliceosome

Splicing:

a) See → DNA splicing.

b) RNA splicing (RNA processing, nuclear processing of RNA, pre-messenger RNA splicing): The → small nuclear RNA catalyzed excision of → introns from a pre-mRNA molecule (→ primary transcript) and the ligation of → exons to create translatable mRNA molecules. This process is part of the → post-transcriptional modification of RNA. RNA splicing is essentially a two-step process. First, cleavage occurs at the 5' splice site (the junction between exon 1 and the intron) to generate intermediates with a free 5' exon and a → lariat form of the intron plus the 3' exon. Second, the 5' and 3' exons are ligated, releasing the mRNA and the fully excised intron lariat. Both reactions are catalyzed by the → spliceosome. See also → abortive splicing, → alternative splicing, → cell-specific splicing, → *cis*-splicing, → constitutive splicing, → cryptic splice site, → differential splicing, → DNA splicing, → guide sequence, → GT-AG rule, → nonsense associated alternative splicing, → protein splicing, → reading frame splicing, → splice junction, → TACTAAC box, → transfer RNA splicing, → *trans*-splicing. Compare → RNA editing, → DNA splicing, → protein splicing.

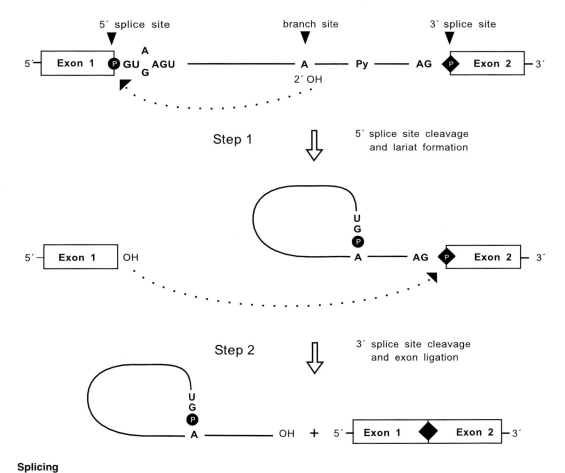

Splicing

Splicing complex: See → spliceosome.

Splicing control element (SCE): Any sequence or genomic region, that is necessary for correct → splicing. For example, the GGG triplet, purine-rich elements or polypyrimidine tracts in the 3'-intronic regions function as such SCEs, which are usually located in the vicinity of a → splice site, but can be away as far as 150 kb. Splicing control elements are found within an → intron or also → exon and at their 5'or 3'ends. SCEs are frequently arranged in repeats.

Splicing enhancer: A purine-rich sequence element in → exons (also → introns) downstream of a → 3' splice site in → pre-mRNAs, that promotes → splicing. Splicing enhancers are address sites for → SR proteins and help to assemble the → spliceosome complex.

Splicing factor 1 (Sp1, branch point-binding protein, BBP): A nuclear protein of the signal transduction and activation of RNA (STAR) family, that specifically recognizes the → intronic branch point sequence (BPS) 5'-UACUAAC–3' in → pre-messenger RNA, and binds to it cooperatively with the U2 auxiliary factor (U2AF). SF1 contains a conserved region, the socalled QUA2 (Quaking homology 2) domain located C-terminal of another domain KH: The 3'-part of the BPS is specifically recognized by a hydrophobic cleft formed by the gly-pro-arg-gly motif and the variable loop of the KH domain, whereas the 5'nucleotides of the BPS are specifically contacted by the QUA2 region. The recognition of the BPS by SF1 is a prime step in the formation of the prespliceosomal complex A.

Splicing junction: See → splice junction.

Splicing silencer: Any sequence in genomic DNA that suppresses → splicing of a particular → messenger RNA. For example, socalled exonic splicing silencers, residing in → exons and comprising from 4 to 74 nucleotides in length, reduce the assembly of early spliceosomal complexes, thereby preventing conventional splicing. Splicing silencers function in → alternative splicing. Intronic splicing silencer sequences also exist. Silencer sequences are binding sites for nuclear proteins (e.g. hnRNPA1, hn RNP I).

Splinker (sequencing primer linker): A synthetic oligodeoxynucleotide that contains an → inverted repeat sequence which forms a double-stranded → stem- and loop structure with a → restriction endonuclease → recognition site located in the double-stranded region. Splinkers serve both as → linkers and → sequencing primers in → direct sequencing of DNA restriction fragments.

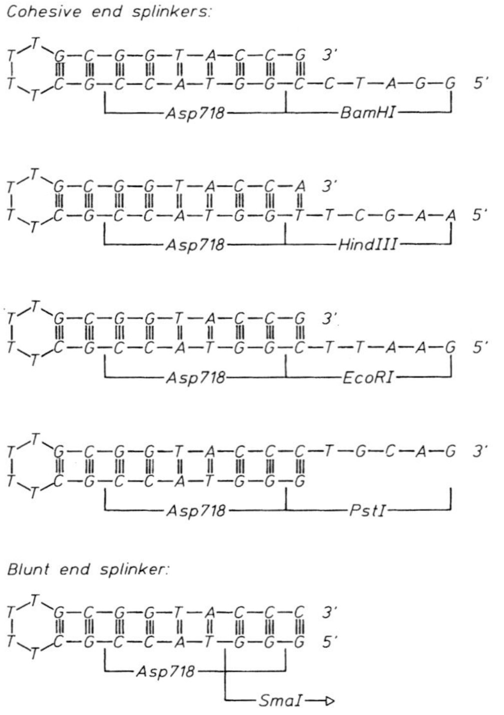

Cohesive end splinkers:

Blunt end splinker:

Splint: See → adaptor.

Split end vector (SEV, SEV system): A plant transformation vector system in which the two → T-DNA border sequences are present on two separate plasmids sharing a region of homology that allows them to cointegrate to form a selectable → disarmed vector. In short, from an → octopine-type → Ti-plasmid of → *Agrobacterium tumefaciens* the → T_L-DNA oncogenic functions, the T_L-DNA right → border and the → T_R-DNA are deleted and replaced by a → kanamycin resistance gene from → transposon Tn 903. This plasmid contains only the T_L-DNA left border sequence and a 1.6 kb region (*left inside homology* region, LIH) that allows recombination with an → intermediate vector (e.g. → pMON 200). Such intermediate vectors carry a chimeric kanamycin resistance gene, a nopaline synthase → reporter gene, a → polylinker for the insertion of foreign DNA, and a T-DNA right border sequence. The LIH sequences allow → cointegrate formation in *Agrobacterium tumefaciens*, which produces the competent plant transformation vector with two T-DNA borders.

Split gene (compound gene, interrupted gene, mosaic gene): A gene that is composed of → exons and → introns. Only the genetic information contained in the exons appears in the mature → messenger RNA. The introns are removed from the primary transcript (see → splicing) and discarded. Split genes are found in → eukaryotes and → archaebacteria.

Split genome: Any → genome, that is divided into two (or more) parts, each containing sequences homologous to sequences of the other part(s). For example, the split genome of the Lyme disease spirochete *Borrelia burgdorferi* consists of a → linear bacterial chromosome of 910 kb, twelve → linear plasmids and nine circular plasmids (altogether 610 kb). The plasmids harbor many sequences that share homology to sequences at the 3' end of the chromosome, allowing → homologous recombination between the different entities.

Split-hybrid system: A variant of the → LexA two-hybrid system, designed to detect protein-protein disruption rather than protein-protein interaction(s) as e. g. in the conventional → two-hybrid or → three-hybrid systems. In the split-hybrid system, the bait-encoding gene (or cDNA) is fused to the LexA → DNA-*b*inding *d*omain (DBD) of *E. coli* and cloned into a DNA-binding domain fusion vector, and the prey-encoding cDNA, fused to the transcriptional → activation domain (AD) of VP16, is inserted into the activation domain fusion vector. Interaction between LexA-bait and VP16-prey fusion proteins leads to the induction of a → tetracycline repressor protein (TetR), which prevents the transcription of the *HIS*3 → reporter gene. This inhibition leaves the yeast host cells unable to grow without histidine. If, however, the protein-protein interaction is disrupted, the ability to grow on histidine-free media is restored (because expression of TetR is blocked). The split-hybrid system can therefore be used to screen for proteins or peptides that interfere with protein-protein interaction(s), to detect mutations in both the prey- or bait-encoding genes, that interfere with this interaction, or to identify genes encoding these proteins or peptides. See → dual-bait two-hybrid system, → interaction trap, → one-hybrid system, → reverse two-hybrid system, → RNA-protein hybrid system, → split-ubiquitin membrane two-hybrid system.
Figure see page 1068.

Split promoter: A → promoter whose sequence elements are not contiguous, but arranged as two (or more) sequence blocks separated by spacer DNA (e.g. in → transfer RNA genes).

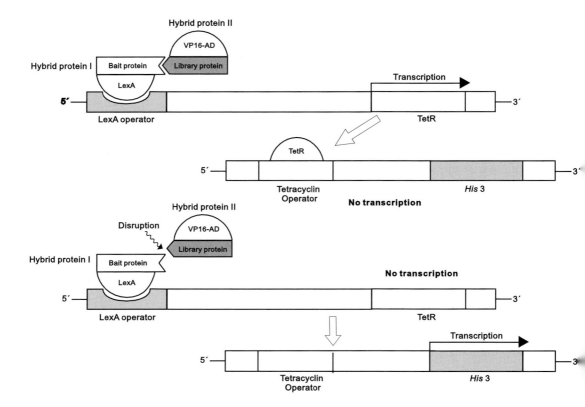

Split-hybrid system

Splitting genes: A misleading term for the *in silico* separation of two genes, that are mistakenly predicted to be one single gene. One of the genes retains the original gene name, and the other gene receives a new name. See → merging genes. Do not confuse with → split gene.

Split-ubiquitin system: See → split ubiquitin two hybrid system.

Split ubiquitin two hybrid system (split ubiquitin system, membrane-based two-hybrid system): A variant of the conventional → two-hybrid system, that allows to identify interactions between integral membrane proteins and to screen for small molecules inhibiting such interactions. In short, the gene encoding one membrane protein is fused to a sequence coding for the *N*-terminal half of → *ubi*quitin (Nub), and a separate fusion is made between the sequences encoding the *C*-terminal half of *ubi*quitin (Cub) and an other membrane protein. Both constructs are transformed in yeast cells and expressed. Should both membrane proteins interact, the two halves of the ubiquitin molecule come into close proximity to form a quasi-native ubiquitin.

This system is therefore based on → ubiquitin, that is attached to the N-terminus of proteins to tag them for → proteasomal degradation. Ubiquitin-tagged proteins are recognized by *ubi*quitin-specific protease*s* (UBPs), resulting in the cleavage and subsequent degradation of the attached proteins. Ubiquitin can be split into an N-terminal (Nub) and a C-terminal (Cub) half. The two parts retain an affinity for each other and spontaneously reassemble into the socalled split-ubiquitin, that is recognized by UBPs. These in turn cleave the peptide bond adjacent to the C-terminal amino acid residue of the Cub (i.e. glycine). Now if a → reporter protein is fused to the C-terminus of the Cub, it will be cleaved upon assembly of the Nub and Cub regions. A point mutation in the N-terminal half of ubiquitin (NubG) completely abolishes the affinity of both halves for each other. Since the separated NubG and Cub parts are not recognized by the UBPs, no cleavage of the attached reporter occurs. However, if the two parts are fused to two interacting proteins, this interaction brings them close enough together to reconstitute split-ubiquitin, resulting in the cleavage of the reporter protein by the UBPs. The reporter protein consists of the DNA-binding domain of LexA fused to the *Herpes simplex* VP16 transactivator. The reporter is fused to the Cub domain, which in turn is fused to a membrane protein X. A second transmembrane protein Y is fused to the NubG moiety. The only prerequisite for the function of this system: both the Cub and NubG have to be located on the cytoplasmic side of the membrane. Now if proteins X and Y interact, Cub and NubG are brought into close proximity, reconstitute split-ubiquitin, resulting in cleavage and release of the LexA-VP16 reporter fusion protein (PLV). The reporter enters the nucleus of the yeast host cell, binds to the UAS, and activates the reporter gene(s), for example *HIS*3 or *lac*Z. Analogous to the two-hybrid system, protein X is the bait, and protein Y is replaced by a → cDNA library, that is screened for interacting proteins. The cDNA libraries are fused to the NubG domain. The membrane-based two-hybrid system allows to analyse membrane proteins or membrane-associated proteins, their interaction with each other and with cytosolic proteins, and the interference of small molecules with these interactions. See → dual-bait two-hybrid system, → interaction trap, → LexA two-hybrid system, → one-hybrid system, → reverse two-hybrid system, → RNA-protein hybrid system, → split-hybrid system.

Spm: See → *s*uppressor-*m*utator *e*lement.

SPM: See → segregation of partly melted molecules.

spm **gene:** See → suppressor gene.

Sp 1 (specificity protein): A phosphorylated and O-glycosylated nuclear sequence-specific → DNA-binding protein that recognizes 5'-ATCGGGGCGGGGC-3' stretches or related GC runs (→ GC boxes) and binds to DNA by interaction of three contiguous Zn(II) finger motifs (see → Zinc finger proteins). The binding of Sp 1 to properly positioned GC boxes in a → promoter activates → RNA polymerase II transcription of linked genes. Sp 1 recognition sequences are found in many cellular and viral promoters in single or multiple copies.

Spontaneous mutation ("background mutation"): Any → mutation which is not experimentally induced but occurs naturally. The spontaneous → mutation rate varies from genome to genome (e. g. in bacteriophages: 7×10^{-5} to 1×10^{-11}; in bacteria: 2×10^{-6} to 4×10^{-10}; in fungi: 2×10^{-4} to 3×10^{-9}; in plants: 1×10^{-5} to 1×10^{-6}; in insects: 1×10^{-4} to 2×10^{-5}; in humans: 1×10^{-5} to 2×10^{-6}).

Spontaneous transformation: Any → transformation which is not experimentally induced (e.g. by chemicals) but occurs naturally.

Spot: A laboratory slang term for any single element on a → microarray, that contains immobilized target molecules such as → oligonucleotides, → cDNAs, DNAs, also peptides, → antibodies, or proteins, and is delivered by → spotting.

Spot blot: See → dot blot.

Spot picker: A laboratory slang term for a robot, that automatically excise a protein spot on a two-dimensional polyacrylamide gel and transfers the excised material into a microtiter plate for further processing. See → spot picking.

Spot picking: The excision of a target protein spot from a → two-dimensional polyacrylamide gel. The separated proteins of a sample are first visualized by either → Coomassie Brilliant Blue, or → silver, or → fluorophore staining. The stained gel is then scanned with appropriate imaging systems, the scan transferred to a database, and the desired spots isolated by a robot ("picking robot", "spot picker") with XY geometry (equipped with a UV source for fluorophore excitation, an emission filter, and a CCD camera).

Spotted array (spotted microarray): Any → microarray, onto which target molecules (e.g. → antibodies, DNAs, → cDNAs, → oligonucleotides, peptides or proteins, but also low molecular weight compounds) are immobilized by → spotting rather than by electromotive forces (see → electronic microarray).

Spotting: The application of minute volumes (e. g. 4-10 nl) of DNA, cDNA or oligonucleotide → probes onto a solid carrier (e. g. a glass or quartz slide, a membrane, or a silicon chip) at extremely high densities (e. g. 1500-3000 spots/cm^2), using either solid pins or quill (split) pins, pin rings, solenoid valves, or piezoelectric or thermal devices. The applied DNA is then stably fixed onto the carrier surface by either chemical or physical crosslinking techniques (e. g. UV irradiation). The resulting chips are prerequisites for → DNA chip technologies.

SPR: See → *surface plasmon resonance*.

SPR imaging: The process of profiling interactions between chip-bound target molecules (e.g. DNAs, RNAs, proteins, or ligands) and interacting partners (e.g. single molecules, mixtures of different molecules, also cell extracts, column fractions) by → *surface plasmon resonance* (SPR). SPR imaging produces a sensogram, in which interacting partners are detected by minimal mass changes induced by either binding or dissociation and can be localized unequivocally. See → surface plasmon resonance biosensor.

S–protein homologue (SPH): Any member of a large gene family of plants with sequence homology to the stigmatic *self*-incompatibility (S) genes of the field poppy (*Papaver rhoeas*). SPH genes encode relatively small (12-20 kDa), basic, secretory proteins with similar secondary structures (mostly based on β-strands), and are clustered in the genome (e. g. of *Arabidopsis thaliana*). The function(s) of the encoded proteins is presently obscure.

SP6 *in vitro* transcription system: An *in vitro* system to generate large amounts of specific, homogeneous, biologically active and radiolabeled RNA. In short, the DNA template is first cloned into the → polylinker of a plasmid → expression vector. This vector contains a phage promoter located 5 – 8 bp upstream of the polylinker, allowing the → transcription of the cloned insert. After linearization of the vector the transcription is mediated by the addition of → SP6 RNA polymerase (or coliphage → T3, T5 or → T7 RNA polymerases), that possess a high degree of specificity for their own (cognate) → promoter *in vitro*.
Frequently two phage promoters in opposite orientation and separated by the multiple cloning site (→ dual promoter vector) are used that allow transcription from either strand of the insert so that e.g. → sense and → anti-sense RNAs may be produced. In the SP6 system SP6 and T7 RNA polymerase → promoters on either side of the polylinker are combined.

Phage promoter consensus sequences:

SP6 +1
ATTTAGGTGACACTATAGAAGNG

T7 +1
TAATACGACTCACTATAGGGAGA

T3 +1
AATTAACCCTCACTAAAGGGAGA
−17   +6

The SP6 *in vitro* transcription system is used to produce single-stranded RNA for the analysis of → post-transcriptional modifications, → *in vitro* translation and → RNA sequencing, and radio-active → probes for nucleic acid → hybridization.

SP6 RNA polymerase (SP6 polymerase; EC 2.7.7.6): A 96 kDa DNA-dependent → RNA poly-merase of *Salmonella typhimurium* cells infected with bacteriophage SP6. The enzyme shows an extremely high affinity for the SP6 → promoter sequences and may be used to synthesize RNA transcripts of any DNA cloned in appropriate plasmid → expression vectors such as pSP 18, 19, 64 or 65 containing the SP 6 promoter. See → SP6 *in vitro* transcription system, → SP6 sequenc-ing primer.

SP6 sequencing primer: The synthetic oligodeoxyribonucleotide 5'-CACATACGATTTAGG-3' that hybridizes to a conserved 20 bp sequence of the bacteriophage → SP6 RNA polymerase → promoter, and allows → Sanger sequencing of double-stranded DNA inserted into vectors containing this promoter without the need of subcloning (into e.g. an → M13 vector).

SP6 vector: Any → cloning vector that contains a → promoter from the bacteriophage SP6, for example → pSP 64.

Spun-column procedure: A method to separate labeled DNA (or RNA) from unincorporated labeled precursors (e.g. α-^{32}P-dNTPs) by centrifuging the → labeling mixture through a Sepha-dex G-50 column contained in a small syringe. During this procedure the high molecular weight DNA is eluted first and can be collected free of nucleotides.

sqrs: See → *simple quadruplet repeat sequence.*

Squash dot hybridization: A variant of the conventional → dot blot hybridization procedure that uses cell or tissue squashes as a source of DNA or RNA. The cells or tissues are directly crushed on → nitrocellulose membranes, fixed, the DNA (or RNA) is denatured, and hybridized to radioactive or non-radioactive → probes.

Squelching: The inhibition of transcription of various genes (lacking GAL4-binding sites) by high levels of the transcriptional activator GAL4 in yeast cells. Transcription factor GAL4 binds to specific sites in the promoter of certain genes encoding enzymes of galactose utilization, and activates these genes. If overexpressed, the high concentrations of GAL4 titrate out a transcrip-tion factor by its → activation domain, so that the level of this transcription factor does no more allow the activation of the corresponding genes (which are then squelched). For example, the transcription of yeast gene *HIS*3, inducible by GAL4 or GCN4, is squelched by trapping the → TATA box binding protein. Special nuclear proteins, the mediators, release the squelch in vitro.

SRAP: See → sequence-related amplified polymorphism.

***Src* gene (sarcoma-inducing gene):** An → oncogene of the Rous sarcoma virus encoding a cytoplasmic tyrosine-specific protein kinase (pp 60^{V-src}) that activates cellular metabolism by phosphorylation of catalytic proteins.

SR-DNA: See → *s*imple *r*epetitive DNA.

***Sre* gene:** See → short response gene.

SRFA: See → *s*elective *r*estriction *f*ragment *a*mplification.

S RNA (small RNA): One of the three linear single-stranded RNAs of the genome of Tospoviruses (family: Bunyaviridae), that is 2.9 kb in length, is associated with the nucleocapsid proteins and forms a quasi-circle (pseudo-circle) due to the presence of 65–70 nucleotide long inverted complementary repeats at its 5' and 3'ends. It encodes a socalled N protein and a nonstructural protein NSs. The S RNA is ambisense, i.e. the RNA contains the sequence for one protein (NSs) in the messenger sense (+), and the other protein (N) in the complementary sense (–). See L → RNA, → M RNA.

sRNA: See → transfer RNA.

S-RNase: A glycoprotein encoded by the so-called *S*-locus (*s*elf-*i*ncompatibility, SI) of various plants with the biological activity of RNase. Expression of the *S*-locus and accumulation of S-RNase may cause degradation of → ribosomal RNA in a growing pollen tube and thus inhibit fertilization (cause self-incompatibility).

SRP: See → *s*ignal *r*ecognition *p*article.

SR protein (serin-arginin-rich protein): Any one of a series of eukaryotic proteins with an → RNA recognition motif that are involved in → pre-mRNA → splicing. The name derives from the presence of serine (S) and arginine (R) clusters in a specific domain of the protein. The N-terminus of SR proteins consists of an *RNA r*ecognition *m*otif (RRM; also RNA-binding domain or RNP consensus sequence) arranged in four antiparallel β sheets and two α helices. The C-terminus harbors an arginine-serine-rich domain (SR domain) that is highly phosphorylated. SR proteins function in splice-site selection and → spliceosome assembly. SR proteins bind to → exon enhancers (or exon silencers) of the pre-messenger RNA via their RNA-binding domains and additionally interact with the RS domains of protein components of the → small nuclear ribonucleoprotein particle (snRNP), e.g. protein U1–70K. This interaction stabilizes the binding of U1 → snRNA to the → donor splice junction (i.e. the 5' splice site) and enhances the recognition of the corresponding exon. Most of the SR proteins are constitutively expressed, but some are synthesized in special tissues only and regulate the selection of specific alternative exons. Moreover, the ratio between SR proteins and → *h*eterogeneous *n*uclear *r*ibo*n*ucleo*p*roteins (hnRNPs) varies from tissue to tissue in an organism, for which reason the usage of alternative

exons is tissue-specific. Also, SR protein and hnRNPs are stored in specific nuclear compartments, socalled "speckles", from which they are released after their phosphorylation. This represents yet another mechanism for the control of → alternative exon selection.

SrRNA: See → spatial development RNA.

SRS: See → simple repetitive sequence.

SSA: See → synthesis-dependent strand-annealing.

S-SAP: See → sequence-specific amplification polymorphism.

SSB: See → single-stranded DNA-binding protein.

SSC:
a) See → short single copy sequence.
b) See → sodium saline citrate buffer.

SSC (saline sodium citrate): A widely used buffer for nucleic acid → hybridization and → blotting
procedures, consisting of 150 mM sodium chloride and 15 mM sodium citrate, pH 7.0.

SSCA: See → single-strand conformation analysis.

sscDNA: Single-stranded circular DNA.

SSCPA: See → polymerase chain reaction single-strand conformation polymorphism analysis.

ssDNA: See → single-stranded DNA.

ssDNA binding protein: See → single-stranded DNA-binding protein.

SSH: See → suppression subtractive hybridization.

SSLLM: See → single-strand linker ligation method.

SSLP: See → simple-sequence length polymorphism.

SSLP fingerprinting: See → simple-sequence length polymorphism DNA fingerprinting.

SSM: See → slipped-strand mispairing.

Ssn6–Tup1 complex (Ssn6–Tup1 repressor): An evolutionary conserved eukaryotic protein complex composed of one Ssn6 (also Cyc8) and four Tup1 proteins, that highly efficiently represses
more than 200 genes (in *Saccharomyces cerevisiae*, yeast), whose encoded proteins function in
many different cellular pathways (e.g. **a**-specific genes, haploid-specific genes, glucose-repressible
genes, DNA damage-inducible genes, osmotic stress-inducible genes, sporulation-specific genes,
meiosis-specific genes, flocculation genes, genes encoding starch degrading enzymes). Ssn6–Tup1
complexes are recruited to the promoters of the genes they repress by so called sequence-specific
DNA-binding proteins, that localize to their recognition sequences and bring the repressor com-

plex in close proximity to the transcription start site. For example, many glucose-repressible genes are recognized by the Mig1 protein, DNA-damage-inducible genes by Crt1, and hyoxia-induced genes by Rox1. The portion of Tup1, that contacts such DNA-binding proteins, consists of seven copies of a repeated amino acid motif (WD40), which folds into a seven-bladed propellor. Ssn6 in turn includes the *tetratricopeptide* type of *repeat* (TPR), which occurs in many different proteins, but only in four proteins recruiting the Ssn6–Tup1 complex. The mechanism of repression by Ssn6–Tup1 complexes is still obscure, but may involve a direct interference with an activator (or activators), a change of the local chromatin structure (e.g. a changed → nucleosome positioning), or an interaction with the → transcriptosome itself. See → active repression, → global repressor, → repression.

SSO: See → sequence-specific oligonucleotide typing.

SSP: See → sequence-specific priming.

SSP-PCR: See → *s*ingle *s*pecific *p*rimer *p*olymerase *c*hain *r*eaction.

SSPT: See → *s*emi-*s*pecific *p*rimer *t*echnology.

SSR: See → *s*imple *s*equence *r*epeat.

ssRAPD: See → *s*ilver-*s*tained *r*andom *a*mplified *p*olymorphic *D*NA.

SSSR: See → *s*elf-*s*ustained *s*equence *r*eplication.

Stab culture: A bacterial culture, derived from a single colony, that is stabbed into → agar with a sterile needle or tooth-pick and can be stored for longer times at room temperature or 4°C.

Stable transfection (permanent transfection): The uptake and stable integration of foreign DNA into the genome of cultured animal or human cells, mediated by → direct gene transfer techniques. See → transfection.

Stable transformation: The permanent modification of the genome of one cell by the transfer of purified, also recombinant DNA from another cell of different genotype. See → transformation.

Stacked gene: Any one of two (or more) genes that have been transferred to an organism via → artificial or → natural gene transfer.

Stacking: The specific arrangement of adjacent nitrogenous bases in a DNA duplex molecule.

Stacking gel: A portion of a protein-, DNA- or RNA-separating gel that is casted on top of the normal separating gel, differs from the latter in concentration, and focusses the molecules at the stacking-gel/separating-gel interface. See → polyacrylamide gel electrophoresis.

Staggered bases: Any two bases set off vertically out of the mean plane of a base pair in a DNA double helix.

Staggered cuts: The cleavage of two opposite strands of a duplex DNA at points close to one another, as e.g. the cleavage pattern of type II → restriction endonuclease producing → cohesive ends.

Staggered extension process (**StEP**): A single-tube technique for the *in vitro* recombination of DNA fragments from different genes to create new combinations ("molecular evolution"), that consists of priming the template genes followed by repeated cycles of denaturation and extremely short annealing-extension times. In each cycle, growing fragments anneal to different templates, if sequence homology exists, and are extended further. Since continuous → template switching occurs, the growing polynucleotides ("recombination cassettes") contain sequence information from the different parental genes. In short, equimolar mixtures of single-stranded plasmid DNAs containing both parental genes are used as templates, and a 5'- and a 3'-primer added for conventional → polymerase chain reaction amplification. The annealing/extension step, though allowing stringent primer annealing, does, however, limit the extension to only a few seconds. Therefore, only very short extension products are generated in each cycle. In the subsequent cycle, these fragments randomly prime the templates ("template switching") and extend further for a few seconds, and so on. This process is repeated, until full-length new recombinants are generated (as compared to the lengths of the wild-type genes). The final products are then amplified by PCR, purified, and tested for new properties. For example, StEP with five thermostable subtilisin E gene variants produced recombinants, that were more thermostable than the parental genes (i.e. had 50 times increased half-life time at 65°C). See → DNA shuffling, → incremental truncation for the creation of hybrid enzymes.

S-tagging: The → fusion of any protein with the socalled S-tag, a 15 amino acids long peptide. Such S-tagged fusion proteins can easily be detected and quantified, if the socalled S-protein (104 amino acids) is added. The strong interaction between the S-tag and S-protein (K_d = 10^{-9}M) generates an active → RNase S, whose activity can be measured with the artificial substrate poly(C). The higher the RNase S activity, the more ribonucleotides are released, and the higher is the absorbance of the medium at 280 nm. By comparison with a known S-protein standard, the molar concentration of the target protein can be determined. See → RNase S.

Stains-all: The chemical compound 1-ethyl-2-3-(1-ethylnaphtho-1,2-d-thiazolin-2-ylidene)-2– methylpropenylnaphtho-1,2-d-thiazolium bromide, that is used to stain DNA, RNA, proteins, and acid polysaccharides in → polyacrylamide and → agarose gels. The dye is prepared as a 0.1% stock solution in 100% formamide, directly used as a 0.005% working solution in 50% formamide, and stains DNA blue, RNA bluish-purple, and protein pink or red.

Standard codon: Any one of the the most frequently used → codons in an organelle (e.g. a mitochondrium or a plastid) or the nucleus of cells of an organism. Standard codons for human nuclear genes:

TTT	Phe	TCT	Ser	TAT	Tyr	TGT	Cys
TTC	Phe	TCC	Ser	TAC	Tyr	TGC	Cys
TTA	Leu	TCA	Ser	TAA		TGA	
TTG	Leu	TCG	Ser	TAG		TGG	Trp
CTT	Leu	CCT	Pro	CAT	His	CGT	Arg
CTC	Leu	CCC	Pro	CAC	His	CGC	Arg
CTA	Leu	CCA	Pro	CAA	Gln	CGA	Arg
CTG	Leu	CCG	Pro	CAG	Gln	CGG	Arg
ATT	Ile	ACT	Thr	AAT	Asn	AGT	Ser

ATC	Ile	ACC	Thr	AAC	Asn	AGC	Ser
ATA	Ile	ACA	Thr	AAA	Lys	AGA	Arg
ATG	Met	ACG	Thr	AAG	Lys	AGG	Arg
GTT	Val	GCT	Ala	GAT	Asp	GGT	Gly
GTC	Val	GCC	Ala	GAC	Asp	GGC	Gly
GTA	Val	GCA	Ala	GAA	Glu	GGA	Gly
GTG	Val	GCG	Ala	GAG	Glu	GGG	Gly

Standard codon

Standard type: See → wild type.

Staphylococcal nuclease: See → micrococcal nuclease.

STAR:
a) See → sequence-characterized amplified region.
b) See → sequence-tagged restriction site.

Star activity: A change in the → recognition site specificity of certain → restriction endonucleases. In general, most restriction endonucleases cleave their recognition sequences with high specificity, but some restriction enzymes relax the stringency of specificity under suboptimal reaction conditions such as high enzyme concentration, high pH, low ionic strength, the presence of glycerol and substitution of magnesium by manganese, copper, cobalt or zinc ions. Relaxed stringency leads to the restriction of more sequence motifs than are recognized under stringent conditions (e.g. *Eco* RI: G↑TAATTC; *Eco* RI*: N↑TAATTN). See the table "Star activity of restriction endonucleases" of the Appendix.

Starch **gel** *electrophoresis* **(SGE; starch inclusion gel electrophoresis):** A technique to separate proteins on the basis of their size and charge by electrophoresing them through a matrix of starch. Starch-degrading enzymes (or → isozymes) can easily be detected after electrophoresis by simply incubating the starch gel at room temperature for a few hours (decomposition of starch) and then staining the whole gel with potassium iodine. The locations of e. g. amylases or phosphorylases then appear as light bands on an otherwise homogeneously stained black background. Superseded by → polyacrylamide gel electrophoresis.

Start: A laboratory slang term for the base pair in a chromosome, at which a specific clone (e.g. → bacterial artificial chromosome clone) starts.

Start codon (initiation codon, initiator codon, initiator, translation initiation site):
a) A trinucleotide in a messenger RNA (prokaryotes: AUG, GUG; eukaryotes: AUG) that initiates polypeptide synthesis (see → translation). The start codon codes for → N-formylmethionine in bacteria and methionine in eukaryotes and sets the → reading frame for translation. N-formylmethionine is often removed post-translationally (→ post-translational modification).
b) The trinucleotide ATG that represents the initiation codon in mRNA (but additionally codes for internal methionine). See also → cap site.

Start point (start site): The first nucleotide of a transcript.

STAT: See → *s*ignal *t*ransducer and *a*ctivator of *t*ranscription.

Stationary phase: The time period, during which a cell culture does no longer grow (i. e. the cell number remains constant, since new cells are produced at the same rate as older cells die). The stationary phase is depicted as a plateau of the growth curve after previous logarithmic growth of the culture.

STC: See → *s*equence-*t*agged *c*onnector.

Stealth pin: Any stainless steel pin with a flat tip, that contains chambers for liquid volumes of 0.1 – 1.0 μ l and serves to take up samples (e.g. oligonucleotides) and to deliver them onto → microarray supports (e.g. glass slides). Direct contact between the liquid and chip surface results in drop deposition. Spots can be sized with diameters of 50 – 300 μm. Stealth pins allow multiple prints per load.

Stem: The fully base-paired, double-stranded part of a → stem and loop structure.

Stem and loop structure (hairpin structure): Any secondary structure in a nucleic acid molecule in which complementary sequences within the same strand anneal, forming a double-stranded stem while nucleotides between the paired regions form an unpaired, single-stranded loop. Such stem-and loop structures in DNA are potential recognition sites for various nuclear proteins. See also → fold-back DNA.

Stem and loop structure

StEP: See → *st*aggered *e*xtension *p*rocess.

Step gradient: See → density gradient.

Step-out *polymerase chain reaction* (step-out PCR): A laboratory slang term for the → polymerase chain reaction-based amplification of → cDNAs with complete 5' ends, exploiting the (minimal) → terminal transferase activity of → *M*oloney *m*urine *l*eukemia *v*irus (MMLV) reverse transcriptase, which leads to the template-independent extension of the 5' end with cytosyl residues. These 5'-overhangs are used to anchor G-rich → primers for an amplification of complete cDNAs.

Steric availability: The specific spatial configuration of a → target molecule spotted onto a → microarray support, that is optimal for its interaction with a → probe molecule. For example, the attachment of the target sequence at one of its ends to the microarray support increases the physical accessibility of the target by the probe and therefore also its steric availability.

Sticky end: See → cohesive end.

Stimulon:
a) A group of → genes or → proteins, whose → transcription or synthesis, respectively, is regulated by the same environmental parameter. For example, the heat stress stimulon of e.g. *Bacillus subtilis* comprises a small group of specific → heat shock protein genes (*GroE*, *DnaK*) and a large group of general stress protein genes (GSPs), that are up-regulated (stimulated) by a temperature increase in the medium (heat shock). Compare → regulon.
b) The entirety of all proteins of a cell, that are either induced (i.e. up-regulated) or repressed (i.e. down-regulated) at a given time. The term is usually used for induced ("stimulated") proteins.

STM: See → *s*ignature-*t*agged *m*utagenesis.

STMS: See → *s*equence-*t*agged *m*icrosatellite *s*ite.

Stoffel fragment: A modification of recombinant → *Thermus aquaticus* DNA polymerase from which the N-terminal 289 amino acids have been deleted, which is more thermostable than the holoenzyme, and which lacks any intrinsic 5' → 3' exonuclease activity. Stoffel fragment is used to amplify templates with complex secondary structures by running the → polymerase chain reaction at higher temperatures than is possible with conventional *Taq* polymerase.

Stoke's shift: The distance in wave-length (measured in nanometers) between the absorption maximum and the emission maximum of a → fluorochrome.

Stop codon (nonsense codon, chain terminating codon, terminator, chain terminator, termination codon): A trinucleotide in a messenger RNA that signals the termination of polypeptide synthesis and the release of the completed polypeptide chain from the → ribosome. Three different stop codons exist: UAG (amber), UGA (opal), and UAA (ochre). None of them corresponds to an → anticodon of a → tRNA molecule. See also → nonsense mutation (→ amber, → ochre, → opal mutation).

Stop codon mutation: Any change in the base sequence of the → stop codon of a → gene that abolishes its transcription termination function. As a consequence, → RNA polymerase II(B) transcribes beyond the original termination signal, generating an abnormally long → messenger RNA. The encoded protein either does not function correctly or loses its function. Stop codon mutations cause human diseases. For example, the thanatophoric dwarfism type 1, the most common form of lethal neonatal chondrodysplasia with shortened limbs and extremities, a narrow thorax, enlarged head and an abnormal lobulation of the brain, is caused by a stop codon mutation in the fibroblast growth *f*actor *r*eceptor 3 (FGFR 3) gene. Therefore, the coding region continues to be transcribed for 423 bp beyond the original stop codon, until another in-frame stop codon is reached. The expressed protein is elongated by 141 amino acids.

Stop unwanted rearrangement events (SURE): An engineered *E.coli* K12 strain carrying a chromosomal → tetracyclin resistance gene and a → kanamycin resistance gene on the → F-factor. Due to targeted mutations in its genes responsible for → recombinations (e.g. UV repair gene *uvrC* and the SOS repair gene *umuC*), the SURE strain is defect in its recombinatorial potential, so that e.g. → repetitive DNA sequences can be cloned without, or only a low number of → rearrangements, and is 10–20 times more stable than in normal *E. coli* strains. See → rec⁻ mutant.

Storage area network (SAN): A data storage system, consisting of storage arrays, fibre channels or gigabit Ethernet switches, host servers and storage management software, that has the capacity to rapidly move (100 MB/sec and faster) contiguous stored data between servers and storage arrays. For an improvement of the system, it can be combined with → network-attached storage.

Stormo rules: The description of the conditions that must be fulfilled for an efficient function of the → Shine-Dalgarno box. One rule determines the length of the SD → spacer (i.e. the sequence between the AGG-triplet of the SD-box and the → initiation codon ATG). The spacer should comprise in between 6 to 9 bp.

Stowaway: A trivial name for a family of plant → transposon-like insertion elements of about 100-300 bp in length, that occur in non-coding sequences of genes of both monocotyledonous and dicotyledonous plants. Stowaways carry subterminal repeats with specificity for potential target sites (5'-TAA-3').

STR:
a) See → *s*hort *t*andem *r*epeat.
b) See → *s*ub*t*elomeric *r*epeat.
c) See → *s*ynthetic *t*andem *r*epeat.

Strain: Any group of related organisms which differ from other groups by specific characteristics. Compare → selection.

Strand break induction by photolysis (SBIP): A technique for the induction and detection of single-strand nicks and → cuts in double-stranded DNA. In short, proliferating cells are first incubated with → BrdU, which is incorporated into newly synthesized DNA. The cells are then stained with → HOECHST 33258, and exposed to UV light to induce DNA strand breaks (photolysis). HOECHST 33258 increases the efficiency of photolysis. The cells are then fixed, and the exposed 3'hydroxyl ends at the strand breaks directly labeled *in situ* with → *t*erminal de-oxynucleotidyl transferase-mediated d*UTP nick-end*labeling assay (TUNEL), in which fluorescent or biotinylated nucleotides are attached to these 3'hydroxyl ends by mammalian → *t*erminal *d*eoxynucleotidyl *t*ransferase (TdT).

Strand displacement: The ability of a → DNA polymerase to displace the 5' template strand of a → double-stranded DNA at the → replication bubble, as it polymerizes a new copy of the displaced → single-stranded DNA.

Strand *d*isplacement *a*mplification (SDA): An isothermal *in vitro* DNA amplification technique that uses a → restriction enzyme to → nick the unmodified strand of a hemiphosphorothioate → recognition site and the → Klenow fragment of DNA polymerase I to initiate replication at the nick and to displace the down-stream non-template strand (strand displacement). In short, target DNA is first heat-denatured and four → primers (P1 – P4) annealed to opposite strands of the target DNA at positions flanking the sequence to be amplified. Two of these primers (P1, P2) carry a restriction enzyme recognition site 5' to their target-binding regions, the other two primers (P3, P4) contain only target-binding sequences. The Klenow fragment extends all primers, using dCTP, dGTP, TTP, and dATPαS (thioate). However, during extension of P3 and P4 the extended portions of P1 and P2 are displaced, and serve as targets for the opposite primers. After multiple rounds of extension and displacement the target fragment is exponentially amplified by coupling → sense and → antisense reactions in which the displaced strands from the sense reaction serve as targets for the antisense reaction, and vice versa. With the exception of heat denaturation of the target DNA, the SDA technique works at a single temperature so that thermal cycling as in conventional → polymerase chain reactions is not necessary.

Strand invasion: The displacement of one strand of a DNA → duplex molecule by an incoming single-stranded DNA, RNA, or PNA molecule. For example, → *p*eptide *n*ucleic *a*cid (PNA) strands containing only thymine and cytosine residues bind to a duplex DNA molecule such that one PNA forms a double helix with one DNA strand by → Watson-Crick hydrogen bonds, whereas another PNA strand binds to the major groove of the PNA-DNA duplex by → Hoogsteen base-pairing. The remaining strand of the original DNA duplex is then displaced as a → displacement loop. Such PNA-DNA-PNA triplexes are extremely stable (T_m > 70° C for a 10mer).

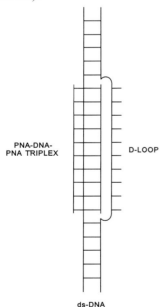

PNA-DNA-PNA TRIPLEX   D-LOOP

ds-DNA

Strand polarity: The polarity of oligo- or polynucleotide chains, characterized by a free or phosphorylated → 3' end and a free or phosphorylated → 5' end.

Strand selection: The property of DNA-dependent $\rightarrow$ RNA polymerases to recognize and select the $\rightarrow$ template strand of a DNA duplex molecule.

Strand-separating gel: A neutral $\rightarrow$ agarose or $\rightarrow$ polyacrylamide gel in which the two strands of a denatured double-stranded DNA molecule can be separated. This technique has been replaced by vectors that allow the generation of single-stranded DNA (or RNA) molecules (see $\rightarrow$ M13).

STR biosensor: See $\rightarrow$ surface plasmon resonance biosensor.

STRE: See $\rightarrow$ *s*tress-*r*esponsive *e*lement.

Streaking (streak plating): The distribution of an inoculum of cells (bacteria) on the surface of a solid medium in a way that during incubation individual colonies of the inoculated organism (e.g. a bacterium) will form.

Streak plating: See $\rightarrow$ streaking.

Strep **tag:** A nine amino acid (H_2N-AWRHPQFGG-COOH) sequence, incorporated into a target protein, that permits the one-step purification of this tagged protein. The gene encoding the target protein is first cloned into an appropriate $\rightarrow$ plasmid vector containing the *strep* tag sequence between a *m*ultiple *c*loning *s*ite (MCS) and a $\rightarrow$ terminator sequence. Expression of the cloned gene then leads to a $\rightarrow$ fusion protein, that can be isolated and purified by $\rightarrow$ streptavidin affinity chromatography. The bound fusion protein can be competitively displaced from the column by low concentrations of the natural ligand biotin or diaminobiotin, the oxydation-insensitive biotin analogon desthiobiotin or other derivatives. The *strep*-tag can also be exploited for the detection of recombinant proteins by $\rightarrow$ enzyme-linked immunosorbent assay or $\rightarrow$ Western blotting. It does not interfere with the biological function of the tagged protein. The immobilized streptavidin can also be optimized for *strep* tag binding. For example, mutagenesis of a flexible peptide loop in streptavidin by combinatorial protein engineering leads to streptavidin derivatives ("streptactins") that show higher affinity to *strep* tags.

Streptavidin **(STV):** An extracellular tetrameric protein (mol. wt: 4×15000) from *Streptomyces avidinii*. As its homologous protein $\rightarrow$ avidin, streptavidin binds up to four molecules of D-biotin with high affinity (dissociation constant $K_D = 10^{-15}$ M), and is also active towards $\rightarrow$ biotin derivatives. The essentially irreversible and specific linkage between streptavidin and biotinylated nucleotides forms the basis for the non-radioactive detection of specific sequences in $\rightarrow$ Southern blotting and other nucleic acid detection techniques. See $\rightarrow$ DNA detection system, $\rightarrow$ streptavidin agarose, $\rightarrow$ streptavidin-conjugated alkaline phosphatase.

Streptavidin agarose: An $\rightarrow$ agarose matrix that contains covalently attached $\rightarrow$ streptavidin. This type of agarose is used for the isolation of biotinylated molecules and complexes containing biotinylated components (e.g. antibodies, antigen-antibody complexes, nucleic acids, and nucleic acid-protein complexes).

Streptavidin-conjugated alkaline phosphatase: A $\rightarrow$ streptavidin molecule that is covalently bound to an $\rightarrow$ alkaline phosphatase. Such conjugates are used for the detection of biotinylated $\rightarrow$ probes in nucleic acid $\rightarrow$ hybridization procedures. After binding of streptavidin to the biotin moieties, the coupled enzyme is detected using the colorless bromochloroindolyl phosphate ($\rightarrow$ X-phos), which is converted into the colored indigo.

Streptolydigin: An → antibiotic that reversibly binds to the β-subunit of bacterial DNA-dependent → RNA polymerase, and inhibits elongation of nascent RNA chains.

Streptomyces: A genus of Gram-positive saprophytic soil bacteria that produce many → antibiotics (e.g. → streptomycin, → tetracycline) and are responsible for the degradation of carbohydrates such as cellulose, starch and proteins.

Streptomycin (Sm): An aminoglycoside → antibiotic produced by *Streptomyces* strains (e.g.
Streptomyces griseus) that binds to the S12 protein of the 30S ribosomal subunit (→ ribosome) of
prokaryotes and inhibits the elongation step during mRNA → translation. See → streptomycin
resistance and → streptomycin sensitivity.

Streptomycin A

Streptomycin resistance (streptomycin suppression; Sm^r): The ability of a prokaryotic organism to
grow in the presence of the → antibiotic → streptomycin. Resistance is encoded by a → plasmid
gene (*str*) whose expression produces streptomycin-inactivating enzyme(s), which phosphorylate, acetylate or adenylate the antibiotic. Some bacterial mutants possess an altered → ribosomal
protein (S12) which lacks the target site for streptomycin and therefore can initiate protein
synthesis in the presence of the antibiotic ("streptomycin suppression"). Compare → streptomycin sensitivity.

Streptomycin sensitivity (Sm^s): The inability of a prokaryotic organism to grow in the presence of the → antibiotic → streptomycin, which binds to the 30S ribosomal subunit and inhibits the elongation step during mRNA translation. Compare → streptomycin resistance.

Streptomycin suppression: See → streptomycin resistance.

Stress response RNA (srRNA): Any → microRNA, that is involved in the response to various biotic and abiotic stresses.

Stress-responsive element (STRE): The conserved pentanucleotide sequence 5'-CCCCT-3' that serves as address site for → transcription factors (e.g. → zinc finger proteins Msn 2 and Msn 4 of yeast) and forms part of the → promoter of → genes induced by various stresses (e.g. heat, oxygen radicals, hyperosmolarity, heavy metals). These genes encode proteins that counteract the detrimental influence(s) of various stresses (e.g. → ß-1.3-glucanase, catalase, chitinase, heat-shock proteins and peroxidase in yeast and plants).

Stringency: The reaction conditions (e.g. salt concentration, temperature, pH) during nucleic acid → hybridization, that allow the formation of DNA-DNA-, DNA-RNA-, and RNA-RNA → duplexes from → single-stranded molecules. The term also describes → primer-template interactions and especially the influence of reaction conditions on the fidelity of annealing. Stringency is an important parameter for all hybridization or annealing reactions. For example, at high stringency, duplexes only form between strands with perfect complementarity, whereas lower stringency also allows annealing of strands with some mismatches. See → high stringency, → low stringency.

Stringent control (tight control): A mechanism whereby a bacterial cell controls the number of → plasmids by strictly coupling their → replication to chromosomal DNA synthesis. Compare → relaxed control, → runaway plasmid, → low copy number plasmid.

Stringent mutant: Any bacterial mutant that abruptly stops RNA synthesis upon addition of a depleted growth medium.

Stringent plasmid: See → low copy number plasmid.

Stringent response: The reduction of → transfer RNA and → ribosome synthesis in bacteria that grow in a poor medium (under stringent conditions).

Stringent washing: See → high stringency, → low stringency.

Stripping: A laboratory slang term for the complete removal of a radioactively labeled probe from its DNA or RNA target bound to a filter (e.g. → nitrocellulose filter). Also used for the removal of primary and secondary antibodies from an antigen-carrying matrix.

Strong overlap: The number of bases matched between the two clones (e.g. → bacterial artificial chromosome clones), determined by using the strictest matching criteria. See → medium overlap, → weak overlap, → total overlap.

Strong positive element: Synonym for → enhancer.

Strong promoter ("high level promoter"): Any → promoter that allows the frequent attachment of DNA-dependent → RNA polymerase with high affinity and concomitant increase in the rate of transcriptional initiation of the adjacent gene.

STRP:
a See → microsatellite polymorphism.
b) See → short tandem repeat polymorphism.

STR technology: The whole repertoire of techniques that aim at detecting → microsatellites (→ short tandem repeats, STRs) in genomic digests (see → DNA fingerprinting), or use micro satellite sequences for the amplification of genomic regions flanked by microsatellites (see → microsatellite-primed PCR, → anchored microsatellite-primed PCR, digested random amplified microsatellite polymorphism, → random amplified microsatellite polymorphism in its RAMP and → RAMPO version, → retrotransposon-microsatellite amplified polymorphism → selective amplification of polymorphic loci.

Structural gene (producer gene, sense gene): Any gene that encodes the amino acid sequence (primary structure) of a protein. In eukaryotic organisms structural genes are transcribed by → RNA polymerase II (B).

Structural genomics: The whole repertoire of techniques to systematically characterize the architecture of → genomes, encompassing sequence-directed domain formations (e. g. → foldback DNA, → stem-and-loop structure), folding of DNA (e. g. in → nucleosomes, → solenoids, → looped domains), condensation of DNA in → chromosomes, and the fine and overall structure of chromosomes or → chromatin. The term is also used for a complete description of the three-dimensional structure of proteins (by X-ray crystallography, NMR spectroscopy, or compositional analysis) or RNAs (by analysis of specific motifs such as → tetraloops, → U-turns, dinucleotide platforms, ribose zippers, S-turns, and other 4–11 nt motifs). Compare → behavioral genomics, → comparative genomics, → environmental genomics, → epigenomics, → functional genomics, → genomics, → horizontal genomics, → integrative genomics, → medical genomics, → metabolome, → nutritional genomics, → pharmacogenomics, → phylogenomics, → proteome, → proteomics, → recognomics, → ribonomics, → structural genomics, → transcriptome, → transcriptomics, → transposomics.

Structural maintenance of chromosomes (SMC): A family of proteins, that catalyse siter chromatid cohesion and chromosome condensation during mitosis. Usually, SMC proteins form heterodimeric complexes (e.g. the SMC1/SMC3, and the SMC2/h-CAP-C complexes).

STS: See → sequence-tagged site.

STS content mapping: A → physical mapping procedure, which uses the presence of → sequence-tagged sites (STSs) on large-insert clones (e.g. → BAC clones, or → YAC clones) to infer → linkage, and to order the STSs.

STS map: See → sequence-tagged sites map.

Stuffer: See → stuffer fragment.

Stuffer fragment (stuffer): An internal region of λ phages coding for recombination (red) and integration (att, int), that is not essential for phage growth. In λ-derived → replacement vectors the stuffer is removed and replaced by foreign DNA.

Subarray: See → subgrid.

Subcellular proteomics: The whole repertoire of techniques to identify the locations, orientations and movements of proteins in a cell or its organelles. See → protein linkage map, → protein-interaction map, proteomics, → topological proteomics.

Sub-clone: A smaller DNA fragment derived from a larger cloned fragment by → restriction endonuclease digestion, which is separately cloned into a vector. See → sub-cloning.

Sub-cloning (subcloning, recloning): A technique to subdivide a large cloned DNA fragment (e.g. 9-25 kb in → genomic libraries) into smaller pieces which are easier to analyze. The original fragment is removed from the vector by digestion with → restriction endonucleases, and then cleaved with other restriction endonucleases. The resulting fragments of sizes up to 5 kb are reinserted either into the same or another vector. Subcloning allows the easy and fast characterization of the subclones (e.g. → sequencing, → *in vitro* mutagenesis, → *in vitro* transcription).

Subculture: Any cell culture that is derived from an original culture ("mother culture") and independently grows on a fresh medium.

Subfunctionalization: The mutation of an ancestral gene and its gene duplicate (see → gene duplication) to an extent, that their total capacity is less than that of the single-copy ancestral gene. See → neofunctionalization, → nonfunctionalization.

Sub-genome: Any subset of → genomic DNA sequences (e.g. genes, or repetitive DNA), that represents a fraction of the complete genome. See → sub-proteome.

Subgenomic library: Any → gene library that contains only parts of a genome of an organism (e.g. one chromosome or part of a chromosome).

Subgenomic messenger RNA **(sg mRNA):** Any one of several → messenger RNAs generated by → discontinuous transcription of the coding part of arterivirus and coronavirus (nidovirus) genomic RNAs [(+)-RNAs].

Subgenomic RNA **(sgRNA):** The → RNA transcribed from the genomic RNA(s) of certain plant viruses and translated into essential viral proteins (e. g. → capsid proteins). For example, the genome of the red clover necrotic mosaic Diantho-virus consists of two RNAs (RNA-1 and RNA-2), of which RNA-1 encodes a viral coat protein, but cannot be recognized by the host transcriptional machinery. For the transcription of the coat protein gene, a 34 nucleotide sequence on RNA-2 is needed. Through interaction of this sequence with RNA-1, the coat protein gene becomes accessible and is transcribed into the subgenomic RNA which in turn can be translated into the protein. Compare → subgenomic messenger RNA.

Subgrid (subarray, block): Any part of a → microarray support (e.g. a glass slide), that has been printed by a single pin. See → spotting.

Sublimon: Any one of a series of rearranged forms of the mitochondrial genome in angiosperm plants, that arose by → homologous recombination(s) between small and large (> 2 kb) repeated sequences. Sublimons ("recombinational isomers") coexist together with the main mitochondrial DNA, and differ from each other only in the relative orientation of the single copy sequences flanking the rapidly recombining repeats.

Submarine electrophoresis: The separation of macromolecules (DNA, RNA, oligonucleotides, peptides, proteins) by electrophoresis in either an agarose or polyacrylamide matrix under a thin layer of electrophoresis buffer (as opposed to electrophoresis in a gel, that is soaked in buffer, but not submerged). See → submarine minigel.

Submarine minigel: Any small → agarose slab gel which is run horizontally under approximately 1 mm of buffer. Minigels are an effective compromise between standard vertical and horizontal slab gels and allow fast runs since they possess a greater conductance.

Sub-proteome: Any subset of proteins and peptides, that represents a fraction of the total → proteome. Frequently, sub-proteomes encircle proteins cooperating in a specific metabolic pathway (e. g. glycolysis, signaling), or originate from a specific cellular compartment (e. g. extracellular space, cell wall, membranes, cytosol, or organelles).

Substitution: Any → mutation in DNA, that leads to the replacement of one base pair by another base pair. See → point mutation, → transition, → transversion.

Substitution editing (modification editing): A variant of → RNA editing, in which one nucleotide in a transcript is converted to another one. See → deletion editing, → insertion editing.

Substitution mutagenesis: Any change in the base sequence of a DNA duplex molecule caused by the replacement of one base with another base (e.g. C → T). See → bisulfite mutagenesis.

Substitution vector: See → replacement vector.

Substrate chip: A solid support (e.g. glass, silicon), onto which hundreds or thousands of peptide and protein substrates for cellular proteases are arrayed. Substrate chips serve to determine the entire functional protease → degradome of a cell at a given time.

Substrate noise: The undesirable contribution of surface reflection, surface coating, pre-treatment of the surface or other surface-related parameters of → microarrays to the fluorescence readings by a fluorescence detector instrument. See → background subtraction, → dark current, → electronic noise, → microarray noise, → optical noise, → sample noise.

Subtracted probe (subtracted cDNA): Any → cDNA sequence that has been isolated by → subtractive hybridization, and is used to screen → gene libraries for cell-specifically expressed genes.

Subtractive cloning: The molecular cloning of sequences expressed in only one of two cell types. These sequences are isolated by e.g. → subtractive hybridisation.

Subtractive hybridization (plus-minus hybridization, differential hybridization): A technique for the detection of sequences expressed in only one of two cell types that is based on the hybridization of cDNAs from mRNAs of one (A) to mRNAs from another cell type (B). Only sequences that are expressed in both cell types can form cDNA:mRNA hybrids which are then separated from single-stranded mRNAs and cDNAs by → hydroxyapatite chromatography. The fraction with single-stranded sequences is then treated with alkali (destruction of RNA), and contains cDNAs from mRNAs expressed in cell type A only. Now a → second strand can be synthesized to complete the cDNA, and the resulting double-stranded molecules can be inserted in a → cloning vector to establish a → subtractive library. Compare → cDNA representational difference analysis, → differential hybridization, → differential display reverse transcription polymerase chain reaction, → electronic subtraction.

Subtractive library (subtraction library): A collection of → cDNA clones representing → messenger RNAs that are expressed in only one of two or more cell types. A subtractive library is established by → substractive hybridization.

Succinate dehydrogenase inhibition test **(SDI test):** A technique for the detection of metabolic activity in viable cells and the quantitation of cell proliferation and cytotoxicity, that uses tetrazolium salts as substrates for dehydrogenases in a → microtiter plate format. Such tetrazolium salts are cleaved only by metabolically active cells. For example, MTT bromide is reduced by the dehydrogenases of viable cells to a colored, water-insoluble formazan. This product can be solubilized, and the quantity of the formazan be estimated in a conventional ELISA plate format at 570 nm. Modified tetrazolium salts (as e.g. XTT or WST–1) are converted to water-soluble formazans by viable cells, and allow to determine cell activation, proliferation and cytotoxicity of certain chemicals.

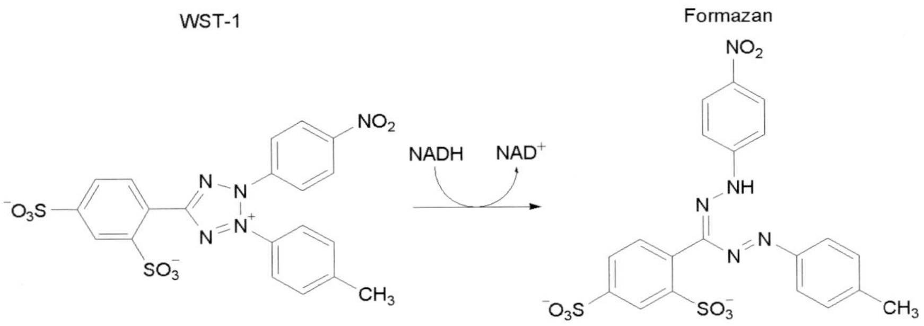

WST-1   Formazan

Sucrose gradient centrifugation (sucrose density gradient centrifugation; sucrose density gradient zonal centrifugation): A method for the separation of macromolecules and subcellular particles which uses centrifugal force. Usually linear or exponential gradients are formed with a gradient former that mixes two sucrose solutions of different concentration (density). Step gradients can be formed – after layering solutions of decreasing sucrose concentration on top of each other – during ultracentrifugation. The particles are separated by their sedimentation through the gradient. The sedimentation velocity in sucrose gradients is largely determined by molecular size and shape, see → sedimentation coefficient. Compare → isopycnic centrifugation, where nucleic acids are separated according to their buoyant density.

Sugar-phosphate backbone: See → backbone.

Suicide gene: Any gene that encodes a protein (e.g. an enzyme) converting a nontoxic substance ("prodrug") into a cytotoxic drug. For example, the prodrug *gancilovir* (GCV) can be mono-phosphorylated to P-GCV by *herpes simplex virus thymidine kinase* (HSV-TK). P-GCV is then phosphorylated to the cytotoxic P_3-GCV by an endogenous thymidine kinase. Suicide genes are potential anti-tumor agents. If, for example, the HSV-TK gene is introduced into a mammalian tumor cell, it confers specific sensitivity to the anti-herpes drug GCV: the cells expressing the gene are therefore killed.

Suicide gene therapy: A variant of the conventional → gene therapy, in which an *E. coli* → suicide gene is transferred into mammalian tumor cells, where it is expressed such that the encoded protein *cytosine deaminase* (CD) converts the non-toxic prodrug 5–*fluorocytosine* (5–FC) into the toxic 5-fluorouracil (5–FU). 5–FU then attacks the tumor cells. The tumor cells carrying the foreign gene are either injected subcutaneously or directly into the manifest tumors (e.g. liver tumors), then 5–FC is administered. In laboratory animals, liver tumor sizes decreased and spread of the cancer cells to other sites stopped.

Suicide plasmid: See → suicide vector.

Suicide *polymerase chain reaction* (suicide PCR): A variant of the conventional → polymerase chain reaction, for which target-specific primer pairs are used only one single time and no positive controls (i. e. amplifications of target DNA) are included. This procedure avoids amplifications from contaminating → templates. For example, *Yersinia pestis* has been identified in the dental pulps of a child and two adults from a 14th century catastrophe mass grave of Black Death victims in Montpellier (France) by suicide PCR. Teeth were extracted from remnants, washed, longitudinally fractured, powdery pulp relics removed from the pulp cavities, DNA isolated and used for suicide PCR with primers targeting at specific *Y. pestis* gene sequences (e. g. the *pla* gene). Each primer pair was used only once and the amplicon was sequenced. The presence of diagnostic *Y. pestis* gene sequences in the dental pulps of presumptive Black Death victims ends the controversy over the causative agent of this medieval pandemic.

Suicide vector (suicide plasmid):
a) Any → cloning vector that contains a gene encoding a function lethal for its host. Transformants receiving such suicide vectors will inevitably be killed ("commit suicide"). Transformants with a suicide vector into which foreign DNA has been inserted, destroying the lethal function, will survive.
b) Any → cloning vector that is used for the → cloning of DNA sequences and their transfer to recipient cells, that undergoes → homologous recombination with → replicons within the host cell and loses vital functions by this process. The disabled vector cannot survive in the recipient and is eliminated.

Sulfonated DNA detection: A method for the → non-radioactive labeling and detection of single-stranded DNA. In short, an antigenic sulfone group is coupled to carbon atom 6 of the cytosine moieties of the single-stranded DNA, stabilized by the substitution of an amino group on carbon 4 of the cytosine with methoxyamine. The sulfone groups can be detected by an immuno-enzymatic reaction (→ sandwich technique). In this reaction a → monoclonal antibody binds the sulfone groups on the modified DNA, and is in turn recognized by an enzyme-conjugated anti-immunoglobulin antibody. The enzyme converts soluble chromogenic substrates (e.g. nitroblue tetrazolium or → X-phos) into a colored, insoluble dye that precipitates at the location of the immune reaction and marks the position of the sulfonated DNA.

Sulphation: The transfer of sulphate from the donor 3'-*p*hospho*a*denosyl–5'-*p*hospho*s*ulphate (PAPS) onto tyrosyl residues of target proteins, catalysed by tyrosylprotein sulphotransferase, a membrane-bound enzyme of the *trans*-Golgi system. Sulphation is the most abundant → post-translational modification of proteins, especially soluble and membrane-passing proteins of the secretory pathway of metazoan cells, as e.g. fibrin and gastrin, and is involved in protein-protein interactions during the intracellular transport of proteins and their secretion.

Sumoylation: The conjugation of the *s*mall → *u*biquitin-like *mo*difier protein (SUMO; also called sentrin, UBL1, PIC1, or GMP1) to target proteins, which changes their characteristics and/or locations or interactions with other proteins. SUMO is first activated by a heterodimeric E1 enzyme (SAE1–SAE2 in man, Uba2–Aos1 in yeast), that adenylates the C-terminal glycine of SUMO using ATP. Then a thioester bond is formed between the C-terminus of SUMO and a cystein in SAE2, releasing AMP. In a transesterification process, SUMO is then transferred from SAE to the E2 SUMO-conjugating enzyme Ubc9. The thioester-linked SUMO-Ubc9 conjugate then catalyses the formation of an isopeptide bond between the C-terminus of SUMO and an amino group of a lysine in the target protein. This process is reversible: SUMO-specific proteases catalyse the cleavage of the isopeptide bond. The function of sumoylation is not clear in each case. However, deletion of *SMT3* and *ULP1* in yeast, encoding a SUMO-specific protease are lethal. Moreover, the trafficking protein RanGAP1 (shuttle for proteins from the cytoplasm to the nucleus) requires sumoylation for its localization to the → nuclear pore.

Supercoil (superhelix): A specific conformation of a double-stranded DNA molecule containing extra twists in the → double helix as a result of either overwinding (positive supercoil) or under-winding (negative supercoil) causing it to coil upon itself. For the maintenance of these extra twists it is necessary that the ends of the duplex are constrained from free rotation. Supercoiled circular DNA can be detected electrophoretically, because it migrates faster than relaxed DNA. See → negative supercoiling, → positive supercoiling. Compare → coiled coil, a superhelical structure of proteins.

Supercoil sequencing: See → plasmid sequencing.

Supercontig: A large fragment of genomic DNA, that is composed of a series of → contigs. Adjacent supercontigs can be combined to larger → ultracontigs. For example, in the mouse genome sequencing project a supercontig had an average length of 16.9 Mb.

Supercycle: An infelicitous term for a thermal cycling program (see → polymerase chain reaction) that is composed of multiple repetitions of one cycle of low, and two cycles of high → stringency, employing one → gene-specific primer and a second → arbitrary primer such as used in → thermally asymmetric interlaced PCR. Such asymmetric cycling allows to amplify only sequence-specific regions of a → template in an exponential way.

Superfluous gene (costly gene, unnecessary gene): An anthropocentric and somewhat arrogant term for any gene, that is apparently supernumerary and has no obvious function for the organism (and can therefore be deleted without any selective disadvantage). For example, selection for high yielding cultivars in traditional plant breeding entailed selection against such unnecessary and costly genes, e.g. resistance genes. In spring wheat, increased yield was coupled to a decrease in the number of → resistance gene analogs.

Supergene family: A category of genes with different chromosomal location and restricted nucleotide sequence homology, that encode proteins with structurally (and functionally) similar → domains (e.g. the immunoglobulin genes). Compare → multigene family.

Superhelix:
a) A tertiary helical structure in DNA molecules, see → supercoil.
b) A tertiary structure of proteins, see → coiled coil.

Superimposed substitution (multiple hit): The occurrence of two (or more) base → substitutions at the same site in a genome. Principally, any nucleotide can be substituted by any other nucleotide with equal probability (Jukes-Cantor model), but there may be some nucleotide sites that substitute more frequently than others (i.e. evolve more rapidly).

Superinfecting phage: A → bacteriophage which infects a bacterial cell that has already been infected by other bacteriophages of the same type. See → superinfection.

Superinfection:
a) A process whereby one bacterial cell is infected by two or more related → bacteriophages.
b) A process whereby a bacterial cell is infected that already harbors a → prophage. See → phage exclusion.

Superinfection immunity: The resistance of a virus-infected cell against an infection by a second virus of the same or similar type (see → phage exclusion). The term is also used for cell-cell interactions, e.g. → *Agrobacterium tumefaciens* and plant cells. If competent plant cells are infected by virulent bacteria, then a secondary infection by the same bacteria is excluded.

Superlinker: See → superpolylinker.

Superlinker vector: See → superpolylinker vector.

Supernatant: That part of a centrifuged solution which cannot be sedimented by the centrifugal force employed, but remains above the precipitate (sediment, pellet).

Superpolylinker (superlinker): A synthetic oligodeoxynucleotide that contains a total of 64 tandemly arranged hexameric → restriction endonuclease → recognition sites with interspersed unique sites for eight-base cutters (e.g. *Not* I, see → rare cutter). Such superpolylinkers serve to add new recognition sites to the termini of any DNA inserted into any of their recognition sites. See also → superpolylinker vector.

Superpolylinker vector (superlinker vector): A → cloning vector that contains a → superpolylinker of up to 64 tandemly arranged hexameric restriction endonuclease recognition sites in addition to an → origin of replication, → selectable marker genes and → promoters for the production of e.g. a → fusion protein of *lacZ* α' and the product of the cloned sequence, or the production of → sense and → antisense RNA from the insert (e.g. T3 and T7 promoters in inverted orientations).

SuperSAGE: A variant of the conventional → serial *a*nalysis of *g*ene *e*xpression (SAGE) technique, that allows the genome-wide and quantitative → gene expression profiling of cells, tissues, organs and organisms. SuperSAGE basically follows the original SAGE protocol, but involves the type III → restriction endonuclease *Eco*P15I, that cleaves the → cDNA template most distantly from its → recognition site. Therefore, the resulting → tags are 26 bp long, and much longer than the tags from traditional SAGE (13 bp) or → LongSAGE (19–21 bp). The advantages of SuperSAGE are twofold. First, the information content of a SuperSAGE tag of 26 bp is higher than the conventional tags and allows to identify a gene directly from the → Genbank databases. Second, the ends of linker-tag fragments generated by SuperSAGE are blunt-ended to ensure random association of the tags to form ditags. SuperSAGE has the additional benefit of discovering host and pathogen messages simultaneously from the same infected material. In short, → messenger RNA is first isolated, → reverse-transcribed into single-stranded → cDNA using a reverse transcription-primer with the sequence:

5'-CTGATCTAGAGGTACCGGATCC**CAGCAG**TTTTTTTTTTTTTTTTT–3'

containing the 5'-CAGCAG–3'→ recognition site for the type III → restriction endonuclease *Eco*P15I from *Escherichia coli* strain TG1. The product is converted to double-stranded cDNA, digested with *Nla*III, and the 3'-end fragments of the cDNAs bound to → streptavidin-coated magnetic beads. The streptavidin-bound cDNA is washed, and divided into two portions in separate tubes. Two → linkers (linker–1E and linker 2E) are labeled with → *f*luoro*isot*hiocyanate (FITC), and the unblocked 5'-termini of linker–1E and linker–2E phosphorylated by → T4 polynucleotide kinase. Both linker–1E and linker–2E harbor the *Eco*P15I recognition sequence (5'-CAGCAG–3'). Linker–1E or linker–2E, respectively, are then added to the two tubes containing cDNA bound to magnet beads and ligated to the cDNA ends by → T4 DNA ligase. Consequently, each cDNA fragment is flanked by two inverted repeats of 5'-CAGCAG–3'.

*Eco*P15I recognizes the asymmetric hexameric sequence 5'-CAGCAG–3' and cleaves the DNA 25 nt (in one strand) and 27 nt (in the other strand) downstream of the recognition site, leaving a 5' overhang of two bases. Two unmethylated and inversely oriented recognition sites in head-to-head configuration (5'-CAGCAG-$N_{(i)}$-CTGCTG–3') are essential for efficient cleavage. Linker-ligated cDNA on the magnetic beads is digested with *Eco*P15I at 37°C for 90 min. Digestion fragments are separated by → poly*a*crylamide *g*el *e*lectrophoresis (PAGE), the ca. 69-bp "linker-tag" fragment (linker: 42 bp, tag: ca. 27 bp) is visualized by FITC fluorescence under UV light, and collected from the gel. "Linker–1E-tag" and "linker–2E-tag" fragments are mixed, their ends blunt-ended by filling-in with DNA → *Thermococcus kodakaraensis* (KOD) polymerase and subsequently ligated to each other. The resulting → ditags were amplified by conventional → *p*olymerase *c*hain *r*eaction (PCR), using biotinylated primers (ditag primer1E: biotin–5'-CTAGGCTTAATACAGCAGCA–3' and ditag primer2E: biotin–5'-TTCTAACGATG-TACGCAGCAGCA–3'), and the ditag PCR products digested with *Nla*III. The digested fragments are then separated on PAGE, and the fragment of ca. 54 bp isolated from the gel. This fragment is concatenated by → ligation. Concatemers larger than 500 bp are size-selected by PAGE, isolated and cloned into a → plasmid vector, that is transformed into electrocompetent *E.coli* cells by → electroporation, and plated on selective LB medium (containing 100 µ-g/ml ampicillin, 20 µg/ml → X-gal and 0.1 mM → IPTG). Plasmid inserts are amplified by → colony PCR, directly sequenced, and the sequences analyzed by the SAGE2000 software package (extraction of the 22-bp tags adjacent to CATG). The resulting 26–27 bp sequence from each cDNA is called "SuperSAGE" tag. The main advantage of using *Eco*P15I over conventional enzymes is the longer tag, that allows better identification of the underlying cDNA (or gene) by → annotation. See → microSAGE, → miniSAGE, → SAGE-LITE, → small *a*mplified *R*NA-SAGE (SAR-SAGE).

Supershift assay: A variant of the conventional → mobility-shift DNA-binding assay for the detection of specific DNA-protein interaction(s). The supershift assay works with an antibody that is directed against the analyte protein, but does not interfere with the DNA-protein interaction. If the antibody can bind, it induces an additional shift in mobility of the DNA-protein complex ("supershift").

Super *Taq* polymerase: The trademark for a highly active thermostable → *Thermus aquaticus* DNA polymerase that has been purified to homogeneity.

Supervirulent vector (Supervirulence vector): A variant of a → binary → Ti plasmid vector, which is composed of a T-DNA-containing Ti plasmid with a supervirulent → vir region and a → helper plasmid with a normal virulence region. The use of this supervirulent vector leads to a significant increase in → transformation efficiency of target plants.

Suppression: See → suppressor mutation.

Suppression hybridization: See → chromosomal *in situ* suppression (CISS) hybridization.

Suppression mutation gene: See → suppressor gene.

Suppression PCR: See → suppression *p*olymerase *c*hain *r*eaction.

Suppression polymerase chain reaction (suppression PCR): A variant of the conventional → polymerase chain reaction for the isolation of → promoters and regulatory elements, for the walking from sequences of cloned cDNAs, for the exact determination of exon-intron boundaries of split genes, and the up- or down-stream walking from → sequence tagged sites in a genome. In short, genomic DNA is separately digested with a series of → restriction endonucleases with six base → recognition sites, that generate blunt ends (e. g. *Dra*I, *Eco* RV, *Pvu*II, *Sca*I and *Ssp*I). Then a special → adaptor is ligated to the blunt ends of the restriction fragments using → T4 DNA ligase. The adaptor possesses a blunt end (to facilitate ligation to the blunt-ended genomic restriction fragments), carries two rare restriction enzyme sites (for e. g. *Not*I and *Srf*I/*Sma*I, generating staggered or blunt ends, respectively, for cloning into appropriately cut vectors), and an amino group on the 3' end of the lower strand. This group blocks the extension of the lower adaptor strand and prevents the formation of a primer binding site (unless a defined distal gene-specific primer extends a DNA strand opposite to the upper strand of the adaptor). After adaptor ligation, an aliquot of the DNA fragments serves as template for a PCR with an adaptor-primer and a gene-specific primer using a mixture of two long-distance thermostable DNA polymerases (→ *Thermus thermophilus* DNA polymerase, and → Vent™ DNA polymerase). The adaptor-primer is shorter than the adaptor and capable of hybridizing to the outer primer-binding site. If any PCR products are generated which contain double-stranded adaptor sequences at both ends (non-specific DNA synthesis), the ends of the individual strands will form panhandle structures due to the presence of inverted terminal repeats. These panhandles are more stable than primer-template hybrids and suppress exponential amplification. If a distal gene-specific primer extends a DNA strand through the adaptor, then the extension product contains the adaptor sequence on one end only (i. e. cannot form a panhandle structure). In this case, amplification proceeds normally. Finally, the amplification products are separated by → agarose gel electrophoresis and → ethidium bromide fluorescence.

Suppression subtractive hybridization (SSH, supressive subtractive hybridization): A variant of the → suppression PCR technique, that allows to selectively amplify differentially expressed target → cDNAs and simultaneously suppress non-target DNA amplification. The technique is based on the selective and efficient suppression of the amplification of undesirable sequences, if long inverted terminal repeats are attached to the cDNAs. Moreover, it combines the → normalization (equalizes the abundance of cDNAs in a target population) and subtraction steps (excludes common sequences between the target and driver population). In short, the tester (contains the differentially expressed cDNAs) and driver cDNAs (contains these cDNAs at very low levels only) are first restricted with a → four-base cutter yielding blunt ends (e. g. *Hae*III or *Rsa*I). The tester cDNA fragments are then divided into two samples (here referred to as tester population 1 and 2) and ligated to two different → adaptors (here referred to as adaptor 1 and 2), resulting in two tester populations. The ends of the adaptors do not contain phosphate groups, so that only the longer strand of each adaptor can be covalently attached to the 5'-termini of cDNA. Then two successive hybridizations are employed. First, an excess of driver is added to each tester population, the samples denatured and allowed to anneal. The single-stranded cDNA tester fraction (A) is normalized (i. e. the concentrations of high and low abundance cDNAs are equalized), because the reannealing process producing homohybrid cDNAs (B) is much faster for the more abundant cDNAs (second order kinetics of hybridization). Also, the ss cDNAs in tester fraction A are enriched for cDNAs from differentially expressed genes. Second, the two samples from the first hybridization are mixed. Only the remaining, normalized and subtracted ss tester cDNAs can reassociate to form B, C, and new E hybrids. Addition of a second portion of denatured driver further enriches fraction E, which is different from all other fractions, because its cDNAs are linked to different adaptor sequences at their 5'-termini (one from sample 1, the other from sample 2). These two different adaptor sequences allow preferential amplification of the subtracted normalized fraction E using conventional → polymerase chain reaction techniques and

primers P1 and P2 directed against the terminal part of adaptor 1, and 2, respectively. This selective amplification requires the fill-in of the 3' sticky ends with DNA polymerase. Exponential amplification can only occur with E type molecules. The type B molecules contain long inverted repeats at their termini and form stable fold-back structures after each PCR step, which are no templates for exponential PCR, since intramolecular annealing of longer adaptor sequences is favoured over the intermolecular annealing of the shorter primers (which is the basis of the suppression effect). Type A and D molecules do not contain primer binding sequences, and type C molecules are amplified at a linear rate only. See → representational difference analysis. Compare → suppression hybridization.
Figure see page 1095.

Suppressive subtractive hybridization (SSH): See → suppression subtractive hybridization.

Suppressor gene:
a) Any gene that reverses the effect of → mutations in other genes (*suppression-mutation gene, spm*).
b) Nonsense suppressor gene, a mutant gene coding for an abnormal → transfer RNA that suppresses → stop codons by reading them as codons encoding amino acids. For example, the *supF* gene of *E. coli* codes for a mutant tyrosyl-tRNA (tRNA^{Tyr}) that reads the stop signal UAG as a tyrosine codon. Consequently, polypeptide synthesis is not terminated at the UAG position, but extends beyond it. The stop codon is said to be suppressed. In gene technology, such suppressor genes are used as → selectable markers, or as constituents of → containment vectors, since their expression allows the suppression of → nonsense mutations in strategic genes of → cloning vectors. Such vectors are not able to replicate in suppressor-less host cells (*sup^0*). See also → suppressor mutation.

Suppressor mutation (suppression, second site mutation): A secondary → mutation that totally or partially restores function(s) lost by a primary mutation in a defined DNA sequence. The site of the secondary mutation is distinctly different from the site of the primary mutation so that a suppressor mutation does not eliminate the original mutation (as is the case in classical → reverse mutations). Frequently a suppressor mutation corrects a previous → reading frame shift mutation. See also → nonsense suppression, → suppressor gene.

Suppressor-mutator element (Spm; suppressor-mutator system): A → transposable element of *Zea mays* that is almost identical to the related → enhancer element (En) and encodes four alternatively spliced transcripts translated into four proteins (TnpA, TnpB, TnpC, TnpD). Complete Spm elements transpose autonomously, and additionally trans-activate internally deleted, transposition-defective derivatives (dSpm). This trans-activation is catalyzed by TnpA and TnpD. The Spm element can be inactivated through methylation at → cytosine residues at its 5' end close to the → transcription start site. Selective TnpA-mediated reactivation of Spm leads to a demethylation of these sequences. TnpA binds to a specific 12 bp recognition sequence that is repeated several times in direct and inverted orientation at the 5' and 3' end of the transcription start site, bringing the ends of Spm together. TnpD acts as specific → endonuclease and generates the transposition complex. In addition to its function in transposition and transactivation of the methylated Spm → promoter, TnpA also acts as repressor of the unmethylated constitutive Spm promoter, but as activator of the methylated promoter.

Suppressor-mutator system (Spm): See → enhancer.

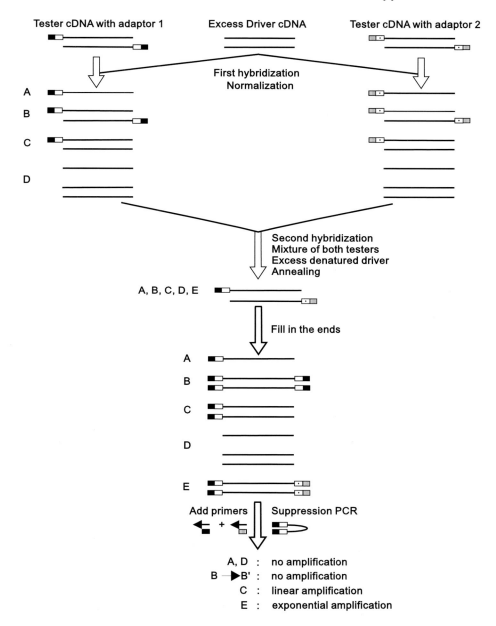

Suppression subtractive hybridization

Suppressor-sensitive mutant: Any → mutant organism that is only viable in the presence, not the absence of a suppressor.

Suppressor *transfer RNA* (suppressor tRNA): An abnormal → transfer RNA that suppresses → stop codons on a → messenger RNA molecule and reads them as codons encoding amino acids. For example, the *sup* F gene of *E. coli* encodes a mutant tyrosyl-tRNA able to read the stop codon UAG as a tyrosine codon. See → suppressor gene.

SURE: See → *s*top *u*nwanted *r*earrangement *e*vents.

Surface *a*coustic *w*ave liquid agitation (SAW liquid agitation): The movement of liquid droplets in the µl (or even nl) range or also solid particles on the surface of a solid support (e.g. a chip), driven by socalled *s*urface *a*coustic *w*aves (SAWs). These waves are generated by applying radio-frequency signals to specially designed socalled transducer electrodes and propagate across the chip. At low power, SAWs induce chaotic streaming within a droplet (effective mixing), at higher amplitudes the droplets move themselves. SAW liquid agitation speeds up → microarray hybridisation, since agitation mixes the reactions more efficiently than simple diffusion.

Surface-*e*nhanced *l*aser *d*esorption/*i*onization (SELDI): A variant of the conventional → MALDI-TOF technique for the simultaneous purification and identification of single peptides or proteins out of a complex mixture, that uses a surface with multiple discrete spots (SELDI chip), where each spot is derivatized with a specific chemical group commonly used for peptide or protein purification (as e.g. SO_3^-, $^+NR_3$, Me [II]). The mixture is therefore first fractionated according to chemical reactivity, the specific reactants (proteins interacting with e.g. hydrophobic groups) are then fractionated by varying the concentration of hydrophobic solvents, and the different fractions finally analyzed by laser desorption/ionization.

Surface *p*lasmon *r*esonance (SPR): An electron charge density phenomenon (excitation), that occurs at the boundary between a metal (gold, silver) and a dielectric medium (air, water) and allows to detect e. g. interactions (binding or dissociation) between a ligand (e. g. an immobilized → antigen) and a biomolecule (e. g. an → antibody) and to measure real-time reaction kinetics between two (or more) molecules (e. g. a ligand and an analyte). It makes use of a 50 nm thin nobel metal layer (e. g. gold), in which surface plasmons, socalled plasmonic surface polaritons (localized regions of oscillating electrons) can be excited by impinging light (e. g. from a monochromatic laser). At a defined wavelength and angle of incidence of the light, and favorable refractive index of the material, light energy is absorbed by the plasmons, and therefore no light is reflected from the surface. A photodetector monitoring the intensity of the reflected light will therefore show minimal reflection intensity. The sensor surface is coated with either a hydrophilic polymer ("hydrocoat") modified with carboxyl or amino groups, or polystyrene, into which the ligands are embedded by covalent coupling to the support, usually via a flexible spacer. Now the analyte (e. g. an → antibody) is added, interacts with the immobilized ligand (e. g. an → antigen) and changes the refractive index at the gold surface, shifting the angle of the intensity minimum. Such optical biosensors do not require labeled analytes.

Surface *p*lasmon *r*esonance biosensor (SPR biosensor): A biochip for the real-time detection of interaction(s) between two unlabeled molecules (e. g. DNA and protein), based on → *s*urface *p*lasmon *r*esonance (SPR). One interaction partner is bound to the chip surface by covalent bonds, adsorption, or spacer molecules. The other interacting partners, which may be single molecule species, but also crude cell extracts, supernatants, or column fractions, are then applied to the sensor surface. Now SPR is used to detect minimal mass changes as a consequence of binding or subsequent dissociation, resulting in a → sensorgram. After regeneration of the chip (by removal of ligands), it may be re-used for another round of SPR measurements. SPR biosensors can be applied for DNA-DNA-, DNA-RNA-, DNA-protein-, RNA-protein-, protein-protein-, and DNA-, RNA- or protein-ligand interaction(s).

Surfactant: Any chemical substance, that disrupts the hydrogen bond interactions between water molecules and thereby reduces the surface tension of a liquid.

Surrogate genetics:
a) A discipline of genetics that is concerned with the expression and processing of DNA and RNA from one donor organism in another acceptor organism. Originally, vertebrate oocytes (e.g. *Xenopus* oocytes) were microinjected with foreign DNA and RNA and the mechanism(s) of gene expression, its regulation, and pathways of processing were elucidated.
b) See → reverse genetics.

Surrogate ligand: Any short peptide that binds to a target protein with high affinity and thereby inhibits its function. In → pharmacogenomics, in vitro assays can be designed, whereby compound libraries are screened for small molecules that competitively displace such peptides and occupy the peptide-binding site on the target protein. Such compounds represent potentially specific drugs. Compare → surrogate marker.

Surrogate marker: Any gene that is deregulated, if another specific essential target gene is inactivated by a low molecular weight molecule (e. g. a drug). Such inactivation can also be brought about by in vivo expression of small peptide inhibitors, or shifting of a → conditional mutant towards non-permissive conditions. Expression profiling or → proteome analysis then allows to identify gene(s) (or encoded protein[s]) that are deregulated as a consequence of target gene inhibition. Therefore, such deregulated genes (proteins) are indicators for the biological activity of the inactivated target gene (protein). Surrogate marker genes can be fused to a → reporter gene, transformed into a target organism, and any deregulation be detected easily as a measure of target gene inhibition. Compare → surrogate ligand.

Suspension array: A misleading term for a suspension of up to 1000 unique polysterene microspheres, each of which carries one special → probe molecule (e.g. a peptide or protein such as → an antibody, an enzyme, a receptor, or an → oligonucleotide). Such beads are carriers, on whose surfaces analytes can be trapped by specific interaction(s) with the probes (e.g. antigen-antibody-, receptor-ligand-, enzyme-substrate-, DNA-DNA-, DNA-RNA-, or DNA-protein- interactions). Since a multitude of differently loaded microspheres can be used simultaneously, and thousands of analytes be pumped through the array synchronously, suspension arrays are platforms for high-throughput analysis, miniaturization and automation, that allow real-time analyses of binding partners and parameters.

S value: See → sedimentation coefficient.

Svedberg constant: See → sedimentation coefficient.

Svedberg unit: See → sedimentation coefficient.

SV 40: See → simian virus 40.

SVT: See → single virus tracing.

Swing-out rotor: A centrifuge rotor containing buckets that swing out horizontally during the centrifuge run. Such rotors are used to generate density gradients or to sediment compact pellets. Compare → fixed-angle rotor.

SWI-SNF complex: A multi-subunit protein assembly that remodels → chromatin at → enhancer (or → silencer) regions thus opening it for the binding of transcriptional activator or repressor proteins.

Switching mechanism at 5' end of RNA template (SMART): A technique for the generation of full-length → cDNAs from either total RNA or → poly(A)$^+$-RNA using → long-distance → polymerase chain reaction methodology. In short, RNA is reverse transcribed into → first-strand cDNA using a modified oligo(dT) oligonucleotide → primer and → Moloney murine leukemia virus reverse transcriptase (M-MLV-RT). Upon reaching the 5' end of → mRNA template, the → terminal transferase activity of M-MLV-RT adds 3-5 deoxycytidine nucleotides to the 3' end of the first strand cDNA. Then a G-rich oligonucleotide primer (SMART primer) is added, that anneals to the C-rich 3' end of the cDNA, forming an extended template. At this point the RT switches the template and replicates the SMART oligonucleotide such that the produced single-stranded cDNA carries the complete 5' end of the mRNA together with the sequence complementary to the SMART oligonucleotide (SMART anchor). This anchor at the 5' end, and the primer sequence at the 3' end of the cDNA serve as universal priming sites for an amplification by → long distance PCR. The resulting double-stranded cDNA is enriched for full-length sequences.

Switching site: A breakpoint within genes where → recombination occurs in gene rearrangements. Incorrectly abbreviated as S.

Swivelase: See → topoisomerase.

Symbiosin: Any one of a series of plant genes that are induced in the early stages of the endosymbiotic plant-microbe interaction necessary for the establishment of *a*rbuscular *m*ycorrhiza (AM) and *r*oot-*n*odule *s*ymbiosis (RNS) of leguminous plants with N$_2$-fixing bacteria. SYMRK (*sym*biosis *r*eceptor-like *k*inase), a receptor-like kinase is encoded by such symbiosin genes and is involved in early symbiotic signal transduction.

Symbiosis island: The complete region of a *Mesorhizobium loti* chromosome that contains the genes for the symbiotic interaction between the bacterium and the host plant.

Symbiosome: The functional unit of symbiotic nitrogen fixation in some higher plants (e.g. most leguminous plants). Symbiosomes are membrane-bounded compartments containing one to several → bacteroids, each of which fixes atmospheric nitrogen. Each cell infected by bacteroids harbors a few thousand symbiosomes (the root nodule of leguminous crops contains hundreds of infected cells). After its fixation as ammonia, the nitrogen moves across the symbiosome membranes and through the cytoplasm of the Symbiosomes are membrane-bounded compartments containing one to several → bacteroids, each of which fixes atmospheric nitrogen. Each cell infected by bacteroids harbors a few thousand symbiosomes (the root nodule of leguminous crops contains hundreds of infected cells). After its fixation as ammonia, the nitrogen moves across the symbiosome membranes and through the cytoplasm of the infected cells into a series of non-infected cells and finally into the vascular system of the plant for transport.

Symmetrical transcription: The complete → transcription of both strands of a → double-stranded DNA molecule such that two RNAs each of the length of the corresponding strand are produced. This type of transcription is rare, but typical for → mitochondria, where the → D-loop region contains two → promoters, one for the transcription of the → H strand, and another for the → L strand of → mitochondrial DNA.

Syn conformation: A conformation of free nucleotides, and nucleotides in nucleic acids, where the C6 of the → pyrimidines and the C8 of the → purines are positioned at the greatest possible distance from the oxygen atom of the → ribose or → deoxyribose. See → anticonformation.

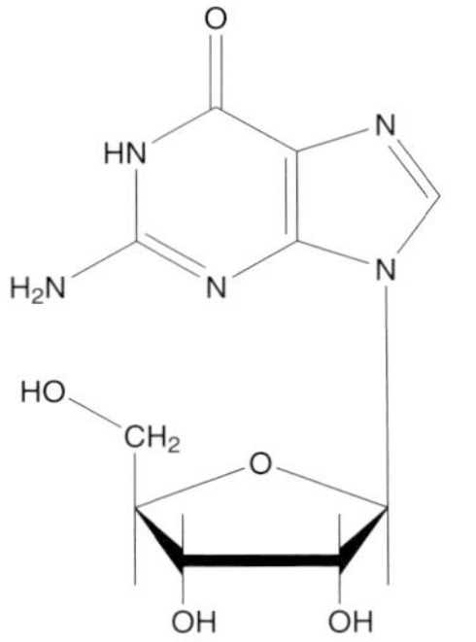

Guanosin: syn conformation

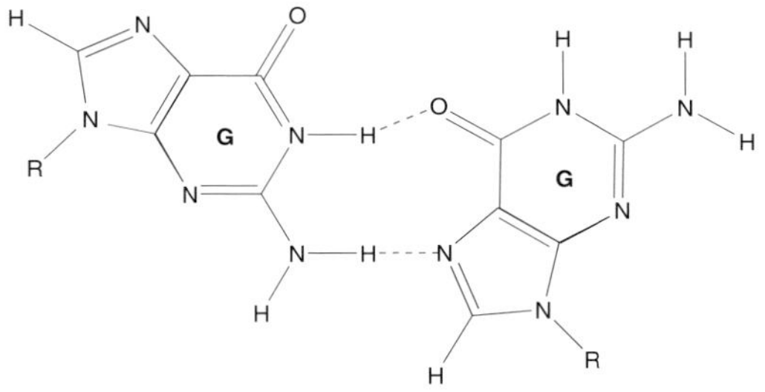

G-anti: G-syn (antiparallel chains)

Syn conformation

Synonymous codons: Two (or more) → codons that encode the same amino acid (e.g. GCA, GCC, GCG and GCU all code for alanine; GGA, GGC, GGG and GGU all code for glycine). Also all → stop codons. See → degenerate code.

Synonymous sequence change: Any alteration in the nucleotide sequence of a coding region, that does not change the amino acid sequence of the encoded protein. See → non-synonymous sequence change.

Synonymous single nucleotide polymorphism (**synonymous SNP, synSNP**): Any → single nucleotide polymorphism, that occurs in an → exon, but does not change the amino acid composition of the encoded protein. See → anonymous SNP, → candidate SNP, → coding SNP, → exonic SNP, → gene-based SNP, → human SNP, → intronic SNP, → non-synonymous SNP, → promoter SNP, → reference SNP, → regulatory SNP.

Syntenic gene: Any gene localized in a genomic region that is conserved over evolutionary times. The order of syntenic genes in different genomes is usually identical or at least similar (→ microsynteny), which can be exploited in → comparative genomics. See → synteny.

Synteny:
a) The localization of two (or more linked or unlinked) gene loci on the same chromosome.
b) The conserved order, sequence, and orientation of → genes in the range of 100 to 500 kb (or more) in the genomes of closely related species. Syntenic regions have e. g. been detected in the genomes of cereals (barley, fox tail millet, rice, rye, sugar cane, and wheat), but also in other plant families. While gene sequences and their map order are highly conserved, neither are the size, sequence, or composition of the DNA in between genes. Therefore most of the genetic diversity between species resides in the → intergenic regions. See → microsynteny.

Synthesis-dependent strand-annealing (**SSA**): A mechanism for the repair of → double-strand breaks (DSBs) in DNA. In short, the 5'-ends of DSBs are resected by 5'→ 3'exonucleases (Rad50/Mre11/Xrs2 proteins), the target helix is unwound, one of the 3'-termini invades the → template and is used as → primer for repair synthesis (Rad51, 52, 54, 55 and 57, and Rdh54 involved). The invading strand displaces one of the target strands, a → D-loop is formed and migrates as repair synthesis proceeds. After the synthesis of a leading and a lagging strand by respective → DNA polymerases, an → endonuclease cleaves the nascent single strand at a specific position, which leads to gap repair without → crossing over. Or, alternatively, it nicks the single-strand of the D-loop, which leads to the formation of a → Holliday junction adjacent to the repaired gap. All the newly synthesized DNA is part of the recipient DNA molecule.

Synthetic enzyme: See → synzyme.

Synthetic gene (artificial gene): Generally, any *in vitro* synthesized DNA sequence that encodes a peptide, protein or RNA. Such synthetic genes are ligated to appropriate → promoter and termination sequences and transferred to target organisms by → gene transfer techniques. Numerous genes have been synthesized *in vitro*. Originally, relatively small genes could be produced (e.g. encoding tRNA [126bp] α-interferon [542bp], secretin [81bp], urogastrone [162bp], thaumatin [650bp], γ-interferon [453bp], eglin C [232bp], proenkephalin [77bp], ATPase [170bp], lysozyme [385bp], c-Ha-ras [576bp], Rnase T1 [324bp], cytochrome b_5 [330bp], hirudin [226bp] and RNase A [375bp]), later on larger genes as well (e.g. encoding rhodopsin [1.057kb] and tissue plasminogen activator, t-PA [1.61kb]). Today, any gene of any length can (at least theoretically) be synthesized *in vitro*. In particular, coding sequences, which do not naturally occur, as for example, an artificial gene that codes for a protein with a high content of essential amino acids (HEAAE-DNA, *h*igh *e*ssential *a*mino *a*cid *e*ncoding DNA). This gene has been transferred to recipient plant cells by → *Agrobacterium* mediated gene transfer and expressed in the host. As a consequence, proteins with a relatively high lysine and methionine content accumulated in the plant (in this case potato). See → designer gene.

Synthetic genome: Any artificial → genome, that is composed of a novel arrangement of either synthetic or naturally occurring genes. For example, the original genome of *Mycoplasma genitalium,* consisting of 517 genes, is removed and a synthetic genome, made of a total of 300 genes in a completely new order is inserted. Threehundred genes are probably the minimum requirement for life. The novel organism containing this minimal genome could then be exploited for specific functions (as e.g. reduction of atmospheric CO_2, binding of hydrogen).

Synthetic messenger RNA: Any → messenger RNA (mRNA) that is synthesized in an → *in vitro* transcription system as opposed to an mRNA synthesized in vivo.

Synthetic nucleic acid (SNA): Any DNA, RNA or oligonucleotide molecule produced *in vitro* by chemical synthesis.

Synthetic promoter: Any → promoter that contains regulatory sequences that have been synthesized *in vitro* (e.g. → TATA boxes, → CAAT boxes → enhancer cores, → negative elements). The term "synthetic promoter" is also frequently used synonymously with → hybrid promoter.

Synthetic receptor: Any → biosensor that combines a fully synthetic receptor molecule (mimicking the recognition properties of a biological receptor) with transducers converting the event of a receptor-ligand interaction into a digital electronic signal. Such sensors are variants of → affinity biosensors. The synthetic receptors are designed by combinatorial or computational chemistry, molecular imprinting, self-assembly, rational design or combinations. For example, so-called *m*olecular *i*mprinted *p*olymera (MIPs) mimic native receptors ("biomimetic sensors") with regard to affinity and specificity, but are more stable, cheaper, and easily prepared on an industrial scale, and therefore can effectively replace natural receptors for experimentation. The interaction between receptor and ligand can be detected optically or electrochemically. In one example, a synthetic socalled "bite-and-switch" receptor, designed by molecular imprinting, allows to detect creatine and creatinine in clinical blood samples. A thioacetale reaction between the synthetic receptor polymer and the amine groups of the analytes creatine and creatinine produces a fluorescent isoindole complex, that can be detected optically.

Synthetic *tandem* *repeat* (STR): A synthetic → oligonucleotide containing short random sequences that are tandemly arranged, and used as a → probe to detect tandemly repeated polymorphic sequences (see for instance → variable number of tandem repeats) in human and other genomes.

Synthetic virus (artificial virus): Any virus, whose genome (DNA or RNA) is synthesized *in vitro*. For example, the 7.741 bases of the RNA genome of the single-stranded poliomyelitis virus can be retrieved from public databases, the information transformed into DNA, fragments of this DNA synthesized *in vitro*, and the various fragments ligated in the correct order. Then the DNA is converted into RNA *in vitro*, the RNA transfected into e.g. human cells or incubated with an extract of such cells, which copy the RNA and produce new poliomyelitis virus particles.

Synzyme (*synthetic enzyme*): A protein with catalytic activity that has been designed (see → protein design) and synthesized *in vitro*.

Systematic *evolution* of *ligands* by *exponential* enrichment (SELEX): A technique for the development of high-affinity RNA ligands (→ RNA aptamers), that recognize and bind specific amino acid → domains of a target peptide or protein or low molecular weight ligands. In short, large repertoire libraries of different single-stranded RNA (or DNA) sequences, completely randomized at specific positions, are generated by → in vitro transcription of (usually synthetic) single-stranded DNA → templates varying at eight positions critical for binding. Theoretically, 65.536 individual RNA species can be created this way. The transcripts are then mixed with the target protein, and RNA-protein complexes retained on nitrocellulose filters. After removing non-bound RNA and protein, the candidate RNAs are eluted from the filters, reverse transcribed into cDNAs using → AMV reverse transcriptase and the cDNAs amplified by *Taq* DNA polymerase in a conventional → polymerase chain reaction. One of the → primers contains a → T7 RNA polymerase → promoter. The double-stranded DNA amplification products are then transcribed in vitro (→ in vitro transcription), and the resulting RNAs used in the next round of selection. Multiple rounds of this selection procedure enrich the RNA population for RNA molecules with the highest affinity for the protein, that can then be isolated, cloned, and sequenced. SELEX can analyze the optimal binding sequences and their sequence context for any → DNA-binding protein (e. g. → transcription factors, → repressors, replication proteins, and translational repressors at ribosome binding sites, to name few). Moreover, interactions of small molecules (e. g. amino acids, nucleotides) with specific RNAs can be studied by binding the effectors to solid supports and partition those RNAs that interact specifically with these substrates. For example, SELEX allows the isolation of protein targets (as e. g. bacteriophage T4 DNA polymerase and HIV reverse transcriptase) or oligonucleotides with affinity for low-molecular weight targets (as e. g. ATP, tryptophan, arginine, theophyllin, caffein, and others. Also, SELEX is a first technical step towards "evolution in a test tube".

Systemic *acquired* *silencing* (SAS): The systemic spread of → gene silencing throughout a → transgenic organism. If gene A is transformed into a → genome of a target cell (usually in multiple copies), then frequently its own expression, or that of the corresponding cellular → homologue is repressed ("silenced"; see → co-suppression). This effect may be transmitted from cell to cell, leading to SAS of all copies of gene A and its homologues in the organism. For example, when a plant leaf expressing the → green fluorescent protein (GFP) is infiltrated with *Agrobacterium tumefaciens* carrying another GFP gene in the → T-DNA of its → Ti-plasmid, then the T-DNA together with the inserted GFP integrates into the nuclear genome of the exposed leaf. As a result, GFP expression is silenced throughout the whole plant although the agrobacterium is restricted to the infiltrated leaf only. The systemically spreading silencing signal is unknown, but may be an → RNA or a → ribonucleoprotein complex.

T

T:

a) Abbreviation for "*t*ransformant". For example, the outcome of a → transformation experiment are T_0 organisms, which are selected from non-transformants by e.g. expression of an → antibiotic resistance gene as → selectable marker. In case of plants, T_0 plants can be selfed to → homozygous T_1 progeny.

b) Large T and small t, see → T antigen.

c) See → *t*wisting number.

d) The single-letter code for threonine, an → amino acid.

e) Abbreviation for → *t*hymine.

TAB linker (*t*wo *a*mino acids *B*arany linker): A hexameric single-stranded oligodeoxynucleotide → linker used for → TAB linker mutagenesis.

TAB linker mutagenesis (*t*wo *a*mino acids *B*arany linker mutagenesis): The introduction of mutations into DNA by the use of hexameric, single-stranded oligodeoxynucleotides (→ TAB linkers) that code for two amino acids and are complementary to → cohesive ends of the restricted target DNA molecule. The TAB linkers are annealed to the target overhangs and ligated. This introduces into the molecule a new → recognition site for a → restriction endonuclease in addition to a mutation (insertion of two new amino acid codons). TAB linker mutagenesis is a technique of → protein engineering. Compare → linker mutagenesis, → linker scanning.
Figure see page 1104.

TAC:

a) See → transcription, → transcription initiation complex.

b) See → *t*ransformation-competent *a*rtificial *c*hromosome vector.

c) See → *t*riplex *a*ffinity *c*apture.

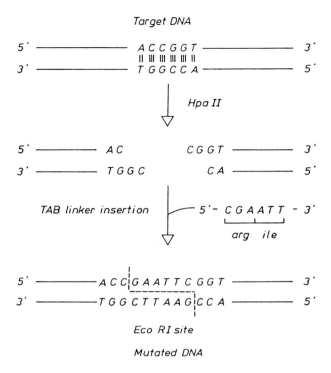

Target DNA

TAB linker mutagenesis

TA cloning (TA overhang cloning): A procedure for the cloning of → *Taq* DNA polymerase-generated amplification products. TA cloning is based on the terminal transferase activity of this thermostable polymerase, i. e. the addition of a single adenosine residue to the 3' end of the otherwise completed amplification product to produce a 3'-A overhang (see → extendase). For the cloning procedure, a special plasmid cloning vector is designed with a single 3'-T overhang such that the insert can directly form A-T base pairs with the ends of the linerearized vector. The ligation can be catalyzed by → DNA ligase *in vitro*, or by the ligases in the host cell *in vivo*. TA cloning circumvents the need for → restriction digestion of the PCR product, → adaptor ligation, or purification of the PCR product prior to ligation.
Figure see page 1105.

***tac* promoter (*tac* promotor):** A → hybrid promoter composed of the -35 box 5'-TTGACA-3' of the tryptophan operon (*trp*) promoter (P_{trp}) and the -10 box 5'-TATAAT-3' of the *lac Z* promoter (P_{lac}). This hybrid promoter can be repressed by the → *lac* repressor and induced by → IPTG, and is stronger than both parent promoters together. Compare → *trc* promoter.

TACTAAC-box: The conserved → consensus sequence 5'-TACTAAC-3' (core: 5'-CTGAC-3') in yeast → heterogeneous nuclear RNA (hnRNA), located about 30 bases upstream of the 3' end of an → intron, and necessary for → lariat formation. The hnRNA is first cut at the 5' end of an intron, yielding a linear exon and an intron-exon molecule whose 5' end binds 5' → 2' to the A of the core sequence CTGAC, forming a lariat. In higher eukaryotes, the analogous box is less conserved (consensus sequence: 5'- PyNPyPyPuAPy-3'), but reacts in the same way. See also → splicing.

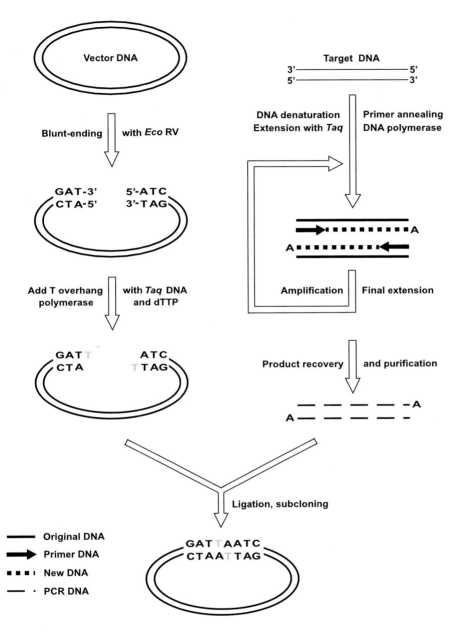

TA cloning

TAE: See → *t*ris-*a*cetate-*E*DTA buffer.

TAF:
a) See → *TATA* box binding protein-*a*ssociated *f*actor.
b) See → *t*ranscription *a*ctivation *f*actor.
c) See → *t*elomere-*a*ssociated chromosome *f*ragmentation.

TAFE: See → *t*ransverse-*a*lternating *f*ield *e*lectrophoresis.

Tag: See → label.

***T*agged random primer *p*olymerase *c*hain *r*eaction (tagged random primer PCR, T-PCR):** An improved variant of the → *primer extension preamplification* (PEP) technique for the sampling of an entire genome that avoids the disadvantage of primer-dimer artifacts characteristic for PEP. In short, a set of all possible nonanucleotide primers (4^9) is synthesized. Each primer is tagged at its 5'-terminus with a constant 17 bp tag (e. g. 5'-GTTTTCCCAGTCACGACN$_9$-3'), and the set used to amplify a target genome in a conventional → polymerase chain reaction. After a few PCR rounds, the amplification products are fractionated by gel filtration (exclusion of small molecules as e. g. primer-dimers). Then the remaining mixture is amplified with tag sequence-specific primers only, producing amplification products that cover up to 75 % of complex genomes.

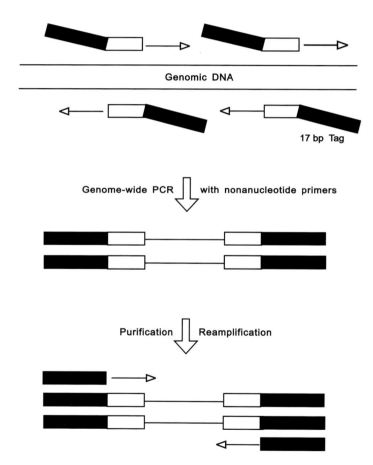

Tagging: See → gene tagging.

Tag sequencing techniques: A series of high-throughput → RNA profiling techniques that are based on the sequencing of short nucleotide stretches ("tags") identifying a specific RNA (e.g. different → messenger RNAs from each other). Tag sequencing is e.g. prerequisite for → *massively* *parallel* *signature* *sequencing* (MPSS) or → *serial* *analysis* *of* *gene* *expression* (SAGE).

Tail:
a) A → protruding terminus in a dsDNA molecule.
b) Single-stranded nucleotide sequences added to linear double-stranded DNA molecules by the enzyme → terminal transferase. See → DNA tailing.
c) See → poly(A) tail.

Tailing:
a) The post-transcriptional attachment of adenine nucleotides to the 3'-terminus of many hnRNA and mRNA molecules by → poly(A) polymerase.
b) See → DNA tailing.

TAIL-PCR: See → *t*hermal *a*symmetric *i*nterlaced *p*olymerase *c*hain *r*eaction.

T-allele: Any → single nucleotide polymorphism, that is caused by the exchange of either an adenine, a cytidine or a guanosine for a thymidine. See → A-allele, → C-allele, → G-allele.

TAMRA: The → fluorochrome *tetra*-*m*ethyl-6-carboxy*r*hod*a*mine, that is used as a marker for → fluorescent primers in e. g. automated sequencing procedures or for labeling in → DNA chip technology. The molecule can be excited by light of 560 nm wave-length, and emits fluorescence light at 582 nm. Since the wave-length of the excitation and emission maxima is pH-dependent, the exact values vary.

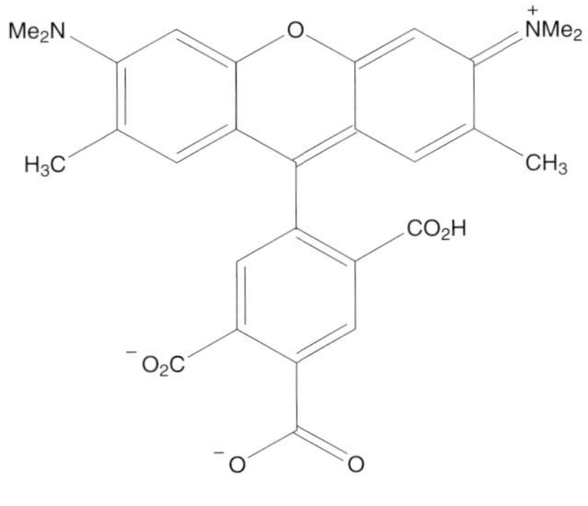

TAMRA

Tandem *affinity* *purification* (TAP) tagging: A technique for the quantitation of proteins in crude cell extracts or in purified samples, in extreme cases in a whole → proteome. In short, a sequence encoding two → protein A modules is inserted into the target gene such that it tags the C-terminus of the encoded protein. If this gene is expressed, i.e. the encoded protein is synthesized, then its TAP tag can be bound to immunoglobulin G with high affinity. Usually the cell extract is transferred to a polyvinylidene difluoride membrane, which is washed and blocked (to minimize unspecific background). Then the target is detected by a one-step immunoaffinity procedure with an anti-peroxidase-IgG conjugate and → chemiluminescence.

Tandem array: See → tandem repeat.

Tandem duplication: See → tandem repeat.

Tandem fluorescent dye conjugate: Any construct, in which two → fluorochromes with identical excitation wavelengths are covalently linked such that they act as donor-acceptor pairs for → *fluorescence resonance energy transfer* (FRET). For example, in a specific tandem conjugate an *R-p*hycoerythrine (R-PE) molecule is excited, the excitation energy transferred to a second fluorochrome without the emission of a photon. The acceptor fluorophore emits a photon at a specific wavelength, which differs from the emission wavelength of R-PE. Therefore, excitation by the same wavelength results in two different fluorescence colors, allowing simultaneous multicolor labeling and detection.

Tandem gene family: Any group of genes whose members are arranged *in tandem*. Such gene families in turn may be tandemly reiterated manifold. For example, the → ribosomal RNA genes in the → nucleolar organizer region are arranged in families, and many such families arranged in tandem (*Drosophila* X chromosome: 250; Y chromosome: 150; human: 250), the → transfer RNA genes are arrayed tandemly at about 50 chromosomal sites (human) with 10-100 copies at each site. Also, → histone genes occur in tandem arrays, which are reiterated manifold. See → dispersed gene family.

Tandem gene repeat: See → tandemly oriented genes.

Tandem lesion: A radiation-induced damage of two adjacent bases in the same DNA strand. Tandem lesions may be generated by a single radical hit (e.g. ˙OH hit). The radical is added to either C5 or C6 of any of the two pyrimidines at the site, then a fast reaction between the generated pyrimidyl radical and molecular oxygen ensues. The resulting peroxyl radical at least partly reacts with a vicinal guanine by intramolecular addition at C8. Subsequently the adduct is rearranged, giving rise to 8–oxo–7,8–dihydroguanine (8–oxoGua) and an oxyl type pyrimidine radical. See → clustered lesion.

Tandemly oriented genes (tandem gene repeat): Any two (or more) genes whose orientation in a genome is identical (i.e. 5'- start codon → stop codon-intergenic space-start codon → stop codon–3') and which are arranged next to each other (*in tandem*). For example, in *Arabidopsis thaliana*, such tandem gene repeats comprise about 15% of all genes. See → convergently oriented genes, → divergently oriented genes.

***Tandem mass spectrometer* (TMS):** A specially designed mass spectrometer that allows to determine the masses of peptide fragments, generated by the extensive fragmentation of isolated proteins by gas wave shocks. As a result of TMS analysis, a spectrum of peptide fragment ions becomes availabe, that can be compared to the theoretically expected fragment ions of the known proteins, or peptide sequence accessions in appropriate data banks, so that proteins and their post-translational modifications can be identified, using software packages as e. g. SEQUEST. See → MALDI post source decay mass spectrometer.

Tandem promoters: A special sequence arrangement where a → promoter is duplicated and the two promoters are localized one after the other. Such tandem promoters are characteristic for → rDNA genes and serve to accumulate → RNA polymerase I molecules for an efficient → transcription of the linked genes (RNA polymerase I trap, RNA polymerase I trapping center). Promoters in tandem array can also be found in → histone genes, and are used in gene technology to ensure high expression of cloned genes.

Tandem repeat (tandem duplication, tandem array):
a) The arrangement of two or more identical sequences within a DNA duplex molecule such that these sequences are close neighbors. There are two orientations possible, namely → direct repeat (head-to-tail-arrangement) or → indirect repeat (head-to-head arrangement).

Direct tandem repeat

5'-CGAATC GTTATCG GTTATCG ACCGT-3'

Indirect tandem repeat

5'-CGAATC GTTATCG GCTATTG ACCGT-3'

Tandem repeats in the coding region of a gene may lead to a tandemly repeated amino acid sequence in the corresponding protein.
b) The arrangement of two or more identical chromosomal segments within a → chromosome such that these segments are close neighbors. Such a tandem repeat may arise by → amplification.

T antigen (large T, T): A 95 kDa phosphoprotein of → simian virus 40 involved in the initiation of viral DNA → replication. The T antigen molecule binds to each of four simple pentameric repeats (5'-GAGGC-3') in the SV 40 → origin of replication to form an organized nucleoproteincomplex competent for initiation. After binding, the T antigen catalyzes the ATP-dependent unwinding of the two DNA strands. This unwinding reaction is the result of a → DNA helicase activity of the T antigen molecule. It establishes two → replication forks and thus generates the substrate for the assembly of proteins required for priming and elongation of nascent strands.

TA overhang cloning: See → TA cloning.

TAP:
a) See → tandem affinity purification.
b) See → *t*obacco *a*cid *p*yrophosphatase.

Tapping mode *atomic force microscopy* (tapping mode AFM): A specific configuration of → atomic force microscopy for the production of high-resolution topography images of individual DNA or protein molecules, that are deposited on a freshly split mica surface. In this mode, a tip with a radius of 20–50 nm is raster-scanned across the surface in question (on which e.g. DNA or protein molecules, or also DNA-protein complexes are deposited). This tip is attached to a cantilever, which oscillates at high frequency (~300 kHz) above the surface. If the tip interacts with the sample, a reduction of the oscillation amplitude ensues, which is used to measure the topography of the surface.

***Taq*Man technique:** See → quantitative polymerase chain reaction.

Taq **polymerase:** See → *Thermus aquaticus* DNA polymerase.

Taquenase™: See → *Thermus aquaticus* DNA polymerase.

TAR: See → transformation-associated recombination cloning.

Target: A more general term for any molecule, cell, or organism under consideration. More specifically, any labeled RNA or cDNA hybridized to a → DNA chip is also called target, but should be called → probe. In → microarray experiments, a target may also mean the conventional probe, but may also be used in its original sense to describe the molecule spotted onto the chip. Even more misleading and confusing for such an RNA is the generic term "sample". To avoid further confusion, it is therefore strongly recommended to use the term "probe" as described (see → probe) and to restrict the term "target" for any molecule tethered to a microarray, that interacts (hybridizes) with a complementary probe. Amen.

Target capture oligonucleotide (TCO): Any oligonucleotide that serves to anchor a target sequence (e. g. a → cDNA) to a → universal capture probe in DNA capture assays.

Target DNA: Any DNA sequence within a genome that
a) can be isolated using a homologous → probe;
b) represents the binding site for hormones, drugs, hormone-receptor – or drug-receptor complexes, or other → DNA-binding proteins;
c) is cloned into a → cloning vector.

Targeted comparative sequencing: The estimation of the base sequences of sections of → genomes from many species, their alignment and the detection of orthologous, mostly non-coding sequences (see → conserved noncoding sequence). For example, the comparison of human sequences to orthologues in other primates will detect differences, but also similarities between them. Targeted comparative sequencing requires the establishment of → bacterial artificial chromosome (BAC) libraries and computer programs for sequence alignment and determination of the extent of matching. First ,BAC libraries are screened for highly conserved regions, the selected clones are sequenced, and the orthologous sequences analysed by software as e.g. PIPmaker (http://bio.cse.psu.edu). PipMaker compares two (or more) genomic sequences of 500 kb to 1.5 Mb in length and records the degree of identity down to the nucleotide level. See → percent identity plot.

Targeted display: A variant of the conventional differential → display technique for the identification and isolation of differentially expressed genes. In short, total RNA is first isolated, and then reverse transcribed into → cDNA using → oligo(dT) primers. The resulting cDNAs are then amplified in a conventional → polymerase chain reaction with specially designed socalled → targeted display primers, the produced fragments separated by → agarose gel electrophoresis and stained with → ethidium bromide. From the multiple banding pattern differentially expressed cDNAs are isolated, cloned and sequenced (and thereby characterized) or used as → probes in → Northern blotting or → RT-PCR analyses to verify the targeted display results. See → adapter-tagged competitive PCR, → enzymatic degrading subtraction, → gene expression fingerprinting, → gene expression screen, → linker capture subtraction, → module-shuffling primer PCR, → preferential amplification of coding sequences, → quantitative PCR, → two-dimensional gene expression fingerprinting. Compare → cDNA expression microarray, → massively parallel signature sequencing, → microarray, → serial analysis of gene expression.

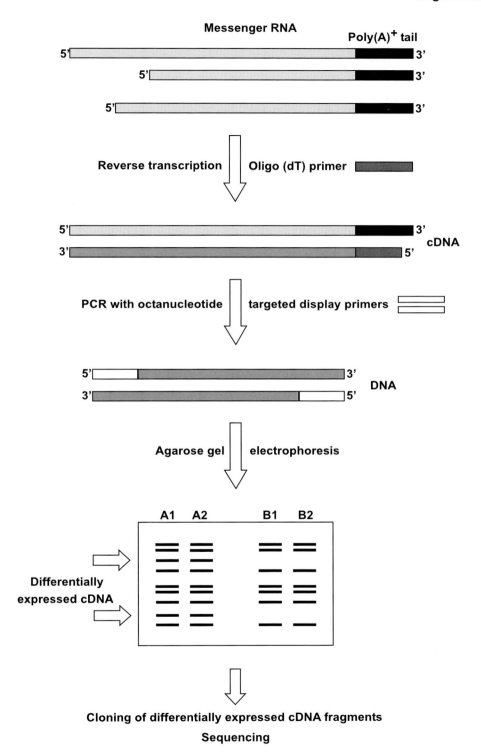

Targeted display

Targeted display polymerase chain reaction (targeted display PCR): A variant of the conventional → polymerase chain reaction, that represents part of the → targeted display technique and allows the amplification of differentially expressed → cDNAs, employing so called → targeted display primers.

Targeted display primer: Any one of a pair of 13bp → primers, designed from a group of about 30 octanucleotides significantly over-represented in most → cDNAs of human origin. Such octanucleotide sequences are mostly present in protein-coding regions (i.e. → exons) of sense DNA strands, and used to derive primer pairs in the → sense and → antisense orientation. Each primer consists of the octanucleotide and additional five base pairs to obtain the length of a → differential display primer. Targeted display primers are employed for → targeted display, and allow to target at least 70% of all mammalian genes.

Targeted gene transfer: See → gene targeting.

Targeted gene walking PCR: See → targeted gene walking polymerase chain reaction.

Targeted gene walking polymerase chain reaction (targeted gene walking PCR, gene walking): A variant of the conventional → polymerase chain reaction (PCR) for the amplification of unknown DNA sequences located → upstream or → downstream of known sequences. In short, first a sequence-specific PCR → primer (target primer, e.g. complementary to an → exon) and a second specific primer (internal primer, hybridizing to the same strand of DNA close to the target primer site) are annealed to their genomic target sites. Then nonspecific so-called walking primers, complementary to sequences either up- or downstream of the target and internal primerbinding sites, are used to amplify the intervening region by → *Taq* DNA polymerase. The technique can be used to isolate sequences adjacent to known regions without cloning and screening procedures and has potential for → DNA fingerprinting, the detection of regulatory regions and the isolation of genes.

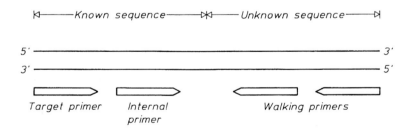

Targeted proteomics: The use of techniques of → proteomics to examine subsets of the → proteome, e. g. only proteins that interact with DNA, or only specifically modified proteins, or only proteins as parts of higher order complexes, or only membrane-bound proteins.

Target excess: The presence of very high → target concentrations on the surface of a → microarray such that the hybridizing probe is not sufficient to react with all target molecules. See → probe excess.

Targeting induced local lesions in genomes (TILLING): A → reverse genetics technique, that allows to introduce a high density of → point mutations into a genome by conventional chemical mutagenesis and a subsequent mutational screening to rapidly detect induced lesions. In short, the target organism is first mutagenized by → *ethylmethane sulfonate* (EMS), which primarily induces CG→TA transitions (of which about 50% are silent, and most of the rest are missense mutations). Then DNA is isolated, and the region of interest (e.g. a gene) is amplified by conventional → polymerase chain reaction techniques with region-specific → primers. The amplified products are denatured and allowed to reanneal to form → heteroduplexes, which are analysed by → *denaturing high pressure liquid chromatography* (DHPLC). DHPLC detects mismatches in heteroduplex molecules, that appear as extra peaks in the chromatogram. The mutant allele can then be sequenced. As an alternative for mutation detection, the plant endonuclease → CEL I from *celery* can be employed. This enzyme recognizes a mismatch and cleaves exactly at the 3'side of the mismatch. Therefore, cutting the mutated DNA by CEL I and subsequent DHPLC or high-resolution → polyacrylamide gel electrophoresis pinpoints the precise position of the mismatch. TILLING is used for → functional genomics (e.g. the proof of the function of a gene of interest by introducing EMS mutations into it).

Target of *unknown function* (TUF): Any DNA sequence of unknown function that is a target for mutation(s).

Target site: The specific sequence which is cleaved by a → restriction endonuclease. Compare → recognition site.

Target *site duplication* (TSD): The 5 bp host sequence element flanking the insertion site of a → transposon. TSDs are generated by staggered → cuts during the integration process.

TAR-RNA: See → *trans-activation response region* RNA.

TAS:
a) See → *telomere-associated sequences.*
b) See → *telomeric association.*
c) See → *transcription-based amplification system.*

TATA-binding protein: See → transcription factor II D.

TATA box (Goldberg-Hogness box, Hogness box): An AT-rich region with the → consensus sequence TATAT/AAT/A (plants: TATAAATA), most frequently located about 15-32 bp (in yeast: 60-120 bp) upstream of the → transcription initiation site of eukaryotic → structural genes. The TATA sequence is analogous to the prokaryotic → Pribnow box and represents the binding site for → transcription factors (e.g. → transcription factor II D). It is essential for accurate initiation of transcription, but is not necessary for quantitative expression. The → promoter of most → house-keeping genes does not contain a TATA box.

TATA box-binding protein: See → transcription factor II D.

TATA box binding protein-associated factor: See → transcription activation factor.

TAT peptide vector: A → vector for the transfer of foreign DNA into target cells, that combines the arginine-rich motif in the N-terminal region of the TAT protein of the AIDS virus HIV–1 with the cationic polymer *polyethylenimine* (PEI) and the DNA to be transferred. This DNA is first precondensed by the TAT peptide vector, then a solution of PEI is added, and the resulting ternary complex transferred into target cells. Since the TAT protein is a transcriptional transactivator, it is actively transported into the nucleus of the infected cell. Nuclear import is mediated by the arginine-rich motif at the N-terminus, which directly interacts with the nucleocytoplasmatic shuttle protein importin b. The presence of this motif (in some cases, as a dimer) facilitates nuclear import and enhances the efficiency of gene transfer.

TA under-representation: The occurrence of the dinucleotide 5'-TpA-3' in a genome at a significantly lower level than inferred from purely statistical distribution. The TA dinucleotide is underrepresented both in pro- and eukaryotes (exceptions: metazoan, fungal and plant → mitochondrial DNA and → chloroplast genomes). TA underrepresentation probably protects the DNA from unwinding and bending of the double helix at too many sites, and instead reduces the occurrences of TA to locations where it is functionally necessary (e.g. at → transcription initiation sites, → TATA boxes, → polyadenylation signals).

TBE: See → *t*ris-*b*orate-*E*DTA buffer.

T-box gene: Any one of a series of vertebrate and invertebrate genes encoding socalled T (T for tail)-box proteins, that contain a 229 amino acid T-box domain. This domain binds to a 20 bp partially palindromic sequence 5'-T(G/C)ACACCTAGGTGTGAAATT–3' in the minor groove of ist target DNA, acting as a → transcription factor. T-box genes are expressed at many stages of embryonic development and in various tissues (e.g. kidney, lungs, mammary gland, muscle, nerve system and skeleton) and the encoded proteins needed for e.g. control of gastrulation, heart development and extremity formation.

T-box protein: Any one of a series of proteins, that contain a highly conserved DNA binding domain, the socalled T-box and function as transcriptional regulators. For example, brachyury ("T protein") is involved in the formation and differentiation of the posterior mesoderm and the axial development of all vertebrates.

TBP-associated factor: See → transcription activation factor.

***Tbr* DNA polymerase:** See →*T*hermus *b*rockianus DNA polymerase.

TBS: See → *t*ransformation *b*ooster *s*equence.

T$_c$: See → *t*etracycline.

TC: See → *t*entative *c*onsensus sequence.

TCC: See → *t*ranscriptionally *c*ompetent *c*omplex.

T-circle: A circular covalently closed → T-DNA intermediate which is formed by joining its ends (the right and left 25bp → T-DNA border sequences). It is produced in → *Agrobacterium tumefaciens* cells after induction of the → Ti-plasmid → *vir* genes by plant phenolics, and represents one of the potential candidates for T-DNA transfer from the bacterium to the host plant. See → T-DNA processing, → T-complex.

TCO: See → *t*arget *c*apture *o*ligonucleotide.

T-complex (*transfer complex*): A stable complex between the → T-strand of the → *Agrobacterium tumefaciens* → Ti-plasmid and proteins encoded by the → *vir*-region genes *E* and *D*. The T-complex mediates T-DNA transfer from *Agrobacterium* to plant cells. Vir E2 proteins possibly protect the → T-strand from endonucleolytic attack during the transfer process, and vir D2 proteins guide the complex from the bacterium to the plant cell nucleus (pilot protein).

TCR:
a) See → transcription chain reaction.
b) See → *t*ranscription-*c*oupled *r*epair.

tc RNA: See → *t*ranslational *c*ontrol RNA.

Td: See → dissociation temperature, → midpoint dissociation temperature.

TDF: See → transcript-derived fragment.

T-DNA (*transferred DNA*): A part of the → Ti-plasmid in virulent → *Agrobacterium tumefaciens*, or the → Ri-plasmid in virulent → *Agrobacterium rhizogenes* that has been transferred from the bacterium to the nuclear genome of plant host cells. There it is stably integrated and expressed causing permanent proliferation of the host cell(s) into tumors (→ crown gall tumors, → hairy root disease). T-DNA is flanked by → direct repeats, the so-called → T-DNA borders. Proliferation is induced by the expression of genes 1, 2 and 4 of the T-DNA. Genes 1 and 2 encode enzymes that convert tryptophan to indoleacetamide (tryptophan monooxygenase) and indole acetamide to indole acetic acid (IAA; indoleacetamidehydrolase). Gene 4 codes for an isopentenyl transferase catalyzing the synthesis of the plant hormone cytokinin of the zeatin-type.
The T-DNA, integrated into the plant genome, differs in tumors induced by different *Agrobacterium tumefaciens* strains with regard to size and copy number as well as the genes encoding enzymes catalyzing the synthesis of different → opines. For example, in so-called nopaline tumors the T-DNA is present as contiguous stretch of about 23 kb encoding at least thirteen genes, among them a → nopaline synthase, that produces the abnormal amino acid → nopaline. In octopine tumors (producing the abnormal amino acid → octopine) the T-DNA is divided into a 12 kb → T_L-DNA (left) and a 7 kb → T_R-DNA (right) with an intervening plant sequence. The T_L-DNA is responsible for tumorigenesis and is transcribed into 8 transcripts, the T_R-DNA codes for 5 distinct transcripts. See also → T-DNA processing, → T-DNA-mediated gene fusion.
Figure see page 1116.

tDNA: See → *t*ransfer RNA genes.

T-DNA border: One of two nearly perfect 25 bp → direct repeat sequences flanking the → T-region of → *Agrobacterium tumefaciens* → Ti-plasmids. These borders are essential for the *vir*-induced excision of the → T-strand after contact of the bacterium with wounded plant cells, because they contain → recognition sites for a border-specific, *vir*-region-encoded endonuclease that nicks the bottom strand within the border.

Border sequence:

5'-TCTTTCTTTAGGTTTACCCGCCAATATATCCTGTC↑AAACACAACA-3'
(excision site)

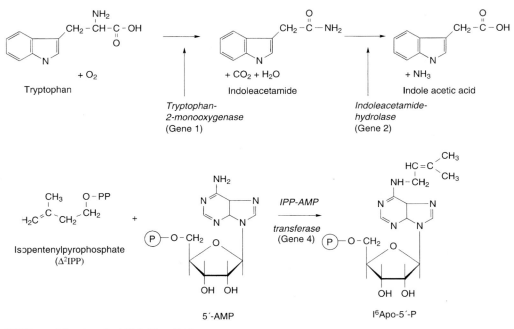

T-DNA encoded auxin and cytokinin biosynthesis

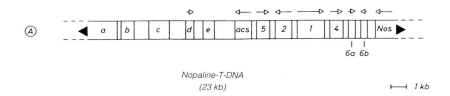

Nopaline-T-DNA
(23 kb)

⊢————⊣ *1 kb*

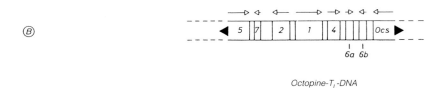

Octopine-T$_L$-DNA
(13.2 kb)

◀ *T-DNA Border* ▶

Sequence: $GGCAGGATATT_{GG}^{CA}G_{G}^{T}TGTAA_{TC}^{AT}$

acs : Agrocinopine synthase gene
Nos : Nopaline synthase gene
Ocs : Octopine synthase gene

Structure of nopaline (A) and octopine T-DNA (B)

T-DNA

Only the right border (RB) in its natural orientation seems to be essential for T-DNA transfer and possibly its integration into the host cell genome. Advanced plant transformation vector systems contain only the border sequences flanking foreign DNA of up to 50 kb which replaces the T-DNA.

T-DNA gene tagging: See → T-DNA tagging.

T-DNA insertion: The integration of the → T-strand of the *Agrobacterium tumefaciens* → Ti-plasmid or any foreign DNA flanked by the → T-DNA borders into the target plant genome. T-DNA insertion is probably catalysed by a plant recombinase and obviously occurs randomly (i.e. has no → hot spots of recombination).

T-DNA insertion mutant (T-DNA insertion line): Any plant, into whose nuclear genome a complete or truncated → T-DNA from the *Agrobacterium tumefaciens* → Ti-plasmid has been inserted. This insertion may be unique (i.e. at only one single locus), or several insertion events occur at different loci in the target genome. T-DNA insertion mutants of e.g. *Arabidopsis thaliana* are available, that exhibit a variety of different phenotypes. The genes underlying these phenotypes can be identified, since the T-DNA can be localized on genomic DNA by either hybridisation with labeled complementary → probes (see → Southern blotting) or by its conventional → polymerase chain reaction amplification, using primers complementary to the T-DNA. See → T-DNA insertion.

T-DNA-mediated gene fusion: The fusion of a transcriptionally and translationally incompetent (i.e. promoter-less) → reporter gene, located at the end of → T-DNA, to a promoter sequence in a plant genome, close to which the T-DNA inserts after → *Agrobacterium*-mediated gene transfer. The reporter gene is first separated from its promoter, cloned into a site at one end of the T-DNA of a specially designed T-DNA vector and transformed into → *Agrobacterium tumefaciens*. This organism is then used to transfer the T-DNA together with the promoter-less reporter gene into competent plant cells where it integrates at a high frequency into numerous locations within the plant's genome. If the initiation codon of the reporter gene is juxtaposed to plant promoter sequences, the reporter gene will be expressed. This procedure allows to screen plant genomes for promoter sequences. Compare → promoter trap vector.

T-DNA processing: The sum of modifications that a → T-strand molecule undergoes, after it has been specifically excised from the → T-region of the → *Agrobacterium tumefaciens* → Ti-plasmid following the induction of at least several *vir*-genes by plant phenolic compounds. One of these modifications is the packaging of the T-strand by vir E proteins into a → T-complex, another is the protection of the T-strand termini from exonuclease attack by vir D1 and vir D2 proteins. All these processing steps serve to protect and guide the T-strand during its transfer into recipient host plant cells.

T-DNA tagging (T-DNA gene tagging): A method to isolate a gene that has been mutated by the → insertion of a → T-DNA sequence. In short, T-DNA is integrated into the genomes of plant → protoplasts, the → transformants regenerated to complete plants, and these plants screened for mutant → phenotypes (e. g. a change in growth behaviour due to → loss-of-function or → gain-of-function of an interesting gene). Then a → genomic library is constructed from a T-DNA-induced mutant, and screened with a radiolabeled T-DNA as → probe. The T-DNA containing clones are sequenced, and the gene, into which the T-DNA inserted, can be isolated directly. See → gene tagging, → transposon tagging.

T-DNA transmission enhancer: See → overdrive.

TD-PCR: See → terminal deoxynucleotidyl transferase-dependent polymerase chain reaction.

TdT: See → terminal transferase.

TE: See → transposon.

TE buffer: A buffer used for suspending nucleic acids, especially DNA. It contains 10 mM Tris-HCl (pH 7.5) and 1 mM ethylene-diamine-tetraacetate (EDTA) as metal chelator. See → TSE buffer.

TEC: See → *t*ernary *e*longation *c*omplex.

tec MAAP: See → *t*emplate *e*ndonuclease *c*leavage *m*ultiple *a*rbitrary *a*mplicon *p*rofiling.

TEL-DNA: See → telomere.

Telobox: A conserved amino acid sequence motif at the C-terminus of → telomere-binding proteins from humans (e. g. orf1, orf2), animals, fungi (Tbf1) and plants (IBP1, BP F1). The telobox comprises 60 amino acids in a bipartite structure with a central region variable in length and sequence, and conserved terminal sequences (e. g. the C-terminal sequence H_2N-VDLKDKWRT-COOH). The telobox is part of the DNA-binding domain of the telomere-binding proteins and is essential for an interaction with the target DNA.

Telomerase (telomere terminal transferase): A *ribonucleo*protein (RNP), that catalyzes the addition of telomeric core sequences (e.g. TTAGGG or TTGGGG) onto the 3'end of single-stranded chromosomal → telomeres, thereby maintaining telomere length and stabilizing chromosome ends. *In vitro*, telomerase from the ciliate *Tetrahymena* elongates single-stranded guanosine-rich DNA primers by adding repeats of the T*etrahymena* type dT_2G_4. Moreover, this ribozyme possesses a 3'→ 5'nucleolytic and non-processive elongation activity. The nuclease activity eliminates mismatches between the DNA primer and the RNA template sequences. The 159 nucleotide *Tetrahymena* telomerase RNA contains a 9 nucleotide sequence complementary to 1.5 repeats of telomeric DNA, 5'-CAACCCCAA–3', which provides the template for the addition of dT_2G_4 repeats *in vitro* and *in vivo*. Compare → terminal transferase.
Figure see page 1119.

Telomerase-immortalized cell: Any cell line that divides indefinitely (is immortal), probably because it stably and strongly expresses the gene for → *t*elomerase *r*everse *t*ranscriptase (TERT). For example, a human *r*etinal *p*igment *e*pithelial (RPE) cell line, engineered to stably express *h*uman *t*elomerase *r*everse *t*ranscriptase (hTERT), does not age (i. e. still divides even after 300 mitoses). In contrast, the original wild-type RPE cell enters senescence after only 50 divisions. hTERT-RPE cells possess normal phenotype (i. e. primary cell morphology) and genotype (i. e. diploid → karyotype), and can be used for studies on aging, telomerase regulation, and the stability of → telomerases after knocking-out the telomerase gene.

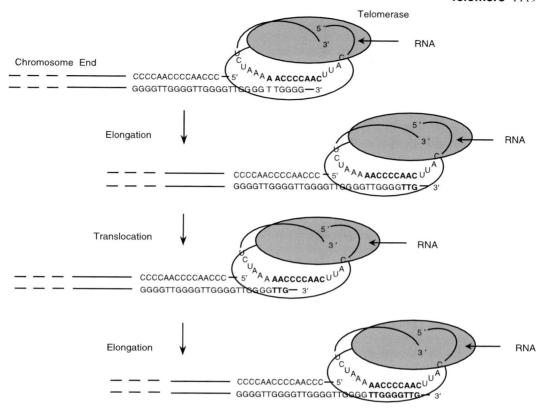

Telomerase

Telomerase PCR ELISA: A variant of the → *t*elomeric *r*epeat *a*mplification *p*rotocol (TRAP) for the measurement of → telomerase activity. In short, the telomerase adds → telomere sequences onto the 3'-end of a synthetic → biotinylated → primer (here P1-TS). The products are amplified in a conventional → polymerase chain reaction, using primer P1-TS and P2-TS (specific for telomeric repeats), and → *Taq* DNA polymerase. An aliquot of the amplification mixture is denatured and hybridized to a specific probe, that is complementary to the telomeric repeats and labeled with → digoxigenin. The resulting hybrid molecule binds to a → streptavidin-coated matrix (e.g. a microtiter plate) and can be detected by a → peroxidase-conjugated antibody directed against digoxigenin. The amount of colored product is directly proportional to the amount of hybridized digoxygenated telomere-specific probe.

Telomerase primer: Any → oligonucleotide, that can be extended by → telomerase. Typically, telomerase primers are telomeric DNA sequence elements.

Telomere (*tel*omere DNA, TEL-DNA, telomeric DNA): The end of a eukaryotic → chromosome, usually consisting of a terminal, densely stained protelomere with→ highly repetitive DNA sequences, and a subterminal, only weakly stained eutelomere. The protelomere part consists of short, tandemly repeated GC-rich sequences of up to 14bp in length with a core → consensus sequence (TTGGGG in *Tetrahymena*; TTAGGG in *Arabidopsis thaliana*; TTAGGG in *Trypanosoma brucei*; TTAGGG and TTCAGGG in *Plasmodium berghei*; TTGGGG and TTTGGG in *Paramecium tetraurelia*; TTTTGGGG in *Oxytricha* and *Stylonychia pustulata*;

TG_{1-8} inDictyostelium; $(TG)_{1-3}TGG(G)$ in yeast; TTAGGG in humans; general arrangement C_n $(A/T)_m$. Telomere length of the same chromosome is variable in different cellular stages (e.g. decreases in human fibroblasts during ageing; increases by 7-10 bp per generation in trypanosomes), and can range from less than 50 bp (*Euplotes*) to more than 100kb (mouse). Elongation of telomere sequences is catalyzed by the → telomerase, which binds specifically to telomere DNA. Telomeres serve principally three functions. First, they preserve a chromosome's linear integrity and block illegitimate recombination of chromosome ends (nontelomeric DNA ends are highly unstable and underly fusions with the ends of other, broken chromosomes, and are exonucleolytically degraded), prevent the loss of internal sequences, and generally chromosome degradation. Second, telomeres facilitate complete replication of the extreme end of chromosomes and interact with different proteins to suppress the expression of adjacent genes (→ telomeric position effect). Third, telomeres maintain the three-dimensional architecture of the chromosomes within the nucleus, and are involved in the attachment of chromosome ends to the nuclear membrane. Loss or mutation of telomeric regions result in high frequency of recombination, failure of chromosome segregation and/or cell division, and loss of expression of telomeric genes.

Organism	Telomere Sequence (5' → 3')	RNA Template Sequence (3' → 5')	RNA Length (Bases)
Tetrahymena	TTGGGG	CAACCCCAA	159
Euplotes	TTTTGGGG	CAAAACCCCAAAACC	190
Oxytricha	TTTTGGGG	CAAAACCCCAAAACC	190
Homo sapiens	TTAGGG	CUAACCCUAAC	450
Mus musculus	TTAGGG	CCUAACCCU	450
S. cerevisiae	TG(1–3)	CACCACACCACAC-AC	~1300
K. lactis	TTTGATTAGG-TATGTGGT-GTACGGA	UCAAAUCCGUACA CCAAUACCUAAU CAAA	~1300

Telomere-*associated* chromosome *fragmentation* (TAF, telomere-directed chromosome fragmentation): A technique for the production of mammalian → minichromosomes, that works with a cloned human telomeric fragment containing 500 bp of $(TTAGGG)_n$ repeats able to fragment mammalian chromosomes. In short, a → vector containing a correctly oriented stretch of this human terminal repeat sequence is first linearized and then electroporated into a Chinese hamster – human bybrid cell line. About 20 % of transfectants generated by random integration of the construct now possess a chromosome, in which one telomere is provided by the introduced DNA sequences. The input DNA is then elongated by several hundred base pairs, leading to cell lines that have a new small chromosome in their complement. TAF is based on the integration of the construct into a double-stranded break. The $(TTAGGG)_n$ sequences are recognized by → telomerase (in rare cases) and elongated, so that a new stable chromosome end is generated. The acentric fragmentation product segregates and is lost from the cell. See → mammalian artificial chromosome.

Telomere-associated sequence (TAS): Any DNA sequence located close to the → telomeres of eukaryotic → chromosomes. TAS basically fall into two broad categories: relatively long (several kb), conserved and moderately repetitive sequences (e.g. X and Y' elements of *Saccharomyces cerevisiae*; He-T elements of *Drosophila melanogaster*), and unique or moderately repeated sequences, frequently found only near telomeres, containing short sequences homologous to corresponding telomeric repeats.

Telomere-binding protein (TP, telomere repeat binding factor, TRF): A protein that binds to → telomeres. It is encoded by a TP gene and serves to suppress unspecific → transcription initiation of this gene.

Telomere capping: The process of assembling a series of specific proteins at the end of chromosomes to form a nucleoprotein complex that protects the → telomere from being degraded. The tandemly reiterated telomeric DNA repeats attract and bind a set of sequence-specific DNA-binding proteins, amongst them → telomerase itself, and other structural and catalytic proteins (e.g. TRF2, Rap1, Ku, and others). Decapping (uncapping) of the telomeres is sensed as DNA damage, and induces DNA repair systems and a re-formation of a telomere cap.

Telomere-directed chromosome fragmentation: See → telomere-associated chromosome fragmentation.

Telomere DNA (TEL-DNA, telomeric DNA): The highly conserved DNA sequence of a → telomere.

Telomere erosion: The loss of telomeric repeats from the end of chromosomes in successive rounds of DNA replication, caused by the inability of DNA polymerase to replicate the ends of the chromosomal DNA. Telomere erosion leads to continuous telomere shortening, unless the cells reach a second proliferative block characterized by chromosomal instability, end-to-end fusion and cell death.

Telomere loop (T loop): A 10-20 kb double-stranded DNA lariat structure at the → telomeres of mammalian chromosomes, formed by looping back of the telomeric DNA on itself. Thereby the 3' G-strand extension invades the duplex telomere repeats, creates a → displacement loop and is thus protected from exonuclease attack. Duplex DNA telomere-binding proteins bind along the T loop. Protein TRF2 stabilizes D-loop formation by binding to the junction of the lariat. *Figure see page 1122.*

Telomere map: The graphical depiction of the order of sequences at the → telomeres of a chromosome. Usually this region is missing in → genetic and → physical maps, because a series of difficult-to-sequence elements are crowded at the termini of the chromosomes (as e.g. → microsatellites). A telomere map is the result of → telomere mapping.

Telomere mapping: The localization of → telomeres on the ends of chromosomes by e.g. → fluorescent *in situ* hybridisation, and the identifation of other sequence elements within the telomeric (and also subtelomeric) region. See → centromere mapping, → telomere map.

Telomere repeat binding factor (TRF): See → telomere-binding protein.

Telomere terminal transferase: See → telomerase.

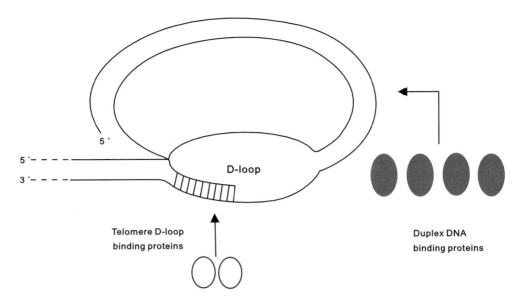

5´

5´- - - -
3´- - - -

D-loop

Telomere D-loop
binding proteins

Duplex DNA
binding proteins

Telomere loop

Telomere transposon: Any one of a series of → transposons, that are an integral part of → telomeres of the chromosomes of *Drosophila melanogaster*, and replace the conventional telomeric repeats. Telomere transposons resemble → retrotransposons (i.e. have poly(A)-stretches at their 3'end, and internal sequences with homology to *gag* regions of retrotransposable elements (and → retroviruses). One of these transposons, the 6kb HeT-A element carries two overlapping → open reading frames (ORFs) flanked by a 5'- and 3'-noncoding region, and a $[dA/dT]^n$ stretch joining each element to the proximal part of the chromosome. Neither ORF encodes a → reverse transcriptase, so that this function must be supplied in trans. Another telomeric transposon, the 10kb, less abundant TART has a similar structure. Both elements form long chains of telomeric repeats. Telomere transposons obviously transpose only to chromosome ends and are responsible for telomere maintenance.

Telomeric association (TAS): The fusion of → telomeres of different chromosomes, representing an → illegitimate chromosomal recombination, and occurring in higher frequency in e. g. tumors (e. g. giant cell bone tumors).

Telomeric DNA: See → telomere.

Telomeric gene: Any gene that is located close to or within the → telomeric repeat region of eukaryotic chromosomes. Telomeric genes probably underly an accellerated mutation rate as compared to genes of other chromosomal regions. See → telomeric transposon.

Telomeric position effect (telomeric silencing): The suppression of the → transcription of otherwise active genes that have been brought into close contact to → telomeres by chromosomal rearrangements or transposition processes. Possibly the telomeres impose an altered → chromatin structure (heterochromatinization) onto the transposed genes, so that they are no longer accessible for regulatory proteins. This change in chromatin conformation is mediated by telomere-binding protein RAP 1 (*r*epressor *a*ctivator *p*rotein 1) and various SIR (*s*ilent *i*nformation *r*egulatory) proteins.

Telomeric *r*epeat *a*mplification *p*rotocol (TRAP): A two-step procedure to detect → telomerase activity in biological samples. In short, the first step involves mixing a cell, tissue or organ homogenate with a synthetic → oligonucleotide containing the → telomeric repeat 5'-TTAGGG-3' (T_S oligonucleotide). The telomerase recognizes this T_S template, and adds further telomeric repeats onto its terminus. This procedure leads to the labeling of the product, since one of the nucleotide-5'-phosphates carried e. g. ^{32}P. For the second step, an oligonucleotide complementary to the telomeric repeat ("CX" oligonucleotide) is then added, and the extended synthetic oligonucleotides amplified in a conventional → polymerase chain reaction. The lengths of the amplification products vary, depending on the number of repeat units incorporated, and on the different phases of a cell'-s life cycle, e.g. is short in normal and long in mitotically active tumor cells. These products are electrophoresed in sequencing gels. The resulting → autoradiograph shows a characteristic DNA ladder, where the bands are 6 bp (telomeric repeat length) apart from each other. See → telomerase PCR ELISA.

Telomics: Another term of the → omics era for a technique to monitor a whole series of metabolic responses of tissues or tissue slices (lat. tela: tissue) to environmental parameters. For example, a specific tissue (e.g. hippocampal tissue slice of 350 μm thickness, consisting of neurons, interneurons and various types of glial cells still capable of synaptic transmission) is cultured and exposed to conditions mimicking a certain disease or otherwise abnormal situation (e.g. an ischemic shock may be simulated by a transient oxygen and glucose deprivation). During or after the exposure a series of metabolic parameters can be measured, as e.g. apoptotic cell death by → propidium iodide fluorescence exclusively in damaged cells, or gene expression by → RT-PCR, the localization, modification and function of proteins by → immunohistochemistry, → Western blotting and activity measurements, and others. Since such tissue slices retain complex intercellular networks, they reflect the potential of *in situ* organs much better than single cells or cell extracts.

Temperate bacteriophage (temperate phage): A → bacteriophage that has the option of existing either in the lytic or the lysogenic state. In the lysogenic state the phage (→ prophage) is integrated into the bacterial chromosome (see → lysogenic bacteriophage, → lysogeny). In the lytic state phage particles are released by cell → lysis after their replication (see → lytic cycle). Compare → virulent bacteriophage.

Temperature *g*radient *g*el *e*lectrophoresis (TGGE): An electrophoretic technique for the analysis of conformational transitions and sequence variations of DNA and RNA, and protein-nucleic acid interactions. The TGGE combines → gel electrophoresis (separation of macromolecules by size, charge and conformation) with a superimposed temperature gradient perpendicular to the electrical field (separation by differing thermal stabilities of the different macromolecules). Compare → temperature sweep gel electrophoresis.
Figure see page 1124.

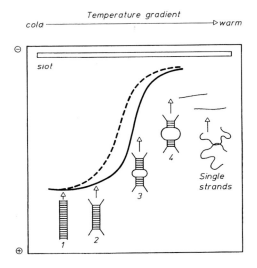

Temperature gradient

cold ─────────────────────→ warm

1 : Fully base-paired duplex migrates fastest at low temperatures
2,3,4 : Local melting of base pairs causes slow-down of electrophoretic migration
--- : Shift of profile as a consequence of altered stability of the duplex (caused by e.g.a mutation)

Temperature *modulated heteroduplex analysis* (TMHA): A technique to scan → genomes for the presence of mutations (as e. g. → single nucleotide polymorphisms), generally sequence → polymorphisms. In short, wild-type and mutant DNA are first amplified in a → touchdown polymerase chain reaction with → *Pfu* DNA polymerase (thus minimizing → misextension by proofreading), using two → primers flanking the region of interest. DNA polymerase is inactivated by EDTA and heating to 95° C. The mixture is then allowed to cool to 25° C over one hour. During this time, a mixture of heteroduplexes (harbouring base mismatches) and homoduplexes is formed that can be resolved by denaturing high performance → capillary electrophoresis as a function of temperature. For example, under non-denaturing conditions (such as necessary for separating DNA fragments), homo- and heteroduplexes have identical retention times. As the temperature increases, the heteroduplexes start to denature in the region flanking the mismatched bases and emerge as separate peaks ahead of the still intact homoduplexes. At still higher temperatures, the homoduplexes also begin to denature, with the A-T wild type homoduplex being earlier denatured than the G-C homoduplex ("temperature modulation"). In TMHA, DNA from homozygous wild-type individuals forms one homoduplex species.
Figure see page 1125.

Temperature-sensitive mutant (*ts* mutant): Any → mutant with an upper limit of temperature tolerance, which is lower than that of the wild-type organism (→ heat-sensitive mutant) or, alternatively, which is inactivated by lower temperatures (→ cold-sensitive mutant). See → temperature-sensitive mutation.

Temperature-sensitive mutation (*ts* mutation): Any conditional lethal → mutation which becomes apparent only above (or below) a certain temperature threshold. The gene affected by such a mutation encodes a protein that is unstable above (or below) a certain temperature. The mutant therefore behaves like the wild type at permissive temperatures but exhibits the mutant phenotype at non-permissive (restrictive) temperature. See → cold-sensitive and → heat-sensitive mutation.

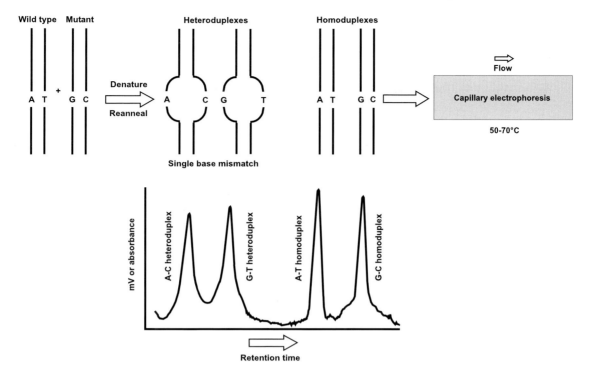

Temperature-modulated heteroduplex analysis

Temperature sweep gel electrophoresis (TSGE): A variant of the → temperature gradient gel electrophoresis (TGGE) technique for the detection of → point mutations in DNA. In contrast to TGGE, the temperature of the gel plate is raised gradually and uniformly from 45 to 63°C during electrophoresis. TSGE is based on the same principles as TGGE which exploits the decrease in electrophoretic mobility of a DNA fragment that occurs, if localized regions begin to melt. The melting point of such regions is shifted by a point mutation so that single base substitutions in these first-denatured regions can be detected.

Template: See → template strand.

Template endonuclease cleavage multiple arbitrary amplicon profiling (tecMAAP): A → MAAP technique that allows the amplification of template nucleic acids restricted with one or more → restriction endonucleases.

Template strand (template):
a) Any nucleic acid strand that serves as a mold for the synthesis of a complementary nucleic acid copy strand (e.g. a DNA strand during → replication of DNA; a DNA strand during → transcription of RNA; an RNA strand during the → reverse transcription of DNA; an RNA strand during replication of RNA).
b) See → antisense strand.

Template switch: The change of → DNA-dependent DNA polymerase from one → template strand to another one, such that the polymerization product does not only contain sequences from the original template. Template switching is usually accompanied by → deletions or re-arrangements.

Temporal expression pattern: The differential expression of multiple genes of an organism as a function of time. This pattern changes continuously in response to internal and external stimuli. See → spatial expression pattern.

Temporal gene expression: Any → transcription of a gene and the → translation of the resulting → messenger RNA into a protein, that is restricted to only a limited time period (e.g. in the development of an organism).

Temporal temperature gradient gel electrophoresis (TTGE): A variant of the conventional → *t*emperature gradient *g*el *e*lectrophoresis (TGGE), that allows to analyze conformational transitions and sequence variations of DNA and RNA, and protein-nucleic acid interactions and to separate DNA fragments differing from each other by → point mutations. TTGE is based on the same principle as e. g. → constant denaturing gel electrophoresis and → denaturating gradient gel electrophoresis (described there), but does not require a chemical denaturant gradient. In short, pre-amplified wild-type and mutant DNA is loaded onto a → polyacrylamide gel containing a constant concentration of 6 M urea. During electrophoresis, the temperature is gradually and uniformly increased (from 62°-68°C), and hence a linear temperature gradient accompanies the entire electrophoresis run. The denaturing environment in TTGE then is formed by a constant denaturant concentration in the gel, combined with a temporal temperature gradient, resulting in good separation of wild-type and mutant sequences differing by → single nucleotide mutations. See → constant denaturing capillary electrophoresis, → constant denaturing gel electrophoresis.

Tentacle resin: A methacrylate resin with bound polyelectrolyte chains ("tentacles"), that are flexible, hydrophilic backbones for the attachment of ion exchange groups, affinity ligands, hydrophobic moieties or other functional groups for → affinity chromatography or other chromatographic techniques. The flexibility of the tentacles exposes more functional groups for the binding of ligands, resulting in an increased resin capacity.

Tentative consensus sequence (TC): A unique virtual transcript, derived from comprehensive → *e*xpressed *s*equence *t*ag (EST) databanks by clustering of the sequences and assembly of cluster elements at high stringency (i.e. removal of low quality, misclustered or chimeric sequences). TCs are generally longer than the individual ESTs, that comprise them, so that TCs can be used more efficiently for functional annotation. See → tentative human consensus.

Tentative human consensus (THC): A → consensus sequence for each putative protein derived from potential protein-coding regions deposited in the databases. Sequences are first grouped together, if they contain at least 40 bases with greater than 95% identity. Then the groups are assembled to generate a THC, and discordant sequences are eliminated.

Terminal deletion: See → deletion.

Terminal deoxynucleotidyl transferase: See → terminal transferase.

Terminal deoxynucleotidyl transferase-dependent polymerase chain reaction (TD-PCR): A variant of the conventional → polymerase chain reaction for the detection of sequence specific DNA damaging by e.g. → cisplatin or DNA adduct formation by UV light, that starts with the isolation of → genomic DNA from a control and treated organism, and the specific linear amplification of a target region (e.g. a gene) with → gene-specific primers. The amplified products are then mixed with → terminal deoxynucleotidyl transferase and rGTP, riboG-tailed and precipitated with ethanol. After rehydration, → linker oligonucleotides are ligated to the ends of the fragments with → T4 DNA ligase, and the linkered DNA molecules amplified with a → primer complementary to the linker and a ^{32}P-labeled → nested gene-specific primer. The labeled fragments are then separated in → sequencing gels and the various bands detected by → autoradiography.

Terminal deoxynucleotidyltransferase-mediated d*U*TP *n*ick-end *l*abeling (TUNEL) assay (*in situ* end *l*abeling, ISEL): A technique for the *in situ* detection of fragmented DNA in apoptotic cells or tissue sections, that uses → *t*erminal *d*eoxynucleotidyl *t*ransferase (TdT) catalyzing the repeated addition of dUTP onto the 3'-OH termini of the DNA fragments to form oligomeric tails. If the dUTP is → fluorescein-labeled (as fluorescein-12-dUTP), it serves as reporter to visualize apoptotic DNA directly by fluorescence microscopy or to quantitate it by → flow cytometry (direct assay). The fragments can also be labeled by → biotinylated dUTP and detected by → streptavidin-coupled fluorochromes, or colorimetrically by streptavidin-coupled → alkaline phosphatase or peroxidase (indirect assay).

Terminal exon: Any → exon that is located at the 3'-terminus of a → multi-exonic gene, preceding the → stop codon or → polyadenylation signal. See → initial exon, → internal exon.

Terminal extendase: See → extendase.

Terminal *i*nverted *r*epeat (TIR; inverted terminal repeat, ITR, ITS): Sequence motifs that flank → transposons and are identical or partly identical and present in inverse orientations. They function as → recognition sites for the excision of the transposon. For example, the → *en*hancer (En) transposition system of *Zea mays* is flanked by a 180 bp terminal region at its 5' end and a 300 bp terminus at its 3' end. These termini consist of several domains (e.g. one 13 bp perfect TIR, and 12 bp sequence motifs present as direct and inverse repetitions; both elements are *cis*-determinants for transposition). See also → fold-back element; compare → inverted repeat.

Terminal loop: Any RNA sequence that folds back on itself and forms a base-paired stem and a non-base-paired loop. See → internal loop.

Terminal redundancy: The occurrence of homopolymeric or redundant sequence elements at both ends of genomic DNA (e.g. a → chromosome).

Terminal repetition: The occurrence of a → direct or → inverted repeat sequence at both ends of a genomic DNA. See for example → terminal inverted repeat.

Terminal *r*estriction *f*ragment (TRF): Any DNA fragment, generated by → restriction of → genomic DNA with one or more → restriction endonucleases, that contains the → telomere array (TTAGGG)$_n$ and a portion of the subtelomeric DNA.

Terminal *r*estriction *f*ragment *l*ength *p*olymorphism (terminal RFLP, T-RFLP): Any variation(s) in the length of DNA fragments produced by a specific → restriction endonuclease from the terminal sequences of → ribosomal DNA or the 16S rDNA genes of two or more individuals of a species. In short, genomic DNA is isolated, and ribosomal DNA amplified with primers an chored at both termini of the rDNA transcription unit. One of these primers is labeled with a → fluorochrome (e. g. carboxyfluorescein or → rhodamine). The rDNA amplicon (bacteria: 1.5 kb) is then digested with a suitable → four-base cutter restriction enzyme, producing restriction fragments that can be separated by → polyacrylamide gel or → capillary electrophoresis. After their separation only those fragments are detected which are labeled at their termini: terminal restriction fragments, T-RFLPs. The number of detectable restriction fragments can be doubled, if the second primer is labeled with a different fluorochrome. Since the distribution of restriction sites is different in the rDNAs from different organisms, different T-RFLP fingerprints are generated that allow the discrimination of the different samples.

Terminal RFLP: See → terminal *r*estriction *f*ragment *l*ength *p*olymorphism.

Terminal transferase (*t*erminal *d*eoxynucleotidyl *t*ransferase, TdT; deoxynucleotidyl transferase; EC 2.7.7.31): An enzyme, usually prepared from calf thymus, catalyzing the template-independent addition of about 10-40 deoxynucleotide-5'-triphosphates to the 3'-OH groups of both termini of duplex DNA, or one terminus of single-stranded DNA acceptor molecules, generating 3'-homopolymer extensions (e.g. polydeoxyadenylate). The enzyme is used for

a) the radioactive labeling of DNA molecules. In this case the enzyme is provided with an α-^{32}P labeled deoxyribonucleotide which is transferred to the acceptor DNA molecule.

b) the → homopolymer tailing of 3' ends of DNA duplex molecules. In this case, one duplex (e.g. a cloning vector) is linearized at a single site, incubated with terminal transferase and tailed with only one kind of deoxyribonucleotide (e.g. GTP). Another duplex (e.g. a DNA to be cloned into the vector) is treated in the same way except that it is tailed with the complementary nucleotide (e.g. CTP). The single-stranded tails are then annealed to each other, and the → gaps are ligated in vitro (by → DNA ligase) or in vivo (by the host bacterium after its transformation). Terminal transferase therefore serves as a tool in → recombinant DNA technology. Compare → telomerase, → DNA polymerase.

Termination:

a) The dissociation of the ternary → elongation complex into → messenger RNA, DNA → template, and → RNA polymerase with concomitant stop of → transcription. In bacteria, → termination sequences are usually composed of an inverted repeat followed by an oligo(T) sequence, so that in RNA they appear as a GC-rich hairpin and a run of several uridine residues at the 3'-terminus. The termination sequence determines the efficiency of termination:

5'-TTTTTATA–3'	Effective
5'-TATATA–3'	Very effective
5'-TACATA–3'	Less effective
5'-TAGTAGTA–3'	Less effective
5'-TAGATATATATGTAA–3'	Less effective
5'-TTTTTTTATA–3'	Little effect

Reverse orientation: No function

b) The dissociation of the newly synthesized peptide chain, the peptidyl-tRNA, messenger RNA and the ribosomal subunits, catalyzed by socalled → termination factors that recognize and bind to one (or more) → stop codons (e.g. UAG, UAA and/or UGA). This process terminates the synthesis of the protein.

Termination analogue: Any 3' modified deoxynucleotide triphosphate, terminating DNA synthesis catalyzed by → DNA-dependent DNA polymerase, that is chemically different from the conventional → dideoxynucleoside-triphosphate (ddNTPs) used in → Sanger sequencing. Such termination analogs are e. g. 3'-0-(2-nitrobenzyl)-dATP, 3'-0-methyl-dATP, or 3'-0-methyl-dTTP and currently employed in → base addition sequencing scheme.

Termination codon: See → stop codon.

Termination factor (*release factor,* RF): A protein that recognizes a → termination codon (e.g. UAG, UAA, and/or UGA) and separates the polypeptide chain, → messenger RNA and → ribosome from each other. This leads to the termination of protein synthesis. For example, RF1 (323 amino acids) recognizes the UAA and UAG, RF2 (329 amino acids) the UAA and UGA stop codons.

Termination sequence (termination site, terminator sequence; terminator):
a) A DNA sequence at the end of a → transcription unit which signals the end of transcription. See also → rho factor-dependent and -independent termination.
b) A DNA sequence downstream of the coding region of pro- and eukaryotic genes that functions as a signal for RNA polymerase to stop transcription.
c) The sequence element 5'-$(GC)_{10}$-$(N)_5$-$(GC)_{10}$-$(A)_{4-8}$-U-3' in → messenger RNAs that is able to form a → stem- and loop structure and serves as a signal for → RNA polymerase to terminate transcription.

Termination signal: A specific sequence element, at which → DNA-dependent RNA polymerase dissociates from the DNA template. See → arrest signal, → pause signal.

Termination site: See → termination sequence.

Terminator:
a) See → dideoxynucleoside triphosphate.
b) See → stop codon.
c) See → termination sequence.

Terminator gene: Any → gene that prevents normal embryo development in higher plants, and can be used to produce embryo-less seeds. Farmers would then be forced to purchase such seeds each year from agrobiotechnology companies that produce them. The terminator technology comprises three genes. Gene I is a repressor gene (encoding a repressor protein), gene II a → recombinase gene with its promoter, that is separated from the gene by an address site for the repressor, and gene III a toxin gene controlled by a promoter ("*late* promoter", LP), that is induced during the later stages of seed development (i. e. the time of embryo formation). Between the late promoter and the toxin gene, a silencing sequence element (the "blocker") is inserted, that prevents an activation of the promoter. Under non-inducing conditions, the repressor binds to its address site and the recombinase gene is silenced. If an inducer (structure not disclosed by companies) is added, it interferes with the interaction of the repressor and its cognate sequence, and the recombinase gene is turned on, producing recombinase protein. Recombinase cuts out the blocker sequence, the LP is activated late in the season, the toxin gene transcribed. Toxin is synthesized and kills the embryo, before the mature seeds are harvested. See → killer gene.

Terminator probe plasmid: A → plasmid → expression vector that contains a single cloning site or a → polylinker 3' downstream of the coding region of a → selectable marker gene. Any DNA segment that is cloned into the cloning site and contains a → termination sequence will lead to → transcription termination. Using a terminator probe vector, the termination sequence and its function can be studied in detail.

Terminator-probe vector (transcription-terminator vector): A bacterial cloning vector that allows the detection of sequences functioning as transcription terminators. Such vectors are designed such that a → selectable marker gene (e.g. a → tetracycline or → chloramphenicol resistance gene) and its promoter are separated by a cloning site so that DNA fragments can be inserted. If such an insert carries terminator sequences then the transcription of the gene beyond the terminator will be stopped (or reduced), leading to a change in the phenotype of the host bacterium. The efficiency of a terminator can also be quantitatively determined, if a gene is included that encodes an easily assayable product (e.g. the galactokinase [gal K] gene), especially in gal K⁻ hosts. In some transcription-terminator vectors two marker genes are placed in tandem, and separated by the cloning site(s). This type of vector is superior to the vector containing only a single marker, because the expression of the first gene represents an internal control for the efficiency of the promoter. If both genes encode assayable products, the ratio of the expression of the second versus the first gene will be a measure for the termination efficiency of the cloned DNA fragment.

Terminator sequence (terminator): See → termination sequence.

Ternary elongation complex (TEC): The relatively stable complex of core → RNA polymerase, the DNA → template and the RNA → transcript.

Terrific broth: A buffered growth medium for the propagation of recombinant *E. coli* strains. It contains 12 g tryptone, 24 g yeast extract, 2.3 g KH_2PO_4 and 12.5 g K_2HPO_4 per liter.

TET: A derivative of the → fluorochrome → fluorescein, a 5–*tet*rachlor-fluorescein, that is used as a marker for → fluorescent primers in e. g. automated sequencing procedures or for labeling in → DNA chip technology. The molecule can be excited by light of 521 nm wave-length, and emits fluorescence light at 538 nm. Since the wave-length of the excitation and emission maxima is pH-dependent, the exact values vary.

TET

Tethering:
a) A process that leads to the approach of two (or more) DNA sequences, that are remote from each other on a linear scale. Tethering is important for e. g. → chromatin structure, → looped domain formation, → enhancer function, → lariat formation.
b) A technique for the identification of low molecular weight ligands that bind with low affinity to specifically modified sites in target proteins. In short, a ligand library consisting of e.g. small compounds (about 250 Da), each one containing an artificially introduced disulfide, is first established by conventional chemistry ("disulfide library"). The purified ligands are then reacted with the target protein under partially reducing conditions that promote rapid thiol exchange (e.g. by the presence of 2-mercaptoethanol). If this protein contains a native or artificially introduced cysteine, a covalent disulfide bond ("disulfide tether") will be formed between ligand and protein, but only if the ligand has even weak affinity for the protein. All the other thousand or hundreds of thousands of ligands that do not possess any affinity to the target, will be removed. The weakly binding ligands can be rapidly optimised by structure-aided design such that the novel ligands bind with much stronger affinity than the precursor compound. The tethered complex can be analysed by e.g.mass spectrometry and/or X-ray crystallography.

Tet NR (*tetranucleotide repeat*): Any tetranucleotide motif (e.g. [GATA]$_n$, [GACA]$_n$, [GTGT]$_n$), that is tandemly reiterated in a → genome (e.g. 5'-GATAGATAGATAGATAGATAGATA-3'). See → microsatellite, → repeat.

Tet-on/tet-off gene expression system (tet-on/tet-off system): A two-plasmid, high-level expression system for mammalian cells, which allows to tune (regulate) the expression of cloned genes by varying the concentration of → *tet*racycline (Tc) or derivatives (such as the more frequently used *dox*ycyclin (Dox). The tet-off version responds to a removal of Tc (or Dox) from the culture medium with activation of the cloned gene. In contrast, gene expression is turned on in the tet-on version upon the addition of the antibiotic. In short, the system is based on two regulatory elements derived from the tetracycline-resistance → operon of the *E-coli* → transposon 10, namely the gene encoding the *tet*racycline *r*epressor protein (Tet R) and the 42 bp *tet*racycline *o*perator sequence (tetO) to which Tet R binds. These two elements are cloned separately into two plasmid vectors, the response plasmid and the regulator plasmid, respectively. The 3.1 kb **response plasmid** contains a → *tet*-responsive *e*lement (TRE) containing seven copies of tetO, driven by a minimal *cyto*megalo*v*irus (CMV) promoter, a *m*ultiple *c*loning *s*ite (MCS) immediately downstream, and an SV 40 poly(A) site at its 3' end. The target gene (whose expression is desired) is cloned into this MCS. The 7.4 kb **regulator plasmid** (pTet-off or pTet-on) harbors the strong immediate early promoter of CMV upstream of a strong *tet*racycline-controlled *t*ranscriptional *a*ctivator (tTA) construct, consisting of a → fusion between the Tet repressor gene and the VP 16 *a*ctivation *d*omain (AD) of herpes simplex virus. The transcribed → fusion protein acts as a tetracycline-responsive transcriptional activator. Both plasmids are transfected into target cells, stably integrated into the host cell genome, and the tTA expressed. The resulting tTA fusion protein binds to the tetO sequences on the response plasmid (pTRE), the strength of binding being dependent on the concentration of the tetracycline. Now, in the tet-off system the fusion protein activates transcription of the cloned gene (in the absence of tetracycline). If tetracycline is added to the culture medium, transcription of the gene is turned off in a dose-dependent kinetics. In the tet-on system, the fusion protein contains a socalled "reverse *TetR* " (rTetR), a mutant tetracycline repressor protein, that differs from its wild-type precursor by four amino acid substitutions. The resulting mutant rtTA binds to its target sequence, but is turned on in the presence of tetracycline and turned off in its absence. Variations of the system include the use of a → dual promoter (that allows the simultaneous expression of two different genes under control of a single TRE), or the expression of a tetracycline-controlled transcriptional → silencer (tTS), e.g. the KRAB-AB domain of the Kid-1 protein (where the fusion protein binds to the tetO sequences in the tetracycline-responsive element on the response plasmid and prevents gene expression. Addition of tetracycline activates this system). Tet-on/tet-off systems allow the stable transformation of mammalian cells with genes encoding toxic peptides or proteins, whose expression can be modulated by the addition or removal of tetracycline.

Tet-on/tet-off system: See → tet-on/tet-off-gene expression system.

Tet^r: See → *tet*racycline *r*esistance.

Tetra-allelic *single nucleotide polymorphism* (tetra-allelic SNP): Any → single nucleotide polymorphism, of which four different → alleles are present in a population. See → A-allele, → C-allele, → diallelic single nucleotide polymorphisms, → G-allele, → T-allele → tri-allelic single nucleotide polymorphism.

Tetracycline (Tc; amethocaine): A member of a group of broad-spectrum → naphthalene antibiotics from *Streptomyces aureofaciens* (chlorotetracycline), *Streptomyces rimosus* (oxytetracycline), and other *Streptomyces* species. Tetracyclines bind to a protein of the 30S subunit of prokaryotic ribosomes and prevent the binding of aminoacyl tRNA to the ribosomal A site, but also affect the P-site, polypeptide initiation and chain termination. See → tetracycline resistance, → tetracycline sensitivity.

Generic	R^1	R^2	R^3
Tetracycline	–H	–H	–H
Rolitetracycline	–H	–H	– CH$_2$ –N
Oxytetracycline	–H	–OH	–H
Chlorotetracycline	–Cl	–H	–H

Tetracycline resistance (tet^r): The ability of a bacterium to grow in the presence of the → antibiotic → tetracycline. The tetracycline resistance genes (*tet*) are plasmid-borne and contained within → transposons and amplifiable DNA sequences. For example, the *tetA* genes are related to → RP1 plasmid and transposon 1721, the *tetB* genes to → transposon 10 and the *tetC* genes to → pSC 101 and its derivative → pBR 322. Tn 10 specifies a repressor as well as a structural tet protein that is incorporated into the cytoplasmic membrane and functions as carrier to mediate active efflux of the antibiotic.

Tetracycline sensitivity (tet^s): The inability of a bacterium to grow in the presence of the → antibiotic → tetracycline. Compare → tetracycline resistance.

Tetraloop: Any 4 nucleotide terminal loop structure in RNAs (e.g. → ribosomal RNAs). Principally, three classes of tetraloops exist, based on the terminal loop motif: the UNCG, GNRA and CUYG categories.

Tetramethylrhodamine: A fluorescent dye ($\rightarrow$ fluorochrome) that is used as a marker for $\rightarrow$ fluorescent primers in e.g. $\rightarrow$ automated DNA sequencing procedures. The fluorochrome is exited by light of a wave-length of 550 nm, and emits fluorescence light at 570 nm.

Tetramethylrhodamine

Tetranucleotide repeat: See $\rightarrow$ Tet NR.

Tetra-octa sequencing: A technique to estimate the base sequence in DNA, using a special $\rightarrow$ phagemid vector with a $\rightarrow$ *multiple cloning site* (MCS) containing the octanucleotide $\rightarrow$ recognition sites for several $\rightarrow$ rare-cutting $\rightarrow$ restriction endonucleases, which allow to linearize the recombinant vector without previous $\rightarrow$ restriction mapping. The phagemid vector does not carry an *Rsa*I site (four base cutter). Therefore, $\rightarrow$ partial digestion of the insert is possible without destruction of the vector sequence, yielding a series of truncated inserts coupled to the vector, which are then circularized. After determination of the respective insert sizes in different subclones, the termination reactions for each subclone (see $\rightarrow$ Sanger sequencing) are separated on $\rightarrow$ sequencing gels, where the samples are arranged with decreasing insert size. Using the *Rsa*I site as marker, the overlapping regions of the target DNA can be inferred by nucleotide sequence reading from the corresponding $\rightarrow$ autoradiograph.

Tetraplex: Any four-stranded (quadruple-stranded) DNA structure (e.g. present at the $\rightarrow$ telomere of eukaryotic chromosomes), where a four-base interaction between G residues in oligonucleotide stretches containing short runs of three or more Gs stabilize the fold-back helical structure, whose folding is inhibited by cytosine methylation. See $\rightarrow$ duplex, $\rightarrow$ triplex.

Tet^s: See $\rightarrow$ *tet*racycline *s*ensitivity.

T-even phage: See $\rightarrow$ T phage.

TEV protease: See $\rightarrow$ *t*obacco *e*tch *v*irus protease.

Texas red: A fluorescent dye ($\rightarrow$ fluorochrome) that is used as a marker for $\rightarrow$ fluorescent primers in e.g. $\rightarrow$ automated DNA sequencing procedures. The fluorochrome is exited by light of a wave-length of 593 nm, and emits fluorescent light at 612 nm.

TF: See → *t*ranscription *f*actor.

***Tfl* DNA polymerase:** See → *Thermus flavus* DNA polymerase.

TFO: See → *t*riplex *f*orming *o*ligonucleotide.

T4 DNA ligase (EC 6.5.1.1): An enzyme from → T4 phage-infected *E. coli* cells, requiring ATP to catalyze the formation of a → phosphodiester bond between neighboring nucleotides of a DNA molecule which expose a 5' phosphate and a 3' hydroxyl group. The ligase serves to repair single-stranded → nicks in DNA duplex molecules, and to join two DNA duplex molecules by either → blunt end or → cohesive end ligation. See → DNA ligase.

T4 DNA polymerase (EC 2.7.7.7): A DNA-dependent → DNA polymerase with potent 3' → 5' exonuclease activity and low → processivity from → T4 phage-infected *E. coli* cells, encoded by gene *43* of the phage. The enzyme is used to produce short single-stranded DNA sequences starting from a → nick, to synthesize → blunt-ended DNA, and to label → 3' recessed termini (filling-in reaction).

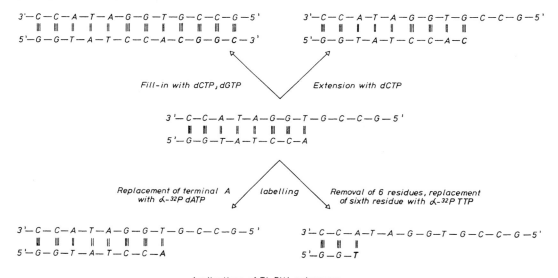

Applications of T4 DNA polymerase

T4 endonuclease V: An enzyme from → T4 phage-infected *E. coli* cells, that has N-glycosylase (→ DNA glycosylase) and *a*purinic/*a*pyrimidinic lyase (→ AP lyase) activities, and binds to *cis-syn c*yclobutane *p*yrimidine *d*imers (CPD; covalently linked photoproducts in DNA, induced by UV light). After binding, the enzyme cleaves the N-glycosylic bond of the 5' pyrimidine of the dimer (pyrimidine dimer DNA glycosylase activity), and then breaks the phosphodiester bond 3' of the resulting abasic site (3' AP lyase activity). The enzyme is used for the detection and repair of DNA damage by UV light and the localization of UV mutational hotspots.

T4-lambda hybrid vector: A → hybrid vector that accomodates more than 160 kb of foreign DNA, can still be packaged *in vitro* into → T4 phage heads and is able to form → plaques on appropriate → *E. coli* host cells.

T4 phage (bacteriophage T4): A → bacteriophage that infects *E. coli* (→ coliphage). Many enzymes encoded by its DNA are used in gene technology (e.g. → T4 DNA ligase, → T4 DNA polymerase, T4 → polynucleotide kinase, → T4 RNA ligase). See also → T4-lambda hybrid vector; → T phages.

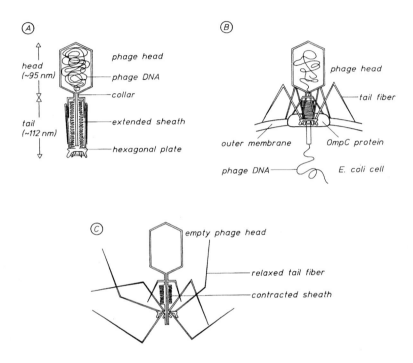

T4 phage
before (A), during (B) and after (C) injection of DNA

T4 polynucleotide kinase: See → polynucleotide kinase.

T4 RNA ligase (EC 6.5.1.3): A 43.5 kDa enzyme from the → T4 phage that catalyzes the ATP-dependent covalent joining of a 5' phosphate end of one single-stranded polynucleotide (donor: e.g. DNA or RNA) to the 3' hydroxyl end of another single-stranded polynucleotide (acceptor). The enzyme is preferentially used in recombinant RNA technology (e.g. for 3' endlabeling of RNA, oligoribonucleotide synthesis, RNA modification, and the site-specific introduction of nucleoside analogs in oligoribonucleotides at the ligation junction).

TF II D: See → *t*anscription *f*actor II D.

***Tfu* DNA polymerase:** A highly thermostable 90 kDa → DNA polymerase with a 3'→ 5' → proofreading exonuclease activity, isolated from the hyperthermophilic marine archaebacterium *Thermococcus fumicolans*. The enzyme is available free from → nickase and → endonuclease activity, and used for high-fidelity amplification by → polymerase chain reaction techniques (especially since it is more heat-stable than → *Taq* DNA polymerase) → site-directed mutagenesis, and correction of misincorporated nucleotides during the polymerase chain reaction.

TGE: See → *t*ransgene *e*xpression.

TGGE: See → *t*emperature gradient *g*el *e*lectrophoresis.

Thaumatin: A highly basic and intensely sweet-tasting member of a small protein family from the arils of the fruit of the African plant *Thaumatococcus daniellii*. These proteins are about 1600 times sweeter than a 10% sucrose solution, and regarded as potential low-caloric substitutes for sugar. They are synthesized as precursor proteins from which 22 mainly hydrophobic amino acids of the amino-terminal, and six mainly acidic amino acids of the carboxy-terminal end are removed during maturation. The genes encoding thaumatins have been isolated and expressed in *E. coli, Saccharomyces cerevisiae, Kluyveromyces lactis*, and other hosts. Their expression can be easily monitored by simple tasting (the human tongue detects thaumatin at concentrations as low as 10^{-3} M).

THC: See → *t*entative *h*uman *c*onsensus.

Theme array: Any → microarray, onto which oligonucleotides, cDNAs, or gene fragments encoding proteins representing a particular metabolic pathway, or the proteins themselves are immobilized, and which can be used to determine the expression patterns of these thematically related genes under a certain experimental or environmental regime.

Therapeutic cloning (*cell nuclear replacement, CNR*): A misleading term for a hybrid technology consisting of → somatic cloning (i. e. the transfer of a nucleus from a differentiated cell into the → cytoplast of an egg cell; the widely discussed "Dolly" has been generated this way) and embryonic stem cell engineering. In short, pluripotent embryonic *s*tem cells (ES) are first isolated from the inner cell population of a blastocyst (i. e. an egg cell 5 – 6 days after fertilization). Given a suitable medium and manipulation, these cells can be differentiated into all the different cell types of an adult organism (theoretically). Therapeutic cloning is based on the use of embryonic material with genomic DNA identical to that of a patient, whose diseased tissue is to be replaced ("substituted") by ES-derived cell clones. Therefore, the embryo is destroyed. On one hand, therapeutic cloning is expected to improve treatment of a series of diseases (e. g. Parkinson's or Alzheimer's diseases, diabetes, osteoporosis, muscular dystrophy, hepatitis, leukemia, even cancer) and prevent the undesirable rejection of transplant. On the other hand, therapeutic cloning is still in its infancy. For example, it is not completely clear, whether the isolated ES cells really are pluripotent (i. e. develop into any desired cell type and tissue), or the tissue will function normally in the acceptor patient, or the ES-derived tissue will not transform into a tumor. Also, the aging process of the ES-derived clones may be different from their "normal" counterparts, or could accumulate mutations at a rate higher than in the corresponding original tissues. The term "therapeutic cloning" wrongly suggests, that only cell clones, but no embryos are involved. In any case, the therapeutic cloning technology has nothing to do with → genetic engineering or → molecular cloning. Compare → reproductive cloning.

Therapeutic DNA repair: The correction of a disease-causing mutation in DNA or RNA by the direct recognition and correct replacement of the faulty bases within the nucleus of a cell. Therapeutic DNA repair starts with the recognition of the mutation, which is based on triple strand formation (a triple strand-forming oligonucleotide carrying a base-modifying activity binds to the mutant region in double-stranded DNA). Then, either → peptide nucleic acids (also carrying base modification activity) are hybridized to the plus or minus strand of the DNA target after strand invasion, or → antisense oligonucleotides are hybridized to the mutated DNA (the antisense oligomer also carries a base-modifying activity). The last step in therapeutic DNA repair is the modification of the aberrant base by enzymatic activity or chemical reaction.

Therapeutic gene: Any gene that encodes a protein with a therapeutic function. See for example → suicide gene.

Thermal asymmetric interlaced (TAIL) polymerase chain reaction (TAIL-PCR): A variant of the conventional → polymerase chain reaction that uses two → primers of different lengths with different thermal stabilities. First, PCR reactions are carried out at relatively high annealing temperatures that favor priming by the longer primer. Then lower temperatures allow both primers to anneal. By switching the amplification cycles from high to low stringency, target sequences detected by the long and sequence-specific primer are amplified preferentially. TAIL-PCR can be used to isolate → promoters of specific genes fairly easily: → gene-specific primers are designed and used in concert with → arbitrary primers to walk → upstream of the gene. Any amplification product will contain promoter sequences. Full promoters can be isolated by repeating this step, i. e. designing promoter-specific primers and using another arbitrary primer to walk still further upstream. Characterization of the promoter can be done by functional analysis (e. g. by → *Bal*31 deletion and → transient expression of → reporter genes with truncated promoter fragments) and sequencing. See → thermally asymmetric PCR.

Thermal cycle sequencing (cycle sequencing, CS): A technique for the sequencing of an amplification product, generated in the → polymerase chain reaction. In short, either single- or double-stranded → template DNA is prepared, and → primers are then annealed to the template in four separate reactions, that contain all ingredients for amplification (e.g. $MgCl_2$ all four deoxynucleotide triphosphates, thermostable → *Taq* polymerase). Additionally, each reaction is supported with one of the → dideoxynucleotides, and → Sanger sequencing is started. After the procedure, → denaturing → polyacrylamide gel electrophoresis and → autoradiography is used to detect the sequence of the amplified product. Cycle sequencing requires only small amounts of template, and a high temperature profile during cycling ensures high specificity of primer annealing and good resolution of intramolecular secondary structures, that otherwise interfere with sequencing.

Thermal DNA ligase: See → DNA ligase.

Thermal hysteresis protein: See → antifreeze protein.

Thermal Klenow fragment: A polypeptide generated by partial proteolytic digestion of *Bacillus stearothermophilus* DNA polymerase, that is used in → Sanger sequencing procedures at elevated temperatures (between 37° and 70°C), at which secondary structures of DNA are largely eliminated.

Thermally asymmetric PCR: See → thermally asymmetric polymerase chain reaction.

Thermally asymmetric *polymerase chain reaction* (thermally asymmetric PCR): A variant of the conventional → polymerase chain reaction, in which the temperature program switches between high and low → primer annealing temperatures. This design allows to use two primers of different sequence (e.g. a → RAPD primer and a → microsatellite primer) to amplify genomic regions flanked by a microsatellite and the address site for the RAPD primer. This type of PCR is employed e.g. for the → random amplified microsatellite polymorphism (RAMP) technique, and → thermal asymmetric interlaced polymerase chain reaction.

Thermo-blot: See → thermoblotting.

Thermoblotting (thermo-blot, hot blot): A variant of the conventional → Southern blotting procedure that uses elevated temperatures to increase the efficiency of the transfer of DNA fragments or RNA molecules from an → agarose gel onto a filter.

***Thermococcus gorgonarius* (*Tgo*) DNA polymerase (*Tgo* DNA polymerase):** A highly thermostable DNA polymerase of the extremely thermophilic sulfur-metabolizing archaebacterium *Thermococcus gorgonarius* from geothermal vents in the Pacific ocean, that possesses a 5'→ 3' polymerase and a 3' → 5' exonuclease activity. It catalyzes the thermostable replication of the archaeon genome with high fidelity (error rate: 3.3 – 2.2 x 10^{-6}). See → *Pfu* DNA polymerase.

***Thermococcus kodakaraensis* DNA polymerase (KOD DNA polymerase):** An enzyme from the extremely thermophilic archaebacterium *Thermococcus kodakaraensis* (strain KOD 1), that polymerizes deoxynucleotides and possesses an integral 3' → 5' exonuclease proofreading activity. KOD DNA polymerase has a high elongation capacity (106–138 bases/second), high processivity (more than 300 bases), good fidelity (mutation frequency: 3.5 x 10^{-3}), and generates blunt-ended PCR fragments. KOD DNA polymerase is also available as the product of a recombinant gene. Compare → *Pyrococcus furiosus* DNA polymerase, → *Pyrococcus woesii* DNA polymerase, → *Thermus aquaticus* DNA polymerase, → *Thermus flavus* DNA polymerase, → *Thermus thermophilus* DNA polymerase.

***Thermococcus litoralis* (*Tli*) DNA polymerase (*Tli* DNA polymerase; Vent™ DNA polymerase; EC 2.7.7.7):** A 90-93 kDa highly thermostable DNA polymerase from the extreme thermophilic archaebacterium *Thermococcus litoralis*, that grows at 98°C in thermal vents on the ocean floor. It polymerizes deoxynucleotides with an integral 3'→ 5' exonuclease proof-reading activity, which is extremely thermostable (optimum temperature: 75° C; still active after exposure to 100 °C). For the → polymerase chain reaction, this enzyme is considered superior to the conventional → *Thermus aquaticus* DNA polymerase, since its 3'→5' proof-reading capacity will correct single-base substitution errors during *in vitro* DNA synthesis. Thus the fidelity of base incorporation is much higher than with → *Taq* polymerase. Vent™ DNA polymerase is also available as the product of a recombinant gene. Compare → *Pyrococcus furiosus* DNA polymerase, → *Pyrococcus woesii* DNA polymerase, → *Thermus aquaticus* DNA polymerase, → *Thermus flavus* DNA polymerase, → *Thermus thermophilus* DNA polymerase.

Thermocycler: An instrument that allows the programmed and rapid heating and cooling of small volumes of reaction mixtures. Such thermocyclers are used in → polymerase chain reaction procedures to set temperatures for thermal denaturation of the target DNA (at around 94°C), for primer annealing (at 37°C), and primer extension (at around 72°C).

Thermolabile DNA polymerase: Any one of a series of → DNA polymerases that are irreversibly inactivated at elevated temperatures (e.g. above 45-50°C), in contrast to heat-resistant → *Thermus aquaticus* DNA polymerase. To this class belong e.g. the → Klenow fragment of *E. coli* → DNA polymerase, → T7 DNA polymerase and → reverse transcriptase.

Thermophilic enzyme: Any enzyme originating from a thermophilic organism, that does not withstand an exposure to denaturing temperatures (e.g. < 70°C). Compare → thermostable enzyme.

Thermophilic *species* DNA polymerase (*Tsp* DNA polymerase): A heat-stable DNA polymerase from a thermophilic bacterium (name not disclosed), catalyzing the template-dependent synthesis of DNA polymers without nontemplated addition of extranucleotides(s), as is the case with e. g. → *Taq* DNA polymerase. See → *Bst* polymerase, → *Pyrococcus woesii* DNA polymerase, → *Thermotoga maritima* DNA polymerase → *Thermus aquaticus* DNA polymerase, → *Thermus brockianus* DNA polymerase, → *Thermus flavus* DNA polymerase, and *Thermus thermophilus* DNA polymerase.

Thermo*sensitive* *alkaline* *phosphatase* (TsAP): A recombinant bacterial enzyme catalyzing the hydrolysis of phosphate monoesters, that is used to remove 5' terminal phosphate groups from DNA molecules prior to their ligation with other DNA molecules. This procedure prevents self-ligation of e. g. → vector DNA in → cloning experiments. TsAP can more easily be inacti vated by heat than its wild-tpye precursor → BAP, which is necessary after the dephosphorylation process.

Thermostable enzyme: Any enzyme originating from a thermophilic organism, that withstands even long exposures to denaturing temperatures (e.g. > 90°C) without appreciable loss of functions). See → *Thermus aquaticus* DNA polymerase. Compare → thermophilic enzyme.

Thermotoga maritima (*Tma*) DNA polymerase (*Tma* DNA polymerase): A 70 kDa enzyme from the anaerobic hyperthermophilic gram-negative eubacterium *Thermotoga maritima*, polymerizing deoxynucleotides and possessing both 5' → 3' and 3'→ 5' exonuclease activity. The enzyme has been engineered to lack 5' → 3' exonuclease activity by truncation at its → amino terminus. Compare → Stoffel fragment, → Klenow fragment.

Thermus aquaticus (*Taq*) DNA polymerase (*Taq* polymerase, Taquenase™; EC 2.7.7.7): An enzyme from the thermophilic eubacterium *Thermus aquaticus*, strain YT 1, or BM, polymerizing deoxynucleotides with little or no 3' → 5' or 5' → 3' exonuclease activity, which is highly thermo stable (optimum temperature: 70-75°C) and allows the selective amplification of any cloned DNA about 10 million-fold with very high specificity and fidelity in the so-called → polymerase chain reaction. *Taq* polymerase can also be used to label DNA fragments either with radioactive nucleotides, or non-radioactively with → biotin or → digoxygenin. Furthermore it is ideal for the → Sanger sequencing of templates with a high degree of secondary structure, since high temperatures will melt out such secondary structures. DNA sequencing with *Taq* DNA polymerase produces uniform band intensities and low background on → sequencing gels. The enzyme is also available as recombinant *Taq* polymerase (Ampli Taq™, → Taquenase™). See also → Stoffel fragment, → super *Taq* polymerase. Compare → *Thermus thermophilus* DNA polymerase.

Thermus brockianus (*Tbr*) DNA polymerase (*Tbr* DNA polymerase): An extremely thermostable → DNA polymerase from the thermophilic bacterium *Thermus brockianus* with a 5' → 3' exonuclease activity. Its extreme heat-stability (half-life time: 2.5 hrs at 96° C) and its very low misincorporation rate recommend *Tbr* DNA polymerase for DNA sequencing, especially of DNA with many secondary structures. See → *Pfu* DNA polymerase, → *Pwo* DNA polymerase, → *Taq* DNA polymerase, → *Tfl* DNA polymerase, → *Tli* DNA polymerase, → *Tma* DNA polymerase, → *Tsp* DNA polymerase, → *Tth* DNA polymerase.

Thermus flavus (*Tfl*) DNA polymerase (*Tfl* DNA polymerase; EC 2.7.7.7): An enzyme from the thermophilic bacterium *Thermus flavus*, polymerizing deoxynucleotides with little or no 3' → 5' – or 5' → 3' – exonuclease activity. *Tfl* polymerase is highly thermostable (optimum temperature: 70-75° C) and used for the sequencing of template DNA with a high degree of secondary structure that will melt out at these high temperatures. It is also employed for → primer extension and thermal cycle sequencing (especially for templates with complex secondary structure).

Thermus thermophilus (*Tth*) DNA polymerase (*Tth* polymerase): A single-subunit 93 kDa enzyme from the thermophilic bacterium *Thermus thermophilus*, polymerizing deoxynucleotides with little or no intrinsic 3' → 5' or 5' → 3' exonuclease activity. The polymerase is optimally active at 75°C and survives temperatures as high as 95°C. It is used for → Sanger sequencing at high temperatures (70°C) where templates have minimal secondary structures that may otherwise disturb the sequencing process. A recombinant *Thermus thermophilus* DNA polymerase (r*Tth*) can also be used and functions as a → reverse transcriptase and a DNA polymerase, depending on the buffer used (see → reverse transcription polymerase chain reaction). Compare → *Thermus aquaticus* DNA polymerase.

Thionucleotide (α-thionucleotide; NTPαS; thiophosphate): Any nucleoside triphosphate, that carries a sulfur atom at the 5'-α (or also 5'-γ) position instead of an oxygen atom. Thionucleotides are more stable against nucleases than their normal counterparts and are used for site-directed mutagenesis. These compounds can also be labeled with ^{35}S in the α- or γ-position, and possess a relatively long half-life time ($\rho_{1/2}$=87.1 days).

2′-Deoxynucleoside-5′-O-(α-thio)-triphosphate

Thionucleotide mutagenesis: A variant of the → oligo-mismatch mutagenesis technique for the introduction of mutations in DNA, that is based on the use of → thionucleotides. In short, the mutagenic oligonucleotide is hybridised to the single-stranded → template DNA, and extended by the → Klenow fragment of DNA polymerase in the presence of a → DNA ligase and a thionucleotide, producing a mutated phosphothio → heteroduplex. The phosphothio strand of the heteroduplex cannot be digested by certain → restriction endonucleases. The corresponding strand can be removed by → exonucleases, whereas the mutated strand remains intact. It then serves as template for the *in vitro* synthesis of a complementary new strand: the mutated duplex DNA is generated.

Thioredoxin fusion protein: A recombinant protein consisting of an *E.coli* thioredoxin oxidore-ductase protein (encoded by *trx*A gene) fused to the N-terminus of another protein. Such → fusion proteins are expressed to high levels in *E.coli* host cells without formation of insoluble inclusion bodies from which proteins would have to be resolubilized and accurately refolded. Instead, the 11.7 kD thioredoxin moiety keeps the fusions readily soluble while constituting only a minor portion of the fusion.

The thioredoxin fusions can be purified by two techniques exploiting two properties of thiore-doxin. First, thioredoxin accumulates in specific areas of *E. coli* cells at the inner membrane, called adhesion zones, which can easily be induced to release their constituents by a single osmotic shock. Second, thioredoxin is heat-stable, and so is the thioredoxin fusion. By simply heating the released proteins up to 85^0 C, the contaminating proteins are denatured and the thioredoxin fusion can be recovered.

3': Three prime. Derived from the 3' carbon atom of ribose or deoxyribose, respectively. The term is used to describe the orientation of nucleic acid molecules. It is synonym to → downstream, and contrary to → 5' and → upstream. See also → 3' end.

3'-deoxyadenosine: See → cordycepin.

Three-dimensional proteomics (3D proteomics): A misleading term for the combination of sub-cellular fractionation, protein chromatography and/or affinity capture of more abundant pro-teins (first dimension) with the two-dimensional → polyacrylamide gel electrophoresis (second and third dimension).

3D proteomics: See → three-dimensional proteomics.

3' end (three prime end, 3' carbon end, 3' carbon atom end, 3'-terminus): One of the ends of a linear DNA or RNA molecule that carries the free hydroxyl group of the 3' carbon of the pentose. Conventionally this end is written to the right in nucleic acid nomenclature.

3' endlabeling: A method to label specifically the 3' end(s) of single- or double-stranded DNA, see → endlabeling. The 3' ends of RNA can be labeled using → T4 RNA ligase.

3'-end processing: The enzymatic reactions leading to the → polyadenylation of → pre-messenger RNA, involving an → endonucleolytic cleavage and the subsequent addition of a → poly(A) tail to the newly generated 3' end. The 3'-end processing is catalyzed by a multiprotein complex. For example, in mammals the endonucleolytic step requires at least five different proteins, the *c*lea-vage and *p*olyadenylation *s*pecificity *f*actor (cPSF), *c*leavage *st*imulation *f*actor (cstF), *c*leavage *f*actors I and II (CF Im, CF IIm), and → poly(A) polymerase. The polyadenylation step is managed by three proteins, the *p*oly(*A*)–*b*inding protein II (PAB II), CPSF, and poly(A) poly-merase. CPSF recognizes and binds the specific → poly(A) addition signal 5'-AAUAAA-3', and recruits the polymerase to the RNA substrate. Once 10-11 adenosine residues are added, PAB II joins the polyadenylation complex by binding to the growing poly(A) tail. The quarternary com-plex between CPSF, PAB II, RNA (plus poly[A] tail), and polymerase allows the processive synthesis of a tail of up to 250 adenosine residues. See → posttranscriptional modification.

3' EST: See → 3' *e*xpressed *s*equence *t*ag.

3' expressed sequence tag (3' EST): Any → expressed sequence tag derived from a → cDNA, which marks the → trailer (i. e. the 3' non-coding region) of a → gene. 3' ESTs are markers for the 3' termini of genes.

3' extension: The single-stranded protrusion at the 3'-terminus of a DNA duplex molecule, generated by specific → restriction endonucleases such as e.g. *Kpn* I:

$$\downarrow$$

5'-GGTACC–3'	5'-GGTAC–3'	5'C–3'
	→	
3'-CCATGG–5'	3'-C	3'-CATGG–5'

$$\uparrow$$

³⁵S: The beta-emitting radioactive isotope of sulfur (half-life: 87.1 days), used either as ^{35}S methionine or as ^{35}S to label proteins (in the latter case by incorporation into sulfur-containing amino acids like methionine or cysteine).

35S promoter: A strong, mostly constitutive → promoter of the → cauliflower mosaic virus, driving the transcription of a 35S RNA (hence the name). This promoter is widely used as → promoter in → plant expression vectors for plant transformation experiments.

3' flanking region: The sequences → downstream of the so-called → trailer of eukaryotic genes. This region contains signals for the precise termination of transcription and the processing of the → 3' end of the transcript (e.g. for → polyadenylation).

Three-hybrid system (yeast three-hybrid system, tribrid system): A variant of the conventional → two-hybrid system, designed for the detection of interaction(s) in ternary protein complexes, that is based on the conventional transcriptional → activation domain fusion vector and a modified → DNA-binding domain fusion vector with two separate → multiple cloning sites, one for the prey-encoding gene, the other one for the third protein. This third protein is conditionally expressed from a methionine promoter (active only in the absence of the amino acid). The three-hybrid system is fully dependent on the function of the third protein, that either links the two other, normally not reactive proteins, or stabilizes a possible weak interaction between both proteins, enzymatically modifies one or both protein(s) as e. g. a phosphokinase, or inhibits protein-protein interaction(s). The three-hybrid system then also allows to identify proteins that inhibit protein-protein interactions. See → dual-bait two-hybrid system, → interaction trap, → LexA two-hybrid system, → one-hybrid system, → reverse two-hybrid system, → RNA-protein hybrid system, → split-hybrid system, → split-ubiquitin membrane two-hybrid system, → two-hybrid system.
Figure see page 1143.

3' hydroxy group (3'-hydroxy residue): Any hydroxy group at the → 3' end of a nucleic acid molecule.

3' hydroxy residue: See → 3' hydroxy group.

3'-mismatched primer: Any → primer oligonucleotide for → polymerase chain reaction experiments, that carries a base → mismatch at its 3'-terminus. The incorporation of such a mismatched primer into an amplification product leads to an → misextension (i.e. a faulty product).

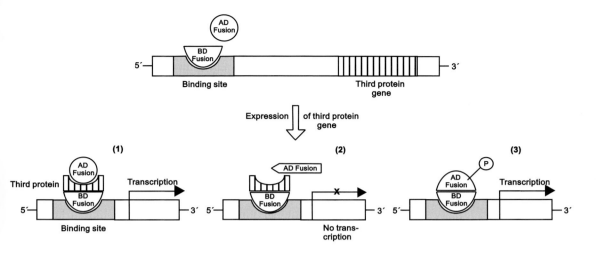

3'-modified oligonucleotide: Any → oligonucleotide, whose 3'-terminus is covalently linked to non-nucleosidic compounds such as amino- or thiol-linkers, adaptors, spacers, or other molecules. See → 5'-modified oligonucleotide.

3' non-coding region: See → trailer.

3' overhang, 3' overhanging end: See → 3' protruding terminus.

3'-PACA: See → anchored polymerase chain reaction.

3' protruding end: See → 3' protruding terminus.

3' protruding terminus (3' protruding end, 3' overhanging end, 3' overhang, 3' extension): The end of a DNA duplex molecule where one strand is longer ("protruding") by some bases than the other which is referred to as "recessed". The last base carries a free hydroxyl group. Compare → recessed 3'-terminus.

3'-RACE: See → rapid amplification of cDNA ends.

3' recessed terminus: See → recessed 3'-terminus.

3' splice site: See → acceptor splice junction.

3SR: See → self-sustained sequence replication.

3'-terminator: A modified base that blocks the 3'-terminus of an → oligodeoxynucleotide and thereby prevents its extension catalyzed by → DNA polymerases or its ligation by → DNA ligases. For example, dideoxy-cytidine is such a 3'-terminator, since it lacks 2'- and 3'-hydroxyl groups.
Figure see page 1144.

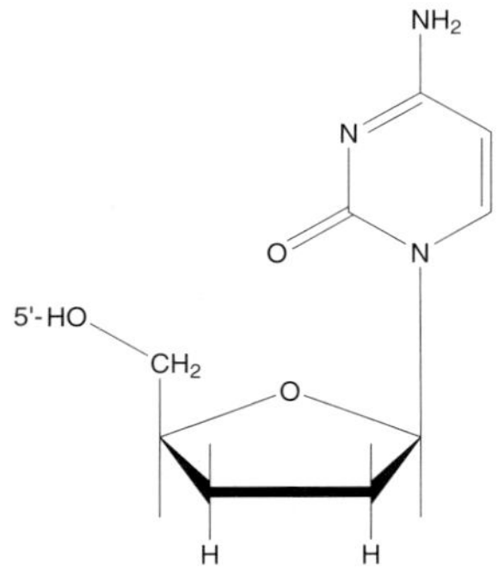

Dideoxycytidine

3' to 5': A term describing the direction from the → 3' end towards the → 5' end of a linear nucleic acid molecule (symbol: 3' → 5'). Compare → 5' to 3'.

^{32}P: A radioactive isotope of phosphorus that emits β-particles and is used in → nick-translation or other nucleic acid → labeling techniques (half life: 14.3 days).

3' untranslated region: See → trailer.

3' untranslated sequence: See → trailer.

3'-UTR: See → trailer.

Threshold cycle (C_t): The cycle of a conventional → quantitative polymerase chain reaction, at which the fluorescence signal is detectable above background.

Thymidine (deoxythymidine, thymine deoxyriboside, 2'-deoxythymidine): A → nucleoside that consists of thymine (→ T) linked to a → deoxyribose molecule.

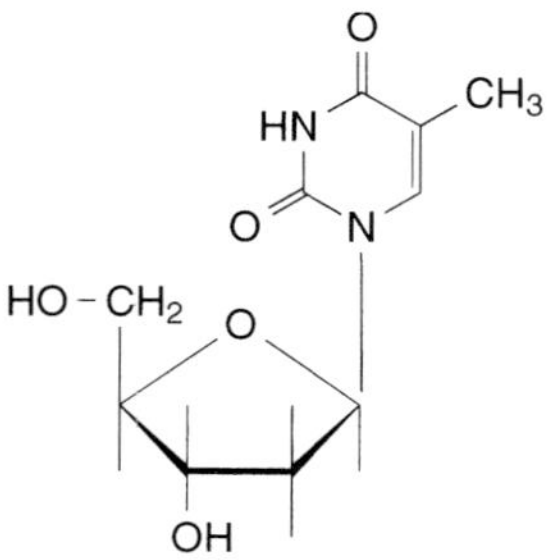

Thymidine

Thymidine-5'-monophosphate (2'-deoxy*t*hymidine-5'-*mono*phosphate, TMP): A → pyrimidine nucleotide with a phosphorous group in deoxyribose-O-phosphoester linkage at the 5' position of → deoxyribose. TMP serves as elementary unit in DNA synthesis.

Thymidine glycol: The → thymidine derivative 5,6-dihydroxydihydrothymidine, that is formed from DNA oxidatively damaged by free radicals or ionizing radiation, excised by repair enzymes and excreted (e.g. in urine). See → thymine glycol, → OH^8dG.

Thymidine *k*inase (tk): An enzyme catalyzing the conversion of → thymidine to deoxythymidine-monophosphate. In gene technology, the gene encoding thymidine kinase is used to tag other genes that are transferred into host cells but cannot be detected by conventional techniques. For this purpose, cells carrying a mutation in the thymidine kinase gene (*tk⁻* mutation) serve as host cells. Such cells do not survive in *h*ypoxanthine-*a*minopterine-*t*hymidine (HAT) growth medium. However, if such mutants are complemented by an intact thymidine kinase gene linked to the gene(s) of interest, they can be selected. Such → revertants (*tk⁺*) usually also contain the gene of interest and transcribe it.

Thymine (T): A → pyrimidine base (2,6-dihydroxy-5-methyl-pyrimidine, 5-methyluracil) charac-teristic for DNA, and absent from RNA. See → thymidine, → thymidine dimer.

Thymine

Thymine dimer: Two cross-linked thymine (→ T) molecules, arising after irradiation of DNA duplex molecules with UV light (<340 nm). Such thymine dimers effectively block DNA → re-plication and are removed by → DNA repair systems, especially the → SOS repair pathway, or light of longer wave length (> 400 nm).

Thymine   Thymine dimer

Thymine glycol: The → thymine derivative 5,6-dihydroxydihydrothymine, that is formed from DNA oxidatively damaged by free radicals or ionizing radiation, excised by repair enzymes, and excreted (e. g. in urine). See → thymidine glycol, → OH^8dG.

TID: See → transposon insertion display.

Tight control: See → stringent control.

Tiled oligonucleotide: Any one of a series of overlapping → oligonucleotides that altogether span a genomic region containing e.g. a mutation (for example, a → single nucleotide polymorphism). Tiled oligonucleotides can be used to produce a → tiling path across the region and neighboring sequences.

Tiling: The process of generating a → tiling path.

Tiling path: The ordered arrangement of → BAC, or → PAC, or → YAC clones using sequence overlaps (see → contig, → sequence-tagged connector) of neighboring clones such that they completely cover the corresponding region of → genomic DNA.

TILLING: See → targeting induced local lesions in genomes.

Time-of-flight (TOF): A special type of mass spectrometer, that allows the separation of fragment ions based on their mass-to-charge ratio (m/z). The traveling time of an ion in the flight tube is correlated to its m/z, lighter ions arriving sooner than heavier ions.

Time-resolved fluorescence (TRF): The temporal separation of the → fluorescence of an experimentally used → fluorochrome from background fluorescence emitted by e.g. biological specimens, plasticware, or buffer components. Since the fluorescence lifetime of most fluorochromes is about 10 nanoseconds, TRF employs fluorochromes with considerably longer lifetimes (100–400 µsec), so that the fluorescence maximum is reached only after the background fluorescence has dissipated. Usually lanthanide chelates are used as long-lived fluorophores. See → time-resolved fluorescence resonance energy transfer.

Time-resolved fluorescence resonance energy transfer (TR-FRET): A technique for the separation of short- and long-lived fluorescence emissions in a sample, that combines → time-resolved fluorescence and → fluorescence resonance energy transfer. For example, for TR-FRET the acceptor → fluorochrome → cyanin 5 (short lifetime) and lanthanide (Ln^{3+}) chelates (long lifetime in the range between 200 and 1500 µsec) as donor fluorochrome are used. In particular, a complex of Eu^{3+} with *t*erpyridine-bis(*m*ethylenamine)*t*etraacetic acid (TMT) is an efficient lanthanide fluorochrome.

Tiny expressed RNA (tx RNA): Any one of a family of conserved small (21 nucleotides) eukaryotic RNAs, that are produced from 70 nucleotides long RNA molecules (containing a helix of 30 bases) by a mechanism similar to → RNA interference.

Tiny RNA: See → microRNA.

Ti-plasmid (*t*umor-*i*nducing plasmid, pTi): A large → conjugative plasmid of about 200 MDa in size which is found in all virulent strains of the Gram-negative soil bacterium → *Agrobacterium tumefaciens*. It contains genes for replication (→ oriV), plasmid transfer (*tra* → genes), → phage exclusion (*Ape*), incompatibility (*Inc*), virulence (→ *vir*-region), root and shoot induction in host plants (*Roi, Shi*), → opine synthesis in host plants (*Nos, Ocs, Ags*), opine catabolism (*Noc, Occ, Agc, Arc*) and catabolism of phosphorylated sugars (*Psc*). The genes for root and shoot induction and opine synthesis are clustered in a specific segment of the Ti-plasmid (→ T-region) and flanked by 25 bp border sequences. These borders are the recognition sites for a Ti-plasmid encoded endonuclease ("border endonuclease") that excises one strand of this region (→ T-strand) which is packaged into a → T-complex and transported into a recipient plant cell. There it is integrated into the nuclear genome and causes the cell's permanent proliferation into a tumor.
Figure see page 1147.

Ti-plasmid mediated gene transfer: See → *Agrobacterium*-mediated gene transfer.

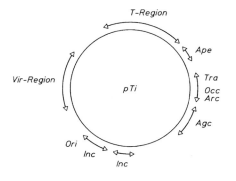

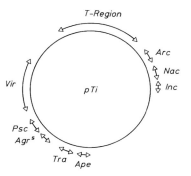

Octopine-Type   Nopaline-Type

Agc	:	Agrocinopine catabolism	Occ	:	Octopine catabolism
Agr^s	:	Agrocin sensitivity	Ori	:	Origin of replication
Ape	:	Phage exclusion	Psc	:	Catabolism of phosphorylated sugars
Arc	:	Arginine catabolism	Tra	:	Transfer genes
Inc	:	Incompatibility region	Vir	:	Virulence region
Noc	:	Nopaline catabolism			

Ti-plasmid

TIR: See → *terminal inverted repeat*.

TIRFM: See → total internal reflection fluorescence microscopy.

TIS: See → cap site.

Tissue array (tissue microarray, TMA): A glass or quartz microscopic slide, onto which tissue specimens of multiple origin (e. g. from different tissues of one organism, or from identical tissues of different organisms) are spotted at medium (e. g. 60 specimens per slide) or high density (e.g. 600 specimens per slide). In short, the production of tissue arrays starts with the removal of tissue samples by surgery or biopsy, their fixation (in e. g. 4 % neutralized formaline for 12 – 24 hrs), dehydration in ethanol, and embedding into paraffin after treatment with xylene. Also, frozen tissue samples can be used. Host DNA, RNA or protein molecules remain intact. Then tissue fragments of 2 mm diameter (~ 100,000 cells) are removed from these samples and positioned on the slide. DNA or RNA can subsequently be detected on the array by → fluorescent in situ hybridization with suitable → probes (e. g. a gene). Proteins can be visualized by immunohistochemical techniques. For DNA or RNA, the slide is de-paraffinized, hydrated and treated with proteinase K, then incubated in various buffers and hybridized to a → digoxygenin-labeled probe (e. g. a gene). Subsequently an anti-digoxygenin antibody is added, and the hybrids detected by e. g. → alkaline phosphatase attached to the antibody. For proteins, the slide is also de-paraffinized and hydrated, thereby liberating the antigens of the samples. Then blocking reagent (preventing unspecific binding) and a primary antibody raised against the target protein are added. The resulting antigen-antibody complex is finally visualized with a biotinylated secondary antibody → avidin coupling and a colorigenic test (e. g. a peroxidase test). Such tissue arrays allow to detect genes, differentially expressed genes, and proteins. Compare → cDNA array, → cDNA expression array, → expression array, → gene array, → microarray, → sequencing array.

Tissue culture: Maintenance of living cells or tissue in a liquid or on a soft gel medium *in vitro*. Three broad categories can be distinguished:
a) → Cell culture: the culture of individual cells isolated enzymatically or physically from tissues.
b) Tissue culture: the culture of a specific tissue or a tissue fragment isolated physically from organs.
c) Organ culture: the culture of a specific organ isolated physically from an organism.
→ Protoplast culture is a special type of cell culture. In this case, cells have been treated with cell wall-degrading enzymes and are cultured as naked protoplasts.

Tissue engineering: A multidisciplinary approach to cultivate monolayers of specific somatic cells with proliferative potential on synthetic or organic media in vitro and to (1) re-implant the multiplied cells into the organism from which their mother cells are derived, to (2) develop biologically functional tissues and (3) whole organs. The ultimate aim of tissue engineering is the replacement of damaged, missing, or functionally defective tissues or organs. Ideally, such cells are taken from the target organism (e.g. a patient), so that the re-implantation does not trigger a rejective immune response ("autologous cells"). Such cells are presently derived from fibroblasts, endothelic cells, chondrocytes, osteoblasts, enterocytes, urothelial cells, less so from cardiomyocytes and hepatocytes. In future, the preferred source will be pluripotent and highly proliferative or adult, tissue-specific embryonic stem cells (ES). The cells are grown on xenogenic matrices (e.g. networks of collagen, fibrin, or other components of the extracellular matrix, synthetic compounds as e.g. PTFE, or short synthetic peptides [PGD peptides] and growth factors). For example, a xenogenic matrix would be an assembly of heart valve cells from beef or sheep, that is used as a support for the growth of human valve cells. The matrix is usually resorbed by the growing cells.
Re-implantation frequently means direct injection of the cells into the target organ of the recipient. For example, mesenchymatic bone marrow cells can be isolated before chemotherapy or ray treatment of a patient and, after the therapy, can be injected back into the bone marrow.
The future potential of tissue engineering is the development and production of complex tissues and complete organs for transplantation (e.g. skin, mucous membranes, heart muscle, blood vessels). Neither tissue engineering nor → therapeutic cloning have anything to do with → gene technology, but are described in some detail, because they are frequently confused with → genetic engineering.

Tissue microarray: The arrangement of DNAs, RNAs, or proteins from many different tissues of an organism on a membrane (e. g. nitrocellulose), glass slide, or silicon chip at a very high density. Such microarrays are used for the high-throughput analysis of either specific DNA or RNA sequences, or proteins in the different tissues. See → cDNA expression array, → cDNA microarray, → DNA chip technology, → expression array, → gene array, → sequencing array.

Tissue print (tissue printing): A technique to detect specific DNA (or RNA) sequences in cells, tissues, or organs. The specimen is simply pressed onto a → nitrocellulose or nylon-based filter which leads to the fixation of the DNA (or RNA) to the filter ("squash blot"). The filters are then processed as in conventional → Southern blot procedures, hybridized to specific radioactive, → biotin- or → digoxygenin-labeled → probes, and subsequently exposed to X-ray films. The resulting → autoradiograph reflects the presence and (in favorite cases) distribution of specific DNA (or RNA) sequences in the specimen. Compare → squash dot hybridization.

Tissue printing: See → tissue print.

Tissue proteomics: A branch of → proteomics that aims at isolating all the proteins and peptides of a given tissue at a given time, characterizing them functionally and structurally, and determining all possible interactions between them.

Tissue-specific gene: See → luxury gene.

Tissue-specific locus control element: Any → locus control region, that directs the tissue-specific expression of a family of genes nearby.

Tissue-specific *m*icrodissection *c*oupled with *p*rotein chip array technology (TMCP): A technique to determine the protein patterns in whole tissue extracts or specific cells microdissected from specific organs. The target organ is first isolated, and specific cell populations are separated from parts of the organ using a joystick-controlled electrical micromanipulator with an extended glass needle and an inverted microscope. The separated cells are then lysed, the cell debris centrifuged out, and the resulting proteins incubated with a → protein chip, that carries an array of potential target proteins. Interactions between tissue-specific and chip-bound proteins can then be detected by → surface-enhanced laser desorption and ionization. The analysis of the retained proteins is then possible with → *m*ass *s*pectrometry (MS) techniques. For an accurate measurement of the molecular mass of the target protein, → *m*atrix-*a*ssisted *l*aser *d*esorption and *i*onization time of-flight mass spectrometry (MALDI-TOF) is commonly employed.

Tissue-specific promoter: Any → promoter that is activated in specific tissues only, and drives the transcription of a → luxury gene.

T_L-DNA: The left part of the → T-DNA of → octopine-type → Ti-plasmids of → *Agrobacterium tumefaciens*. Upon infection of plant cells it is transferred independently from the right part, the → T_R-DNA, and integrated into the nuclear genome of the host cell. This 12 kb T_L-DNA carries 8 genes, encoding two enzymes for auxin, for cytokinin, and one for → octopine biosynthesis. Expression of these genes causes tumorigenesis.

Tli DNA polymerase: See → *Thermococcus litoralis* DNA polymerase.

T loop: See → telomere loop.

TLS: See → *t*rans*l*esion *s*ynthesis.

T_m: See → *m*elting *t*emperature.

TMA: See → *t*ranscription-*m*ediated *a*mplification.

TMCP: See → tissue-specific microdissection coupled with protein chip array technology.

Tma DNA polymerase: See → *Thermotoga maritima* DNA polymerase.

TMHA: See → *t*emperature *m*odulated *h*eteroduplex analysis.

TMP: See → *t*hymidine-5'-*m*ono*p*hosphate.

tmRNA (tRNA-mRNA hybrid, 10 Sa RNA): A stable bacterial RNA of at least 300 nucleotides in length, that combines features of both → messenger RNA and → transfer RNA, and functions to tag polypeptides translated from defective mRNAs (e. g. lacking a → stop codon). In *E. coli*, tmRNA is encoded by the *ssr*A gene, occurs in about 1,000 copies per cell, and contains a tRNA-like region (that can be charged with alanine), a domain encoding the decapeptide H_2N-ANDENYALAA-COOH, and a series of → pseudoknots. The decapeptide tags the carboxy termini of truncated (i. e. faulty) proteins destined for degradation. The charged alanyl-tmRNA rescues ribosomes stalled at the 3'terminus of truncated mRNAs lacking a stop codon and donates its alanine to the stalled peptide. Then the tmRNA-encoded decameric tag is added to the carboxy terminus of the nascent polypeptide chain by → trans-translation of the ribosome from the faulty mRNA to the internal sequence in tmRNA (final sequence: H_2N-AAN-DENYALAA-COOH). Normal termination at the end of the terminal 10 codons allows the ribosome to recycle. The 11-amino acid tag guides the aberrant protein for degradation. See → riboregulation.

Tn: Symbol for → *t*ranspos*on*.

TNA: See → α-Threofuranosyl- (3'→2') nucleic acid.

Tobacco acid pyrophosphatase **(TAP):** An enzyme catalyzing the removal of the → cap from eukaryotic → messenger RNA, leaving a 5' monophosphate.

Tobacco etch virus protease **(TEV protease):** A 49 kDa highly specific thiol protease of the tobacco etch virus that catalyzes the cleavage of the heptapeptide EXXYXQ↓S/G after the conserved glutamine residue (Q: glutamine; E: glutamic acid; X: any amino acid; Y: tyrosine; S: serine; G: glycine). The natural function of TEV protease is the cleavage of a → polyprotein transcribed from the 9.5 kb → single-stranded RNA → genome of the virus. First, the protease excises itself out of the polyprotein, and then processes the other polypeptides.
The specificity of TEV protease for its cleavage site can be exploited for *in vivo* studies. The corresponding nucleotide sequence of the heptapeptide target can be inserted into any gene using directed → mutagenesis or → transposon integration. The protein expressed from the mutated gene will then be selectively inactivated by a concomitantly expressed TEV protease in vivo, from which the cellular function of the mutated gene can be inferred.

T-odd phage: See → T phages.

TOGA: See → *t*otal *g*ene expression *a*nalysis.

Top-down mapping: A technique to establish long-range → genetic maps. Chromosomes are first separated with suitable electrophoretic methods (e.g. → pulsed-field gel electrophoresis), transferred onto hybridization membranes, immobilized, and hybridized to a radioactively or non radioactively labeled DNA → probe. After autoradiography the probe can be assigned to a specific chromosome. This mapping procedure can also be used for subchromosomal fragments.

TOP gene: See → 5' terminal oligopyrimidine gene.

Topo cloning: See → topoisomerase I cloning.

Topoisomer (*topo*logical *isomer*): Any molecule that differs from otherwise identical molecules in its topological characteristics. For example, a distinct DNA duplex molecule adopts various topological conditions: linear, relaxed circular, or supercoiled circular. See → DNA topology.

Topoisomerase (DNA topoisomerase, swivelase): An enzyme that catalyzes the interconversion of → topoisomers of DNA duplex molecules. The enzyme interconverts knotted or catenated forms (see → catenate) of DNA or changes the → linking number of circular DNA duplexes. Basically two types of topoisomerases are known, see → DNA topoisomerase I and → DNA topoisomerase II.

Topoisomerase I cloning (topoisomerase TA cloning, topo cloning): A technique for the rapid and efficient → cloning of → *Taq* DNA polymerase-amplified PCR fragments, that uses vaccinia virus → topoisomerase I instead of → DNA ligase for the recombination of → vector and → insert DNA. The technique exploits the ability of topoisomerase I to bind to duplex DNA, to cleave the phosphodiester backbone of one strand, and to rejoin the DNA strands. The virus-encoded enzyme cleaves at a consensus pentapyrimidine element (5'-CTCCTT-3'), more specifically 3' of the final T of this element. In the cleavage reaction, bond energy is conserved in the formation of a covalent adduct between the 3'-phosphate of the cleaved strand and a tyrosyl residue on the enzyme. Therefore, the engineering of the cleavage motif CTCCTT within 10 bp of the 3'-terminus of duplex DNA allows to form a stable, highly recombigenic topoisomerase-DNA complex with a 5' single-stranded tail (with a T at the end), that religates to an acceptor DNA with a 5'-OH-tail complementary to the end of the donor molecule (here, an A). Since the downstream portion of the cleaved strand spontaneously dissociates from the complex, ligation occurs predominantly with a heterologous acceptor DNA.
The design of a → plasmid vector with a → multiple cloning site flanked by CTCCTT elements in inverted orientation allows to linearize such that the topoisomerase I sites are located at both vector ends. Once covalently activated by topoisomerase I, the vector can be directly used for cloning. In short, the target sequence is first amplified by conventional → polymerase chain reaction employing Taq DNA polymerase, the amplification product with a 3' A-overhang (see → extendase) mixed with the linearized and activated topoisomerase I cloning vector (covalently bound to the enzyme). The vector contains a → kanamycin and → ampicillin resistance marker, a Col E1 → origin of replication, and a → multiple cloning site (MCS) inserted, into a *lac*Z α gene. The MCS is flanked by a T7 promoter and an M13 reverse priming site, and the T-overhangs at the cloning site are bracketed by *Eco*R I restriction recognition sequences. A similar vector for topo I cloning in eukaryotic cells harbors the same antibiotic resistance markers, a Col E1 origin, an MCS with flanking T 7 promoter for in vitro transcription of the sense RNA and sequencing of the insert, a C-terminal V5 → epitope tag for easy detection of expressed recombinant proteins (using anti-V5 antibodies), a C-terminal → histidine (6 × histidine) tag for expressed protein purification using → immobilized metal affinity chromatography and subsequent detection with an anti-histidine tag antibody, and one topoisomerase I molecule covalently bound to each T-overhang. The vector is also equipped with a strong *cyto*megalo*v*irus (CMV) promoter for high-level expression of the insert, a → polyadenylation signal from SV 40 and a transcription termination sequence from *b*ovine growth *h*ormone (BGH) gene. The ligase activity of topoisomerase I now inserts the DNA fragment covalently into the vector, and the enzyme dissociates from the construct, which can immediately be used for → transformation, Transformants can be selected on kanamycin or ampicillin and visualized by → blue/ white screening. The insert can be transcribed into the corresponding RNA by the T 7 promoter, and be easily excised from the vector by *Eco* RI. Topo cloning then works without DNA ligase (and associated difficulties), vector preparation, restriction enzymes, or selective precipitation or purification of the amplification product.

Topoisomerase I recognition site

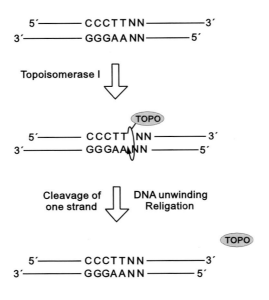

Reaction Mechanism of Topoisomerase I

Topoisomerase I recognition sites

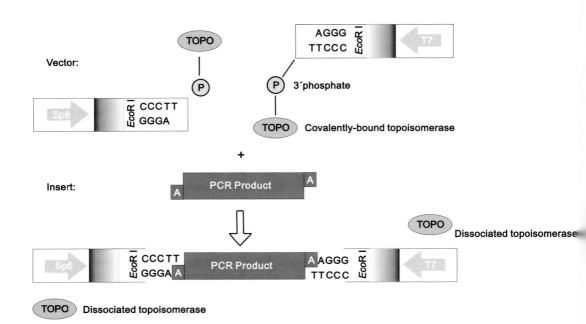

Topoisomerase TA cloning: See → topoisomerase I cloning.

Topological isomer: See → topoisomer.

Topological proteomic fingerprint: The specific three-dimensional distribution of specific protein fractions (e.g. membrane proteins, signalling proteins) in a cell at a given time. The topological proteomic fingerprint is continuously changing during the life of a cell, and can be used to discriminate between normal and aberrant cell types (e.c.cancer cells). See → topological proteomics, → toponome.

Topological proteomics: The whole repertoire of techniques to decipher the → toponome, i.e. the three-dimensional protein networks in a cell or an organelle. See → protein linkage map, → protein-protein interaction map, → compare → chemical proteomics, → proteomics, → subcellular proteomics.

Topological winding number: α, see → linking number.

Toponome **(topological proteome):** The complete three-dimensional arrangement and location of all proteins of a cell. See → topological proteomics.

Total gene expression analysis **(TOGA):** A technique for the automated high-throughput analysis of the expression of nearly all genes in a given cell, tissue, or organ. The method is based on the fact, that almost all → messenger RNAs can be identified by an 8 nucleotide sequence and the distance of this sequence from the → poly(A)-tail. In short, poly(A)$^+$-mRNA is first isolated and double-stranded → cDNA synthesized by → reverse transcription, using a pool (e.g. 48) of equimolar *Not*I-containing 5'-biotinylated anchor → primers, degenerate in their 3' ultimate three positions (e.g. 5'-T$_{18}$VNN [V = A,C or G; N = A,C,G or T]). One primer of this primer mixture initiates synthesis at a fixed position at the 3' end of all copies of each mRNA species in the sample (defining a 3' endpoint for each species). Then the cDNAs are cleaved with *Msp*I (recognition site: 5'-CCGG–3'), the 3'-fragments isolated by → streptavidin bead capture and released from the beads by *Not*I digestion. *Not*I cleaves at an 8 nucleotide sequence within the anchor primers (but rarely within the mRNA-derived part of the cDNAs). The resulting *Not*I-*Msp*I fragments are then directionally cloned into a *Cla*I-*Not*I-cleaved → expression vector in an → antisense orientation to its → T3 RNA polymerase promoter, and the constructs transformed into an *E. coli* host. The plasmids are then isolated, the insert-containing vectors linearized with *Msp*I, that cleaves at several sites within the vector but not in the cDNA inserts or the T3 promoter (insert-less plasmids are concomitantly inactivated), and anti-sense cRNA transcripts of the cloned inserts produced with T3 RNA polymerase. These transcripts contain known vector sequences ("tags") abutting the *Msp*I and *Not*I sites. These cRNAs, after removal of the plasmid DNA template with RNase-free DNase, serve as substrates for reverse transcriptase using a primer complementary to the vector sequences. Then the resulting cDNA is amplified with a primer extending across the non-reconstituted *Msp*I/*Cla*I site (with either A,C,G or T) and a universal 3'-primer in a conventional → polymerase chain reaction. A subsequent PCR with a fluorescent 3'-primer and each of the 256 possible 5'-primers extending 4 bases into the inserts (each one in a separate reaction) generates products that are separated on denaturing → sequencing gels, and the bands detected by laser-induced fluorescence. Each final PCR product carries an identity tag, a combination of an 8 nucleotide sequence (in case of *Msp*I: CCGGN$_1$N$_2$N$_3$N$_4$) and its distance from the 3' end of the mRNA (also a known vector-derived sequence added during TOGA processing). See → differential display, → RNA arbitrarily primed PCR. Compare → serial analysis of gene expression.

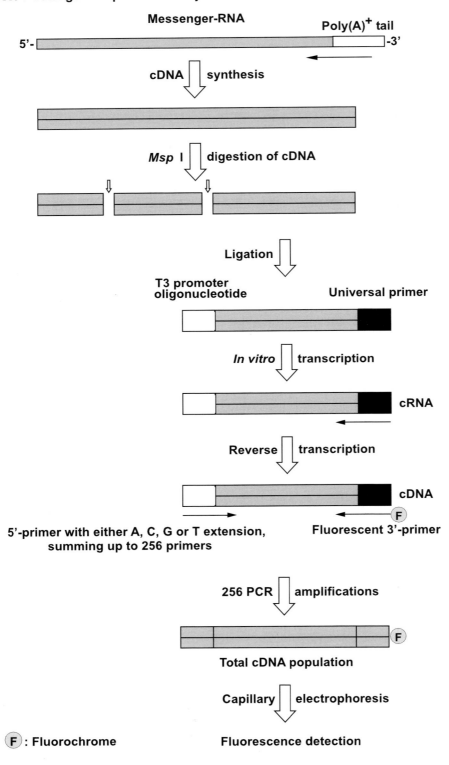

Total gene expression analysis (TOGA)

Total *internal* reflection *fluorescence* *microscopy* (TIRFM; *evanescent* *wave* *microscopy*, EWM, *attenuated* *total* reflection microscopy, ATR microscopy): A microscopic technique for the visualization of fluorescent molecules in fixed or living cells. In short, a target cell is first infiltrated with → fluorochromes attached to a reactive molecule (e.g. an → antibody), fixed or grown on a microscope cover slide, and laser light focused onto the cell through an objective. Within the cover slide glass a "total" internal reflection of the laser light takes place. However, if a medium with lower refractive index is in close vicinity of the cover slide (as e.g. the cell), a small quantum of light "leaks" into the neighboring optically less dense medium. This evanescent light penetrates the cell up to a depth of 100 nm and excites the fluorochromes, whose fluorescence light is emitted and collected through the objective. TIRF microscopy allows to visualize interactions between the cell membrane and penetrating molecules (e.g. DNA or RNA).

Total noise: The sum of all undesirable contributions to the reading of a detection instrument, that obscure the real data. For example, in → microarray experiments, the → dark current, → electronic noise, → microarray noise, → optical noise, → sample noise and → substrate noise all add to the total noise and therefore have to be reduced, eliminated or at least corrected for.

Total overlap: The sum of → medium, → strong, and → weak overlap matches.

Touchdown PCR: See → touchdown *polymerase* *chain* *reaction*.

Touchdown *polymerase* *chain* reaction (touchdown PCR): A variant of the conventional → polymerase chain reaction in which the annealing temperature is initially set very high. This prevents any unspecific → priming, but allows only a small fraction of the → primers to anneal, so that the yield of the amplification product is low. Subsequently the annealing temperature is reduced stepwise to increase the amplification efficiency. Since the concentration of the primary, highly specific amplification product is increased during the following cycles, it will be amplified preferentially.

Touch gene (TCH gene): A misleading term for any one of a series of plant genes (TCH 1,2,3,4, and so on), that are induced by mechanical stimulation ("touch") of cells, tissues, or organs. Mechanical stress translates to an increased influx of Ca^{2+} with concomitant activation of signal networks finally leading to the transcription of the touch genes.

Tourist: A trivial name for a large family of plant → transposon-like insertion elements of about 130 bp in length, that occur in → exons, → introns or flanking sequences of maize and barley genes. The tourist elements have conserved 14 bp → terminal inverted repeats, a subterminal pentamer repeat 5'-GGATT-3', are flanked by a 3 bp target site direct repeat, and possess insertion site specificity, i.e. prefer the target sequence 5'-TAA-3'. Tourist sequences own the potential to form a → hairpin structure. The copy number of these elements in the maize genome is very high (1 tourist per 30 kb maize DNA).

T-overhang vector: See → T vector.

Toxicogenomics: A variant of → genomics, which uses → microarrays on glass or polystyrene chips (called ToxiChip, or similar) with thousands of spotted genes to probe potentially toxic or genotoxic (i.e. mutagenic) chemical compounds. To that end, a cell, tissue or organ is first exposed to the toxicant, the RNA extracted, reverse-transcribed into → cDNA, which is then labeled with an appropriate → fluorochrome and then hybridized to the ToxiChip. Hybridizations per se indicate presence of a particular → messenger RNA, the extent of hybridization reflects the → abundance of the mRNA in the sample. Differences in hybridization signals between control and toxin-treated cells are indications for toxin-induced expression of a gene.

Toxicoproteomics: The whole repertoire of techniques for the identification and characterization of proteins induced by toxic substances. Compare → toxicogenomics.

Toyocamycin: A → nucleoside antibiotic of *Streptomyces* species, that is used as → base analogue by nucleolar → RNA polymerase I (A), and thereby incorporated into the → pre-ribosomal RNA, where it prevents normal → processing. See → tubercidin.

Toyocamycin

TP:
a) See → *t*elomere-binding *p*rotein.
b) See → *t*ransgenic *p*lant.
c) See → *t*ransport *p*rotein.

T-PCR: See → *t*agged random primer *p*olymerase *c*hain *r*eaction.

Tp53 **(tumor suppressor gene 53):** A 20,303 kb human gene, containing 11 exons and located on the short arm of chromosome 17 (17p13.1), that encodes a protein of 393 amino acids involved in various metabolic pathways (e.g. control of the cell cycle, regulation of DNA repair, commitment in→ apoptosis, and tumor angiogenesis). Mutations in certain exons of this gene frequently are associated with either pre-cancerogenesis or tumorigenesis. For example, 90% of all mutations in this gene are found in exons 5–8 (which altogether encode only 181 amino acids, but essential functions of the p53 phosphoprotein). These mutations are clustered in socalled hot spots, about 18 codons with high mutability. The frequency with which these hot spots occur, is different for different tumors. For example, in Aflatoxin B1–induced liver carcinoma, the → transversion G:C → T:A (arginine to serine exchange) occurs specifically in the third base position of codon 249 in exon 7. Tp53 mutations are better indicators for tumorigenesis in some carcinomas (e.g. the prostate carcinoma, where it is a better prognostic factor as compared to the *p*rostate-*s*pecific *a*ntigen PSA). Cells with a mutated *Tp53* gene frequently show higher Tp53 expression levels. Mutations in this gene can be detected by → temperature-gradient gel electrophoresis (and verified by subsequent sequencing).

T phages: A series of → bacteriophages infecting *E. coli* (→ coliphages) with a linear, double stranded DNA genome of about 40 (e.g. T7 phage) to 165 kb in length (e.g. → T4 phage). The T phages are arbitrarily categorized into T-even phages (e.g. T2, T4, T6) and T-odd phages (e.g. T1, T3, T5, T7). These phages are valuable sources for a series of enzymes used in gene technology (e.g. → T3 RNA polymerase, → T4 DNA ligase, → T4 DNA polymerase, → T4 polynucleotide kinase, → T4 RNA ligase, → T7 DNA polymerase, → T7 RNA polymerase, and others).

T-pilus: A flexible filamentous protrusion of → *Agrobacterium tumefaciens* cells, formed after contact of the bacteria with wounded plant cells, which induces the → vir region of the → Ti plasmid. T-pili consist of virB2, virB5 and virB6 proteins, and probably tether the plant host cell to the bacterium and provide the conduct for → T-complex transport.

T primer: An oligonucleotide → primer containing an oligo(dT) tail of about 5–20 residues, that serves as a → reverse primer in → differential display reverse transcription polymerase chain reaction in combination with a → P primer of arbitrary sequence.

TPS (*triisopropylsulfonyl chloride*): A → coupling reagent used for → chemical DNA synthesis.

TψC loop: A → stem- and loop structure close to the 3' acceptor terminus of → transfer RNA molecules that interacts with → ribosomal RNA. This loop contains the modified base → pseudouridine (ψ) between a T and a C residue (TψC).

Tracking dye: Any organic compound that allows the visualization of the running progress during an electrophoretic run (→ gel electrophoresis). Tracking dyes are included in the sample buffer (e.g. → bromophenol blue, that migrates with DNA fragments of about 10-100 bp in length, bromocresol green, methyl green, and → xylene cyanol, that migrates with DNA fragments of about 5 kb). In most electrophoresis systems both bromophenol blue and xylene cyanol are used.

***tra* genes (*transfer genes*):** A set of about 12 → plasmid genes necessary for the transfer of a plasmid from one bacterial cell (donor) to another (acceptor, recipient) during → conjugation. The *tra* system of the → F factor, for instance, harbors about a dozen genes required for → pilus formation (e.g. *traA*), other genes involved in DNA processing (e.g. *traY* and *traZ*), in transfer and → replication (*traM*) and unwinding of the duplex DNA prior to transfer of one strand (*traI*). Other genes encode proteins of the inner (*traS*) and outer (*traT*) cell membrane which act as entry exclusion proteins (i.e. they reduce the ability of the cell to act as a recipient for conjugation, thereby preventing unproductive mating between cells carrying the same or a related plasmid). Plasmids can be categorized as → conjugative or → non-conjugative depending on whether or not they carry *tra* genes. See also → *mob*.

Trailer:
a) 3' *un*translated *r*egion, 3'-UTR, 3' untranslated sequence, untranslated sequence: Sequences of eukaryotic → messenger RNAs flanked by the coding part to the 5' and the → poly(A) tail to the 3' side. This region contains the → poly(A) addition signal, which is located about 5 to 30 nucleotides → upstream of the poly(A) sequence and serves as signal sequence both for an → endonuclease to cleave the mRNA at a site 14-20 bases downstream of it and for a → poly(A) polymerase to add a → poly(A) tail to the 3'-terminus of the molecule. The length of the trailer increases with evolutionary complexity of the organisms: 200 (fungi, plants), 300 (invertebrates), 400 (cold-blooded vertebrates), 420 (rodents) and 500 nucleotides (humans). See → leader sequence, → messenger RNA circularization.
b) 3' non-coding region: The sequences at the 3' end of eukaryotic genes that do not code for proteins but are also transcribed and contain important signal sequences, e.g. the sequence

AATAAA coding for the mRNA → poly(A) addition signal which may occur in more than one copy per message. Furthermore a consensus sequence YGTGTTYY ("GT box"), located about 30 bases downstream of the poly(A) signal, is possibly involved in transcription termination and polyadenylation of the message. The DNA of the 3' non-coding region thus encodes the mRNA 3' untranslated region.

Trait: A phenotypically detectable character or property of an organism.

Trans: A prefix denoting 'on the other side', contrary to → *cis*. Used:
a) in biochemistry, for a molecule having certain atoms or groups of atoms on the other side of the molecule (in trans) as viewed from reference atoms or groups of atoms.
b) in molecular biology, for a DNA sequence (e.g. a gene) that is located on one → chromosome as viewed from a reference sequence on another chromosome.
c) in molecular biology, for a protein encoded by sequences on one chromosome but acting positively (*trans*-activating protein) or negatively (*trans*-silencing protein) on the expression of reference sequences (genes) on another chromosome. See → transcription factor, → activator, → repressor.

Transa-Bind™: The trademark for a chemically modified cellulose membrane filter, that is used for → blotting and → hybridization procedures.

Trans-acting factor: See → transcription factor.

Trans-acting protein: See → transcription factor.

Trans-acting RNA (DsrA): An 87-nucleotide untranslated regulatory RNA of *E. coli*, that contains sequence complementarity to messenger RNAs from at least five different genes (i.e. *hns*, *arg*R, *ilv*IH, *rpo*S, and *rbs*D), interacts *in trans* with the messages from genes *hns* and *rpo*S via → RNA-RNA interactions and regulates them differently. Whereas DsrA enhances turnover of *hns* mRNA by blocking H-NS translation, it stimulates the translation of RpoS through a stabilization of the *rpo*S mRNA. RpoS represents the stationary phase and stress-responsive sigma factor, H-NS is an abundant nucleoid-structuring protein with global transcription repressor functions. In the case of *hns* mRNA, DsrA interacts with the start and stop codon regions and forms a coaxial stack, whereas it forms base pairs with the translational operator of the *rpo*S RNA, opens a stable stem-loop of the RNA such that the → Shine-Dalgarno sequence becomes accessible, finally enhancing translation. DsrA consists of three stem-loops, the first one carries an *rpo*S complementary motif, the centrally located second stem-loop harbors an *hns* RNA complementary region, and the third stem-loop is the transcription terminator of DsrA. Therefore, one single RNA regulates the activities of at least two transcriptional regulators in an opposite way.

Transactivation response region RNA (TAR-RNA): A 59 base stem-loop structure at the 5'end of all nascent HIV-1 transcripts, that interacts with specific regions of the HIV-1 Tat protein (e.g. amino acid residues $_{48}$GRKKRRQRRR$_{57}$) and *trans*-activates HIV-1 gene expression.

Trans-cleaving ribozyme: Any → ribozyme, that binds to a target and substrate RNA by basepairing, cleaves it and releases the cleavage products.

Transcomplementation: The → complementation of a mutation in a gene of organism A by a wildtype gene from organism B, that has been transferred into the genome of A by gene transfer.

***Trans*-conjugant:** The bacterial cell (recipient) that receives DNA from another bacterium (→ donor cell) during the process of bacterial → conjugation.

Transcribed spacer: The DNA segments separating individual genes in a → transcription unit that are transcribed into a → primary transcript (e.g. → pre-ribosomal RNA, also → histone gene transcripts) but excised during the formation of mature and functional RNA.

Transcript: The single-stranded RNA molecule produced by RNA polymerase I (A) on → rDNA (transcript: → ribosomal RNA), by RNA polymerase II (B) on → structural genes (transcript: → messenger RNA), and by RNA polymerase III (C) on tDNA (transcript: → transfer RNA). See → RNA polymerase, → transcription.

Transcript array: See → cDNA expression array.

Transcriptase: Synonym for → RNA polymerase.

Transcript cluster: Any one of a series of groups of homologous sequences from different genomes (i.e. different organisms) deposited in different databases(e.g. → cDNAs, → expressed sequence tags), that harbor potential genes, which, however, are not yet proven to be functional.

***Transcript-derived fragment* (TDF):** Any sequence derived from a → transcript, that is generated by → restriction of the corresponding → cDNA. Usually, → messenger RNA is first isolated, reverse-transcribed into a double-stranded cDNA, the cDNA restricted with appropriate → restriction endonucleases (e.g. *Eco* RI and *Mse* I), *Eco* RI- and *Mse* I-complementary → adaptors ligated to the restriction fragment, and the fragment amplified by conventional → polymerase chain reaction, using adaptor-specific → primers. By necessity, TDFs represent only parts of mRNAs.

Transcript heterogeneity: The production of multiple transcripts from one particular gene, that differ in sequence and function(s). For example, tissue-specific transcripts from the SA gene appear in different organs of the rat. Transcript heterogeneity in this case is brought about by → exon repetition. Other mechanisms give rise to transcript heterogeneity. At the DNA level, the use of multiple → transcriptional start sites, at the RNA level an → alternative splicing of the → primary transcript and variability in the site and length of the 3'poly(A)-tail may also be responsible for transcript heterogeneity.

Transcript imaging (transcriptome imaging): The visualization of – ideally all – transcripts (i. e. → messenger RNAs) of a cell at a given time. A complete imaging can only be achieved by high throughput techniques as e. g. → cDNA microarray screening.

Transcription: The synthesis of an RNA molecule on a DNA or RNA → template, catalyzed by DNA-dependent or RNA-dependent → RNA polymerases, respectively. See also → *in vitro* transcription, → run-off transcription. For further details see → transcription initiation complex, → transcription initiation site, → transcription initiation unit. Compare → reverse transcription, → replication.

***Transcription activation factor* (TAF;** *TATA* **box binding** *protein* **associated factor, TBP-associated factor):** Any one of a series of nuclear transcription co-factors that associate with the → transcription factor II D, to form a large heterogeneous multisubunit protein complex (pre-initiation complex), that is a prerequisite for the formation of a → transcription initiation complex, and mediates activation (or repression) signals from regulatory proteins bound to → enhancer (or → silencer) sequences.

Transcriptional control: The regulation of the expression of a particular gene through controlling the number of transcripts per unit time, as opposed to → translational control.

Transcriptional enhancer: See → enhancer.

Transcriptional fusion: The ligation of two protein-encoding genes or parts of them to form a → fused gene, in which all protein-coding sequences are derived from one (the → reporter gene), and the regulatory sequences from the other gene (the controller). See also → sandwiched gene.

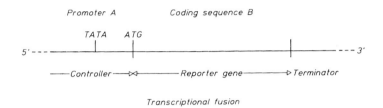

Transcriptional fusion

***Transcriptional gene silencing* (TGS):** The complete suppression of the transcription of an endogenous gene or a → transgene. TGS may be brought about by different mechanisms. For example, → chromatin structure may impose silencing (a transgene inserted into → heterochromatin adopts this condensed state and is silenced), the presence of endogenous repetitive sequences may recruit chromatin components (e.g. proteins) that induce silencing of neighboring transgenes, a → paramutation may be induced in one allele, leading to TGS of the corresponding allele, or the integration of multiple copies of a transgene in a special spatial arrangement leads to cytosine methylation and TGS in both *cis* and *trans* (i.e. transgene loci silence → ectopic transgenes driven by homologous promoters). Moreover, *trans*-TGS can be mediated by an aberrant RNA (e.g. truncated or non-polyadenylated), or even by DNA viruses (e.g. Cauliflower mosaic virus produces an aberrant RNA, that impedes transcription through DNA-RNA interactions). TG-silenced transgenes are hypermethylated, and hypermethylation attracts nuclear proteins (e.g. MeCP2) that assemble the local chromatin into a repressive heterochromatic complex. Obviously many genes are involved in TGS. In *Arabidopsis thaliana*, genes *hog*1, *sil*1, *som/ddm*1 and *mom*1 encode proteins that fix the suppressed state, whereas mutations in these genes relax it (by reverting cytosine methylation). Compare → posttranscriptional gene silencing.

Transcriptional ground state: The inherent activity of a → promoter (i.e.core transcription machinery) *in vivo* in the absence of specific regulatory sequences (and therefore absence of activators and repressors).

***Transcriptionally competent complex* (TCC):** A complex of about 17 distinct proteins and a molecular mass of → 2 Mda, which is the absolute minimal requirement for → transcription initiation.

Transcriptional noise: The presence of spurious, non-spliced, or truncated → transcripts in the → transcriptome of a cell.

Transcriptional silencer: See → negative element.

Transcriptional synergy: The cooperative synergistic effect of specific combinations of → transcription factors on the → transcription of a specific → gene. This synergism leads to a significantly higher level of → transcription than the sum of the effects of all individual factors.

Transcriptional terminator: See → terminator sequence.

Transcription and mRNA export **(TREX):** A conserved protein complex coupling the transcription of a specific gene with the export of the resulting → messenger RNA (mRNA) into the cytoplasm. The nascent mRNA is first bound by the socalled THO complex, consisting of the proteins Tho2p, Hpr1p, Mft1p and Thp2p (and others). The messenger RNA export proteins Sub2p, Yra1p and Tex1p also interact with this THO complex, upon which Yra1p recruits the heterodimer Mex67p-Mtr2p. This in turn dissociates Sub2p from the mRNP. The Mex67p-Mtr2p messenger exporter mediates the transport of the mRNP through the nuclear pore via a direct interaction with the nucleoporins. On the cytoplasmic side of the nuclear membrane the mRNA is separated from the Mex67p-Mtr2p complex and is then translatable by ribosomes. The export factors then re-enter the nucleus and become engaged in a new round of mRNA export.

Transcription attenuation: See → attenuation, definition b.

Transcription-based *amplification* *system* **(TAS):** A technique for the isothermal amplification of nucleic acid molecules. In short, total RNA is first isolated and reverse transcribed into cDNA using a → primer complementary to the target sequence at its 3' end and carrying a → T7 RNA polymerase → promoter at its 5' end. A subsequent transcription of the cDNA with T7 RNA polymerase results in multiple RNA copies of the target cDNA, which are again converted to cDNA with → reverse transcriptase. These cDNAs can again be transcribed by T7 RNA polymerase. Compare → nucleic acid sequence-based amplification, → transcription-mediated amplification.

Transcription *chain* *reaction* **(TCR):** A solid-phase RNA amplification technology, that allows to amplify a single RNA template several million-fold. In short, total RNA is first isolated from the organism of interest, the polyadenylated → messenger RNA captured by an oligo(dT)-tail linked to a → T7 (or T3) promoter sequence, which in turn is covalently bound to a solid surface, and → cDNA synthesis started by the addition of → reverse transcriptase. The resulting double-stranded cDNA is then copied by a → DNA polymerase. Subsequently, a socalled TCR tag is ligated to the cDNAs, and T7 (or T3) RNA polymerase used to amplify these fragments in several rounds to very high concentrations (one T7 promoter produces from 100–1,000 copies of anti-sense or sense RNAs).
Figure see page 1162.

Transcription complex: See → transcription initiation complex.

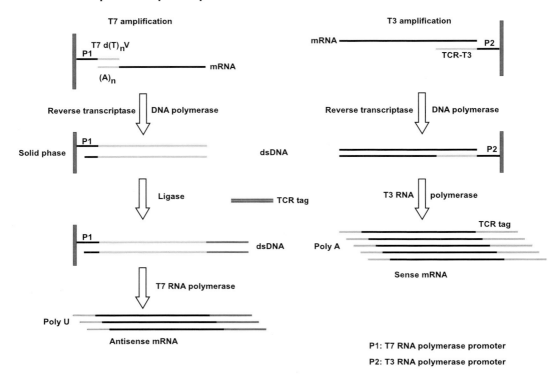

Transcription chain reaction

Transcription-*c*oupled *r*epair (TCR): A somewhat misleading term for the repair of DNA damage (e.g. UV-induced → cyclobutane dimers, → thymine glycols, → 8–oxo-guanine, or other mutated bases) involving the recognition of the damaged site by the transcriptional complex. During the transcription of a gene, the elongating DNA-dependent RNA polymerase arrests at an injury in the → template strand, causing to stall the transcription complex and leading to a transcriptional collapse. The stalled RNA polymerase II complex harbors, among others, CSA, CSB, transcription factor IIH, XPG, and probably other proteins, that process the arrested RNA polymerase, thereby making the damaged site accessible for repair. Processing involves → ubiquitination and subsequent degradation of the polymerase. The damage itself is then repaired by lesion-specific or lesion-independent repair systems.

Transcription cross-talk: A laboratory slang term for the use of an → RNA polymerase II → promoter by RNA polymerase III, or vice versa. Transcription cross-talk is normally impossible.

Transcription *f*actor (TF; *trans*-acting factor, *trans*-acting protein, nuclear factor): Any one of a class of nuclear → DNA-binding proteins that interacts with its recognition sequence (binding site), and facilitates the initiation of → transcription by eukaryotic DNA-dependent → RNA polymerase. Transcription factors may bind to → *u*pstream *r*egulatory *s*equences, to the → TATA box or also to sequences within the coding region (e.g. in the case of → class III genes). See for example → SP 1, → transcription factor II D. Compare → activator, → repressor.

Transcription factor binding site (transcription factor address site, transcription factor recognition element, transcription factor recognition site): Any one of short (4–25 bp) conserved sequence elements in → promoters, that function as address modules for the specific recognition and binding by the corresponding → transcription factors. Such sites are also present in different promoters. For example, the octamer element 5'-ATGCAAAT–3'is part of immunoglobulin gene → enhancers, but also present in promoters of → house-keeping genes such as e.g. → histone genes and → snRNA genes. At least two → helix-turn-helix transcription factors recognize this element, one (OTF–1) is ubiquitous, the other one (OTF–2) is lymphoid-specific. Moreover, their acidic → activation domains are different as is their capacity to interact with other proteins of the pre-initiation complex.

Transcription factor family: A group of regulatory proteins involved in → transcription, that either recognize similar or identical target sequences (see → transcription factor binding site) as e.g. transcription factors NF1, CTF, NFY, or CBF, all recognizing the motif 5'-CCAAT–3', or contain similar protein → domains (e.g. → helix-turn-helix, → leucine zipper, or zinc fingers).

Transcription factor recognition element: See → transcription factor binding site.

Transcription factor recognition site: See → transcription factor binding site.

Transcription factor II A **(TFIIA):** A nuclear protein complex (α: 37 kDa; β: 19 kDa; γ: 13 kDa) that contacts → transcription factor II D and stabilizes its interaction with the → TATA box. See → transcription initiation complex.

Transcription factor II B **(TFIIB):** A nuclear 35 kDa zinc finger protein that binds to → transcription factor II D and spans about 30 base pairs to determine the → cap site. It recruits → RNA polymerase II (B) and → transcription factor II F into the → pre-initiation complex. See → transcription initiation complex.

*Transcription factor II*B *recognition* e*lement* **(TFIIB-RE; BRE):** A → core promoter element (consensus sequence: 5'-G/C-G/C-G/ACGCC-3') in addition to the → TATA box, the → initiator element, and the → downstream promoter element, that represents the binding site for general → transcription factor IIB. This protein, after being bound to the BRE via a canonical → helix-turn-helix motif, affects the assembly of the → preinitiation complex and the transcription of the adjacent gene.

Transcription factor II D **(TFIID, TATA-binding protein, TATA box-binding protein):** A complex of at least ten different nuclear proteins, one of which recognizes the → TATA box and binds to this motif via a highly conserved carboxyterminal domain (TATA-box-binding protein, TBP, 38 kDa). Other proteins of this complex possess histone acetyltransferase activity (250 kDa TAF II 250), or similarity to → histone H2B (20 kDa TAF II 20), histone H3 (31 kDa TAF II 31), and histone H4 (80 kDa TAF II 80). Binding of TFIID to its recognition sequence is a prerequisite for the assembly of the → transcription initiation complex.

Transcription factor II E **(TFIIE):** A nuclear zinc finger protein complex that recruits → transcription factor II H into the → pre-initiation complex and additionally modulates helicase and kinase activities of TFIIH. See → transcription initiation complex.

Transcription factor II F **(TFIIF):** A nuclear protein complex (26 kDa RAP 30 and 58 kDa RAP 74 proteins) that recruits → RNA polymerase II (B) into the → pre-initiation complex, and prevents erroneous transcription initiation. See → transcription initiation complex.

Transcription factor II H (**TFII H**): A nuclear protein complex of at least 8 subunits with helicase and kinase activities that is recruited by → transcription factor II E into the → pre-initiation complex, where it is necessary for → transcription initiation. TFIIH also functions in DNA repair. Mutations in TFIIH subunits therefore can lead to repair disorders (e. g. Xeroderma pigmentosum). Two subunits (the 37 kDa cyclin H and the 40 kDa cdk 7 proteins) act in concert to catalyze the phosphorylation of the C-terminal domain of RNA polymerase II (B).

Transcription-free cloning: The → cloning of DNA fragments into a → plasmid vector (e.g. derived from → pUC), that contains transcription terminator sequences flanking the → multiple cloning site (MCS) and hence the cloned → insert. The terminators prevent (or at least reduce) undesirable transcription out of the insert (and also into it). In conventional cloning vectors, the MCS is contained within the indicator gene (e.g. → ß-galactosidase), and any insert cloned into this site can be transcribed. If the transcript encodes a toxic protein, then expression selects against such inserts. Such bias is avoided by transcription-free cloning vectors.

Transcription initiation: The start of the → transcription of a gene into the corresponding → messenger RNA, which presupposes the formation of the → RNA polymerase → holoenzyme (in prokaryotes) or a → transcription initiation complex, consisting of various → transcription factors and DNA-dependent → RNA polymerase (in eukaryotes). The transcription initiation site is located downstream of the → TATA box and upstream of the → translation initiation site in eukaryotes (→ cap site). See → transcription termination.

Transcription initiation codon: See → start codon.

Transcription initiation complex (*transcription complex, TC*): The complex between more than 100 different → *transcription factors* (TFs) and DNA-dependent → RNA polymerase II (B), that assembles at the → cap site of eukaryotic genes, has a molecular weight of more than 2 MDa, and directs the polymerase to the → start codon for *correct* initation of gene transcription. A first step in the assebmly process in animal → class II gene → promoters is the binding of → transcription factor II D (TF II D), consisting of the *TATA box* binding *protein* (TBP) and a series of so-called *TBP-associated factors* (TAF), to the → TATA box. In → footprinting experiments this leads to protection of a region spanning base pairs –42 to –17 from → DNase I attack. The bound TF II D nucleates the core complex, to which transcription factors TF II A and TF II B (and possibly other proteins, e.g. so-called coactivators) bind serially. This multi-protein complex extends from –80 to +10 and is recognized by RNA polymerase II (B) and TF II E, that binds → downstream of the complex and extends the protected region to +30. Then TF II F, TF II H and TF II J join in. The whole complex interacts with other proteins bound to the → GC box (e.g. SP 1), the → CAAT box (e.g. CTF), and other sequence motifs in distal promoter regions (e.g. → enhancers). These proteins interfere positively (→ enhancers) or negatively (→ silencers) with the activity of the transcription initiation complex. See also → scanning model.
Figure see page 1165.

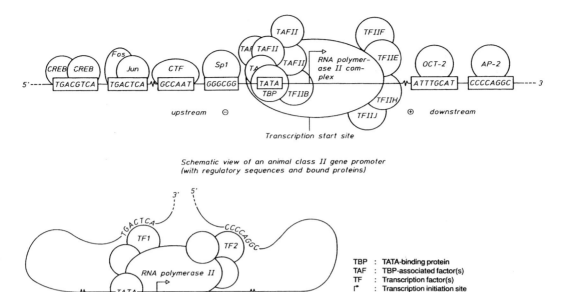

Schematic view of an animal class II gene promoter
(with regulatory sequences and bound proteins)

Transcription initiation complex

TBP	:	TATA-binding protein
TAF	:	TBP-associated factor(s)
TF	:	Transcription factor(s)
⌐→	:	Transcription initiation site

Transcription initiation sequence (transcription initiator sequence; *initiator*, Inr; initiator element; initiator box, transcription initiation site, mRNA initiation site, transcription start site, cap site) A 17 bp sequence element (consensus sequence: 5'-CTCA-3', 5'-PyPyCA(Py)₅-3'(animals), and 5'-PyPyCA(Py)ₙ-3' (plants), more generally 5'-PyPyCAPyPyPyPyPy-3' (where A marks + 1) of → promoters of → RNA polymerase II (B) genes, that is located at + 1 to + 11 and contains the → transcription start site. It is necessary to nucleate the assembly of the various proteins of the → transcription initiation complex, and drives accurate basal transcription initiation of various genes. The initiator represents the simplest functional promoter and is present in many → TATA-box containing promoters. It also may substitute the TATA sequence in TATA-less promoters of e. g. → housekeeping genes. Inr elements are address sites for a series of proteins (e.g. CIF, E2F, USF, → *transcription factor* IIB, and → RNA polymerase II).

Transcription initiation site: See → cap site.

***Transcription-mediated amplification* (TMA):** A variant of the conventional → polymerase chain reaction technique for the amplification of target sequences (e. g. → cDNAs), using transcription as intermediate step. In short, → messenger RNA is first isolated and annealed to a primer that carries a → promoter sequence from → bacteriophage T7. Then this primer is extended by → reverse transcriptase, and a double-stranded cDNA synthesized that contains a T7 promoter sequence. This promoter allows transcription of the DNA template, generating up to 1000 transcripts, which are now amplified with the promoter-primer and a second primer complementary to the RNA transcript, leading to millions of copies after a few rounds of transcription-amplification. Compare → transcription-based amplification system.
Figure see page 1166.

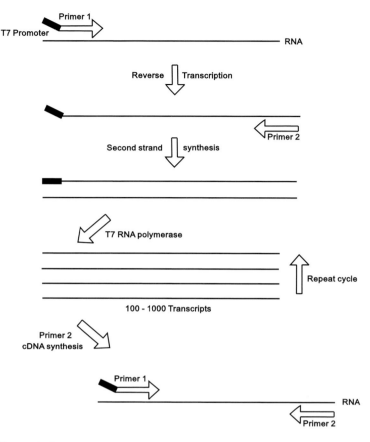

Transcription-mediated amplification

Transcription profiling: The determination of all expressed (transcribed) genes of a cell, a tissue, an organ or organism at a given time. The process of transcription profiling produces an → expression profile. See → transcript profiling, → expression profiling.

Transcription-regulating sequence (TRS): A short conserved pentanucleotide sequence in the RNA genome of mammalian nidoviruses (arteriviruses, coronaviruses), where it precedes every → transcription unit (so-called body TRS), and additionally is part of the 5'leader sequence of the genome (so-called leader TRS). In equine arteritis virus (EAV) TRS, the conserved sequence, 5'-UCAAC-3', is located in a loop region of a → hairpin structure formed by base pairing of the EAV genome. Both the former (so-called body TRS) and the latter (so-called leader TRS) interact by base-pairing to initiate synthesis of → subgenomic messenger RNAs (sgmRNAs) by → discontinuous transcription.

Transcription start site: See → cap site.

Transcription termination: The stop of the → transcription of a gene into the corresponding → messenger RNA. In prokaryotes, → rho factor-dependent and rho factor-independent termination occurs. The transcription termination process in eukaryotes is largely unknown, but seems to involve different signals for the different DNA-dependent → RNA polymerases. RNA polymerase I recognizes a termination sequence of 18 bp to which an auxiliary protein is bound, RNA polymerase II possibly leaves the → template strand after contact with a specific secondary structure at the termination site which has a specific sequence (5'-TTTTTATA-3'), and RNA polymerase III terminates transcription at a U_4-sequence embedded in a GC-rich region. In yeast, transcription termination of polymerase II genes occurs at AT-rich sequences, e.g. a 38 bp region including the sequences 5'-TTTTTATA-3', 5'-TATATA-3' (very effective), 5'-TACATA-3', and 5'-TAGTAGTA-3'. None of these sequences functions as transcriptional terminators in the reverse orientation. See → transcription initiation.

Transcription unit (transcripton): The complete DNA sequence between the transcription initiation (→ cap site) and transcription termination sites recognized by DNA-dependent → RNA polymerase. The transcription unit may comprise only one or more than one gene. In the latter case transcription produces a → polycistronic message which in prokaryotes may be translated as such, (giving rise to a → polyprotein which has to be cleaved during → post-translational modification), whereas in eukaryotes the precursor molecule is processed (e.g. the → pre-ribosomal RNA that is cleaved to generate 5.8 S, 18S and 28S rRNA).

Transcription vector: See → expression vector.

Transcript map: See → expression map.

Transcriptome: The entirety of all expressed → genes of a → genome. Also called "expressed genome".

Transcriptome-derived marker (transcriptome marker): Any transcript band detected by → cDNA-AFLP or other transcript profiling techniques, that allows to identify → polymorphisms between parents (and the progeny from a cross between these parents) and can therefore be mapped (see → genetic map). Transcriptome markers differ from conventional → molecular markers: they are not anonymous, but derived from active genes.

Transcriptome imaging: See → transcript imaging.

Transcriptome map: A misleading term for a compilation of co-regulated genes. Compare → transcript map.

Transcriptomics: The whole repertoire of techniques to analyze and characterize the → transcriptome of an organelle or a cell, including → RNA isolation, → messenger RNA isolation, → reverse transcription into → complementary DNA, → agarose or → polyacrylamide gel electrophoresis, → Northern blotting, → cDNA array techniques, isolation of specific transcripts, their → sequence analysis, and use in → transgenics.

Transcripton: See → transcription unit.

Transcriptosome: The multi-protein complex consisting of → RNA polymerase and → transcription factors that is actively transcribing a → gene. See → transcription initiation complex.

Transcript profile: See → expression profile.

Transcript profiling: See → expression profiling.

Transcript sequencing: A method for the determination of → messenger RNA sequences. First, an → oligo (dT) primer is annealed to the → poly(A) tail of isolated polyadenylated mRNA, and used to synthesize a cDNA strand, catalyzed by → reverse transcriptase. This cDNA strand can then be sequenced using the → Sanger sequencing procedure.

Transduced element: The fragment of the chromosome of a bacterial donor cell which becomes incorporated into a → transducing phage and is subsequently transferred into a recipient cell.

Transducing phage: A → bacteriophage particle, which has packaged host DNA associated with the phage DNA. See → transduction.

Transducing retrovirus: A → retrovirus that has integrated host cell genes (generally, DNA sequences) into its genome, but usually suffered deletions of own sequences. These deletions do no longer allow viral replication, which is only possible if a → helper virus supplies replicative functions in → trans. See also → transduction.

Transductant: The bacterial cell (recipient) that receives DNA from another bacterium (donor) in the process of bacteriophage → transduction.

Transduction: The transfer of DNA from one bacterium (donor cell) to another bacterium (recipient cell) with the aid of → temperate or → virulent bacteriophages as vectors. The transduced DNA is derived from host sequences flanking the site of integration of the phage DNA (e.g. DNA flanking the → attachment sites of the → lambda phage in *E. coli*). The *complete* or *recombinative* transduction leads to the covalent integration of the donor DNA into the acceptor DNA, where as in abortive transduction the bacterial DNA carried by the phage is transferred to the recipient cell (and may be expressed there), but is not replicated (see → transient expression). Principally two variants of complete transduction exist:
a) The non-specific or *generalized transduction* (gt). In this case the phage can integrate at almost any position of the host chromosome, and therefore almost any host gene can be incorporated into the transducing phage and subsequently be transferred to the recipient bacterium.
b) The specialized (special, restricted, restrictive) transduction. In this case the phage can integrate only at a specific position of the host chromosome, and therefore only host gene(s) close to this position can be incorporated into the transducing phage and subsequently be transferred to the recipient bacterium.
The transducing phages are defective, since part of their genome has been substituted by the transduced DNA, and can only replicate in the recipient cell if complemented by a wild-type → helper phage. See also → cotransduction.

Transfection (DNA transfection; *trans*formation-in*fection*):
a) The uptake of viral nucleic acid by bacterial cells or → spheroplasts, resulting in the production of a complete virus (e.g. a lytic → bacteriophage).
b) The integration of foreign DNA into the genome of cultured animal or human cells via → direct gene transfer (e.g. by → calcium phosphate precipitation).

Transfer: The process of replacing an exhausted medium by a freshly prepared medium in tissue and cell culture, or the movement of the cells or tissues from the depleted medium to a new medium (especially for solid media).

Transfer DNA: See → transfer RNA genes.

Transfer operon (tra operon): An operon in → conjugative plasmids containing the transfer (→ tra) genes that encode functions needed for conjugal transfer of a plasmid from one bacterial cell (donor) to another (acceptor, recipient).

Transferred DNA: See → T-DNA

Transferrinfection: A method for the introduction of foreign DNA into animal cells that exploits a natural ion-uptake mechanism for DNA transport. In short, conjugates of apotransferritin (an iron-delivery protein) or its analogue from chicken (conalbumin) and a polycationic peptide (e.g. poly-lysine or protamine) are synthesized and ligated through a disulfide bridge. These conjugates are recognized by specific transferrin receptors located in cellular membranes, and imported by receptor-mediated endocytosis. Since polycations such as poly-lysine bind DNA strongly, any DNA (of up to 14 kb) can be transferred into the target cell together with the transferrin molecule by an endocytotic process.

Transfer RNA (tRNA; formerly: soluble RNA, sRNA; 4S RNA; adaptor RNA; amino acid acceptor RNA): Any one of a large number of structurally similar, low molecular weight single-stranded RNAs (4S; about 75-80 nucleotides in length) that transfers an amino acid to a growing polypeptide chain during the process of → translation. The tRNA molecule adopts a characteristic "clover leaf" configuration through maximal intrastrand → base pairing. This structure consists of a stem with a 3'-terminal CCA (the amino acid attachment site), the → TψC loop with 7 unpaired bases, the → anticodon loop containing the codon-recognition site that is specific for a particular tRNA molecule and responsible for the correct recognition of the corresponding → codon in an mRNA molecule, and the dihydrouridine loop. All loops are involved in protein, → messenger RNA or → ribosome recognition processes.

Transfer RNAs contain several purine and pyrimidine residues that are not ubiquitous (→ rare bases), such as ψ(→ pseudouridine), T (ribothymidine), U^d (dihydrouridine), G^m (methyl guanosine), G^d (dimethyl guanosine), I (→ inosine), and I^m (methyl inosine).

Figure see page 1170.

Transfer-RNA directed integration (tRNA-directed integration): The preferential insertion of → retrotransposons into the vicinity of → transfer RNA genes in the genome of the slime mold *Dictyostelium discoideum* (position-specific integration, see → tRNA-directed retrotransposon). Similar integration preferences are characteristic for the → *Ty* elements of *Saccharomyces cerevisiae*. A total of seven such positions-specific retrotransposon sites exist in the *D. discoideum* genome.

Transfer RNA editing (tRNA editing): The post-transcriptional exchange of distinct bases in → transfer RNA. This process occurs in mitochondria of e.g. the amoeboid protozoon *Acanthamoeba castellanii* and the primitive fungus *Spizellomyces punctatus*, and concerns 13 of a total of 16 tRNAs encoded by the mtDNA of these organisms. RNA editing affects several bases. In Marsupialia, as e.g. the opossum *Didelphis virginiana*, mature tRNA molecules are modified by C→U changes catalyzed by deaminases. In tRNA^{Asp}, the RNA editing changes the specificity of a codon: the unedited tRNA contains a GCC codon (specifying the amino acid glycine), which is replaced by GUC (specifying aspartate). Both tRNAs are present in mitochondria in about the same concentration. In essence, tRNA editing allows to synthesize two different tRNAs from a single gene.

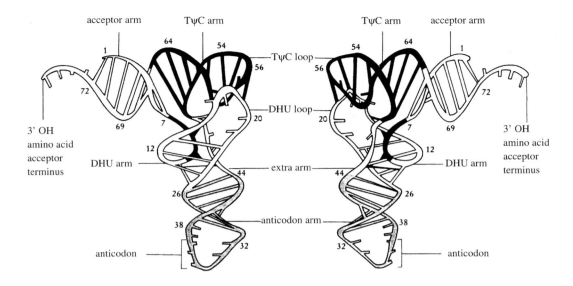

Schematic tertiary configuration of tRNA^Phe of yeast

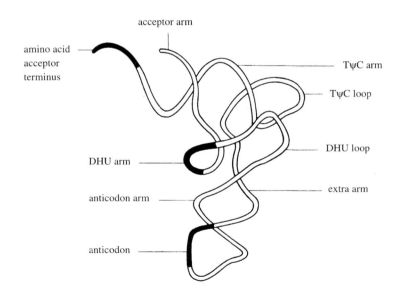

Transfer RNA

Scheme of three-dimensional tRNA structure. The three regions of contact with aminoacyl tRNA synthetase are indicated in black.

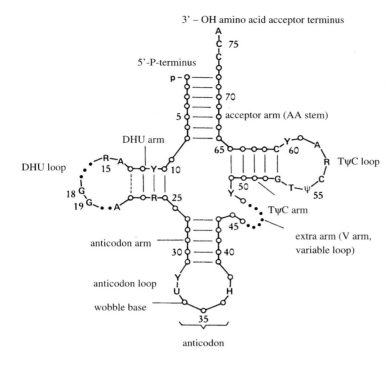

3' – OH amino acid acceptor terminus

5'-P-terminus

DHU arm

DHU loop

acceptor arm (AA stem)

TψC loop

TψC arm

anticodon arm

extra arm (V arm, variable loop)

anticodon loop

wobble base

anticodon

**Schematic „clover leaf"
configuration of tRNAs.**

Dotted line: number of nucleotide
changes from tRNA to tRNA
R: purine base
Y: pyrimidine base
DHU: dihydrouracil
ψ: pseudouridine
H: hypermodified purine
p: 5'-terminal phosphate
T: ribothymidine

Transfer RNA gene (tRNA gene, tDNA): Any gene that codes for a → transfer RNA. In euka-ryotes, tRNA genes are transcribed by DNA-dependent → RNA polymerase III, and are present in multiple copies with average repetition frequencies for each tRNA gene from 5 (yeast), 10 (*Drosophila*) and 15 (humans) to over 200 (*Xenopus*) per haploid genome.
Transfer RNA genes also serve other functions:
a) On compact mitochondrial genomes of mammals, tRNA genes separate other genes and mark positions for → processing.
b) tRNA gene-like structures on retroviral genomes act as → primers to copy the genetic infor-mation of these viruses.
c) In *Saccharomyces cerevisiae*, tRNA genes are frequently associated with → transposons (e.g. sigma elements).
d) In *Dictyostelium discoideum*, tRNA genes offer the preferential integration sites for → trans-posons.

Transfer RNA isoacceptor: See → isoacceptor tRNA.

Transfer RNA splicing (tRNA splicing): A complex process by which → introns are removed from → *precursor* tRNA (pre-tRNA) to produce the functional → transfer RNA. The mechanism of tRNA splicing is best known in yeast (*Saccharomyces cerevisiae*), which contains 272 tRNA genes. Of these, 59 carry introns of 14-60 bp in length, interrupting the → anticodon loop im-mediately 3' to the anticodon. Intron removal is catalyzed by the concerted action of (1) the low-abundant, nuclear membrane-bound tRNA splicing endonuclease, consisting of four differ-ent subunits (Sen 15, Sen 34, Sen 44, Sen 54, where Sen 34 and Sen 44 catalyze cleavage), (2) the tRNA splicing ligase and (3) a 2' phosphotransferase. The endonuclease cleaves the pre-tRNA at the 5' and 3' splice sites to release the introns, producing two tRNA half-molecules with a 2' phosphate remaining at the splice junction, which is transferred to NAD$^+$ by a 2' phosphotransferase. See → transfer RNA. Compare → mRNA splicing.

Transfer RNA suppressor (tRNA suppressor): A → mutation in a → transfer RNA gene that changes its → anticodon to a sequence complementary to a → termination codon. This mutation allows the suppression of amino acid chain termination. See → suppressor gene.

Transfer RNA synthetase recognition site: A sequence in → transfer RNA that serves as recognition and binding site for tRNA synthetase that catalyzes the covalent attachment of a specific amino acid to the 3'-CCA-terminus of the tRNA.

Transformant: A cell that has undergone a → transformation.

Transformation:
a) An alteration in cell morphology and/or cell characteristics (e.g. loss of contact inhibition, neoplastic growth), occurring after integration of nucleic acid from → oncogenic viruses into the cell genome, after exposure to specific chemical carcinogens, or spontaneously (oncogenic transformation).
b) The directed modification of the genome of a cell by the external application of purified, also recombinant DNA from another cell of different → genotype leading to its uptake and integration in this cell's genome (genetic transformation). See also → natural transformation, → transformation frequency.

Transformation-*a*ssociated *r*ecombination cloning (TAR): A technique for the cloning of large → inserts of → genomic DNA, that relies on a 5.7 kb → shuttle vector for bacteria and yeast. This plasmid vector consists of a broad host range bacterial → origin of replication for stable maintenance of large insert clones, a → kanamycin resistance selectable marker gene (selection in bacteria), a *HIS*–3 as selectable marker and *CEN*6 for stability in yeast. A yeast origin of replication is provided in the cloned fragment. The use of a broad host range bacterial origin of replication permits replication also in → *Agrobacterium*, which makes the vector suitable for plant transformation. TAR exploits → homologous recombination for cloning and manipulation of inserts, and can be employed for the isolation and comparison of identical genomic regions in different individuals of an organism and different species, for bridging gaps in → physical maps, for construction of chromosome-specific libraries, and the construction of mutations in large inserts.

Transformation *b*ooster *s*equence (TBS): A genomic DNA segment from *Petunia hybrida* that increases → transformation frequencies if present on the transforming DNA, but exerts no influence on its integration pattern. TBS contains elements with sequence homology to → consensus sequences in → scaffold-associated regions.

Transformation-competent *a*rtificial *c*hromosome vector (TAC vector): A plant transformation vector for the cloning of large → genomic DNA fragments, that is stable in both *E.coli* and → *Agrobacterium tumefaciens* and can be efficiently mobilized from *Agrobacterium tumefaciens* into plant target cells to produce → transgenic plants. The TAC vector consists of the pRiA4 replicon of the Ri-plasmid, the P1 bacteriophage replicon (both responsible for a single plasmid copy in the host cells, which stably maintains foreign DNA), a plant → selectable marker gene (e.g. a → hygromycin phosphotransferase gene driven by the → nopaline synthase promoter and terminated by the nopaline synthase 3'-termination sequence, located at the right rather than left → T-DNA border, so that hygromycin resistant plants can be expected to contain the complete → T-DNA), a → multiple cloning site with various → rare cutter recognition sequences (e.g. *Asc* I, *Bam*HI, *Fse*I, *Hind*III, I-*Sce*I, *Not* I, *Sal*I, *Sfi*I and *Srf*I), a → kanamycin resistance gene, an → overdrive sequence and a → left border sequence. Genomic DNA is ligated into the *Hind*III site of the vector, transformed into *E. coli*, mobilized into *Agrobacterium tumefaciens* by → triparental mating, and then used to transfer the passenger DNA into recipient plant cells (→ gene transfer).

Transformation frequency: The effectiveness with which a microbial host takes up foreign DNA and thereby acquires new properties (as e.g. resistance against an → antibiotic). The transformation frequency is expressed as number of transformants per µg of foreign DNA. The term is also used in → transformation experiments with eukaryotic cells.

Transformation-infection: See → transfection.

Transformation vector: A → cloning vector whose transfer and expression in the host cell leads to the → transformation of this cell, that is to its permanent proliferation and tumorigenesis. See for example → promoter trap vector, → disarmed vector, binary vector, → intermediate vector, → pMon, → retroviral vector.

Trans–4–*hydroxy-L-proline* peptide *nucleic acid* (HypNA): An artificial, negatively charged → peptide nucleic acid (PNA) with high affinity to → ribonucleic acid. It is composed of alternating monomers of trans–4–hydroxy-L-proline and → phosphono PNA monomers, and lacks → polarity. It therefore can bind to target RNA both in parallel and antiparallel orientation. If modified to carry e.g. → thymidyl residues, the resulting poly(T) HypNA invades double-stranded RNA (e.g. in the regions of stem-loop structures at the 3'-end of poly(A)$^+$-mRNAs) and displaces one strand to form a stable → D-loop. HypNA is not degradable enzymatically and therefore stable in biological systems and hybridizes to poly(A)$^+$-mRNA also in vivo. In combination with an → oligo(T)-PNA, HypNA is used to isolate mRNA with short oligo(A)-tails and extensive secondary structure.

Transgene: Any gene that originates from one and has been transferred to a second organism (cell).

Transgene chip: Any → DNA chip containing spotted oligonucleotides complementary to genes or → promoters used in → gene transfer experiments. For example, in plant transformation experiments frequently → 35S promoter and → nos terminator sequences are used for constitutive and correct expression of → transgenes. The transgene chip allows to detect such sequences in → transgenic plants by → hybridization of restricted plant DNA to the chip.

Transgene coplacement: The transfer of two genes into a target organism and their insertion at exactly the same position in the recipient genome. Transgene coplacement is exploited for the study of → lineage-specific position effects.

Transgene DNA (tg DNA): Any DNA, that originates from one, and has been inserted into the genome of another organism. See → transgene, → transgenesis, → transgenic organism.

Transgene expression (heterologous gene expression): The → transcription of a foreign gene in a → transgenic organism and the → translation of the resulting → messenger RNA into a protein. Heterologous gene expression in bacterial hosts, notably *E. coli*, is exploited for the production of pharmacologically interesting proteins (e. g. insulin, human growth hormone, blood coagulation proteins), and necessitates the use of a bacterial promoter. Transgene expression in eukaryotic host cells encounters a series of problems. For example, very AT-rich gene sequences are hardly or not at all expressed. However, many signal sequences such as e. g. → poly(A) addition signals, transcriptional → termination sequences, messenger RNA destabilizing sequences and also some → introns are AT-rich. Also, the transgene may not be expressed (or at a very low level only), if it lacks introns. The underlying mechanism is not clear, but involves post-transcriptional processes. Some introns (e. g. immunoglobulin gene introns) also contain transcriptional → enhancers. Even if the transgene is fully expressed, the protein may be toxic to the host cell, so that transgene expression cannot be exploited.

Transgene genetics: A branch of genetics, that focuses on the design of vectors for gene transfer, the gene transfer process itself, the identification of the position and copy number of the transgene in the host genome, its expression and regulation, stability and inheritance.

Transgene inactivation: See → co-suppression.

Transgene-induced gene silencing: See → co-suppression.

Transgene-induced mutation: Any → mutation that is caused by the → insertion of a → transgene into target DNA.

Transgene rescue (transgenic rescue): The substitution of a mutated gene in an organism by a wild-type → transgene (or a → BAC or → YAC clone carrying the correct gene) such that the original → phenotype is restored. For example, the reversal of a mutant phenotype in mice can be achieved by transferring a BAC or YAC clone with the transgene into the germ line, subsequent crossing of the transgenic mouse with mutant mice, and selection for a complemented phenotype. See → plasmid rescue.

Transgene silencing: See → co-suppression.

Transgenesis: The transfer of genes (or, more generally, DNA) into a cell or organism that does not naturally contain them.

Transgenic animal: An animal whose nuclear → genome contains foreign DNA (e. g. genes) that has been transferred by → transfection or → direct gene transfer. See → knock-out mouse, → transgenic plant.

Transgenic founder: The first-generation transgenic organism arising from a non-transgenic precursor into which foreign DNA (e. g. a gene) has been transferred by standard → gene transfer techniques. See → transgenic line.

Transgenic line: The direct progeny derived from a → transgenic founder. A transgenic line contains the transferred DNA (e. g. a gene) as stably inherited genetic element.

***Transgenic mitigator gene* (TM):** A gene used in genetic → transformation experiments, that is neutral or beneficial for a target organism, but deleterious for a non-target organism. For example, natural fertilization of a wild ("weedy") plant by pollen from a transgenic relative will lead to a progeny, some of which may express the transgene (the proverbial "superweed"). The TM technology employs two transgenetic mitigator genes flanking the gene conferring the desired trait (e.g. insect resistance), that are deleterious only in the weedy transgenic plant.

Transgenic organism: Any organism into which foreign DNA has been introduced, see → transgenic animal, → transgenic plant.

Transgenic plant: A plant into which foreign DNA has been transferred either by → direct, → *Agrobacterium*-mediated, or virus-mediated gene transfer. See → transgenic animal.

Transgenic rescue: See → transgene rescue.

Transgenome: Any → genome that carries one or more foreign genes (generally, DNA sequences).

Transgressive segregation: The appearance of phenotypes in the progeny of a cross, that are either more or less pronounced than the parent phenotype. For example, if in the progeny of a cross between a virulent and an almost avirulent parent some segregants show higher virulence than the virulent parent, and others lower virulence than the almost avirulent parent, they are collectively called transgressive segregants.

Transient: A laboratory slang term for any temporary gradient of molecules (e.g. peptides, proteins, transcripts, also low molecular weight substances such as e.g. hormones) across a cell or also across a cell membrane.

Transient expression (transient gene expression):
a) Temporary → transcription of a gene in a → cloning vector which has been introduced into a non-permissive host system. Since the vector is kept from replicating, its copy number is reduced with each division of the host and so is the extent of expression of its genes. Compare → transient expression vector.
b) The expression of foreign genes that have been introduced into cells, → spheroplasts or → protoplasts by → direct gene transfer, but are not covalently integrated into cellular DNA. These genes are nevertheless transcribed until they are degraded by cytoplasmic and/or nuclear → nucleases so that their expression is only transient. Transient expression assays are preferentially used to test the functionality of gene constructs in host cells, especially their → promoter strength, and their compatibility with → transcription factors. They also serve to optimize DNA delivery into the host cell.

Transient expression vector: An → expression vector that allows the transient → overexpression of cloned foreign DNA sequences in suitable host cells. The expression of the foreign gene(s) is transient only, because the host cell's transcription-translation machinery is soon inactivated, and the host cell dies after a few days. Such → transient expression vectors are used to monitor cloned genes for their function(s) and to detect the effect(s) of → promoter or gene mutations on gene expression.

Transient transfection: See → abortive transfection.

Transilluminator: A UV-light source which serves to induce the emission of fluorescent light from dyes (e.g. → ethidium bromide) intercalated between the two strands of a DNA duplex or double stranded RNA. It serves to visualize nucleic acid molecules separated by → gel electrophoresis. In a transilluminator the UV lamp is positioned beneath an appropriate support for the gel, and UV irradiation is directed towards the experimentor (protection of eyes is necessary).

Transition (transition mutation, transitional mutation, base pair substitution): The replacement of one → purine base by another purine, or a → pyrimidine base by another pyrimidine in a DNA duplex molecule, leading to a transition mutation. The result of a transition finally is the exchange of a G-C pair with an A-T pair, or vice versa. See → point mutation, compare → transversion.

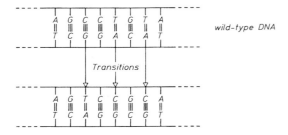

Transition

Transitional mutation: See → transition.

Transition mutation: See → transition.

Transition *stringency* **(TS):** The → stringency, at which a distinct → probe-target DNA duplex dissociates. The lower the TS, the less sequence complementarity between probe and target exists, and *vice versa*. TS is a parameter to determine the reliability of → DNA chip hybridisation experiments, e. g to exclude non-specific cross-hybridizations. See → specificity assessment from fractionation experiments.

Transition temperature: The temperature at which a double-stranded nucleic acid dissociates into single strands. See → denaturation.

Transitive interference RNA (transitive RNAi, transitive RNA): Any double-stranded RNA newly synthesized from a template → messenger RNA by an RNA-dependent RNA polymerase in *Caenorhabditis elegans,* that is usually shorter than the template RNA and act as a → small interfering RNA (i. e. lead to the specific silencing of the corresponding message). Transitive RNAi appears in normal → RNA interference, ensures a drastic RNAi response, and also silence other messenger RNAs, that share regions of homology with the initial target message.

Transit peptide: See → signal peptide.

Transit peptide coding sequence: The DNA sequence coding for the N-terminal domain of a nuclear-encoded organellar protein. This domain is involved in post-translational import of this protein into the organelle (e.g. the chloroplast in plants, the mitochondrium in plants and animals).

Translation: The de-coding of a specific → messenger-RNA (mRNA) by → ribosomes and the translation of this code into a protein. In short, this process starts once the 5' leading terminus of an mRNA molecule is bound to the ribosome. The mRNA then moves through the ribosome (→ translocation, definition a) and serves as a → template for the assembly of amino acids into a protein. As soon as the 5' leading terminus of the mRNA has moved through the first ribosome, it can be bound by a second ribosome and the de-coding of the message and the synthesis of an identical protein is repeated (see → translational amplification). As soon as the 3' → poly(A) tail of the mRNA molecule emerges from the first ribosome the newly synthesized protein is released, and this ribosome may start a second round of translation. The assembly of amino acids into the protein starts at the amino terminus (→ N-terminal end) and ends at the carboxyl terminus (→ C-terminal end) and is mediated by → transfer RNAs. Two binding sites for tRNAs exist on the ribosome: the P site and the A site. The → P (*p*eptidyl-tRNA binding site) binds the tRNA molecule that is attached to the end of the nascent polypeptide chain. The A (*a*minoacyl-tRNA binding site) binds the incoming tRNA that carries the next amino acid. The binding of each tRNA to this site allows base-pairing of its → anticodon with the juxtaposed → codon of the mRNA molecule that progresses through the ribosome. See also → aminoacyl-tRNA-synthetase. Compare → *in vitro* translation.

Translational amplification: The repeated usage of the same → messenger RNA molecule for the production of large amounts of a specific protein. See also → translation.

Translational bypassing: The overriding of noncoding disruptive sequences in a → messenger RNA molecule such that they are not included in the mature protein. For example, the mRNA transcribed from bacteriophage T4 gene 60 (encoding a → topoisomerase subunit) contains a UAG → stop codon after codon 46, followed by 47 nucleotides that are skipped (translationally bypassed) by the ribosome. Synthesis continues at a CUU leucine codon for amino acid 47 and progresses until the complete protein is produced. Translational bypassing is comparatively rare in nature. See → recoding.
Figure see page 1178.

Translational control: The regulation of gene expression by modulation of the rate of → translation of a specific → messenger RNA into a polypeptide, as opposed to → transcriptional control. For example, the 15fold increase in the rate of protein synthesis after fertilization of sea urchin eggs is an example for *quantitative* translational control. Or, the suppression of host cell protein synthesis by certain viruses (e.g. polio- or adeno-virus) with concomitant synthesis of viral proteins is an example for *qualitative* translational control. Translational control generally is determined by such diverse processes and factors as mRNA degradation, mRNA modification, subcellular localization of mRNA, availability of mRNA for competent ribosomes, or → translational control RNA and RNA-binding proteins.

Translational *c*ontrol RNA (tcRNA): Any member of basically two classes of eukaryotic RNAs that are part of large nuclear RNA molecules (→ heterogeneous nuclear RNA) and modulate the → translation of → messenger RNA. The tcRNA from one class, isolated from → *m*essenger ribonucleoprotein particles (mRNP-tcRNA) inhibits the translation of mRNP-mRNA, but has no effect on polysomal mRNA. Polysome-tcRNA, on the other hand stimulates the translation of polysomal RNA but has no effect on mRNP-mRNA. The interaction may be mediated by poly(U) tracts on the tcRNA that form hybrids with the → poly(A) tail of mRNA.

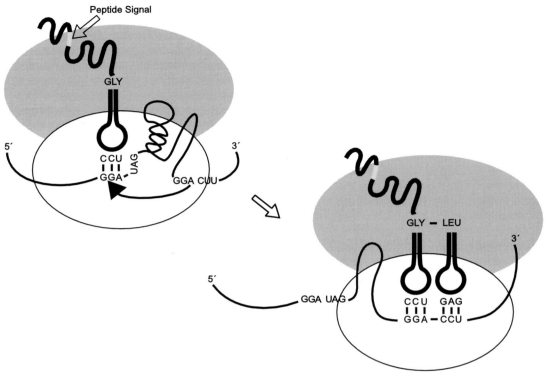

Translational bypassing

Translational fusion: The → ligation of two protein-encoding genes or parts of them to form a hybrid gene (→ fused gene), in which the information for the synthesis of a polypeptide (fused protein) originates from both the reporter and the controller. The activity of the reporter can easily be measured and serves to detect the fused protein in e.g. → expression gene libraries (→ protein tagging).

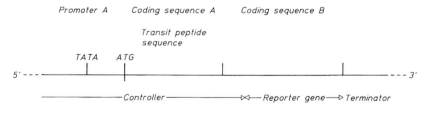

Translational fusion

***T**ranslational **r**epression **a**ssay **p**rocedure* (TRAP): A technique for the study of RNA-protein interactions and the effect of pharmacological compounds on RNA-protein interactions *in vivo*, the cloning of RNA-binding proteins and the characterization of RNA sequence or protein domains essential for binding. TRAP is based on the reduced translation of a reporter → messenger RNA to an indicator protein, if a protein binds to its cognate binding site, artificially introduced into the 5'-UTR of the mRNA. The protein-mRNA complex inhibits the stable association of the small ribosomal subunit and thereby represses the translation of the message. In short, yeast cells are transformed with (1) a plasmid for the expression of an RNA-binding protein or a cDNA expression library, driven by a galactose-inducible → promoter and (2) a plasmid for the expression of a reporter protein driven by a constitutively active promoter. Preferred reporter protein is the → green fluorescent protein or its derivatives. The binding site for the RNA-binding protein of interest is cloned into the 5'-UTR of the mRNA encoding the reporter protein. Expression of the reporter is turned off in glucose medium (the promoter driving the expression of the RNA-binding protein is silenced): the phenotype is GFP$^+$. After induction of RNA-binding protein expression by the replacement of glucose by galactose in the growth medium, the translation of the reporter mRNA is repressed: the phenotype of the cells is now GFP$^-$.

Translational synergy: The synergistic cooperation of the → cap structure and → poly(A) tail of → messenger RNA, that leads to a pseudo-circularization of the molecule and an enhancement of translational competence beyond the additive effect of each modification separately. Translational synergy requires the attachment of the poly(*A*)*binding protein* (PABP; in yeast: Pab 1p) to the poly(A) tail, which recruits the 40S ribosome subunit to the mRNA. The Pab 1p reacts with the N-terminal part of the heterotrimeric, multi-functional → initiation factor eIF 4G (yeast), forming a circular complex, to which other proteins (as e.g. eIF 4E, eIF 4A, eIF3, eIF 4E kinase Mnk–1) bind. This → protein machine recruits the ribosomal subunits for an effective translation of the message. See → translation, → translational control.

Translation efficiency: The percentage of → messenger RNAs in a cell that are bound to, and efficiently translated by → ribosomes. Translation efficiency is usually estimated by the ribosome loading profile, i.e. the distribution of mRNAs between free → *messenger ribonucleoprotein* (mRNP) particles or the cytoplasm (ribosome-free mRNAs), and → polysome-bound mRNAs. In a given cell, each mRNA species is translated with a characteristic efficiency, which may vary more than 100-fold, depending on cellular states (resting, active, stimulated, blocked, proliferative, differentiating).

Translation initiation site: The → start codon (prokaryotes: AUG, GUG; eukaryotes: AUG) in a → messenger RNA molecule, at which polypeptide synthesis is initiated. See → Kozak consensus translation initiation sequence.

***Translation state array analysis* (TSAA):** A technique for the estimation of the → messenger RNA translation state, i.e. the number of messenger RNAs in a given cell at a given time, that are actually translated into proteins. In short, total ribosomes are first isolated, and mRNP particles (mostly containing translationally inactive mRNAs) and monosomes (i.e. one ribosome attached to one undertranslated message) on one hand, and polysomes (actively translating the mRNAs) on the other separated in → sucrose density gradients. The RNA from the various fractions is then extracted, reverse-transcribed into ^{32}P-labeled → cDNA probes, and these probes hybridized to → microarrays either containing fragments of cDNAs of known functions or oligonucleotides representing characteristic sequences of known genes. The resulting hybridization pattern is then analyzed by → autoradiography or → phosphorimaging and appropriate software.

Translation termination efficiency: The precision with which → translation termination is executed at the → termination codon of a → messenger RNA. Translation termination efficiency is an important parameter of gene expression: more efficient translational termination results in better expression rates, i. e. higher levels of correctly synthesized protein. Most organisms prefer a specific sequence context encompassing the termination codon. For example, the base immediately downstream of this codon (*E. coli*, yeast, insects: UAA; mammals and monocotyledonous plants: UGA) exerts a strong influence on translation termination efficiency (with decreasing efficiency: G > U, A > C for UAA; U, A > C > G for UAG). For *E. coli*, the translation termination efficiency varies from 80 % (for UAAU) to only 7 % (for UGAC). For high expression of a particular protein in → cell-free translation assays it is therefore necessary to optimize the context around the stop codon, otherwise a significant increase in translational → read-through and thus reduced protein expression will result. For example, in → rabbit reticulocyte lysates, translational read-through of the UGAC context can be as high as 10 %. If the fourth base is A, G or U, however, the read-through is less than 1 %.

Translatome: The complete set of peptides and proteins expressed in a cell, a tissue, an organ or an organism at a given time. Basically identical to → proteome.

Translesion synthesis (TLS): A replication mechanism, that does not avoid or excise faulty bases in DNA (e.g. created by → mutagens), but tolerates them and replicate through the mutated site. TLS is therefore error-prone and potentially mutagenic in itself (depending on the nucleotide that is incorporated across from the mutated site). Compare → damage avoidance mechanism.

Translocase: See → elongation factor.

Translocation:

a) The stepwise, codon-to-codon advance of a → ribosome along a → messenger RNA with simultaneous transfer of the peptidyl-tRNA from the A site to the P site of the ribosome. Each translocation step exposes an mRNA → codon for base-pairing with its specific tRNA → anticodon.

b) Any change in the position of a specific → chromosome segment within the same chromosome (intrachromosomal translocation, "shift"), or from one chromosome to another nonhomologous chromosome (interchromosomal translocation). Though translocations may involve whole chromosome arms ("whole-arm translocations"), they do not change the overall number of genes per genome. See also → insertional translocation.

***trans* marker:** Any→ molecular marker, designed and informative for one genome (where it is linked to e. g. a disease-causing gene), that can be used successfully to tag the corresponding gene in another genome. See → *cis* marker.

Transmembrane domain (TM): A → domain of membrane-bound proteins, that anchors the protein in the cellular membrane. For example, in receptor proteins, the extracellular domain interacts with the corresponding ligand, and the cytoplasmic domain mediates the recruitment and activation of signal-transducing proteins. The TM bridges both domains, but additionally modulates the efficiency of signal transduction through self-interaction. For example, the erythropoietin receptor (EpoR) exists as a homodimer. Even without the ligand erythropoietin, but more pronounced after its binding by the extracellular domains (representing a kind of molecular scissors), the two TMs interact with each other and bring the cytoplasmic domains closer to each other. This process is transient, unless stabilized by the interaction with erythropoietin, which allows binding of the socalled *Ja*nus *k*inase (JAK–2). This in turn leads to the phosphorylation of a strategic tyrosine residue of the kinase, which is activated and phosphorylates a tyrosine residue of the receptor, a prerequisite for the recruitment of the signal-transducing proteins.

Transmembrane helix: An α-helical domain of 20–30 amino acids in membrane proteins, that spans the lipid bilayer of the membrane. Such domains are usually rich in hydrophobic amino acids (isoleucine, leucine, valine, alanine, and phenylalanine), whose side chains are oriented towards the surface of the proteins and interact with the lipids of the transmembrane section. See → β-barrel, → multi-pass protein, → single-pass protein.

trans-NAT: Any two → messenger RNAs that form → sense-antisense complexes and are transcribed from different genomic loci. Many trans-NATs encode → ribosomal proteins.

Transoplex: A complex (*"trans*fection *complex"*) between cationic *s*olid *l*ipid *n*anoparticle*s* (SLNs), 100 nm in diameter, and DNA (e.g. plasmid DNA), in which the DNA is electrostatically bound. Such complexes protect the DNA from chemical degradation, are physiologically inert (or at least tolerated), and are used for → transfection of foreign DNA into target cells. In addition to cationic liposomes ("lipoplexes") or polycationic polymers ("polyplexes") transoplexes are potent gene delivery systems, whose efficiency can be increased by helper substances such as chloroquine, that facilitates the release of transfection complexes from endosomes into the cytoplasm (inducing endosome lysis).

Transplacement: See → gene replacement.

Transplacement vector (DNA transplacement vector): See → gene replacement.

Transplastome: The entirety of genetically modified → chloroplast genomes of a cell, tissue, organ or plant. Modifications are usually introduced by → transplastomic transformation. See → transplastomic plant.

Transplastomic plant: Any plant that underwent a → transplastomic transformation (i.e. the integration of foreign genes into the plastids of all cells).

Transplastomic transformation: The transfer of foreign genes into chloroplasts, their stable integration into → chloroplast DNA by → homologous recombination, and their expression over numerous generations. Usually, the genes are transferred by → particle bombardment. Transplastomic plants have potential for high expression of the recombinant protein, because the ploidy level of the plastic genome is very high. For example, in a mature leaf cell, 100 chloroplasts are present with up to 100 pDNA molecules each, making up a total of ~ 10,000 genome copies per cell. Additionally, plastids are only rarely transmitted through pollen, i. e. are safe gene containments. Also, transplastomic proteins do not undergo as many different post-translational modifications as e. g. proteins encoded by nuclear genes and synthesized *in nucleo*.

Transport *protein* (TP; *movement protein*; MP): A → virus-encoded protein that facilitates the spread of a viral genome from a primarily infected plant host cell to neighbouring cells by changing the permeability of plasmodesmata, the cytoplasm-filled channels between adjacent cells. The constitutive expression of defective TP genes in → transgenic plants confers resistance to viral infection by interfering with the cell-to-cell transport of viral genetic material.

Transposable element: See → transposon.

Transposable *exon* (trexon): A short → inverted repeat segment in a genome resembling a → transposon, that encodes an α-helix and a short turn or loop → domain of a protein (the galactose-glucose-binding protein or neurophysin–3, for example). Such trexons or similar sequence elements are occurring in many protein-coding genes (e.g. of *Rickettsia conorii*, where they are called *Rickettsia palindromic elements* (RPEs) and probably function to create new protein sequences.

Transposase: An enzyme encoded by a gene of class II → transposons that catalyzes the excision, transfer and insertion of the transposable element which carries its gene. See → transposition.

Transposition: The process of excision of a → transposon from its original site and its transfer to and integration at another site in the recipient DNA. Before transposition of prokaryotic transposons can take place, the transcriptional block exerted at the *res* site by the *tnp* R gene product ("resolvase") has to be released. After expression of the *tnp* A gene into the → transposase protein the transposition process begins. In a first step the transposase catalyzes the formation of → nicks in one strand of the → inverted repeats flanking the transposon. Two nicks are then introduced into the recipient DNA at the target site, and → site-specific recombination occurs. In the case of class I transposons replication of the elements occurs at this stage. After filling-in of single stranded regions and ligation the resulting cointegrate contains two → replicons, separated by two directly repeated copies of the transposon. The second step of the transposition involves site-specific cleavage and recombination, and separates the two copies of the transposon. This step is called resolution and requires the action of the *tnp* R product ("resolvase"). See also → transposition immunity.

Transposition immunity: The reduced capability or inability of a → transposon to transpose into a → plasmid which already contains a copy of it. In case of Tn 3 the transposition immunity is extremely specific. If one copy of Tn 3 is already present in a → replicon, the insertion of a second Tn 3 copy is excluded, whereas the insertion of a closely related transposon (e.g. Tn 501) is permitted.

Transposome (α transposome): A transient synaptic complex between a → transposase protein (e.g. from transposon 5) and a → transposon sequence (e.g. → transposon 5), containing a → selectable marker gene (e.g. an → antibiotic resistance gene), that functions to stably integrate the transposon into the target DNA. Transposed cells can then be selected on the basis of their resistance phenotype.

Transposomics: The use of a → transposome to transfer a → transposon or → foreign DNA integrated between its *outer ends* (OEs), into a bacterial host cell and to insert it randomly into the host cell's DNA. Though transposomes are only formed during transposition, stable transposomes can be isolated, if Mg^{2+} is excluded. These stabilized transposomes are incubated with the target DNA, and an aliquot of this mixture is electroporated into bacterial or yeast host cells, where they are activated (by e. g. cellular Mg^{2+}), and randomly insert the transposon into the genome of the cell. Moreover, mutations in the → transposase gene and changes in the 19 bp transposase recognition sequence lead to a hyperactive transposome, so that the insertion frequency is high. Screening for antibiotic resistance allows to select clones of any desired phenotype. Transposomics is therefore used to randomly generate insertion mutants (as e. g. → gene knock-outs or → gene knock-ins), to facilitate sequencing without cloning (if the transposome contains appropriate sequencing primer sites), to map specific sequences, and to insert any foreign DNA sequence (e. g. tags, promoters, control elements, genes) into any target DNA. See → *in vitro* transposition.

Transposon **(Tn; *transposable element*, TE; mobile element, mobile genetic element; "jumping gene"; selfish genetic element, SGE):** The usage of the term "transposon" is partly contradictory. In a strict sense it designates only prokaryotic transposable elements, while eukaryotic transposable elements are called "transposon-like elements". It is, however, also used synonymously to "transposable element", as name for both eukaryotic and prokaryotic mobile sequences. Accordingly the following definitions are used:

a) Generally, all segments of DNA that can change their location within a genome. Transposable elements are flanked by short → inverted repeat sequences and encode enzymes that catalyze their excision from their original site, and their transfer to and insertion into a new site (→ transposition). Transposons can be used for the construction of → transposon-based cloning vectors, for → transposon mutagenesis and for → transposon tagging. A number of transposons are described in this book in more detail, see → activator-dissociation system, → enhancer (definition b), both of *Zea mays*; → fold-back element, → hobo element, → *P* element of *Drosophila*, → *Ty* element of yeast. See also a list of prokaryotic transposons, following definition b, → conjugative transposon, and compare → insertion sequence, → retrosequence.

b) Mobile DNA sequences of bacteria, → bacteriophages, or → plasmids, flanked by terminal repeat sequences and typically harboring genes for transposition functions (→ transposase, resolvase). They insert at random and independently of the host cell recombination system into plasmids or → chromosomes. Transposons can be broadly categorized into compound or class I, and complex or class II transposons (see → composite transposon). Class I transposable elements are characterized by a drug resistance gene flanked by → insertion sequences either as → direct or, more frequently, as → inverted repeats. The IS elements provide the transposition functions and transpose the intervening drug resistance gene(s) in concert. The IS elements may also transpose independently. Examples for class I transposons are *Tn* 5, and *Tn* 10 (see below). Class II transposons contain genes encoding transposition functions and drug resistance(s) flanked by short, inverted repeats (IR) of 30-40 bp in length. A copy of the transposon is retained at the original location. The transposon's insertion at a new target site generates a direct repeat of 3-11 bp within the target DNA borders. A class II transposon is for example *Tn* 3. Transposons of both classes are used as source for drug resistance genes and as → mutagens. Some of them are listed below.

Class I transposon (e.g. Tn 5 or Tn10) :

IS   Drug resistance gene(s)   IS

Class II transposon (e.g. Tn1 or Tn3) :

IR   Transposition   Drug resis-   IR
      functions   tance gene(s)

Transposon-based cloning vector: A → cloning vector that contains a → transposon into which a foreign gene (generally, DNA sequence) can be cloned. After its transfer into the genome of a host cell, the transposon together with the foreign sequence may transpose to various loci in the recipient genome.

Transposon display: See → transposon insertion display.

Transposon 5 (Tn 5): A 5.818 kb bacterial transposon, flanked by IS 50 terminal inverted repeats, and encoding a → neomycin phosphotransferase (npt) that catalyzes the inactivation of the → aminoglycoside antibiotics → kanamycin, → neomycin and → geneticin. The *npt* gene from *Tn* 5 (*npt II*) is now routinely used for the construction of effective → cloning vectors for bacteria, plants and animals, because most of these organisms are sensitive against these antibiotics. Expression of the *npt II* gene in the host cell confers resistance against the antibiotic so that → transformants can be easily selected. In addition, *Tn* 5 encodes genes for its own transposition (e.g. a → transposase, protein 1).

Transposon insertion display (TID, transposon display): A high-resolution technique for the simultaneous visualization of individual members of → transposable element families in high copy number lines, the analysis of element copy numbers, → insertion frequencies and transpositional activities of the elements, and the isolation of → transposon-tagged genes and sequences flanking transposable elements in plants (e.g. *En/Spm*, *Mu*1 and the non-LTR retrotransposon *Cin*4 in *Zea mays*). Transposon display is a variant of the conventional → amplified fragment length polymorphism technique. In short, genomic DNA is first isolated, restricted with a frequently cutting → restriction enzyme (e.g. *Bbs*I and *Mse*I for *Cin*4TID, *Bst*XI and *Mse*I for *Mu*1–TID), then → splinkerette-like linkers ligated to the fragments, and biotinylated → primers complementary to the linkers used to amplify the fragments in a conventional → polymerase chain reaction (PCR). The amplified products are then purified from the primers and bound to → streptavidin-coated beads. Then a second PCR with a nested 5'- or 3'-end transposon-specific and labeled primer (e.g. labeled with ^{32}P or ^{33}P) and a linker-primer is performed, the products electrophoretically separated on a denaturing → polyacrylamide gel and detected by → autoradiography (display of only transposon-specific fragments). Fragments of interest are then isolated and amplified for sequencing. See → expression transposon insertion display. Compare → differential display, → phage display, → protein complementation assay, → random peptide display, → ribosome display.

Transposon internal eliminated sequence: Any one of a highly repetitive family of → internal eliminated sequences from the → micronucleus of hypotrich ciliate protozoa (e.g. *Oxytricha nova, O. fallax, O. trifallax, Stylonychia lemnae, S. pustulata*, and *Euplotes crassus*), that represent → transposons. Since many such elements contain telomeric repeat sequences $(5'-CA-3')_n$ within the distal 17 bp of the 77–78bp → inverted terminal repeats, they are called → *telomere-bearing element* (TBE; do not confuse with → Tris-borate-EDTA). For example, the *O. fallax* TBE1–fal is 4.1 kb in length, bounded by 3bp direct repeat target sequences (ANT), and harbors three → open reading frames, of which one encodes a 42 kDa → transposase. About 2,000 TBE1s exist within the micronuclear genome, and all are eliminated during macronuclear development. See → chromosome fragmentation, → gene scrambling, → nuclear dimorphism, → short internal eliminated sequence.

Transposon-like element: See → transposon.

Transposon mutagenesis: A method to introduce → insertion mutations at random within a target DNA, using → transposons. Basically, two approaches can be followed. In the non-targeted approach the transposon inserts at random, in the targeted approach it inserts into a particular gene sequence. See also → interposon.

Transposon 9 (Tn 9): A 2.638 bp bacterial transposon, flanked by → IS 1 terminal direct repeats, and encoding a → chloramphenicol acetyltransferase (CAT) gene that catalyzes the inactivation of the antibiotic → chloramphenicol. The CAT gene from Tn 9 is routinely used for the construction of → cloning vectors and serves as a → selectable marker and → reporter gene in → transformation and → transfection experiments.

Transposon 7 (Tn 7): A bacterial transposon encoding a → dihydrofolate reductase that is highly tolerant to the bacteriostatic action of → trimethoprim. Provided its DHFR gene is driven by a → promoter compatible with the host (e.g. a *tk* promoter for animal cells, a *nos* promoter for plant cells), *Tn* 7 may be used as a → selectable marker. It has also been used for → transposon mutagenesis.
The transposition mechanism involves 5 transposition genes (*tns A, B, C, D* and *E*) that mediate transposition via two pathways, differing in their choice of the target site for *Tn* 7 integration. Both pathways require the activity of *tnsA*, *tnsB* and *tnsC*, where *tnsB* encodes a specific → DNA binding protein that recognizes sequences within the termini of Tn 7. In one pathway, *tnsA, B, C* and *D* promote transposition to target sites like *att Tn 7*, to which *Tn* 7 transposes with high frequency, and *pseudo-att Tn 7*, to which *Tn* 7 transposes at low frequency. In the second pathway, *tnsA, B, C* and *tnsE* promote low frequency transposition of *Tn* 7 to many different sites that have no obvious relationship to *att Tn 7*.

Transposon silencing: The prevention of the transposition of a → transposable element by e.g. → RNA interference. Transposon silencing is a defense mechanism of organisms to protect the genome from the accumulation of transposons.

Transposon 601 (Tn 601): A bacterial transposon encoding a → neomycin phosphotransferase (npt) protein that catalyzes the inactivation of the → aminoglycoside antibiotics → kanamycin, → neomycin and → geneticin. In contrast to the npt II gene from *Tn* 5, the gene from *Tn* 601 is designated as neomycin phosphotransferase gene I (*npt I*). In addition, *Tn* 601 encodes genes for its transposition (e.g. a → transposase).

Transposon tagging: A method to isolate a gene that has been mutated by the → insertion of a → transposon. Basically, two approaches can be followed. In the non-targeted (random) approach the transposon is randomly integrated into the target → genome. Insertion into, or very close to the gene of interest may alter the → phenotype of the host organism, which makes selection of interesting → transformants easier. The targeted approach is based on the insertion of a transposon into a cloned gene, and the transformation of a suitable host with this construct. The host incorporates the construct into its chromosome by → homologous recombination (compare also → gene targeting). In both cases can the target DNA be identified by hybridization of genomic restriction digests or → genomic clones to a radiolabeled → probe completely, or only partly identical to the transposable element. See → gene tagging, → T-DNA tagging.

Transposon 10 (Tn 10): A 9.3 kb bacterial transposon, flanked by IS 10 elements in opposite orientation, and encoding proteins that mediate high-level → tetracycline resistance in *E. coli* and other enteric bacteria. Basically two genes of *Tn* 10, *tet A* (encoding the resistance protein) and *tet R* (encoding a → repressor protein) are involved in tetracycline resistance, both being tightly controlled at the level of transcription by the *tet R* gene product. The resistance against the antibiotic, which enters the cell via passive diffusion, is based on membrane-bound *tet A* protein that functions as an excretion carrier for the energy-dependent export of tetracycline from the cytoplasm. See also → multicopy inhibition.

Transposon 3 (Tn 3): A 4.957 kb bacterial transposon, flanked by 38 bp terminal inverted repeats, and encoding a → β-lactamase protein that converts → penicillin into penicilloic acid. The gene for β-lactamase has been integrated into → pBR 322 (Ap^r, → ampicillin resistance) and functions as → selectable marker gene. In addition, *Tn* 3 encodes genes for its transposition (e.g. a → transposase gene, *tnA* and resolvase gene, *tnpR*).

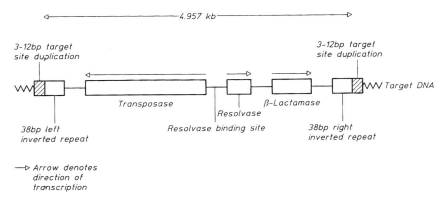

Schematic structure of Tn 3.

Transposon walking: A technique for the sequencing of unknown DNA, that contains long stretches of A or T (which are difficult to sequence), in which → transposons are inserted into the target DNA at random and exploited as starting points for sequencing using transposon-specific primers.

Transprimer: A → *trans*posable element that can be mobilized *in trans* by a → transposase, randomly inserts into a target DNA molecule, and is used as a → primer-binding site for the sequencing of the target DNA. Such transprimers are e. g. employed in → genome priming, where they are mobilized in vitro from a donor plasmid into the target DNA cloned into an acceptor plasmid. *In vitro* mobilization can be achieved by purified Tns ABC transposase.

***Trans*-reporter:** An *in vitro* technique for the detection of an indirect or direct involvement of a target protein in a signal transduction pathway. In short, first three → constructs are made, the socalled reporter plasmid (containing a GAL 4 → *upstream activation site* [UAS] fused to a → luciferase gene), the expression plasmid transcribing a gene of interest (driven by e.g. the cytomegalovirus [CMV] promoter) and a pathway-specific fusion trans-activator plasmid (containing a chimeric GAL 4 binding domain fused to the specific activation domain, driven by an SV40 promoter). All three plasmids are cotransfected into mammalian cells. Now, the fusion *trans*-activator plasmid expresses the chimeric fusion *trans*-activator protein, the unknown gene expresses a protein that directly or indirectly phosphorylates the activation domain of the fusion *trans*-activator protein. The phosphorylated fusion *trans*-activator protein binds as a dimer to the GAL 4 UAS of the reporter: the reporter gene is activated, and its activity detected and quantified with a luciferase assay. Should the unknown protein not be part of a signal cascade, then no reporter gene activity can be detected. See → *cis*-reporter.

trans **ribozyme:** Any → ribozyme that catalyzes the cleavage of single-stranded RNA substrates, but does not cleave itself. See → *cis* ribozyme.

Trans-silencing protein: See → repressor.

Trans-**splicing (*trans*-RNA splicing, pre-mRNA *trans*-splicing):** The ligation of → exons from two (or more) different → messenger RNA (mRNA) molecules to form one mature message with a new combination of coding sequences. *Trans*-splicing falls into two categories. First, the spliced leader type of *trans*-splicing, characteristic for protozoa (e.g. trypanosomes) and lower invertebrates (e.g. nematodes) results in the addition of short, capped 5'-noncoding sequences to the mRNA. Second, the discontinuous group II intron types in chloroplasts of algae and higher plants (and plant mitochondria as well) involves the joining of independently transcribed coding sequences through unusual interactions between intronic RNA stretches. Both types of *trans*-splicing processes probably accelerate the evolution of novel proteins. See → splicing.

Transvection (transvection effect): The inactivation of a specific gene by the physical proximity of its homologous allele. For example, in *Drosophila* the aberrant → phenotypes of certain bithorax gene mutants (the bithorax complex belongs to the class of the → homeotic genes) can be altered by → rearrangements leading to the disruption of the alignment of the corresponding homologous chromosomes. Such transvection effects have also been found in the *zeste, white, decapentaplegic*, glue protein gene *Sgs-4, notch*, and *cubitus interruptus* loci of *Drosophila*.

Transvection effect: See → transvection.

Trans-vector: See → binary vector.

Transverse-alternating field **electrophoresis (TAFE):** A method to separate DNA molecules in the size range from 50 kb to over 1000 kb in → agarose gels by subjecting the molecules alternately to two electric fields oriented transversely to the agarose gel. This modification of the conventional → pulsed field gel electrophoresis avoids the characteristic "bent" lanes of other variations of PFGE. See → gel electrophoresis.

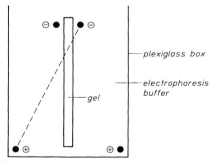

Electrode configuration of TAFE

Transversion (transversion mutation, base pair substitution): The substitution of a → purine by a → pyrimidine, or a pyrimidine by a purine base in duplex DNA. The result of a transversion finally is the exchange of an A-T pair with a T-A pair or C-G pair. Transversions are less frequent than transitions. See → point mutation, compare → transition. See also → rotational base substitution.

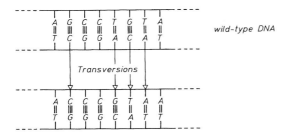

wild-type DNA

Transversion

Transversion mutation: See → transversion.

Transwitch: The expression of pieces of target genes in the → sense orientation which interferes with the correct processing of the target-gene RNA and with the expression of the target-gene trait. For example, the block of genes for the pigment-producing flavonoid synthetic pathway by transwitch techniques leads to the conversion of Petunia's normal purple flowers into white ones.

TRAP:
a) See → *t*elomeric *r*epeat *a*mplification *p*rotocol.
b) See → *t*ranslational *r*epression *a*ssay procedure.

TR aptamer: Any RNA → transcript derived from an → aptamer-encoding DNA that is cloned into the → cloning site of a → T7–RNA expression cassette in a → TR vector and whose transcription is driven by a → T7 promoter. The transcript binds its peptide or protein target and thereby interferes with its function(s). For example, an RNA aptamer directed against the intracellular domain of the ß2 integrin LFA–1 (a transmembrane protein mediating cell adhesion to *i*nter*c*ellular *a*dhesion *m*olecule 1 [ICAM–1]) reduced or abolished cell adhesion to TCAM–1.

Trash DNA sequencing ("trash sequencing"): The estimation of the nucleotide sequence of genes (e. g. encoding → ribosomal RNA) directly from crude samples (e. g. sediment, soil), that contain a mixed population of organisms, without prior isolation, cultivation and identification of the individual organisms. The heterogeneous DNAs are simply amplified in a conventional → polymerase chain reaction using e. g. rDNA-specific primers, the various amplification products sequenced, and the sequences aligned. This approach allows to detect new species (e. g. of archaebacteria living in hot springs), that otherwise escape detection, because they cannot be cultivated yet. See → environmental DNA, → environmental gene, → environmental gene cloning, → environmental genomics.

TR cassette: See → T7–RNA expression cassette.

trc promoter: An improved → *tac* promotor in which the distance between the -35 box of the tryptophan (*trp*) operon promoter and the -10 box of the *lac* Z promoter has been optimized for efficient transcription of linked genes.

T$_R$-DNA:
a) The right part of the → T-DNA of → octopine producing → Ti-plasmids of → *Agrobacterium tumefaciens*. Upon infection of plant cells it is transferred independently from the left part, the → T$_L$-DNA, and integrated into the nuclear genome of the host cell in variable copy numbers. It encodes five different transcripts.

b) The TATAA box regulating *his 3* → transcription in yeast (*Saccharomyces cerevisiae*).

T-region: A part of the → Ti-plasmid of → *Agrobacterium tumefaciens* or the → Ri-plasmid of → *Agrobacterium rhizogenes* that is transferred from the bacterium to the nuclear genome of plant host cells. Compare → T-DNA.

TREX: See → transcription and mRNA export.

Trexon: See → transposable exon.

TRF:
a) See *t*elomere *r*epeat binding *f*actor.
b) See → *t*erminal *r*estriction *f*ragment.

T-RFLP: See → *t*erminal *r*estriction *f*ragment *l*ength *p*olymorphism.

Tri-allelic *s*ingle *n*ucleotide *p*olymorphism (tri-allelic SNP): Any → single nucleotide polymorphism, of which three different alleles exist in a population. See → A-allele, → C-allele, → diallelic single nucleotide polymorphism, → G-allele, → T-allele, → tetra-allelic *s*ingle *n*ucleotide *p*olymorphism.

Tribrid protein: A polypeptide encoded by a → fused gene composed of three individual parts ("three-part gene") from different origin.

Tribrid system: See → three-hybrid system.

Tricistronic message: Any → messenger RNA that carries the information for three ("tri") proteins and is transcribed from a common → promoter. Compare → bicistronic message. See → cistron, → operon.

Trifluoroleucine: A leucine derivative, that can be incorporated into proteins and confer improved stability against denaturation without changing the overall structural characteristics of the protein.

***Trimethoprim* (Tp; 5–[3,4,5-trimethoxyphenyl]-methyl-2,4-pyrimidine-diamine):** An antibacterial folic acid analogue. It blocks the enzyme → dihydrofolate reductase and inhibits the synthesis of tetrahydrofolic acid which serves as a coenzyme for the transfer of hydroxymethyl and formyl groups (in e.g. DNA and RNA synthesis). Also active in eukaryotes, but with very low effectivity. See → trimethoprim resistance.

***Trimethoprim* resistance (*Tp^r*):** The ability of an organism to grow in the presence of the bacteriostatic compound → trimethoprim. Resistance is encoded by a → plasmid gene whose expression produces a → dihydrofolate reductase protein that is almost resistant against trimethoprim. Compare → trimethoprim sensitivity.

Trimethoprim sensitivity (Tp^s): The inability of an organism to grow in the presence of the bacteriostatic compound → trimethoprim which is a competitive inhibitor of the enzyme → dihydrofolate reductase. Compare → trimethoprim resistance.

Trimming: The → post-transcriptional modification(s) of primary transcripts to functional → messenger RNA, → ribosomal RNA, or → transfer RNA. For example, in transfer RNA precursors the intron-encoded sequences (especially in archaebacteria and eukaryotic organisms) are enzymatically removed and some bases chemically modified. See → rare bases.

Trinitrophenyl nucleotide (TNP nucleotide): A non-fluorescent nucleotide derivative, that becomes fluorescent after its binding to a protein. Such modified nucleotides are used to study interactions between the nucleotide (e. g. ATP) and its target protein (e. g. a kinase, ATPase).

Tri NR (*trinucleotide repeat*): Any trinucleotide motif (e.g. [ATA]$_n$, [CGG]$_n$, [GAC]$_n$), that is tandemly reiterated in a → genome (e.g. 5'-CGGCGGCGGCGGCGG-3'). See → DNR, → TetNR, → microsatellite, → repeat.

Trinucleotide repeat: See → Tri NR.

Triparental cross: The simultaneous infection of one bacterial host cell by three different → phages which repeatedly recombine to produce a triparental recombinant phage within the infected cell. This process involves two successive crosses, first between two parents, and then between a third parent and the biparental recombinant.

Triparental mating: A method to transfer genes (generally, DNA sequences) from an *E. coli* host into an unrelated second host (e.g. *Agrobacterium tumefaciens*) using a mediating *E. coli* strain which provides a wide host range → origin of transfer. In short, triparental mating starts with the mixing of *E. coli* cells containing a → plasmid cloning vector that carries the DNA to be transferred, and *E. coli* cells containing a broad host range plasmid. This mixture is then added to a suspension of freshly grown → *Agrobacterium tumefaciens* cells containing a helper → Ti-plasmid (to provide the → vir functions needed for gene transfer into plants). The three types of cells are filtered out and the filter is placed on a rich agar medium to allow the three-way conjugation to occur. The wide host range plasmid is transferred into *E. coli* cells containing the vector plasmid. In these *E. coli* cells, it provides transfer proteins (→ tra genes) and other factors that act on the origin of transfer of the vector plasmid and mobilizes it into *Agrobacterium tumefaciens* cells. These can then be used to transform susceptible plants (see → *Agrobacterium*-mediated gene transfer).

Triparental recombinant: Any progeny → phage that contains genes from each of three different parental phages which simultaneously infected a bacterial host cell.

Triple gene block: A set of three overlapping genes in the genomes of Potexvirus, Carlavirus, Furovirus and Hordeivirus groups, encoding proteins essential for the cell-cell movement of virus particles in the host plant via plasmodesmata (cytoplasmic connections between neighbouring plant cells). The 5'-most gene of the triple gene block resembles genes encoding RNA helicases, and all three genes possess sequences encoding transmembrane domains. Genetic engineering for virus resistance makes use of mutated genes of the triple gene block, that are transferred to target plants, expressed there, and interfere with the movement of the virus from host cell to host cell.

Triple helix: See → H-DNA.

Triple helix cosmid vector: A 7.6 kb → cosmid cloning vector that contains two different → H DNA forming sequences ("triplex sites", THBS) rich in purine on one and pyrimidine residues on the other strand, in addition to duplicate → cos sites, an → origin of replication, → selectable marker genes, → T3 and T7 RNA polymerase → promoters flanking a *Bam* HI cloning site, and other specific → restriction endonuclease → recognition sites. Triple helix cosmid vectors may be used for the construction of → genomic libraries. In short, chromosomal DNA is partially digested with either restriction endonuclease *Mbo* I or *Sau* 3A to obtain the optimal insert size of 30-45 kb. The inserts are dephosphorylated with → calf intestinal alkaline phosphatase and ligated into the *Bam* HI site of the linearized and dephosphorylated cosmid vector. After → packaging, the cosmid library is plated out, and screened for insert-containing positive plaques. The DNA is isolated from single plaques, completely restricted with *Not* I (a → rare cutter restriction endonuclease), cutting externally to the triplex sites, and subjected to partial digestion with another restriction enzyme of interest. The THBS at the end of the generated fragment are then hybridized to the corresponding radiolabeled complementary probes to form a comparatively stable triple helix, the different fragments are separated by → agarose gel electrophoresis, and the dried gel is exposed to an X-ray film. The acquired data allow the → restriction mapping of the clones and the T3- or T7-promoter-driven generation of → contigs for → chromosome walking and ordering of overlapping clones. The triple helix cosmid vector cannot be used for the cloning of human DNA, since it contains cross-hybridizing sequences.

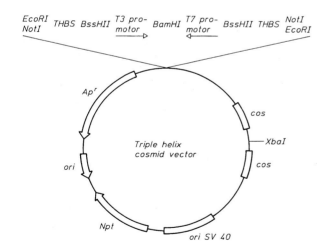

Triple helix cosmid vector

Simplified map of a triple helix cosmid vector.

Ap^r	:	Ampicillin resistance gene
cos	:	Phage λ cohesive site
Npt	:	Neomycin phosphotransferase gene
ori	:	Plasmid origin of replication
ori SV 40	:	SV 40 origin of replication
THBS	:	Triple helix-forming sequence

Triple helix-directed DNA cross-linking: The interstrand cross-linking in DNA double helices by → psoralen-oligonucleotide conjugates. The oligonucleotide first binds to its complementary target sequence, and forms a → triple helix structure with it. Upon irradiation with UV-light, the psoralen-oligonucleotide is cross-linked (*in vitro* or *in vivo*) to both strands of the target DNA. The cross-linked site inhibits → transcription and → replication. Basically, the oligonucleotide serves to guide the conjugate to a specific complementary target, and the psoralen-mediated cross-linking stabilizes the triple helix. Psoralen is usually linked to the oligonucleotide via a hexamethylene spacer arm. The technique is used to specifically silence targeted genes *in vivo* ("antigene strategy"). Compare → antisense oligonucleotide.

Triple splice acceptor site: A sequence within an → intron that regulates the → splicing at either the first, the second, or the third base in a → codon. Usually, splice site 1 is used by the → spliceosome, but the triple splice acceptor site can also direct to splice splice site 2 or 3.

Triplet: A → codon.

Triplet code: A → code in which a set of three consecutive → nucleotides specifies a distinct amino acid.

Triplet repeat expansion: See → microsatellite expansion.

Triplex: Any triple-stranded DNA, DNA-RNA or RNA complex. See → H-DNA.

Triple-stranded DNA

Triplex affinity capture **(TAC):** A method to isolate specific → triple helix (triplex)-forming clones from a → genomic library. In short, plasmid DNA from a library is pooled and incubated with a → biotinylated homopyrimidine oligonucleotide under mildly acidic conditions to form an intermolecular triplex between the target DNA and the oligo probe. Then → streptavidin-coated → magnetic beads are used to capture the biotin-streptavidin complex and separate it from bulk DNA. Double-stranded target DNA is recovered by treating the magnetic beads under mildly alkaline conditions that selectively destabilize the triple helix but not the normal Watson-Crick double helix.

Triplex-forming oligonucleotide **(TFO):** Any → oligodeoxynucleotide or its modified analogue, that recognizes its complementary → Watson-Crick base pairs in the → major groove of double-stranded DNA and forms a → triple-helix by Hoogsteen or reverse Hoogsteen bonds to already bonded bases. The TFO then forms the third strand, if all purine bases of the target sequence are on the same strand of the double helix. Therefore, triplex formation is mainly restricted to oligopyrimidine · oligopurine sequences. TFOs can be used to e. g. inhibit → transcription of specific genes. For example, the TFO can bind to promoter sequences and prevent the attachment of → transcription factors, can induce a conformational change in DNA, precluding the assembly of the → transcription initiation complex, or can recognize sequences within the gene and inhibit the elongation of the nascent → messenger RNA by → RNA polymerase. TFO-mediated gene silencing ("anti-gene strategy") is also possible in living cells. For example, a TFO with a sequence complementary to a 15 bp oligopyrimidine-oligopurine stretch in the promoter of the α-subunit of the interleukin–2 receptor (IL–2Rα), is cloned into a plasmid and transfected into lymphocytic cell lines. After its transcription, the TFO inhibits transcription of e. g. → reporter genes downstream of the IL–2Rα promoter. Alternatively, the TFO can be linked covalently to a → psoralen moiety (Pso-TFO). After its binding to target DNA, the psoralen is UV-photoinduced and crosslinks with the DNA strands at the triplex site, permanently blocking the binding of proteins to the target sequence, or preventing activation of the underlying gene sequence. This strategy allows to block the *polypurine tract* (PPT) of HIV–1 (provirus) in native chromatin of the infected cell.

Triplex polymerase chain reaction (triplex PCR): A variant of the conventional → polymerase chain reaction procedure, in which three different → primers are used to amplify two target sequences simultaneously in one single reaction. For example, in the → retrotransposon-based insertion polymorphism technique two primers complementary to sequences flanking the retrotransposon are used to amplify a specific genomic locus, and a retrotransposon-specific primer in combination with one flanking primer is additionally employed to prove the presence (or absence) of a retrotransposon at this locus. Compare → duplex polymerase chain reaction, → simplex polymerase chain reaction.

Tris-acetate-EDTA buffer (TAE): A buffer commonly used for the electrophoretic separation of DNA or RNA molecules in → agarose or → polyacrylamide gels (40 mM Tris, pH 8.0; 20 mM sodium acetate; 2 mM EDTA).

Tris base (tris-[hydroxymethyl]-aminomethane): A methane derivative used for buffer solutions in the pH range from 7 to 9 (pK$_a$ at 25°C: 8).

Tris-borate-EDTA buffer (TBE): A buffer commonly used for the electrophoretic separation of DNA or RNA molecules in → agarose or → polyacrylamide gels (89 mM Tris, pH 8.0; 89 mM boric acid; 8 mM EDTA).

Trisomy: The presence of three instead of two chromosomes of one type within a diploid genome.

Tritium (^{3}H): A radioactive isotope of hydrogen that is relatively long-lived (→ half time : 12.26 years) and emits weak beta particles.

Triton X-100: The viscous non-ionic detergent octylphenylpolyethyleneglycol ether, that is used to solubilize membranes by replacing the membrane phospholipids. Triton X-100 is usually added to extraction buffers in order to destroy membranes that may entrap macromolecules such as DNA or RNA, and to solubilize proteins or protein complexes.

tRNA: See → transfer RNA.

tRNA-directed retrotransposon (TRE): Any → retrotransposon, that inserts preferentially (or exclusively) into the vicinity of → transfer RNA genes in the genome of the slime mold *Dictyostelium discoideum*. See → transfer-RNA directed integration.

tRNA editing: See → transfer RNA editing.

tRNA genes: See → transfer RNA genes.

TRNA-*m*RNA hybrid (tmRNA): A hybrid molecule of prokaryotes, consisting of a → transfer RNA and a → messenger RNA molecule, that recognizes an empty A site on ribosomes.

tRNA splicing: See → transfer RNA splicing.

tRNA suppressor: See → transfer RNA suppressor.

***trp* promoter (*tryptophan* promoter):** A → promoter that controls the expression of the genes involved in trytophan synthesis (tryptophan synthesis operon) of *E. coli*. As a strong promoter it is used for the construction of → expression vectors. Its -35 sequence forms part of the → *tac* promoters.

TRS: See → *t*ranscription-*r*egulating *s*equence.

Truncated gene: See → pseudogene.

TR vector (*transfer* [TR] vector): An → expression vector for → aptamer sequences, which contains a → T7–RNA expression cassette, a → multiple cloning site with *Hpa* I, *Sph* I, *Xba* I, *Spe* I, *Bal* I, *Bgl* II, *Pac* I, *Sal* I and *Not* I recognition sequences and an → aptamer sequence inserted between the *Xba* I and *Pac* I sites. The vector is used to transfect cells, in which the aptamer sequence is to be transcribed.

TR vector cassette: See → T7–RNA expression cassette.

Tryptone: A tryptophan-rich protein hydrolysate (e.g. a pancreatic digest of casein).

***Tryptophan* promoter:** See → *trp* promoter.

TSA: See → *t*yramide *s*ignal *a*mplification.

TSAA: See → *t*ranslation *s*tate *a*rray *a*nalysis.

TSA-AP: See → *t*yramide *s*ignal *a*mplification using *a*lkaline *p*hosphatase.

TsAP: See → *t*hermosensitive *a*lkaline *p*hosphatase.

TSD: See → *t*arget *s*ide *d*uplication.

TSE buffer: A buffer used for suspending nucleic acids, especially → oligonucleotides. It contains 25 mM *T*ris-HCl (pH 8.0), 100 mM *s*odium chloride, and 0.1 mM *E*DTA as metal chelator. See → TE buffer.

T7 DNA polymerase: An enzyme of bacteriophage T7 that catalyzes the 5' → 3' polymerization of deoxyribonucleoside-triphosphates into duplex DNA. The polymerase complex consists of an 84,000 Da subunit derived from T7 gene 5, and a 12,000 Da protein (thioredoxin) derived from host gene *trxA*. The gene 5 protein itself has a low → processivity only (it dissociates from the template after incorporation of 1-50 nucleotides). In a complex with the accessory thioredoxin, its interaction with the primer-template is stabilized, and its processive activity is enhanced (incorporation of more than 1000 nucleotides without dissociation from the primer). The enzyme also possesses a potent 3' → 5' single and double-strand exonuclease activity (form I enzyme). The 3' → 5' exonuclease activity can be removed almost completely (form II enzyme) without affecting the polymerase activity. This form II enzyme is used to sequence long tracts of DNA with the → Sanger sequencing technique, especially since the modified T7 polymerase does not discriminate between different deoxynucleotide analogs. The form II enzyme is also produced by recombinant DNA techniques and is commercially available (→ Sequenase™).

T7 endonuclease I: An enzyme of the bacteriophage T7 that recognizes and cleaves non-perfectly matched DNA (e. g. → cruciform DNA, → Holliday junction, → heteroduplexes, or → nicked circular DNAs).

T7 *in vitro* transcription system: An *in vitro* system to generate large amounts of specific, homogeneous, biologically active and labeled RNA. In short, total RNA is first isolated from the target cell, reverse transcribed with → reverse transcriptase using oligo(dT)-primers, and the template RNA removed by → RNase H. The resulting single-stranded antisense cDNA is then exposed to a mixture of random → primers, that carry a → T7 promoter sequence at their 5' terminus. The resulting double-stranded cDNA then harbors a functional T7 promoter at one end. After denaturation, another T7 random primer is added and amplification again leads to a double-sranded cDNA with a T7 promoter sequence. After purification, → T7 RNA polymerase allows to amplify the sense strand *in vitro* to high quantities. The resulting sense mRNA therefore possesses defined sequences at both ends, that can be used as primer binding sites for further amplification. The T7 *in vitro* transcription system is a linear isothermal amplification procedure, that allows to increase otherwise limiting amounts of mRNAs for e.g. microarray experiments. See → SP6 *in vitro* transcription system.

T7–RNA expression cassette (TR, TR cassette, TR vector cassette): A synthetic → expression cassette for the cloning of → aptamer sequenes, consisting of a → T7 RNA polymerase promoter, located upstream of a → multiple cloning site, which in turn is flanked by stabilizing → hairpin-loop structures on each side ("TΦ-terminator" at the 3' side), proper termination codons, and *Hpa* I and *Not* I → restriction recognition sites at the 5' and 3' termini, respectively. The target DNA (e.g. the aptamer) is directionally cloned into the *Xma* I and *Pac* I sites by → forced cloning and can be transcribed by induction of the T7 promoter. The TR cassette is part of the socalled → TR vector, and serves to transcribe the aptamer-DNA in target cells.

T7 RNA polymerase (EC 2.7.7.6): A 98 kD → DNA-dependent RNA polymerase encoded by bacteriophage T7 that catalyzes the synthesis of RNA on double-stranded DNA molecules containing the T7 → promoter. This highly conserved, 23 bp promoter sequence is not recognized by *E. coli* RNA polymerase but has a high affinity to T7 RNA polymerase. Therefore it is used in T7-directed → *in vitro* transcription systems. T7 RNA polymerase is almost unaffected by the presence of non-T7 transcription terminators and thus produces complete transcripts from almost any insert DNA linked to a T7 promoter. The enzyme may also be used in conjunction with a variant version of the → polymerase chain reaction for → genomic amplification with transcript sequencing.

T7 sequencing primer: The synthetic oligodeoxyribonucleotide 5'-ATCGAAATTAATACG-3' that hybridizes to a conserved 20 bp sequence of the bacteriophage → T7 RNA polymerase → promoter, and allows → Sanger sequencing of double-stranded DNA inserted into vectors containing this promoter. Thus any subcloning (e.g. into an → M13 vector) is superfluous. Compare → sequencing primer.

TSGE: See → *t*emperature *s*weep *g*el *e*lectrophoresis.

***ts* mutant, *ts* mutation:** See → temperature-sensitive mutant/mutation.

***Tsp* DNA polymerase:** See → *t*hermophilic *sp*ecies DNA polymerase.

T-strand: A linear, single-stranded DNA molecule arising through a strand-specific → nick within the so-called → border sequences flanking the → T-region of → *Agrobacterium tumefaciens* → Ti-plasmids. The excision of the T-strand is catalyzed by a → *vir*-region encoded → endonuclease that is induced in a cascade of events following the contact of an *Agrobacterium* with wounded plant cells. In most plants, wounding induces the synthesis of phenolic compounds, some of which (e.g. acetosyringone) serve as signal substances for Agrobacteria (i.e. are recognized by specific *vir*-encoded membrane chemoreceptors, vir A proteins, and are transmitted to the *vir*-region through specific *vir*-encoded transfer or activator proteins, vir G proteins). After the excision of the T-strand it is packaged into a → T-complex, a protective coat of vir E2 proteins (single-strand specific → DNA-binding proteins) and transported into the recipient plant cell by a vir D2 protein (pilot protein).

TTGE: See → *t*emporal *t*emperature *g*radient *g*el *e*lectrophoresis.

***Tth* DNA polymerase:** See → *Thermus thermophilus* DNA polymerase.

***Tth* polymerase:** See → *Thermus thermophilus* DNA polymerase.

T3 RNA polymerase (EC 2.7.7.6): A DNA-dependent → RNA polymerase encoded by bacteriophage T3 that catalyzes the synthesis of RNA on double-stranded DNA molecules containing the T3 promoter. The enzyme has a high affinity for T3 promoter sequences, and is used for the synthesis of specific RNA as a hybridization → probe for → Northern and → Southern blotting, for → in vitro translation studies, exon and intron mapping of → genomic DNA, and the analysis of → post-transcriptional modifications of RNA.

T-transporter (T-complex transporter): A complex of 11 proteins, encoded by the → *vir*ulence (vir) operon (*vir*B operon) of the → Ti-plasmid of → *Agrobacterium tumefaciens,* and the virD4 protein. The T-transporter catalyzes the transfer of the → T-complex through the bacterial envelope into the plant recipient cells, the energy for this complex process being derived from ATP, which is hydrolyzed by the three ATPase proteins virB4, virD4 and virB11 associated with the T-transporter. T-transporters belong to the typeIV secretion system family.

Tubercidin: A → nucleoside antibiotic of *Streptomyces* species with cytostatic effect. See → toyocamycin.

***Tub* polymerase:** A thermostable DNA polymerase from the thermophilic bacterium *Thermus flavus*, that catalyzes the amplification of DNA targets of more than 10 kb with a PCR regime of short extension times and low enzyme concentrations.

TUF: See → *t*arget of *u*nknown *f*unction.

Tumor virus: See → oncogenic virus.

TUNEL assay: See → terminal deoxynucleotidyltransferase-mediated dUTP nick-endlabeling assay.

Turbid plaque: Any → plaque that appears turbid in opposite light, because some of the infected bacterial cells in this area withstand lysis and grow. The growing cells have most probably suffered a → mutation that renders them resistant against bacteriophage infection. Alternatively some phages have become lysogenic and fail to lyse the bacterial cells.

T/U vector: Any one of a series of → cloning vectors, that contains either a T or a U → overhang at a cloning site, into which a PCR product (generated with e.g. a proof-reading DNA polymerase with terminal transferase activity and containing an A overhang) can directly be inserted ("direct cloning"). See → TA cloning.

T-vector (T-overhang vector): A → plasmid → cloning vector, either cut with the → restriction endonuclease *Xcm* I or tailed enzymatically such that it contains 3' terminal thymidine nucleotide overhangs. Such overhangs allow easy cloning of DNA fragments ending with 3' A-overhangs ("T/A cloning"). Many thermostable → DNA polymerases, such as → *Thermus aquaticus* DNA polymerase or → *Thermus thermophilus* DNA polymerase, add single dA nucleotides to the 3' end of amplification products (→ extendase activity), without corresponding T residues on the template strand. The presence of this A-overhang prevents direct cloning. T-vectors are specifically equipped with single T nucleotide overhangs at the cloning sites to match these 3' extensions and to facilitate → ligation.

Tween-20: The non-ionic detergent poly(oxyethylene)$_n$-sorbitan-monolaurate, that is used for → enzyme-linked immunosorbant assays.

Twinning: An undesirable disorder in a protein crystallization process, which leads to the formation of distinct domains with different orientations. For example, two crystals are linked to each other, but do not face the same direction. The diffraction then represents two different crystal orientations.

Twin promoter: See → dual promoter.

Twintron: A hybrid → intron composed of a group II intron into which a second group II intron has been inserted. Twintron → splicing proceeds sequentially, i.e. the internal intron is removed prior to the excision of the external intron, and involves → lariat intermediates. Twintrons are possibly created by intron → transposition.

Twin vector: See → dual promoter vector.

***T*wisting number (T):** The number of base pairs in a relaxed closed circular double-helical DNA molecule, divided by the number of base pairs per turn of the DNA helix, as a measure of DNA supercoiling.

Two amino acids Barany linker: See → TAB linker.

2-aminophenyl thioether paper: See → APT paper.

Two-bait interaction trap: See → dual-bait system.

Two-bait system: See → dual-bait system.

Two-bait two-hybrid system: See → dual-bait system.

Two-color overlay: A composite graphical image in → microarray experiments. Overlays consist of two originally separate images of one and the same microarray probed with two differently labeled target molecules (e.g. two oligonucleotides labeled with → cyanin3 and → cyanin 5, respectively), that are merged *in silico*. For example, a green-red overlay is produced by subtracting microarray image values from separate channels of a microarray scanner and superimposing the resulting images to create a composite image. The green spots in that image represent sequences present in higher concentration in the test sample, whereas red spots correspond to sequences more abundant in the reference sample. Two-color overlays of whatever combinations allow the easy identification of up- and down-regulated genes in a sample.

Two-component transposon system: An artificial → transposon system, that consists of two separate elements: an element expressing a transposase gene, but lacking characteristic → terminal inverted repeats (TIRs), and a second transposase-less element with TIRs. The transposase element mobilizes the transposase-less element *in trans*.

2D-electrophoresis: See → two-dimensional gel electrophoresis.

2D-GEF: See → two-dimensional gene expression fingerprinting.

Two-dimensional difference gel electrophoresis (2–D DIGE): A variant of the conventional → two-dimensional gel electrophoresis, that allows the simultaneous separation of proteins from two samples and the identification of proteins specific for only one sample. In short, the proteins from e.g. two different tissues are differentially labeled with the → fluorochromes → cyanin 3 and → cyanin 5, respectively. Then the pre-labeled protein samples are mixed and co-separated by standard two-dimensional SDS polyacrylamide gel electrophoresis. The gel is imaged for the two fluorophores, and the resulting, perfectly overlapping image analysed and spots of interest (e.g. proteins only present in one sample) excised from the gel, in-gel digested with trypsin, and the tryptic fragments analysed by → mass spectrometry.
See color plate 8.

Two-dimensional DNA fingerprinting: A technique for high-resolution → genotyping, that produces two-dimensional restriction fragment fingerprints. In short, genomic DNA is first restricted (with e.g. *Hae* III or *Hinf* I), and the restriction fragments separated by electrophoresis in 6% neutral → polyacrylamide gels. The relevant region is cut out of the gel(e.g. the region of fragment sizes from 0.3 – 3.0 kb, or from 1.0 – 10 kb) and applied to a 6% polyacrylamide gel containing a 10–75% linear gradient of denaturant (10%: 0.7 M urea, 4% formamide). This combination of neutral and denaturing gradient gel electrophoresis leads to a resolution of up to 1000 (or more) spots per DNA sample on the stained denaturing gel. Specific sequences (e.g. genes, or repetitive sequences as → mini- or → microsatellites) can then be detected by → Southern blotting, hybridization of radiolabeled → probes to the blots and → autoradiography. Two-dimensional DNA fingerprinting allows to measure mutation frequencies in genes, to associate genetic variation with a → trait (e.g. a disease), to detect genomic instability in cancer and aging, to establish linkages, and to map target genes (e.g. disease genes). See → DNA fingerprinting.

Two-dimensional fingerprint: The highly specific pattern of restriction fragments separated by (1) electrophoresis in neutral → polyacrylamide gels and (2) subsequent electrophoresis in a linear gradient of the denaturants urea and formamide. See → two-dimensional DNA fingerprinting.

Two-*d*imensional gel electrophoresis (2D-electrophoresis, O'Farrell gel electrophoresis): A method to separate proteins on the basis of two of their characteristics: isoelectric point (pI) and molecular weight. Usually the protein mixture is first subjected to electrofocussing in a → polyacrylamide matrix containing an ampholines-stabilized pH-gradient (see → ampholyte). This serves to separate the various proteins according to their pI. Then the first-dimension gel is polymerized onto a second-dimension → SDS-polyacrylamide slab gel in which the proteins are separated according to their molecular mass (O'Farrell gel).

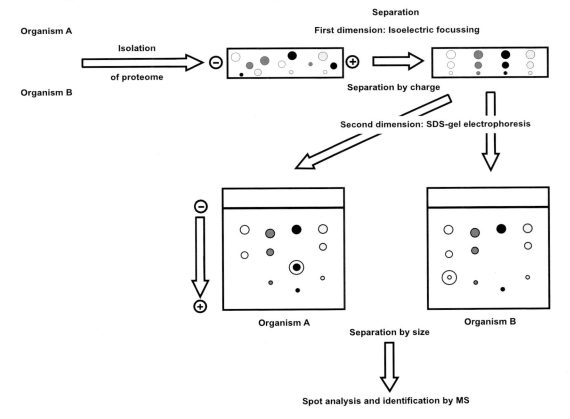

Two-dimensional gel electrophoresis

Two-dimensional gel map (2D gel map): A graphical depiction of the positions of all peptides and proteins separated by → two-dimensional polyacrylamide gel electrophoresis and visualized by either protein staining or → autoradiography (if e.g. newly synthesized proteins are labeled with ^{35}S-methionine).

Two-dimensional *gene* expression *fingerprinting* (2D-GEF): A variant of the → *gene* expression *fingerprinting* (GEF) technique for the visualization and characterization of (preferably) all → messenger RNAs of a cell at a given time, that combines the original GEF procedure with the resolving power of → two-dimensional polyacrylamide gel electrophoresis. In short, a first step leads to a set of → cDNA fragments, generated by the same procedure as in conventional GEF. This primary cDNA fragment population is then resolved according to size in the first dimension by → denaturing polyacrylamide gel electrophoresis. For a transfer to the second dimension, the resulting gel is subdivided into 96 fractions, and each fraction eluted into a well of a 96-well → microtiter plate. Eluted single-stranded fragments are captured by → streptavidin-coated beads, a second strand synthesized with → *Taq* DNA polymerase and a ^{32}P-labeled → adapter-primer, and the double-stranded fragments sequentially restricted with several → restriction endonucleases (selected as to minimize variations in the number of cDNA fragments liberated after each round of restriction). The cDNA fragments are then resolved by two-dimensional polyacrylamide gel electrophoresis and subsequent → autoradiography. The complex patterns can be analysed by appropriate software, or interesting bands be isolated, cloned, sequenced, and compared to database entries. See → adapter-tagged competitive PCR, → enzymatic degrading subtraction, → gene expression screen, → linker capture subtraction, → module-shuffling primer PCR, → preferential amplification of coding sequences, → quantitative PCR, → targeted display. Compare → cDNA expression microarray, → massively parallel signature sequencing, → microarray, → serial analysis of gene expression.

2D/3D biochip: Any glass, metal or teflon support with an ultrathin hydrophobic or lyophobic surface of modified silicone (not to be moistened by oleophilic liquids), that is interrupted by an ordered array of hydrophilic islands ("hydrophilic anchors") 100 mm in diameter. Such anchors either are simple holes in the hydrophobic surface layer or small areas of gold depositions. This configuration represents the 2D format of the chip. If such a chip is exposed to analyte solutions (e.g. DNAs, RNAs, or proteins), they localize exclusively above the hydrophilic islands, whose dimensions define the volume of the analyte (usually droplets of a few nanoliters). These droplets represent the third dimension on the chip (3D format) and serve as individual reaction chambers for biochemical or molecular experiments (e.g. measurement of enzymatic activities, enzyme-catalyzed syntheses, or cleavage of proteins). The 3D chip can be kept in a moist chamber (for reactions) or be exposed to high temperatures (for the evaporation of solutes, so that the samples are concentrated, in extreme cases are crystallized in the hydrophilic anchors).

Two-fold rotational symmetry: The symmetry of → palindromes.

Two-genome model: See → bimodal DNA replication.

Two-hybrid system (*yeast two-hybrid system*, YTH, Y2H): A technique for the detection of even relatively weak and transient protein-protein interactions *in vivo* and the identification of genes encoding interacting proteins, that is based on the dual modular composition of many eukaryotic transcriptional activators (e. g. *GAL*4 protein of yeast). These activators contain two discrete, physically separable, functionally independent molecular domains, a target-specific *DNA-binding domain* (DBD) that binds to a specific → promoter sequence, and a target-independent → *activation domain* (AD). The DBD serves to target the transcription factor to specific promoter sequences (e. g. → *upstream activation sequences* [UAS] in yeast), whereas the AD directs the → RNA polymerase II(B) transcription complex to transcribe the gene downstream of the DNA-binding site. Both domains are required for transcriptional activation, and neither domain alone can activate transcription. However, a non-covalent interaction of two independent hybrid proteins containing a DBD and an AD, respectively, leads to a reconstituted (i. e. active) transcription factor, triggering the expression of a → reporter gene by the specific UAS for the DBD. Therefore, the system exploits the interaction of proteins expressed from two hybrid genes, that are constructed in vitro and the independently transferred into and maintained in a yeast cell on two separate, but compatible plasmids. In short, the socalled bait hybrid gene consists of a fusion of the coding region of a → DNA-binding protein and a DNA segment encoding a socalled bait protein. The other hybrid gene is composed of a cDNA or a genomic DNA fragment (prey), whose interaction with the bait protein has to be tested, and a fused transcriptional activation domain. Both constructs are on separate high-expression plasmid vectors (DNA-binding domain fusion vector, and activation domain fusion vector) co-resident in the same transformed yeast cell ("reporter strain"), which contains one (or more) → reporter genes with upstream binding sites for specific DNA-binding proteins. Both protein domains can be expressed individually from the recombinant vectors, and are then targeted to the nucleus. Transformants are selected for growth, due to expression of a reporter gene that encodes an essential enzyme.

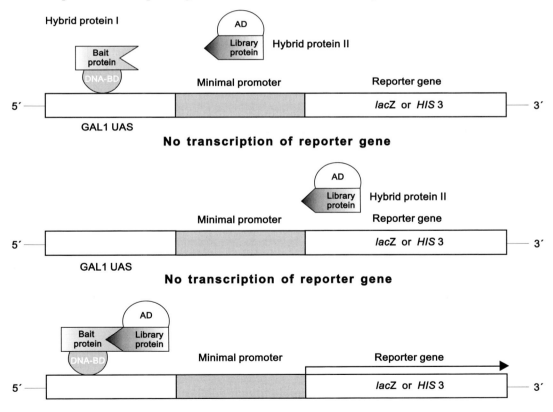

A DNA-binding domain/bait hybrid localizes to the reporter gene, but does not activate its transcription. If, however, protein-protein interaction between bait and prey occurs, the activation domain is brought into close proximity to the DNA-binding site (functional reconstitution of the transcription factor), which induces transcription of the reporter gene. This reporter gene activation is then taken as indication for an interaction between bait and prey proteins.

Variations of the system make use of different DNA-binding domains, activation domains, and reporter genes. Some variants include two different integrated reporter genes that share minimal sequence overlap in their promoters. For example, a convenient reporter combination uses the yeast biosynthetic gene *HIS*3 (nutritional selection for histidine, i. e. yeast grows in the absence of histidine), and the bacterial *lacZ* gene (colorigenic assay for β-galactosidase activity with the chromogenic substrate → X-gal or use of the → green fluorescent protein gene). Other systems exploit other selectable marker genes in the bait plasmid (e. g. resistance towards → zeocin). The two-hybrid system originally developed for yeast cells can also be adapted to higher eukaryotes (e. g. mammals) by e. g. the exchange of the *GAL4*-AD for the activation domain of herpes simplex virus type 1 VP16 protein, a firefly luciferase reporter gene and the strong cytomegalo-virus promoter.

The two-hybrid system allows to detect, which proteins can be in contact with each other and is a prerequisite for the establishment of → protein maps. Moreover, genes can be mutagenized to reveal protein domain(s) or amino acid residues necessary for protein-protein interaction(s). See → dual-bait two-hybrid system, → interaction mating, → interaction trap, → LexA two-hybrid system, → one-hybrid system, → reverse two-hybrid system, → RNA-protein hybrid system, → split-hybrid system, → split-ubiquitin two-hybrid system, → two bait system.

Two micron circle (2 μ circle, 2 μ plasmid; o-micron DNA; *Saccharomyces cerevisiae* plasmid, scp; yeast 2 μm plasmid): A nuclear → plasmid of yeast (*Saccharomyces cerevisiae*) that has a contour length of 2 μm (corresponding to 6318 bp), is packaged into → nucleosomes and is present in about 30-200 copies per cell, but has no known function for the host. Its six genes (*FLP, REP1, REP2, RAF,* and two genes of unknown function) code for amplification (*FLP* for a site-specific recombinase), plasmid → replication and → partitioning (*REP1, REP2, STB* locus), and transcriptional activation of the *FLP* gene (*RAF*). Its replication is controlled by nuclear genes. The → origin of replication (containing an → autonomously replicating sequence) of the 2 μ circle has been used for the construction of → yeast cloning vectors (e.g. YEp, → yeast episomal plasmid).

2 μ circle: See → two micron circle.

2 μ plasmid: See → two micron circle.

2'-ACE: See → 2'-bis(acetoxyethoxy)methylether.

2'-bis(acetoxyethoxy)methylether (2'-ACE): A silylether, that protects the 5'-hydroxyl group (5'-silyl) in combination with an acid-labile orthoester protection group. This 2'-ACE is attached during conventional → chemical DNA or RNA synthesis.

2'-Fluorine-modified RNA (fluorine-modified RNA): Any ribonucleic acid, that is synthesized by a mutant → T7 RNA polymerase able to incorporate 2'-fluorine CTP or 2'-fluorine UTP (as well as ATP and GTP). Fluorine-modified RNA is completely resistant to ubiquitous → RNase A and related RNases, but is still sensitive to → RNase T1 and → RNase H. Therefore, the fluorine-modified RNA is not attacked by RNases present on human skin ("finger nucleases"),whereas unmodified RNAs are degraded after contact with human skin.

2'-O-*aminopropyl*-RNA (AP-RNA): Any RNA in which the 2'carbon of the ribose moiety is modified by the introduction of a 2'-O-aminopropyl residue. AP-RNAs are highly resistant towards 3'-exonucleases, and the longer the alkyl chain of the substituent, the more efficient the protection. The unusual stability of AP-RNAs in vivo makes them versatile compounds for → antisense therapy.

2'-O-amino-RNA (2'-O-NH2–RNA): Any RNA, that contains riboses with an amino group at the 2'position. Such RNAs are more stable against nucleases. See → 2'-O-fluoro-RNA, → 2'-O-methyl-RNA.

2'-O-fluoro-RNA (2'-O-F-RNA): Any RNA, that contains riboses with a fluorine atom at the 2'position. Such RNAs are more stable against nucleases. See → 2'-O-amino-RNA, → 2'-O-methyl-RNA.

2'-O-fluoro-RNA

2'-O-methyl adenosine: An → adenosine nucleoside in which the 2'-hydroxyl group has been replaced by a methoxy group. 2'-O-methyl adenosine is incorporated into oligoribonucleotides to increase their stability when bound to RNA targets in → antisense technology. See → 2'-O-methyl cytosine, → 2'-O-methyl guanosine, → 2'-O-methyl uridine.

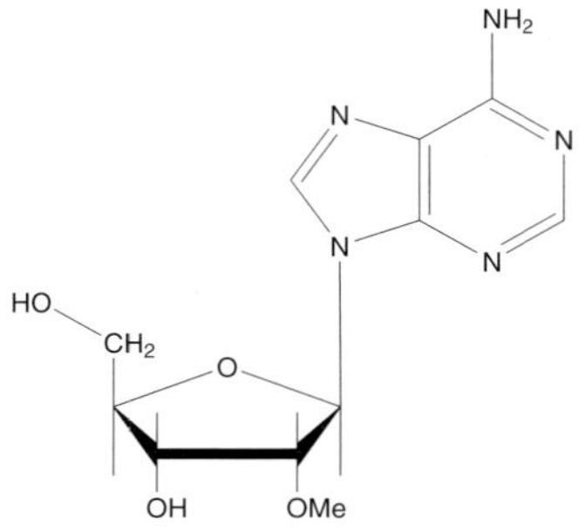

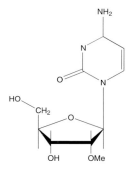

2'-O-Methyl Adenosine

2'-O-methyl cytidine: A → cytidine nucleoside in which the 2'-hydroxyl group has been replaced by a methoxy group. 2'-O-methyl cytidine is incorporated into oligoribonucleotides to increase their stability when bound to RNA targets in → antisense technology. See → 2'-O-methyl adenosine, 2'-O-methyl guanosine, → 2'-O-methyl uridine.

2'-O-Methyl Cytidine

2'-O-methyl guanosine: A → guanosine nucleoside in which the 2'-hydroxyl group has been replaced by a methoxy group. 2'-O-methyl guanosine is incorporated into oligoribonucleotides to increase their stability when bound to RNA targets in → antisense technology. See → 2'-O-methyl adenosine, → 2'-O-methyl cytidine, → 2'-O-methyl uridine.
Figure see page 1205.

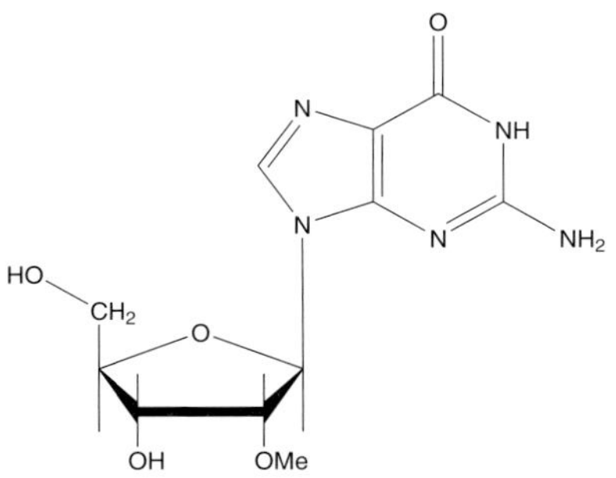

2'-O-Methyl Guanosine

2'-O-methyl-RNA (2'-O-Me-RNA): Any RNA, that contains riboses with an OCH_3-group at the 2'position. Such RNAs are more stable against nucleases. See → 2'- O-amino-RNA, → 2'-O-fluoro-RNA.

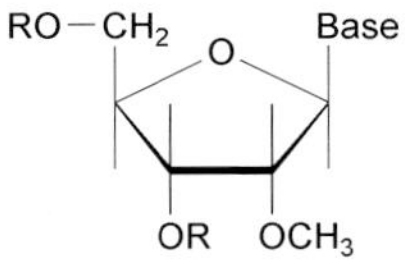

2'-O-methyl uridine: A → uridine nucleoside in which the 2'-hydroxyl group has been replaced by a methoxy group. 2'-O-methyl uridine is incorporated into oligoribonucleotides to increase their stability when bound to RNA targets in → antisense technology. See → 2'-O-methyl adenosine, → 2'-O-methyl cytidine, → 2'-O-methyl guanosine.

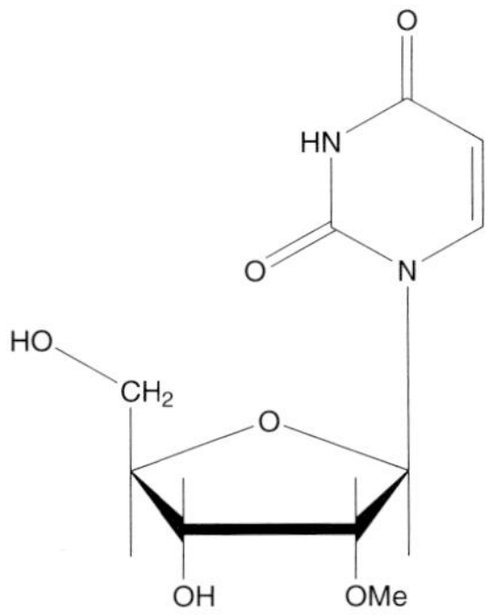

2'-O-Methyl Uridine

Two R hypothesis (2R hypothesis): The still hypothetical postulate, that the human genome experienced two rounds of duplication, one occurring about 550 million years, the recent one about 80 million years ago. The first duplication then coincides with the divergence of vertebrates from their immediate ancestors (as e.g. the lancelet *Amphioxus*), whereas the second round preceded the divergence of amphibians, reptiles, birds, and mammals from the bony fishes. Other organisms (e.g. yeast) also experienced genome duplication(s) in evolutionary times.

Two-stage *polymerase chain reaction* (two-stage PCR): A variant of the conventional → polymerase chain reaction that can be used to amplify two → exons separated by an → intron, or to prove the linkage of particular exons in members of a → multigene family. In short, genomic DNA is first amplified using two outside primers corresponding to the 5' and 3' ends of neighboring exons (or genes). After removal of free primers the PCR product is repaired by the → Klenow fragment and phosphorylated by kinase. The DNA is then circularized with → T4 DNA ligase and then amplified using two inside primers corresponding to the 3' and 5' ends of the neighboring exons (or genes), but with opposite orientation to the first-round outside primers. The relatively short second-round PCR product can then be analysed by → agarose gel electrophoresis and → Southern hybridization.

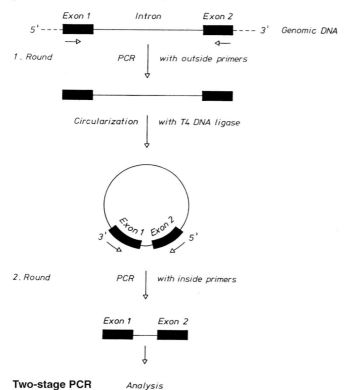

Two-stage PCR   *Analysis*

Two-step gradient *polymerase chain reaction* (two-step gradient PCR): A variant of the conventional → polymerase chain reaction, which combines the traditionally separated steps of → primer annealing and → elongation into one single step, so that only two steps (denaturation of → template DNA, and primer annealing/elongation remain. Usually a temperature gradient in between 48 and 60,8°C is gradually established, so that initially the primer(s) can bind, and subsequently the primer(s) can be extended. Two-step gradient PCR therefore saves both time and reagents.

TX-primer (TX unidirectional primer): Any synthetic single-stranded oligodeoxyribonucleotide that contains one or more → restriction endonuclease → recognition site(s) immediately adjacent to a homopolymeric oligo(dT) sequence. This primer is used for the → forced cloning of → cDNA. First, the TX-primer is annealed to the → poly(A) tail of mRNA, → first strand synthesis is achieved with → reverse transcriptase and → second strand synthesis with *E. coli* → DNA polymerase I and *E. coli* → DNA ligase. When the overhangs are filled in with → T4 DNA ligase, a functional restriction (X) site is generated at one end of the cDNA (e.g. a *Not* I site; in this case the primer is a T *Not* primer). If non-palindromic → linkers containing a different restriction site are added to the other end of the cDNA, → concatemerization is avoided and the cDNA can be inserted into an appropriate vector in only one orientation. See → unidirectional primer.

TX RNA: See → tiny expressed RNA.

TX unidirectional primer: See → TX-primer.

Ty (*transposon yeast*) element: A series of transposable elements of *Saccharomyces cerevisiae*. There are usually 30-35 copies of Ty elements present per haploid genome. Each → transposon has roughly a length of 5.6 kb and consists of a central region containing two → open reading frames (Ty A and Ty B) and 250 bp flanking → terminal direct repeats (∂ sequences). Insertion of Ty elements into target sequences (preferentially → promoter regions, e.g. 5' sequences of → transfer RNA genes) is accompanied by a 5 bp duplication of its target DNA. New copies of the transposon arise by a replicative → transposition process in which the Ty transcript is converted to a progeny DNA molecule by a Ty-encoded → reverse transcriptase (*tyB* gene product). Ty 1 sequences are also ubiquitous components of plant genomes. See → retroposon.

Ty 1 (*transposon yeast 1*): A → Ty element of yeast.

Type I locus: Any one of an immense number of DNA sequences, generally → genes, whose sequence is conserved across species. See → synteny, → type II locus.

Type II locus: Any one of a series of highly variable, → minisatellite- or → microsatellite-containing genomic regions. See → type I locus.

***T*yramide *s*ignal *a*mplification (TSA; *c*atalyzed *r*eporter *d*eposition, CARD):** A technique for the enhancement of signal strength in immunoblotting, immunocytochemistry, → fluorescent *in situ* hybridization on chromosomes and → *in situ* hybridization for the detection of → messenger RNAs (see → tyramide signal amplification using alkaline phosphatase). TSA employs → *h*orse*r*adish *p*eroxidase (HRP) to catalyze the deposition of previously → biotin or → fluorochrome-labeled tyramide onto chromosome spreads, cell preparations, or tissue sections (paraffin-embedded or semi-thin sections in methacrylate plastic), whose surfaces are coated with proteins. The binding of the labeled tyramides to the surface is covalent, because it reacts with tyrosine residues of surface-bound endogenous proteins. The deposition of numerous biotin molecules occurs close to the enzyme, and can be visualized with → *s*treptavidin (SA) conjugated either to HRP (SA-HRP) or → alkaline phosphatase (SAAP) and chromogenic detection ("indirect TSA"). Tyramides covalently linked to → fluorochromes (e. g. → fluorescein [green], coumarin [blue], cyanin 3 [red] or → tetramethylrhodamine [red]) can also be used, and their localization be detected by direct fluorescence microscopy ("direct TSA").
Figure see page 1208.

Tyramide signal amplification using alkaline phosphatase (TSA-AP): A variant of the → tyramide signal amplification technique for the detection of → low abundancy messenger RNAs, that visualizes the hybridization of a → biotin-labeled → probe to an RNA target by → streptavidin-alkaline phosphatase conjugates which catalyze the deposition of coloured products at the site of enzyme action. TSA is used to amplify the chromogenic signal.

U

U: Abbreviation for *u*racil (2,4-dihydroxy-pyrimidine), a pyrimidine base characteristic for RNA, and not found in DNA.

UA cloning (UA hybridization): A variant of the → TA cloning procedure for the cloning of → *Taq* DNA-polymerase-generated amplification products, that uses a UA instead of a TA → overhang. Non-proofreading DNA polymerases such as *Taq* DNA polymerase add an extra A to the 3' termini of their amplicons. Therefore TA cloning vectors are designed to exploit these overhangs for cloning. However, the T also hybridizes to non-complementary bases (i.e. C, G, or T), so that the vectors may self-anneal, or primers are cloned. The UA cloning procedure avoids this problem by replacing T for U, which has only a low tolerance for non-specific base-pairing such that the UA cloning is both more efficient and faster than comparable techniques (e.g. → blunt-end or → sticky end cloning, → topoisomerase I cloning).

UA hybridization: See → UA cloning.

UAP: See → *u*niversal *a*dapter *p*rimer.

UAS:
a) See → *u*pstream *a*ctivation *s*ite.
b) *U*pstream *a*ctivator *s*equence: An incorrect term for → enhancer.

uAUG: See → upstream AUG.

UBF: See → *u*pstream *b*inding *f*actor.

Ubiquitin: A ubiquitously occurring, phylogenetically highly conserved, globular and compact 85 kDa acidic eukaryotic protein of 76 amino acids, which can be conjugated to a variety of cellular and nuclear proteins (e. g. → histones H2A and H2B, actin, the lymphocyte homing receptor, the platelet-derived growth factor receptor, the growth hormone receptor and ribosomal proteins) via an isopeptide bond between its C-terminal glycine and the ε-amino group of a lysine residue in the acceptor protein. This reversible conjugation to cellular proteins – an ATP-requiring process – mediated by three different enzyme classes (the ubiquitin-activating enzyme E1, the ubiquitin-conjugating enzymes known as E2 proteins, and the ubiquitin-protein ligase E3). Two different gene families encode ubiquitins, the ubiquitin-fusion genes, which consist of a single *ub*iquitin-encoding sequence (Ub) fused to a *c*arboxy-terminal *e*xtension *p*rotein-encoding sequence (CEP), and the polyubiquitin genes, that encode from 2-60 head-to-tail repeats of ubiquitin genes (human: 3-9 repeats). The Ub-CEP genes maintain the basal ubiquitin level in the cell, whereas the polyubiquitin genes function during acute cellular stress (e. g. heat stress). The addition of multiple ubiquitin adducts to a protein acts as signal for its degradation. Polyubiquitinylation starts with the addition of one ubiquitin molecule, and the addition of a second one to the lysine residue K48 of the first one, and so on. The tagged proteins are then degraded by the non-lyso-somal protease functions of the 26S multi-subunit → proteasome. Selection of the proteins for degradation makes use of their structure at the N-terminus ("N-end rule"). For example, a protein with an N-terminal AG has a 10^3 times higher life-time than a protein with an N-terminal RF. After degradation, the ubiquitin is liberated to continue the cyclic process. Dramatic increases in ubiquitin and ubiquitin-protein conjugates are observed in neurons during a variety of neurodegenerative disorders, as e. g. Alzheimer's and Parkinson's disease. Also patients suffering from the autoimmune disease Systemic Lupus Erythematosus generate antibodies reacting with ubiquitin and ubiquitinated histones.

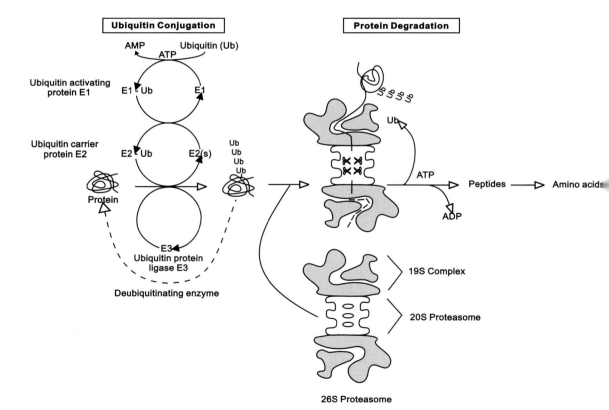

UCL: See → *u*niversal *c*DNA *l*ibrary.

UCO: See → *u*nequal *c*rossing-*o*ver.

UCP: See → *u*niversal *c*apture *p*robe.

UDG: See → *u*racil *D*NA *g*lycosylase.

U5: An 80-100 bases long sequence element from the 5' end of the genomic RNA of → retroviruses.

UHD: See → *u*ltra-*h*igh *d*ensity map.

uHTS: See → ultra-high throughput screening.

***uid* A:** The gene encoding → ß-glucuronidase in *E. coli*.

Ultracentrifugation: The process of sedimenting cells, subcellular particles or molecules in extreme gravity fields (e.g. more than $500000 \times g$) through → density gradients.

Ultracentrifuge: An instrument that drives rotors of various designs (e.g. fixed angle rotor, swing out rotor, vertical rotor) up to 100000 revolutions per minute, rpm (corresponding to gravity fields of more than $500000 \times g$). These gravitational forces serve to separate cells, subcellular particles or molecules on the basis of either their density or their size. Preparative ultracentrifuges are used to prepare cells, subcellular particles or molecules for their subsequent analysis. Analytical ultracentrifuges allow the visualization of sedimenting particles in the spinning rotor through quartz windows and by means of UV or Schlieren optics.

Ultracontig: A large fragment of genomic DNA composed of a series of adjacent → supercontigs, that in turn consist of neighboring → contigs. For example, in the mouse genome sequencing project, ultracontigs had a length of 50 Mb, so that 88 ultracontigs covered the whole genome.

Ultra-*h*igh *d*ensity map (UHD): A → genetic map which consists of a very dense arrangement of molecular markers (e. g. 5-10 → AFLPs or → STMS per centiMorgan).

Ultra-high density microarray: Any one of a series of → microarrays, that contains millions of different immobilized probe molecules. For example, a socalled → nanoarray, produced by → *d*ip *p*en *n*anolithography (DPN), may harbor 50,000 dots ("low resolution DPN") or even 100 millions of dots ("high resolution DPN") instead of a single dot of a conventional → microarray.

Ultra-*h*igh *t*hroughput *s*creening (uHTS): A robust routine process which allows to perform more than 10^5 assays per day in a working volume of 1 to 10 µl per assay, using high density platforms (e.g. 1536 well assay plates).

Ultrasound-mediated gene transfer ("ultrasonic gene transfer"): The introduction of genes (generally, DNA sequences) into recipient genomes of target cells, which were briefly exposed to ultrasonic waves. This procedure facilitates DNA uptake by the cells, its stable integration into the target genome and its expression. See → calcium phosphate or → DEAE dextran precipitation, → direct gene transfer, → electrophoretic transfection, → electroporation, → particle gun technique. See also → chromosome-mediated gene transfer, → fragment transfer, → hypo-os-motic shock loading, → irradiation and fusion gene transfer, → lipofection, → microcell-mediated gene transfer, → microinjection, → polybrene transformation, → protectifer, → scrape-loading, → sonication loading, → transferrinfection. Compare → indirect gene transfer.

Ultra-zoom gel (narrow range immobilized pH gradient strip, IPG strip): A gel strip for the separation of proteins by → isoelectric focusing (IEF), that comprises only one to two pH units instead of six to nine units, but has the same length as the conventional IEF strip. Therefore, ultra-zoom gels allow to separate more proteins (IEF strip of pH 3–10: only 1,000 to 3,000 proteins; five ultra-zoom gels comprising the same pH range: 15,000 to 20,000 proteins).

UMP: See → uridine-5'-monophosphate.

UMS: See → upstream mouse sequence.

Unassigned reading-frame (URF, unidentified reading-frame): A gene-like nucleotide sequence with proper → start and → stop codons but without any known function, usually derived from inspection of DNA sequences. Compare → reading frame.

Unbalanced genome: Any → genome that suffered → deletions, → duplications, → inversions, → translocations, generally → rearrangements during → recombination with a second alien genome, producing newly arranged sequence contexts, which influence the viability or fitness of the resident organism.

Uncharacterized open reading frame (uncharacterized ORF): Any → open reading frame (ORF) detected by sequencing a genome from one species, supposed to be real, because → orthologs exist in one or more other species, which are, however, not verified experimentally (i.e. no gene product is found yet). See → verified open reading frame.

Uncharacterized protein family (UPF): Any protein family that does not contain biochemically characterized members, but is present in at least three taxonomically distinct species, or the three major kingdoms (archaea, eubacteria, and eukaryotes).

Uncoating: The release of viral nucleic acid from a → virus particle during the process of viral infection of a cell, leaving the coat proteins outside the cell (in the case of → bacteriophages). Uncoating may also occur at the cell membrane or within the cytoplasm (animal viruses).

Underwinding: See → negative supercoiling.

Unequal crossing-over (UCO): The non-reciprocal exchange of segments of homologous chromosomes that are not precisely paired. UCO leads to the loss of sequences on one, and the concomitant gain of sequences on the other chromosome, so that chromosomes of unequal length are generated. Such unequal cross-over events are enhanced in chromosomal regions, where clusters of tandemly repeated sequences are located. See for example → magnification. Compare → homologous recombination, → cross-over.

Unfolded protein response (UPR): A cellular stress response of the endoplasmic reticulum (ER) as a consequence of an accumulation of wrongly folded proteins. Normally, the processing of membrane-bound or secretory proteins in the intraluminal space of the ER is terminated by the correct folding of the proteins, assisted by molecular → chaperons. If the folding becomes increasingly imprecise, the ER-bound transmembrane kinase/nuclease Ire1p (yeast) or Ire1 and Ire1 m (mammals) identifies the imbalance between the proteins to be folded and the proteins incorrectly folded and transmits a signal to the cytoplasm or nucleus. In the normal state, Ire1 is inactive (monomeric), but is activated by oligomerization. This process is catalyzed by the chaperon BiP, that forms a stable complex with Ire1 in resting cells. Now, incorrectly folded proteins compete with Ire1 for BiP, the latter is blocked, and the oligomerization of Ire1 takes place. This leads to an autophosphorylation of the transmembrane kinase, inducing the endonuclease function of the protein, which in turn splices the messenger RNA of the → transcription factor Hac1p. Hac1p binds to the socalled *unfolded protein response elements* (UPREs) in promoters of BiP and a total of 208 other genes, enhancing their transcription (and finally expression).

UNG: See → uracil DNA glycosylase.

Ungapped sequence: Any (usually genomic) DNA sequence stretch, that is not interrupted by → gaps.

Unidentified reading-frame: See → *u*nassigned *r*eading-*f*rame.

Unidirectional cloning: See → forced cloning.

Unidirectional primer: A synthetic oligodeoxyribonucleotide that contains a homopolymeric (dT)-tail at its 3' end and one or more → restriction sites in close proximity. This type of → primer is designed to prime → cDNA synthesis from poly(A)⁺-mRNA while generating a specific restriction site at the 3' end of the cDNA. After restriction at this site the cDNA can be cloned into an appropriately linearized vector in a predictable orientation (→ forced cloning). See for example → TX-primer.

Unidirectional replication: A specific mode of DNA → replication, in which the → replisome moves in only one direction along the DNA template.

Uniformity: The consistency of fluorescence signals on a → microarray (hybridized to a fluorescently labelled → probe) at different locations across the array support (e.g. a glass slide). Preferably, the uniformity at the different spots of a chip warrants that no bias is introduced to the generation of reliable hybridisation data.

Unigene:
a) A set of non-redundant → complementary DNAs representing a cluster of → expressed sequence tags in eukaryotic genomes.
b) Any → *e*xpressed *s*equence *t*ag (EST), that is generated by the alignment of many EST sequences of an EST library sharing more than 95% homology. The unigene represents the most likely candidate for a → cDNA of a particular gene. Do not confuse with → UniGene.

UniGene: A database consisting of GenBank sequences organized into non-redundant sets of gene clusters, each of which probably represents only one unique gene. UniGene additionally contains pertinent informations as e.g. the type of tissue in which the gene is expressed, and hundreds of thousands of novel → *e*xpressed *s*equence *t*ags (ESTs). The database is subdivided into separate databases for e.g. human, mouse and rat sequences. Do not confuse with → unigene.

Unigene EST: See → unigene *expressed sequence tag*.

Unigene *expressed sequence tag* (unigene EST): Any → expressed sequence tag derived from a gene that occurs only once in a genome.

Uniplex DNA sequencing: A technique for the directional determination of the sequence of bases in DNA. In short, the foreign DNA is inserted into an appropriate → cloning vector. The strands of this vector are then separated, and a biotinylated → primer annealed to the selected strand. Then the → Sanger sequencing reactions are performed, the DNA oligomers separated by → electrophoresis, the fragment pattern transferred to a membrane, cross-linked and detected by → chemiluminescence.

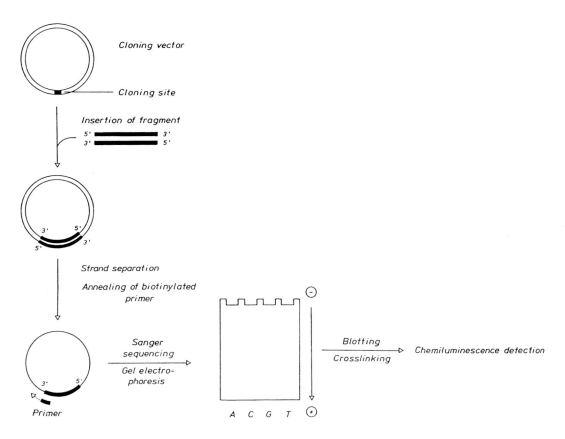

Uniplex DNA sequencing

Unique DNA (single-copy DNA): A fraction of denatured → genomic DNA that forms duplexes very late in a → C_0t analysis (i.e. reassociates at high → C_0t values) because its sequences occur only once in the genome. This fraction comprises virtually all → structural genes in prokaryotes and eukaryotes. Compare → repetitive DNA.

Unique sequence: A DNA sequence present only once per haploid genome. See → unique DNA.

Unitig: A laboratory slang term for a set of different genomic sequence reads assembled into a single contiguous sequence such that no fragment in this unitig overlaps any fragment not present in the unitig. Compare → contig. See → U-unitig.

Universal *adapter-primer* (UAP): An → adapter-primer that is complementary to sequences conserved among widely divergent species. It can therefore be used to prime the amplification of sequences common to a variety of different organisms in the → polymerase chain reaction.

Universal array (universal microarray, "one-chip-for-all"): Any → gene array, which contains all possible genes of the living world in the form of 10–16 nucleotides long, chemically synthesized oligonucleotides (universality requires one million spots of 10-mers per array, or 64 million spots per array for 13-mers). Universal arrays would allow to globally determine the gene expression patterns ("expression signatures") of any organism.

Universal base: Any nucleic acid base that forms weak base-pairs with dA, dC, dG, or dT residues in a target DNA. For example, → inosine, 2'-deoxynebularine and 3-nitropyrrol are such universal bases.

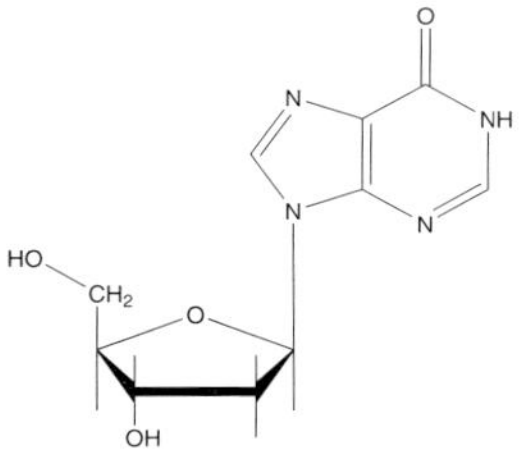

Deoxyinosine 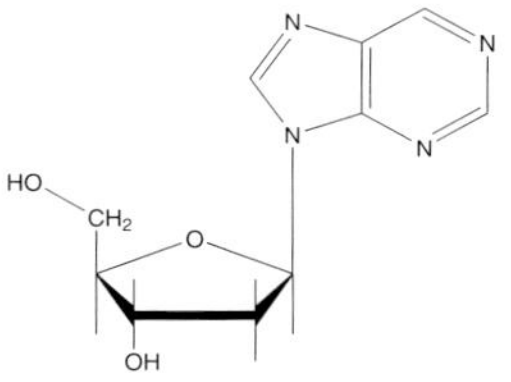   DeoxyNebularine 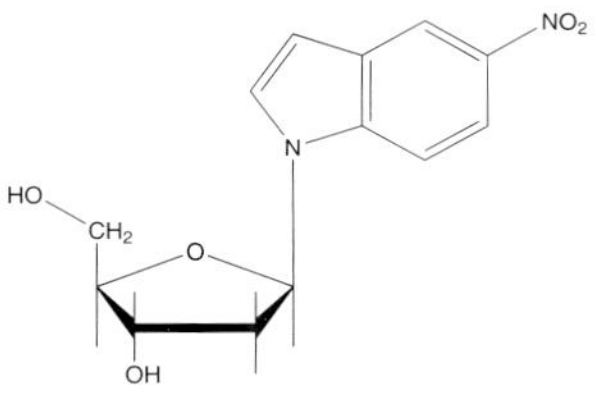   5-Nitroindole 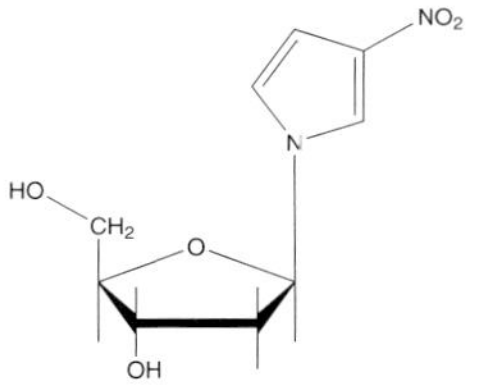   3-Nitropyrrole

Universal *capture probe* (UCP): Any chimeric oligonucleotide composed of a sequence complementary to a socalled → target capture oligonucleotide and a universal sequence. UCPs serve to anchor the target capture oligonucleotide to the solid support (e. g. the well of a microtiter plate) in → DNA capture assays.

Universal *cDNA library* (UCL): A comprehensive → cDNA library that contains the → messenger RNA transcripts of many different cells, tissues, or organs of an organism. Normally these universal superlibraries are extensively normalized to reduce the over-representation of → abundant messages. The clones can be arrayed on supports (e. g. → nitrocellulose filters, → microchips) and serve to detect specific cDNAs of a sample by → hybridization procedures. See → normalization.

Universal code: The → genetic code that is identical in most organisms. It is, however, slightly different in e.g. mitochondria of specific organisms where AGG and AGA (normally codons for arginine) are → stop codons, and UGA (normally a → stop codon) stands for tryptophan.

Universally *primed polymerase chain reaction* (UP-PCR): A technique for the detection of sequence → polymorphisms between two (or more) genomes, that is similar to the → *random amplified polymorphic DNA* (RAPD) method, but uses specific → primer features (e. g. primers of 15-22 bases; RAPD: 10 bases), high primer annealing temperatures (e. g. 52-65° C; RAPD: 33-38° C), fast ramping and an initially long primer annealing time (2 min). These parameters improve the reproducibility of the technique, so that it is superior to RAPD fingerprinting. Do not confuse with → unpredictably primed PCR.

Universal microarray: See → universal array.

Universal primer: A synthetic oligonucleotide complementary to sequences that are conserved among widely divergent species. It may be used in the → polymerase chain reaction as → primer for the amplification of a particular nuclear or organellar gene fragment from nearly all members of a major taxonomic group (e.g. fungi, plants). Compare → universal probe.

Universal probe: Any nucleic acid sequence that detects homologous sequences in a whole variety of different organisms (e.g. → ribosomal RNA or → rDNA which has been conserved during evolution). See also → universal primer.

***Universal protein array* (UPA):** A nitrocellulose membrane onto which hundreds or thousands of highly purified proteins are spotted, that serve as anchors to detect protein-protein, protein-DNA, protein-RNA, or protein-ligand interactions. In short, the protein array is prepared by spotting highly purified proteins (or protein fractions), preferably in functionally related groups (e.g. → transcription factors together with activators and coactivators and RNA processing enzymes, or → replication proteins, → translation proteins, or signal transduction proteins). The various proteins are first overexpressed in *E.coli* or → baculovirus, and purified to near-homogeneity, retaining their full activity. Then either protein → probes (labeled with ^{32}P using the catalytic subunit of Ca^{2+}-independent bovine heart muscle protein kinase A and [γ^{32}P]-ATP), or double-stranded DNA probes (labeled at the 3'end with → Klenow fragment and [^{32}P]dCTP), or single-stranded DNA probes (labeled at the 5'end with → T4 polynucleotide kinase and [γ^{32}P]ATP), or RNA probes (synthesized from a plasmid by → SP6 RNA polymerase and ^{32}P-UTP), or ligand probes (e.g. L-3,5,3'-[^{125}I]triiodothyronine) are hybridized to the UPA. After hybridization under non-denaturing conditions, the array is washed, the resulting signals visualized by → autoradiography and quantified by a densitometer. The same UPA can be re-used for hybridizations with either probe, after the previous hybridization probe has been removed by high salt and denaturant (1M urea) stripping. UPA allows to map protein interaction domains and DNA- or RNA-binding domains of a protein.

Universal reference RNA: A mixture of DNA-free total RNAs from several tissue types of an organism (e.g. ten different human tissues or cell lines as e.g. B-lymphocytes, brain, cervix, glia, kidney, liver, macrophages, mammary glands, skin and testis) in equal quantities, representing preferably all expressed genes of these tissues/organs. These pooled RNAs are used as internal standard in expression array experiments. For example, universal reference RNA allows to compare multiple expression → arrays for homogenous spotting and hybridisation.

Unknome: A laboratory slang term for all genes (more generally, DNA sequences), whose sequence is completely known, but their function not yet deciphered.

Unnecessary gene: See → superfluous gene.

***Unpredictably primed polymerase chain reaction* (UP-PCR):** A technique to walk from any known genomic locus into the unknown flanking sequences. In short, genomic target DNA is amplified in two successive steps. First, a → walking primer is annealed to denatured genomic target DNA at low → stringency and extended by → *Taq* DNA polymerase. Second, the resulting amplification product is then denatured, and specifically re-amplified using a → *sequence-specific prim-er* (SP; "outer SP") complementary to the known sequence from which the walk should begin, with its 3' end towards the unknown region. This amplification occurs at high stringency. Third, the resulting amplicon is now more specifically amplified with a → nested primer (*inner specific primer*, iSP; i. e. a primer complementary to the known sequence) and a *short walking primer* (sWP; partly identical to the initially used walking primer). The final product then contains part of the known sequence (complementary to the iSP) and the unknown flanking region extending to the sequence complementary to the sWP. Compare → thermally asymmmetric interlaced polymerase chain reaction.

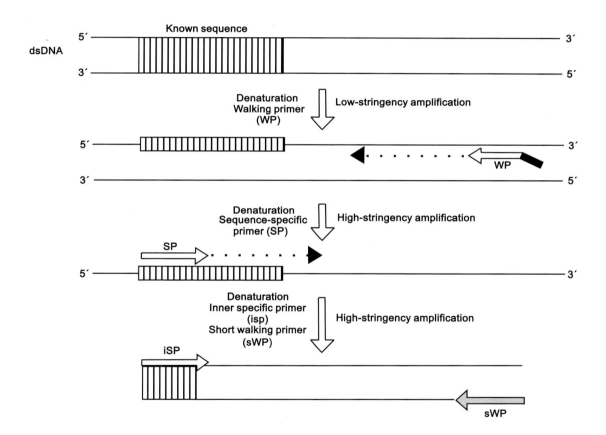

Unstable mutation: Any → mutation with a high frequency of reversion (see → reverse mutation; e.g. a mutation caused by the → insertion of a → transposon that has a high frequency of → transposition).

Untranslated RNA: A generic term for all → ribonucleic acids, that are not translated into proteins (as e.g. → messenger RNAs are). For example, → non-coding RNA, → ribosomal RNA, → short interfering RNA, → small endogenous RNA, → small non-messenger RNA, → small regulatory RNA, → tiny RNA, → transfer RNA (among others) are such untranslated RNAs.

Untranslated sequence (UTR): See → leader sequence (b) and → trailer (a).

Untwisting enzyme: See → DNA topoisomerase II.

Unweighted pair group method with arithmetic means (UPGMA): A clustering algorithm for the analysis of → microarray data, that calculates the average Euclidean distance between each point in a cluster and all the points in a neighboring cluster. The two clusters closest to each other (i.e. with the smallest average distance) are connected to a higher order cluster. The resulting data are depicted as a socalled dendrogram (dendrograph, tree-like graph), with the closest branches of the tree representing genes with similar expression patterns. UPGMA analysis can also be applied to the estimation of genetic diversity in a population.

Unwinding protein: See → DNA helicase.

uORF: See → upstream open reading frame.

UPA:
a) See → universal protein array.
b) See → *u*niversal *p*rotein *a*ssay.

UPE: See → upstream regulatory sequence.

UPGMA: See → unweighted pair group method with arithmetic mean.

Up mutation: See → promoter-up mutant.

UP-PCR:
a) See → *u*niversally *p*rimed PCR.
b) See → *u*npredictably *p*rimed PCR.

Up-promoter mutation: See → promoter-up mutant.

UPR: See → unfolded protein response.

Upregulation: A generic term for any regulated increase in the activity of a peptide, protein (e.g. enzyme) or gene. See → downregulation.

Upstream: A term to describe sequences in a linear DNA or protein molecule, preceding a given site in the opposite direction from gene expression or protein synthesis, respectively (e.g. on the → 5' side of any site in DNA or RNA, and on the free amino terminus side of any site in a protein). See → downstream.

Upstream activation site (UAS, UPS, upstream activation sequence): Any one of a series of short sequence elements 5' upstream of the initiation sites of yeast (*Saccharomyces cerevisiae*) genes that is functionally similar to → enhancer elements of higher eukaryotes and required for optimal expression of adjacent genes. It is, however, non-functional if placed downstream of a → promoter.

Upstream activator sequence: See → enhancer.

Upstream AUG (uAUG): Any AUG codon within the 5'-untranslated → leader sequence of a → messenger RNA encoded by certain classes of genes as e.g. → oncogenes and other genes innvolved in the control of cellular growth and differentiation. In some cases, uAUGs also control → translation.

Upstream binding factor (UBF): A eukaryotic → transcription factor that interacts with a → TATA box-binding protein complex (SL1) to establish a stable pre-initiation complex on the → promoter of → rDNA genes. UBF is essential for → RNA polymerase I-directed transcrip tion of → ribosomal RNA genes. The UBF protein carries a cluster of → high mobility group (HMG) box-specific DNA-binding domains (e. g. 5 in the UBF of *Xenopus laevis*, that interact with irregular DNA structures (e. g. → cruciform and → supercoiled DNA), and can bind two or more DNA → duplexes simultaneously to form crossover junctions. A 107 amino acids long N-terminal dimerization domain allows homodimer formation. The C-terminus contains mostly acidic residues. The UBF and redundant rDNA repeats are assembled in the → nucleolus to support optimal ribosomal gene transcription.

Upstream enhancer: See → enhancer.

Upstream mouse sequence (UMS): A DNA sequence element that contains a transcription ter- mination signal, isolated from the → promoter of the murine c-*mos* gene and used to reduce uninduced background transcription of cloned → reporter genes.

Upstream open reading frame (uORF): Any → open reading frame in a → messenger RNA, that is located → upstream of the main open reading frame, recognized by a scanning 40S ribosome subunit and associated initiation factors and translated, before it reinitiates translation at an AUG codon further downstream. Translation of uORFs normally blocks or reduces translation from a downstream UAG by e.g. the peptide encoded by the uORF, or by ribosome stalling on the mRNA. For example, translation of a basic helix-loop-helix transcriptional activator protein encoded by Lc, a member of the maize R/B gene family functioning in the anthocyanin pathway, is repressed by a 38-codon uORF in the 5'leader. Repression results from inefficient re-initiation of ribosomes.

Upstream primer (UP): Any → oligonucleotide → primer that binds to its homologous sequence → upstream of a target DNA (e. g. a primer with homology to → promoter sequences will bind upstream of the → cap site).

Upstream promoter: One of a pair of → promoters, both driving the expression of a particular gene, that is located 5' → upstream of the gene. The other promoter (→ downstream promoter) lies at the 3' end of the gene. For example, transcription of the human *RCC 1* gene is initiated at two different promoters about 9 kb apart. Initiation at the downstream promoter produces a pre-mRNA, in which a 5'terminal single → noncoding exon is spliced to downstream exons encoding the RCC 1 protein. Initiation at the upstream promoter leads to the synthesis of a transcript containing four short noncoding exons spliced to the coding part of the mRNA.

Upstream promoter element: See → enhancer, → upstream regulatory sequence.

Upstream regulatory sequence (URS; upstream promoter element, UPE): A short *cis*-acting sequence upstream of the → TATA box in eukaryotic → promoters. URSs are binding sites for → transcription factors and effect their optimal positioning for the formation of a stable and active → transcription initiation complex (up-regulation, see e.g. → enhancer element). They may, however, also cause reduction of expression of the adjacent genes (down-regulation, see → negative element). See also → weak positive element. Compare → enhancer.

Upstream stimulatory activity (USA): The property of a complex nuclear mixture of activating and inhibiting proteins, that stimulate → *in vitro* transcription. For example, one of the many components of USA, the activator p15 (synonym: PC 4), binds to single- and double-stranded DNA, stimulates the transactivation of different → transcription factors, and initiates transcriptional activation during → transcription factor II A-TF II D-promoter complex formation.

Uracil: See → U.

Uracil DNA glycosylase (UDG; uracil N–glycosylase, UNG): An enzyme from *E. coli* that catalyzes the excision of deoxyuracil (dU) from dU-containing double- or single-stranded DNA by cleaving the N-glycosidic bond between the uracil base and the deoxyribose phosphate backbone. RNA and normal dT-containing DNA are no substrates for the enzyme, which is employed for → polymerase chain reaction carry-over prevention and → base excision sequence scanning. See → DNA glycosidase, → uracil glycosylase inhibitor.

Uracil glycosylase inhibitor (UGI): A thermostable, small (9.5 kDa) protein of the PBS1 phage of *Bacillus subtilis*, that inhibits the → uracil-DNA glycosylase of *E. coli* by reversible binding to the enzyme. UGI is used to suppress residual uracil-DNA glycosylase activity after heat treatment, so that subsequent DNA degradation is prevented.

Uracil interference assay (uracil interference): A technique for the isolation of a DNA sequence, that serves as target site for a → DNA-binding protein. In short, an oligonucleotide or → restriction fragment containing a protein-binding sequence is first amplified by conventional → polymerase chain reaction using an unlabeled forward → primer and a 5' labeled reverse primer, and the four dNTPs as well as dUTP. Deoxyuracil is randomly substituting thymine on both → template strands. This modified DNA is now incubated with the DNA-binding protein, and sequences with deoxyuracil that do not interfere with protein-binding, are selected (separation of unbound DNA and DNA-protein complexes). The selected DNA is now treated with → uracil-DNA gylcosylase, that generates → apyrimidinic sites susceptible to piperidine. The reaction products are finally separated by denaturing → polyacrylamide gel electrophoresis.

Uracil N-glycosylase: See → uracil DNA glycosylase.

Uranylacetate: An uranium salt that is very electron-dense and used in electron microscopy to stain nucleic acid-containing structures in thin sections.

Uranyl photocleavage (uranyl cleavage): The introduction of single-strand breaks into DNA by uranyl(VI) ions. UO_2^{2+}, which binds strongly to DNA, is excited by long wavelength UV light, and – as very strong oxidant and coordinated to a phosphate group – oxidizes proximal sugars via a direct electron transfer mechanism. Uranyl phosphocleavage reflects phosphate accessibility in DNA and can therefore be employed in e.g. Studies of protein-DNA interactions (via phosphate-protein contacts).

URF: See → *u*nassigned *r*eading-*f*rame.

Uridine (3-β-D-ribofuranosyluracil): A nucleoside that consists of → uracil linked to a D-ribose molecule. A specific component of RNA.

Uridine-5'-*mono*phosphate (5'-UMP, UMP): A → pyrimidine nucleotide with a phosphorous group in ribose-O-phosphoester linkage at the 5' position of ribose.

Uridine-5'-*tri*phosphate (UTP): A → pyrimidine nucleotide with an energy-rich triphosphate group in ribose-O-phosphoester linkage at the 5' position of the ribose. UTP serves as elementary unit in RNA synthesis.

U-RNA (U-snRNA): A member of a family of eukaryotic, uridine-rich, small (100-300 nt) and strongly conserved nuclear RNAs, designated as U1, U2, U3.... U-RNAs are synthesized by → RNA polymerase II and capped at their 5' end. U-RNAs are a sub-family of → small nuclear RNAs, and are constituents of the → small nuclear ribonucleoprotein particles.

URS: See → *u*pstream *r*egulatory *s*equence.

USA: See → *u*pstream *s*timulatory *a*ctivity.

U-snRNA: See → U-RNA.

U-snRNP: See → *s*mall *n*uclear *r*ibo*n*ucleoprotein particle.

U3: A 170-1250 bases long sequence element from the 3' end of the genomic RNA of → retroviruses.

UTP: See → *u*ridine-5'-*tri*phosphate.

UTR: See → *un*translated sequence.

U-turn: A → terminal loop motif of RNA molecules, especially the → anticodon loops of → transfer RNA. U-turns contain the hyperabundant socalled terminal loop motif UNRN (N= any nucleotide; R= any purine).

U-unitig: Any → unitig that consists of *u*nique DNA (probably genes).

UV partial: A mixture of fragments arising from partial → digestion of DNA that has been irradiated with UV light. This treatment produces pyrimidine dimers. If they occur at or near the → recognition site of certain → restriction endonucleases, these enzymes will not be able to cleave there. For example, *Eco* RI (or *Rsr* I) normally cleaves the sequence 5'-GAA TT C-3', but does not recognize it when a thymidine dimer has been formed (5'-GAATTC-3').

***uvr* genes:** A set of genes that are involved in the repair of DNA damaged by *u*ltraviolet *r*adiation.

UV shadowing: The visualisation of nucleic acid molecules separated by → agarose gel electrophoresis by placing the gel onto a fluorescent support (e.g. a thin layer chromatography plate with a UV indicator), and irradiating it with short wave UV light (254 nm). Since RNA and DNA absorb UV, the location of these molecules can easily be seen as nonfluorescent bands ("shadows") on a fluorescent background.

V

V: Abbreviation for any nucleotide in a DNA sequence except → thymidine (i. e. → adenine, → cytosine, and → guanosine).

Vacuum blotting: A variant of the conventional capillary transfer (→ Southern transfer) used for the blotting of DNA fragments from 1 kb up to the size of whole chromosomes from an → agarose gel to → nitrocellulose or nylon-based matrices employing controlled → vacuum. Vacuum blotting can also be applied to proteins separated on a → polyacrylamide gel. See also → blotting.

Variable number of dinucleotide repeats (VNDR): The occurrence of different numbers of dinucleotide → microsatellites (e.g. $[CA]_n$, $[GA]_n$) at identical genomic loci of two organisms. VNDR polymorphism can be detected by → DNA fingerprinting techniques. See → variable number of tandem repeats (VNTR).

Variable number of tandem repeats (VNTR): A set of tandemly repeated, short (11-60 bp) oligonucleotide sequences with a conserved → core sequence 5'-GGGCAGGAXG-3' The number of these repeats within a given DNA region of the human genome varies from one individual to another. The diminution or amplification of the number of such tandem repeats may be due to a high frequency of → unequal crossing-over events at the VNTR recombinational → hot spots. VNTRs thus are responsible for considerable DNA sequence → polymorphisms in the human genome (e.g. since the length of a restriction fragment carrying VNTRs is a function of the copy number of tandem repeats present within the fragment, → RFLPs may be due to the presence of VNTRs). VNTRs can be detected for example with → ligated oligonucleotide probes. Compare also → hypervariable region, → microsatellite, → minisatellite. See → variable number of dinucleotide repeats.

Variable region: The part of → antibody heavy chain and light chain molecules which is specific for each individual antibody → clone. It is responsible for antigen recognition and binding and shows little or no sequence conservation between different antibodies. Compare → conserved region.

Variant repeat unit: A member of a → minisatellite repeat family whose sequence differs from that of adjacent members. For example, in certain human minisatellite repeats (e.g. D1S8), a → transition of a specific A to a G leads to the disappearance of a *Hae* III → restriction endonuclease → recognition site. Since all repeat units are still cut by restriction endonuclease *Hinf* I, → partial digests with both enzymes allow each repeat unit to be scored for *Hae* III restriction sites (or, in other words, for single transitions).

VCR: See → *Vibrio cholerae repeat*.

V(D)J recombination: A programmed → DNA recombination process in developing B and T cells, whereby V (variable), D (diversity) and J (joining) segments are assembled into functional immunoglobulin and T-cell receptor genes. Recombination is absolutely dependent on the expression of the recombination activating genes *RAG1* and *RAG2*, that recognize and cleave DNA at the border between a conserved recombination signal sequence and the flanking coding DNA segment. The DNA bending proteins HMG1 and/or HMG2 assist in this process. Then the cleaved recombination signals are ligated to form a signal joint, and the deletion and addition of nucleotides to the coding DNA forms a coding joint. This reaction is mediated by ubiquitous proteins involved in the repair of DNA double-strand breaks. This combinatorial process, combined with the imprecise joining of the gene segments, generates a huge variability in the → antigen recognition part of receptors. V(D)J recombination probably evolved from a transposable element, whose transposase is now represented by the RAG proteins ("RAG transposon"), about 450 millions of years ago (after the divergence of jawless and jawed vertebrates). See → class switch recombination.

V-DNA: A special form of dsDNA, generated by the annealing of two complementary single stranded cccDNA molecules, in which any region of right-handed double-helical conformation has to be compensated by → negative supercoiling and/or by region(s) of left-handed conformation. V-DNA exists *in vitro* only. See → A-DNA, → B-DNA, → C-DNA, → D-DNA, → E DNA, → ε-DNA, → G-DNA, → G4-DNA, → H-DNA, → M-DNA, → P-DNA, → Z-DNA.

Vector: See → amplicon vector, → autocloning vector, → baculovirus vector, → bicistronic expression vector, → bidirectional vector, → binary vector, → broad host range vector, → BRP vector, → bacterial release protein vector, → CAT vector, → *cis* vector, → cloning vector, → cointegrate vector, → dicistronic vector, → directional vector, → direct selection vector, → disarmed vector, → display vector, → double cos-site vector, → dual expression vector, → dual promoter vector, → dual selection marker vector, → episomal expression vector, → epitope tagging vector, → excretion vector, → exon trap vector, → expression cloning vector, → expression phagemid vector, → expression-secretion vector, → expression vector, → expression shuttle vector, → gene replacement vector, → gutless vector, → inducible expression vector, → integration vector, → integrative vector, → intermediate vector, → linearized vector, → Lorist vector, → minimal domain vector, → minimal vector, → modular vector, → non-palindromic vector, → Okayama-Berg cloning vector, → open reading frame vector, → open reading frame expression vector, → ORF vector, → pEX vector, → phage cloning vector, → plant expression vector, → plasmid cloning vector, → plasmid vector, → positive selection vector, → prey vector, → pUC vector, → pUR expression vector, → replacement vector, → replicative vector, → reporter vector, → retroviral vector, → runaway plasmid vector, → secretion vector, → self-inactivating expression vector, → sequence replacement vector, → shuttle vector, → SI expression vector, → SIN vector, → siRNA expression vector, → split end vector, → SP6 vector, → substitution vector, → suicide vector, → supervirulent vector, →TAT peptide vector, → terminator probe vector, → transformation-competent artificial chromosome vector, → transformation vector, → transient expression vector, → transposon-based cloning vector, → *trans*-vector, → triple-helix cosmid vector, → TR vector, → TU vector, → T vector, → vigilant vector, → viral vector, → yeast cloning vector.

Vector arm primer: See → vector primer.

Vector cassette: A construct containing all sequence elements necessary for → stable transformation of cells, e. g. a → selectable marker gene, and a → reporter gene. It also carries a → multiple cloning site into which a target gene can be cloned.

Vectorette: A synthetic → oligodeoxynucleotide that contains a mismatched central region (partially single-stranded), and two flanking double-stranded DNA regions. Vectorettes are ligated to fragments of → genomic DNA, using DNA ligase, and serve as complementary address sequences for the annealing of a → vectorette PCR primer in → vectorette PCR techniques.

Vectorette cloning: See → vectorette polymerase chain reaction.

Vectorette library: A collection of → restriction fragments from a target DNA (e. g. a → genome), to which → vectorette PCR primers are ligated to prepare them for → vectorette PCR.

Vectorette PCR: See → vectorette *polymerase chain reaction*.

Vectorette PCR primer: A synthetic → oligodeoxynucleotide, which is complementary to the bottom strand of the mismatched region in → vectorettes.

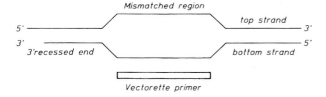

Vectorette *polymerase chain reaction* (vectorette PCR, vectorette cloning, "chemical genetics"): A variant of the conventional → polymerase chain reaction for the amplification, cloning and sequencing of unknown DNA sequences adjacent to a known sequence. In short, one or several → vectorette libraries are established by digestion of the target DNA with only one or a variety of restriction enzymes and ligation of an excess of vectorettes to the restriction fragments catalyzed by → T4 DNA ligase. Vectorettes are synthetic oligonucleotide duplexes consisting of two complementary strands containing an internal mismatched sequence and either a blunt or sticky end complementary to the overhang generated by the chosen restriction endonuclease(s). Then a locus-specific → primer (initiating primer, IP) is used in a first PCR cycle to amplify a new complementary strand to which the vectorette primer (identical to the internal mismatched sequence of the vectorette's bottom strand) can anneal in the second PCR cycle. In all subsequent cycles, priming occurs from both the locus-specific and vectorette primer, thus including sequences from the unknown region. Priming from other vectorette-flanked fragments, that do not contain the known sequence, is therefore impossible. The amplified fragment can either be cloned (and the insert used as hybridization probe for → Southern analyses) or sequenced with a sequencing primer identical to the 3' end of the vectorette's bottom strand. After sequencing, a new primer can be designed that allows further walking into flanking unknown DNA, using a new aliquot of the vectorette library. Vectorette PCR is used for → chromosome walking in a specific direction, characterization of sequences flanking inserted DNA (e. g. → transgenes, → transposons, or → viruses) or amplification of terminal sequences of e. g. → bacterial or → yeast artificial chromosome clones.
Figure see page 1226.

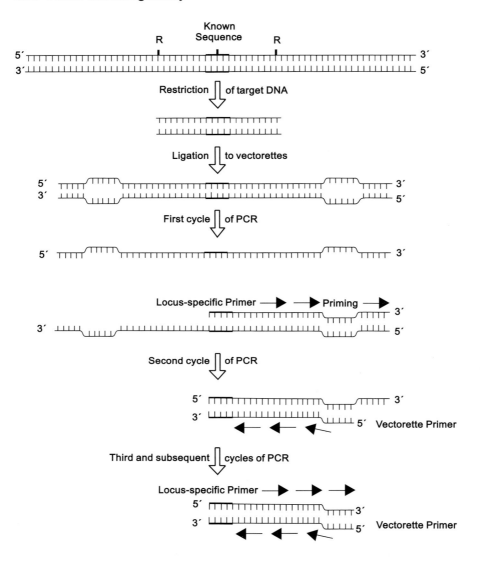

Vectorette PCR

Vector immunogenicity: The activation of a host's immune system by → sectors used to transfer genes. For example, the widely used → baculovirus expression vector, that allows to transfer genes of interest into mammalian target cells and to express them under the control of appropriate → promoters (e.g. *Rous sarcoma virus* [RSV] promoter), is recognized by the host cells and attacked by about 20 different proteins of the complement system. A socalled *membrane attack complex* (MAC) leads to the destruction of the viral surfaces and consequently to the elimination of the vector. Vector immunogenicity can be modified by engineering a → fusion product between a *decay-accellerating factor* (DAF; but see → DNA amplification fingerprinting), that speeds up the degradation of complement complexes and delays the unspecific immune response, and a baculoviral coat glycoprotein (gp64). This construct is then inserted into the → baculovirus expression system.

Vector-insert PCR: See → vector-insert *polymerase chain reaction*.

Vector-insert *polymerase chain reaction* (vector-insert PCR): A variant of the conventional → polymerase chain reaction, that allows the amplification of a specific → cDNA from → cDNA libraries by using → primers complementary to → vector and → insert sequences. In short, the cDNA is first amplified and enriched, employing a cDNA-specific, nested, → biotinylated primer. The specifically amplified cDNA is then isolated by → streptavidin-coated → paramagnetic beads. Then anchors are added to the 3' ends, either by ligation of a synthetic oligonucleotide or by → terminal transferase-catalyzed addition of → homopolymeric sequences. Now vector-specific and 3' anchor-complementary primers are used to amplify the cDNA.

Vectorless gene transfer: See → direct gene transfer.

Vector primer (VP; vector arm primer): A synthetic → oligodeoxynucleotide that is complementary to part of a → vector (e. g. a → plasmid), and serves as a → primer for the amplification of the → insert cloned into this vector, using conventional → polymerase chain reaction techniques. For example, the use of the left VP 5'-CATGGTGTCCGACTTATGCCC-3' (VPl) and the right VP 5'-CAGCGCACATGGTACAGCAAG-3' (VPr) allows to amplify any insert cloned into → lambda phage cloning vehicles such as e.g. → EMBL vectors. Some vectors are designed such that the cloned insert can be transcribed from both ends, because adjacent to the cloning site a → T7 promoter (upstream) and an → SP6 promoter (downstream) are located. The RNA transcript, either in its → sense or → antisense polarity, can then be used as specific → probe in e. g. hybridization experiments. Moreover, the T7 and SP6 promoter sequences may serve as binding sites for primers with complementary sequences, that allow to amplify the insert in between.

Vector priming: A special technique for → cDNA cloning that allows the synthesis of the first cDNA strand attached to a cloning vector. See → Okayama-Berg cloning and → Heidecker Messing method.

Vector retention: The presence of → vector sequences in the genome of → transgenic cells, tissues, organs, or organisms. Since all → transgenes have to be cloned into such vectors, and in some cases vectors are necessary for → gene transfer (as e. g. → *Agrobacterium tumefaciens*-mediated gene transfer in plants), vector sequences are cointegrated into recipient genomes together with the transgenes. However, vector retention is not desirable, since the vector sequences have no function(s) in the recipient genome, but have potential to disturb replication and/or transcription processes.

Vehicle:
a) Any → host cell that allows the replication and/or expression of cloned foreign genes.
b) Any → cloning vector.

Venn diagram: A graphical description of the chemical similarities between the 20 more common amino acids, based on their physico-chemical properties which determine protein structure. The Venn diagram shows overlapping groups of amino acids (e.g. polar, hydrophobic, small).

Vent™DNA polymerase: See → *Thermococcus litoralis* (*Tli*) DNA polymerase.

Verified *o*pen *r*eading *f*rame (verified ORF): Any → *o*pen *r*eading *f*rame (ORF), for which a gene product exists in a target organism. Generally, verified ORFs have orthologs in one or more other related species. See → uncharacterised open reading frame.

Vertical rotor: Any ultracentrifuge rotor in which the wells for the centrifuge tubes are orientated parallel to the rotational axis and perpendicular to the centrifugal field. In such rotors, density gradients rapidly form across the width of the tube, so that centrifugation times are comparatively low.

Vertical transmission: The transfer of DNA from an individual cell or organism to its progeny via mitosis or meiosis.

Vesicle-mediated gene transfer: See → lipofection.

V-gene: See → virulence gene.

***Vibrio cholerae* *r*epeat (VCR):** Any one of about 60-100 highly repeated sequences in the *Vibrio cholerae* genome, each being 123-126 bp long with an imperfect dyad symmetry starting with an inverse core-site 5'-RYYYAAC-3' and ending with its analogue 5'-GTTRRRY-3' further downstream. All VCRs are in the same orientation to each other, abutting at least one → open reading frame. VCRs share sequence homology with the 59 bp sequence element associated with the *Pseudomonas* antibiotic resistance gene *bla* P3 (encoding carbenicillinase CARB-4), that is part of an → integron.

Vigilant vector: A eukaryotic → expression vector containing a socalled → therapeutic gene under the control of an → inducible, tissue- or cell-specific promoter, that is tranfected into target cells or organs. Under inducible (usually pathological) conditions, the therapeutic gene is transcribed. The encoded protein exerts a protective action against the otherwise deleterious external stimulus. For example, a cardioprotective vigilant vector consists of a stable non-pathological vector sequence (e.g. the *a*deno-*a*ssociated *v*irus, AAV), a heart-specific promoter, reduced to its core size (e.g. the *m*yosin *l*ight *c*hain, MLC–2v, promoter), a *h*ypoxia-*r*egulatory *e*lement (HRE), which is activated by transactivating *h*ypoxia-*i*nducible *f*actor (HIF–1) in response to a reduced oxygen level, and a theurapeutic gene, whose product could protect the heart tissue during ischemia (e.g. the gene encoding angiotensin II type 1 receptor, or hemoxygenase, or superoxide dismutase). This construct is injected into the target cell or animal systemically (direct organ injection or via the jugular vein), where it is inserted into the DNA of every organ, but specifically expressed only in the heart tissue. Ischemic conditions now reduce the oxygen level in the heart tissue, which prevents the degradation of HIF–1α subunit by the → proteasome and thereby induces HRE. HRE in turn activates the MLC–2v promoter about 10fold in a matter of 6 hours, and the therapeutic gene is transcribed. Vigilant vectors can be constructed for a constant protection against diabetes type 1, stroke, or cancer, to name few.

Viral ghost: A viral → capsid, that does not contain any genetic material (RNA or DNA). Such ghosts can be filled with DNA or RNA, and serve as → transformation vectors.

Viral-like element: See → retrotransposon.

Viral-like retroposon: See → retrotransposon.

Viral *oncogene* (*v-onc*): An → oncogene (see → cellular oncogene) that has been integrated into the genome of a virus. In specific virus-host combinations its expression leads to the → transformation of the host (i.e. the formation of tumor cells), because it is no longer subjected to its proper cellular control in its new genetic environment.

Viral piracy: The acquisition of host genes by an invading virus. For example, the *human herpes virus* (HHV), especially HHV–8 (Karposi's sarcoma associated herpes virus) contains a series of human genes, that the virus has integrated into its genome in the course of evolution.

Viral vector: Any → vector for the transfer of foreign genes into eukaryotic cells, that is constructed from viral sequences. For example, → adenovirus vectors contain viral functions that import the gene(s) efficiently, but do not integrate them into the genome of the target cell, so that they are lost over time. Adenovirus vectors therefore are employed in therapeutic strategies, that require the therapeutic gene to be active only for a short time. In addition, adenovirus vectors may cause undesirable inflammatory reactions of the host. For long-term gene expression, → retroviral vectors are preferred, that are based on *murine leukemia virus* (MLV) sequences, because transferred genes are integrated into the target chromosomes. Viral vectors are used in → gene therapy.

Vir box: The dodecameric consensus sequence 5'-TNCAATTGAAAPy-3' located in the upstream region of the six → *vir* region → operons of the → Ti plasmid of → *Agrobacterium tumefaciens*. These vir boxes serve as address sites for the binding of the phosphorylated → vir G protein, which induces the activation of the vir genes with subsequent excission of the → T-strand, its packaging and transport shaping, and its transfer to the wound-activated plant recipient cell. See → crown gall.

Virion: Synonym for the inert, complete → virus particle.

Virocidal: The property of causing the inactivation of a virus. Compare → virostatic.

Viroid (Vd): A single-stranded covalently closed circular protein-free RNA molecule of about 270-380 nucleotides, extensively base-paired with itself, which causes diseases in angiosperm plants (e. g. spindle tuber disease in potato). The viroid RNA is folded into five different domains (the two terminal loops T1 and T2, a central conserved region, a variable domain, and a pathogenic domain), of which the bend of the pathogenic domain, containing an (A) $_{5-6}$ stretch, determines the disease and its severity (the more pronounced the bend, the more pronounced the disease symptoms). Viroids do not integrate into the host cell genome, but are targeted to the nucleus – possibly by proteins with a → nuclear localization signal – and replicated there by a → rolling circle replication mechanism, catalyzed by host RNA polymerase II (B). Some viroids are chimeric, i. e. consists of pieces of two or more other viroids. The mechanism leading to these chimeric forms is not clear, but the generation of such composite viroids occurs in planta. The viroids can be broadly categorized as HH viroids (containing socalled *hammer head* structures) and *retroviroids* (Rvd; reverse transcribed into a DNA copy, that resides *in nucleo* as an extra-chromosomal entity). Viroid RNA is never translated into protein(s). The mechanisms by which viroids cause a disease are not yet clear. Possibly they involve interference with the processing or → pre-ribosomal RNA in host cells. Viroids bear some homology to group I → introns, suggesting that viroids may have evolved from them.

Viropexis: The pinocytotic engulfment and uptake of → virus particles by cells.

Viroplasm (viroplast): The site of → virus replication and/or assembly in a virus-infected cell.

Viroplast: See → viroplasm.

Virostatic: The property of interfering with viral replication. Compare → virocidal.

***vir*-region (*virulence* region):** A 35 kb region of the → Ti-plasmid of → *Agrobacterium tumefaciens*, that is responsible for the excision and transfer of the → T strand from the bacterium into the recipient plant cell. Its function supplements the action of several constitutively expressed chromosomal loci (e.g. *chvA*, *chvB*, *pscA* or *exoC*) whose products are involved in the → attachment process (see → crown gall). The different *vir* genes *pinF*, *vir A*, *B*, *C*, *D*, *E*, *F*, and *G* are under tight control. *vir A* and *G* genes are constitutively expressed, the other genes are inducible. Inducers are plant phenolic compounds (e.g. acetosyringone, AS; α-hydroxyacetosyringone, OH-AS, and many others) that are synthesized by plant cells during the process of wound healing (e.g. lignin formation, phytoalexin synthesis). The vir A protein is a membrane-bound chemoreceptor for plant phenolics that bind to its amino-terminal sensory domain. After binding this membrane spanning domain undergoes a conformational change, that also changes the conformation of the C-terminal domain, thereby inducing its kinase activity. After autophosphorylation of vir A and transphosphorylation of the vir G protein, the latter adopts a DNA-binding conformation and acts as a transcriptional → activator for the *vir* genes generally. The activation of these genes leads to the production of several proteins involved in site-specific excision of the T-strand (vir D1 and D2 proteins), the protection of the T-strand from exo- and endonucleolytic attack (e.g. by vir D2 proteins at the ends of the T-strand; vir E2 and E3 proteins along the T-strand), the piloting of the T-strand through the bacterial and possibly the plant membrane system (vir B proteins), and its nuclear targeting (vir D2 protein containing a nuclear targeting peptide).

Virtual cloning (*in silico* cloning): The use of appropriate software packages to design and virtually perform a complete → cloning procedure. Such software ("gene construction kit") also allows to document any step in → recombinant DNA experiments.

Virtual screening (*in silico* screening): A technique to identify compounds with desirable properties (e.g. binding to a target receptor protein) by computational analysis of hugh real or virtual chemical or genetic databases. Screening of literally millions of candidate molecules with potential structural affinity to biological targets (based on the tree-dimensional structure of this target) results in a collection of lead compounds, from which high-affinity ligands can in turn be rationally designed using programs like GRID, FlexX, LUDI, and others.

Virulence (pathogenicity): The capacity of a pathogenic organism to cause a disease. See also → virulence plasmid, → *vir*-region.

Virulence gene (V-gene): Any gene encoding a protein, that determines the degree of pathogenicity of a virus, bacterium or fungus. V-genes may be either dominant (V-gene) or recessive (v-gene). See → resistance gene.

Virulence plasmid: A → plasmid which increases the pathogenicity of bacteria.

Virulence region: See → *vir*-region.

Virulent bacteriophage (virulent phage): Any bacteriophage that causes the → lysis of its host cell. Compare → temperate bacteriophage.

Virulent phage: See → virulent bacteriophage.

Virus: An infectious complex of either an RNA (RNA virus) or a DNA molecule (→ DNA virus) enclosed in a protein coat (→ capsid). Since viruses do not contain ribosomes or cell organelles (e.g. mitochondria), they rely on the host cell's metabolism for their → replication. The host cell may or may not be destroyed in the process of viral replication and release. Viruses pathogenic for bacteria are called → bacteriophages. See also → cryptic virus, → helper virus, → latent virus, → oncogenic virus, → provirus, → replication-defective virus. Viruses described in some detail in this book are the DNA-containing → adenovirus, → baculovirus, → caulimo virus, → gemini virus, → Simian virus 40; and the RNA-containing → bromovirus, → positive strand RNA virus, → retrovirus (also → transducing retrovirus). See also → lambada virus, which has not yet been classified satisfactorily.

Virus-induced gene silencing (VIGS): The specific blockage of the expression of genes of a virus after its entry into a host cell as a defense mechanism. Many plant viruses produce double-stranded RNA (dsRNA) intermediates during their replication, and thereby trigger VIGS. For example, the infection of a plant with a recombinant virus containing a sequence from a host nuclear gene may lead to the silencing of that gene. Or, double-stranded RNA (as indicator for a potential viral invasion) in the cytoplasm of mammalian cells triggers a socalled interferon response. The dsRNA (of more than 30 bp) binds to and activates protein kinase PKR and a 2', 5'-oligoadenylate synthetase (2', 5'-AS). The activated PKR phosphorylates the translation initiation factor eIF2ß and thereby stalls → translation, whereas the activated 2', 5'-AS causes messenger RNA degradation by 2', 5'-oligoadenylate-activated ribonuclease L. All these reactions are not dependent on any particular RNA sequence.

Virus-like particle (VLP): A → baculovirus-derived particle, that does not contain viral DNA or RNA and is used to transfer foreign DNA into insect (and other animal) cells. In short, insect cells in culture are first infected with a recombinant baculovirus carrying human polyoma virus (JCV)-VP1 sequences. VP1 is expressed, the VP1 monomers assemble into pentamers, the pentamers form VLPs spontaneously, which are secreted into the medium. These VLPs can be isolated from the medium and dissociated into monomers by EDTA. Then the foreign DNA is mixed with these monomers, Ca^{2+} ions are added, and reassociation leads to the formation of VLPs, that engulf the foreign DNA. Since VLPs have very little immunogenicity, they may be employed in gene transfer to mammalian cells.

Visigel™**:** The trademark for a fully transparent agarose-based matrix for the separation of small DNA fragments in the range of 60-1000 bp. The → ethidium bromide-stained fragments appear more focussed and fluoresce more intensively as compared to agarose-separated DNA. However, no blotting and → freeze-squeeze techniques can be used with Visigel™.

Visual mapping: See → optical mapping.

Visual marker: Any genetic → marker that can be identified by simple visual inspection. For example, if a dominant gene inducing fur and eye pigmentation (a tyrosinase-encoding minigene driven by a strong promoter) and an unrelated clone carrying the gene of interest are coinjected into one-cell stage albino mouse embryos, both constructs are cointegrated into the genome and expressed in the embryo. Therefore, embryos transgenic for the tyrosinase gene (i. e. animals able to synthesize pigment) can easily be identified as early as 12 days after fertilization, when pigment becomes visible in the epithelial cells of the retina. See → visual transformation marker.

Visual transformation marker: Any → genetic marker that allows simple visual identification of → transformation events (e.g. → transformants). See → visual marker.

Vitamin H: See → biotin.

VLP: See → virus-like particle.

VNDR: See → *v*ariable *n*umber of *d*inucleotide *r*epeats.

VNTR: See → *v*ariable *n*umber of *t*andem *r*epeats.

Von Heijne rule (-3/-1 rule): The prediction of the site of cleavage between a → signal sequence and the mature secreted protein on the basis of amino acid sequence → homology. The transient N-terminal → signal peptide of most secretory proteins, that functions to initiate export across the inner membrane (prokaryotes) or the endoplasmic reticulum (eukaryotes), contains three structural and functional regions. First, the basic N-terminal region (n-region), second, the central hydrophobic region (h-region, correponding to positions -13 to -6), and a more polar C-terminal region (c-region, corresponding to residues -5 to -1, but extending to +1 or +2). The structural determinants reside in the n- and h-regions, with positions -3 and -1 relative to the cleavage site being the important ones for the cleavage process ("-3/-1 rule"). This rule also defines the amino acids tolerated in the different regions without abolishing cleavage: the -1 residue must be small (e. g. an alanine, serine, glycine, cysteine, threonine, or glutamine), the -3 residue must **not** be aromatic (e. g. phenylalanine, histidine, tyrosine, or tryptophan), charged (e. g. aspartic acid, glutamic acid, lysine, or arginine), or large and polar (e. g. asparagine, glutamine). Proline must be absent from positions -3 through +1.

Voxelation: A technique for the three-dimensional expression profiling of thousands of genes in an organ (e.g. brain). The organ is first cut into multiple cubes ("voxels"), mRNA from each cube isolated, reverse transcribed into cDNAs, and these cDNAs independently hybridised to expression chips loaded with thousands of genic sequences. Chip readers then localize hybridisation events and allow to establish an expression image of each particular cube. This way, a series of volumetric maps of gene expression for the target organ are established, which altogether image gene expression patterns in three dimensions. Voxelation is used to detect differences in gene expression patterns between e.g. brains of normal mice and mice treated with methamphetamine (as a pharmacological model for Parkinson's disease). In this case, the expression of distinct genes proved to be different especially in the striatum.

VP: See → *v*ector *p*rimer.

VP22 translocation: The efficient transfer of a protein from one cell to all cells of a cell culture by fusing this protein at its N- or C-terminus to the VP22 protein of the herpes simplex virus type 1. First, → plasmid vectors, containing the VP22 encoding DNA sequence together with a strong → promoter for its high-level expression, a suitable → polyadenylation signal and transcription termination sequence as well as a → multiple cloning site are used to fuse the DNA sequence of the target protein to the VP22 sequence. Secondly, this construct is transfected into a target cell, where the fusion protein is synthesized, then exported from the transfected cell and translocated into the surrounding cells with subsequent accumulation in the nucleus. The target protein exerts its function in all cells of the cell culture.

W

W:
a) Abbreviation for "*weak*", i.e. the relatively weak interaction(s) between hydrogen-bonded A = T base pairs in DNA. Used to symbolize either A or T.
b) Abbreviation for *w*yosine, the → nucleoside of base Y, see → rare bases.

***Walking primer* (WP):** Any synthetic → oligodeoxynucleotide of a defined sequence (e. g. T_mG_n; as 5'-$T_{11}GT_3GT_2GTG_5TGT$-3'), that can be used as a → primer in → unpredictably primed PCR. The walking primer is employed in a first low → stringency amplification of anonymous regions of a genome, yielding products that are subsequently specifically amplified with → sequence-specific primers (e. g. a → gene-specific primer) at high stringency. The products of this second amplification are re-amplified with a nested ("inner") specific primer and a *short walking primer* (sWP) complementary to part of the original WP. The final product then contains part of the known sequence (complementary to the iSP) and the unknown flanking region extending to the sequence complementary to the short walking primer.

Watson-Crick base-pairing: See → base-pairing.

Watson-Crick helix: See → double helix.

Watson strand: The strand in a → double helix that runs from → 5' to → 3'. Compare → Crick strand.

Wavelength-shifting molecular beacon: A variant of the conventional single-stranded → molecular beacon, that contains two separate → fluorophores, a so-called harvester fluorophore (located within the 5'arm of the oligonucleotide, absorbing strongly in the wavelength range of the employed monochromatic light source) and a so-called emitter fluorophore (located at the distal end of the 5'arm, separated from the harvester fluorophore by a spacer several nucleotides in length, emitting fluorescence light of the desired spectral quality) in addition to a non-fluorescent quencher (located at the end of the 3'arm). In the non-hybridized wavelength-shifting molecular beacon, which forms a → hairpin structure, the harvester fluorophore is in close proximity to the quencher, i.e. the molecular beacon does not emit any fluorescence light. If hybridized to a complementary target DNA, the molecular beacon undergoes a conformational change, separating the 5'arm from the 3'arm (hence the quencher from the harvester). Now the harvester can absorb light, but does not emit itself. Instead, the absorbed energy is transferred to the emitter fluorophore through → *f*luorescence *r*esonance *e*nergy *t*ransfer (FRET). The emitter in turn emits fluorescence light of its own characteristic wavelength. Wavelength-shifting molecular beacons allow the detection of many hybridization events at the same time, since each single hybridization has its own characteristic fluorescence.

WC: See → Watson-Crick base pairing. Do not confuse with another WC.

Weak overlap: The number of bases matched between two clones (e.g. → bacterial artificial chromosome clones), that are not matched using the strictest or less strictest criteria, but are matched only by the least strict criteria. See → medium overlap, → strong overlap, → total overlap.

***Weak positive element* (WPE):** A *cis*-acting DNA sequence motif of 20-100 bp in → promoters of eukaryotic genes that enhances their expression. If this motif is deleted, transcription from the resulting mutant promoter is slightly reduced. The WPEs are probably target sites for the binding of specific → transcription factors.

Weak promoter (low level promoter): Any → promoter that does not allow the frequent attachment of DNA-dependent → RNA polymerase so that the adjacent gene can only be transcribed at a low frequency.

Weak splice site (weak splice junction): Any → splice junction, that is only used, if a → splicing enhancer is present. Weak splice sites are underlying several → alternative splicing events.

Web cloning: A laboratory slang term for the design of a cloning experiment using relevant informations deposited in the Internet (web), with vector selection (http://www.invitrogen.com/vectors.html or http://www.stratagene.com/vectors/index.htm), type of cloning (e.g. PCR cloning: http://www.premierbiosoft.com), transformation (http://www.genome.ou.edu/protocol_book/protocol_adxF.html), and clone sequencing (http://www. gatc.de).

Weigant halo: The arrangement of extended DNA fibers around a partially preserved nucleus, from which they exploded after a special treatment (deproteinization). In such Weigant halos the DNA (normally 3 Å/bp) is decondensed severalfold. Therefore the DNA of such halos can serve as substrate for → extended fiber FISH.

Western blot (protein blot): Any solid support (e.g. a nitrocellulose or nylon membrane), onto which proteins are transferred from a polyacrylamide gel by → electroblotting. The immobilized proteins can then be probed with e.g. specific antibodies (and antibody-antibodies labeled with e.g. → fluorochromes) to detect the presence of specific proteins. See → microarray Western, → Northern blot, → Southern blot, → Western blotting.

Western blotting (Western transfer, immunoblotting, ligand blotting, affinity blotting): A technique for the detection of specific proteins, that are separated by → polyacrylamide gel electrophoresis and then transferred ("blotted") onto a solid support (e.g. a nitrocellulose or nylon membrane). This membrane (→ Western blot) can then be probed with specific radioactively labeled, fluorescence-conjugated or → enzyme-conjugated antibodies ("immuno probes"). The term "Western" is a phantasy name, and leans against the analogous → "Southern" (named after Ed Southern, the inventor of the technique) and "Northern" transfer techniques, where Northern again is phantasy. Compare → microarray Western.

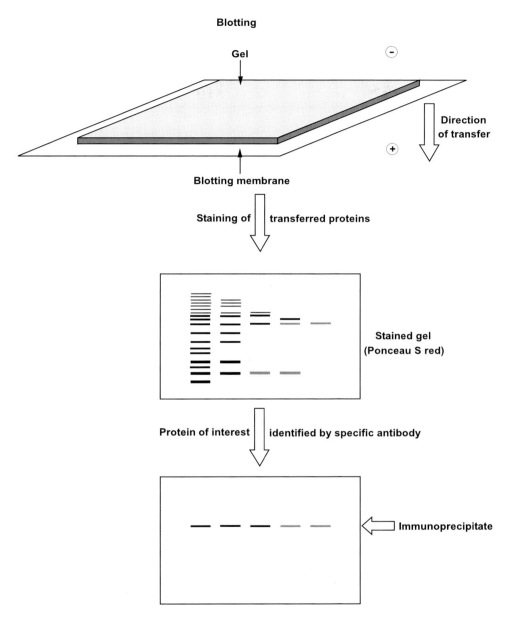

Western blotting

Western transfer: See → Western blotting.

Wheat germ system: A complete → *in vitro* translation system from wheat embryos, including ribosomes, a full complement of tRNAs, aminoacyl tRNA-synthetases, amino acids, enzymes, initiation, elongation and termination factors, K^+ and Mg^{2+}, ATP and GTP, and an energy-re-generating system (e.g. creatine phosphokinase). This system is used to translate heterologous mRNAs into proteins *in vitro*. Since the wheat germ system is practically free from endogenous mRNA, any translational activity is fully dependent on exogenous messages. However, large mRNAs (coding for proteins of more than 80 kDa) are only inefficiently translated, because their synthesis is frequently not completed. Compare → rabbit reticulocyte lysate.

White biotechnology: A laboratory slang term for the application of the methodological repertoire of → biotechnology to microorganisms (predominantly bacteria). Compare → blue biotechnology, → green biotechnology, → grey biotechnology, → red biotechnology.

Whole cell transcription system: A concentrated and dialyzed whole-cell extract prepared from eukaryotic cells (e.g. HeLa cells) that contains endogenous → RNA polymerase II and is used to initiate → messenger RNA synthesis from exogenous → templates *in vitro*.

Whole chromosome shotgun sequencing (WCS sequencing): A technique for the sequencing of isolated DNA from single → chromosomes, that requires separation of all → chromosomes of a target organism with e.g. → pulsed field gel electrophoresis or → flow cytometry (chromosome sorting), the isolation of single chromosomal bands from the pulsed field gel, → shearing of the selected chromosome into fragments of defined length, cloning of the fragments into a suitable → cloning vector (e.g. → a plasmid) and subsequent sequencing of the → inserts. Usually as many clones are sequenced as is necessary for a 5–10 times → coverage of the chromosome. The sequences are ordered via overlapping fragments, and the complete chromosomal DNA sequence reconstructed. See → whole genome shotgun sequencing.

Whole genome amplification (WGA): A general term for a series of different techniques, that altogether aim at using → polymerase chain reaction amplification with random or degenerate oligonucleotide primers to amplify whole genomes or substantial parts of them. For example, → degenerate oligonucleotide-primed PCR, → multiple displacement amplification, or → rolling circle amplification are such WGA methods. Compare → whole genome polymerase chain reaction.

Whole genome duplication: The amplification of a whole genome (or also parts of it) during evolution, such that the present-day genomes of many organisms are consisting of duplicated regions. For example, the duplicated regions of the *Arabidopsis thaliana* genome constitute 68 Mb, or 60% of the genome. The number of homologous genes in the duplicated regions vary considerably, ranging from 20–50%, which is either a consequence of tandem duplications or gene losss after segmental duplication. Whole genome duplications are a cause for polyploidization.

Whole genome oligonucleotide array: A somewhat misleading term for any → microarray, that consists of thousands of oligonucleotide probes, that are regularly spaced along a genomic DNA, but do not fully cover this genomic DNA (as the name suggests). Such arrays serve e.g. to discover transcribed sequences that are not detectable by → annotations of known exons.

Whole genome PCR: See → whole genome polymerase chain reaction.

Whole genome polymerase chain reaction (whole genome PCR): A variant of the conventional → polymerase chain reaction which allows to select and amplify specific genomic fragments. In short, genomic DNA is either sonicated or cleaved with the → restriction endonuclease *MboI* (→ recognition sequence 5'-GATC-3'), blunt-ended with the → Klenow fragment of *E. coli* DNA polymerase I, and the blunt-ended fragments ligated to → linkers ("catch linkers"). Then the linkered fragments are cleaved with another restriction enzyme (e.g. *Xho* I; recognition sequence 5'-CTCGAG-3') and the desired fragments selected (by e.g. radioactive → probes, such as proteins, if DNA-protein interactions are to be studied). Primers complementary to linkers are then used in PCR amplification of the selected sequence.

Whole genome shotgun sequencing (*whole genome sequencing, WGS*): The technology to determine the sequence of large genomes (e. g. the human genome with its 2.8×10^9 base pairs) without the traditional → BAC, → PAC, or → YAC cloning, subcloning, sequencing, and overlap screening. The shotgun approach starts with the mechanical fragmentation of the whole genome into overlapping pieces of about 5000 bases each, the sequencing of the ends of these fragments, and their assembly into a complete genome by powerful computer programs (e. g. the base-calling program PHRED, that assists in unequivocal conversion of peaks in a sequence fluorogram to the correct bases, the program PHRAP for the assembly of many short fragments into longer sequences exploiting overlaps, the program PRIMO for the detection of gaps in the sequences and the design of primers to close the gaps, and FINISH, also for filling-in the gaps and calling for more data, if needed). For a better management of this direct genome sequencing technique, multicapillary gene sequencers are used, that employ → capillary electrophoresis and fluorescence detection by an argon ion laser beam. See → whole chromosome shotgun sequencing.

Whole-mount *in situ* hybridization (whole-mount ISH): A method to identify specific sequences in cells, tissues or organs of a whole plant or animal or cross-sections of them (e.g. an embryo) by hybridization with radioactively labeled complementary nucleic acid → probes. Usually the organism is prepared for ISH by dehydration – rehydration, hydrogen peroxide treatment to decrease background, proteinase K digestion to increase the accessibility of mRNAs for the probe, and refixation. The hybridization procedure is identical to that used in → *in situ* hybridization. See → fluorescent *in situ* hybridization, → primed *in situ* labeling, → genomic *in situ* hybridization.

Whole proteome microarray: Any → microarray, onto which preferably all proteins of a cell, a tissue, organ, or organism are spotted. Such arrays are used for the detection of protein-protein interactions on a proteome scale and allow to establish → protein interaction maps.

Wild type (standard type):
a) The most frequently occurring → phenotype (strain, organism) in natural breeding populations.
b) The genetic constitution of an organism at the onset of recombinant DNA experiments with its genome or its plasmid(s). Thus, this "wild type" is an arbitrarily specified → genotype used as a basis for comparison in genetic programs. See also → wild type gene.

Wild type allele: Any → allele of a gene, that was first sequenced.

Wild type gene: Any gene that is commonly occurring in nature, or any gene sequence that serves as original before being modified in genetic engineering experiments. See → wild type.

Winged *h*elix-*t*urn-*h*elix protein (winged HTH protein): Any protein containing a variant of the → helix-turn-helix motif, that harbors a third → α-helix and an adjacent β-sheet structure. Both are components of the DNA-binding motif. The recognition helix of winged HTH proteins binds to DNA in the same way as HTH proteins, whereas the additional secondary structures (→ α-helix and β-sheet) mediate additional contacts with the DNA backbone.

Wobble base: The third base in degenerate codons. See → wobble hypothesis.

Wobble base pairing: The hydrogen-bonding of the third base of an → anticodon with more than one complementary → "wobble" bases (e. g. G may pair with U or C, or I with U, C, or A). See → wobble hypothesis.

Wobble hypothesis: A hypothesis that principally aims at explaining why one species of → transfer RNA (tRNA) may recognize more than one → codon (see → degenerate code). It is a firm rule that the first two bases of the codon of an mRNA molecule pair with the first two bases of the → anticodon via Watson-Crick → base-pairing interactions. The third base of the anticodon, however, does not strictly follow → this base-pairing but has a limited range of play ("wobble"). This wobble allows it to pair unconventionally with a variety of bases at the third position of different codons. For example, uracil in the third position of the anticodon pairs with either adenine or guanine, and guanine with uracil or cytosine in the third position of the codon. Inosine, resembling guanine in its physico-chemical properties, may pair with uracil, cytosine or adenine.

Guanine-Uracil — Codon or anticodon / Codon or anticodon

Inosine-Uracil — Anticodon / Codon

Inosine-Cytosine — Anticodon / Codon

Inosine-Adenine — Anticodon / Codon

Wobbles: A laboratory slang term for a series of oligonucleotides that contain equimolar mixtures of two (or more) different bases at a given position. Wobbles are used as → probes to detect an unknown gene, that encodes a known protein sequence, by hybridization.

Working draft genomic sequence: A set of DNA sequences representing about 90% of the genome in question. For example, in the human genome sequencing effort, mapped clones are first shotgun-sequenced, the shotgun data assembled in a "draft sequence" (covering most of the region of interest, but still containing ambiguities and gaps). Then the gaps are filled and a → finished genomic DNA sequence produced. The working draft of the human genome combines both finished genomic DNA sequence and draft sequence, where the latter has an average → contig length of about 15 kb with about one sequencing error per 5000 bases.

WP: See → *walking primer*.

WPE: See → *weak positive element*.

Writhing **number (W):** The number of times with which the axis of a closed circular double-helical DNA molecule crosses itself by supercoiling (→ supercoil).

Wun **promoter:** A plant → promoter that is induced by *wounding*.

X

X: Synonymous with → N.

Xanthine (2,6-dihydroxy purine): A metabolite of purine biosynthesis and turnover (e.g. deamination of guanine).

Xanthosine: A nucleotide derivative of → xanthine, that is used as analogue of G-nucleotides in signal transduction experiments.

Xanthosine–5'-triphosphate

Xaptonuon (e*xapted* *nuon*): Any → potonuon, that is actually converted ("exapted") to an element with a new function. See → naptonuon, → nuon, → retronuon.

X-CHIP: See → formaldehyde fixation (X) and chromatin immunoprecipitation (X-Chip).

Xenogenomics: A somewhat misleading term for the whole repertoire of techniques for the discovery and analysis of genes in native and exotic organisms (e.g. plants), that are uniquely adapted to environmental stresses (in plants, especially adaptation to drought, salinity, aluminum toxicity, low nutrients).

Xenology (Greek: xenos-strange, foreign): The molecular description of genes in one species, that were transferred from another species by → horizontal gene transfer.

X-gal (5-bromo-4-chloro-3-indolyl-β-D-*gal*actopyranoside): A colorless chromogenic substrate for → β-galactosidase which is converted into a blue indolyl derivative (indigo) through cleavage.

X-gal

5-Bromo-4-chloro-indoxyl

5,5′-Dibromo-4,4′-dichloro-indigo

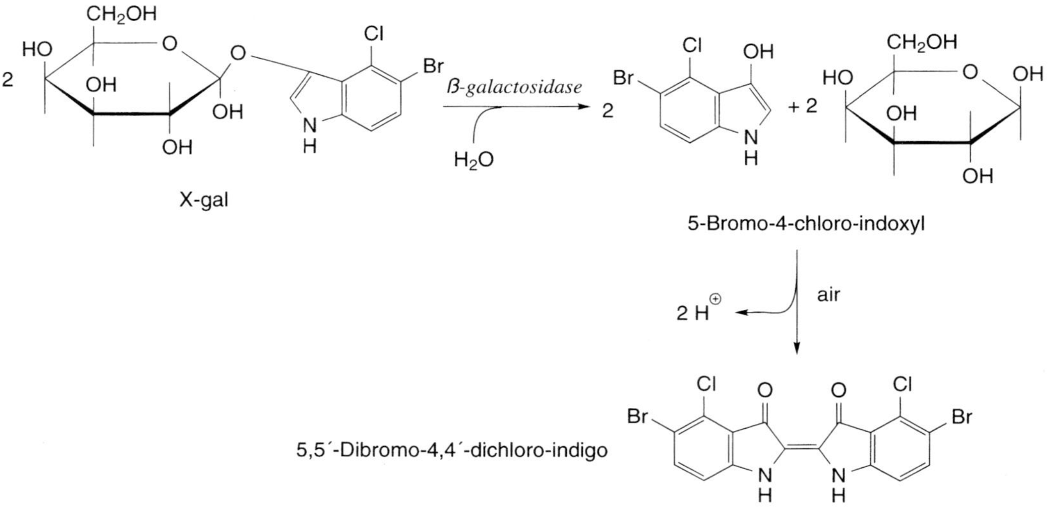

X-gluc (5-bromo-4-chloro-3-indolyl-β-D-*glu*curonide): A colorless chromogenic substrate for → β-glucuronidase which is converted into a blue indolyl derivative through cleavage. The derivative can be used to quantify GUS activity and to localize GUS action in histochemical preparations. The blue indolyl compound is also secreted from GUS-containing organisms (e.g. transgenic plants) into e.g. → agar or → agarose so that it can be monitored without destroying the living cells.

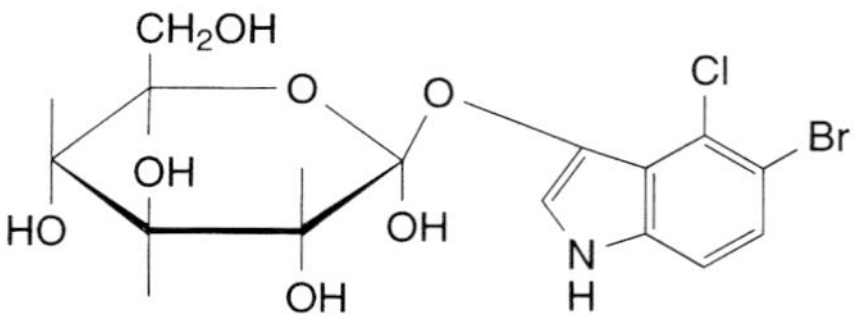

Xist-RNA: A regulatory RNA molecule responsible for the inactivation of one of the two X chromosomes in somatic cells of female mammals.

X-phos (5-*b*romo-4-*c*hloro-3-*i*ndolyl*p*hosphate, BCIP): A colorless chromogenic substrate for → alkaline phosphatase which is converted into a blue indolyl derivative (indigo) through cleavage. BCIP (in combination with nitroblue tetrazolium chloride, NBT) is used as chromogen for the immunologic detection of proteins or nucleic acids in biotinylation or digoxigenation procedures (see → biotinylation of nucleic acids; → digoxigenin-labeling).
Figure see page 1245.

X-ray crystallography: A technique for the determination of the three-dimensional structure of atoms or molecules in a crystal, using the diffraction patterns produced by X-ray scattering.

Xylene cyanol (xylene cyanol FF): An organic dye, used as electrophoresis marker in → agarose gel electrophoresis of DNA. It migrates with DNA fragments of about 5 kb (see → tracking dye).

Xylene cyanol FF

XYZ arm (XYZ robotic arm): A part of a laboratory robot, that moves in three dimensions (x-y-z) and allows to e.g. take up liquid (or media, bacteria, clones) in one microtiter plate, transfer them into the wells of another microtiter plate (where e.g. a reaction takes place), and dispense the resulting mixture into the wells of yet another microtiter plate. See → XYZ robotic workstation.

XYZ robotic workstation: An instrument for high-throughput experimentation, that usually consists of a stage (holding pipetting heads, that can automatically be exchanged during operation, pipette tips, reagent containers, temperature-controlled blocks, in which e.g. reactions are performed under controlled temperature regimes, and filtration blocks with filtering capacity in a 96-well format, or higher), and one or more movable arms with e.g. pipetting heads. The stage can move in the X direction, while the arm and the pipetting head can move in the Y and Z direction, respectively. This robotic workstation can be programmed, runs automatically, and has the potential to transfer reagents to and from 96-well (or larger) microtiter plates, mix them, incubate them, and dispense them. In some instruments, the heads can be exchanged, and the workstation can then be modified to spot proteins or cDNAs onto glass slides. XYZ robotic workstations are time-saving and continuously working parts of genomics laboratories.

Y

Y: A grammalogue for → pYrimidine (→ C or → T), used in sequence data banks.

YAC: See → *yeast artificial chromosome.*

YAC end primer: Any → primer oligonucleotide complementary to sequences flanking the → cloning site of a → *yeast artificial chromosome* (YAC). YAC end primers allow to amplify YAC ends, an essential step for YAC identification via → YAC fingerprinting. See → BAC end primer.

YAC fingerprinting: The establishment of a → DNA fingerprint by → hybridization of → microsatellite motifs (e.g. [GGC]$_5$) to → yeast artificial chromosome (YAC) clones, that were restricted by one or several → restriction endonucleases, electrophoresed and blotted onto a membrane. YAC fingerprinting allows to detect similar or identical sequences in different YAC clones and allows to either exclude doublets or to identify → contigs.

YAC library: A collection of randomly cloned, about 600 kb long fragments of → genomic DNA of an organism, inserted into → yeast artificial chromosome plasmid cloning vectors, and representing either the DNA of one single chromosome or the entire genome.

YAC map: The ordered alignment of → yeast artificial chromosome clones such that a → physical map of the genome is constructed. See → BAC map.

Yatalase: A mixture of different enzymes from culture supernatants of *Corynebacterium* sp. OZ–2, mainly consisting of chitinase, chitobiase, chitosanase, protease, and β–1,3–glucanase, that is used to lyse cell walls of filamentous fungi and to prepare fungal protoplasts.

Y base: Any modified → guanine base in → transfer RNA. See → rare base.

YCp: See → *yeast centromere plasmid.*

Yeast (baker's yeast): The trivial name for various unicellular ascomycetes (protoascomycetes) that are of interest for biotechnology, and gene technology in particular. For example, gene technology profits from yeast as → host for a series of versatile → cloning vectors such as → yeast centromere plasmid, → yeast episomal plasmid, → yeast integrative plasmid, → yeast replicative plasmid, → yeast artificial chromosome, that allow the cloning and expression of foreign, especially eukaryotic → split genes. These genes are correctly transcribed, the transcripts efficiently and precisely processed, and the encoded proteins synthesized with high fidelity. Moreover, the proteins are comparatively stable in the heterologous host due to its low content of proteolytic enzymes, which renders their isolation in pure form relatively easy. Furthermore, yeasts are a rich source for useful genes, regulatory and other useful DNA sequences, which can be isolated without extensive screening procedures (because of the haploid genome's size of 1.45×10^7 bp). Yeast does not produce toxins or otherwise hazardous compounds for humans so that it is the host of choice for the production of pharmaceutically interesting molecules. The biotechnological capacities of yeasts have been exploited for many centuries (e.g. for brewing, baking, wine production). One of the most prominent yeasts is *Saccharomyces cerevisiae* (but also *Schizosaccharomyces pombe*, *Kluyveromyces lactis*, and others).

Yeast *artificial* chromosome (YAC, pYAC): A high-capacity 11.5 kb → plasmid cloning vector, replicating both in *E. coli* and → yeast, and containing a single *Sma*I → cloning site within a gene whose interruption is phenotypically visible (e.g. an → ochre suppressor gene of a tyrosyl tRNA, *sup 4*), an → autonomously replicating sequence, a → centromere, and → selectable markers on both sides of the centromere (e.g. the N-[5'-phosphoribosyl anthranilate isomerase gene *trp 1* and orotidine-5'-phosphate decarboxylase gene *ura 3*). For cloning, this vector is digested so that a left and right chromosome arm and a → stuffer region are generated. The phosphatased arms are then ligated onto large (350-620 kb) exogenous DNA fragments, and the ligation product is transformed into yeast → spheroplasts, where it is maintained as an → artificial chromosome (i.e. is replicated once per cell cycle, and segregated at mitosis). Compare → yeast cloning vectors. See → bacterial artificial chromosome, → human artificial chromosome, → mammalian artificial chromosome, → pBeloBac 11, → plant artificial chromosome, → P1 cloning vector, → *Schizosaccharomyces pombe* artificial chromosome, → transformation-competent artificial chromosome vector.
Figure see page 1249.

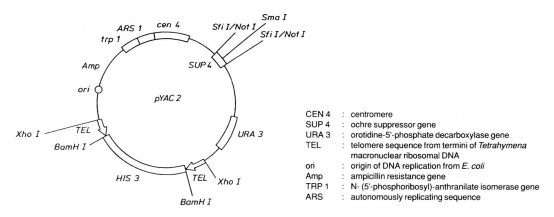

CEN 4	:	centromere
SUP 4	:	ochre suppressor gene
URA 3	:	orotidine-5'-phosphate decarboxylase gene
TEL	:	telomere sequence from termini of *Tetrahymena* macronuclear ribosomal DNA
ori	:	origin of DNA replication from *E. coli*
Amp	:	ampicillin resistance gene
TRP 1	:	N- (5'-phosphoribosyl)-anthranilate isomerase gene
ARS	:	autonomously replicating sequence

Simplified map of a YAC vector

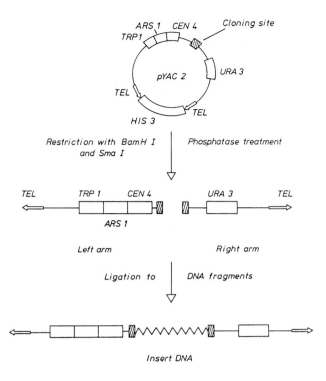

Cloning into YAC vectors

***Yeast* *c*entromere *p*lasmid (yeast centromeric plasmid, YCp, pYC):** A → low copy number plasmid cloning vector that contains a yeast → centromere sequence, an → autonomously replicating sequence, one or more → yeast chromosomal markers, and an *E. coli* → origin of replication. This plasmid is used to transform yeast cells (transformation frequency: $10^2 – 10^3$ transformants per µg DNA), and generally segregates as a → minichromosome at meiosis.

Yeast centromeric plasmid: See → yeast centromere plasmid.

Yeast chromosomal marker (yeast marker): Any → yeast chromosome gene that serves as a → selectable marker gene in yeast transformation experiments. Yeast markers are central genes in e.g. amino acid (e.g. *arg 8*, *his 3*, *leu 2*) or uracil biosynthesis pathways (*ura 3*). If transferred into a specific auxotrophic yeast cell as part of a → yeast cloning vector, the gene complements the deficient genome. In consequence, the cell becomes prototrophic and can be easily selected.

Yeast cloning vector: Any one of a series of specially designed → plasmid cloning vectors that contain regulatory elements functioning in → yeast cells. See for example → yeast centromere plasmid, → yeast episomal plasmid, → yeast integrative plasmid, → yeast replicative plasmid, → yeast artificial chromosome, → yeast expression plasmid, → yeast linear plasmid, → yeast promoter plasmid.

Yeast episomal plasmid **(YEp, pYE):** A → yeast cloning vector that contains the → yeast 2 μm plasmid, → yeast chromosomal markers (e.g. URA 3), selectable markers for bacteria (e.g. ampicillin resistance gene, Ap) and *E. coli* vector sequences (from e.g. pMB9) together with a tetracycline resistance gene. This gene is, however, separated from its promoter by the URA 3 sequence. These constructs replicate both in *E. coli* (pBR 322 origin of DNA replication) and *Saccharomyces cerevisiae* (replication sequence from yeast 2 μ circle plasmid), have a high copy number in yeast (25-100 copies per cell) and transform yeast at a high frequency ($3 \cdot 10^3$ to $5 \cdot 10^4$ transformants per viable cell). Compare → yeast artificial chromosome.

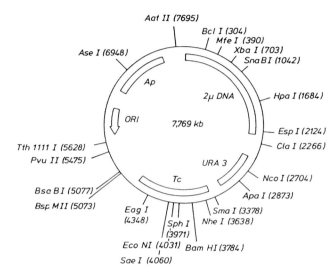

Simplified map of yeast episomal plasmid
(with unique restriction sites)

Ap (bla)	: Ampicillin resistance gene
Tc (tet)	: Tetracycline resistance gene
ORI	: Origin of DNA replication from pBR 322
URA 3	: Orotidine-5'-phosphate decarboxylase gene
2 μ DNA	: Yeast 2 micron circle plasmid
1	: Nucleotide number 1, identical to the first G in the unique Eco RI recognition site 5'-GAATTC-3' at the junction between the Ap and the 2 μ circle sequence

Yeast episomal plasmid

Yeast expression plasmid **(YXp, pYX):** A → yeast plasmid vector that contains strong → promoters ligated → in frame to foreign DNA sequences which are overexpressed in the host cell.

Yeast hybrid plasmid **(YHp, pYH):** Any one of a series of engineered → yeast cloning vectors that contains hybrid gene sequences.

Yeast integrating plasmid: See → yeast integrative plasmid.

Yeast integrative plasmid **(yeast integrating plasmid, YIp, pYI):** A → plasmid cloning vector that contains a → yeast chromosomal marker (e.g. URA 3) cloned into an *E. coli* plasmid (e.g. → Col E1, → pBR 322) and the → ampicillin (Ap) and tetracycline (Tc) resistance genes together with an → origin of replication that functions in *E. coli*. In yeast, this vector can only be stably maintained if integrated into a yeast chromosome, since it does not carry an → autonomously replicating sequence (ARS). It is used to transform yeast cells (low transformation frequency: 1-10 transformants per μg DNA), but transformants are usually stable. See also → yeast cloning vectors. *Figure see page 1251.*

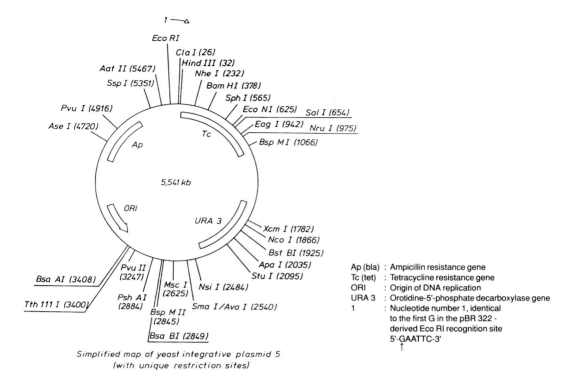

Simplified map of yeast integrative plasmid 5
(with unique restriction sites)

Ap (bla)	: Ampicillin resistance gene
Tc (tet)	: Tetracycline resistance gene
ORI	: Origin of DNA replication
URA 3	: Orotidine-5'-phosphate decarboxylase gene
1	: Nucleotide number 1, identical to the first G in the pBR 322 - derived Eco RI recognition site 5'-GAATTC-3'

Yeast integrative plasmid

Yeast *l*inear *p*lasmid (YLp, pYL): A yeast → plasmid cloning vector that contains yeast telomeric sequences (see → telomere DNA) in addition to a → centromere and a chromosomal → autonomously replicating sequence. See also → yeast cloning vectors.

Yeast marker: See → yeast chromosomal marker.

Yeast one-hybrid system: See → one-hybrid system.

Yeast *o*pen *r*eading *f*rame (yORF): Any → open reading frame detected in the Yeast Genome Sequencing Project.

Yeast *p*romoter *p*lasmid (YPp, pYP): A yeast → plasmid cloning vector that contains fusions of various yeast promoters to *E. coli* genes such as → *lac Z*, *cat* or *gal K*. The expression of these genes in yeast leads to easily detectable products. See also → yeast cloning vectors.

Yeast replicating plasmid: See → yeast replicative plasmid.

Yeast replication plasmid: See → yeast replicative plasmid.

Yeast *r*eplicative *p*lasmid (yeast replication plasmid, yeast replicating plasmid, YRp, pYR): A low-copy number 7 kb yeast 2 μm plasmid cloning vector that contains the URA 3 and TRP 1 genes from yeast as selectable markers, a 1.4 kb → yeast chromosomal DNA fragment with an → autonomously replicating sequence inserted into the *Eco* RI site of → pBR 322, or, alternatively, an → origin of replication e.g. from the → yeast 2 μm plasmid, a replication origin (ORI) from pBR 322, and ampicillin and tetracycline resistance genes (Ap, Tc). This plasmid is used to transform yeast cells (transformation frequency: 10^3 transformants per μg DNA), but transformants are usually extremely unstable. See also → yeast cloning vectors.

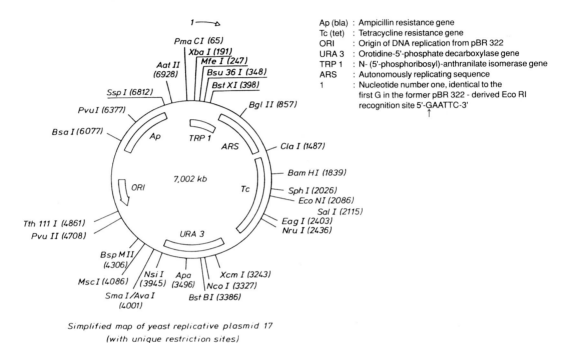

Ap (bla) : Ampicillin resistance gene
Tc (tet) : Tetracycline resistance gene
ORI : Origin of DNA replication from pBR 322
URA 3 : Orotidine-5'-phosphate decarboxylase gene
TRP 1 : N- (5'-phosphoribosyl)-anthranilate isomerase gene
ARS : Autonomously replicating sequence
1 : Nucleotide number one, identical to the first G in the former pBR 322 - derived Eco RI recognition site 5'-GAATTC-3'

Simplified map of yeast replicative plasmid 17
(with unique restriction sites)

Yeast replicative plasmid

Yeast three-hybrid system: See → three-hybrid system.

Yeast two-hybrid system: See → two-hybrid system.

Yeast 2 μm plasmid: See → two micron circle.

Yeast vector: See → yeast cloning vectors.

Yeast whole cell extract (YWCE): An extract from yeast cells, that contains (preferably) all cellular proteins. Usually, target cells are broken mechanically, the debris removed by centrifugation (300,000 x g)and the proteins precipitated by $(NH_4)_2SO_4$. After neutralization, the proteins are collected by centrifugation, resuspended and dialyzed extensively. Aliquots can be stored at –70°C.

YEp: See → *yeast episomal plasmid*.

YHp: See → *y*east *h*ybrid *p*lasmid.

YIp: See → *y*east *i*ntegrative *p*lasmid.

YLp: See → *y*east *l*inear *p*lasmid.

yORF: See → *y*east *o*pen *r*eading *f*rame.

YPp: See → *y*east *p*romoter *p*lasmid.

YRp: See → *y*east *r*eplicative *p*lasmid.

YRY: An intramolecular triple helix formed by specific pyrimidine-purine-pyrimidine stretch interactions. The normal → Watson-Crick base pairs in a duplex DNA are symbolized by dashes, and the base pairs involving the third strand by colons (e. g. T:A-T).

Y2H: See → two hybrid system.

YXp: See → *y*east *e*xpression *p*lasmid.

Z

ZBP: See → RNA localization.

Z-DNA (zig-zag DNA): A specific zig-zag conformation of the Watson-Crick double helix in which the antiparallel running DNA strands are wound around each other in a left-handed orientation (as opposed to the right-hand winding of → A- or → B-DNA) with about 12 residues per turn. B-DNA adopts the Z-configuration in regions with alternating purines and pyrimidines, e.g.

5'...GCGCGCGCGC...3'
3'...CGCGCGCGCG...5'

Z-DNA (or similarly bizarre conformations of DNA) can be detected by monoclonal anti-Z DNA antibodies (→ monoclonal antibody), and possibly offers address sites for specific nuclear proteins though its occurrence in vivo has not been proven unequivocally. See → C-DNA, → D-DNA, → E-DNA, → ε-DNA, → G-DNA, → G4-DNA, → H-DNA, → M-DNA, → P-DNA, → V-DNA.

Zeocin: A copper and sulfur containing → bleomycin → antibiotic that intercalates into and cleaves → double-stranded DNA, and is effective against bacteria, fungi (including yeast), plant and mammalian cells. Resistance to zeocin is conferred by the *Sh ble* (Stretoalloteichus *h*industanus *ble*omycin) gene product, a 14 kD protein, which binds the antibiotic stoichiometrically and thereby prevents its binding to DNA and its strand-cleaving activity. This gene can be used as → selectable marker gene for transformed bacteria, yeast, plant or mammalian cells, which are selected with zeocin.

Zero-cycle artifact: Any one of the non-specifically primed amplification products that arises from → *Taq* polymerase-catalyzed, low stringency priming events prior to the first amplification cycle in a conventional → polymerase chain reaction.

Zero integrated field electrophoresis (ZIFE): A variant of the → field inversion gel electrophoresis (FIGE), that is capable of resolving very large DNA molecules, but is very slow as compared to FIGE. See → contour-clamped homogeneous electric field gel electrophoresis, → field inversion gel electrophoresis, → orthogonal-field alternation gel electrophoresis, → programmable autonomously controlled electrodes gel electrophoresis, → pulsed-field gel electrophoresis, → pulsed homogeneous orthogonal-field gel electrophoresis, → rotating gel electrophoresis, → secondary pulsed field gel electrophoresis, → transverse alternating field electrophoresis.

Zero time binding DNA ("zero time fraction"): A fraction of denatured → genomic DNA that forms duplexes at the start of a → C_0t analysis because it contains a high proportion of → repetitive DNA.

Zero time fraction: See → zero time binding DNA.

Zeta Bind^R: A nylon-based membrane for → blotting procedures.

Zetaprobe^R: A nylon-based membrane for → blotting procedures.

ZIFE: See → *z*ero *i*ntegrated *f*ield *e*lectrophoresis.

Zig-zag DNA: See → Z-DNA.

Zimmermann cell fusion (electrofusion): A technique for the → fusion of two individual cells to form one → hybrid cell employing low-level, high-frequency electric fields. In short, two cells are oriented in an electric field until they touch each other. Then a current impulse is given that opens micropores in adjacent cell membranes. This allows mixing of the two cytoplasms and finally fusion of the cells. This technique (developed by U. Zimmermann) can also be used to fuse one cell with subcellular particles (e.g. → karyoplasts) of another cell. See also → cell fusion, → protoplast fusion.

Zinc cluster protein: A member of a class of nuclear proteins containing two zinc atoms that form a zinc cluster with six cysteine residues. This specific cluster domain interacts with DNA (e.g. in the → transcription factor GAL 4 of → *Saccharomyces cerevisiae*, that regulates the expression of gen*e*s encoding *gal*actose-metabolizing enzymes). Compare → zinc finger protein, → zinc twist protein.

Zinc finger protein (ZFP): A member of a class of nuclear proteins containing from two to nine imperfect → tandem repeats of the thirty amino acid-sequence (Phe, Tyr)-Xaa-Cys-$(Xaa)_{2(4)}$ Cys-$(Xaa)_3$-Phe-$(Xaa)_5$-Leu-$(Xaa)_2$-His-$(Xaa)_3$-His-$(Xaa)_5$, where Xaa stands for any amino acid. This region may fold into an independent structural → domain organized by a tetrahedrally coordinated Zn^{2+} ion which binds the two cys and the two his residues ("zinc finger": his-his-loop, cys-cys-loop). Zinc fingers interact with DNA in such a way that the α-helix of each domain lies in the → major groove, where sequence-specific contacts are made through residues of the finger motif (lys^{13}, arg^{18}, arg^{21}, and lys^{24}) and the polar side chains on the surface of the helix (glu^{12}, ser^{14}, ser^{17}, and gln^{20}). Alternatively to the cys-cys-loop or his-his-loop type of zinc fingers (e.g. TF III A-type), the Zn^{2+} ion may also be bound by four cys ligands (in e.g. steroid receptors and yeast → transcription factors). Close to 700 human genes encode zinc finger proteins, and the *Arabidopsis thaliana* genome harbors 85 such genes. Compare also → zinc cluster protein, → zinc twist protein. Compare → helix-loop-helix, → helix-turn-helix, → leucine zipper.

Zinc finger protein binding region (ZIP): A sequence element of → promoters (e.g. the human interle*u*kin 2 promoter) that serves as binding site for → zinc finger proteins functioning as → transcription factors (e.g. → SP1).

Zinc finger protein transcription factor (ZFP TF, TF_{ZF}): Any one of a series of → transcription factors, that contain one or more → zinc finger protein motifs. Most natural ZFP TFs have three fingers, but some possess as many as 37 such motifs, arranged one after the other such that they can contact multiple adjacent base triplets along the DNA double helix. ZFP TFs can also be assembled *in vitro,* producing novel combinations of the basic motifs, that can bind to virtually any gene or → promoter sequence in the → genome and thereby either activate or inactivate the corresponding gene.

Zinc ribbon: A conserved domain of yeast and plant DNA-binding proteins, that is similar to a → zinc finger.

Zinc twist protein: A member of a class of nuclear proteins containing two zinc atoms, each of which is bound by four cysteine residues. The amino acid chain in between these two zinc atoms forms a helical DNA recognition and binding domain (e.g. in the rat glucocorticoid receptor, that regulates the expression of a series of steroid-responsive genes by binding to so-called glucocor-ticoid-responsive elements [GRE] 5' upstream of the → promoters of these genes).

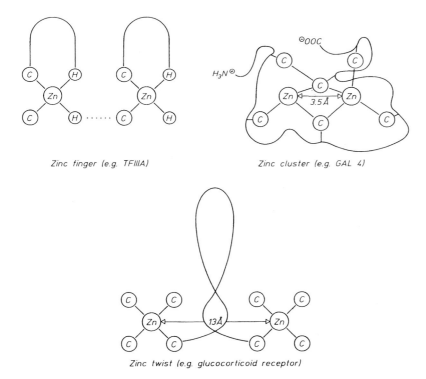

Zinc finger (e.g. TFIIIA)   *Zinc cluster (e.g. GAL 4)*

Zinc twist (e.g. glucocorticoid receptor)

Zinc finger, Zn²⁺-Cluster and Zn²⁺-twist protein

ZIP: See → *zinc finger protein* binding region.

Zippering: The process of successive reannealing of more and more → complementary bases of two single-stranded DNA or RNA molecules, in which the initial formation of hydrogen bonds between one or a few base pairs has an enhancing effect on the formation of the following, and so on (positive cooperativity).

ZOE: The fluorochrome 5-carboxy–2', 4', 5', 7'-tetrachlorofluorescein, that is used as marker for → fluorescent primers in e.g. automated sequencing procedures, as a dye terminator or as a label of probes in → DNA chip technology. The molecule can be excited by light of 495 nm wavelength, and emits fluorescence light at 545 nm. Since the wave-length of the excitation and emission maxima is pH-dependent, the exact values vary.

Zoo blot: A laboratory term for a → Southern blot onto which several restricted → genomic DNAs from different organisms have been transferred and which serves to detect sequences common to all species by hybridization to either → oligonucleotides or DNA → probes.

ZOO-FISH: See → zoo fluorescent *in situ* hybridization.

ZOO *fluorescent in situ hybridization* (ZOO-FISH): The localization of fluorescence-labeled → probes (e. g. → BAC clones, → genes) on panels of metaphase chromosomes from different, yet related animals. See → physical mapping, → synteny.

Zoom gel: A laboratory slang term for a gel strip for → isoelectric focusing, that provides high resolution power over a narrow pH range and accepts much higher protein loads than broad pH range strips.

ZsGreen: A variant of the → green fluorescent protein, that originates from the reef coral *Zoanthus* sp., and has an absorption maximum at 493 nm and an emission maximum at 505 nm. The gene encoding ZsGreen is used as a → reporter gene.

Appendix

1 Units and Conversion Factors

1.1. Metric Prefixes

Symbol	Prefix	Factor	Symbol	Prefix	Factor
Y	yotta	10^{24}	c	centi	10^{-2}
Z	zetta	10^{21}	μ	micro	10^{-6}
E	exa	10^{18}	n	nano	10^{-9}
P	peta	10^{15}	p	pico	10^{-12}
T	tera	10^{12}	f	femto	10^{-15}
G	giga	10^{9}	a	atto	10^{-18}
M	mega	10^{6}	z	zepto	10^{-21}
k	kilo	10^{3}	g	guaco	10^{-23}
h	hecto	10^{2}	y	yocto	10^{-24}
m	milli	10^{-3}			

1.2. Conversions of Nucleic Acids and Proteins

1 pg = 1000 Mbp = 1 Gbp (Number of bp = pg $\times$ 0.9869 $\times 10^{9}$)
1 kb = 1000 bases or 1000 base pairs of single- or double-stranded nucleic acid

1 kb double-stranded DNA (sodium salt)	= 6.6×10^{5} Da
1 kb single-stranded DNA (sodium salt)	= 3.3×10^{5} Da
1 kb single-stranded RNA (sodium salt)	= 3.4×10^{5} Da
1 MDa double-stranded DNA (sodium salt)	= 1.52 kb

1.3. Conversion Bases → Mass

Kilobase pairs (duplex DNA)

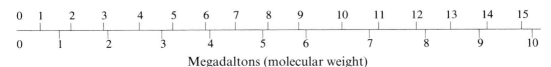

Megadaltons (molecular weight)

dNMP (average mass) = 330 Da
1 bp dNMP   = 660 Da

Nucleotide Triphosphates

Compound	Molecular weight	λ_{max} (at p_H = 7.0)
ATP	507.2	259
CTP	483.2	271
GTP	523.2	253
UTP	484.2	262
dATP	491.2	259
dCTP	467.2	271
dGTP	507.2	253
dTTP	482.2	267

1.4. Nucleic Acids and Proteins

1 µg/ml nucleic acid	= 3 µM phosphate
1 µg of a 1 kb DNA fragment	= 1.5 pmol
1 pmol of a 1 kb DNA	= 0.66 µg
1 mol of linear pBR 322 DNA	= 3.2×10^6 g
1 pmol of 5' ends of linear pBR 322 DNA	= 1.6 µg
1 µg pBR 322	= 0.36 pmol DNA
1 pmol pBR 322	= 2.8 µg

$1\ OD_{260\ nm}$ of double-stranded DNA = 50 µg/ml = 0.15 mM (in nucleotides)

$1\ OD_{260\ nm}$ of single-stranded DNA = 33 µg/ml = 0.10 mM (in nucleotides)

$1\ OD_{260\ nm}$ of single-stranded RNA = 40 µg/ml = 0.11 mM (in nucleotides)

$1\ OD_{260\ nm}$ of linear λ-DNA = 1.6×10^{-12} mol/ml = 9.8×10^{11} molecules/ml = 1.6 nM = 3.2 nM 5' ends

$1\ OD_{260\ nm}$ of linear pBR 322 DNA = 1.8×10^{-11} mol/ml = 1.1×10^{13} molecules/ml = 18.0 nM = 36.0 nM 5' ends

$1\ OD_{260\ nm}$ of an 8–mer oligolinker = 9.8×10^{-9} mol/ml = 5.9×10^{15} molecules/ml = 10.0 µM = 20.0 µM 5' ends

Molecular mass of a double-stranded DNA fragment
= (number of bp) × (660 Da/bp)
Molecular mass of a single-stranded DNA fragment
= (number of bp) × (330 Da/bp)
Moles of ends of a double-stranded DNA fragment
= 2× (g of DNA)/[molecular mass of DNA(Da)]
= 2× (g of DNA)/[(number of bp) × (660 Da/bp)]
Moles of ends produced by a restriction endonuclease
a) circular DNA : 2 × (moles of DNA) × (number of sites)
b) linear DNA : 2 × (moles of DNA) × (number of sites) + 2 × (original ends)

Exact molecular weight of a dephosphorylated oligonucleotide:
MW (g/mol) = [(A × 312.2)+(G × 328.2)+(C × 288.2)+(T × 303.2)–61.0]
Exact molecular weight of a phospho-oligonucleotide:
MW (g/mol) = [(A × 312.2)+(G × 328.2)+(C × 288.2)+(T × 303.2)+17.0]

Length and Molecular Weights of Common Nucleic Acids

Nucleic acid	Number of nucleotides	Molecular weight
Lambda DNA	48,502 (circular, dsDNA)	3.1×10^7
pBR322 DNA	4,363 (dsDNA)	2.8×10^6
28S rRNA	4,800	1.6×10^6
23S rRNA	3,700	1.2×10^6
18S rRNA	1,900	6.1×10^5
16S rRNA	1,700	5.5×10^5

| 5S rRNA | 120 | 3.6×10^4 |
| tRNA (*E. coli* | 75 | 2.5×10^4 |

1 kb coding capacity	= 333 amino acids = 37,000 Da protein
10,000 Da protein	= ~ 270 bp DNA
30,000 Da protein	= ~ 810 bp DNA
50,000 Da protein	= ~ 1.35 kb DNA
100,000 Da protein	= ~ 2.7 kb DNA
Amino acid (average mass)	= 120 Da

100 pmoles of 10,000 Dalton protein = 1 µg
100 pmoles of 50,000 Dalton protein = 5 µg

1.5. RNAs and Genomes

Sizes and molecular weights of various RNAs

RNA	Bases	Molecular weight (daltons)
E. coli		
tRNA	75	3.4×10^4
5S rRNA	120	4.1×10^5
16S rRNA	1541	5.2×10^5
23S rRNA	2904	9.9×10^5
Drosophila		
18S rRNA	1976	6.7×10^5
28S rRNA	3898	1.3×10^6
Mouse		
18S rRNA	1869	6.4×10^5
28S rRNA	4712	1.6×10^6
Rabbit		
18S rRNA	2366	8.0×10^5
28S rRNA	6333	2.2×10^6
Human		
18S rRNA	1868	6.4×10^5
28S rRNA	5025	1.7×10^6

Size and molecular weights of various genomic DNAs

DNA	Base pairs per haploid genome	Molecular weight (daltons)
pBR322 DNA	4,363	2.8×10^6
SV40	5,243	3.4×10^6

ΦX174	5,386	3.6×10^6
Adenovirus 2	35,937	2.3×10^7
Lambda phage	48,502	3.2×10^7
Escherichia coli	4.7×10^6	3.1×10^9
Saccharomyces cerevisiae	1.5×10^7	9.8×10^9
Dictyostelium discoideum	5.4×10^7	3.5×10^{10}
Arabidopsis thaliana	7.0×10^7	4.6×10^{10}
Caenorhabditis elegans	8.0×10^7	5.2×10^{10}
Drosophila melanogaster	1.4×10^8	9.1×10^{10}
Gallus domesticus (chicken)	1.2×10^9	7.8×10^{11}
Mus musculus (mouse)	2.7×10^9	1.8×10^{12}
Rattus norvegicus (rat)	3.0×10^9	2.0×10^{12}
Xenopus laevis	3.1×10^9	2.0×10^{12}
Homo sapiens	3.3×10^9	2.1×10^{12}
Zea mays	3.9×10^9	2.5×10^{12}
Nicotiana tabacum	4.8×10^9	3.1×10^{12}

1.6. Radioisotopes

1 µCi	$= 2.2 \times 10^6$ disintegrations per minute
	$= 3.7 \times 10$ becquerels
1 becquerel	$= 1$ disintegration per second

Physical properties of some beta-emitting radionuclides

Radionuclide	Half-life time	Specific activity (mCi/mmol)	Daughter nuclide (stable)
Tritium	12–43 years	10^2–10^5	helium–3
Carbon–14	5,730 years	1–10^2	nitrogen–14
Sulphur–35	87.4 days	1–10^6	chlorine–35
Phosphorus–32	14.3 days	10–10^6	sulphur–32
Phosphorus–33	25.5 days	10–10^4	sulphur–33
Chlorine–36	3.01×10^5 years	10^{-3}–10^{-1}	argon–36

2 Restriction Endonucleases

2.1 Isoschizomers: few examples

Enzyme	Recognition Site 5'→3'	Termini	Isoschizomer	Termini
Acc II	TCCGGA		Bsp M II	same
Asu II	TTCGAA	2 base 5' extension	Mla I	same
Ava I	CPyCGPuG	4 base 5' extension	Nsp7524 III	same
Ava II	GG$\frac{A}{T}$CC	3 base 5' extension	Afl I, BamN$_x$,I, Cla II, Fdi I, HgiB I, HgiC II, HgiE I, HgiH III	same
BamH I	GGATCC	4 base 5' extension	Bst I	2 base 5' extension
Bbe I	GGCGCC	4 base 3' extension	Nar I, Nda I, Nun II	same
Bcn I	CC$\frac{C}{G}$GG	1 base 5' extension	Cau II	same
BstE II	GGTNACC	5 base 5' extension	AspA I, BstP I, Eca I, FspA I	same
Bvu I	GPuGCPyC	4 base 3' extension	Ban II, HgiJ II	same
Cfr I	PyGGCCPu	4 base 5' extension	Eae I	1 base 5' extension
EcoR II	CC$\frac{A}{T}$GG	5 base 5' extension	Aor I, Apy 1, BstN I, TaqX I	same
Hae III	GGCC	Blunt	BspR I, BsuR I. Clt I, FnuD I, Sfa I	same
HgiC I	GGPyPuCC	4 base 5' extension	HgiH I	same
HgiD I	GPuCGPyC	2 base 5 ' extension	Acy I, Aos II, AstW I, Asu III, HgiG I, HgiH II	same
Hha I	GCGC	2 base 3' extension	FnuD III	2 base 5' extension
			HinP$_1$ I, SciN I	same
Hind II	GTPyPuAC	Blunt	HinJC 1	same
Hind III	AAGCTT	4 base 5' extension	Hsu I	same
Hinf I	GANTC	3 base 5' extension	FnuA I	same
Hpa II	CCGG	2 base 5' extension	Hap I, Mno I, Msp I	same
Mst I	TGCGCA	Blunt	Aos I, Fdi II	same
Mst II	CCTNAGG	3 base 5' extension	Cvn I, Sau I	same
Nsp7524 II	G$\overset{C}{\underset{T}{A}}GC\overset{G}{\underset{T}{A}}$C	4 base 3' extension	NspH I	same
Pst I	CTGCAG	4 base 3' extension	SalP I, Sfl I	same
Pvu I	CGATCG	2 base 3' extension	Nbl I, Rsh I, Xor II	same
Sal I	GTCGAC	4 base 5' extension	HgiC III, HgiD II Nop I	same
Sau3A I	GATC	4 base 5' extension	FnuC I, FnuE I, Mbo II, Pfa I Dpn I (methylated adenine)	blunt same
Sau96 I	GGNCC	3 base 5' extension	Asu I, Nsp7524 IV	4 base 5' extension
Sma I	CCCGGG	Blunt	Xma I	same
Sst I	GAGCTC	4 base 3' extension	Sac I	same
Sst II	CCGCGG	2 base 3' extension	Csc I, Sac II	same
Stu I	AGGCCT	Blunt	Gdi I	same
Tha I	CGCG	Blunt	FnuD II	same
Xho I	CTCGAG	4 base 5' extension	Blu I, PaeR7 I, Sla I, Xpa I	
Xho II	PuGATCPy		Mfl I	

Key: N : Any nucleotide   Pu : Purine base   Py : Pyrimidine base

2 Restriction Endonucleases

2.2 Methyl-sensitivity: few examples

Panel 1

Enzyme	Sequence cleaved	Sequence not cleaved
Aat I	AGG↓CCT	AGG$\overset{m}{C}$TGG
Acc I	GT↓$^{AT}_{CG}$AC	GT$^{AT}_{CG}$$\overset{m}{A}$G
Acc III	T↓CCGGA	TCCGG$\overset{m}{A}$TC
Alu I	AG↓CT	$\overset{m}{A}$GCT
		AG$\overset{m}{C}$T
Apa I	GGGCC↓C	GGG$\overset{m}{C}$CC
Apy I	CC↓^A_TGG	$\overset{m}{C}$C^A_TGG
Ava I	C↓^T_CCG^A_GG	C$^T_C$$\overset{m}{C}G^A_G$G
		C^T_CCG$\overset{m}{A}$G
Ava II	GG↓^A_TCC	GG$^A_T$$\overset{m}{C}C^A_T$GG
		GG$^A_T$$\overset{m}{C}$CGG
		GG$^A_T$$\overset{m}{C}C\overset{m}{C}$GG
Bal I	TGG↓CCA	TGG$\overset{m}{C}$CA
BamH I	G↓GATCC	GGAT$\overset{m}{C}$C
		GGATC$\overset{m}{C}$GG
Ban I	GG$^{TA}_{CG}$CC	GG$^{TA}_{CG}$CC^{m}
Ban II	G^A_GGC^C_TC	GAGCTC
		GGGC$\overset{m}{C}$
Bcl I	T↓GATCA	TGATC$\overset{m}{A}$
Bcn I	CC↓^G_CGG	CC$\overset{m}{^G_C}$GG
		C$\overset{m}{C}$CGG
		CC$\overset{m}{G}$GG
Bgl II	A↓GATCT	AGATCT^{m}
Cfr 13 I	G↓GNCC	GGNC$\overset{mA}{C_T}$GG
		GGNC$\overset{m}{C}$GG
		GGNC$\overset{m}{C}$$\overset{m}{C}$GG
Cla I	AT↓CGAT	ATCGAT^{m}
		ATCGATC
Dde I	C↓TNAG	AG$\overset{m}{C}$TNAG
Dpn II	GA↓TC	G$\overset{m}{A}$TC

Panel 2

Enzyme	Sequence cleaved	Sequence not cleaved
Eac I	^T_CGGCC^A_G	^T_CGGC$\overset{m}{C}$AGG
		^T_CGGC$\overset{m}{C}$GG
EcoR I	G↓AATTC	G$\overset{m}{A}$ATTC
		GAATT$\overset{m}{C}$CGG
Eco R II	↓CC^A_TGG	C$\overset{m}{C}$$^A_T$GG
EcoR V	GAT↓ATC	TCG$\overset{m}{A}$TATC
Hae II	^A_GGCGC↓^T_C	^A_GG$\overset{m}{C}$GC^T_C
Hae III	GG↓CC	GG$\overset{m}{C}$C
		GG$\overset{m}{C}$CGG
Hha I	GCG↓C	GCG$\overset{m}{C}$
		GCG$\overset{m}{C}$CGG
Hinc II	GT^T_C↓^A_GAC	GT$^{TA}_{CG}$A$\overset{m}{}$
		GTCGAC
Hind III	A↓AGCTT	$\overset{m}{A}$AGCTT
		AAG$\overset{m}{C}$TT
Hpa I	GTT↓AAC	GTTAA$\overset{m}{C}$
Hpa II	C↓CGG	$\overset{m}{C}$CGG
Mbo I	↓GATC	G$\overset{m}{A}$TC
		ATCG$\overset{m}{A}$TC
		TCG$\overset{m}{A}$TC
Mfl I	A↓^A_GGATC^T_C	^A_GG$\overset{m}{A}$TC^T_C
Msp I	C↓CGG	$\overset{m}{C}$CGG
		GG$\overset{m}{C}$CGG
		GGATC$\overset{m}{C}$GG
Nci I	CC↓^G_CGG	CC$\overset{mG}{C}$GG
Nco I	C↓CATGG	GG$\overset{m}{C}$CATGG
Nru I	TCG↓CGA	TCGCG$\overset{m}{A}$TC

Panel 3

Enzyme	Sequence cleaved	Sequence not cleaved
Pst I	CTGCA↓G	CTGC$\overset{m}{A}$G
		AG$\overset{m}{C}$TGCAG
Pvu I	CGAT↓CG	CGAT$\overset{m}{C}$G
Sal I	G↓TCGAC	GTCG$\overset{m}{A}$C
		GT$\overset{m}{C}$GAC
Sau3A I	↓GATC	GAT$\overset{m}{C}$
		GAT$\overset{m}{C}$CGG
		GGAT$\overset{m}{C}$C
Sau96 I	G↓GNCC	GGNC$\overset{mA}{C_T}$GG
		GGNC$\overset{m}{C}$GG
		GGG$\overset{m}{C}$C
		GGNC$\overset{m}{C}$CGG
Sma I	CCC↓GGG	CC$\overset{m}{C}$GGG
Taq I	T↓CGA	TCG$\overset{m}{A}$
		TCG$\overset{m}{A}$TC
		ATCG$\overset{m}{A}$T
TthHB8 I	T↓CGA	TCG$\overset{m}{A}$
		TCG$\overset{m}{A}$TC
		ATCG$\overset{m}{A}$T
Xba I	T↓CTAGA	TCTAG$\overset{m}{A}$
		T$\overset{m}{C}$TAGA
Xho I	C↓TCGAG	CTCG$\overset{m}{A}$G
		CT$\overset{m}{C}$GAG
Xho II	A↓^A_GGATC^T_C	^A_GGATC$\overset{m}{C}$GG
Xma I	C↓CCGGG	$\overset{m}{C}$CCGGG
Xma III	C↓GGCCG	CGG$\overset{m}{C}$CG
		CGG$\overset{m}{C}$CGG
Xmn I	GAAN$_2$N$_2$TTC	TCG$\overset{m}{A}$AN$_4$TTC

Key: ↓ : Arrow indicates cleavage site   m : Methyl group (−CH$_3$)   N : Any nucleotide

2 Restriction Endonucleases

2.3 Star activity

Restriction enzyme	Recognition sequence	Relaxed specificity	Inducing conditions
Aat II	GACGT↓C	—	DMSO
*Aor*13H I	T↓CCGGA	—	High glycerol concentration Alkaline pH DMSO
Ava I	C↓PyCGPuG	—	High endounclease concentration DMSO
Ava II	G↓GWCC	—	High glycerol concentration DMSO
Avi II	TGC↓GCA	—	Alkaline pH DMSO
*Bam*H I	G↓GATCC	GGNTCC GGANCC GPuATCC	High glycerol concentration Mn^{2+} DMSO Low ionic strength
Ban II	GRGCY↓C	—	High glycerol concentration DMSO Low ionic strength
Bgl I	GCCNNNN↓NGGC	—	Low ionic strength
Bgl II	A↓GATCT	—	DMSO
*Bsp*T 104 I	TT↓CGAA	—	High glycerol concentration DMSO
Bst I	G↓GATCC	NGATCN	High endonuclease concentration High glycerol concentration
*Bst*P I	G↓GTNACC	—	High glycerol concentration Low ionic strength
Bst 1107 I	GTA↓TAC	—	Low ionic strength
*Bsu*R I	GG↓CC	NGCN	High endonuclease concentration High glycerol concentration High pH
Dde I	C↓TNAG	—	High pH DMSO
Eam 1105 I	GACNNN↓NNGTC	—	High glycerol concentration Alkaline pH Low ionic strength

2 Restriction Endonucleases

2.3 Star activity (continued)

Restriction enzyme	Recognition sequence	Relaxed specificity	Inducing conditions
*Eco*O 65 I	G↓GTNACC	—	DMSO Low ionic strength
*Eco*R I	G↓AATTC	NAATTN	High glycerol concentration Mn^{2+} DMSO Low ionic strength
*Eco*R V	GAT↓ATC	PuATATC GNTATC GANATC GATANC GATNTPy	DMSO
*Eco*T22 I	ATGCA↓T	—	Low ionic strength 2-mercaptoethanol
Fba I	T↓GATCA	—	High glycerol concentration Alkaline pH DMSO Low ionic strength
Hae III	GG↓CC	—	High endonuclease concentration High glycerol concentration
Hha I	GCG↓C	—	High endonuclease concentration High glycerol concentration DMSO
Hinc II	GTY↓RAC	—	DMSO
Hind III	A↓AGCTT	PuAGCTT A^A_GGCTT AA^G_TCTT AAGCNT AAGCTPy	Mn^{2+} DMSO
Hpa I	GTT↓AAC	—	High glycerol concentration DMSO
Kpn I	GGTAC↓C	—	DMSO
Mun I	C↓AATTG	—	High glycerol concentration Low ionic strength
Nco I	C↓CATGG	—	High glycerol concentration DMSO

2 Restriction Endonucleases

2.3 Star activity (continued)

Restriction enzyme	Recognition sequence	Relaxed specificity	Inducing conditions
Nhe I	GĊTAGC	—	High glycerol concentration Alkaline pH DMSO Low ionic strength
Pae R7	ĊTGGAG	—	High glycerol concentration Low ionic strength
Psp 1406 I	AAĊGTT	—	DMSO
Pst I	CTGCAˇG	—	High glycerol concentration DMSO
Pvu I	GAGˇCTC	CCG↓CTG CAT↓CTC CAG↓ATG CAG↓NTG CAG↓GNG CAG↓CGG	High endonuclease concentration High glycerol concentration DMSO
Pvu II	CAGˇCTG	NAGCTG CNGCTG CAGNTG CAGNTG CAGCNG CAGCTN	High glycerol concentration DMSO
Sac I	GAGCTˇC		High glycerol concentration DMSO
Sal I	GˇTCGAC	—	High glycerol concentration DMSO
*Sau*3A I	ˇGATC	SATC GMTC GAKC GATS	High glycerol concentration DMSO
Sca I	AGTˇACT	—	Mn^{2+} Alkaline pH Low ionic strength
Sfi I	GGCCNNNNˇNGGCC	—	Mn^{2+} DMSO
Spe I	AˇCTAGT	—	DMSO Low ionic strength

2 Restriction Endonucleases

2.3 Star activity (continued)

Restriction enzyme	Recognition sequence	Relaxed specificity	Inducing conditions
Ssp I	AAT↓ATT	—	DMSO High glycerol concentration Low ionic strength Alkaline pH
Sst I	GAGCT↓C	—	High endonuclease concentration High glycerol concentration DMSO
Sst II	CCGC↓CGG	—	High endonuclease concentration High glycerol concentration
Swa I	ATTT↓AAAT	—	DMSO Low ionic strength
Taq I	T↓CGA	—	High glycerol concentration Alkaline pH Low ionic strength
Tth 111 I	GACN↓NNGTC	NACNNNGTC GNCNNNGTC GANNNNGTC GACNNNNTC GACNNNGNC GACNNNGTN	Mn^{2+} Alkaline pH High ionic strength
Van 91 I	CCANNNN↓NTGG	—	High glycerol concentration
Vpa K11B I	G↓GWCC	—	High glycerol concentration Alkaline pH DMSO Low ionic strength
Xba I	T↓CTAGA	—	High glycerol concentration DMSO

3 Databases (relevant for gene technology, genomics, proteomics)

Database	URL (Uniform Resource Locator)	Description
Universal Sequence Repositories		
DNA Data Bank of Japan (DDBJ)	www.ddbj.nig.ac.jp	Nucleotide and protein sequences
EMBL Nucleotide Sequence Database	www.ebi.ac.uk/embl.html	Nucleotide and protein sequences
GenBank	www.ncbi.nlm.nih.gov/web/genbank/ www.ncbi.nlm.nih.gov/BLAST/blast_overview.html www.ncbi.nlm.nih.gov/Genbank/GenbankOverview.html	Nucleotide and protein sequences
Genome Sequence Database (GSDB)	www.ncgr.org/gsdb	Nucleotide and protein sequences
Kyoto Encyclopedia of Genes and Genomes (KEGG)	www.genome.ad.jp/kegg	Genome maps and gene catalogs
TIGR Gene Indices	www.tigr.org/tdb/tdb.html	Genes and gene clusters
UniGene	www.ncbi.nlm.nih.gov/unigene/	Genes and gene clusters
Genetic Maps		
DRESH	www.tigem.it/local/drosophila/dros.html	Human cDNA clones homologous to *Drosophila* mutant genes
GB4–RH	www.sanger.ac.uk/rhserver/rhserver.shtml	Genebridge4 (GB4) human radiation hybrid maps
GDB	www.gdb.org	Human genes and genomic maps
GenAtlas	www.citi2.fr/genatlas	Human genes, markers & phenotypes
GeneMap '99	www.ncbi.nlm.nih.gov/genemap/	Human gene map (mapping consortium)
HuGeMap	www.infobiogen.fr/services/hugemap	Human genome map data
IXDB	ixdb.mpimp-berlin-dahlem.mpg.de	Maps of chromosome X
Radiation Hybrid Database	www.ebi.ac.uk/rhdb	Radiation hybrid map data

Genomic Databases

ACeDB	www.sanger.ac.uk/HGP/ www.sanger.ac.uk/Projects/C_elegans/	Sequences: *C. elegans, S. pombe*, human
AMmtDB	bio-www.ba.cnr.it:8000/biowww/#ammtdb	Metazoan mtDNA sequences
Arabidopsis Database (AtDB)	genome-www.stanford.edu/arabidopsis	*Arabidopsis thaliana* genome
CropNet	synteny.nott.ac.uk	Genome mapping in crop plants
CyanoBase	www.kazusa.or.jp/cyano/mutants	*Synechocystis* sp. genome
EcoGene	bmb.med.miami.edu/ecogene/ecoweb	*E. coli* K-12 sequences
EMGlib	pbil.univ-lyon1.fr/emglib/emglib.html	Sequences of bacterial and yeast genomes
FilGenNet	www.neb.com/fgn/filgen1.html	Nematode parasites of humans, genomics
FlyBase	www.fruitfly.org	*Drosophila* sequences
GOBASE	megasun.bch.umontreal.ca/gobase/gobase.html	Organelle genome database
Human BAC Ends Database	www.tigr.org/tdb/humgen/bac_end_search/bac_end_intro.html	Human BAC end sequences
INE	www.dna.affrc.go.jp:82/giot/ine.html	Rice maps and sequence data
Mendel Database	jiio6.jic.bbsrc.ac.uk	Plant EST & STS sequences
MitBASE	www3.ebi.ac.uk/research/mitbase/mitbase.pl	Mitochondrial genomes
MitoDat	www-lecb.ncifcrf.gov/mitodat	Mitochondrial proteins
MITOMAP	www.gen.emory.edu/mitomap.html	Human mitochondrial genome
MITONUC/MITOALN	bio-www.ba.cnr.it:8000/srs6	Nuclear genes for mitochondrial proteins
MITOP	websvr.mips.biochem.mpg.de/proj/medgen/mitop	Mitochondrial proteins and genes
Mouse Genome Database (MGD)	www.informatics.jax/org	Mouse genetics and genomics
Munich Info Center for Protein Seqs (MIPS)	www.mips.biochem.mpg.de	Protein and genomic sequences
NRSub	pbil.univ-lyon1.fr/nrsub/nrsub.html http://pbil.univ-lyon1.fr/emglib/emglib.html utmmg.med.uth.tmc.edu/sphaeroides	*Bacillus subtilis* genome
RsGDB		*Rhodobacter sphaeroides* genome
Saccharomyces Genome Database (SGD)	genome-www.stanford.edu/saccharomyces www.mips.biochem.mpg.de/proj/yeast	*Saccharomyces cerevisiae* genome

TIGR Microbial Database	www.tigr.org/tdb/mdb/mdb.html	Microbial genomes and chromosomes
ZFIN	www.tigr.org/tdb/hgi/searching/hgi_reports.html zfish.uoregon.edu/zfin	Zebrafish genetics & development
ZmDB	zmdb.iastate.edu	Maize genome database

Comparative Genomics

Clusters of Orthologous Groups	www.ncbi.nlm.nih.gov/cog	Protein classification from 30 complete genomes

Gene Expression

ASDB	cbcg.nersc.gov/asdb	Proteins & expression patterns of alternatively spliced genes
Axeldb	www.dkfz-heidelberg.de/abt0135/axeldb.htm	Gene expression in *Xenopus*
BodyMap	bodymap.ims.u-tokyo.ac.jp	Human & mouse gene expression
Expressed Gene Anatomy Database (EGAD)	www.tigr.org/tdb/egad/egad.html	Non-redundant human (HT) and non-human transcript (ET) sequences
FlyView	pbio07.uni-muenster.de	*Drosophila* genetics
Gene Expression Database (GXD)	www.informatics.jax.org/gxdindex.html	Mouse gene expression
Mouse Atlas and Gene Expression Database	genex.hgu.mrc.ac.uk	Gene expression data
TRIPLES	Ycmi.med.yale.edu/ygac/triples.htm	Transposon – insertion phenotypes in *Saccharomyces*

Gene Structure

Ares Lab Intron Site	www.cse.ucsc.edu/research/compbio/yeast_introns.html	Yeast spliceosomal introns
COMPEL	compel.bionet.nsc.ru/funsite.html	Composite regulatory elements
CUTG	www.kazusa.or.jp/codon	Codon usage tables

EID	mcb.harvard.edu/gilbert/eid/	Protein-coding, intronic genes
EPD	www.epd.isb-sib.ch	Eukaryotic POL II promoters
ExInt	intron.bic.nus.edu.sg/exint/exint.html	Exon-intron gene structure
IDB/IEDB	nutmeg.bio.indiana.edu/intron/index.html	Intron sequence and evolution
PLACE	www.dna.affrc.go.jp/htdocs/place	Plant *cis*-acting elements
PlantCARE	sphinx.rug.ac.be: 8080/plantcare	Plant *cis*-acting elements
TransTerm	biochem.otago.ac.nz: 800/transterm/homepage.html	Codon usage, start & stop signals
TRRD	wwwmgs.bionet.nsc.ru/mgs/dbases/trrd4	Regulatory regions of eukaryotic genes
YIDB	www.embl-heidelberg.de/externalinfo/seraphin/yidb.html	Yeast nuclear and mitochondrial intron sequences

Plant Genome

AAtDB	http://genome-www.stanford.edu/Arabidopsis	*Arabidopsis*
Alfagenes	http://naaic.org/	Alfalfa (*Medicago sativa*)
Bean Genes	http://probe.nalusda.gov:8300/cgi-bin/browse/beangenes	*Phaseolus* and *Vigna*
ChlamyDB	http://probe.nalusda.gov:8300/cgi-bin/browse/chlamydb	*Chlamydomonas reinhardtii*
Cool Genes	http://probe.nalusda.gov:8300/cgi-bin/browse/coolgenes	Cool season food legumes
Cotton DB	http://probe.nalusda.gov:8300/cgi-bin/browse/cottondb	*Gossypium* species
Grain Genes	http://probe.nalusda.gov:8300/cgi-bin/browse/graingenes	Wheat, barley, rye and relatives
Maize DB	http://www.agron.missouri.edu/	Maize
Millett Genes	http://jiio5.jic.bbsrc.ac.uk:8000/cgi-bin/ace/search/millet	Pearl millett
Patho Genes	http://probe.nalusda.gov:8300/cgi-bin/browse/pathogens	Fungal pathogens of small-grain cereals
Rice Genes	http://genome.cornell.edu/rice/	Rice
Rice Genome Project	http://www.staff.or.jp	Rice
Sol Genes	http://genome.cornell.edu/solgenes/welcome.html	*Solanaceae*
SorghumDB	http://probe.nalusda.gov:8300/cgi-bin/browse/sorghumdb	*Sorghum bicolor*
Soybean	http://129.186.26.94/	Soybean
Tree Genes	http:/dendrome.ucdavis.edu/index.html	Forest Trees

Protein Databases

DAtA	luggagefast.stanford.edu/group/arabprotein	*Arabidopsis* coding sequences
DExH/D Family Database	www.columbia.edu/~ej67/dbhome.htm	DEAD-box, DEAH-box, and DExH-box proteins
Endogenous GPCR List	www.biomedcomp.com/gpcr.html	G protein-coupled receptors
GenProtEC	dbase.mbl.edu/genprotec/start	*E. coli* genes, gene products, and homologs
Histone Sequence Database	genome.nhgri.nih.gov/histones	Histone and histone fold sequences and structures
Homeobox Page	copan.bioz.unibas.ch/homeo.html	Information relevant to homeobox proteins, classification, and evolution
Homeodomain Resource	genome.nhgri.nih.gov/homeodomain	Homeodomain sequences
HUGE	www.kazusa.or.jp/huge	Large human proteins and cDNA sequences
IMGT	www.ebi.ac.uk/imgt/hla	Immunoglobulin, T cell receptor, and MHC sequences
InBase	www.neb.com/neb/inteins.html	Inteins and other motifs
Kabat Database	immuno.bme.nwu.edu	Immunologically interesting proteins
Membrane Protein Database	biophys.bio.tuat.ac.jp/ohshima/database	Membrane and transmembrane peptides
ooTFD	www.ifti.org	Transcription factors
PhosphoBase	www.cbs.dtu.dk/databases/phosphobase	Protein phosphorylation sites
PKR	delphi.phys.univ-tours.fr/prolysis	Protein kinase sequences
PPMdb	sphinx.rug.ac.be:8080/ppmdb/index.html	*Arabidopsis* plasma membrane protein sequences and expression data
Prolysis	delphi.phys.univ-tours.fr/prolysis/	Proteases and inhibitors
PROMISE	bioinf.leeds.ac.uk/promise	Prosthetic centers and metal ions in protein active sites
Protein Information Resource (PIR)	www.nbrf.georgetown.edu/pir	Non-redundant protein sequences
Receptor Database (RDB)	impact.nihs.go.jp/rdb.html	Receptor protein sequences
SENTRA	wit.mcs.anl.gov/wit2/sentra	Sensory signal transduction proteins
SWISS-PROT/TrEMBL	www.expasy.ch/sprot	Curated protein sequences
TRANSFAC	transfac.gbf.de/transfac/index.html	Transcription factors & binding sites

Protein Sequence Motifs

BLOCKS	www.blocks.fhcrc.org	Protein sequence motifs
Pfam	www.sanger.ac.uk/software/pfam	Multiple sequence alignments and protein domains
PIR-ALN	www-nbrf.georgetown.edu/pirwww/dbinfo/piraln.html	Protein sequence alignments
PRINTS	www.biochem.ucl.ac.uk/bsm/dbbrowser/prints/printscontents.html	Protein sequence motifs and signatures
ProDom	www.toulouse.inra.fr/prodom.html	Protein domain families
PROSITE	www.expasy.ch/prosite	Protein patterns and profiles
ProtoMap	www.protomap.cs.huji.ac.il	Hierarchical classification of SWISS-PROT proteins
SBASE	www2.icgeb.trieste.it/~sbaseserv	Annotated protein domains
SMART	coot.embl-heidelberg.de/smart	Signalling domain sequences
SYSTERS	www.dkfz-heidelberg.de/tbi/services/clusters/systersform	Protein clusters

Proteome Resources

AAindex	www.genome.ad.jp/dbget	Physico-chemical properties of peptides
REBASE	rebase.neb.com/rebase/rebase.html	Restriction enzymes and methylases
SWISS–2DPAGE	www.expasy.ch/ch2d	2D-PAGE images
Yeast Proteome Database (YPD)	www.proteome.com/ypdhome.html	*Saccharomyces cerevisiae* proteome

RNA Sequences

ACTIVITY	wwwmgs.bionet.nsc.ru/systems/activity	Functional DNA/RNA site sequences
Collection of mRNA-like noncoding RNAs	www.man.poznan.pl/5sdata/ncrna	Non-protein-coding RNA transcripts

Name	URL	Description
Database on the structure of LS rRNA	rrna.uia.ac.be	Alignment of large subunit ribosomal RNA sequences
Database on the structure of SS rRNA	rrna.uia.ac.be/ssu	Alignment of small subunit ribosomal RNA sequences
5S Ribosomal RNA Databank	www.rose.man.poznan.pl/5sdata/5srna.html	5S rRNA sequences
Guide RNA Database	www.biochem.mpg.de/~goeringe	Guide RNA sequences
Intronerator	www.cse.ucsc.edu/~kent/intronerator	RNA splicing & gene structure in *C. elegans*
Non-canonical Base Pair Database	prion.bchs.uh.edu/bp_type	RNA structures containing rare base pairs
PLMLTRNA	bio-www.ba.cnr.it:8000/srs	Plant mitochondrial tRNAs and tRNA genes
Pseudobase	wwwbio.leidenuniv.nl/~batenburg/pkb.html	RNA pseudo-knots
Ribosomal Database Project (RDP)	www.cme.msu.edu/rdp	rRNA sequences
RNA Modification Database	medlib.med.utah.edu/rnamods	Naturally modified nucleosides in RNA
SRPDB	psyche.uthct.edu/dbs/srpdb/srpdb.html	Signal recognition particle RNA, protein & receptor sequences
tRNA sequences	www.uni-bayreuth.de/departments/biochemie/trna	tRNA & tRNA gene sequences
UTRdb	bigarea.area.ba.cnr.it:8000/embit/utrhome/	5' & 3' UTRs of eukaryotic mRNAs
Viroid and Viroid-Like RNA Database	www.callisto.si.usherb.ca/~jpperra	Viroid and viroid-like RNA sequences

Structure

Name	URL	Description
IMB Jena Image Library	www.imb-jena.de/image.html	Visualization & analysis of three-dimensional biopolymer structures
LPFC	www-smi.stanford.edu/projects/helix/lpfc	Library of protein family core structures
MODBASE	guitar.rockefeller.edu/modbase	Comparative protein structure models
NDB	ndbserver.rutgers.edu/ndb/ndb.html	Nucleic acid-containing structures

Transgenics

Cre Transgenic Database	www.mshri.on.ca/nagy/cre.htm	Cre transgenic mouse lines
Transgenic/Targeted Mutation Database	tbase.jax.org	Transgenic animals and targeted mutations

Intermolecular Interactions

Database of Ribosomal Crosslinks (DRC)	www.mping-berlin-dahlem.mpg.de/~ag_ribo/ag_brimacombe/drc/	Ribosomal crosslinking data
DIP	dip.doe-mbi.ucla.edu	Catalog of protein-protein interactions
DPInteract	arep.med.harvard.edu/dpinteract/	*E. coli* DNA-binding proteins

Mutation Databases

ALFRED	fondue.med.yale.edu/db2	Allele frequencies & DNA polymorphisms
Atlas of Genetics and Cytogenetics in Oncology	www.infobiogen.fr/services/chromcancer	Chromosomal abnormalities in cancer
Database of Germline p53 Mutations	www.lf2.cuni.cz/homepage.html	p53 gene mutations
dbSNP	www.ncbi.nlm.nih.gov/dbsnp	Single nucleotide polymorphisms
HGBASE	hgbase.interactiva.de	Intragenic sequence polymorphisms
HIV-RT	hivdb.stanford.edu/hiv	HIV reverse transcriptase & protease sequence variation
Human Gene Mutation Database (HMGD)	uwcm.web.cf.ac.uk/uwcm/mg/hgmd0.html	Gene lesions for human inherited diseases
HvrBase	www.eva.mpg.de/hvrbase	Primate mtDNA control region sequences

KinMutBase	www.uta.fi/imt/bioinfo/kinmutbase/	Disease-causing protein kinase mutants
MmtDB	www.ba.cnr.it/~areamt08/mmtdbwww.htm	Mutations & polymorphisms in metazoan mtDNA
Mutation Spectra Database	info.med.yale.edu/mutbase	Gene mutations
Online Mendelian Inheritance in Man	www.ncbi.nlm.nih.gov/omim	Human genetic disorders
RB1 Gene Mutation Database	home.kamp.net/home/dr.lohmann	Human retinoblastoma (RB1) gene mutation

Biomedical Databanks

Molecular Probe Database	srs.ebi.ac.uk	Synthetic oligonucleotides, probes, and PCR primers
MPDB	www.biotech.ist.unige.it/interlab/mpdb.html	Synthetic oligonucleotides
Tree of Life	phylogeny.arizona.edu/tree/phylogeny.html	Phylogeny & biodiversity
Vectordb	vectordb.atcg.com	Characterization & classification of nucleic acid vectors

4 Scientific Journals

Advances in Cell and Molecular Biology
Academic Press

Advances in Enzymology
John Wiley and Sons

Advances in Experimental Medicine and Biology
Plenum

Advances in Genetics
Academic Press

Advances in Genome Biology
JAI Press

Advances in Human Genetics
Plenum

Advances in Protein Chemistry
Academic Press

American Journal of Human Genetics
University of Chicago Press

American Journal of Medical Genetics
Wiley-Liss

Animal Cell Biotechnology
Academic Press

Animal Genetics
(formerly *Animal Blood Groups and Biochemical Genetics*)
Blackwell Scientific

Annales de Génétique
Expansion Scientifique Francaise

Annals of Human Genetics
Cambridge University Press

Annual Review of Biochemistry
Annual Reviews

Annual Review of Biophysics and Biomolecular Structure
Annual Reviews

Annual Review of Cell Biology
Annual Reviews

Annual Review of Genetics
Annual Reviews

Annual Review of Medicine
Annual Reviews

Annual Review of Microbiology
Annual Reviews

Annual Review of Neuroscience
Annual Reviews

Annual Review of Plant Physiology and Plant Molecular Biology
Annual Reviews

Applied Cytogenetics
(Journal of the Association of Cytogenetic Technologists)

Archives of Biochemistry and Biophysics
Academic Press

Archives of Microbiology
Springer-Verlag

Archives of Virology
Springer-Verlag

Biochemical and Biophysical Research Communications
Academic Press

Biochemical Genetics
Plenum

Biochemistry and Cell Biology
National Research Council of Canada

Biochimica et Biophysica Acta
Elsevier / North Holland

Bioinformatics
Oxford University Press

Biopolymers
John Wiley and Sons

Biotechniques
Eton Publishing

Biotechnology and Bioengineering
John Wiley and Sons

Biotechnology and Genetic Engineering News
Intercept

Brookhaven Symposia in Biology
Brookhaven National Laboratory

Canadian Journal of Microbiology
National Research Council of Canada

Cancer Genetics and Cytogenetics
Elsevier / North Holland

Cell
Cell Press

Cellular and Molecular Biology
Pergamon Press

Chromosoma
Springer-Verlag

Clinical Genetics
Munksgaard Forlag

Cold Spring Harbor Symposia on Quantitative Biology
Cold Spring Harbor Laboratory

Comparative and Functional Genomics
John Wiley and Sons

CRC Critical Reviews in Biochemistry and Molecular Biology
Chemical Rubber Company Press

CRC Critical Reviews in Microbiology
Chemical Rubber Company Press

Current Genetics
Springer-Verlag

Current Opinion in Biotechnology
Current Biology

Current Opinion in Genetics and Development
Current Biology

Current Topics in Cellular Regulation
Academic Press

Current Topics in Microbiology and Immunology
Springer-Verlag

Cytogenetics and Cell Genetics
S. Karger

Developmental Genetics
Wiley-Liss

DNA – A Journal of Molecular and Cellular Biology
Mary Ann Liebert Inc. Publishers

DNA and Protein Engineering Techniques
Wiley-Liss

DNA Repair Reports
Elsevier / North Holland

DNA Sequence: The Journals of DNA Sequencing and Mapping
Harwood Academic

ELECTROPHORESIS
Wiley-VCH

EMBO Journal
IRL Press at Oxford University Press

European Journal of Human Genetics
S. Karger

Functional and Integrative Genomics
Springer Verlag

Fungal Genetics Newsletter
(formerly *Neurospora Newsletter*)
Fungal Genetics Stock Center

Gene
Elsevier / North Holland

Gene Function & Disease
Wiley-VCH

Genes and Development
Cold Spring Harbor Laboratory

Genes and Function
Blackwell Scientific

Genes and Immunity

Genes, Chromosomes and Cancer
Wiley-Liss

Genes to Cells
Blackwell Scientific

Gene Therapy
Stockton Press

Genetica
Kluwer Academic

Genetical Research
Cambridge University Press

Genetic Engineering
Academic Press

Genetic Engineering: Principles and Methods
  Plenum Press

Genetics
  Genetics

Genome
  (formerly *Canadian Journal of Genetics and Cytology*)
  National Research Council of Canada

Genome Research

Genomics
  Academic Press

Heredity
  Blackwell Scientific

Human Gene Therapy
  Mary Ann Liebert Inc. Publishers

Human Genetics
  (formerly *Humangenetik*)
  Springer-Verlag

Human Heredity
  (formerly *Acta Genetica et Statistica Medica*)
  S. Karger

Human Molecular Genetics
  Oxford University Press

Hybridoma
  Mary Ann Liebert Inc. Publishers

Immunogenetics
  Springer-Verlag

International Journal of Biochemistry and Cell Biology
  Elsevier / North Holland

International Review of Cytology
  Academic Press

Japanese Journal of Genetics
  Japan Publications Trading Co.

Japanese Journal of Human Genetics
  Institute of Medical Genetics

Journal of Animal Breeding and Genetics
  Paul Parey Scientific Publishers

Journal of Bacteriology
  American Society for Microbiology

Journal of Biochemistry
  Japanese Biochemical Society

Journal of Cell Biology
  (formerly *Journal of Biophysical and Biochemical Cytology*)
  Rockefeller University Press

Journal of Cell Science
  (formerly *Quarterly Journal of Microscopical Science*)
  Company of Biologists

Journal of General and Applied Microbiology
  Journal Press

Journal of General Microbiology
  Cambridge University Press

Journal of General Virology
  Cambridge University Press

Journal of Genetic Counseling
  Plenum

Journal of Heredity
  Oxford University Press

Journal of Histochemistry and Cytochemistry
  Elsevier / North Holland

Journal of Medical Genetics
  BMJ Publishing Group

Journal of Molecular Biology
  Academic Press

Journal of Molecular and Cellular Immunology
  Springer-Verlag

Journal of Molecular Evolution
  Springer-Verlag

Journal of Molecular Medicine
  Springer-Verlag

Journal of Molecular Microbiology and Biotechnology
  Karger
  www.karger.com/mmb

Journal of Molecular Modeling
  Springer-Verlag

Journal of Virology
  American Society for Microbiology

Mammalian Genome
Springer-Verlag

Medical Genetics
Pergamon Press

Methods in Cell Biology
Academic Press

Methods in Cell Physiology
Academic Press

Methods in Enzymology
Academic Press

Methods in Immunology and Immunochemistry
Academic Press

Methods in Molecular and Cellular Biology
Wiley-Liss

Methods in Virology
Academic Press

Microbiological Reviews
American Society for Microbiology

Molecular and Cellular Biochemistry
(formerly *Enzymologia*)
Kluwer Academic

Molecular and Cellular Biology
American Society for Microbiology

Molecular and Cellular Probes
Academic Press

Molecular Biology and Evolution
University of Chicago Press

Molecular Biology and Medicine
Academic Press

Molecular Biology Reports
Kluwer Academic

Molecular Ecology
Blackwell Scientific

Molecular Genetic Medicine
Academic Press

Molecular Genetics and Genomics
(formerly *Zeitschrift für Vererbungslehre*)
Springer-Verlag

Molecular Immunology
Pergamon Press

Molecular Medicine
Blackwell Scientific

Molecular Microbiology
Blackwell Scientific

Molecular Microbiology and Medicine
Academic Press

Molecular Pharmacology
Academic Press

Molecular Phylogenetics and Evolution
Academic Press

Monographs in Human Genetics
S. Karger

Mouse Genome
(formerly *Mouse Newsletter*)
Oxford University Press

Mutagenesis
IRL Press at Oxford University Press

Mutation Research
Elsevier / North Holland

Nature
Nature Publishing

Nature Biotechnology
Nature Publishing

Natur Cell Biology
Nature Publishing

Nature Genetics
Nature Publishing

Nature Medicine
Nature Publishing

Nature Structural Biology
Nature Publishing

Nucleic Acids Research
IRL Press at Oxford University Press

Pharmacogenetics
Chapman and Hall

Physiological Genomics
http://www.physiolgenomics.physio-logy.com
The American Physiological Society

Plant Molecular Biology
Kluwer Academic

Plasmid
  Academic Press

*Proceedings of the National Academy of
  Sciences of the United States of America*
  National Academy of Sciences

Progress in Biophysics and Molecular Biology
  Elsevier / North Holland

*Progress in Nucleic Acid Research and
  Molecular Biology*
  Academic Press

Protein Science
  Cambridge University Press

Proteins: Structure, Function and Genetics
  Wiley-Liss

PROTEOMICS
  Wiley-VCH

RNA
  Cambridge University Press

Science
  American Association for the Advancement
  of Science

*Sequence: The Journal of DNA Mapping and
  Sequencing*
  Harwood Academic

Somatic Cell and Molecular Genetics
  Plenum

Stadler Genetics Symposia
  University of Missouri Press

Sub-Cellular Biochemistry
  Plenum

Theoretical and Applied Genetics
  (formerly *Der Züchter*)
  Springer-Verlag

Transgenic Research
  Chapman and Hall

Transgenics
  Harwood Academic

Trends in Biochemical Sciences
  Elsevier Trends Journals

Trends in Biotechnology
  Elsevier Trends Journals

Trends in Cell Biology
  Elsevier Trends Journals

Trends in Genetics
  Elsevier Trends Journals

Trends in Microbiology
  Elsevier Trends Journals

Virology
  Academic Press

Virus Research
  Elsevier / North Holland

Yeast
  John Wiley and Sons

5 Journal Homepages

Advances in Cell and Molecular Biology
www.elsevier.com

Advances in Enzymology
http://www.wiley.com

*Advances in Experimental Medicine and
Biology*
http://www.kluweronline.com/

Advances in Genetics
http://www.elsevier.com

Advances in Human Genetics
http://www.wkap.nl/prod/s/AHUG

Advances in Protein Chemistry
http://www.elsevier.com

American Journal of Human Genetics
http://www.journals.uchicago.edu/AJHG/
home.html

American Journal of Medical genetics
http://www3.interscience.wiley.com/
cgi-bin/jhome/33129

Animal Cell Biotechnology
http://www.elsevier.com/inca/publications/
store/6/7/8/8/4/6/

Animal Genetics
http://www.blackwellpublishing.com/
journal.asp?ref=0268–9146

Annales de Genétique
http://www.expansionscientifique.com

Annals of Human Genetics
http://journals.cambridge.org

Annual Review of Biochemistry
http://arjournals.annualreviews.org/loi/
biochem

*Annual Review of Biophysics and Biomolecular
Structure*
http://arjournals.annualreviews.org/loi/
biophys

Annual Review of Cell Biology
Publication End Year: 1994
http://arjournals.annualreviews.org

Annual Review of Genetics
http://arjournals.annualreviews.org/loi/
genet

Annual Review of Medicine
http://arjournals.annualreviews.org/loi/
med

Annual Review of Microbiology
http://arjournals.annualreviews.org/loi/
micro

Annual Review of Neuroscience
http://arjournals.annualreviews.org/loi/
neuro

*Annual Review of Plant Physiology and
Molecular Biology.*
Publication End Year: 2001
http://arjournals.annualreviews.org

Applied Cytogenetics
(Journal of the Association of Cytogenetic
Technologists)
http://www.agt-info.org/

Archives of Biochemistry and Biophysics
http://www.elsevier.com/locate/
issn/0003–9861

Archives of Microbiology
http://www.springerlink.com

Archives of Virology
http://www.springerlink.com

*Biochemical and Biophysical Research
Communications*
http://www.elsevier.com/locate/
issn/0006–291X

Biochemical Genetics
http://www.kluweronline.com/
issn/0006–2928

Biochemistry and Cell Biology
http://pubs.nrc-cnrc.gc.ca/cgi-bin/rp/
rp2 jour e

BIOCHIMICA ET BIOPHYSICA ACTA
http://www.elsevier.com/gej-ng/29/50/
show/

Biopolymers
http://www3.interscience.wiley.com/
cgi-bin/jhome/28380

Biotechniques
http://www.biotechniques.com/

Biotechnology and Bioengineering
http://www3.interscience.wiley.com/
cgi-bin/jhome/71002188

Biotechnology and Genetic Engineering News
http://www.intercept.co.uk/gb/
rechrev.html

Brookhaven Symposia in Biology
http://www.bnl.gov/world/

Canadian Journal of Microbiology
http://pubs.nrc-cnrc.gc.ca/cgi-bin/rp/
rp2 jour e

Cancer Genetics and Cytogenetics
http://www.elsevier.com/inca/publications/
store/5/0/5/7/5/2

Cell
http://www.cell.com/

Cellular and Molecular Biology
Publication End Year: 1992
http://www.pergamonpress.com

Chromosoma
http://www.springerlink.com

Clinical Genetics
http://www.blackwellpublishing.com/
journal.asp?ref=0009–9163

Cold Spring Habor Symposia on Quantitative Biology
http://www.cshl.org

Comparative and Functional Genomics
http://www3.interscience.wiley.com/
cgi-bin/jhome/77002016

Critical Reviews in Biochemistry and Molecular Biology
http://www.crcjournals.com/ejournals/
issues/issue archive.asp?section=1062

Critical Reviews in Microbiology
http://www.crcjournals.com/ejournals/
issues/issue archive.asp?section=1050

oder von
http://www.crcjournals.com/default.asp
weiter zu allen critical reviews

Current Genetics
http://www.springerlink.com

Current Opinion in Biotechnology
http://www.current-opinion.com

Current Opinion in Genetics & Development
http://www.elsevier.nl/inca/publications/
store/6/0/1/3/0/2/

Current Topics in Cellular Regulation
www.elsevier.com

Current Topics in Microbiology and Immunology
http://www.springerlink.com

Cytogenetics and Cell Genetics
umbenannt in Cytogenetic and Genome Research
http://content.karger.com

Developmental Genetics
http://www3.interscience.wiley.com/
cgi-bin/jhome/33773

DNA – A Journal of Molecular and Cellular Biology
http://www.liebertpub.com/

DNA and Protein Engineering Techniques
www.interscience.wiley.com

DNA Repair Reports

DNA Repair – Responses to DNA Damage and other Aspects of Genomic Stability
http://www.elsevier.com/inca/publications/
store/6/2/2/2/7/6

DNA Sequence : the Journal of DNA Sequencing and Mapping.
http://www.tandf.co.uk/journals/
titles/10425179.html

Electrophoresis
http://www3.interscience.wiley.com/
cgi-bin/jhome/10008330
http://www.wiley-vch.de/publish/en/
journals/

EMBO Journal
http://emboj.oupjournals.org/

European Journal of Human Genetics
http://www.nature.com/ejhg/

Fungal Genetics Newsletter
http://www.fgsc.net/newslet.html

Gene
http://www.elsevier.nl/inca/publications/
store/5/0/6/0/3/3

Gene Function & Disease
http://www3.interscience.wiley.com/
cgi-bin/jhome/72500531

Genes and Development
http://www.genesdev.org/

Genes and Function
Publication End Year: 1997
http://www.blackwellpublishing.com

Genes, Chromosomes and Cancer
http://www.wiley.com/WileyCDA/
WileyTitle/productCd-GCC.html

Genes to Cells
http://www.blackwellpublishing.com/
journal.asp?ref=1356–9597

Gene Therapy
http://www.nature.com/gt/

Genetica
http://www.kluweronline.com/
issn/0016–6707

Genetical Research
http://titles.cambridge.org/journals/
journal catalogue.asp?mnemonic=grh

Genetic Engeneering: Principles and Methods
http://www.wkap.nl/prod/s/GEPM

Genetics
http://www.genetics.org/

Genome
http://pubs.nrc-cnrc.gc.ca/cgi-bin/rp/
rp2 jour e

Genome Biology
http://genomebiology.com/home/

Genomics
http://www.elsevier.com/locate/
issn/0888–7543

Heredity
http://www.nature.com/hdy/

Human Gene Therapy
http://www.liebertpub.com/hum/

Human Genetics
http://www.springerlink.com

Human Heredity
http://content.karger.com

Human Molecular Genetics
http://hmg.oupjournals.org/

Hybridoma
http://www.liebertpub.com/

Immunogenetics
http://www.springerlink.com

*The International Journal of Biochemistry &
Cell Biology*
http://www.elsevier.nl/inca/publications/
store/3/9/5/

International review of cytology
http://www.elsevier.com

Japanese Journal of Genetics
(jetzt: Genes & Genetic Systems)
Publication End Year: 1995
http://www.cib.nig.ac.jp/GGS/GGS.html

Japanese Journal of Human Genetics
Publication End Year: 1997

Journal of Animal Breeding
http://www.blackwellpublishing.com/
journals/jbg/

Journal of Bacteriology
http://jb.asm.org/

The Journal of Biochemistry
http://jb.bcasj.or.jp/
Journal of Cell Biology
http://www.jcb.org/

Journal of Cell Science
http://jcs.biologists.org/

Journal of General and Applied Microbiology
http://www.iam.u-tokyo.ac.jp/JGAM/
general.htm

Journal of General Microbiology
Publication End Year: 1993
http://www.socgenmicrobiol.org.uk/

Journal of General Virology
  http://vir.sgmjournals.org/

Journal of Genetic Counseling
  http://www.kluweronline.com/
  issn/1059–7700

Journal of Heredity
  http://jhered.oupjournals.org/

Journal of Histochemistry and Cytochemistry
  http://www.jhc.org/

Journal of Medical Genetics
  http://jmg.bmjjournals.com/

Journal of Molecular Biology
  http://www.elsevier.com/locate/
  issn/0022–2836

Journal of Molecular and Cellular Immunology
  Publication End Year: 1990
  www.springerlink.com

Journal of Molecular Evolution
  http://www.springerlink.com

Journal of Virology
  http://jvi.asm.org/

Mammalian Genome
  http://www.springerlink.com

Medical Genetics
  http://medgen.genetics.utah.edu

Methods in Cell Biology
  http://www.ascb.org/publications/
  mcb/mcb.html

Methods in Cell Physiology
  www.elsevier.com

Methods in Enzymology
  www.elsevier.com

Methods in Immunology and Immunochemistry
  www.elsevier.com

Methods in Molecular and Cellular Biology
  http://www.interscience.wiley.com/
  jpages/0898–7750/

Methods in Virology
  www.elsevier.com

Microbiological Reviews
  Publication End Year: 1996
  http://journals.asm.org/

Molecular and Cellular Biochemistry
  http://www.kluweronline.com/
  issn/0300–8177

Molecular and Cellular Biology
  http://mcb.asm.org/

Molecular and Cellular Probes
  http://www.elsevier.com/locate/
  issn/0890–8508

Molecular and General Genetics
  http://www.springerlink.com (see
  Molecular Genetics and Genomics)

Molecular Biology and Evolution
  http://mbe.oupjournals.org/

Molecular Biology and Medicine
  Publication End Year: 1991
  www.elsevier.com

Molecular Biology Reports
  http://www.kluweronline.com/
  issn/0301–4851/contents

Molecular Ecology
  http://www.blackwellpublishing.com/
  journal.asp?ref=0962–1083

Molecular Genetic Medicine
  Publication End Year: 1994
  www.elsevier.com

Molecular Genetics and Genomics (formerly
  Molecular and General Genetics)
  http://www.springerlink.com

Molecular Immunology
  http://www.elsevier.nl/locate/molimm

Molecular Medicine
  http://www.hum-molgen.de/journals/MM/

Molecular Microbiology
  http://www.blackwellpublishing.com/
  journal.asp?ref=0950–382X

Molecular Microbiology and Medicine
  www.elsevier.com

Molecular Pharmacology
  http://molpharm.aspetjournals.org/

Molecular Phylogenetics and Evolution
  http://www.elsevier.com/locate/
  issn/1055–7903

Monographs in Human Genetics
Jetzt: Key issues in human genetics
http://content.karger.com

Mouse Genome
Publication End Year: 1997
http://www.oup-usa.org

Mutagenesis
http://mutage.oupjournals.org/

Mutational Research
DNA Repair – Responses to DNA Damage and
other Aspects of Genomic Stability
http://www.elsevier.com/inca/publications/
store/6/2/2/2/7/6/

Nature
www.nature.com

nature biotechnology
http://www.nature.com/nbt/

nature genetics
http://www.nature.com/ng/

Nature Medicine
http://www.nature.com/nm/

Nature Structural Biology
http://www.nature.com/nsb/

Nucleic Acids Research
http://nar.oupjournals.org/

Pharmacogenetics
http://www.jpharmacogenetics.com/

Plant Molecular Biology
http://www.kluweronline.com/
issn/0167–4412

Plasmid
http://www.elsevier.com/inca/publications/
store/6/2/2/9/3/3/

Proceedings of the National Academy of
Sciences of the United States of America
http://www.pnas.org/

Progress in Biophysics & Molecular Biology
http://www.elsevier.com/locate/
pbiomolbio

Progress in Nucleic Acid Research and
Molecular Biology
http://www.harcourt-international.com/
serials/nucleicacid/

Protein Science
http://www.proteinscience.org/

Proteins: Structure, Function, and Genetics
http://www3.interscience.wiley.com/
cgi-bin/jhome/36176

PROTEOMICS
http://www3.interscience.wiley.com/
cgi-bin/jhome/76510741
http://www.wiley-vch.de/publish/en/
journals/

RNA
http://www.rnajournal.org/

Science
http://www.sciencemag.org/

Sequence: the Journal of DNA Mapping and
Sequencing
http://www.tandf.co.uk/journals/
titles/10425179.html

Somatic Cell and Molecular Genetics
http://www.kluweronline.com/
issn/0740–7750

Stadler Genetics Symposia
http://www.wkap.nl/prod/s/SGSS

Sub-Cellular Biochemistry
http://www.wkap.nl/prod/s/SCBI

Theoretical and Applied Genetics
http://www.springerlink.com

Transgenic Research
http://www.kluweronline.com/
issn/0962–8819

Transgenics
http://www.oldcitypublishing.com/
Transgenics/Transgenics.html

Trends in Biochemical Sciences
http://www.elsevier.com/locate/tibs

Trends in Biotechnology
http://www.elsevier.com/inca/publications/
store/4/0/5/9/1/7

Trends in Cell Biology
http://www.elsevier.nl/inca/publications/
store/4/2/2/5/5/2/

Trends in Genetics
  http://www.elsevier.nl/locate/tig

Trends in Microbiology
  http://www.elsevier.nl/locate/tim

Virology
  http://www.elsevier.com/inca/publications/
  store/6/2/2/9/5/2/

Virus Research
  http://www.elsevier.nl/locate/virusres

Yeast
  http://www3.interscience.wiley.com/
  cgi-bin/jhome/3895

6 Acknowledgements

Illustrations have been kindly provided by (in alphabetical order):

Dr. Monika Blank, Cellzome AG, Heidelberg (Germany): **"Protein-protein interaction map".** This photo is reprinted with permission from Nature 415:141–147 (2002), MacMillan Publishers Ltd. (www.nature.com/nature)

Dr. Ilse Chudoba, MetaSystems GmbH, Robert-Bosch-Straße 6, 68804 Altlussheim (Germany): **"High resolution multicolour banding"** and **"Multicolor fluorescence *in situ* hybridisation".**

Dr. Joachim Häse, Amersham Pharmacia Biosciences Europe GmbH, Am Dachsberg 43, 60435 Frankfurt am Main (Germany): **"Two-dimensional difference gel electrophoresis"**

Dr. Peter R. Hoyt, Molecular Imaging Group, The Oak Ridge National Laboratory, Oak Ridge, Tennessee (USA): **"Atomic force microscopic imaging"**(bottom). This photo is reprinted with permission from Genomics 41: 379–384 (1997).

Dr. Khalid Meksem, Southern Illinois University at Carbondale, Department of Plant, Soil and General Agriculture, Carbondale IL 62901–4415 (USA): **"BAC fingerprint".**

Dr. Stefan Thalhammer, Institute of Crystallography and Mineralogy, Ludwig-Maximilians-University, Munich, Germany): **"Atomic force microscopic imaging"** (top) and **"Chromosome microdissection".**

READER RESPONSE FORM

Günter Kahl: The Dictionary of Gene Technology (ISBN 3-527-30100-3)

1) Terms to add in the next edition (*please supply your definition, or a reference*):

2) Scientific names or common names to add in the next edition (*please supply your definition, or a reference*):

3) Errors or inadequate definitions (*please explain*):

4) Errors in scientific names or common names (*please explain*):

5) Others:

Please give your name and adress:

Please photocopy page and fax or mail it to:

Günter Kahl
Johann Wolfgang Goethe-Universität
Biocentre, Plant Molecular Biology
Marie-Curie-Str. 9
D–60439 Frankfurt am Main
Germany
Fax: +49–(0)69–798–29268
Email: kahl@em.uni-frankfurt.de

2-9-06